Problem-Solving Workbook with Selected Solutions

to accompany

Chemistry: Atoms First

Second Edition

Julia Burdge
University of Idaho

Jason Overby
College of Charleston

Prepared by
Amina El-Ashmawy and Dawn Richardson

Collin College

and

diacriTech

Mc Graw Hill Education

PROBLEM-SOLVING WORKBOOK WITH SELECTED SOLUTIONS TO ACCOMPANY
CHEMISTRY: ATOMS FIRST, SECOND EDITION

Published by McGraw-Hill Education, 2 Penn Plaza, New York, NY 10121. Copyright © 2015 by McGraw-Hill Education. All rights reserved. Printed in the United States of America. Previous edition © 2013. No part of this publication may be reproduced or distributed in any form or by any means, or stored in a database or retrieval system, without the prior written consent of McGraw-Hill Education, including, but not limited to, in any network or other electronic storage or transmission, or broadcast for distance learning.

Some ancillaries, including electronic and print components, may not be available to customers outside the United States.

This book is printed on acid-free paper.

1 2 3 4 5 6 7 8 9 0 QVS/QVS 1 0 9 8 7 6 5 4

ISBN: 978-0-07-764644-8
MHID: 0-07-764644-4

All credits appearing on page or at the end of the book are considered to be an extension of the copyright page.

The Internet addresses listed in the text were accurate at the time of publication. The inclusion of a website does not indicate an endorsement by the authors or McGraw-Hill Education, and McGraw-Hill Education does not guarantee the accuracy of the information presented at these sites.

www.mhhe.com

Table of Contents
Problem-Solving Workbook with Selected Solutions
to accompany *Chemistry: Atoms First, Second Edition*

*Chapter 25 of the Workbook and Selected Solutions can be found on the Atoms First, Second Edition Online Learning Center (www.mhhe.com/burdgeoverby).

Chapter 1
Chemistry: The Science of Change

PROBLEM-SOLVING STRATEGIES AND TUTORIAL SOLUTIONS

TYPES OF PROBLEMS

Problem Type 1: Classification of Matter.
- **A.** Pure Substances.
- **B.** States of Matter.
- **C.** Mixtures.

Problem Type 2: The Properties of Matter.
- **A.** Physical and Chemical Properties.
- **B.** Extensive and Intensive Properties.

Problem Type 3: Scientific Measurement.
- **A.** SI Base Units.
- **B.** Mass.
- **C.** Temperature.
- **D.** Derived Units: Volume and Density.

Problem Type 4: Uncertainty in Measurement.
- **A.** Significant Figures.
- **B.** Calculations with Measured Numbers.
- **C.** Accuracy and Precision.

Problem Type 5: Using Units and Solving Problems.
- **A.** Conversion Factors.
- **B.** Dimensional Analysis – Tracking Units.

PROBLEM TYPE 1: CLASSIFICATION OF MATTER.

A. Pure Substances.

There are many different ways to categorize matter. One of the most straightforward methods is to determine if the sample is a *pure substance* (atom, element, molecule or compound) or a *mixture* (two or more pure substances mixed).

B. States of Matter.

All matter exists primarily in three states: **solid, liquid and gas**. When a substance changes state, the only thing changing is how the particles are arranged relative to each other. In the solid state, the particles are close together with little motion. In the liquid state, the particles are close but have more motion, which allows them to conform to the shape of their container. In the gas state, the particles are very far apart and have a relatively high degree of motion. Gases take both the shape and the volume of their container.

C. Mixtures.

A **mixture** can be classified as either **homogeneous** (all of the components are equally distributed throughout the sample and are in the same state or phase, uniform composition throughout) or **heterogeneous** (the components are not uniform throughout and can be in different phases). Mixtures may be separated using physical processes.

STUDY QUESTION 1.1
Classify the following as either a pure substance or a mixture.

    a.   **Beach sand**
    b.   **The mercury in a thermometer**
    c.   **Air**

Solution: Beach sand contains not only sand (silicon dioxide) but salt from the water, bits of sea plants and sea life. Therefore, beach sand is a mixture because it contains many different types of matter.

Inside of a thermometer, there is only liquid Mercury. Mercury is a metallic element, as we will see in chapter 2, therefore it is a pure substance.

Air contains approximately 78% nitrogen, 21% oxygen and 1% other gases. Air is a mixture since it contains many elements in their gaseous states.

STUDY QUESTION 1.2
What is the process occurring in each of the following?

    a.   **Heating a pot full of water**
    b.   **Putting a water-filled ice cube tray in the freezer**
    c.   **Moisture forming on your windshield in cold weather**
    d.   **Leaving dry ice out at room temperature**

Solution: When heating water it converts from liquid to gas, which is *vaporization*. Placing water in the freezer changes it from liquid to solid, which is *freezing*. Water vapor in the air turns to liquid water when it comes in contact with the cold windshield. This process is called *condensation*. Now, for the dry ice process, first it is good to know that dry ice is solid carbon dioxide. When placed at room temperature, the dry ice goes directly to gas, which is called *sublimation*.

STUDY QUESTION 1.3
Classify the following as either a homogeneous or heterogeneous mixture.

    a.   **Air**
    b.   **Beach sand**
    c.   **Fruit Loops® cereal**
    d.   **Gasoline**
    e.   **Blood**

Solution: Being a gas, air particles move to fill their container. They will achieve uniform appearance and consistency throughout the sample. Air is a homogeneous mixture. Beach sand and Fruit Loops® are both solid mixtures that do not have uniform appearance or consistency throughout the sample. They are heterogeneous mixtures. Gasoline is a liquid mixture containing various organic compounds. It has the same consistency throughout and is a homogeneous mixture. Blood looks homogenous on a macroscopic scale. However, if we look at blood under a microscope we find that there are different types of cells in the blood that are not equally distributed throughout the sample. Blood is a heterogeneous mixture.

PRACTICE QUESTION 1.1
Classify the following as either a pure substance or a mixture.

    a.   **Smarties® candy**
    b.   **The neon gas in glowing signs**
    c.   **Fingernail polish**

PRACTICE QUESTION 1.2
What is the process occurring in each of the following scenarios?

    a.   Putting a chocolate bar in your pocket
    b.   Wet clothes in the dryer

PRACTICE QUESTION 1.3
Classify the following as either a homogeneous or heterogeneous mixture.

    a.   Orange juice with pulp
    b.   Unopened bottle of soda pop
    c.   Maple syrup

PROBLEM TYPE 2: THE PROPERTIES OF MATTER.

A. Physical and Chemical Properties.

Substances are identified by their quantitative (involving numbers) and qualitative (not involving numbers) properties. **Physical properties** are those that can be determined without changing the identity of the matter in question. A **physical change** is one in which the identity of the matter involved does not change during a process. Therefore a chemical interaction with another substance is not needed to determine a physical property. A **chemical change** is one in which the identity of the reacting matter does change during the chemical reaction (a new form of matter is produced as a result of a chemical reaction). **Chemical properties** are determined only as the result of a chemical change or chemical process, in which the original substance is converted to a different substance.

STUDY QUESTION 1.4
Classify each as either a physical or chemical property.

    a.   Bluish-purple color of iodine crystals
    b.   Potassium metal reacting violently with water
    c.   Water freezing at 0 °C

Solution: Color and freezing point are physical properties as they describe the physical existence of the substance. Both of these properties can be determined directly without a chemical reaction occurring. Reactivity is a chemical property because it describes how the substance changes as the potassium reacts with water in a chemical reaction.

STUDY QUESTION 1.5
Classify each as either a physical or chemical change.

    a.   Fireworks exploding
    b.   Getting wet in the rain
    c.   Erasing the board
    d.   Bleaching your hair
    e.   Chewing gum

Solution: As we discussed above, a *physical change* is one where there is no change in identity or composition of the substance and is generally reversible. A *chemical change* involves a substance reacting or changing its identity or composition.

Fireworks explode as a result of a *chemical change* or reaction that burns the chemicals contained within. The burning chemicals give off color. Each chemical has its characteristic color when burned. Bleaching hair involves a chemical reaction that changes the pigment particles in the hair strand. It is a *chemical change.*

Getting wet in the rain is a *physical change* that can be reversed by the evaporation of the water. Erasing the board is also a *physical change* because the chalk or marker is simply moved off the surface of the board and is not changing in composition.

Chewing gum is a bit complicated as it involves several things. Initially, you change the shape of the gum, which is a *physical change*. Your saliva begins extracting the sugar and flavors, which is a *physical change*. Subsequently, the enzymes in your saliva begin breaking down or digesting the sugar, which is a *chemical change*.

PRACTICE QUESTION 1.4
Classify each as either a physical or chemical property.

  a. **The smell of coffee**
  b. **The hardness of diamonds**
  c. **The flammability of natural gas**

PRACTICE QUESTION 1.5
Classify each as either a physical or chemical change.

  a. **Dry ice subliming at room temperature**
  b. **Growing a plant**
  c. **Frying an egg**
  d. **Ice cream melting**
  e. **Table salt dissolving in water**
  f. **Pop Rocks® popping on our tongue**

B. Extensive and Intensive Properties.
All properties of matter are either extensive or intensive. **Extensive properties** vary with the amount of matter being considered. For example, mass depends on the amount of matter; the more material there is, the larger the mass. **Intensive properties** are independent of the amount of material. For example, the color of a small piece of pure gold is the same as the color of a large piece of gold.

STUDY QUESTION 1.6
Classify each as either an extensive or intensive property.

  a. **Melting point**
  b. **Heat needed to vaporize**
  c. **Volume**
  d. **Density**

Solution: Whether you have one gram or one kilogram of a substance, the temperature at which melting will occur does not change. Hence, melting point is an *intensive property*.

If you need to vaporize a glass of water versus a tub full of water, you will need a different amount of heat. Heat, in general, is an *extensive property*. Volume is the amount of space taken up by matter. The more matter you have, the more space it will occupy. Volume is also an *extensive property*.

As we will soon learn later in this chapter, density is determined by the ratio of mass to volume (mass/volume). We know that both mass and volume are extensive properties. As the mass increases, the volume also increases. Therefore, when mass is divided by volume, the dependency on amount cancels out, and the ratio is independent of the amount. Density is therefore an *intensive property*.

PRACTICE QUESTION 1.6
Classify each as either an intensive or extensive property.

a. **Boiling point**
b. **Temperature**
c. **Odor**
d. **Solubility**

PROBLEM TYPE 3: SCIENTIFIC MEASUREMENT.

A. SI Base Units.

Scientists and engineers have adopted a uniform, international (language independent) system of units for labeling observations. The International System of Units (SI) is a metric system built on a foundation of base units for fundamental phenomena such as mass, length (distance) and time. Rather than report observations using very small or very large numbers, SI units are scaled (multiplied) by decimal prefixes. Your textbook lists the prefixes used with metric units to scale up or down the numbers associated with them (see Table 1.2). You need to memorize the prefix and the definition of each. Units for other phenomena, such as volume and density, can be derived (calculated) from the base units.

B. Mass.

The mass of an object is a measure of the amount of matter it contains. The SI base unit for mass is the kilogram (kg). The terms mass and weight are often used interchangeably, although they actually refer to distinct properties. The mass of an object is the same everywhere in the universe. The weight of an object is determined by the force exerted on the object by gravity and will vary depending on the relative masses of the object, the source of gravity and the object's distance from the source of gravity.

C. Temperature.

Temperature is a measure of the amount of motion of the particles that make up a piece of matter. There are three temperature scales, which were devised based on the melting point and boiling point of water at sea level: Celsius, Fahrenheit, and Kelvin. The SI unit for temperature is the Kelvin, which is also known as the absolute temperature scale. At 0 K, the motion of particles within a piece of matter will cease.

STUDY QUESTION 1.7
The freezing point of a 50-50 mixture of antifreeze (ethylene glycol) and water is –36.5 °C. Convert this temperature to degrees Fahrenheit.

Solution: First, write the equation for temperature conversion:

$$^\circ C = (^\circ F - 32\ ^\circ F) \times \frac{5}{9}$$

Next, substitute –36.5 °C and solve for degrees Fahrenheit:

$$-36.5\ ^\circ C = (^\circ F - 32\ ^\circ F) \times \frac{5}{9}$$

$$-36.5\ ^\circ C \times \frac{9}{5} = (^\circ F - 32\ ^\circ F)$$

$$^\circ F = -65.7\ ^\circ F + 32\ ^\circ F = -33.7\ ^\circ F$$

STUDY QUESTION 1.8
Convert –36.5 °C to Kelvin.

Solution: First, write the equation for temperature conversion:

$$K = °C + 273.15$$

Next, substitute –36.5 °C and solve for Kelvin:

$$K = -36.5 °C + 273.15 = 236.65 \text{ K}$$

As we will see later, we cannot have more decimals than the least number of decimals used in the addition/subtraction operation. We would round the answer to only one decimal, or 236.6 K.

> *PRACTICE QUESTION 1.7*
> **Body temperature is 98.6 °F. Convert this temperature to degrees Celsius and to Kelvin.**

> *PRACTICE QUESTION 1.8*
> **If a person's body temperature is elevated by 3 °C, how much higher is it in kelvin? Would their body temperature be less than, the same or greater than 3 °F higher?**

D. Derived Units: Volume and Density

The volume of an object is a measure of the space it occupies. The larger the volume, the more space the object requires. The SI unit for volume is the cubic meter (m^3) and it is a derived unit. It is simple to understand why we have a base unit, like the meter, raised to the third power to imply volume. Remember that volume is a measure of space and to define space we need three coordinates (hence, the third power). The density of an object is the ratio of the mass to the volume, or the mass per unit volume, so it also has a derived unit. The ratio of the fundamental units for mass and volume is kg/m^3, but this unit is rarely used. Densities of liquids and solids are usually reported in g/mL or g/cm^3. Densities of gases are usually reported in g/L.

STUDY QUESTION 1.9
A flask filled to the 25.0 mL mark contained 29.97 g of a concentrated salt–water solution. What is the density of the solution?

Solution: The density (d) of an object or a solution is defined as the ratio of its mass (m) to its volume (V). Substitute the given quantities:

$$d = \frac{\text{mass}}{\text{volume}} = \frac{29.97 \text{ g}}{25.0 \text{ mL}} = 1.20 \text{ g/mL}$$

> *PRACTICE QUESTION 1.9*
> **Calculate the volume of a brick that is 34 cm long by 7.0 cm wide by 14 cm high.**

> *PRACTICE QUESTION 1.10*
> **A certain metal ingot has a mass of 3951 g, and measures 10.2 cm by 8.2 cm by 4.2 cm. Calculate the density of the metal.**

PRACTICE QUESTION 1.11
A person with almost no body fat has to work much harder to float than someone with a high percentage of body fat. Explain why that is.

PROBLEM TYPE 4: UNCERTAINTY IN MEASUREMENT.

A. Significant Figures.

Significant figures are used to specify the uncertainty in a measured number or in a number calculated using measured numbers. Significant figures must be carried through calculations such that the implied uncertainty in the final answer is reasonable. Measured numbers are inexact. Numbers obtained by counting or that are part of a definition are exact numbers (they have an infinite number of significant figures and do not influence the accuracy of a calculation). A set of empirical rules have been devised to assign the number of significant figures in a number. Pay special attention to the rules for the zeroes, since whether they are significant or not will depend on their position within the number. Another important skill to remember is how to express numbers in scientific notation.

STUDY QUESTION 1.10
Determine the number of significant figures in each.

a.	6.02
b.	0.012
c.	1.23×10^7
d.	1.5400

Solution:

a. 3 significant figures. Recall that zeros between nonzero digits are significant.

b. 2 significant figures. Zeros to the left of the first nonzero digit are not significant.

c. 3 significant figures. Scientific notation implies that significant figures are shown.

d. 5 significant figures. When a number contains a decimal point and has trailing zeros, these zeros are significant.

PRACTICE QUESTION 1.12
Determine the number of significant figures expressed in each of the following.

a.	0.609
b.	1.0×10^3
c.	0.0000222
d.	238.0
e.	1.030×10^{-2}

PRACTICE QUESTION 1.13
Round each to the number of significant figures indicated.

a.	0.60945 to three significant figures
b.	1.012×10^3 to two significant figures
c.	0.00022174 to three significant figures
d.	237.95 to four significant figures
e.	1.303 to two significant figures

B. Calculations with Measured Numbers.

In many cases, the results we seek are calculated from the measurements we make. The number of significant figures is limited by the least precise measurement.

To determine the number of significant figures in a calculation use the following rules:

1. The number of significant figures in a product ($\times$) or quotient ($\div$) is the number of significant figures in the least accurate factor, which is the original number that has the least amount of significant figures.
2. The number of significant figures in a sum (+) or difference (–) cannot have more digits to the right of the decimal point than any of the original numbers being added or subtracted.

If adding two numbers, which are in scientific notation but have different exponents, the numbers must be converted to a common exponent. For example,

$(4.5 \times 10^{-5}) + (1.5 \times 10^{-6})$	Convert one of the numbers to match the other's exponent. I will convert the –6 to –5
$(4.5 \times 10^{-5}) + (0.15 \times 10^{-5})$	Now, add the coefficients (i.e., the numbers in front)
$(4.5 + 0.15) \times 10^{-5} = 4.65 \times 10^{-5}$	Use your rounding rules (least number of decimal places)
4.7×10^{-5}	This is YOUR ANSWER to the correct significant figures.

STUDY QUESTION 1.11
Carry out the following operations, rounding the answer to the correct number of significant figures.

   a. $287.12 - 95.333 =$
   b. $7.25 \times 10^{12} \div 92 =$
   c. $(6.0 + 5.21) \div 4781 =$

Solution:
a. First, subtract the numbers giving: 191.787. Only two digits to the right of the decimal are significant, so the answer is 191.79.
b. The exponent plays no part in significant figures when multiplying or dividing. Look at total significant figures in your original numbers. Your answer must have the same as the one with the least (i.e., two) significant figures. Answer: 7.9×10^{10}.
c. When 6.0 and 5.21 are added, the answer is 11.21. However in this addition, you only keep one place to the right, giving 11.2. The number 11.2 has 3 significant figures and it is divided by 4781 which has 4 significant figures. This quotient can only have three significant figures. Answer: 2.34×10^{-3}.

PRACTICE QUESTION 1.14
Carry out the following operations and express the result to the correct number of significant figures.

   a. 12×2143.1
   b. $3.09 \div 7$
   c. $(2.2 \times 10^{-3})(1.40 \times 10^{6})$
   d. $12.70 + 1.222$
   e. $595.2 \times (24.33 - 16.271)$

PRACTICE QUESTION 1.15
Why are the number of eggs reported exactly, irrespective of the number of significant figures?

C. **Accuracy and Precision.**

There are two types of error associated with measurements: random and systematic. With random error there is a 50% chance of the measurement being high and a 50% chance of the measurement being low relative to the actual value. Systematic error is associated with instrumentation where all measured values will err in the same direction, either all too high or all too low. Precision refers to how closely reproducible measurements are to one another. It expresses the degree of random error in your measurements. Accuracy refers to how close measurements are to a true value. It expresses the degree of systematic error in your measurements.

STUDY QUESTION 1.12
Elsa and Tim were asked to weigh a 5.00 g steel bar. The data gathered by both students follows.

Tim	Elsa
4.78	5.03
5.01	4.99
4.89	4.97

Which set of data is most precise? Most accurate?

Solution: Remember what accuracy and precision mean. In order to evaluate the precision of each set of repeated measurements we need to be able to "see" how close they are to one another. This is easily accomplished if we order them from large to small or vice versa (i.e., by increasing or decreasing magnitude).

Tim	Elsa
5.01	5.03
4.89	4.99
4.78	4.97

If we look at the difference between the largest and smallest value in each set, we notice that Tim's values spread over a larger range than Elsa's values. This is an indication that Tim's measurements are more scattered. There is more random error and thus least precise.

Accuracy refers to how close the measured values are to a true, accepted value. The steel bar has a mass of 5.00 g. To evaluate the accuracy of each set as a whole, calculate the average value of each set and compare it to the true value. The average weight for Tim's data is 4.89 g while that of Elsa's data is 5.00 g. Elsa's data is more accurate.

PRACTICE QUESTION 1.16
Evaluate the relative accuracy and precision of the following sets of data on the measurement of the melting point of gallium. The melting point of gallium is reported to be 29.76 °C.

Set #1	Set #2
28.56	30.85
28.60	31.29
28.57	30.14

PROBLEM TYPE 5: USING UNITS AND SOLVING PROBLEMS.

A. Conversion Factors.

A conversion factor is a fraction in which the numerator and denominator are the same quantity expressed in different units. Multiplying by a conversion factor is unit conversion.

In general, conversion factors are simple ratios. Any time units of a given quantity are a ratio (as in m/s, g/mL, etc.) a corresponding conversion factor can be written. For example if the density of a substance is given a 1.78 g/mL, we can write the density as the following conversion factor:

$$\frac{1.78 \text{ g}}{1 \text{ mL}}$$

Any expression that is written as an equality can be turned into a conversion factor. For example, the definition of the centimeter, $1 \text{ cm} = 10^{-2}$ m, can be written as the ratio $\frac{1 \text{ cm}}{10^{-2} \text{ m}}$ or $\frac{10^{-2} \text{ m}}{1 \text{ cm}}$, and can be used as a conversion factor.

STUDY QUESTION 1.13

A single molecule weighs approximately 3.0×10^{-22} g. What is the conversion factor obtained from the above statement?

Solution: Using symbols to represent the words, we get 1 molecule $= 3.0 \times 10^{-22}$ g. From this equation we can write the conversion factor as

$$\frac{1 \text{ molecule}}{3.0 \times 10^{-22} \text{g}}$$

PRACTICE QUESTION 1.17

Write the conversion factor corresponding to the following statement. A sample containing 6.022×10^{23} molecules of water weighs 18 g.

PRACTICE QUESTION 1.18

Using an estimated number, what is the conversion factor for stating the number of eyebrow hairs in a classroom containing 45 people?

B. Dimensional Analysis – Tracking Units.

Dimensional analysis is a series of unit conversions used in the solution of a multistep problem. The key to dimensional analysis is to consider the units as an essential part of any quantity. We should not think of a quantity without its units. The great benefit of dimensional analysis is that when quantities labeled with the correct units are subjected to calculations, the desired unit is correctly calculated. Additionally, analyzing the units in your conversion factors helps you decide whether to multiply or divide to make the desired conversion.

Conversion factors can be used according to the following general form:

$$\text{given value} \times \frac{\text{\# desired unit}}{\text{\# given unit}} = \text{desired value}$$

The ratio $\dfrac{\text{\# desired unit}}{\text{\# given unit}}$ is the conversion factor obtained from known equalities as seen in the previous section.

STUDY QUESTION 1.14
Convert 2.5 gallons to liters.

Solution: First, we need to understand what units are involved in this problem. We are given gallons, which is an English unit of volume, and we are wanting to determine the volume in liters (L), which is a metric unit of volume. The conversion factor is 1 gal = 3.7854 L.

$$2.5 \text{ gal} \times \frac{3.7854 \text{ L}}{1 \text{ gal}} = 9.5 \text{ L}$$

STUDY QUESTION 1.15
Convert 257 milligram to micrograms.

Solution: We are given milligram, which is a metric unit of mass, and we are wanting to determine micrograms, which is also a metric unit of mass. Since we know how many mg are in a gram and how many μg are in a gram, we can convert mg to grams and then grams to μg: mg → g → μg

$$\frac{257 \text{ mg}}{} \times \frac{1 \times 10^{-3} \text{ g}}{1 \text{ mg}} \times \frac{1 \text{ μg}}{1 \times 10^{-6} \text{ g}} = 2.57 \times 10^5 \text{ μg}$$

STUDY QUESTION 1.16
How many seconds (s) are there in one week?

Solution: This problem involves converting from week to seconds. Both are units of time. The conversions we know are 60 sec = 1 min, 60 min = 1 hr, 24 hr = 1 day, and 7 days = 1 week.

$$1 \text{ week} \times \frac{7 \text{ days}}{1 \text{ week}} \times \frac{24 \text{ hr}}{1 \text{ day}} \times \frac{60 \text{ min}}{1 \text{ hr}} \times \frac{60 \text{ sec}}{1 \text{ min}} = 604{,}800 \text{ sec}$$

PRACTICE QUESTION 1.19
Write the conversion factor corresponding to the following statement. A sample containing 6.022×10^{23} atoms of gold weighs 196.97 g.

PRACTICE QUESTION 1.20
How many picometers are there in 1.0 mile?

PRACTICE QUESTION 1.21
What is the mass in grams of a lead (Pb) brick that measures 20.5 cm × 9.7 cm × 62 mm? The density of Pb is 11.4 g/cm^3?

PRACTICE QUIZ

1. Write the following amounts in scientific notation in terms of the base SI unit.
   a. 7 µg
   b. 8.0 nm
   c. 0.14 ML
   d. 1.0 ks

2. Calculate the density of mercury given that a spherical droplet of mercury with a radius of 0.328 cm has a mass of 2.00 g. The volume of a sphere is $\frac{4}{3}\pi r^3$.

3. How many cubic centimeters are equal to 5 m^3?

4. The Voyager 2 mission to the outer planets of the solar system transmitted spectacular photographs of Neptune to Earth by radio. Radio waves, like light waves, travel at 3.00×10^8 m/s. If Neptune was 2.75 billion miles from Earth during these transmissions, how many hours were required for radio signals to reach Earth from Neptune?

5. The density of a 26% salt solution is 1.199 g/mL. What is the volume, in mL, occupied by 20.0 g of this solution?

6. The displacement volume of a certain automobile engine is 350 in^3. How many liters is this?

7. Make the following conversions of metric lengths.
   a. 12.5 cm = _____ m
   b. 8.0×10^{-8} m = _____ nm
   c. 445 cm = _____ km
   d. 32.5 mm = _____ km
   e. 5.73×10^3 nm = _____ mm

Chapter 2
Atoms and the Periodic Table

PROBLEM-SOLVING STRATEGIES AND TUTORIAL SOLUTIONS

TYPES OF PROBLEMS

Problem Type 1: Atomic Number, Mass Number, and Isotopes.

Problem Type 2: Patterns of Nuclear Stability.

Problem Type 3: Average Atomic Mass.

Problem Type 4: Periodic Table.

Problem Type 5: The Mole and Molar Masses.
     A. The Mole.
     B. Molar Mass.
     C. Interconverting Mass, Moles, and Number of Atoms.

PROBLEM TYPE 1: ATOMIC NUMBER, MASS NUMBER, AND ISOTOPES.

The **atomic number, Z**, tells us the number of protons in the nucleus of an atom. In a later chapter, we will see that this refers to nuclear charge. Atomic number determines the identity of the atom. The **mass number, A**, is the sum of the protons and neutrons in the nucleus. The difference between the mass number and atomic number determines the number of neutrons for that atom. Protons and neutrons are referred to, collectively, as nucleons. Atoms of an element may have varying numbers of neutrons. Each atom is considered a specific **isotope** of that element. The symbol for an atom is written as follows.

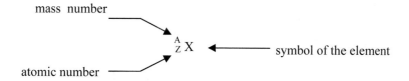

$$\text{mass number} \quad\longrightarrow\quad {}^{A}_{Z}X \quad\longleftarrow\quad \text{symbol of the element}$$
$$\text{atomic number} \quad\longrightarrow$$

STUDY QUESTION 2.1
How many protons, neutrons and electrons are in $^{46}_{20}$Ca?

Solution: The atomic number, subscript preceding the symbol, tells us the number of protons. For a neutral atom, there is the same number of protons as electrons. The number of neutrons is the difference between the mass number and atomic mass. Therefore, there are 20 protons, 20 electrons, and 26 neutrons in a $^{46}_{20}$Ca (calcium-46) atom.

STUDY QUESTION 2.2
How many protons, neutrons and electrons are in ^{106}Pd?

Solution: The atomic number is not given with this symbol. Remember that the atomic number is synonymous with the element. So, looking at the periodic table, we find that the atomic number is 46. There are 46 protons, 46 electrons, and 60 neutrons in a palladium-106 atom.

STUDY QUESTION 2.3
Write the symbol for an atom that is made up of 34 protons, 34 electrons and 41 neutrons.

Solution: The number of protons is the same as the atomic number and defines the element. Looking at the periodic table, we find that atomic number 34 is selenium, Se. The number of electrons is the same as the number of protons in a neutral atom. The mass number is the sum of protons and neutrons, which is 75. The symbol is $^{75}_{34}$ Se.

STUDY QUESTION 2.4
Write the symbol for an atom that is made up of 126 neutrons and 83 electrons.

Solution: The number of protons is not given. However, the number of electrons has to be the same as the number of protons in an atom, so the atomic number is 83, bismuth. The mass number is $(126 + 83) = 209$. The mass number is 83. The symbol is $^{209}_{83}$ Bi. This is a bismuth-209 atom.

STUDY QUESTION 2.5
Two isotopes of thalium have mass numbers of 203 and 205, respectively. How many neutrons are present in each?

Solution: Thalium has atomic number 81, which means there are 81 protons in each of the isotopes. Thalium-203 has $(203 - 81) = 122$ neutrons. Thalium-205 has $(205 - 81) = 124$ neutrons.

PRACTICE QUESTION 2.1
How many neutrons are in tungsten-183?

PRACTICE QUESTION 2.2
Write the symbol for the atom that is made up of 102 neutrons and 68 protons.

PRACTICE QUESTION 2.3
How many electrons are in the atom represented in Practice Question 2.2?

PRACTICE QUESTION 2.4
Two stable isotopes of europium exist. One has a mass of 151 amu and the other 153 amu. How many neutrons are present in each?

PROBLEM TYPE 2: PATTERNS OF NUCLEAR STABILITY.

Little is known about the forces that hold a nucleus together. However, some interesting facts emerge if we examine the numbers of protons and neutrons found in stable nuclei. Nuclei can be classified according to whether they contain even or odd numbers of protons and neutrons. Table 2.2 in your textbook lists the number of stable isotopes with respect to whether the number of protons and neutrons are odd and/or even.

The following rules are useful in predicting nuclear stability:

---

1. All isotopes of elements after bismuth (Z = 83) are radioactive.
2. Nuclei that contain certain specific numbers of protons and neutrons ensure an extra degree of stability. These so–called **magic numbers** for protons and for neutrons are 2, 8, 20, 28, 50, 82, and 126.
3. Nuclei with even numbers of protons or neutrons are generally more stable than those with odd numbers of these particles.

The *magic numbers* correspond to nuclear structures that are stable much like the numbers 2, 8, 18, 32, … (number of electrons to fill each shell) are for electrons.

STUDY QUESTION 2.6
Using the stability rules, rank the following isotopes in order of increasing nuclear stability.

$$^{39}_{20}Ca \qquad ^{39}_{20}Ca \qquad ^{30}_{15}P$$

Solution: Phosphorus–30 should be the least stable because it has odd numbers of both protons (15) and neutrons (15). Calcium–39 has an even number of protons (20) and an odd number of neutrons (19). With an even number of protons and a magic number at that, it should be more stable than P–30. Of the three isotopes, Calcium–40 should be the most stable. It has an even number of protons (20) and of neutrons (20). Both numbers are magic numbers.

$$^{30}_{15}P \; < \; ^{39}_{20}Ca \; < \; ^{39}_{20}Ca$$

The principal factor for determining whether a nucleus is stable is the neutron to proton ratio. Figure 20.1 in your textbook shows a plot of the number of neutrons versus the number of protons in various isotopes. The stable nuclei are located in an area of the graph known as the **belt of stability**. In this figure, we see that at low atomic numbers, stable nuclei possess a neutron to proton ratio of about 1.0. When Z > 20, the number of neutrons always exceeds the number of protons in stable isotopes. The n : p ratio increases to about 1.5 at the upper end of the belt of stability.

If you were given the symbol of a radioisotope, without any experience, it would be impossible to tell its mode of decay. But with knowledge of the belt of stability, you can make accurate predictions of the expected mode of decay.

Isotopes with too many neutrons lie above the belt of stability. The nuclei of these isotopes decay in such a way as to lower their n : p ratio. For example, one neutron may decay into a proton and a beta particle.

$$^{1}_{0}n \; \rightarrow \; ^{1}_{1}p \; + \; ^{0}_{-1}\beta$$

The proton remains in the nucleus, and the beta particle is emitted from the atom. The loss of a neutron and the gain of a proton produces a new isotope with two important properties. It has a lower n : p ratio, and thus is more likely to be stable. Also, the daughter product has a nuclear charge that is one greater than the decaying isotope, due to the additional proton. Consider the decay of carbon–14, for example (Carbon–14 is continually produced in the upper atmosphere by the interaction of cosmic rays with nitrogen). Carbon–14 has a higher n : p ratio than either of carbon's stable isotopes (C–12 and C–13), and decays by beta decay.

$$^{14}_{6}C \; \rightarrow \; ^{14}_{7}N \; + \; ^{0}_{-1}\beta$$

Note that the product isotope $^{14}_{7}N$ has nuclear charge greater by one than carbon. Also, it is stable. Its n : p ratio is 1.0.

Isotopes with too many protons have low n : p ratios and lie below the belt of stability. These isotopes tend to decay by positron emission because this process produces a new isotope with a higher n : p ratio. During positron emission a proton emits a positron, $_{+1}^{0}\beta$, leaving behind a neutron.

$$_{1}^{1}p \rightarrow {}_{0}^{1}n + {}_{+1}^{0}\beta$$

The neutron remains in the nucleus, and the positron is emitted from the atom. Thus, the newly produced nucleus will contain one less proton and one more neutron than the parent nucleus. The n : p ratio increases due to positron decay.

Electron capture accomplishes the same end, that is, a higher n : p ratio. Some nuclei decay by capturing an atom's orbital electron.

$$_{1}^{1}p + {}_{-1}^{0}e \rightarrow {}_{0}^{1}n$$

Lanthanum–138, a naturally occurring isotope with an abundance of 0.089 percent, decays by electron capture.

$$_{57}^{138}La + {}_{-1}^{0}e \rightarrow {}_{56}^{138}Ba + h\nu$$

Electron capture is accompanied by X–ray emission, represented in the above reaction as $h\nu$ (the equation for the energy of a photon).

STUDY QUESTION 2.7
The only stable isotope of sodium is sodium–23. What type of radioactivity would you expect from sodium–25?

Solution: Sodium–23 has 11 protons and 12 neutrons and is in the belt of stability. Sodium–25 has two more neutrons, and so has a higher n : p ratio than the stable isotope. Sodium–25 will decay by $_{-1}^{0}\beta$ emission.

PROBLEM TYPE 3: AVERAGE ATOMIC MASS.

Atomic mass is the mass of an atom in atomic mass units. The periodic table lists the average atomic mass (sometimes called the atomic weight) of each element. The average atomic mass is determined by the relative abundance of each isotope of that element. It is calculated by the sum of (abundance X isotopic mass) all the isotopes. Remember that a value given as a percent would be used mathematically as a decimal. For example, 52.6% is 0.526.

STUDY QUESTION 2.8
Calculate the atomic mass for Si given the following data.

Isotope	Relative Abundance (%)	Mass (amu)
Silicon-28	92.23	27.9769
Silicon-29	4.67	28.9765
Silicon-30	3.10	29.9738

Solution: The average atomic mass, or atomic mass, for Si is
$(0.9223 \times 27.9769 \text{ amu}) + (0.0467 \times 28.9765 \text{ amu}) + (0.0310 \times 29.9738 \text{ amu}) = 28.085485$ amu or 28.09 amu.

STUDY QUESTION 2.9
Identify the element with the following isotopic data.

Isotope	Relative Abundance (%)	Mass (amu)
#1	78.99	23.9850
#2	10.00	24.9858
#3	11.01	25.9826

Solution: The average atomic mass, or atomic mass, is
$(0.7899 \times 23.9850 \text{ amu}) + (0.1000 \times 24.9858 \text{ amu}) + (0.1101 \times 25.9826) = 24.3050$ amu or 24.31 amu.
According to the periodic table, the element with atomic mass 24.31 amu is Mg.

PRACTICE QUESTION 2.5
Determine the atomic mass and the identity of the element based on the following data.

Isotope	Relative Abundance (%)	Mass (amu)
#1	82.58	87.9056
#2	7.00	86.9089
#3	9.86	85.9093
#4	0.56	83.9134

PROBLEM TYPE 4: PERIODIC TABLE.

The periodic table arranges the elements in columns, called **groups**, and rows, called **periods**. Elements in the same group exhibit similar properties. Figure 2.10 in the textbook shows the **metals** below and to the left of the orange-shaded elements. The orange-shaded elements are the **metalloids**. The **nonmetals** are above and to the right of the metalloids. Metals are good conductors of electricity and heat; nonmetals are not good conductors of electricity or heat, otherwise called insulators. Metalloids have properties intermediate between those of metals and nonmetals. Some of the groups have special names including **alkali metals** (Group 1A, except hydrogen), **alkaline earth metals** (Group 2A), **chalcogens** (Group 6A), **halogens** (Group 7 A), **noble gases** (Group 8A), and **transition metals** (Group 1B and Groups 3B-8B).

STUDY QUESTION 2.10
Bromine, Br, belongs to which group and period? What type of element is bromine?

Solution: Br is in group 7A and period 4. It is a halogen and a nonmetal.

STUDY QUESTION 2.11
To which group and period does an atom with 80 neutrons and 56 electrons belong?

Solution: First, we must determine which element this is. Having 56 electrons infers the atomic number is 56. This is barium, Ba. Barium is in group 2A and period 6. It is an alkaline earth metal.

> *PRACTICE QUESTION 2.6*
> **I am a period 4 metalloid chalcogen. Who am I?**

> *PRACTICE QUESTION 2.7*
> **I am the only liquid in Group 2B. Who am I?**

PROBLEM TYPE 5: THE MOLE AND MOLAR MASSES.

A.   The Mole.
A mole is the amount of a substance that contains 6.0221418×10^{23} [Avogadro's number (N_A)] of elementary particles (such as atoms, molecules or ions). We should view it simply as a number, just like a dozen is 12.   We use Avogadro's number to convert between the macroscopic scale and the atomic scale. Avogadro's number has units of particles/mol. Since particles are unitless, the number ends up with the unit of inverse moles (1/mol), and is used specifically to convert between moles and numbers of atoms (or any particles).

STUDY QUESTION 2.12
How many moles of doughnuts are in 5 dozen doughnuts?

Solution: First, we need to know what number of doughnuts (particles) 5 dozen is. This is $(5 \times 12) = 60$ doughnuts. Now, to determine the number of moles,

$$60 \text{ doughnuts} \times \frac{1 \text{ mol doughnuts}}{6.0221418 \times 10^{23} \text{ doughnuts}} = 9.9632327 \times 10^{-23} \text{ mol}$$

> *PRACTICE QUESTION 2.8*
> **How many eggs are in 3.25 moles of eggs?**

> *PRACTICE QUESTION 2.9*
> **Without using a calculator, determine which contains more shrimp, 2.25 moles of shrimp or 1.244×10^{24} shrimp?**

B.   Molar Mass.
Molar mass (M) is the mass of one mole of a substance, usually expressed in grams. The molar mass of an element in grams turns out to be the same as its atomic mass in amu. For example, an atom of osmium, Os, has mass 90.23 amu while a mole of osmium has mass 90.23 g.

STUDY QUESTION 2.13
What is the molar mass of York Peppermint Patties® if one patty weighs 43 g?

Solution: There are 6.0221418×10^{23} miniature cups in a mol.

$$1 \text{ mol} \times \frac{6.0221418 \times 10^{23} \text{ cups}}{1 \text{ mol}} \times \frac{43 \text{ g}}{1 \text{ cup}} = 2.6 \times 10^{25} \text{ g}$$

PRACTICE QUESTION 2.10
A Reese's Peanut Butter Cups® miniatures weigh 8.8 g. How many miniature cups are in 0.25 moles?

C. Interconverting Mass, Moles, and Number of Atoms.
Molar mass and Avogadro's number can be used to interconvert mass, moles, and number of atoms (particles). While the molar mass of different substances is different, each will contain the same number of particles (Avogadro's number).

STUDY QUESTION 2.14
Vanadium's molar mass is 50.94 g. How many atoms are present in 0.750 g of vanadium?

Solution: We need to convert the 0.750 g to moles using molar mass, then we can use Avogadro's number to convert the moles to atoms.

$$0.750 \text{ g} \times \frac{1 \text{ mol}}{50.94 \text{ g}} \times \frac{6.0221418 \times 10^{23} \text{ atoms}}{1 \text{ mol}} = 8.87 \times 10^{21} \text{ atoms}$$

STUDY QUESTION 2.15
How many g would 3.011×10^{22} atoms of rhodium, Rh, weigh?

Solution: We must convert the atoms to moles using Avogadro's number, then convert the moles to grams using the molar mass of Rh.

$$3.011 \times 10^{22} \text{ atoms} \times \frac{1 \text{ mol}}{6.0221418 \times 10^{23} \text{ atoms}} \times \frac{102.905 \text{ g}}{1 \text{ mol}} = 5.145 \text{ g}$$

PRACTICE QUESTION 2.11
What is the molar mass of an element if 1.598×10^{22} atoms weigh 2.36 g?

PRACTICE QUESTION 2.12
How many atoms are present in 1.450 g of platinum, Pt?

PRACTICE QUESTION 2.13
The diamond (pure C) in Suzie's ring weighs 0.1520 g while that in Erica's ring contains 3.81×10^{22} atoms of C. Whose ring has the larger diamond?

PRACTICE QUIZ

1. Complete the following isotope table:

Name	Symbol	Number of protons	Number of electrons	Number of neutrons	Mass Number
sodium	^{23}Na	11			
	^{40}Ar			22	
arsenic					75
lead				120	

2. What is the total number of protons, neutrons, and electrons in an atom of $^{56}_{26}$ Fe?

3. What is the mass number of a copper atom that has 35 neutrons?

4. List the number of protons, neutrons, and electrons in atoms of the following.
    a. $^{17}_{8}$ O   b. $^{107}_{47}$ Ag   c. $^{222}_{86}$ Rn

5. Which of the following are isotopes of element X?
    $^{46}_{20}$ X, $^{20}_{46}$ X, $^{43}_{20}$ X, $^{46}_{43}$ X

6. Rank the following in order of decreasing stability.
    ^{46} Sc, ^{238} U, ^{60} Ni, ^{209} Po

7. What is the mass, in grams, of 1.5×10^{22} atoms of P?

8. What is the mass, in grams, of 10^{13} carbon atoms?

9. What is the mass of a single atom of fluorine in grams?

Chapter 3
Quantum Theory and the Electronic Structure of Atoms

PROBLEM-SOLVING STRATEGIES AND TUTORIAL SOLUTIONS

TYPES OF PROBLEMS

Problem Type 1: Properties of Waves.

Problem Type 2: Quantization of Energy.

Problem Type 3: Bohr's Theory of the Hydrogen Atom.
        A. Atomic Line Spectra.
        B. The Line Spectrum of Hydrogen.

Problem Type 4: The De Broglie Hypothesis.

Problem Type 5: Quantum Numbers.
        A. Principal Quantum Number (n).
        B. Angular Momentum Quantum Number (ℓ).
        C. Magnetic Quantum Number (m_ℓ).
        D. Electron Spin Quantum Number (m_s).

Problem Type 6: Atomic Orbital Energies and Electron Configurations.
        A. Energies of Orbitals in One-Electron Systems.
        B. Energies of Orbitals in Many-Electron Systems.
        C. The Aufbau Principle.
        D. Hund's Rule.
        E. General Rules for Writing Electron Configurations.

PROBLEM TYPE 1: PROPERTIES OF WAVES.

What we commonly refer to as "light" is actually the visible portion of the electromagnetic spectrum. Light travels in waves. There are three fundamental properties that are used to characterize waves: wavelength (λ), frequency (ν), and amplitude. The distance between the maxima in the waves is called the **wavelength**, λ (lambda). Remember, the SI unit for distance is the meter (m), though it is more common for wavelengths to be reported in nanometers (nm). The **frequency**, ν (nu), of a wave is the number of crests (or troughs) that pass a fixed point per second. The unit of frequency is cycles per second, which is written as 1/s or s^{-1}. The SI unit for frequency is the hertz (Hz). The product of the wavelength and the frequency of any wave is the velocity or speed of the wave, usually in meters per second. All electromagnetic waves travel at the same speed through a vacuum. This speed, known as the *speed of light*, is 3.00×10^8 m/s and has the symbol c.

$$c = \lambda \nu$$

Long wavelength radiation, such as radio waves, has low frequency; short wavelength radiation, such as X rays, has high frequency. Visible radiation covers a very small range of wavelengths (400 – 700 nm) in the middle of the infinite electromagnetic spectrum.

STUDY QUESTION 3.1
Domestic microwave ovens generate microwaves with a frequency of 2.45 GHz. What is the wavelength of this microwave radiation?

Solution: The equation relating the wavelength to the frequency is $c = \lambda v$, where c is the speed of light. Rearranging and substituting, we get:

$$\lambda = \frac{c}{v} = \frac{3.00 \times 10^8 \text{ m/s}}{2.45 \text{ GHz}} \times \frac{1 \text{ GHz}}{10^9 / \text{s}} = 0.122 \text{ m}$$

PRACTICE QUESTION 3.1
What is the wavelength of electromagnetic radiation having a frequency of 5.0×10^{14}/s?

PRACTICE QUESTION 3.2
What is the frequency of light that has a wavelength of 750 nm?

PROBLEM TYPE 2: QUANTIZATION OF ENERGY.

The work of Max Planck in 1900 successfully described the entire wavelength range of radiation emitted by hot solids by constraining energy to be absorbed and emitted in discrete packets called **quanta** (plural of **quantum**). A single quantum of energy is defined as proportional to the frequency, v (nu), of the radiation.

$$E = hv$$

In this equation, h is Planck's constant and it has a value of 6.63×10^{-34} J·s.

STUDY QUESTION 3.2
The orange light given off by a sodium vapor lamp has a wavelength of 589 nm. What is the energy of a single quantum of this radiation?

Solution: The energy of a quantum is proportional to its frequency. Since wavelength is given here, substitute c/λ into Planck's equation for v.

$$E = hv = \frac{hc}{\lambda}$$

Substitute values for Planck's constant and the speed of light into the equation. The units will not cancel unless the wavelength is converted from nanometers into meters.

$$E = \frac{(6.63 \times 10^{-34} \text{ J·s})(3.00 \times 10^8 \text{ m/s})}{(589 \times 10^{-9} \text{ m})} = 3.38 \times 10^{-19} \text{ J}$$

PRACTICE QUESTION 3.3
What is the energy of a quantum of radiation having a frequency of 6.2×10^{14}/s?

PRACTICE QUESTION 3.4
The red line in the spectrum of lithium occurs at 670.8 nm. What is the energy of a quantum of this light? What is the energy of 1 mole of these quanta?

PROBLEM TYPE 3: BOHR'S THEORY OF THE HYDROGEN ATOM.

Scientists had known for years that each element emits a unique spectrum (series of radiation frequencies), but the source of the emission was not understood until 1913 when Niels Bohr predicted the emission spectrum of the hydrogen atom. An emission spectrum is the light given off by an object when it is excited thermally; it may be continuous, including all the wavelengths within a particular range, or it may be a line spectrum, consisting only of certain discrete wavelengths. Bohr kept the popular solar system-inspired picture of an electron making a circular orbit around the nucleus, but restricted the electron to a finite set of **orbits** (n), each associated with a specific radius and energy.

A. **Atomic Line Spectra.**
    When energy is absorbed by the atom, the electron must jump to a higher energy orbit. The lowest energy for the atom occurs when the electron is closest to the nucleus, when $n = 1$. This is called the **ground state**. Hydrogen atoms that have an electron in a higher energy orbit are called **excited state** atoms. Bohr was able to calculate the radii of the allowed orbits and their energies. The energies of the H atom are given by:

$$E_n = -R_H \left(\frac{1}{n^2} \right)$$

where R_H is the Rydberg constant, which has a value equal to 2.18×10^{-18} J. The integer n is a label for each electron orbit. The negative sign in the equation means that the energy of the hydrogen atom is lower than that of a completely separated proton and electron for which the force of attraction is zero.

B. **The Line Spectrum of Hydrogen.**
    Bohr attributed the lines of the hydrogen emission spectrum to the radiation emitted by the atom as the electron drops from a higher to a lower energy orbit. In order for energy to be conserved, the energy lost as radiation must equal the difference in the energies of the initial and final electron orbits. For the emission process, E_i and n_i (i stands for initial) represent the higher atomic energy and larger orbit radius; n_f and E_f (f stands for final) represent the lower energy and smaller radius. Substituting the expression for each energy level, the change in energy of the atom ΔE_{atom} is

$$\Delta E_{atom} = E_f - E_i = -R_H \left(\frac{1}{n_f^2} - \frac{1}{n_i^2} \right)$$

Keep in mind that this equation is only good for the hydrogen atom. The energy of the photon is always positive. Once the energy of a photon is known, the wavelength and frequency can be determined from the equations we discussed previously. It is also important to notice that the energy difference between successive orbits decreases as they are farther away from the nucleus.

STUDY QUESTION 3.3
What amount of energy, in joules, is lost by a hydrogen atom when an electron transition from $n = 3$ to $n = 2$ occurs in the atom?

Solution: The energy of the atom is quantized and depends on the orbit of the electron. Each orbit is assigned a principal quantum number n:

$$E_n = -R_H \left(\frac{1}{n^2} \right)$$

For this transition $n_i = 3$ and $n_f = 2$, the change in energy of the electrons can be determined using:

$$\Delta E_{atom} = E_f - E_i = -R_H \left(\frac{1}{n_f^2} - \frac{1}{n_i^2} \right), \text{ where } n_i = 3 \text{ and } n_f = 2.$$

Substitution gives:

$$\Delta E_{atom} = -2.18 \times 10^{-18} \text{ J} \left(\frac{1}{2^2} - \frac{1}{3^2} \right) = -3.03 \times 10^{-19} \text{ J}$$

STUDY QUESTION 3.4
What is the wavelength of the light emitted by the transition described in Study Question 3.3?

Solution: The energy lost by the atom appears as a photon of radiation with its respective frequency and wavelength. Since wavelengths are positive quantities, we take the absolute value of the change in energy for the atom just calculated.

$$|\Delta E_{atom}| = E_{photon} = \frac{hc}{\lambda}$$

$$\lambda = \frac{hc}{\Delta E_{atom}} = \frac{hc}{3.03 \times 10^{-19} \text{ J}}$$

$$\lambda = \frac{(6.63 \times 10^{-34} \text{ J} \cdot \text{s})(3.00 \times 10^8 \text{ m/s})}{3.03 \times 10^{-19} \text{ J}}$$

$$\lambda = 6.56 \times 10^{-7} \text{ m} = 656 \text{ nm}$$

PRACTICE QUESTION 3.5
Consider the first four orbits in a H–atom, $n = 1$ through $n = 4$.

    a.   **How many emission lines are possible as a result of electron transitions?**
    b.   **Which transition produces photons of the greatest energy?**
    c.   **Which transition produces the emission line with the longest wavelength?**

PRACTICE QUESTION 3.6
What wavelength of radiation will be emitted when an electron in a hydrogen atom jumps from the $n = 5$ to the $n = 1$ principal energy level? Name the region of the electromagnetic spectrum corresponding to this wavelength.

PRACTICE QUESTION 3.7
Which is higher energy, 5.0×10^{14}/s or 750 nm?

PROBLEM TYPE 4: THE DE BROGLIE HYPOTHESIS.

Bohr's theory works for H atoms but fails to predict spectra of any other elements. Moreover, no one, including Bohr, could justify the quantization of electron orbits or atomic energy levels. In 1924 Louis de Broglie hypothesized that if photons can have the properties of particles as well as waves, electrons could have the properties of waves as well as particles! De Broglie deduced that the wavelength λ of a particle depends on its mass, m, and velocity, u:

$$\lambda = \frac{h}{mu}$$

STUDY QUESTION 3.5

When an atom of Th-232 undergoes radioactive decay, an alpha particle which has a mass of 4.0 amu is ejected from the thorium nucleus with a velocity of 1.4×10^7 m/s. What is the de Broglie wavelength of the alpha particle?

Solution: The de Broglie wavelength depends on the mass and velocity of the particle:

$$\lambda = \frac{h}{mu}$$

Because Planck's constant has units of J · s, and 1 J = 1 kg · m²/s², the mass of the alpha particle should be expressed in kg. If 1 alpha particle has a mass of 4.0 amu, then 1 mole of alpha particles has a mass of 4.0 g. You need the mass of one alpha particle in kg and this is calculated as follows:

$$\frac{4.00 \text{ g}}{1 \text{ mol}} \times \frac{1 \text{ kg}}{1000 \text{ g}} \times \frac{1 \text{ mol}}{6.022 \times 10^{23} \text{ particles}} = 6.64 \times 10^{-27} \text{ kg/particle}$$

The wavelength of this alpha particle is:

$$\lambda = \frac{h}{mu} = \frac{(6.63 \times 10^{-34} \text{ J} \cdot \text{s}) \times \left(\dfrac{1 \text{ kg} \cdot \text{m}^2/\text{s}^2}{1 \text{ J}} \right)}{(6.64 \times 10^{-27} \text{ kg})(1.4 \times 10^7 \text{ m/s})}$$

Notice that all the units cancel except meters. The wavelength is 7.1×10^{-15} m.

> ### PRACTICE QUESTION 3.8
> **Calculate the wavelength, in nanometers, of a proton (mass = 1.6725×10^{-27} kg) that is moving at 10% of the speed of light.**

PROBLEM TYPE 5: QUANTUM NUMBERS.

Erwin Schrödinger developed a theory for computing electron energies based on the idea that we don't have to specify the position or path of the electron in order to compute its energy and predict spectra. The solutions (wave functions) of the Schrödinger equation specify the energy levels and spatial distributions of the electron in the hydrogen atom. In contrast to the circular orbits of the Bohr solar model, the wave function specifies an atomic orbital, a representation of the volume an electron can occupy. The Schrödinger equation has an exact solution for the hydrogen atom, but can only be solved using an approximation method for atoms with more than one electron. The approximation method uses the Schrödinger equation solutions for the hydrogen atom as approximations to describe electrons in larger atoms. The probability that an electron can be found in a particular region around an atom is called **orbital** or **electron density**.

Each solution of the Schrödinger equation is specified by the values of three variables called **quantum numbers**. We use the quantum numbers as symbols of the wave functions, the atomic orbitals they represent or the electrons that reside in them. There is a hierarchy, or "chain-of-command" among the quantum numbers. The values that we can assign to some of them will be dependent on the value of the other quantum numbers.

A. **Principal quantum number, (n.)**
   The principal quantum number can take any positive integer value: $1 \geq n > \infty$. The principal quantum number reflects the average distance of an electron from the nucleus. The collection of orbitals having the same value of n is known as an orbit (same as from the Bohr model), level or shell. All the known elements have n between 1 and 7.

B. **Angular momentum quantum number, (ℓ.)**
   The angular momentum quantum number can take integer values between zero up to a maximum of $(n-1)$. It is very easy to remember that the value that ℓ can take depends on the value of n and that ℓ can never be equal to or larger than n. The angular momentum quantum number reflects the shape of the electron distribution. (The boundaries of an electron cloud need not be well defined for an orbital to have an overall shape.)

   The following table correlates the value of ℓ to the letter designation and shape.

ℓ	letter	shape
0	s	spherical
1	p	dumbbell or figure 8
2	d	clover-leaf*
3	f	multi-lobed

   *Most but not all *d*-orbitals are clover-leaf shaped. The d_z^2 orbital is a different shape.

C. **Magnetic quantum number, (m_ℓ.)**
   The magnetic quantum number can have integer values between $-\ell$ and $+\ell$. The number of m_ℓ values is important because it designates the number of orbitals within a subshell. For example, when $\ell = 1$ (*p*-orbital), there are three m_ℓ values, -1, 0, and $+1$. Each m_ℓ value corresponds to a different *p*-orbital. The three *p*-orbitals all have the same energy (**degenerate**), but have different orientations in space. These would be designated as p_x, p_y, and p_z.

D. **Electron Spin Quantum Number, (m_s.)**
   The fourth quantum number comes from an extension of applying the Schrödinger equation to atoms with multiple electrons. The **Pauli Exclusion Principle** states that *no two electrons in an atom can have the same set (all four) quantum numbers*. This principle limits the number of electrons in each atomic orbital to two. The electron spin quantum number, m_s, designates one of two possible spin directions for electrons. (Remember that spinning charges induce magnetic fields.) The two possible values of m_s are $+1/2$ and $-1/2$.

STUDY QUESTION 3.6
If the principle quantum number of an electron is $n = 2$, what are allowed values for the following quantum number?

   a. ℓ

   b. m_ℓ

   c. m_s

Solution:
a. Recall that the angular momentum quantum number ℓ has values that depend on the value of the principal quantum number n. $\ell = 0, 1, 2$, on up to the highest value $(n - 1)$, which can never be equal to or greater than n. When $n = 2$, then $\ell = 0, 1$.
b. The magnetic quantum number m_ℓ depends on the value of ℓ where $m_\ell = -\ell, -\ell +1, ..., 0, ..., \ell + 1, \ell$. There are two ℓ values and so two sets of m_ℓ values. When $\ell = 1$, then $m_\ell = -1, 0, 1$; and when $\ell = 0$, then $m_\ell = 0$.
c. The m_s quantum number for a single electron can be either $+1/2$ or $-1/2$.

STUDY QUESTION 3.7
List all the possible types of orbitals associated with the principle energy level $n = 4$.

Solution: The type of orbital is given by the angular momentum quantum number. When $n = 4$; $\ell = 0, 1, 2$, and 3. These correspond to subshells (orbitals) of $s, p, d,$ or f.

STUDY QUESTION 3.8
For the following sets of quantum numbers, indicate which quantum numbers (n, ℓ, m_ℓ) could not occur and state why.

    a.   3, 2, 2
    b.   2, 2, 2
    c.   2, 0, –1

Solution:
a. When $n = 3$, then $\ell = 0, 1, 2$. When $\ell = 2$; then $m_\ell = -2, -1, 0, 1, 2$. Set (a) can occur.
b. When $n = 2$, then $\ell = 0, 1$. The set in (b) cannot occur because $\ell \neq 2$ when $n = 2$. This means that the $n = 2$ principle level cannot have d orbitals.
c. When $n = 2$, $\ell = 0, 1$. When $\ell = 0$, $m_\ell = 0$. The set in (c) cannot occur because $m_\ell \neq -1$ when $\ell = 0$.

STUDY QUESTION 3.9
What are the possible values of m_ℓ for a

    a.   **3d orbital?**
    b.   **2s orbital?**

Solution: The values of m_ℓ depend only on ℓ. First, convert each orbital designation to the corresponding value of ℓ, then list allowed values for m_ℓ.
a. For a d orbital $\ell = 2$; therefore, possible m_ℓ values are $-2, -1, 0, 1,$ and 2.
b. For an s orbital $\ell = 0$; therefore, $m_\ell = 0$.

> ### PRACTICE QUESTION 3.9
> **Which of the following sets of quantum numbers is/are not allowed for describing an electron in an orbital?**
>
> | | n | ℓ | m_ℓ | m_s |
> |---|---|---|---|---|
> | a. | 3 | 2 | –3 | 1/2 |
> | b. | 2 | 3 | 0 | –1/2 |
> | c. | 2 | 1 | 0 | –1/2 |
>
> ### PRACTICE QUESTION 3.10
> **How many orbitals in an atom can have the following designations?**
>
>   a. 2s b. 3d c. 4p d. $n = 3$
>
> ### PRACTICE QUESTION 3.11
> **How many electrons in an atom can have the following designations?**
>
>   a. 2s b. 3d c. 4p d. $n = 3$

PROBLEM TYPE 6: ATOMIC ORBITAL ENERGIES AND ELECTRON CONFIGURATIONS.

As we have seen, atomic orbitals come in different sizes, shapes, and orientations about a set of coordinates which are dependent upon the values of n, ℓ, and m_ℓ. There are several factors that affect the energies of the orbitals within a level or shell. Some of these factors include: the charge of the nucleus, the ability of the orbital to lie closer to the nucleus (penetration), the presence of other electrons in orbitals that have a smaller size and are closer to the nucleus (shielding), the presence of electrons occupying the orbitals.

A. Energies of Orbitals in One-Electron Systems.

The energies of all the orbitals in a level or shell are the same for a hydrogen atom. The electron density distributions in the same shell are different for s and p orbitals, but they have the same energy. In other words, the energies of the orbitals increase <u>only</u> with n:

$$1s < 2s = 2p < 3s = 3p = 3d < 4s = 4p = 4d = 4f < 5s \cdots$$

We can write an atomic formula that reflects the energy and distribution for every electron in every element on the periodic chart. We will see directly that the similarity in chemical and physical properties associated with the elements in a group arise from similarities in the arrangement of the electrons around the nucleus of those atoms. We call this arrangement the electron configuration. The ground state electron configuration of H is $1s^1$, as illustrated.

B. Energies of Orbitals in Many-Electron Systems.

When an atom contains more than one electron, the orbital energies in a shell are no longer equal. This inequality of orbital energies is due to electron repulsion and electrostatic shielding. Interestingly, the energy associated with orbitals in different shells starts blending where a higher energy orbital (d– or f–orbital) in a smaller shell can have higher energy than a lower energy orbital (s–orbital) of a larger shell. For example, the energy of an atom is lower when the $4s$ orbital is filled before the $3d$. Consequently, the orbital energies depend on <u>both</u> n and ℓ. The order of the energy levels in many electron atoms becomes:

$$1s < 2s < 2p < 3s < 3p < 4s < 3d < 4p < 5s < 4d < 5p < 6s < 4f < 5d \ldots$$

C. The Aufbau Principle.

The electron configurations of many-electron atoms are constructed by adding an electron to the next (lowest energy) empty orbital. This is called the **Aufbau Principle** or building up principle. So, the electron configuration of He is $1s^2$.

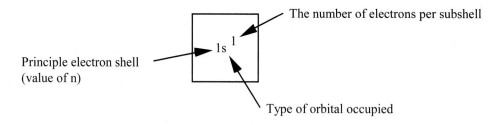

The number of electrons per subshell

Principle electron shell (value of n)

Type of orbital occupied

STUDY QUESTION 3.10
Write the electron configuration for a potassium atom.

Solution: Potassium has 19 electrons. Looking at the position of potassium in the Periodic Table we see that it is an element in Group 1A and it is in period 4. Group 1A elements will have their last electron in an *s*–orbital. Since potassium is in period 4, we know that the last subshell occupied will be the 4*s* orbital. Now, let's take a look at the sequence of orbitals in terms of energy given above.

$$1s < 2s < 2p < 3s < 3p < 4s < 3d < 4p < 5s < 4d < 5p < 6s < 4f < 5d < \ldots$$

A ground-state electron configuration should have all of the subshells of lower energy filled to their maximum capacity. This means we will have 2 electrons in every *s* subshell of lower energy, 6 electrons in every *p* subshell of lower energy, and if any *d* subshells were filled they would contain 10 electrons. Since the last subshell filled for K is the 4*s* subshell, each subshell that is lower in energy will be fully occupied. The electron configuration of potassium is:

$$\text{K} \qquad 1s^2 2s^2 2p^6 3s^2 3p^6 4s^1$$

You can always check that the sum of the superscripts written in the electron configuration correspond to the number of electrons in the neutral atom or ion you are writing it for.

STUDY QUESTION 3.11
Which of the following electron configurations would correspond to ground states and which to excited states?

  a. $1s^2 2s^2 2p^1$
  b. $1s^2 2p^1$
  c. $1s^2 2s^2 2p^1 3s^1$

Solution:
a. In this configuration, electrons occupy the lowest possible energy shells. It corresponds to a ground state.
b. In this configuration electrons do not occupy the lowest possible energy shells. The 2*s* orbital which lies lower than the 2*p* is vacant, and the last electron is in the 2*p* orbital. This is an excited state.
c. In this case the last electron is in the 3*s* orbital while the lower energy 2*p* subshell is not completely filled. This is an excited state.

> ### PRACTICE QUESTION 3.12
> **Write the ground-state electron configuration for the following atoms: Sb, V, Pb.**

As you can see, the electron configurations for the large elements are long and cumbersome. The electron configuration of the preceding noble gas can be used as an abbreviation when writing electron configurations.

STUDY QUESTION 3.12
Write the abbreviated configuration for the following atoms.

  a. Sn
  b. P
  c. Sg

Solution:
a. $[Kr]5s^2 4d^{10} 5p^2$
b. $[Ne]3s^2 3p^3$
c. $[Rn]7s^2 5f^{14} 6d^4$

> ### PRACTICE QUESTION 3.13
> **Write the abbreviated configuration for the following atoms.**
>
>   a. Ti   b. Os   c. Br

D. Hund's Rule.

A particularly useful way to represent the arrangement of electrons into orbitals for any atom or ion is done through an orbital diagram. An **orbital diagram** groups boxes that represent individual orbitals, to designate subshells. An arrow pointing up ($\uparrow$) stands for an electron spinning in one direction, $m_s = +1/2$, and an arrow pointing down ($\downarrow$) stands for an electron spinning in the opposite direction, $m_s = -1/2$. Electrons are placed into orbitals according to **Hund's rule**, which states that electrons fill degenerate orbitals (multiple orbitals of same energy) singly to the maximum extent with the same spins before they begin pairing up. For example, carbon has a single electron in two of the three (degenerate) 2p orbitals, rather than two electrons paired in the same 2p orbital.

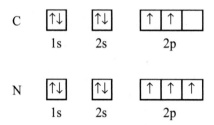

Similarly, Hund's rule predicts that nitrogen atoms have the electrons in the 2p orbitals distributed with one each of the three 2p orbitals. The orbital diagram indicates the number of unpaired electrons in an atom. The presence, or absence, of unpaired electrons is indicated experimentally by the behavior of an element when placed in a magnetic field. **Paramagnetic** elements are attracted to a magnetic field because the atoms of paramagnetic substances contain unpaired electrons. **Diamagnetic** elements are repulsed by a magnetic field because the atoms of diamagnetic substances contain only paired electrons.

STUDY QUESTION 3.13

    a. Write the electron configuration for arsenic.
    b. Draw its orbital diagram.
    c. Are arsenic atoms paramagnetic or diamagnetic?

Solution:

a. From the atomic number of As we see there are 33 electrons per arsenic atom. The order of filling orbitals is 1s, 2s, 2p, 3s, 3p, 4s, etc. Placing electrons in the lowest energy orbitals until they are filled, the first 18 electrons are arranged as $1s^2 2s^2 2p^6 3s^2 3p^6$, corresponding to an Ar core. The next two enter the 4s, and the next ten enter the 3d. This leaves 3 electrons for the 4p subshell. The electron configuration of As is:

$$\text{As} \qquad [\text{Ar}] 4s^2 3d^{10} 4p^3$$

b. All of the orbitals are filled except for the 4p orbitals. The electrons must be placed into the 4p orbitals according to Hund's rule. The diagram for the outer orbitals is:

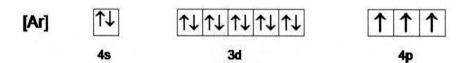

c. Arsenic atoms are paramagnetic because they contain 3 unpaired electrons.

E. General Rules for Writing Electron Configurations.

1. Each shell (principle level) contains n subshells and a maximum total of $2n^2$ electrons.
2. Each subshell contains $2\ell+1$ orbitals.
3. Electrons fill degenerate (equal energy) orbitals 1 at a time.
4. Each orbital can hold up to 2 electrons.
5. Pairs of electrons in same orbital have opposite spins.
6. The symbol of the preceding noble gas can be used to abbreviate the electron configuration of the core electrons.

In most cases, you should be able to write the electron configuration of any element on the periodic table. Transition metals, lanthanides and actinides often exhibit non-sequential ordering of subshells; however, the rules above will give correct electron configurations in most cases.

> **PRACTICE QUESTION 3.14**
> **What third period element has atoms in the ground state with three unpaired electrons?**

> **PRACTICE QUESTION 3.15**
> **How many fourth period elements have atoms in the ground state with three unpaired electrons?**

PRACTICE QUIZ

1. Calculate the wavelength of a quantum that has energy of 2.05×10^{-19} J.

2. Silver bromide (AgBr) is the light-sensitive compound in most photographic films. Assume, when film is exposed, that the light energy absorbed dissociates the molecule into atoms. (The actual process is more complex.) If the energy of dissociation of AgBr is 100 kJ/mol, find the wavelength of light that is just able to dissociate AgBr.

3. A hydrogen emission line in the ultraviolet region of the spectrum at 95.2 nm corresponds to a transition from a higher energy level n to $n = 1$. What is the value of n for the higher energy level?

4. The second line in the Balmer series (transitions to $n = 2$) occurs at 486.1 nm. What is the energy difference between the initial and final energy levels involved in the electron transition?

5. Which choice is a possible set of quantum numbers for the highest energy electron in a gallium (Ga) atom in its ground state?

	n	ℓ	m_ℓ	m_s
a.	4	2	0	$-1/2$
b.	4	1	0	$1/2$
c.	4	2	-2	$-1/2$
d.	3	1	$+1$	$1/2$
e.	3	0	0	$-1/2$

6. Which of the following is/are incorrect designations for an atomic orbital?
    a. 3f b. 4s c. 2d d. 4f

7. For each of the following, give the subshell designation, the m_ℓ values, and the number of possible orbitals.
    a. $n = 3$ $\ell = 2$
    b. $n = 4$ $\ell = 3$
    c. $n = 5$ $\ell = 1$

8. For each of the following subshells, give the n and ℓ values, and the number of possible orbitals.
    a. 2s b. 3p c. 4f

9. What element has atoms with the electron configuration $[Xe]6s^2 4f^{14} 5d^{10} 6p^2$?

10. Write the complete and abbreviated electron configurations for rubidium, Rb.

11. What fourth period element has atoms in the ground state with six unpaired electrons?

12. How many unpaired electrons do sulfur atoms have?

Chapter 4
Periodic Trends of the Elements

PROBLEM-SOLVING STRATEGIES AND TUTORIAL SOLUTIONS

TYPES OF PROBLEMS

Problem Type 1: Classification of Elements.

Problem Type 2: Effective Nuclear Charge.

Problem Type 3: Periodic Trends in Properties of Atoms.
        A. Atomic Radius.
        B. Ionization Energy.
        C. Electron Affinity.
        D. Metallic Character.

Problem Type 4: Electron Configuration of Ions.
        A. Ions of Main group Elements.
        B. Ions of *d*–Block Elements.

Problem Type 5: Ionic Radius.
        A. Comparing Ionic Radius with Atomic Radius.
        B. Isoelectronic Series.

PROBLEM TYPE 1: CLASSIFICATION OF ELEMENTS.

The periodic table can be divided into the main group elements (also known as the representative elements) and the transition metals. It is further divided into smaller groups or columns of elements that all have the same configuration of valence electrons. The 18 columns of the periodic table are labeled 1A through 8A (*s*–and *p*–block elements) and 1B through 8B (*d*–block elements), or by the numbers 1 through 18.

The arrangement of the modern periodic table corresponds to the type of subshells being filled by the atom. Groups 1A through 7A (the representative or main group elements) of the periodic table include elements that have incompletely filled sets of *s* or *p* orbitals. The noble gases have completely filled outer *ns* and *np* subshells, ns^2np^6, except for helium. The **valence electrons** are those that are in the outermost shell of the atom, which determine how the atoms will interact with another atom. The number of valence electrons for a main group element is the same as its group number.

Transition metal atoms of elements in Groups 3B–8B and 1B have incompletely filled *d* subshells. The elements Zn, Cd, and Hg (Group 2B) are often not considered true transition elements, because their *d* subshells are completed.

STUDY QUESTION 4.1
Without consulting the periodic table, write the ground state-electron configuration and block designation for an atom with:

    a.   **20 electrons.**
    b.   **39 electrons.**
    c.   **49 electrons**

Solution:

Using the Aufbau principle and Figure 3.23, we arrive at:
   a. $1s^2 2s^2 2p^6 3s^2 3p^6 4s^2$, s-block, alkaline earth metal.
   b. $1s^2 2s^2 2p^6 3s^2 3p^6 4s^2 3d^{10} 4p^6 5s^2 4d^1$, d-block, transition metal
   c. $1s^2 2s^2 2p^6 3s^2 3p^6 4s^2 3d^{10} 4p^6 5s^2 4d^{10} 5p^1$, p-block, main group element

PRACTICE QUESTION 4.1
Name an element with valence configuration of $ns^2 np^1$.

PRACTICE QUESTION 4.2
Write the ground-state electron configuration for gallium.

PROBLEM TYPE 2: EFFECTIVE NUCLEAR CHARGE.

Nuclear charge (Z) is the number of protons in the nucleus of an atom. Effective nuclear charge (Z_{eff}) is the nuclear charge that is "felt" by the valence electrons. The outermost electrons in an atom do not feel the full positive charge of the nucleus. Z_{eff} is usually lower than the nuclear charge due to shielding by the core electrons. Electrons in inner shells, lying between the nucleus and the outermost electrons, tend to shield the outermost electrons from the nuclear charge. In a sodium atom, for example, the inner 10 electrons ($1s^2 2s^2 2p^6$) shield the outer 3s electron from the positive charge of 11 protons. The Z_{eff} experienced by the 3s electron is only about +1. The Z_{eff} is equal to the nuclear charge Z minus the shielding constant σ. The shielding constant has a value between 0 and Z. Therefore, $Z_{eff} = Z - \sigma$.

STUDY QUESTION 4.2
Predict which atom in each pair would have greater effective nuclear charge, Z_{eff}.
   a.   **Li or Rb**
   b.   **F or Cl**
   c.   **Na or S**

Solution: To determine which atom would have a greater Z_{eff}, we must consider the valence shell, number of shielding electrons and the nuclear charge. As we step down a group, the number of core shells increases. We know that as the number of core shells increases, Z_{eff} decreases. As we move from left to right within a period, the amount of shielding from core electrons stays the same, but the nuclear charge increases. Therefore, Z_{eff} increases from left to right on the periodic table.
a. Li has only 2 shielding electrons and valence shell of $n = 2$. Rb has 36 shielding electrons and valence shell of $n = 5$. Li has greater Z_{eff}.
b. F has only 2 shielding electrons and valence shell of $n = 2$. Cl has 10 shielding electrons and valence shell of $n = 3$. Cl has greater Z_{eff}.
c. Na and S both have 10 shielding electrons and valence shell of $n = 3$. Na has nuclear charge of +11 while S has nuclear charge of +16. S has greater Z_{eff}.

PRACTICE QUESTION 4.3
Predict which atom in each pair would have greater effective nuclear charge, Z_{eff}.

   a.   **Li or Ar**
   b.   **Ru or O**
   c.   **C or P**

PROBLEM TYPE 3: PERIODIC TRENDS IN PROPERTIES OF ATOMS.

A. Atomic Radius.

Atomic radius is the distance between an atom's nucleus and its valence shell. The atomic radius of a metal atom is defined as the metallic radius, which is one-half the distance between adjacent, identical nuclei in a metal solid. The atomic radius of a nonmetal is defined as the covalent radius, which is one-half the distance between adjacent, identical nuclei in a substance. In general, atomic radii increase from top to bottom down a group and decrease from left to right across a period of the periodic table. Thus, the alkali metals have the largest atoms and the noble gases the smallest.

Atomic radius is greatly influenced by Z_{eff}. The greater the Z_{eff}, the more pull on the valence electrons. In turn, the valence electrons are drawn closer to the nucleus making the atom smaller.

STUDY QUESTION 4.3
Which has the smallest atomic radius, Li, Na, Be, or Mg?

Solution: Atomic radii increase from top to bottom in a group, therefore, Li atoms are smaller than Na atoms, and Be atoms are smaller than atoms of Mg. Next, compare Be and Li. The atomic radius decreases from left to right within a period of the periodic table, thus, Be atoms are smaller than Li atoms, as well as the other choices given.

STUDY QUESTION 4.4
Arrange the following atoms in order of increasing atomic radius: Na, Cl, Ne, S, and Li.

Solution: Increasing order means we should start with the smallest and end with the largest atomic radius. Atomic radius increases from top to bottom in a group. We have already determined that Li is smaller than Na. Within the same period, atomic radius decreases from left to right. In the second period, Ne is smaller than Li. In the third period, Cl is smaller than S, which is smaller than Na.

$$Ne < Li < Cl < S < Na$$

PRACTICE QUESTION 4.4
Which of the following is a smaller atom? Na, Ar, Ru, O

PRACTICE QUESTION 4.5
Without looking up atomic radii, arrange the following atoms in order of decreasing atomic radius.
      Ar   Li   Se   Y

PRACTICE QUESTION 4.6
Na is a larger atom than Tl, atomic radii of 186 pm and 170 pm, respectively. What is a reasonable explanation for this fact?

B. Ionization Energy.

Ionization energy (IE) is the minimum energy required to remove an electron from the ground state of an atom in the gas phase. The magnitude of the ionization energy is a measure of how strongly the outermost electron is held by an atom. The greater the ionization energy, the more strongly/tightly the electron is held.

The first ionization energy (IE) is smaller than subsequent ionization energies [e.g., second (IE_2), third (IE_3), and so on]. The first ionization of any atom removes a valence electron. An atom with a net positive charge is a called a **cation**. Ionization energies increase dramatically when core electrons are being removed. Ionization energy is a chemical property, and ionization is a chemical process.

First ionization energies (IE_1 values) tend to increase left to right across a period and decrease down a group. Exceptions to this trend can be explained based upon the electron configuration of the element. The alkali metals have the lowest ionization energy and the noble gases the highest.

An easy way to remember the periodic trends discussed so far is that as the atoms become smaller, their first ionization energy increases. In atoms of smaller size, their outermost electrons experience a greater attraction towards the nucleus making them "harder" to remove.

STUDY QUESTION 4.5
Which has the highest first ionization energy, K, Br, Cl, or S?

Solution: Ionization energy increases from left to right within a period. Thus, the value for Cl is greater than for S, and the value for Br is greater than for K. In comparing Cl and Br, Cl has the higher ionization energy value because ionization energy decreases from top to bottom within a group.

STUDY QUESTION 4.6
The first, second, and third ionization energies for calcium are:

$$IE_1 = 590 \text{ kJ/mol} \qquad IE_2 = 1145 \text{ kJ/mol} \qquad IE_3 = 4900 \text{ kJ/mol}$$

Explain why so much more energy is required to remove the third electron from Ca, as compared to removal of the first and second electrons.

Solution: The successive ionization energy values follow the usual trend $IE_3 > IE_2 > IE_1$. However, we must explain the very large difference between IE_2 and IE_3. The first two electrons are valence electrons and are removed from the $3s$ orbital. But the third electron must be removed from the inner $2p$ subshell. The $n = 2$ principle energy level lies much closer to the nucleus than the $n = 3$ energy level; therefore, its electrons are held much more strongly. Consequently, $IE_3 \gg IE_2$.

PRACTICE QUESTION 4.7
Based on the periodic trends, which atom has the lowest first ionization energy? Second ionization energy?

PRACTICE QUESTION 4.8
For silicon atoms, which ionization energy value (IE) will show an exceptionally large increase over the preceding ionization energy value? IE_2, IE_3, IE_5, IE_6

C. Electron Affinity.

Electron affinity (EA) is the energy released when an atom in the gas phase accepts an electron $[A(g) + e^- \rightarrow A^+(g)]$. An ion with a net negative charge is called an **anion**. Electron affinities tend to increase from left to right in a period. As with first ionization energies, one can explain exceptions to the trend based on the electron configuration of the element. Electron affinity is also a chemical property.

As with atomic radius and ionization energy, the trend in electron affinity is related to effective nuclear charge. It is related as well to the energy of the orbital that the electron will enter. The Group 6A and 7A elements have the highest electron affinities. These elements have high Z_{eff} and the added electron enters the valence shell. The noble gases with even higher Z_{eff} have no affinity for an additional electron. With filled s and p subshells, noble gas atoms have no tendency to add an electron because the next available orbital is in a higher energy level beyond the valence shell.

STUDY QUESTION 4.7

Some anomalous behavior in the trend of electron affinity is observed for Group 2A elements. Be and Mg have negative electron affinities while the rest of the elements have small, but positive, electron affinities. Using arguments similar to our discussion for the noble gases, give a reasonable explanation for the anomalies.

Solution: Group 2A elements share a common electron configuration for their outermost energy level of ns^2. We notice that for Be and Mg, addition of a new electron involves using orbitals in a higher energy subshell. Furthermore, it results in losing the stability of a complete subshell (neutral Be: $[He]2s^2$ while Be$^-$: $[He]2s^22p^1$). Thus, it seems that the decrease in stability happens not only when adding electrons to a new energy level like for the noble gases. It also occurs when adding an electron to a new energy subshell, which makes the resulting anion less stable than the neutral atom, thus, a negative electron affinity results. For Ca, Sr, Ba the values are positive although rather small compared to other elements in their vicinity. For these atoms, adding the "extra" electron involves using $(n-1)d$ subshells, which lie closer to the nucleus. It thus seems that the destabilization brought about by adding electrons to the $(n-1)d$ subshell is slightly counterbalanced by the Z_{eff} that those electrons experience.

> *PRACTICE QUESTION 4.9*
> **Which one of the following elements has both low ionization energy and positive electron affinity, K, Ne, Br, Fe, or N?**

D. Metallic Character.

Metals tend to be shiny, lustrous, malleable, ductile, and conducting (for both heat and electricity). Metals typically lose electrons when combining with nonmetals to form cations. **Metallic character** is a periodic trend that relates to the chemical behavior or reactivity of the elements and to some common physical properties. Within a group or column, metallic character increases as we move from top to bottom. For metals, this increase in metallic character is accompanied by an increase in their reactivity. A reactive metal will more readily lose its outermost electrons when reacting with nonmetals. In general, metallic character decreases across a period and increases down a group of the periodic table. Metals have low ionization energies, low electron affinities, and relatively large atomic radii. All metallic elements combine with nonmetals such as oxygen and chlorine to form salts. The most reactive metals are at the left of the periodic table. The transition metals are less reactive than the Group 1A and 2A metals.

Nonmetals vary in color, tend to be brittle, and are generally not good conductors (for either heat or electricity). They can gain electrons to form anions. Eighteen of the elements are nonmetals. These elements are on the right side of the periodic table. At 25 °C, eleven are gases, one (bromine) is a liquid, and the rest are brittle solids. Typically, their densities and melting points are low. Atoms of nonmetallic elements have high ionization energies and high electron affinities. Nonmetals combine with metals to form ionic compounds, and with other nonmetals to form molecular or covalent compounds, as we will see in a later chapter.

Metalloids are elements with properties intermediate between metals and nonmetals. Many periodic tables show a zig–zag line separating the metals from nonmetals. The elements that border this line on both sides are metalloids (except Al, which is a metal). They include boron, silicon, germanium, arsenic, antimony, and tellurium. The metalloids have ionization energies and electron affinity values intermediate between metals and nonmetals.

STUDY QUESTION 4.8
Predict the missing values.

a.

Element	Density (g/cm^3)
Ca	1.55
Sr	?
Ba	3.5

b.

Element	Atomic Radius (nm)
Cl	0.099
Br	0.114
I	?

Solution: Within a group of the periodic table, the physical properties vary in a regular fashion as you read down, in the group.

a. From the positions of these elements in Group 2A of the periodic table, the density of Sr could be estimated to be half-way between the values for Ca and Ba. The average of 1.55 and 3.5 g/cm^3 is 2.5 g/cm^3. The observed value is 2.6 g/cm^3.

b. In this case, we must extrapolate rather than interpolate as above because I is farther down in Group 7A than Cl and Br. We could assume the change in the radius from Br to I will be the same as the change from Cl to Br. The difference between Cl and Br is 0.015 nm. Adding 0.015 to 0.114 gives the estimated radius for I as 0.129 nm. The observed radius of I is 0.133 nm.

> *PRACTICE QUESTION 4.10*
> **Estimate the melting point of Br$_2$(s) given that the melting points of Cl$_2$(s) and I$_2$(s) are –101 °C and 114 °C, respectively.**

PROBLEM TYPE 4: ELECTRON CONFIGURATION OF IONS.

A. Ions of Main Group Elements.

Ions of main group elements are isoelectronic with noble gases. Writing electron configurations for their ions requires knowing which orbital is the last one filled in the neutral atom. For cations of main group elements, the electron removed comes from the outermost orbital (the orbitals with the largest value of n and ℓ). For example, to write the electron configuration for Ca^{2+} we need to remember that the last orbital with electrons in the neutral Ca atom is $4s$ (which has 2 electrons in it). Thus, the two electrons are removed from this orbital.

Ca: $1s^2 2s^2 2p^6 3s^2 3p^6 4s^2$
Ca^{2+}: $1s^2 2s^2 2p^6 3s^2 3p^6$

Notice that the $4s$ orbital is the one with the highest n value. If we had an $ns^2 np^x$ electron configuration, the electrons in the np orbital would be removed first (from the same n value, these are the ones with the largest ℓ value).

When forming anions of main group elements, we just need to note of how many electrons are in the last orbital filled and whether or not we can place additional electrons in it. For example, in an atom of oxygen there are 4 electrons in the $2p$ orbital, which is the last orbital filled. To form the oxide anion, the two "extra" electrons can be placed in the $2p$ sublevel since it can hold a maximum of 6 electrons. The two configurations look like

O: $1s^2 2s^2 2p^4$
O^{2-}: $1s^2 2s^2 2p^6$

STUDY QUESTION 4.9
Write the electron configurations of the following ions.

    a. Mg^{2+}
    b. S^{2-}

Solution: We must first write the electron configuration of the neutral atoms then either remove electrons from (for cations) or add electrons to (for anions) the highest energy orbital.

a. Mg: $1s^2 2s^2 2p^6 3s^2$ or $[Ne]3s^2$. Now, we remove two electrons from the outermost energy level, which is $3s$, to form the 2+ ion. The configuration of Mg^{2+} is: $1s^2 2s^2 2p^6$ or [Ne].

b. S: $1s^2 2s^2 2p^6 3s^2 3p^4$. The sulfur atom becomes the negative S^{2-} ion by gaining two electrons. Adding two electrons to the outermost subshell gives S^{2-}: $1s^2 2s^2 2p^6 3s^2 3p^6$ or [Ar].

PRACTICE QUESTION 4.11
Write the ground-state electron configuration for:

    a. K^+
    b. I^-
    c. Sr^{2+}
    d. Se^{2-}

PRACTICE QUESTION 4.12
Arrange the following ions in order of increasing radius.
K^+ I^- Sr^{2+} Se^{2-}

B. Ions of d-Block Elements.

When a d-block element loses one or more electrons, it loses them first from the shell with the highest principle quantum number (e.g., electrons in the $4s$ orbital are lost before electrons in the $3d$ orbital). Be sure to write the ground-state electron configuration of the metal and then remove the electrons from the highest principle quantum number (n). This means, remove the electron from the outer s–orbital before removing the inner shell d–orbital electron(s).

Mn $[Ar]4s^2 3d^5$ Mn^{2+} $[Ar]3d^5$
Cr $[Ar]4s^1 3d^5$ Cr^{3+} $[Ar]3d^3$
Cu $[Ar]4s^1 3d^{10}$ Cu^+ $[Ar]3d^{10}$

STUDY QUESTION 4.10
Write the electron configurations of the following ions.

    a. Rh^{2+}
    b. Rh^{3+}

Solution: As before, we first write the electron configuration of the neutral atom then remove electrons from the highest energy orbital.

a. Rh: $[Kr]5s^24d^7$. Now, we remove two electrons from the *outermost* energy level, which is $4s$, to form the 2+ ion. The configuration of Rh^{2+} is $[Kr]4d^7$.

b. We need to remove one more electron from the Rh^{2+} ion. That give us Rh^{3+} $[Kr]4d^6$.

NOTE:

The stable ions of all but a few representative elements are isoelectronic to a noble gas. Keep in mind that most transition metals can form more than one cation, and that for the most part, these ions are not isoelectronic with the preceding noble gases.

> *PRACTICE QUESTION 4.13*
> **Write the electron configuration for:**
> _____
> a. Ru^{3+}
> b. Ni^{2+}
> c. Co^{3+}
> _____

PROBLEM TYPE 5: IONIC RADIUS.

A. Comparing Ionic Radius with Atomic Radius.

Ionic radius is the distance between the nucleus and valence shell of a cation or an anion. Because atoms and ions of the same element have different numbers of electrons, we expect atomic radii and ionic radii to be different. The radii of cations are smaller than those of the corresponding neutral atoms because positive ions are formed by removing one or more electrons from the outermost shell. Since these electrons are furthest from the nucleus, their absence will make the cation smaller than the original atom. In addition, loss of an electron causes a decrease in the amount of electron-electron repulsion which also causes the cation to be smaller than the neutral atom.

In contrast, the radius of an anion is larger than that of the corresponding neutral atom. When an electron is added to an atom to form an anion, there is an increase in the electron-electron repulsion. This causes electrons to spread out as much as possible, and so anions have a larger radius than the corresponding atoms.

B. Isoelectronic Series.

An **isoelectronic series** consists of one or more ions and a noble gas, all of which have identical electron configurations. Within an isoelectronic series, the greater the nuclear charge, the smaller the radius.

Na	$[Ne]3s^1$	F	$[He]2s^22p^5$
Na^+	$[Ne]$	F^-	$[He]2s^22p^6 = [Ne]$

Notice that Na^+ and F^- are isoelectronic; both have the electron configuration of neon. We can also compare radii within an isoelectronic series.

Species	S^{2-}	Cl^-	Ar	K^+	Ca^{2+}
Radius (pm)	219	181	154	133	99

Each species has 18 electrons arranged in an argon configuration, and so the electron–electron repulsion is about the same in each ion. The reason for the decrease in radius is that the nuclear charge increases steadily within this series. This causes the electrons to be attracted more strongly toward the nucleus, and the ionic radius to contract as the nuclear charge increases.

STUDY QUESTION 4.11
Choose the ion with the largest ionic radius in each pair.

   a. K^+ or Na^+
   b. K^+ or Ca^{2+}
   c. K^+ or Cl^-

Solution:

a. K^+ and Na^+ are in the same group in the periodic table. The outer electrons in K^+ occupy the third principle energy level, and those in Na^+ occupy the second. K^+ has the greater ionic radius.
b. K^+ and Ca^{2+} belong to an isoelectronic series; both have 18 electrons. Calcium ions with their greater nuclear charge attract their electrons more strongly than K^+ ions and so are smaller. K^+ is larger.
c. Again K^+ and Cl^- are isoelectronic species. The ion with the greater nuclear charge (K^+) will be smaller. Cl^- has the greater ionic radius.

STUDY QUESTION 4.12
Identify the isoelectronic pairs.
$Kr \quad Cl^- \quad Rb^+ \quad K^+ \quad Cu^{2+} \quad Se^{2-} \quad Zn^{2+}$

Solution: Isoelectronic means that they have the same number of electrons. So, if we count the number of electrons in each, we can determine the isoelectronic species.

K has 19 electrons.
Cl^- has $(17 + 1) = 18$ electrons.
Rb^+ has $(37 - 1) = 36$ electrons.
K^+ has $(19 - 1) = 18$ electrons.
Cu^{2+} has $(29 - 2) = 27$ electrons.
Se^{2-} has $(34 + 2) = 36$ electrons.
Zn^{2+} has $(30 - 2) = 28$ electrons.

Cl^- and K^+ are isoelectronic with 18 electrons each. Rb^+ and Se^{2-} are isoelectronic with 36 electrons each.

STUDY QUESTION 4.13
Arrange the following in order of decreasing radius,
$Cu^{2+} \quad K^+ \quad Rb^+ \quad Zn^{2+}$.

Solution: We should consider number of electrons, nuclear charge and valence shell for each.

Cu^{2+}: 27 electrons, nuclear charge of $+29$, and valence shell of $n = 3$.
K^+ has 18 electrons, nuclear charge of $+19$, and valence shell of $n = 3$.
Rb^+ has 36 electrons, nuclear charge of $+37$, and valence shell of $n = 4$.
Zn^{2+} has 28 electrons, nuclear charge of $+30$, and valence shell of $n = 3$.

We see that Rb^+ is the largest because it has electrons in the 4th shell compared to the 3rd shell for the other cations. For the other ions, the $+2$ cations will be smaller than the $+1$ cation due to reduced electron–electron repulsion. Of the two $+2$ cations, the one with greater nuclear charge will be smaller. For decreasing order, we should rank the ions from largest to smallest.

$$Rb^+ \; < \; K^+ \; < \; Cu^{2+} \; < \; Zn^{2+}$$

PRACTICE QUESTION 4.14
Arrange the following in order of increasing radius, F^-, Ne, Na^+, Mg^{2+}, Al^{3+}.

PRACTICE QUIZ

1. What group of elements in the periodic table have:
   a. the highest ionization energies?   b. the lowest ionization energies?

2. Based on periodic trends, which one of the following elements has the lowest ionization energy?
   K, S, Se, Li, Br

3. Based on periodic trends, which one of the following elements has the largest ionization energy?
   Cl, K, S, Se, Br

4. Define electron affinity. What group of elements has negative electron affinities?

5. Which of the following atoms has both high ionization energy and a large positive electron affinity? K, Ne, Br, Fe, N

6. Write the ground–state electron configurations of the following ions:
   a. Na^+ b. N^{3-} c. Ba^{2+} d. Br^- e. Li^+ f. Al^{3+}

7. Write the ground–state electron configurations of the following ions.
   a. Sn^{2+} b. Sn^{4+} c. Cu^{2+} d. Cu^+ e. Ti^{2+} f. Ti^{4+}

8. Write the ground-state electron configurations of the following ions.
   a. Sc^{3+} b. V^{5+} c. Pb^{2+} d. Pb^{4+}

9. Identify any isoelectronic pairs among the following.
   Na^+, Ar, Cl^-, Ne, Se^{2-}

10. Which of the following has the largest radius?
    S^{2-}, Ar, Se^{2-}, O^{2-}, Al^{3+}

11. Which species is larger:
    a. Co^{2+} or Co^{3+}
    b. S or S^{2-}

Chapter 5
Ionic and Covalent Compounds

PROBLEM-SOLVING STRATEGIES AND TUTORIAL SOLUTIONS

TYPES OF PROBLEMS
Problem Type 1:   Lewis Dot Symbols.

Problem Type 2:   Ionic Compounds and Bonding.
      A.   Lattice Energy.

Problem Type 3:   Naming Ions and Ionic Compounds.
      A.   Formulas of Ionic Compounds.
      B.   Naming Ionic Compounds.

Problem Type 4:   Covalent Bonding and Molecules.
      A.   Lewis Theory of Bonding.
      B.   Molecules.
      C.   Molecular Formulas.
      D.   Empirical Formulas.

Problem Type 5:   Naming Molecular Compounds.
      A.   Specifying Numbers of Atoms.
      B.   Compounds Containing Hydrogen.
      C.   Organic Compounds.

Problem Type 6:   Covalent Bonding in Ionic Species.
      A.   Polyatomic Ions.
      B.   Oxoacids.
      C.   Hydrates.

Problem Type 7:   Molecular and Formula Masses.

Problem Type 8:   Percent Composition of Compounds.

Problem Type 9:   Molar Mass.
      A.   Determining Molar Mass.
      B.   Interconverting Mass, Moles, and Numbers of Particles.
      C.   Determination of Empirical Formula and Molecular Formula from Percent Composition.

PROBLEM TYPE 1: LEWIS DOT SYMBOLS.

Electron configurations can be used to write the Lewis dot symbols of the representative elements.
A Lewis dot symbol of an element consists of the chemical symbol with one or more dots placed around it. Each dot represents a valence electron. The orbital diagram and the Lewis symbol for the fluorine atom are shown on next page. Fluorine has 7 electrons in its outermost principle energy level ($n = 2$), and therefore, has 7 dots in its Lewis symbol.

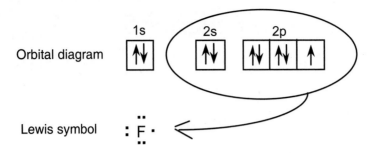

For representative elements, the number of valence electrons is the same as the group number.

The representative metals form ions by losing all of their valence electrons. Lewis symbols can be used to represent the formation of positive magnesium and aluminum ions as follows:

$$\cdot Mg \cdot \longrightarrow Mg^{2+} + 2e^-$$

$$\cdot \overset{\cdot}{Al} \cdot \longrightarrow Al^{3+} + 3e^-$$

The nonmetal elements form negative ions by acquiring electrons until they are isoelectronic with a noble gas atom. Ions of nonmetal have 8 electrons in their valence shells.

$$: \overset{\cdot}{\underset{\cdot}{S}} : + 2e^- \longrightarrow \left[: \overset{\cdot\cdot}{\underset{\cdot\cdot}{S}} : \right]^{2-}$$

$$: \overset{\cdot\cdot}{Cl} \cdot + e^- \longrightarrow \left[: \overset{\cdot\cdot}{\underset{\cdot\cdot}{Cl}} : \right]^-$$

STUDY QUESTION 5.1
Write Lewis symbols for the following elements:

a. **Ca**
b. **O**
c. **S**
d. **Se**

Solution: The Lewis symbol of an element consists of the element symbol surrounded by between 1 and 8 dots, representing valence electrons.

Ca is in Group 2A. It has two electrons in the outermost energy level. These are its valence electrons. The Lewis symbol is

$$\cdot Ca \cdot$$

O, S, and Se are all in the same group, 6A. Each has 6 valence electrons.

$$\cdot \overset{\cdot\cdot}{\underset{\cdot\cdot}{O}} \cdot \qquad \cdot \overset{\cdot\cdot}{\underset{\cdot\cdot}{S}} \cdot \qquad \cdot \overset{\cdot\cdot}{\underset{\cdot\cdot}{Se}} \cdot$$

STUDY QUESTION 5.2
Write Lewis symbols for the following ions:

a. Ca^{2+}
b. Se^{2-}

Solution:

a. Removing two electrons (dots) from ·Ca· gives simply Ca^{2+} as the Lewis symbol of a calcium ion. No electrons are shown around the Ca^{2+} symbol because this ion has lost its valence shell electrons.

b. Adding two electrons to ·Se· gives $\left[: \ddot{Se} : \right]^{2-}$ as the Lewis symbol for a selenide ion.

> **PRACTICE QUESTION 5.1**
> **Write Lewis dot symbols for atoms of the following elements:**
>
> a. Mg
> b. Al
> c. Br
> d. Xe

> **PRACTICE QUESTION 5.2**
> **Write Lewis dot symbols for the atom and predominate ion for a generic element (X) that is found in:**
>
> a. Group 1A
> b. Group 7A

> **PRACTICE QUESTION 5.3**
> **Write Lewis dot symbols for the following ions:**
>
> a. K^+
> b. S^{2-}
> c. N^{3-}
> d. I^-
> e. Sr^{2+}

PROBLEM TYPE 2: IONIC COMPOUNDS AND BONDING.

It is important to identify any properties of atoms that affect their ability to form ionic compounds. In the formation of ions from atoms, the metal atom loses an electron and the nonmetal atom gains an electron.

$$Li(g) \rightarrow Li^+(g) + e^-$$
$$e^- + F(g) \rightarrow F^-(g)$$

The overall change is:

$$Li(g) + F(g) \rightarrow Li^+(g) + F^-(g)$$

One factor favoring the formation of an ionic compound is the ease with which the metal atom loses an electron. A second factor is the tendency of the nonmetal atom to gain an electron. Thus, ionic compounds tend to form between elements of low ionization energy and those of high electron affinity. The alkali metals and alkaline earth metals have low ionization energies. They tend to form ionic compounds with the halogens and Group 6A elements, both of which have high electron affinities.

A third factor is the lattice energy. Electrostatic attraction of the ions results in large amounts of energy being released when two kinds of gaseous ions are brought together to form a crystal lattice.

$$Li^+(g) + F^-(g) \rightarrow LiF(s)$$

A. Lattice Energy.

The force that gives rise to the ionic bond is the electrical attraction existing between a positive ion and a negative ion. Chemical bonding lowers the energy of two interacting atoms (or ions). Coulomb's law states that the potential energy (E) of interaction of two ions is directly proportional to the product of their charges and inversely proportional to the distance between them:

$$E = k\frac{Q_{cation}Q_{anion}}{r}$$

where Q_{cation} and Q_{anion} are the charges of the two ions, r is their distance of separation, and k is proportionality constant (its value will not be needed). When one ion is positive and the other negative, E will be negative. Bringing two oppositely charged particles closer together lowers their energy. The lower the value of the potential energy, the more stable is the pair of ions. Energy would need to be added to separate the two ions.

The factors that govern the stability of ion pairs are the magnitude of their charges, and the distance between the ion centers. The distance between ionic centers (r) is the sum of the ionic radii of the individual ions.

$$r = r_{cation} + r_{anion}$$

The energy of attraction between two oppositely charged ions depends:

- directly on the magnitude of the ion charges. The greater the ion charges, the stronger the attraction. The interaction of Mg^{2+} and O^{2-} ions is much greater than that of Na^+ and Cl^- ions.

- inversely on the distance between ion centers. The distance between ions depends on the sizes of the ions involved. As the sum of the ionic radii increases, the interaction energy decreases.

The lattice energy provides a measure of the attraction between ions and the strength of the ionic bond. The lattice energy is the energy required to separate the ions in 1 mole of a solid ionic compound into gaseous ions. For example, the lattice energy of NaF(s) is equal to 908 kJ/mol. This means 908 kJ are required to vaporize one mole of NaF(s) and form one mole of Na^+ ions and one mole of F^- ions in the gas phase.

$$NaF(s) \rightarrow Na^+(g) + F^-(g) \qquad \text{lattice energy} = 908 \text{ kJ}$$

The lattice energy is related to the stability of the ionic solid; the larger the lattice energy, the more stable the solid. The lattice energies are reflected in the melting points of ionic crystals. During melting the ions gain enough kinetic energy to overcome the potential energy of attraction, and they move away from each other. The higher the melting point, the more energy the ions need to separate from one another. Therefore, as the lattice energy increases, so does the melting point of the compound.

In general, the smaller the ionic radius and the greater the ionic charge, the greater the lattice energy.

STUDY QUESTION 5.3
Consider NaCl, NaBr and NaI. Which compound would you expect to have the greatest lattice energy?

Solution: Since all of the compounds contain Na^+, and all the anions have the same ionic charge, -1, the distance between the ions will determine the relative lattice energies. Remember that as the distance between the ions decreases, the lattice energy and the attractive force between the oppositely charged ions will increase. The distance between the ions will be minimized between the pair of ions with the smallest ionic radii. Recall that ionic size increases as you go down a group. Since Cl^- will have the smallest ionic radius, NaCl is expected to have the greatest lattice energy.

> *PRACTICE QUESTION 5.4*
> **Write a chemical equation for the process that corresponds to the lattice energy of MgO.**

> *PRACTICE QUESTION 5.5*
> **Which member of the pair will have the higher lattice energy?**
>> a.   **NaCl or CaO**
>> b.   **KI or KCl**

PROBLEM TYPE 3: NAMING IONS AND IONIC COMPOUNDS.

A. Formulas of Ionic Compounds
Ionic compounds are combinations of anions and cations held together by electrostatic forces. Sodium chloride (table salt, NaCl) is an example of an ionic compound. The formulas of ionic compounds are generally written as the empirical formula (formula in which elements have smallest whole number ratio) because these compounds do not exist as discrete molecular units. These compounds exist as three-dimensional networks of ions in the solid phase or as ions surrounded by water in the aqueous phase.

B. Naming Ionic Compounds
The principal feature of these compounds is that the negative and positive charges must balance resulting in an electrically neutral combination. Consequently, ionic compounds of equally charged ions, such as LiF and MgS, have equal numbers of cations and anions in their formulas. Ionic compounds composed of cations and anions of different charge have ratios that equalize the positive and negative charge. In fact, the number of cations (cation subscript) is often equal to the value of the charge on the anion and the number of anions (subscript) is equal to the value of the charge on the cation. (Remember, the lack of a subscript means one atom). Some metals form more than one ion and there are two ways to name them. The modern method uses the Stock convention in which the charge on the ion is denoted using Roman numerals. Metals which do not need the Roman numeral are the Alkali Metals (1A), the Alkaline Earth Metals (2A), as well as Silver (Ag), Cadmium (Cd), Zinc (Zn) and Aluminum (Al) because these metals always have but one charge in an ionic compound.

STUDY QUESTION 5.4
Write the formulas of the following compounds.
>> a.   **Magnesium chloride**
>> b.   **Magnesium oxide**

Solution:

a. The ions in magnesium chloride are Mg^{2+} and Cl^-. Two chloride ions are needed for each magnesium ion. [1 cation (+2)] + [2 anions (–1)] = 0. The formula is $MgCl_2$.

b. The ions in magnesium oxide are Mg^{2+} and O^{2-}. In this case just one cation and one anion will give electrical neutrality. (+2) + (–2) = 0. The formula is MgO.

STUDY QUESTION 5.5
Name the following compounds according to the Stock system:

    a.   **CuBr**
    b.   **CuS**

Solution: In the Stock system the symbol of the metallic element in a compound is followed by a Roman numeral derived from the charge of the cation.

a. Since Cu and Br combine in a 1:1 ratio and the bromide ion has charge of –1, Cu must be a +1 ion. Therefore, CuBr is named copper(I) bromide.

b. Since the sulfide ion has a –2 charge and Cu combines in a 1:1 ratio with S, copper must be a +2 ion and the compound is copper(II) sulfide.

PRACTICE QUESTION 5.6
Which of the following compounds are likely to be ionic?

    a.   **KCl**
    b.   **CH_4**
    c.   **$AlCl_3$**
    d.   **SO_2**
    e.   **MgO**
    f.   **CCl_4**

PRACTICE QUESTION 5.7
Write formulas of the following binary compounds.

    a.   **barium chloride**
    b.   **magnesium nitride**
    c.   **iron(III) oxide**
    d.   **iron(II) fluoride**

PRACTICE QUESTION 5.8
Name the following compounds.

    a.   **K_3N**
    b.   **BaF_2**
    c.   **Al_2O_3**
    d.   **AgCl**
    e.   **$CoCl_2$**

PROBLEM TYPE 4: COVALENT BONDING AND MOLECULES.

A. Lewis Theory of Bonding.
Molecules are held together by bonds resulting from the sharing of electrons between two atoms in a manner that is consistent with the octet rule. A simple covalent bond is formed when two atoms in a molecule share a pair of electrons. The octet rule states that when forming bonds, most atoms of the representative elements tend to gain, lose, or share electrons until they have eight electrons in the valence shell, hence, they end up with the same number of valence electrons and the nearest noble gas.

Lewis Structures

The formation of a covalent bond in hydrogen chloride can be represented with Lewis structures:

$$\text{H} \cdot + \cdot \overset{\cdot\cdot}{\underset{\cdot\cdot}{\text{Cl}}} : \rightarrow \text{H} : \overset{\cdot\cdot}{\underset{\cdot\cdot}{\text{Cl}}} : \text{ or } \text{H}—\overset{\cdot\cdot}{\underset{\cdot\cdot}{\text{Cl}}} :$$

where the dash represents a covalent bond, or a pair of electrons shared by both the H atom and the Cl atom. By sharing the electron pair, the Cl atom has eight valence shell electrons. The stability of this bond results from both atoms acquiring noble gas configurations. Notice that hydrogen is an exception to the octet rule. Rather than achieving an octet, it needs only two electrons to achieve a filled outer energy level. The H atom becomes **isoelectronic** (same number of electrons) with helium. The electron pairs on the Cl atom that are not involved in bonding are called lone pairs, unshared pairs, or nonbonding electrons.

B. Molecules.

A molecule is a discrete particle composed of atoms that are bonded together in a fixed arrangement. A molecule can consist of atoms of the same element, such as O_2 or S_8. If a molecule consists of atoms of different elements, the molecule is the smallest particle that has the properties of that compound. Diatomic molecules contain two atoms, while polyatomic molecules contain more than two atoms. (The prefix "poly" is Greek for many). One of the most important facts to remember is that seven of the naturally occurring elements on Earth exist as diatomic molecules. These elements are hydrogen (H_2), nitrogen (N_2), oxygen (O_2), fluorine (F_2), chlorine (Cl_2), bromine (Br_2), and iodine (I_2).

C. Molecular Formulas.

One of the biggest challenges for new chemistry students is mastering the rules for constructing molecular formulas and chemical names. Chemists use chemical formulas to represent elements and compounds in terms of their chemical symbols. The chemical formula reflects the number of the different types of atoms in the substance. Molecular models and structural formulas are used to represent elements and compounds pictorially. The structural formula depicts the spatial arrangement of the atoms in the substance.

D. Empirical Formulas.

For molecular compounds we may write their formulas expressing the smallest whole–number ratio between the atoms in the compound. This representation is known as an empirical formula. Two or more compounds that have different molecular formulas may share a common empirical formula. Thus, for a chemist, an empirical formula, albeit useful, does not provide enough information to distinguish between compounds. Empirical formulas may be determined by looking at the subscripts following each symbol for the elements in the formula. If the subscripts are divisible by the greatest common factor, then the formula is not an empirical formula. When each subscript is divided by this greatest common–factor, the empirical formula is generated.

STUDY QUESTION 5.6
What are the empirical formulas of the following compounds?

a. B_2H_6
b. P_4O_{10}

Solution: To find the smallest whole–number ratio of atoms, reduce the numbers as if reducing a fraction, i.e., divide both by the largest number which give whole numbers.

a. The simplest ratio of H to B is 6 to 2 which reduces to 3 to 1. The empirical formula is BH_3.

b. The simplest ratio of O to P is 10 to 4 which reduces to 5 to 2. Notice that you must keep a whole–number ratio. An atom ratio of 2.5 to 1 would not be physically reasonable. The empirical formula is P_2O_5.

PRACTICE QUESTION 5.9

In the following generic formulas, which would represent an empirical formula?

a. M_3X_2
b. M_2X_4
c. M_2X_2
d. MX_3
e. M_2X_3

PRACTICE QUESTION 5.10

Styrene has the molecular formula C_8H_8. What is its empirical formula?

PRACTICE QUESTION 5.11

What are the empirical formulas of the following compounds?

f. N_2O_4
g. C_4H_8
h. $AlCl_3$
i. Fe_2O_3
j. S_2F_{10}

PROBLEM TYPE 5: NAMING MOLECULAR COMPOUNDS.

A. Specifying Numbers of Atoms.

The rules for naming compounds depend on which category the compound is in. Compounds can be ionic or molecular (covalent).

- Ionic compounds, as we discussed above, are composed of cations and anions bound in a three dimensional network (often containing a metal and a nonmetal).
- Molecular (covalent) compounds are composed of discrete combinations of atoms (typically atoms are all nonmetals).

For our discussion, there are two types of molecular compounds: inorganic compounds and organic compounds. Molecular inorganic compounds are usually combinations of nonmetal elements. The first element is the one that is furthest to the left of the periodic table. Compounds containing different numbers of the same elements are distinguished using the numerical prefixes in Table 5.5 in your textbook.

Study the table below to learn how to name binary molecular compounds.

Table 5.1 Naming Binary Molecular Compounds

Class	General Formula	Nomenclature Rules	Example	Structural Formula
Molecular Compounds	X_pY_q	p# prefix + element 1 name + q# prefix + root of element 2 + 'ide'	CO_2 carbon dioxide	O=C=O
			N_2O_4 dinitrogen tetraoxide	

Some molecular compounds, especially those containing hydrogen, are named with their common, not systematic names.

STUDY QUESTION 5.7
What are the empirical formulas of the following compounds?

    a. SO_3
    b. N_2O_5
    c.   tetraphosphorus hexoxide
    d.   carbon tetrachloride

Solution: Although you were told that these compounds were molecular, you should always first check the combination of atoms present in the formula. Remember that compounds between nonmetallic elements are usually molecular, while compounds between metals and nonmetals are usually ionic.

a. The Greek prefix for three is tri–, therefore this compound is sulfur trioxide.

b. The Greek prefixes for two and five are di– and penta–, respectively. Therefore, the name for this compound is dinitrogen pentaoxide. Sometimes, you will find that the names of the prefixes are "contracted" when two vowels follow each other. It is also acceptable to name this compound as dinitrogen pentoxide.

c. Tetra– and hexa– are the prefixes for 4 and 6, respectively. The formula for this compound is P_4O_6.

d. The formula for this compound is CCl_4.

> *PRACTICE QUESTION 5.12*
> **Provide names or formulas for the following binary molecular compounds.**
>
>   a. PCl_3
>   b. Cl_2O_7
>   c.   carbon disulfide
>   d.   diboron trioxide
>   e. P_4O_{10}
>   f.   dinitrogen monoxide
>   g. NO_2

B. Compounds Containing Hydrogen.

Some molecular compounds, especially those containing hydrogen, are named with their common, not systematic names.

Table 5.2 Common Names & Formulas of Molecular Inorganic Compounds

H_2O	water	H_2S	hydrogen sulfide
NH_3	ammonia	PH_3	phosphine

Binary hydrogen–containing compounds are particularly "tricky" to name because the rules we employ will depend on the phase of the compounds. When binary hydrogen–containing covalent compounds are part of an aqueous solution, they behave as acids and we name them using a different set of rules. Take a look at Table 5.7 in your textbook. There is a very systematic approach to naming those binary acids. It goes like this:

hydro + root name of the second element in the formula + ic acid

Once you are introduced to the names of other acids you will notice that a very easy way to recognize these binary acids (compounds that contain two elements, one of them being hydrogen, which is usually written first) is by the way their name begins. When an acid name begins with "hydro" you should immediately think "binary acid."

STUDY QUESTION 5.8
What are the names of the following binary acids?

    a. HCl(*aq*)
    b. H$_2$S(*aq*)

Solution: It is important to first look at the compound that you are trying to name and first decide if it is ionic or covalent. Since both of these compounds contain only nonmetallic elements, they are both covalent molecules. Upon further inspection, we see that each of these species contains only two elements and includes an (*aq*) notation. This is a visual clue that we are dealing with binary acids. Remembering that binary acids have names that begin with "hydro–" and have "ic acid" endings. We can name the two compounds as hydrochloric acid and hydrosulfuric acid, respectively.

PRACTICE QUESTION 5.13
Provide the formulas for the following binary acids.

      a.   hydrobromic acid
      b.   hydrofluoric acid

C. Organic Compounds.

Hydrocarbons are molecular compounds that contain only carbon and hydrogen. They are the simplest type of organic compounds. Table 5.8 in your textbook gives an exhaustive list of the simplest hydrocarbons and shows the three dimensional ball and stick model for their molecules. Note how the H's in the structural formulas are directly attached to the carbon which they follow. Notice also that their formulas follow the pattern C$_n$H$_{2n+2}$, where n is the number of carbon atoms. These hydrocarbons are collectively known as alkanes. Below is a list of some organic compounds which are best known by their common names. We will have a more detailed discussion of organic compounds in chapter 24.

Table 5.3 Names of Common Organic Compounds

Formula	Name	Formula	Name
CH$_2$O	formaldehyde	CH$_3$OH	methanol
CH$_3$CH$_2$OH	ethanol	CH$_3$COOH	acetic acid
C$_6$H$_6$	benzene	C$_6$H$_{12}$O$_6$	glucose

PROBLEM TYPE 6: COVALENT BONDING IN IONIC SPECIES.

A. Polyatomic Ions.

When a group of atoms acquires a net charge, it is called a polyatomic ion. The poison cyanide is a common example. Cyanide is an anion of carbon and nitrogen, CN$^-$ (the 1 is not written in the formula of ions with a net charge of $\pm$1.) You should become familiar with Table 5.10 in your textbook. Check with your instructor if you will be required to learn some of these polyatomic anions. With so many polyatomic ions, let's look at some easy ways to remember those formulas and names. Look at all of the anions on Table 5.10 in your textbook and also at the following Periodic Table.

Anions ending in -ate

													CO_3^{2-}	NO_3^-			

Period 2 — CO_3^{2-} NO_3^-

Period 3 — PO_4^{3-} SO_4^{2-} ClO_4^-

All of these polyatomic anions end in –ate and for all of them, except for ClO_4^- , the name just follows the pattern: root of nonmetal name + –ate. Notice also that for all of them the number of oxygen atoms is just the "nometal's period number + 1". This pattern should make it easy to remember that NO_3^- is *nitr*ate (nitrogen is on period 2, the anion has 3 oxygen atoms) and that SO_4^{2-} is *sulf*ate (sulfur is on period 3, the anion has 4 oxygen atoms). Notice also the progression of the negative charges which become more negative as we move to the left, just as in the number line. Using this very simple "pattern" we could predict that the oxyanion made by boron should be BO_3^{3-} and be named borate.

In addition, we notice that the –ite ending anions differ from the corresponding –ate ending anions by having one less oxygen atom in their formulas. Notice also that not all of the –ate ending anions listed form –ite ending anions. Pay attention to the naming of oxyanions that contain halogen atoms, which follow a slightly different pattern.

STUDY QUESTION 5.9
Write the formula for magnesium phosphate.

Solution: The ions in magnesium phosphate are Mg^{2+} and PO_4^{3-} . In this case 2 phosphate anions would have a –6 charge and it will take 3 magnesium cations to have the needed +6 charge.
[3 cations (+2)] + [2 anions (–3)] = 0. The formula is $Mg_3(PO_4)_2$. Notice that we had to put PO_4 in parentheses before adding the subscript 2 in order to avoid it appearing as though we are saying one P and 42 O's. This way, it is clear that we have 3 Mg and 2 units of PO_4.

STUDY QUESTION 5.10
Name CuSO₄ according to the Stock system.

Solution: Since the sulfate ion has a –2 charge, copper must be a +2 ion and the compound is copper(II) sulfate.

PRACTICE QUESTION 5.14
Write the formulas for the following compounds.

    a.   ammonium chloride
    b.   sodium phosphate
    c.   potassium sulfate
    d.   calcium carbonate
    e.   potassium hydrogen carbonate
    f.   magnesium nitrite
    g.   sodium nitrate
    h.   strontium hydroxide
    i.   copper(II) cyanide

PRACTICE QUESTION 5.15
Name the following compounds.

    a. Ag_2CO_3
    b. $Mg(OH)_2$
    c. $NaCN$
    d. NH_4I
    e. $Fe(NO_3)_2$

B. Oxoacids.
The rules for naming acids are based on the type of anion present in the acid. We have already discussed the naming of binary acids; binary compounds containing hydrogen (normally written first in the formula) plus a monoatomic anion. Monoatomic anions are named by keeping the root of the name of the nonmetal and adding the "–ide" ending. We can therefore establish that the acids derived from anions whose names end in "–ide" will be named using the pattern "hydro + root of name of nonmetal + –ic acid." We can see why the acid derived from cyanide (CN^-) is named hydrocyanic acid. The other two endings that we find for anions (usually for polyatomic anions) are –ate and –ite. Acids derived from –ate ending anions are named using the pattern "root of the name of the polyatomic anion + –ic acid" while acids derived from polyatomic anions ending in –ite are named following the pattern "root of the name of the polyatomic anion + –ous acid."

STUDY QUESTION 5.11
Provide names and formulas for the acids derived from the following anions.

    a. ClO_2^-
    b. PO_4^{3-}
    c. I^-

Solution: The two things that will allow you to succeed in naming acids are: (1) knowing the name of the anion present in the acid, and (2) paying attention to the charge of the anion.

a. The name of the anion is chlo**rite**, thus applying the pattern given above the name is chlor**ous acid**. Since compounds are neutral (whether they are molecular or ionic), and acids always contain H^+ as cation, we need one hydronium ion to cancel the charge on the chlorite anion. The formula for the acid is $HClO_2$.

b. The name of the anion is phosph**ate** so the name of the acid is phosphor**ic acid**. The charge of the anion is –3, thus three hydronium ions are required to form the neutral compound, H_3PO_4.

c. The name of the anion is iod**ide** and it is a monoatomic anion. Following the pattern for binary acids, the name is **hydro**iod**ic acid** and the formula is HI.

PRACTICE QUESTION 5.16
Name the following acids.

a. HNO_3
b. HNO_2
c. HCN
d. $HClO_4$

C. Hydrates.
Some ionic compounds have specific numbers of water molecules bonded to them, even in the solid state. The number of water molecules is denoted by Greek numerical prefixes.

STUDY QUESTION 5.12
What is the name of $CaSO_4 \cdot 2H_2O$?

Solution: From the formula we see that there are two molecules of water attached to this ionic salt. We will name the ionic salt following the rules that we previously learned as, calcium sulfate. Next, we need to specify how many water molecules are involved using Greek prefixes. The prefix for "2" is "di–." Putting these pieces together we arrive at the name, calcium sulfate dihydrate.

PRACTICE QUESTION 5.17
Write the formula of the following hydrates.

a. **nickel(II) nitrate hexahydrate**
b. **copper(II) sulfate pentahydrate**

PROBLEM TYPE 7: MOLECULAR AND FORMULA MASSES.

Molecules are composed of a number of atoms bonded together in a fixed arrangement. By the law of conservation of mass, the molecular mass is the sum of the atomic masses of the atoms in the molecular formula. For example, the molecular masses of two nitrogen oxides NO_2 and N_2O_5 are as follows:

$$\text{molecular mass of } NO_2 = \text{atomic mass of N} + 2(\text{atomic mass of O})$$

$$= 14.01 \text{ amu} + 2(16.00 \text{ amu})$$

$$= 46.01 \text{ amu}$$

$$\text{atomic mass of } N_2O_5 = 2(\text{atomic mass of N}) + 5(\text{atomic mass of O})$$

$$= 2(14.01 \text{ amu}) + 5(16.00 \text{ amu})$$

$$= 108.02 \text{ amu}$$

For ionic compounds, instead of using the term "molecular mass", the sum of the masses of all of the atoms in a unit is called a "formula mass." Although both terms are mathematically equivalent, that is, they are calculated by adding the masses of all the atoms present in the compound, the main difference lies in the "type of unit" to which they refer.

STUDY QUESTION 5.13
Calculate the molecular mass of carbon tetrachloride.

Solution: The molecular mass is the sum of the atomic masses of all the atoms in the molecule:
molecular mass $CCl_4 = (\text{atomic mass of C}) + 4(\text{atomic mass of Cl}) = (12.01 \text{ amu}) + 4(35.45 \text{ amu})$
molecular mass $CCl_4 = 153.81$ amu

PRACTICE QUESTION 5.18
Calculate the molecular mass of the following.

    a. octane, C_8H_{18}
    b. formaldehyde, H_2CO
    c. xenon difluoride, XeF_2

PRACTICE QUESTION 5.19
Calculate the molecular mass of the following compounds.

    a. Na_2SO_4, sodium sulfate
    b. $FeCl_3$, iron(III) chloride
    c. $Ba(OH)_2$, barium hydroxide
    d. SO_3, sulfur trioxide

PROBLEM TYPE 8: PERCENT COMPOSITION OF COMPOUNDS.

The percent composition of a compound is the percentage by mass of each element in the compound. It measures the relative mass of an element in a compound. The formula for the percent composition of an element is:

$$\% \text{ composition of element} = \frac{n \times \text{element molar mass}}{\text{compound molar mass}} \times 100\%$$

where n is the number of atoms of the element of interest in the compound. Keep in mind that percent is always calculated as the part per total times 100. In this case, the part is the grams of the individual element; the total is the grams of the entire compound.

Consider sodium chloride (NaCl) as an example. The formula mass, 58.44 amu, is the sum of the mass of 1 atom of Na, 22.99 amu, and the mass of 1 atom of Cl, 35.45 amu. The percentage of Na by mass is

$$\% \text{Na} = \frac{1 \times 22.99 \text{ amu Na}}{58.44 \text{ amu NaCl}} \times 100\% = 39.33\%$$

The percentage of Cl by mass is

$$\% \text{Cl} = \frac{1 \times 35.45 \text{ amu Cl}}{58.44 \text{ amu NaCl}} \times 100\% = 60.66\%$$

The percentages sum to 99.99% rather than 100% because the atomic masses and molar mass of NaCl were rounded to two decimal places.

STUDY QUESTION 5.14
Calculate the percent compostion of glucose ($C_6H_{12}O_6$).

Solution: First, percent composition means the percentage of each element in the compound. The molecular mass of glucose must be calculated using the atomic masses of each element.

$$\text{molecular mass of } C_6H_{12}O_6 = \left[6 \text{ atoms C} \times \frac{12.01 \text{ amu}}{1 \text{ atom C}} \right] + \left[12 \text{ atoms H} \times \frac{1.008 \text{ amu}}{1 \text{ atom H}} \right] +$$

$$\left[6 \text{ atoms O} \times \frac{16.00 \text{ amu}}{1 \text{ atom O}} \right]$$

$$= 72.06 \text{ amu} + 12.10 \text{ amu} + 96.00 \text{ amu} = 180.16 \text{ amu}$$

Then, the percent of each element present is calculated by dividing the total mass of each element in 1 molecule of glucose by the total mass of 1 molecule of glucose.

$$\text{mass percent of C} = \frac{\text{mass of C in 1 molecule } C_6H_{12}O_6}{\text{mass of 1 molecule } C_6H_{12}O_6} \times 100$$

$$\%C = \frac{6 \text{ atoms C} \times (12.01 \text{ amu/atom C})}{180.16 \text{ amu } C_6H_{12}O_6} = \frac{72.06 \text{ amu}}{180.16 \text{ amu}} \times 100\% = 40.00\% \text{ C}$$

$$\%H = \frac{12 \text{ atoms H} \times \left(\dfrac{1.008 \text{ amu}}{\text{atom H}} \right)}{180.16 \text{ amu } C_6H_{12}O_6} = \frac{12.10 \text{ amu}}{180.16 \text{ amu}} \times 100\% = 6.716\% \text{ H}$$

$$\%O = \frac{6 \text{ atoms O} \times \left(\dfrac{16.00 \text{ amu}}{\text{atom O}} \right)}{180.16 \text{ amu } C_6H_{12}O_6} = \frac{96.00 \text{ amu}}{180.16 \text{ amu}} \times 100\% = 53.29\% \text{ O}$$

Thus, glucose ($C_6H_{12}O_6$) contains 40.0% C, 6.72% H, and 53.3% O by mass.

NOTE:
A good way to check the results is to sum the percentages of the elements, because their total must add to 100%. Here we get 100.02%.

> **PRACTICE QUESTION 5.20**
> **Calculate the percent composition by mass of the following compounds.**
>
> a. CO_2
> b. H_3AsO_4
> c. $CHCl_3$
> d. $NaNO_2$
> e. H_2SO_4

> **PRACTICE QUESTION 5.21**
> **Calculate the mass of hydrogen present in 355 g of acetone, C_3H_6O.**

PROBLEM TYPE 9: MOLAR MASS.

A. Determining Molar Mass.

The atomic mass scale is useful for small numbers of atoms or molecules, but macroscopic amounts of elements and compounds are too large to be measured conveniently on the atomic mass scale. Macroscopic amounts of elements and compounds are measured in grams, but to measure out equal *numbers* of atoms of two elements, say carbon and oxygen, we cannot simply weigh out equal masses of the two elements. Instead, we must measure a gram ratio of the two that is the same as the mass ratio of one C atom to one O atom. The atomic masses of C and O are 12.01 amu and 16.00 amu, respectively, so any amounts of C and O that have a mass ratio of 1.0:1.33, will contain equal numbers of C and O atoms. Therefore, 16.00 g O contains the same number of atoms as 12.01 g C. The number of C atoms in 12.01 g C is 6.022×10^{23}, called Avogadro's number. The quantity of a substance that contains Avogadro's number of atoms or other entities is called a mole (abbreviated mol).

The concept of a mole is a simple concept but often difficult for students to grasp. "Mole" is a word which represents a number, the number being very large: 6.022×10^{23}.

The molar mass of an element or compound is the mass of one mole of its atoms or molecules. The molar mass of an element or molecule is also equal to its atomic or molecular mass expressed in grams rather than atomic mass units. For example, the atomic mass of Na is 22.99 amu, or we can say that there are 22.99 grams of Na in one mole, that is, 22.99 grams/mol.

The term *mole* can be used in relation to any kind of particle, such as atoms, ions, or molecules. For clarity, the particle must always be specified. We say 1 mole of O_3 (ozone), or 1 mole of O_2 (diatomic oxygen), or 1 mole of Na^+ (sodium ions).

STUDY QUESTION 5.15
Calculate the molar mass of iron (III) hydroxide, $Fe(OH)_3$.

Solution: Molar masses are calculated using the same concept as for a molecular or formula mass; the difference being that we calculate the mass in grams for 1 mole of the compound. Therefore, in 1 mol of $Fe(OH)_3$ there are:

> 1 mole of Fe atoms
> 3 moles of H atoms
> 3 moles of O atoms

Adding up all of these masses we get:

$$\frac{55.85 \text{ g}}{\text{mol}} + \left(\frac{3 \times 1.008 \text{ g}}{\text{mol}} \right) + \left(\frac{3 \times 16.00 \text{ g}}{\text{mol}} \right) = 106.87 \text{ g/mol}$$

STUDY QUESTION 5.16
Calculate the mass of carbon in 10.00 g of glucose ($C_6H_{12}O_6$).

Solution: In Study Question 5.13, we calculated that glucose contains 40.00% carbon. Therefore,

$$\text{mass C} = 40.00\% \text{ of } 10.00 \text{ g } C_6H_{12}O_6 = \frac{40.00 \text{ g C}}{100 \text{ g } C_6H_{12}O_6} \times 10.00 \text{ g } C_6H_{12}O_6 = 4.000 \text{ g C}$$

This calculation would have to be done from scratch if we did not already know the percentage of carbon. The plan is:

$$\text{mass of glucose} \rightarrow \text{mol glucose} \rightarrow \text{mol carbon} \rightarrow \text{g carbon}$$

$$10.00 \text{ g } C_6H_{12}O_6 \times \frac{1 \text{ mol } C_6H_{12}O_6}{180.16 \text{ g } C_6H_{12}O_6} \times \frac{6 \text{ mol C}}{1 \text{ mol } C_6H_{12}O_6} \times \frac{12.01 \text{ g C}}{1 \text{ mol C}} = 4.000 \text{ g C}$$

PRACTICE QUESTION 5.22
What is the mass of 1 mole of each of the following?

　　a.　**Cu**
　　b.　**CuO**
　　c.　**$CuSO_4$**

   d. $CuSO_4 \cdot 5H_2O$

B.   Interconverting Mass, Moles, and Numbers of Particles.
   We can make conversions between mass, moles and number of particles by remembering that:
   1.   The molar mass provides a convenient "link" between mass and moles
   2.   Avogadro's number represents the numerical relationship (definition) of the number of particles in one mole of anything.
   The following examples will serve to illustrate all of these relationships.

STUDY QUESTION 5.17
How many molecules of N_2O are present in 0.245 mol of N_2O? How many N atoms? How many O atoms?

Solution: Since N_2O is a molecular compound, the unit contained in a mole of this compound is the molecule. Within each molecule of N_2O, there are two nitrogen atoms and one oxygen atom.

$$0.245 \text{ mol } N_2O \times \frac{6.022 \times 10^{23} \text{ molecules } N_2O}{1 \text{ mol } N_2O} = 1.48 \times 10^{23} \text{ molecules } N_2O$$

To calculate the number of atoms of each element, all we need is to use the ratios given by the subscripts in the molecular formula:

$$\frac{2 \text{ N atoms}}{1 \text{ molecule } N_2O} \quad \text{and} \quad \frac{1 \text{ O atom}}{1 \text{ molecule } N_2O}$$

Thus;

$$1.48 \times 10^{23} \text{ molecules } N_2O \times \frac{2 \text{ N atoms}}{1 \text{ molecule } N_2O} = 2.96 \times 10^{23} \text{ N atoms}$$

$$1.48 \times 10^{23} \text{ molecules } N_2O \times \frac{1 \text{ O atom}}{1 \text{ molecule } N_2O} = 1.48 \times 10^{23} \text{ O atoms}$$

STUDY QUESTION 5.18
What is the mass of 2.25 moles of iron (Fe)?

Solution: To convert moles of Fe to grams we use the molar mass as the conversion tool. The molar mass of Fe is 55.85 g/mol or

$$? \text{ g Fe} = 2.25 \text{ mol Fe} \times \frac{55.85 \text{ g}}{1 \text{ mol Fe}} = 126 \text{ g Fe}$$

STUDY QUESTION 5.19
How many zinc atoms are present in 20.0 g Zn?

Solution: Using tools we know we can do the following: g Zn → mol Zn → atoms Zn. Step one uses molar mass of Zn (65.39 g/mol), step two uses Avogradro's number.

Step 1: Starting with the given quantity: 20.0 g Zn, convert to moles:

$$? \text{ Zn mol} = 20.0 \text{ g Zn} \times \frac{1 \text{ mol Zn}}{65.39 \text{ g}} = 0.3059 \text{ mol Zn}$$

Step 2: Here we will carry one more digit than the correct number of significant figures. To convert moles of zinc to atoms of zinc, we use Avogadro's number:

$$? \text{ Zn atoms} = 0.3059 \text{ mol Zn} \times \frac{6.022 \times 10^{23} \text{ Zn atoms}}{1 \text{ mol Zn}} = 1.84 \times 10^{23} \text{ Zn atom}$$

•Comment
Instead of calculating the number of moles separately, we could have strung the conversion factors together into one calculation.

$$\frac{1 \text{ mol Zn}}{65.39 \text{ g Zn}} \times \frac{6.022 \times 10^{23} \text{ Zn atoms}}{1 \text{ mol Zn}}$$

$$? \text{ Zn atoms} = 20.0 \text{ g Zn} \times \frac{1 \text{ mol Zn}}{65.39 \text{ g}} \times \frac{6.022 \times 10^{23} \text{ Zn atoms}}{1 \text{ mol Zn}} = 1.84 \times 10^{23} \text{ Zn atoms}$$

STUDY QUESTION 5.20
How many molecules of ethane (C_2H_6) are present in 50.3 g of ethane? How many atoms each of H and C are in this sample?

Solution: Plan: g C_2H_6 → mol C_2H_6 → molecules C_2H_6. Step one uses the molar mass of C_2H_6, step two uses Avogadro's number.

The molecular mass of C_2H_6 is:

$$2(12.01 \text{ amu}) + 6(1.008 \text{ amu}) = 30.07 \text{ amu or } 30.07 \text{ g/mol}$$

Step 1:

$$? \text{ mol } C_2H_6 = 50.3 \text{ g } C_2H_6 \times \frac{1 \text{ mol } C_2H_6}{30.07 \text{ g } C_2H_6} = 1.67 \text{ mol } C_2H_6$$

Step 2:

$$? \, C_2H_6 \text{ molecules} = 1.67 \text{ mol } C_2H_6 \times \frac{6.022 \times 10^{23} \text{ molecules}}{1 \text{ mol } C_2H_6} = 1.01 \times 10^{24} \, C_2H_6 \text{ molecules}$$

Plan: molecules C_2H_6 → atoms C; molecules C_2H_6 → atoms H. Both of these steps require you to look at the subscripts in the formula, giving the following conversion factors.

$$\frac{6 \text{ H atoms}}{C_2H_6 \text{ molecule}} \quad \text{and} \quad \frac{2 \text{ C atoms}}{C_2H_6 \text{ molecule}}$$

Multiplying the number of C_2H_6 molecules by the number of atoms of each kind per molecule gives the number of each kind of atom present.

$$1.01 \times 10^{24} \text{ molecules} \times \frac{6 \text{ H atoms}}{C_2H_6 \text{ molecule}} = 6.06 \times 10^{24} \text{ H atoms}$$

$$1.01 \times 10^{24} \text{ molecules} \times \frac{2 \text{ C atoms}}{C_2H_6 \text{ molecule}} = 2.02 \times 10^{24} \text{ C atoms}$$

PRACTICE QUESTION 5.23
How many moles of…

---

a. Silver are in 5.00 g of Ag?
b. Sodium are in 5.00 of Na?

PRACTICE QUESTION 5.24
How many…

---

a. Silver atoms are in 5.00 G of Ag?
b. H_2 molecules are there in 4.0 g of H_2?
c. Molecules are in 25.0 g of methane (CH_4)?

PRACTICE QUESTION 5.25
Determine the masses of the following scenarios.

---

a. What is the mass of 2.0 moles of H_2?
b. What is the mass of 2.79×10^{22} atoms of Ag?
c. What is the average mass in grams of one atom of aluminum?

PRACTICE QUESTION 5.26
How many moles of…

---

a. $NaNO_3$ are in 8.72 g of $NaNO_3$?
b. O atoms are in 8.72 g of $NaNO_3$?

PRACTICE QUESTION 5.27
Which one contains more moles of formula units?

---

a. 95 g of $LiNO_3$ or 95 g of $NaSO_4$?
b. 150 g of MgI_2 or 125 g of $BaBr_2$?

PRACTICE QUESTION 5.28
The density of silver is 10.5 g/cm^3. How many Ag atoms are present in a silver bar that measures 0.10 m x 0.05 m x 0.01 m?

PRACTICE QUESTION 5.29
The pesticide malathion has the chemical formula $C_{10}H_{19}O_6PS_2$. What is the molar mass of malathion? The dose that is lethal to 50 percent of a human population is about 1.25 g per kilogram of body mass. How many molecules are in a dose lethal to an adult male weighing 70 kg? To an adult female weighing 58 kg?

C. **Determination of Empirical Formula and Molecular Formula from Percent Composition.**
Percent composition can be calculated directly from the formula of the compound or can be determined by chemical analysis if the formula is not known. You might expect, then, that given the percent composition you could calculate the molecular formula. This is almost true. Remember, the percent composition is a measure of the relative contribution of an element to a compound. So we can calculate the *empirical formula*, reflecting the simplest whole-number ratio of the different kinds of atoms in the compound, from the percent composition.

General procedure for determining the empirical formula from percent composition:
- Assume a 100 g sample of the compound and treat the percents as grams. (If the percent composition of a substance is 30.43% N and 60.56% O, and there are 100.00 grams of the sample, 30.43 grams is nitrogen and 60.56 grams is oxygen)
- Convert the mass to moles. Remember, an empirical formula is a ratio of the number of atoms so we must leave the "mass world" and enter the "number world", and moles is a number.
- Divide by the smallest number of moles to get the mole ratio.
- If this gives fractions (like 1/2 or 3/4) multiply by whatever number is necessary to remove the fraction. Empirical formulas have the smallest whole number ratio of atoms.

To determine the molecular formula from the empirical formula, one more piece of information must be known: the molar mass. The molecular formula is a whole number multiple of the empirical formula. To determine the molecular formula, divide the molar mass by the molar mass of the empirical formula.

STUDY QUESTION 5.21
An elemental analysis of a new compound reveals its percent composition to be: 50.7% C, 4.23 % H, and 45.1% O. Determine the empirical formula.

Solution: The empirical formula gives the relative numbers of atoms of each element in a formula unit. To start, assume you have exactly 100 g of compound, and then determine the number of moles of each element present. Then, find the ratios of the number of moles.

$$50.7 \text{ g C} \times \frac{1 \text{ mol C}}{12.01 \text{ g C}} = 4.22 \text{ mol C}$$

$$4.23 \text{ g H} \times \frac{1 \text{ mol H}}{1.008 \text{ g H}} = 4.20 \text{ mol H}$$

$$45.1 \text{ g O} \times \frac{1 \text{ mol O}}{16.00 \text{ g O}} = 2.82 \text{ mol O}$$

This gives the mole ratio for C: H: O of 4.22: 4.20: 2.82.

Dividing by the smallest of the molar amounts gives:

$$\frac{4.22 \text{ mol C}}{2.82 \text{ mol O}} = 1.50 \text{ mol C to } 1.00 \text{ mol O and } \frac{4.20 \text{ mol H}}{2.82 \text{ mol O}} = 1.49$$

$$\cong 1.5 \text{ mol C and H to } 1.00 \text{ mol O}$$

Therefore, $C_{1.5}H_{1.5}O_{1.0}$ is a possible formula.

This is not an acceptable formula because it does not contain all whole numbers. To convert these fractions to whole numbers, multiply each subscript by the same number, in this case a 2. The empirical formula therefore is $C_3H_3O_2$.

STUDY QUESTION 5.22
Mass spectrometer experiments on the compound in Study Question 5.19 show its molecular mass to be about 140 amu. What is the molecular formula?

Solution: The molar mass of the empirical formula $C_3H_3O_2$ is 71 g. Molar mass ÷ empirical formula mass gives 140/71 = 2. Therefore, the molecular formula must have twice as many atoms as the empirical formula.

$$\text{molecular formula} = C_6H_6O_4$$

PRACTICE QUESTION 5.30
Determine the empirical formula of each compound from its percent composition.

    a. **46.7% N, 53.3% O**
    b. **63.6% N, 36.4% O**
    c. **55.3% K, 14.6% P, 30.1% O**
    d. **26.6% K, 35.4% Cr, 38.1% O**

PRACTICE QUESTION 5.31
When 2.65 mg of the substance responsible for the green color on the yolk of a boiled egg is analyzed, it is found to contain 1.42 mg of Fe and 1.23 mg of S. What is the empirical formula of the compound?

PRACTICE QUESTION 5.32
Cyclohexane has the empirical formula CH_2. Its molecular mass is 84.16 amu. What is its molecular formula?

PRACTICE QUIZ

1. Arrange the following ionic compounds in order of increasing lattice energy: RbI, MgO, $CaBr_2$.

2. What is the empirical formula of each of the following compounds?
   a. $C_6H_8O_6$
   b. C_2H_2
   c. Hg_2Cl_2
   d. H_2O_2
   e. $C_2H_2O_4$
   f. $MgCl_2$

3. Write the formulas for the following compounds:
   a. calcium hypochlorite
   b. mercury(II) sulfate
   c. barium sulfite
   d. zinc oxide
   e. dinitrogen oxide
   f. sodium carbonate
   g. copper(II) sulfide
   h. lead(IV) oxide

4. Name the following compounds:
   a. Na_2HPO_4
   b. HI (gas)
   c. P_4O_6
   d. $LiNO_3$
   e. HI (solution)
   f. $Sr(NO_2)_2$
   g. $NaHCO_3$
   h. K_2SO_3
   i. Na_3PO_4
   j. $Al(OH)_3$

5. Name the following acids:
   a. H_2SO_3
   b. HClO
   c. $HClO_3$
   d. H_3PO_3
   e. H_2S

6. Explain the difference in the meaning of the symbols O_3 and $3O$.

7. Why are the chemical formulas of ionic compounds the same as the empirical formulas?

8. What is the mass, in grams, of 1.5×10^{22} molecules of PCl_5?

9. What is the mass, in grams, of 10^{12} carbon atoms?

10. What is the mass of a single atom of fluorine in grams?

11. How many O atoms are in each of the following amounts?
    a. 1 mol ozone, O_3
    b. 1 mol CuO
    c. 20 g $CuSO_4 \cdot 5 H_2O$

12. What mass of $Ca_3(PO_4)_2$ would contain 0.500 mol of Ca?

13. a. How many moles of $CaSO_4$ are there in 600 g of $CaSO_4$?
    b. How many moles of oxygen are there in 600 g of $CaSO_4$?
    c. How many oxygen atoms are in 600 g of $CaSO_4$?

14. How many antimony atoms are there in 5.00 mol of Sb_2O_5?

15. Answer the following questions concerning sulfur trioxide.
    a. What is the mass of 17.5 mol of SO_3?
    b. What is the mass of 7.5×10^{20} SO_3 molecules?

16. Calculate the percent composition by mass of the following compounds:
    a. Al_2O_3
    b. HNO_3
    c. CCl_2F_2

17. Calculate the percentage of fluorine by mass in the refrigerant HFC–134a. Its formula is $C_2H_2F_4$.

18. The empirical formula of styrene is CH, and its molar mass is 104 g/mol. What is its molecular formula?

19. Chromic oxide is a compound of chromium and oxygen. Determine the empirical formula of chromic oxide given that 2.00 g of the compound contains 1.04 g of Cr.

Chapter 6
Representing Molecules

PROBLEM-SOLVING STRATEGIES AND TUTORIAL SOLUTIONS

TYPES OF PROBLEMS

Problem Type 1: Lewis Structures and Multiple Bonds.
        A. Lewis Structures.
        B. Multiple Bonds.

Problem Type 2: Electronegativity and Polarity.
        A. Electronegativity.
        B. Dipole Moment, Partial and Percent Ionic Character.

Problem Type 3: Drawing Lewis Structures.

Problem Type 4: Lewis Structures and Formal Charge.

Problem Type 5: Resonance.

Problem Type 6: Exceptions to the Octet Rule.
        A. Incomplete Octets.
        B. Odd Number of Electrons.
        C. Expanded Octets.

PROBLEM TYPE 1: LEWIS STRUCTURES AND MULTIPLE BONDS.

A. Lewis Structures

Lewis structures are drawn to represent molecules and polyatomic ions, showing the arrangement of atoms and the positions of all valence electrons. Lewis structures represent the shared pairs of valence electrons either as two dots, or as a single dash,–. Any unshared electrons are represented as dots.

The formation of a covalent bond in hydrogen chloride can be represented with Lewis structures:

$$H \cdot + \cdot \overset{\cdot\cdot}{\underset{\cdot\cdot}{Cl}} \colon \;\rightarrow\; H \colon \overset{\cdot\cdot}{\underset{\cdot\cdot}{Cl}} \colon \quad \text{or} \quad H\!-\!\overset{\cdot\cdot}{\underset{\cdot\cdot}{Cl}} \colon$$

where the dash represents a covalent bond, or a pair of electrons shared by both the H atom and the Cl atom. By sharing the electron pair, the Cl atom has eight valence shell electrons. The stability of this bond results from both atoms acquiring noble gas configurations. Notice that hydrogen is an exception to the octet rule. Rather than achieving an octet, it needs only two electrons to achieve a filled outer energy level. The H atom becomes isoelectronic with helium. The electron pairs on the Cl atom that are not involved in bonding are called **lone pairs**, **unshared pairs**, or non–bonding electrons.

B. Multiple Bonds

In some cases two or three pairs of electrons are shared by two atoms in order to reach an octet. In these molecules, *multiple bonds* exist. A *double bond* is a covalent bond in which two pairs of electrons are shared between two atoms, as between C and O in formaldehyde.

H_2CO (formaldehyde):

$$H : \overset{\cdot\cdot}{C} :: \overset{\cdot\cdot}{\underset{\cdot\cdot}{O}} \qquad \text{or} \qquad H - C = \overset{\cdot\cdot}{\underset{\cdot\cdot}{O}}$$

In general, atoms joined by a double bond lie closer together than atoms joined by a single bond. The $C = O$ bond length is shorter than the C—O bond length. Nitrogen molecules (N_2) contain a triple bond.

$$: N \vdots \vdots N : \qquad \text{or} \qquad : N \equiv N :$$

STUDY QUESTION 6.1

How many bonding and non-bonding electrons are in the Lewis structure of NO_2?

$$\overset{\cdot\cdot}{\underset{\cdot\cdot}{O}} = \overset{\cdot}{N} - \overset{\cdot\cdot}{\underset{\cdot\cdot}{O}} :$$

Solution: There are three dashes shown between the N and two O atoms. Each dash represents one pair of shared or bonding electrons. The dots represent non-bonding electrons. There are 6 bonding and 11 non-bonding electrons in the Lewis structure of NO_2.

> *PRACTICE QUESTION 6.1*
> **Determine the number of bonding and non-bonding electrons in formaldehyde, H_2CO.**
>
> $$H : \overset{\cdot\cdot}{C} :: \overset{\cdot\cdot}{\underset{\cdot\cdot}{O}}$$

PROBLEM TYPE 2: ELECTRONEGATIVITY AND POLARITY.

A. Electronegativity.

Electronegativity is an atom's ability to draw shared electrons toward itself. Linus Pauling developed a method for determining the relative electronegativity of the elements. Pauling assigned the value 1.0 to Li and 4.0 to F. Electronegativity values exhibit periodic behavior. In general, electronegativity increases from left to right across a period, and decreases within a group from top to bottom.

STUDY QUESTION 6.2

Using the trends within the periodic table, arrange the following elements in order of increasing electronegativity: As, Se, and S.

Solution: Se and As are in the same period, and so the one further to the right has the higher electronegativity. That one is Se. Now, compare S and Se. They are in the same group. The one nearer the top of the group has the greater electronegativity, which is sulfur. Therefore, the ordering should be: As < Se < S.

> **PRACTICE QUESTION 6.2**
> **Which atom is the most electronegative?**
> Li   Cs   P   As   Ge

B. Bond Polarity and Partial Charges.

Chemical bonds are rarely pure covalent or completely ionic. Rather, most bonds exhibit some characteristics of both. Bonds in which electrons are not shared equally are **polar** and are referred to as **polar covalent bonds,** or simply **polar bonds**. Bonds between elements of widely different electronegativities ($\Delta \geq 2.0$) are ionic. The "2.0 rule" is an approximation and does not apply to all ionic compounds. Covalent bonds between atoms with different electronegativities ($2.0 \geq \Delta \geq 0.5$) are polar. Bonds between atoms with very similar electronegativities ($\Delta < 0.5$) are considered nonpolar. Bonds between homonuclear atoms have $\Delta = 0$ and are referred to as pure covalent bonds.

In the HCl molecule, for instance, the electronegativity of Cl is 3.0, and for H it is 2.1. The electronegativity difference is 0.9. The chlorine atom with its higher electronegativity attracts the electron pair more strongly. This makes the Cl atom slightly negative and H slightly positive.

$$\overset{\delta+}{H} - \overset{\delta-}{Cl}$$

Here, δ denotes a partial charge, that is, a charge less than 1.0, as it would be in an ion.

STUDY QUESTION 6.3

For the following pairs of elements, label the bonds between them as ionic, polar covalent, or pure covalent.

> a.   Rb and Br
> b.   S and S
> c.   C and N

Solution: Bond type depends on electronegativity differences.

a. The difference in electronegativities for a Rb—Br bond is 2.0. The bond between Rb and Br is ionic.

b. For S—S, the difference in electronegativity is zero, and so the bond is a pure covalent bond.

c. In the periodic table, carbon and nitrogen are adjacent to each other in period 2. The one on the right is N; it is more electronegative so we expect a polar covalent bond. Electronegativity values give a difference of 0.5.

STUDY QUESTION 6.4

For the following pairs of elements, label the bonds between them as ionic, polar covalent, or pure covalent.

> a.   Ca and F
> b.   H and P
> c.   S and Cl

Solution: Bond type depends on electronegativity differences.

a. The difference in electronegativities for a Ca—F bond is 3.0. The bond between Ca and F is ionic.

b. For H—P, the difference in electronegativity is zero, and so the bond is a pure covalent bond.

c. In the periodic table, sulfur and chlorine are adjacent to each other in period 3. The one on the right is Cl; it is more electronegative so we expect a polar covalent bond. Electronegativity values give a difference of 0.5.

C. Dipole Moment and Percent Ionic Character.

When atoms of elements having different electronegativities interact forming a polar covalent bond, a dipole ("two poles", meaning some separation of charges) is created. Dipoles are vectorial quantities; they have both a magnitude and a direction. The magnitude of the dipole is referred to as the dipole moment (μ). The dipole moment is a quantitative measure of the degree of polarity of a bond. Nonpolar covalent bonds and completely ionic bonds represent extreme situations in bonding. To refer to a bond as being "ionic" or "covalent" is an oversimplification. Sometimes, the term "*percent ionic character*" is used to describe the polar nature of a bond. A purely covalent bond has zero percent ionic character. Bonds that have electronegativity differences of 2.0 or more have at least 50% ionic character and are called ionic. There are no 100% ionic bonds. Percent ionic character quantifies the polarity of a bond, and is determined by comparing the measured dipole moment to the one predicted by assuming that the bonded atoms have discrete charges.

Dipole moment is calculated by

$$\mu = Q \times r$$

where μ is dipole moment, Q is the magnitude of the charge at either end of the molecular dipole, and r is the distance between charges. Dipole moment is usually expressed in debye units, $1\ D = 3.336 \times 10^{-30} C \cdot m$.

STUDY QUESTION 6.5
Arrange the following bonds in order of increasing ionic character: C—O, C—H, and O—H.

Solution: As the electronegativity difference increases, the bond becomes more polar and its ionic character increases. We can determine the electronegativity differences. With greater the difference in electronegativity, the bond will have more ionic character.

$$\begin{array}{lll}
\text{for C—O} & 3.5 - 2.5 = 1.0 \\
\text{for C—H} & 2.5 - 2.1 = 0.4 \\
\text{for O—H} & 3.5 - 2.1 = 1.4
\end{array}$$

The ionic character increases in the order C—H < C—O < O—H.

NOTE:
The electronegativity of hydrogen is unlike that of the other elements of Group 1A. In terms of electronegativity, hydrogen is similar to the nonmetal elements boron and carbon. The bonds of H to nonmetal atoms are polar covalent, rather than ionic (such as the bonds in LiCl and NaCl).

STUDY QUESTION 6.6
Use the following information to determine the percent ionic character of the C≡O bond in carbon monoxide. Assume charges of +2 and –2 on the carbon and the oxygen.

Dipole Moment	Bond Length
0.112 D	113 pm

Solution: In order to calculate the theoretical dipole moment we need to know the values of Q (in C · m) and r (in meters).

$$Q = 2 \times 1.6022 \times 10^{-19}\ C = 3.2044 \times 10^{-19}\ C$$

$$r = 113\ \text{pm} \times \frac{1 \times 10^{-12}\ m}{1\ \text{pm}} = 1.13 \times 10^{-10}\ m$$

$$\mu = \left(3.2044 \times 10^{-19}\ \text{C}\right)\left(1.13 \times 10^{-10}\ \text{m}\right) = 3.62 \times 10^{-29}\ \text{C} \cdot \text{m}$$

All we need to do now is convert the theoretical dipole to debye units.

$$3.62 \times 10^{-29} \text{C} \cdot \text{m} \times \frac{1\ \text{D}}{3.336 \times 10^{-30} \text{C} \cdot \text{m}} = 10.9\ \text{D}$$

The percent ionic character of the $C \equiv O$ bond is:

$$\% \text{ ionic character} = \frac{0.112\ \text{D}}{10.9\ \text{D}} \times 100 = 1.03\%$$

PRACTICE QUESTION 6.3
Arrange the following bonds in order of decreasing ionic character: C—N, C—H, and N—H.

PRACTICE QUESTION 6.4
Using the dipole moment and bond length given for the CO bond in Study Question 6.5, determine the partial charges on the carbon and oxygen atoms in CO.

PROBLEM TYPE 3: DRAWING LEWIS STRUCTURES.

Lewis structures represent the covalent bonding and location of unshared electron pairs within molecules and polyatomic ions. Lewis structures can be drawn using the following step–by–step procedure:

1. Use the molecular formula to draw the skeletal structure. Remember that the central atom will be the least electronegative atom in the molecule.

2. Count the total number of valence electrons, adding electrons to account for a negative charge and subtracting electrons to account for a positive charge.

3. Subtract two electrons from the total number of valence electrons for each bond in the skeletal structure.

4. Distribute the remaining valence electrons to complete octets, completing the octets of the more electronegative atoms (peripheral atoms) first.

5. Place any remaining electrons on the central atom.

6. Include double or triple bonds, if necessary, to complete the octets of all atoms. This is done by moving a lone pair from a peripheral atom and making it a bonding pair with the central atom.

STUDY QUESTION 6.7
Draw the Lewis structure for hydrazine, N_2H_4. How many unshared electron pairs (lone pairs) are there on each N atom?

Solution:
1. Arrange the atoms in a reasonable skeletal form. H atoms form only one bond and so must be located on the outside of the atom.

H N N H

H H

2. Count the valence electrons. Each N atom has 5 valence electrons and each H atom has 1. There are $2(5) + 4(1) = 14$ valence electrons.

3. Connect the atoms with single bonds:

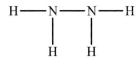

Normally we would add unshared pairs to complete all octets of surrounding atoms, but in this case the H atoms only need the two electrons shared in the bond to the N atom. Counting the number of electrons used; 5 pairs = 10 valence electrons. Now add unshared pairs to complete the octets of the N atoms.

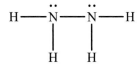

4. Count the electrons: 7 pairs = 14 valence electrons. This is the same number as given in step 2.

PRACTICE QUESTION 6.5
How many lone pairs are on the underlined atoms in the following compounds?

    a. $\underline{P}H_3$
    b. $\underline{S}Cl_2$
    c. $H_2C\underline{O}$

PROBLEM TYPE 4: LEWIS STRUCTURES AND FORMAL CHARGE.

Formal charge is a way of keeping track of the valence electrons in a species. Formal charges should be consistent with electronegativities and can be used to determine the best arrangement of atoms and electrons for a Lewis structure. The formal charge is the charge that an atom seems to have in a Lewis structure and it is calculated under the assumption of equal sharing of the bonding electrons. When determining the formal charge, all non–bonding electrons count as belonging entirely to the atom with which they are drawn. Bonding electrons are split evenly between the two atoms sharing them. The formula for the formal charge of an atom in a particular Lewis structure is:

$$\begin{array}{c}\text{formal}\\\text{charge}\end{array} = \left(\begin{array}{c}\text{number of}\\\text{valence electrons}\end{array}\right) - \left(\text{associated electrons}\right)$$

$$\text{associated electrons} = \left(\begin{array}{c}\text{number of}\\\text{nonbonding electrons}\end{array}\right) + \frac{1}{2}\left(\begin{array}{c}\text{number of}\\\text{bonding electrons}\end{array}\right)$$

It will be good to keep in mind that the formal charge is a property of a structural formula. Formal charges do not indicate actual charges on atoms in the real molecule. There are a few guidelines on the use of formal charges to decide which, among several Lewis structures, is most plausible for any given compound.

1. A Lewis structure in which there are no formal charges is preferred over one where formal charges are present.
2. When formal charges are assigned in a Lewis structure, the sum of the formal charges must be zero in a neutral molecule. For a polyatomic ion the formal charges must add up to the charge of the ion.
3. The most plausible Lewis structures will be those with negative formal charges on the more electronegative atoms. A structure should not have similarly charged atoms next to each other.

STUDY QUESTION 6.8

Assign formal charges to the atoms in the following Lewis structures.

a. 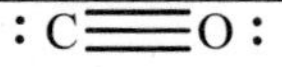 $: C \equiv O :$

b. 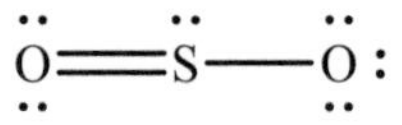 $O = S — O :$

Solution:

The formula used to calculate the formal charge of an atom is:

$$\begin{matrix} \text{formal} \\ \text{charge} \end{matrix} = \left(\begin{matrix} \text{number of} \\ \text{valence electrons} \end{matrix} \right) - \left(\begin{matrix} \text{number of} \\ \text{nonbonding electrons} \end{matrix} \right) - \frac{1}{2} \left(\begin{matrix} \text{number of} \\ \text{bonding electrons} \end{matrix} \right)$$

a. The carbon atom: formal charge = $4 - 2 - 1/2\ (6) = -1$
   The oxygen atom: formal charge = $6 - 2 - 1/2\ (6) = +1$

$$: \overset{-}{C} \equiv \overset{+}{O} :$$

b. The sulfur atom: formal charge = $6 - 2 - 1/2\ (6) = +1$
   The oxygen atom on the right: formal charge = $6 - 6 - 1/2\ (2) = -1$
   The oxygen atom on the left: formal charge = $6 - 4 - 1/2\ (4) = 0$

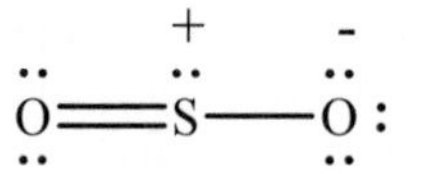

$$O = \overset{+}{S} — \overset{-}{O} :$$

PRACTICE QUESTION 6.6

Assign formal charges to the atoms in the following Lewis structures:

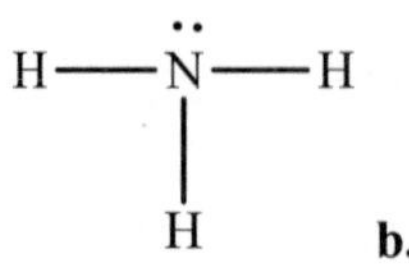

a. $H — N — H$ with H below N

b. 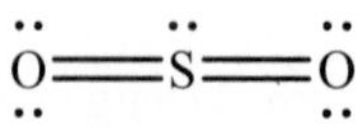 $O = S = O$

PROBLEM TYPE 5: RESONANCE.

Resonance structures are two or more equally correct Lewis structures that differ in the positions of the electrons but not in the positions of the atoms. Different resonance structures of a compound can be separated by a resonance arrow, ↔. For the nitrite ion NO_2^-, for instance, the following structure shows the correct number of valence electrons and satisfies the octet rule. The brackets are used to indicate that the −1 charge belongs to the entire nitrite ion, and not to just one atom in the structure.

$$\left[\ :\overset{\cdot\cdot}{\underset{\cdot\cdot}{O}}\text{——}\overset{\cdot\cdot}{N}\text{==}\overset{\cdot\cdot}{\underset{\cdot\cdot}{O}}\ \right]^{-}$$

However, the structure does not accurately represent what is known about the bond lengths of the N—O bonds in NO_2^-. Both bond lengths are known to be the same, whereas, according to the structure, we expect the double bond to be shorter than the single bond. An N—O single bond length should be about 136 pm, and an N==O double bond length about 122 pm. However, the two bond lengths in NO_2^- are actually equal and are intermediate between these two values.

It turns out to be impossible to draw a single satisfactory Lewis structure for NO_2^-. Situations like this can be handled by using the concept of resonance. First, draw two structures for NO_2^- that reflect different choices of electron arrangements.

$$\left[\ :\overset{\cdot\cdot}{\underset{\cdot\cdot}{O}}\text{——}\overset{\cdot\cdot}{N}\text{==}\overset{\cdot\cdot}{\underset{\cdot\cdot}{O}}\ \right]^{-} \longleftrightarrow \left[\ \overset{\cdot\cdot}{\underset{\cdot\cdot}{O}}\text{==}\overset{\cdot\cdot}{N}\text{——}\overset{\cdot\cdot}{\underset{\cdot\cdot}{O}}:\ \right]^{-}$$

These structures have the same placement of atoms but different positions of electron pairs. **The electron pairs do not actually move from one position to another within the molecule.** The electron pair is actually **_delocalized,_** which is impossible to show with the Lewis Structures. The nitrite ion is not adequately represented by either Lewis structure. The best representation we have is to describe the molecule through the use of a composite (average) of these structures. This composite structure cannot be drawn using the rules for writing Lewis structures. Each of the structures that contribute to the composite structure is called a contributing or resonance structure.

STUDY QUESTION 6.9
Draw three resonance structures for N_2O. The skeletal structure is N—N—O.

Solution:
1. Arrange the atoms.

$$N\text{—}N\text{—}O$$

2. Count the valence electrons.

$$2(5) + 6 = 16 \text{ electrons}$$

3. Add unshared pairs to the terminal atoms:

$$:\overset{\cdot\cdot}{\underset{\cdot\cdot}{N}}\text{——}N\text{——}\overset{\cdot\cdot}{\underset{\cdot\cdot}{O}}:$$

4. Count the electron pairs used: 8 electron pairs = 16 electrons. There are not enough electrons to add any to the central atom.

5. Move one of the unshared pairs from the terminal N atom to make a double bond, and move an unshared pair from the terminal O atom to make a double bond.

$$: \overset{..}{N} = N = \overset{..}{O} :$$

This is a satisfactory structure because it completes the octets of all three atoms and uses the correct number of electrons. Additional resonance structures can be generated by moving electron pairs in such a way that the octet rule for each atom is always satisfied. The positions of the atoms in N_2O cannot be altered.

Two possibilities are

$$: N \equiv N - \overset{..}{\underset{..}{O}} : \quad \text{and} \quad : \overset{..}{\underset{..}{N}} - N \equiv O :$$

NOTE:
The three structures for N_2O are called resonance structures. The actual bonding in N_2O is a composite of these three structures.

PRACTICE QUESTION 6.7

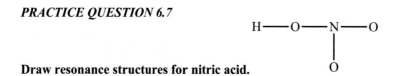

Draw resonance structures for nitric acid.

_____

PROBLEM TYPE 6: EXCEPTIONS TO THE OCTET RULE.

A. Incomplete Octets.
The boron halides BX_3 are well–known examples of molecules in which the central atom has an incomplete octet. The boron atom has only six valence electrons. The boron atom has an incomplete octet. This situation is typical in boron chemistry. Revisit the discussion on formal charge for an explanation of this fact.

B. Odd Numbers of Electrons.
A molecular species with an odd number of valence electrons is known as a **free radical**. Since an even number of electrons is required for complete pairing, the octet rule cannot be satisfied. Two common oxides of nitrogen, NO and NO_2, have odd numbers of electrons. Two additional free radicals that are known to exist in our atmosphere for very short periods of time are OH (hydroxyl radical) and HO_2 (hydroperoxyl radical).

$$\cdot \overset{..}{\underset{..}{O}} - H \qquad H - \overset{..}{\underset{..}{O}} - \overset{..}{\underset{..}{O}} \cdot$$

C. Expanded Octets.
Nonmetal atoms with valence shell of $n \geq 3$ have d–orbitals available in their valence, which provides enough room for these atoms to accommodate more than 8 electrons when bonding. Second period elements never exceed the octet rule. In compounds of P and Cl, PCl_3 obeys the octet rule while the P atom in PCl_5 has 10 electrons in its valence shell. This is called an **expanded octet**.

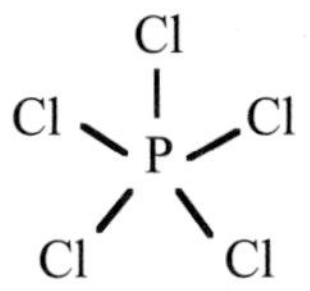

STUDY QUESTION 6.10
Draw Lewis structures for:

a. GaI_3
b. NO_2 (all bonds are equivalent)
c. ClF_3

Solution:

a. Gallium is in Group 3A of the periodic table, a group well known for its electron deficient elements. GaI_3 has $3 + 3(7) = 24$ valence electrons.

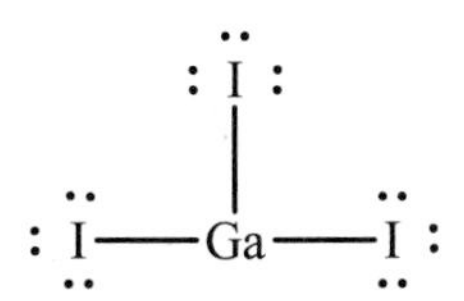

This structure shows 24 electrons, the correct number. Ga, with three electron pairs, is electron deficient. And while it is possible to follow the rules for writing Lewis structures and draw one that obeys the octet rule (moving one of the lone pairs on I and giving a Ga=I double bond), it is not likely to exist because it puts a negative formal charge on the central atom (a metal) and a positive formal charge on Iodine. This is not probable. Draw the structure and calculate the formal charges on the atoms to verify this for yourself.

b. NO_2 has 17 valence electrons. With an odd number of electrons, it cannot obey the octet rule. The best we can do is start with 18 valence electrons as in NO_2^- and then remove one from the nitrogen atom (because it is the unique atom). Two contributing structures are necessary.

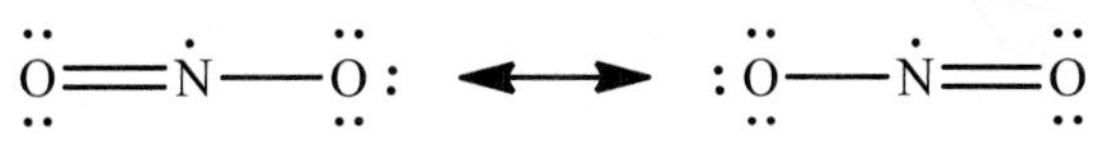

c. ClF_3 has 28 valence electrons. 26 electrons are required to complete the octets of the four atoms. The remaining two electrons are placed on the central Cl atom because chlorine is in the third period, and has vacant $3d$ orbitals that can hold electrons in addition to an octet. Chlorine is said to have an expanded octet.

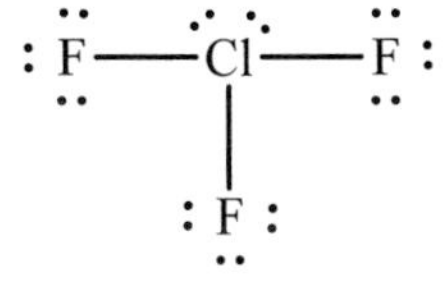

PRACTICE QUESTION 6.8
Draw the Lewis structures for the following molecules.

a. SO_3
b. SF_4
c. NO

PRACTICE QUESTION 6.9
Which of the following molecules is most polar? SO_3 SF_4 NO

PRACTICE QUIZ

1. List the following bonds in order of increasing polarity:
   N—O, Na—O, O—O, S—O

2. Classify the O—H bond in CH_3OH as ionic, polar covalent, or nonpolar covalent.

3. Which of the following is a nonpolar covalent bond (pure covalent)?
   H—Cl   Li—Br   Se—Br   Br—Br

4. Classify the following bonds as ionic, polar covalent, or nonpolar covalent.
   Se—Cl   Al—Cl   K—F   Cl—Cl

5. Write the Lewis structures for the following species.
   a. NH_4^+ b. NCl_3 c. CF_2Cl_2 d. CHCCl

6. Assign formal charges to the atoms in the following possible Lewis structures.

   a. $:O\!\!\equiv\!\!C\!\!—\!\!O$ b. $F\!\!=\!\!Be\!\!=\!\!F$

   c. Draw alternative Lewis structures in which the formal charges are minimized.

7. Do formal charges represent actual charges? What do they represent?

8. Draw resonance structures for bicarbonate ion, HCO_3^-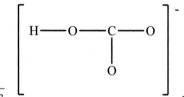

Chapter 7
Molecular Geometry, Intermolecular Forces, and Bonding Theories

PROBLEM-SOLVING STRATEGIES AND TUTORIAL SOLUTIONS

TYPES OF PROBLEMS
Problem Type 1: Molecular Geometry.
    A. The VSEPR Model.
    B. Electron–Domain Geometry and Molecular Geometry.
    C. Deviation from Ideal Bond Angles.
    D. Geometry of Molecules with More than One Central Atom.

Problem Type 2: Molecular Geometry and Polarity.

Problem Type 3: Intermolecular Forces.

Problem Type 4: Valence Bond Theory.

Problem Type 5: Hybridization of Atomic Orbitals.
    A. Hybridization of *s* and *p* Orbitals.
    B. Hybridization of *s*, *p*, and *d* Orbitals.

Problem Type 6: Hybridization in Molecules Containing Multiple Bonds.

Problem Type 7: Molecular Orbital Theory.
    A. Bonding and Antibonding Molecular Orbitals.
    B. σ Molecular Orbitals.
    C. Bond Order.
    D. π Molecular Orbitals.
    E. Molecular Orbital Diagrams.

Problem Type 8: Bonding Theories and Descriptions of Molecules with Delocalized Bonding.

PROBLEM TYPE 1: MOLECULAR GEOMETRY.

A. The VSEPR Model.
The major features of the geometry of molecules and polyatomic ions can be predicted by applying a simple principle: "The valence shell electron pairs surrounding an atom are arranged such that they are as far apart as possible." This statement provides the basis for the Valence–Shell Electron–Pair Repulsion (VSEPR) theory, a particular molecular geometry results from the orientation of electron pairs around a central atom so as to minimize their interactions (i.e., repulsions). The theory depicts the shapes of molecules based on 5 basic geometrical shapes that are designated based on the number of electron pairs (or electron domains) around the central atom.

B. Electron–Domain Geometry and Molecular Geomtery.
In order to successfully assign a shape to a molecule, the first step is to draw a Lewis structure showing all valence electrons. Once the Lewis structure is drawn, we look at the central atom and the number of electron–domains around it. An electron–domain includes all of the bonding electron pairs and the

non-bonding electron pairs. The five basic geometrical shapes are shown below with a table summarizing their major features. These five basic shapes will always be used to describe the electron–domain geometry for any molecule. However, they will only apply to molecules in which there are no lone pairs on the central atom when assigning molecular geometries.

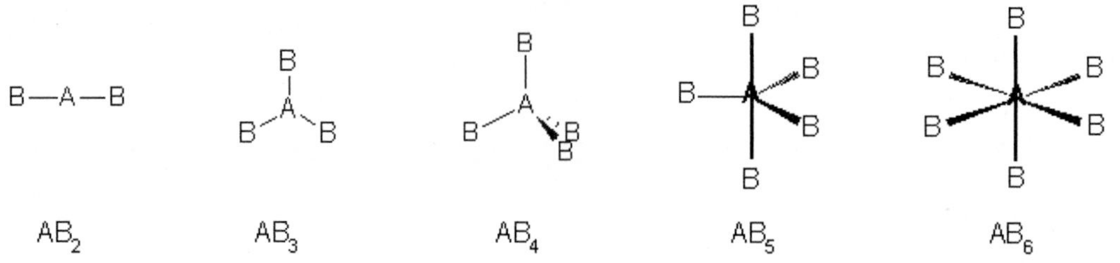

B—A—B

AB_2 AB_3 AB_4 AB_5 AB_6

Number of Electron–Domains around Central Atom	Bond Angles	Shape of Molecule (with no lone pairs)
2	$180°$	Linear
3	$120°$	Trigonal planar
4	$109.5°$	Tetrahedral
5	$120°, 90°$	Trigonal bipyramidal
6	$90°$	Octahedral

When the central atom contains lone pairs (non–bonding pairs) in addition to bonding pairs, the situation becomes slightly more involved. Molecular geometries are assigned based on the position of nuclei, which means lone pairs are "invisible." This gives rise to molecular geometries that are derived from the five basic shapes. Let's take a look at the NH_3 molecule. There are 4 electron pairs in the valence shell of the central N atom; 3 atoms connected to the central atom plus one lone pair. The electron domain geometry considers all 4 electron pairs on the central atom and we thus assign a tetrahedral electron domain geometry to NH_3. However, when describing the shape of a molecule, we consider *only the positions of the atoms*. Its geometry is described as trigonal pyramidal because the N atom sits at the top of a three-sided pyramid.

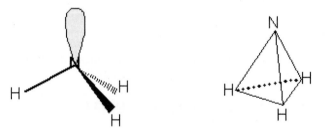

When determining the molecular geometry of any molecule, follow the following procedure:

1. Draw the Lewis structure.
2. Count the number of bonding regions around the central atom. Single, double and triple bonds count as only one bonding domain. So the bottom line is, you count the number of atoms connected to the central atom and the number of lone pairs.
3. Place the lone pairs in the locations which have the most room (this is important for structures derived from a trigonal bipyramidal arrangement). In cases where there are more than one lone pair, place lone pair electrons as far away from each other as possible.
4. After all atoms and lone pairs are located around the central atom, by looking only at the atoms, determine the name by looking at the total electron domains and the number of

them that are lone pairs. Names are assigned as in Table 7.2. Become familiar with the
names and what the shapes look like. Once you have drawn on paper where the atoms
and lone pair are located, the name will be evident.

5.   Repeat for each central atom if there is more than one central atom in a molecule.

STUDY QUESTION 7.1
Predict the geometrical shapes and bond angles for the following compounds using the VSEPR theory:

   a. $SnCl_4$
   b. O_3
   c. IF_5
   d. XeF_2

Solution:

a. First, we determine the number of valence shell electrons for the central Sn atom. A tin atom has four
   valence electrons, and each Cl atom contributes seven electrons giving a total of 32 valence electrons. The
   valence shell electrons are arranged with four electron pairs around the Sn atom. The Lewis structure is:

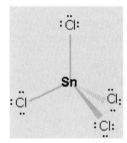

   This structure shows that there are no lone pairs about the Sn atom. Thus, the four bonding electron pairs
   and the Cl atoms, as well, are oriented at the corners of a tetrahedron. $SnCl_4$ is an AB_4 type molecule, and
   has a tetrahedral structure. The Cl—Sn—Cl bond angles are 109.5°.

b. The Lewis structure of ozone (O_3) is given below. Note that there are multiple resonance structures that
   can be drawn for ozone. You only need to draw one Lewis structure in order to determine the geometry.

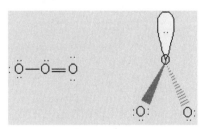

   We look at the central atom (the middle O). There are two bonded atoms and lone pair, for a total of
   3 electron groups. These three electron domains are oriented at the corners of a triangle. However, one of
   these pairs is a lone pair. The positions of the three O atoms are described as bent. The O—O—O bond
   angle should be close to 120°. However, the presence of lone–pair vs. bonding–pair repulsion
   compresses the bond angle. The observed angle in ozone is 117°.

c. Now that we have had some practice we will abbreviate the explanation. To determine the shape of IF_5,
   count the valence shell electrons around the central I atom.

from the I atom	= 7
from the F atoms (5 @ 7 e⁻)	= 35
total valence shell electrons	= 42

The Lewis structure is:

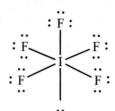

valence shell electron pairs = 6
number of lone pairs = 1
number of bonded atoms = 5
molecular shape: square pyramid
The F—I—F bond angles are 90°.

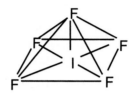

d. To predict the shape of XeF$_2$. First, count the valence shell electrons around Xe.

from the Xe atom = 8
from the F atoms (2 @ 7) = 14
total valence shell electrons = 22

The Lewis structure is:

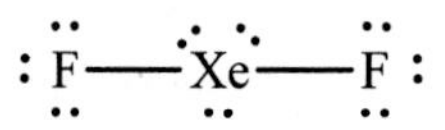

$$: \ddot{F} \!\!-\!\!-\!\! \ddot{Xe} \!\!-\!\!-\!\! \ddot{F} :$$

valence shell electron pairs = 5
number of lone pairs = 3
number of bonded atoms = 2
molecular shape: linear

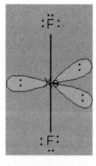

The F—Xe—F bond angles are 180°.

PRACTICE QUESTION 7.1
Predict the shape of the following molecules:

a. BeCl$_2$
b. NF$_3$
c. SF$_4$
d. ClF$_3$
e. KrF$_4$

PRACTICE QUESTION 7.2
Predict the shape of the following ions:

    a. ClO_2^-

    b. CO_3^{2-}

    c. $ZnCl_4^{2-}$

PRACTICE QUESTION 7.3
Which of the following molecules and ions have a tetrahedral geometry?

    a. CCl_4
    b. SCl_4
    c. $AlCl_4^-$

PRACTICE QUESTION 7.4
Predict the following bond angles.

    a.   O–C–O in CO_2
    b.   F–C–F in CF_4
    c.   H–O–O in H_2O_2

C. **Deviation from Ideal Bond Angles.**
VSEPR theory also explains why many observed bond angles are not quite equal to their predicted angles. For instance, a true tetrahedral angle, as in CH_4, is 109.5°, but the measured H—N—H bond angle in NH_3 is 106.7°, and in water the H—O—H bond angle is 104.5°.

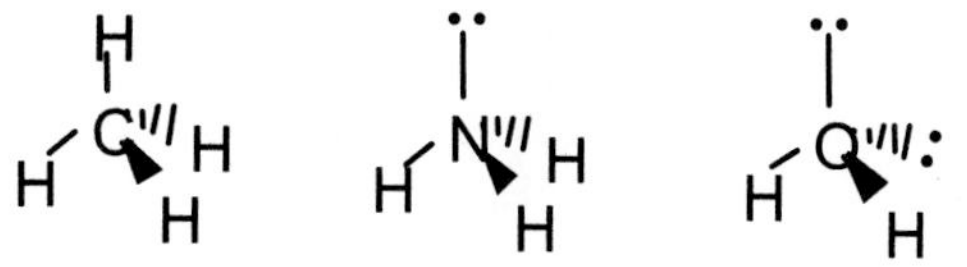

        109.5°   106.7°   104.5°

The deviation from 109.5° arises because there are three types of electron pair repulsions represented: (1) repulsion between bonding pairs, (2) repulsion between lone pairs, and (3) repulsion between a bonding pair and a lone pair. The force of repulsion is not the same among these three, but decreases as follows:

    lone–pair vs. lone–pair > lone–pair vs. bonding–pair > bonding–pair vs. bonding–pair
        repulsion   repulsion   repulsion

In other words, lone pairs require more space than bonding pairs and tend to push the bonding pairs closer together. Bonding electron pairs are attracted by two nuclei at the same time. These pairs require less space than lone electron pairs.

In ammonia, the lone pair on the nitrogen atom repels the bonding pairs more strongly than the bonding pairs repel each other. The result is that the H—N—H bond angle is several degrees less than a true tetrahedral angle.

In water, the oxygen atom has two bonding pairs and two lone pairs. The greater lone pair versus bonding pair repulsion causes the two O—H bonds to be pushed in toward each other. The H—O—H angle in water should be less than the H—N—H angle in ammonia.

In general, when non-bonding electrons are present on the central atom, the shape will be close to, but not exactly, as predicted in Table 7.2.

PRACTICE QUESTION 7.5
Which molecule represents a species in which there would be a deviation from ideal bond angles.

      a. SH_2 or CO_2
      b. BF_3 or PCl_3

D. Geometry of Molecules with More than One Central Atom.
In nature, complex molecules with more than six or seven atoms are quite common. In such cases, predicting an overall shape is more complicated so for now we will concentrate our efforts on being able to predict the shapes about each "central" atom. Let's look at a simple example first.

STUDY QUESTION 7.2
Predict the shape around each nitrogen atom in a molecule of hydrazine, N_2H_4:

Solution: In this molecule, either nitrogen could be considered as a central atom (remember that hydrogen is never a central atom). Therefore, our skeleton will consist of two nitrogen atoms sharing a bonding electron pair at the center. Let's determine the Lewis structure:

from the nitrogen atoms (2 @ 5 e⁻) = 10 electrons
from the hydrogen atoms (4 @ 1 e⁻) = 4 electrons
total   = 14 electrons

The Lewis structure is:

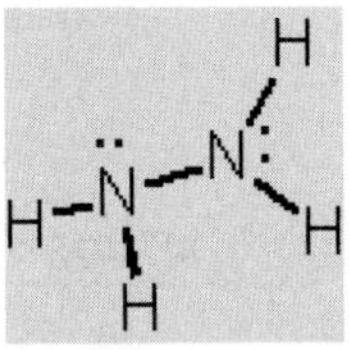

Around each nitrogen atom there are 3 bonding electron pairs and one lone pair. While the electron domain geometry is tetrahedral, the molecular shape (or geometry) observed around each nitrogen atom is trigonal pyramidal.

        PRACTICE QUESTION 7.6
        Predict the molecular shape around each central atom in propane, C_3H_8.

        PRACTICE QUESTION 7.7
        Study the following skeletal structure and assign the shapes around each central atom.

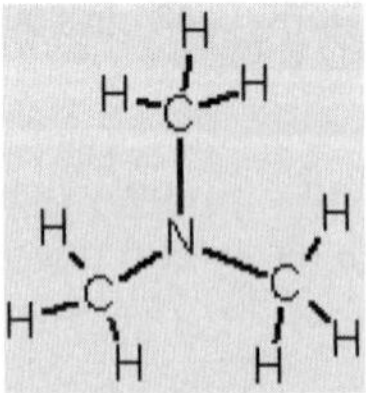

PROBLEM TYPE 2: MOLECULAR GEOMETRY AND POLARITY.

Polar bonds are those made between atoms of different electronegativities. The resulting charge separation is called a dipole. A molecule with an electric dipole is said to be polar and to possess a dipole moment.

Molecules without a dipole moment are called nonpolar molecules. In order for molecules to be classified as polar, two conditions must be fulfilled: (1) they must contain polar bonds, and (2) the orientation of the individual bond dipoles must not lead to their cancellation. All homonuclear diatomic molecules are nonpolar. In general, heteronuclear diatomic molecules are polar.

As mentioned, the dipole moment of a molecule containing more than one bond depends both on bond polarity and molecular geometry. In a polyatomic molecule the vectors representing polar bonds may add together to yield a polar molecule with a resultant dipole moment μ. Conversely, the vectors may cancel each other when added, producing a zero resultant dipole moment.

Although the best way to determine the dipole of a molecule is by vector addition, we can devise a set of empirical rules to predict if a molecule will be polar or nonpolar.

STUDY QUESTION 7.3
Predict whether the following molecules are polar or nonpolar:

  a. **CO**
  b. **H_2CO**
  c. **CCl_4**

Solution: Polarities of diatomic molecules depend on the electronegativity differences of the atoms.
a. In CO, oxygen is more electronegative than carbon. As a result, the $C \equiv O$ molecule is polar with partial positive and negative charges on C and O, respectively.

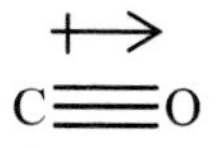

$$C \equiv O$$

b. In H_2CO, the dipolar nature of a polyatomic molecule depends on both the bond polarity and molecular geometry. Formaldehyde is a planar molecule with a polar $C \equiv O$ bond and two C—H bonds of very low polarity. Neglecting the low polarity C—H bond dipoles, the charge distribution is:

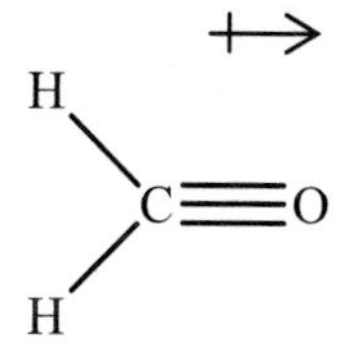

The net effect of all bond dipoles is that the center of positive charge is located near the carbon atom and the center of negative charge lies near the oxygen atom. H_2CO is a polar molecule.

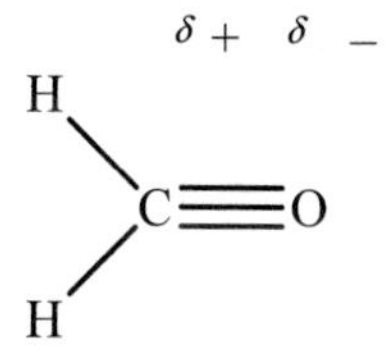

c. Chlorine is more electronegative than carbon. Therefore, CCl_4 has four bond dipoles. However, these polar bonds are arranged in a symmetric tetrahedral fashion about the central carbon atom.

In this situation the centers of positive and negative charge are both on the carbon atom. The bond dipoles have canceled each other and the molecule as a whole does not possess a dipole moment. CCl_4 is nonpolar.

PRACTICE QUESTION 7.8
What two requirements must be met in order for a molecule to be polar?

PRACTICE QUESTION 7.9
Which of the following molecules are polar?
  a. BCl_3
  b. SO_3
  c. PF_3
  d. SF_6

PRACTICE QUESTION 7.10
Which of the following molecules are polar?
  a. BeH_2
  b. CO_2
  c. SO_2
  d. NO_2

PROBLEM TYPE 3: INTERMOLECULAR FORCES.

Each phase or state, solid, liquid and gas, can be described from a molecular viewpoint in terms of four characteristics: distance between molecules, attractive forces between molecules, the motion of molecules, and the orderliness of the arrangement of molecules.

We saw previously the forces that hold atoms together, such as covalent bonds, which exist *within* molecules (intramolecular). Now, we will learn about the types of forces that hold different molecules close together as in a solid or a liquid. Forces of attraction between molecules are called **intermolecular forces (IMF)**, or **van der Waals forces**. These include **dipole–dipole interactions**, **hydrogen bonding**, and **London dispersion forces**, simply called dispersion forces. **Hydrogen bonding** can be viewed as a subset of dipole–dipole interactions.

Molecules that have a dipole moment, μ, are referred to as **polar molecules**. **Dipole–dipole interactions** are attractive forces between polar molecules. In a polar substance, molecules tend to become aligned with the positive end of one molecule directed toward the negative ends of neighboring molecules' dipoles. The attraction is electrostatic: *the larger the dipole moment, the stronger the force of attraction.*

Hydrogen bonding is a special type of dipole–dipole interaction. Hydrogen bonding is limited to compounds containing hydrogen bonded directly to fluorine, oxygen or nitrogen (small, highly electronegative atoms). The hydrogen atom in the polar bond is fairly stressed since the atom it is bonding to pulls the bonding electrons away from hydrogen, which results in a large dipole. Consequently, the H atom interacts with a high electron density atom, F, O, or N, in another molecule so that the hydrogen atom ends up increasing electron density around itself. Hydrogen bonds are the strongest of the intermolecular forces, with energies on the order of 10–40 kJ/mol.

In water, for instance, the attractions are stronger than just the attractions of one dipole for another. Each hydrogen atom, with its partial positive charge, is attracted to one of the lone electron pairs of an oxygen atom of a neighboring molecule.

Dispersion forces exist between nonpolar molecules (or atoms of the noble gases). These forces cause deviations from ideal gas behavior and low freezing points. Although these molecules (or atoms) have no permanent dipole moments, their electrons are in constant motion. If for an instant the electrons should move to the same side of the molecule (or nucleus for an atom), a short–lived dipole will exist. In an instant the electrons will continue their motion and the dipole will be gone, but a new one will be formed. This *instantaneous dipole* can polarize a neighboring molecule, thereby producing an *induced dipole*. The two dipoles will tend to stick together.

The strength of dispersion forces depends on the polarizability of the molecule and can be as large as, or larger than, dipole–dipole forces. **Polarizability** is the tendency of an electron cloud to be distorted by the presence of an electrical charge such as that of an ion or the partial charge of a dipole. In general, the polarizability increases as the total number of electrons in a molecule increases. Since molecular mass and number of electrons are related, the polarizability of molecules and the strength of dispersion forces increase with increasing molecular mass. All molecules, whether polar or nonpolar, exhibit dispersion forces.

The electrostatic attraction between an ion and a polar molecule is called an **ion–dipole force.** When an ionic compound dissolves in water, its cations and anions are attracted to water molecules. Water molecules are polar; they have a negative end and a positive end. The cations attract the negative end of the water molecules, and the anions attract the positive end of the water molecules. As the charge of an ion increases, it attracts polar molecules more strongly.

STUDY QUESTION 7.4
Which of the following substances should have the strongest intermolecular attractive forces: N_2, Ar, F_2, or Cl_2?

Solution: Note that none of these molecules (or atoms, in the case of Ar) is polar and that there is no chance for H bonding to occur. The only intermolecular forces existing in these molecules then are dispersion forces. Dispersion forces increase as the polarizability of the molecule increases, while polarizability increases with molecular mass. The most polarizable molecule will be Cl_2 (70.9 amu).

STUDY QUESTION 7.5
Indicate all the different types of intermolecular forces that exist in each of the following substances.

    a. $CCl_4(l)$
    b. $HBr(l)$
    c. $CH_3OH(l)$

Solution: The types of intermolecular forces (IMF) depend primarily on the polarity of the molecule and the electronegativity of atoms.
a. CCl_4 is nonpolar. The only type of IMF between nonpolar molecules are dispersion forces.
b. HBr is a polar molecule. The types of IMF are dipole–dipole and dispersion forces. There is no hydrogen bonding in HBr as the Br atom is large and has low electronegativity.
c. CH_3OH is polar and has a hydrogen atom bonded to an oxygen atom. The IMF are hydrogen bonding and dispersion forces.

STUDY QUESTION 7.6
The dipole moment μ in HCl is 1.03 D, and in HCN it is 2.99 D. Which one should have the higher boiling point?

Solution: The larger the dipole moment, the stronger the IMF. The stronger the IMF, the higher the temperature needed to provide molecules of the liquid with enough kinetic energy to overcome these attractive forces. The boiling point of HCN should be greater than that of HCl. Observed values are b.p. (HCl) = –85 °C, and b.p. (HCN) = 26 °C.

PRACTICE QUESTION 7.11

Indicate the types of intermolecular forces that exist between molecules (or basic units) in each of the following.

a. PCl_3
b. CO_2
c. Cl_2
d.   ICl
e.   KCl

PRACTICE QUESTION 7.12

Which member of each pair should have the higher boiling point?

a. O_2 or CO
b. Br_2 or ICl
c. H_2O or H_2S
d. PH_3 or AsH_3

PROBLEM TYPE 4: VALENCE BOND THEORY.

The Valence Bond theory is one of two quantum mechanical descriptions of chemical bonding currently in use. The other is the molecular orbital theory which will be discussed in the next section. A covalent bond forms when the orbitals from two atoms overlap and a pair of electrons occupies the region between the atoms. This overlap produces a region between the two nuclei where the probability of finding an electron is greatly enhanced. The presence of greater electron density between the two atoms tends to attract both atoms and bonding occurs.

STUDY QUESTION 7.7

Using unhybridized atomic orbitals, describe what atomic orbitals used in the following molecules to form covalent bonds:

a. F_2
b. H_2S

Solution:

a. From the orbital diagram of fluorine we can see that an F atom has one half–filled $2p$ orbital that can be used for bonding. If this orbital overlaps with the half–filled $2p$ orbital of another F atom, a pair of electrons (one from each atom) spend most of the time between the two atoms and bonding occurs.

$$F \quad \boxed{\uparrow\downarrow} \quad \boxed{\uparrow\downarrow \,|\, \uparrow\downarrow \,|\, \uparrow}$$
$$\quad\quad 2s \quad\quad\quad 2p$$

b. Hydrogen atoms have a $1s^1$ electron configuration. Therefore, they can use the half–filled $1s$ orbital for bonding. The orbital diagram of the valence electrons in an S atom is

$$S \quad \boxed{\uparrow\downarrow} \quad \boxed{\uparrow\downarrow \,|\, \uparrow \,|\, \uparrow}$$
$$\quad\quad 3s \quad\quad\quad 3p$$

Overlap of the two half–filled $3p$ orbitals of a sulfur atom with the half–filled $1s$ orbitals of the two H atoms results in the formation of two covalent bonds.

NOTE: The directional properties of two p–orbitals (say the p_y and p_z) are such that the orbitals project out from the S atom at right angles to each other. The H—S—H bond angle should be 90°. The measured angle is 92.2°.

PRACTICE QUESTION 7.13
Using only atomic orbitals, describe the bond formation in:

    a.   **Cl_2**
    b.   **HBr**

PROBLEM TYPE 5: HYBRIDIZATION OF ATOMIC ORBITALS.

The concept of hybridization is used to account for bonding and geometry in molecules with trigonal planar, tetrahedral, and more complex shapes. The process of orbital mixing is called hybridization, and the new atomic orbitals are called **hybrid orbitals**.

A. Hybridization of *s* and *p* Orbitals.

We can illustrate the main features of hybridization by examining the linear $BeCl_2$ molecule. To start with, an isolated Be atom has the ground state electron configuration $1s^2 2s^2$. In a Be atom, the 2*s* electrons are already paired into an orbital and, in principle, the Be atom should have just a small tendency to form covalent bonds. However, the 2*p* orbitals lie quite close in energy to the 2*s* orbital. Orbitals on the same atom that lie close together in energy, have an ability to combine with one another to form *hybrid orbitals*. The presence of two Be—Cl bonds can be accounted for if, prior to bond formation, a 2*s* electron in a Be atom is promoted into an empty 2*p* orbital.

Be   ↑   ↑ | |

           2s   2p

Now there are two unpaired electrons that could participate in two covalent bonds to two chlorine atoms. Since one electron is in an *s*–orbital and one is in a *p* orbital, we would expect two types of Be—Cl bonds. However, both Be—Cl bonds are observed to be the same. This equivalence is explained by hybridization or "mixing" of the 2*s* and the 2*p* orbitals to create two new equivalent orbitals called *sp* hybrid orbitals.

Be   ↑ | ↑ | |

         2sp   2p

B. Hybridization of *s*, *p*, and *d* Orbitals.

For elements on Period 3 and beyond, it is possible to expand the octet by mixing *s*–, *p*–, and *d*–orbitals. Let's look at the formation of the covalent bonds in SF_6. The Lewis structure for this compound is shown below.

The central sulfur atom has 6 valence electrons arranged in the 3*s* and 3*p* sublevels as follows:

↑↓   ↑↓ | ↑ | ↑

    3s   3p

Looking at this orbital diagram, we may think that sulfur can only share those two electrons that are unpaired in the 3*p* orbitals. How can we make 6 covalent bonds? The answer to this question is once again provided by Valence Bond theory and the concept of hybridization. Sulfur has empty 3*d* orbitals on the third energy level that lie quite close in energy. By promoting some of the electrons to the *d* orbitals, it can maximize its bonding capability. If we mix, one *s*–orbital with three *p*–orbitals and two *d*–orbitals ($1 + 3 + 2 = 6$) we must get as a result 6 orbitals of equal energy, as shown below:

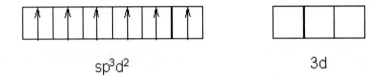

With this arrangement, sulfur can form six equivalent covalent bonds. Notice that we can easily predict the types of hybrid orbitals used by a central atom from a Lewis structure. The structure shown for SF_6 has six electron domains on the central sulfur atom, all of them being bonding electron pairs. In order to "accommodate" six electron pairs, six orbitals of equal energies are needed. Therefore, by mixing the six orbitals, 1 "*s*" + 3 "*p*" + 2 "*d*" = 6 orbitals of $s^1p^3d^2$ hybridization. Notice that the sum of the superscripts equals the number of orbitals mixed (the superscript "1" is normally omitted).

We can summarize the rules of hybridization as follows:

1. Hybridization is a process of mixing orbitals on a single atom in a molecule.
2. Only orbitals of similar energies can be mixed to form hybrid orbitals.
3. The number of hybrid orbitals obtained always equals the number of orbitals mixed together.
4. Hybrid orbitals are identical in shape and they project outward from the nucleus with characteristic orientations in space.
5. To determine the hybridization, determine the arrangement of the electron pairs (all atoms and lone pair included) about the central atom. The hybridization must show the number of orbitals needed to accommodate the number of electron pairs about the central atom. The number of orbitals that should be hybridized equals the number of electron domains on the central atom.

Table 7.4 in your textbook shows several of the more important hybrid orbitals discussed in the textbook and their characteristic geometrical shapes.

STUDY QUESTION 7.8
Using orbital diagrams, outline the steps used to describe the formation of a set of sp^2 hybridized orbitals on a carbon atom.

Solution: The ground state orbital diagram for the *valence electrons* of a carbon atom is:

$$C \quad \boxed{\uparrow\downarrow} \quad \boxed{\uparrow \,|\, \uparrow \,|\, }$$
$$2s \qquad\qquad 2p$$

First, promote one $2s$ electron to the empty $2p$ orbital. Mixing or hybridization of the $2s$ orbital with two of the $2p$ orbitals creates three equivalent sp^2 hybrid orbitals. These hybrid orbitals point out from the carbon atom to the corners of a triangle. One electron remains in an unhybridized $2p^2$ atomic orbital.

$$C \quad \boxed{\uparrow \,|\, \uparrow \,|\, \uparrow} \quad \boxed{\uparrow}$$
$$2sp \qquad\qquad 2p$$

STUDY QUESTION 7.9
Describe the hybridization of the central atom in:

    a. **PH_3**
    b. **PCl_5**

Solution: First, use VSEPR theory to predict the geometry. Then, match the geometry to the corresponding hybridization.

a. According to VSEPR theory, the PH_3 molecule has trigonal pyramidal geometry and a lone pair. There are four electron pairs around the P atom that occupy the corners of a tetrahedron.

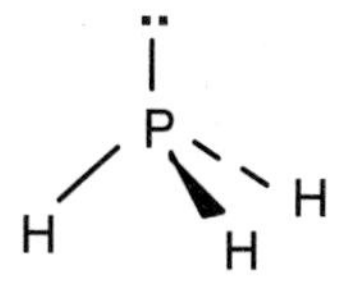

Therefore, four equivalent orbitals project out from the central P atom at 109° from each other. The bonding orbitals of the P atom must be sp^3 hybrid orbitals.

b. According to VSEPR theory, PCl_5 has the geometrical shape of a trigonal bipyramid.

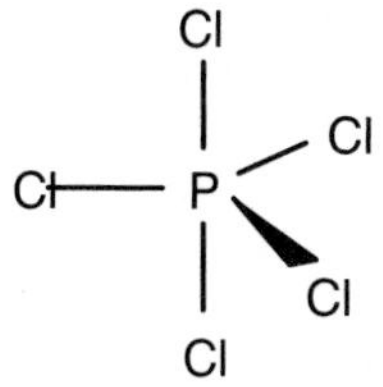

The type of hybridization that gives five equivalent orbitals that point to the corners of a trigonal bipyramid is sp^3d.

PRACTICE QUESTION 7.14
What is the angle between the following two hybrid orbitals on the same atom?

    a.   Two sp orbitals
    b.   Two sp^2 orbitals
    c.   Two sp^3 orbitals

PRACTICE QUESTION 7.15
What hybrid orbital set is used by the underlined atom in the following molecules?

    a. $\underline{C}S_2$
    b. $\underline{N}O_2^-$
    c. $\underline{B}F_3$
    d. $\underline{Cl}F_3$
    e. $H_2\underline{C}O$

PRACTICE QUESTION 7.16
What hybrid orbital set is used by the underlined atom in the following molecules?

    a. $\underline{S}F_2$
    b. $\underline{S}F_4$
    c. $\underline{Xe}F_4$

PRACTICE QUESTION 7.17
The selenium atom in selenium hexachloride is sp^3d^2 hybridized. What is the geometry of $SeCl_6$?

PROBLEM TYPE 6: HYBRIDIZATION IN MOLECULES CONTAINING MULTIPLE BONDS.

Double and triple bonds can also be understood in terms of overlap of atomic orbitals. The C=C double bond in the ethylene molecule (see Figure 7.7 in Section 7.5 of the textbook), consist of one sigma (σ) bond and one pi (π) bond.

A sigma bond is a covalent bond in which the electron density is concentrated in a cylindrical pattern around the line of centers between the two atoms. In the C=C bond, the σ bond results from overlap of an sp^2 hybrid orbital from one of the carbon atoms with a sp^2 hybrid orbital from the other carbon atom.

In a pi bond the electron density is concentrated above and below the C—C internuclear axis, but not cylindrically as in a sigma bond (Figure 7.7b in your textbook). The pi bond has two regions of electron density, but it should be kept in mind that it is only *one* bond. The π bond in ethylene results from sideways overlap of the unhybridized $2p_z$ atomic orbitals of the two carbon atoms.

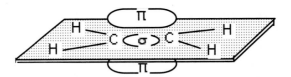

STUDY QUESTION 7.10
Account for the bonding in HCN. State the types of bonds (σ or π) and the orbitals used in their formation.

Solution: First, let's take a look at the Lewis structure for HCN.

On the central carbon atom, there is a total of two electron domains (remember that multiple bonds count as one). The number of orbitals that need to mix in order to form hybrid orbitals that will overlap to form sigma bonds is also two. We can therefore mix one s and one p orbital from the second energy level of a carbon atom. Carbon uses sp orbitals to form the sigma bonds. These are the bonds made between carbon and hydrogen, plus one of the bonds formed with nitrogen. The sp hybrid orbital on carbon overlaps with the $1s$ orbital from hydrogen forming a bond in which the shared electron density lies between the region defined by the internuclear axis. Similarly, overlap of the sp hybrid orbital on carbon and a $2p$ orbital on nitrogen leads to the formation of the second sigma bond.

In order to form a triple bond, a sigma bond and two pi bonds must be used. The formation of the π bonds can be accounted for by using the remaining unhybridized orbitals on both carbon and nitrogen (overlap of $2p$ unhybridized orbital on carbon and a $2p$ unhybridized orbital on nitrogen placing most of the shared electron density in a region above and below the internuclear axis).

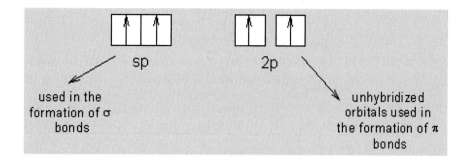

PRACTICE QUESTION 7.18
How many sigma bonds and pi bonds are there in a molecule of CO_2?

PROBLEM TYPE 7: MOLECULAR ORBITAL THEORY.

Until now we have considered covalent bonds in terms of the *Valence Bond theory,* where the electrons in a *molecule* occupy atomic orbitals of the individual *atoms*. Those orbitals, whether hybrids or not, are still considered to be localized on each atom. Covalent bonds are visualized in terms of the overlap of two half–filled atomic orbitals. This overlap increases the electron density between both atomic centers.

The **Molecular Orbital** *model* has a different approach. In this model, the available atomic orbitals are combined to form orbitals that belong to the *entire molecule*. Like atomic orbitals, these molecular orbitals have different shapes, sizes, and energies. Once the molecular orbitals have been derived, electrons are allotted to them in a manner analogous to allotting electrons to the orbitals of atoms.

A. Bonding and Antibonding Molecular Orbitals.
According to the Molecular Orbital (MO) theory, the result of the interaction of two $1s$ orbitals is the formation of two molecular orbitals. One of these is a bonding molecular orbital, which has a lower energy than the original $1s$ orbitals. The other molecular orbital is an antibonding orbital, which is of higher energy than the original $1s$ orbitals. It is the greater stability of the electrons in the bonding MO that results in covalent bond formation. In the bonding MO, electron density is concentrated between the two atoms. While the antibonding MO actually has a node (a region where ψ^2 approaches zero) in the electron density midway between the two nuclei, there is no contribution to bonding when electrons enter antibonding MOs. Take a look at Figure 7.11 in your textbook to visualize these characteristics of bonding and antibonding orbitals.

B. σ Molecular Orbitals.
The antibonding and bonding MOs that result from combining $1s$ orbitals are examples of sigma molecular orbitals. The electron density of sigma MOs is distributed symmetrically about the line of centers between the two atoms. The sigma bonding MO and sigma antibonding MO resulting from combination of $1s$ atomic orbitals are designated σ_s and σ_{1s}^*, respectively. Sigma MOs also result from the combination of certain other atomic orbitals. For instance, the $2p_x$ orbitals on different atoms combine to form σ_{2p} bonding and σ_{2p}^* antibonding orbitals as shown in Figure 7.15 of the text.

C. Bond Order.
When comparing the stabilities of molecules, we use the concept of bond order.

$$\text{bond order} = \frac{1}{2}\left(\begin{array}{l}\text{number of electrons} \\ \text{in bonding MOs}\end{array} - \begin{array}{l}\text{number of electrons} \\ \text{in antibonding MOs}\end{array}\right)$$

The value of the bond order is a measure of the stability of the molecule. For H_2:

$$\text{bond order} = \frac{1}{2}(2 - 0) = 1$$

and for He_2:

$$\text{bond order} = \frac{1}{2}(2 - 2) = 0$$

A bond order of 1 represents a single covalent bond, and a bond order of 0 means the molecule is not stable. Bond orders of 2 and 3 result for molecules with double bonds and triple bonds, respectively.

D. π Molecular Orbitals.

Two $2p_y$ orbitals and two $2p_z$ orbitals on different atoms can approach each other sideways as shown in Figure 7.16 of the text. In the resulting molecular orbital, the electron density is concentrated above and below the line joining the two bonded atoms, rather than around the line of centers. Such an MO is called a pi molecular orbital. The symbol π_{2p} stands for a bonding pi orbital formed by combination of two $2p$ atomic orbitals. The pi antibonding orbital is designated, π_{2p}^{*}.

E. Molecular Orbital Diagrams.

For most homonuclear diatomic molecules containing atoms of second period elements with $Z \leq 7$, an approximate order of molecular energy levels is:

$$\sigma_{1s} < \sigma_{1s}^{*} < \sigma_{2s} < \sigma_{2s}^{*} < \pi_{2p_y} = \pi_{2p_z} < \sigma_{2p_x} < \pi_{2p_y}^{*} = \pi_{2p_z}^{*} < \sigma_{2p_x}^{*}$$

The electron configurations of diatomic molecules of the second period elements and some of their known properties are summarized in Figure 7.18 in the textbook. This figure shows the bond order of these molecules as predicted by MO theory and the corresponding bond enthalpies and bond lengths. The number of unpaired electrons also correlates with the magnetic properties of these molecules. The MO theory satisfactorily predicts the magnetic properties of molecules, something that the valence bond theory does not do. Notice that for atoms with $Z > 7$, there is a reversal of the ordering of the σ_{2p_x} and the π_{2p_y} and π_{2p_z} MOs. This is due to an increase in the interaction of the s and p orbitals with increasing nuclear charge.

In summary:

1. The number of molecular orbitals formed is always equal to the number of atomic orbitals combined.
2. The more stable the bonding molecular orbital, the less stable is the corresponding antibonding molecular orbital.
3. In a stable molecule, the number of electrons in bonding molecular orbitals is always greater than that in antibonding molecular orbitals.
4. As with atomic orbitals, each molecular orbital can accommodate up to two electrons with opposite spins, in accordance with the Pauli's exclusion principle.
5. When electrons are added to molecular orbitals having the same energy, the most stable arrangement is that predicted by Hund's rule. That is, electrons occupy these orbitals singly, and with parallel spins, rather than in pairs.
6. The number of electrons in the molecular orbitals is equal to the sum of all the electrons on the atoms.

STUDY QUESTION 7.11

Represent bonding in the O_2^- ion by means of a molecular orbital diagram. What is the bond order?

Solution: The O_2 molecule has 16 electrons. The O_2^- ion will have 17 electrons. The energy level diagram follows. The 4 electrons in the σ_{1s} and σ_{1s}^* orbitals are not shown because they are not involved in bonding.

$$
\begin{array}{cc}
\underline{} & \sigma_{2p}^* \\[4pt]
\underline{\uparrow\downarrow}\quad\underline{\uparrow} & \pi_{2p}^* \\[4pt]
\underline{\uparrow\downarrow}\quad\underline{\uparrow\downarrow} & \pi_{2p} \\[4pt]
\underline{\uparrow\downarrow} & \sigma_{2p} \\[4pt]
\underline{\uparrow\downarrow} & \sigma_{2s}^* \\[4pt]
\underline{\uparrow\downarrow} & \sigma_{2s}
\end{array}
$$

The electron configuration of O_2^- ion is

$$O_2^- : (\sigma_{1s})^2 (\sigma_{1s}^*)^2 (\sigma_{2s})^2 (\sigma_{2s}^*)^2 (\sigma_{2p_x})^2 (\pi_{2p_z})^2 (\pi_{2p_y})^2 (\pi_{2p_y}^*)^2 (\pi_{2p_z}^*)^1$$

$$\text{bond order} = \frac{1}{2}\left(\begin{array}{c}\text{number of electrons} \\ \text{in bonding MOs}\end{array} - \begin{array}{c}\text{number of electrons} \\ \text{in antibonding MOs}\end{array}\right)$$

$$= \frac{1}{2}(8-5) = 1.5$$

STUDY QUESTION 7.12
Which of the following species are paramagnetic and which are diamagnetic?

    a. N_2

    b. N_2^+

    c. B_2

Solution:
a. The electron configuration of N_2 is:

$$N_2 : (\sigma_{1s})^2 (\sigma_{1s}^*)^2 (\sigma_{2s})^2 (\sigma_{2s}^*)^2 (\pi_{2p_y})^2 (\pi_{2p_z})^2 (\sigma_{2p_x})^2$$

All orbitals contain pairs of electrons. Because of the lack of unpaired electrons, N_2 is diamagnetic.

b. N_2^+ will have one less electron than N_2. Therefore, it has one unpaired electron and is paramagnetic.

c. The electron configuration of B_2 should have 10 electrons:

$$B_2 : (\sigma_{1s})^2 (\sigma_{1s}^*)^2 (\sigma_{2s})^2 (\sigma_{2s}^*)^2 (\pi_{2p_y})^1 (\pi_{2p_z})^1$$

Because of the unpaired electrons, B_2 is paramagnetic.

PRACTICE QUESTION 7.19
Which has the higher energy, a bonding molecular orbital or its corresponding antibonding molecular orbital? A bonding molecular orbital or the atomic orbitals from which it is created?

PRACTICE QUESTION 7.20
Compare the bond order in Be_2 to that in Be_2^+?

PRACTICE QUESTION 7.21
Give the electron configurations for Li_2 and Li_2^+. Which has the stronger Li-Li bond?

PRACTICE QUESTION 7.22
Which of the following diatomic species are paramagnetic?

    a. O_2

    b. O_2^-

    c. Li_2

PRACTICE QUESTION 7.23
Both N_2 and O_2 can be ionized to form N_2^+ and O_2^+, respectively. Explain why the bond order of N_2 is greater than that of N_2^+, but the bond order of O_2 is less than that of O_2^+.

PROBLEM TYPE 8: BONDING THEORIES AND DESCRIPTIONS OF MOLECULES WITH DELOCALIZED BONDING.

Recall the discussion of resonance? To understand the bonding in molecules that exhibit resonance using MO theory, we must discuss delocalization of electrons. If a pair of electrons is not confined between two adjacent atoms, but is shared between three or more atoms, the bond formed involves a delocalized molecular orbital. To demonstrate the concept, we will examine NO_2^-. The resonance structures are as follows:

To describe the bonding, look at the central atom (of either resonance structure). The nitrogen has two bonded atoms and one lone pair for three electron domains. This is a trigonal planar geometry and, thus, the nitrogen has sp^2 hybridization. Two of the sp^2 hybrid orbitals form the sigma bonds with the oxygen atoms and one houses the lone pair. The pi–bond, is shown in the resonance structure to alternate between the two oxygen atoms. In delocalized molecular orbital theory, the two electrons which make up the pi bond are delocalized about the three atoms, by existing in the shared space formed by the overlap of the *p*–orbitals of each atom, which lie perpendicular to the plane formed by the sigma bonds. This gives electron density above and below the plane of the O–N–O bond.

The two electrons which make up the pi–bond are able to move about in the grey area depicted in the bottom diagram. In other words, the two electrons making the double bond (shown in the resonance structure) are now delocalized about the whole molecule. Anytime multiple resonance structures can be drawn for a molecule or ion, delocalization molecular orbitals can be used to describe the pi bonds in the molecule.

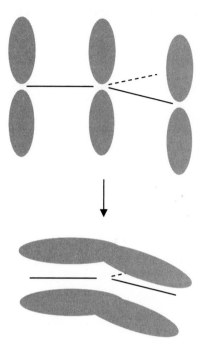

STUDY QUESTION 7.13
Which of the following species will contain delocalized molecular orbitals?

    a. SO_2
    b. CO

Solution: First draw the Lewis structure, including any resonance structures
a. For SO_2, the resonance structures are

$$:\ddot{O}\!-\!\ddot{S}\!=\!\ddot{O} \longleftrightarrow \ddot{O}\!=\!\ddot{S}\!-\!\ddot{O}:$$

The fact that resonance structures can be drawn determines that delocalized molecular orbitals exist. The two electrons which make up the pi bond are delocalized in orbitals formed by the side-ways overlap of the p orbitals on all three atoms.
b. For CO, the Lewis structure is:

$$:C\!\equiv\!O:$$

Because there is only one way to draw this Lewis structure, there are no resonance structures; the molecule does not contain delocalized molecular orbitals.

    PRACTICE QUESTION 7.24
    What orbitals overlap to form a delocalized π bond in ozone (O_3)?

PRACTICE QUIZ

1. Use VSEPR theory to predict the shapes of the following molecules:
   a. BrF_5 b. HCN
   c. BF_3 d. SO_2
   e. SCl_2

2. What type of hybrid orbital is used by the central atom of each of the following molecules?
   a. BrF_5 b. HCN
   c. BF_3 d. SO_2
   e. SCl_2

3. Use VSEPR theory to predict the shapes of the following polyatomic ions:
   a. BeF_3^- b. AsF_4^-
   c. ClF_4^- d. NO_3^-
   e. SO_4^{2-}

4. What type of hybridization is used by the central atom of each of the following polyatomic ions?
   a. BeF_3^- b. AsF_4^-
   c. ClF_4^- d. NO_3^-
   e. SO_4^{2-}

5. Which of the following should have the strongest intermolecular forces?
   a. CH_4 b. Cl_2 c. CO   d. CS_2

6. Which of the following molecules would be tetrahedral?
   a. SO_2 b. SiH_4
   c. SF_4 d. BCl_3
   e. XeF_4

7. Predict the approximate bond angles in
   a. $GeCl_2$ b. IF_4^- c. $TeCl_4$

8. Which of the following species are capable of hydrogen bonding with another like molecule?
   a. CH_3F b. HF   c. CH_3CH_2OH d. CH_3NH_2 e. CH_4

9. What type of forces must be overcome in order to:
   a. vaporize water?   b. dissociate H_2 into H atoms?   c. boil liquid O_2?

10. CCl_4 is a perfect tetrahedron, but $AsCl_4^-$ is a distorted tetrahedron. Explain.

11. $BeCl_2$ and $TeCl_2$ are both covalent molecules, yet $BeCl_2$ is linear while $TeCl_2$ is nonlinear (bent). Explain.

12. What types of hybrid orbitals can be formed by elements of the third period that cannot be formed by elements of the second period?

13. When we describe the formation of hybrid orbitals on a central atom, must all the available atomic orbitals enter into hybridization?

14. Which molecule should have the larger dipole moment, HBr or HI?

15. Which one of the following molecules has a dipole moment?
    a. CCl_4 b. H_2S
    c. CO_2 d. BCl_3
    e. Cl_2

16. Which of the following molecules have no dipole moment?
    a. ClF   b. BCl_3
    c. $BeCl_2$ d. Cl_2O
    e. H_2CO

17. Choose the best answer. For the water molecule:
    a. The bonds are polar, and the molecule is nonpolar.
    b. The bonds are nonpolar, and the molecule is polar.
    c. The bonds are polar, and the molecule is polar.
    d. The bonds are nonpolar, and the molecule is nonpolar.

18. Hydrogen peroxide (H_2O_2) has a dipole moment of 2.1 D. Which of the bonds in H_2O_2 are polar? Is the molecule linear?

19. The compound calcium carbide, CaC_2, contains the acetylide ion C_2^{2-}.

    a. Write molecular orbital electron configurations for C_2 and C_2^{2-}.

    b. Compare the bond orders in C_2 and C_2^{2-}.

20. The bond length in N_2 is 109 pm and in N_2^+ it is 112 pm. Explain why the bond lengths differ in this way.

21. In molecules having the trigonal bipyramidal geometry why does a lone pair occupy an equatorial position rather than an axial position?

22. Use the concept of delocalized molecular orbitals to explain the bonding in the oxalate anion, $C_2O_4^{2-}$.`

Chapter 8
Chemical Reactions

PROBLEM-SOLVING STRATEGIES AND TUTORIAL SOLUTIONS

TYPES OF PROBLEMS

Problem Type 1:   Chemical Equations.
- **A.** Interpreting and Writing Chemical Equations.
- **B.** Balancing Chemical Equations.
- **C.** Patterns of Chemical Reactivity.

Problem Type 2:   Combustion Analysis.
- **A.** Determination of Empirical Formula.

Problem Type 3:   Calculations with Balanced Chemical Equations.
- **A.** Moles of Reactants and Products.
- **B.** Mass of Reactants and Products.

Problem Type 4:   Limiting Reactants.
- **A.** Determining the Limiting Reactant.
- **B.** Reaction Yield.

Problem Type 5:   Periodic Trends in Reactivity of the Main Group Elements.
- **A.** General Trends in Reactivity.
- **B.** Reactions of the Active Metals.
- **C.** Reactions of Other Main Group Elements.
- **D.** Comparison of Group 1A and Group 1B Elements.

PROBLEM TYPE 1: CHEMICAL EQUATIONS.

A. Interpreting and Writing Chemical Equations.

A chemical reaction is a process in which one or more chemical substances are changed into one or more new substances. A chemical equation is symbolic shorthand for representing a chemical reaction. The chemical formulas of the reactants (starting materials) are written on the left side of the equation, and the formulas of the products on the right side. Equations are written according to the following format:

a.   Reactants and products are separated by an arrow. The arrow is read as "produces" or "yields."

b.   Plus signs are placed between reactants and between products. A plus sign between reactants means that all the reactants are required for reaction. A plus between products implies that a mixture of two or more products is formed by the reaction. The plus sign is read as "plus" or "and."

c.   Abbreviations are sometimes included in parentheses to indicate the physical states of the reactants and products.

d.   Balanced equations reflect the law of conservation of mass. That is, the number of atoms of a certain element appearing in the reactants must be equal to the number of atoms of that element appearing in the products.

e.   Differences in the relative amounts of reactants consumed or products generated are reflected by stoichiometric coefficients. A coefficient is a number placed before a chemical formula in an equation that is a multiplier for the formula. For example, $3H_2O$ means three molecules of water, a total of 6 hydrogen atoms and three oxygen atoms. Subscripts, 2 in H_2O, represent the number of atoms in a compound. Absence of a coefficient or element subscript is understood to mean one.

B. Balancing Chemical Equations.

When the hydrocarbon pentane (C_5H_{12}) is burned or combusted (reacted with O_2), carbon dioxide and water are produced. The unbalanced chemical equation representing the reaction is:

$$C_5H_{12} + O_2 \rightarrow CO_2 + H_2O$$

Eventually, you will not need a procedure to balance chemical reactions. The basic idea is to change the coefficients of the reactants and products until the numbers of each type of atom are the same on each side of the reaction. It is important that you <u>NEVER</u> change the subscripts of the reactants or products; changing the subscripts changes the chemical compound. For now, use the following rules to balance equations, or develop your own. For simplicity, the physical state subscripts are omitted during this process.

1. Count and list the number of each type of atom on each side of the arrow.
2. Look for elements that occur once on both sides of the equation; change the coefficients so that these elements are balanced, if necessary.
3. Change the coefficient of any remaining components to accommodate the changes made in step 2.

Check the balanced equation to be sure the numbers of each atom are equal on both sides of the equation.

Consider the combustion (burning) of pentane, a petroleum product:

Step #	Equation	# reactant atoms			# product atoms			Balanced?
1	$C_5H_{12} + O_2 \rightarrow CO_2 + H_2O$	5 C	12 H	2 O	1 C	2 H	3 O	No
2	$C_5H_{12} + O_2 \rightarrow 5CO_2 + H_2O$	5 C	12 H	2 O	5 C	2 H	11 O	No
3	$C_5H_{12} + O_2 \rightarrow 5CO_2 + 6H_2O$	5 C	12 H	2 O	5 C	12 H	16 O	No
4	$C_5H_{12} + 8O_2 \rightarrow 5CO_2 + 6H_2O$	5 C	12 H	16 O	5 C	12 H	16 O	Yes

STUDY QUESTION 8.1

Balance the following reaction. In this displacement reaction, zinc displaces H_2 from chemical combination with chlorine.

$$Zn(s) + HCl(aq) \rightarrow ZnCl_2(aq) + H_2(g)$$

Solution:
Step 1: Identify elements that occur in only two compounds in the equation. In this case, each of the three elements Zn, H, and Cl occurs in only two compounds.
Step 2: Of these, balance the element with the largest subscript. Both Cl and H have the subscript 2, so balance either one of them first. Multiplying HCl by 2 to balance H:

$$Zn + 2HCl \rightarrow ZnCl_2 + H_2$$

Counting atoms on both sides:

Reactants	Products
H (2)	H (2)
Cl (2)	Cl (2)
Zn (1)	Zn (1)

At this point, both H and Cl are balanced. Further inspection reveals that Zn is also balanced. Thus, the equation is balanced!

STUDY QUESTION 8.2
The reaction of a hydrocarbon with oxygen is an example of a combustion reaction. Balance the equation for the reaction of hexyne with oxygen gas:

$$C_6H_{10} + O_2 \rightarrow CO_2 + H_2O$$

Solution: Following step 1 given in the previous example, we note that both H and C appear in only two compounds. According to step 2 (balancing the element with the highest subscript) H has the largest subscript. Balance H first by placing the coefficient 5 in front of H_2O. Next balance the carbon atoms by placing a 6 in front of CO_2.

$$C_6H_{10} + O_2 \rightarrow 6CO_2 + 5H_2O$$

Taking stock:

Reactants	Products
C (6)	C (6)
H (10)	H (10)
O (2)	O (17)

Finally, the oxygen atoms can be balanced. The products already have their coefficients determined, and so the reactants must be adjusted to supply 17 O atoms. Multiplying the O_2 by 17/2 gives us 17 O atoms in the reactants side. However, we should not have fractional coefficients. To make all coefficient whole numbers, we double all the coefficients to get rid of the fraction in front of O_2. This gives us

$$2C_6H_{10} + 17O_2 \rightarrow 12CO_2 + 10H_2O$$

Reactants	Products
C (12)	C (12)
H (20)	H (20)
O (34)	O (34)

The equation is now balanced.

STUDY QUESTION 8.3
Balance the following decomposition reaction:

$$NaClO_3(s) \rightarrow NaCl(s) + O_2(g)$$

Solution: According to steps 1 and 2, O should be balanced first. There are 3 O atoms on the reactant side and 2 on the product side. The previous example showed us how to balance an equation starting with fractional coefficients then multiplying through to make all coefficients whole numbers. Another approach can be taken. When an element that occurs in only two substances in the equation and the number of atoms of that element is even on one side and odd on the other side of the equation, use two coefficients that increase the number of atoms on each side of the equation to the least common multiple.

$$NaClO_3 \rightarrow NaCl + O_2$$

Reactants	Products
O (3)	O (2)

The least common multiple for balancing O is $3 \times 2 = 6$. The two coefficients are 2 and 3.

$$2NaClO_3 \rightarrow NaCl + 3O_2$$

Reactants	Products
O (6)	O (6)

Balancing the Na and Cl atoms gives:

$$2NaClO_3 \rightarrow 2NaCl + 3O_2$$

STUDY QUESTION 8.4

Balance the following precipitation reaction in which sulfuric acid reacts with barium chloride in aqueous solution to yield hydrochloric acid and a precipitate of barium sulfate.

$$H_2SO_4(aq) + BaCl_2(aq) \rightarrow HCl(aq) + BaSO_4(s)$$

Solution: Here, it helps to recognize certain groups of atoms that maintain their identity during the reaction. In this equation the sulfate ion, SO_4^{2-}, appears as a unit.

$$H_2SO_4 + BaCl_2 \rightarrow HCl + BaSO_4$$

Reactants	Products
H (2)	H (1)
SO_4^{2-} (1)	SO_4^{2-} (1)
Ba (1)	Ba (1)
Cl (2)	Cl (1)

Either H or Cl could be balanced first. Balance chlorine by multiplying HCl by 2:

$$H_2SO_4 + BaCl_2 \rightarrow BaSO_4 + 2HCl$$

A check of the H atoms shows they are now balanced along with the other elements.

PRACTICE QUESTION 8.1
Rust (Fe_2O_3) forms readily when iron is exposed to air. Write a balanced chemical equation for the formation of rust.

PRACTICE QUESTION 8.2
Write a balanced chemical equation for the reaction between hydrogen gas and carbon monoxide to yield methanol (CH_3OH).

PRACTICE QUESTION 8.3
The balanced chemical equation for the reaction between hydrogen gas and oxygen gas follows.
$$2 H_2 + O_2 \rightarrow 2 H_2O$$
Explain how this equation is balanced when the total number of moles in the reactant side is 3 and that in the product side is only 2.

C. Patterns of Chemical Reactivity.

Three of the commonly encountered types of reactions are *combination, decomposition, and combustion*. Combination is a reaction where two or more reactants all combine to form one product. A generic form of the reaction is

$$AB \rightarrow A + B.$$

Decomposition, opposite of combination, is a reaction where two or more reactants are formed from only one reactants. A generic form of the reaction is

$$M + Q \rightarrow MQ.$$

Combustion is the burning of an organic compound (contains C, H, and maybe O), considered the fuel. The reaction involves the fuel reacting with oxygen to produce CO_2 and H_2O. Combustion of methane, CH4, is written as

$$CH_4 + 2O_2 \rightarrow CO_2 + 2H_2O.$$

PROBLEM TYPE 2: COMBUSTION ANALYSIS.

A. Determination of Empirical Formulas.

Elemental analysis is a relatively cheap and simple way to gather data to determine the empirical formulas of compounds. One way to perform an elemental analysis is through the complete combustion of the unknown. For organic compounds, the amounts of carbon and hydrogen in the unknown sample may be determined from the amounts of CO_2 and H_2O collected from the combustion.

STUDY QUESTION 8.5

When a 0.761 g sample of a compound of carbon and hydrogen is burned in a C–H combustion apparatus, 2.23 g CO_2 and 1.37 g H_2O are produced.

> a. **Determine the percent composition of the compound.**
> b. **Determine the empirical formula.**

Solution:

a. Here we want the grams of carbon and hydrogen in the original compound. All the carbon is now present in 2.23 g of CO_2. All the hydrogen is now present in 1.37 g of H_2O. How many grams of C are there in 2.23 g of CO_2, and how many grams of H are there in 1.37 g of H_2O? We know there is 1 mole of C atoms (12.01 g) in every mole of CO_2, or 44.01 g.

The solution plan is:

$$g\ CO_2 \rightarrow mol\ CO_2 \rightarrow mol\ C \rightarrow g\ C$$

$$?\ g\ C = 2.23\ g\ CO_2 \times \frac{1\ mol\ CO_2}{44.01\ g\ CO_2} \times \frac{1\ mol\ C}{1\ mol\ CO_2} \times \frac{12.01\ g\ C}{1\ mol\ C} = 0.608\ g\ C$$

$$g\ H_2O \rightarrow mol\ H_2O \rightarrow mol\ H \rightarrow g\ H$$

$$?\ g\ H = 1.37\ g\ H_2O \times \frac{1\ mol\ H_2O}{18.0\ g\ H_2O} \times \frac{2\ mol\ H}{1\ mol\ H_2O} \times \frac{1.008\ g\ H}{1\ mol\ H} = 0.153\ g\ H$$

The percentage composition is:

$$\%H = \frac{0.153\ g\ H}{0.761\ g\ compound} \times 100\% = 20.1\%\ H$$

$$\%C = \frac{0.608\ g\ C}{0.761\ g\ compound} \times 100\% = 79.9\%\ C$$

b. To determine the empirical formula, we need to calculate the number of moles of each element then take a ratio.

$$? \text{ mol C} = 2.23 \text{ g } CO_2 \times \frac{1 \text{ mol } CO_2}{44.0 \text{ g } CO_2} \times \frac{1 \text{ mol C}}{1 \text{ mol } CO_2} = 0.0507 \text{ mol C}$$

$$? \text{ mol H} = 1.37 \text{ g } H_2O \times \frac{1 \text{ mol } H_2O}{18.0 \text{ g } H_2O} \times \frac{2 \text{ mol H}}{1 \text{ mol } H_2O} = 0.152 \text{ mol H}$$

The ratio of H:C is

$$\frac{0.152 \text{ mol H}}{0.0507 \text{ mol C}} = 2.998$$

This tells us that there are 3 mol of H for each mol of C. The empirical formula is CH_3.

PRACTICE QUESTION 8.4
Ascorbic acid is a compound consisting of three elements: C, H, and O. When a 0.214 g sample is burned in oxygen, 0.320 g of CO_2 and 0.0874 g of H_2O are formed. What is the empirical formula of ascorbic acid?

PRACTICE QUESTION 8.5
An unknown element M combines with oxygen to form a compound with a formula MO_2. If 25.0 g of the element combines with 4.50 g of oxygen, what is the atomic mass of M?

PRACTICE QUESTION 8.6
A reaction chamber was filled with equal moles of an unknown hydrocarbon and oxygen gas. After igniting the mixture, 80% of the hydrocarbon was left over. What are the two possible molecular formulas of this hydrocarbon?

PROBLEM TYPE 3: CALCULATIONS WITH BALANCED CHEMICAL EQUATIONS.

A. Moles of Reactants and Products.
Balanced chemical equations contain information about the relative numbers of reactants and products involved in the reaction. Remember that a mole is a number. Consequently, the balanced equation tells us the relative mole ratio of reactants and products.

The quantitative relationship between substances in chemical reactions is a part of chemistry called stoichiometry. In all stoichiometry problems, the balanced chemical equation provides the "bridge" that relates the amount of one reactant to the amount of another; and the amounts of products to reactants. Mole ratios are conversion factors that relate the number of reactants consumed to each other or to the number of product moles formed.

Once you have a balanced equation, stoichiometry problems are solved in three steps:

1. Convert the amounts of given substances into moles, if necessary.
2. Use the appropriate mole ratio constructed from the balanced equation to calculate the moles of the needed or unknown substance.
3. Convert the moles of the needed substance into units specified in the problem, if required.

STUDY QUESTION 8.6
Given the following reaction

$$2Al(OH)_3 + 3H_2SO_4 \rightarrow Al_2(SO_4)_3 + 6H_2O$$

calculate the number of moles of aluminum sulfate and water that will be produced when 3.22 moles of sulfuric acid react with an excess of aluminum hydroxide. How many molecules of water are produced?

Solution: The stoichiometric coefficients from the balanced chemical equation provide all the information (mole ratios) we need to perform these calculations.

$$3.22 \text{ mol } H_2SO_4 \times \frac{1 \text{ mol } Al_2(SO_4)_3}{3 \text{ mol } H_2SO_4} = 1.07 \text{ mol } Al_2(SO_4)_3$$

$$3.22 \text{ mol } H_2SO_4 \times \frac{6 \text{ mol } H_2O}{3 \text{ mol } H_2SO_4} = 6.44 \text{ mol } H_2O$$

The last calculation only requires that we remember that a mole consists of a fixed number of particles, 6.022×10^{23}.

$$6.44 \text{ mol } H_2O \times \frac{6.022 \times 10^{23} \text{ molecules } H_2O}{1 \text{ mol } H_2O} = 3.88 \times 10^{24} \text{ molecules } H_2O$$

PRACTICE QUESTION 8.7
How many molecules of hydrogen can be produced from the complete reaction of 1.55 mol of magnesium with HNO_3?

B. Mass of Reactants and Products.
We can use the molar masses of the reactants and products to convert the moles of reactants and products to their masses. Therefore, in the reaction, we can determine how many grams of one substance will be needed to react with a given mass of another, or how many grams of product can be produced by the reaction of a specific mass of a reactant.

STUDY QUESTION 8.7
How many grams of iron are produced when 25.0 g Fe_2O_3 reacts with aluminum?

Solution: First, we must write and balance the chemical equation that represents the reaction taking place in the problem. We know from the problem that the reactants are Fe_2O_3 and Al, and Fe is a product. This means the Al is displacing Fe in the compound with O.

$$Fe_2O_3 + Al \rightarrow Fe + Al_2O_3$$

Balancing the equation we get

$$Fe_2O_3 + 2Al \rightarrow 2Fe + Al_2O_3$$

Now that we have a balanced equation, we can follow the three steps outlined previously.

Step 1:

$$n_{Fe_2O_3} = 25.0 \text{ g Fe}_2O_3 \times \frac{1 \text{ mol Fe}_2O_3}{159.70 \text{ g Fe}_2O_3} = 0.157 \text{ mol Fe}_2O_3$$

Step 2:

$$n_{Fe_2} = 0.157 \text{ mol Fe}_2O_3 \times \frac{2 \text{ mol Fe}}{1 \text{ mol Fe}_2O_3} = 0.314 \text{ mol Fe}$$

Step 3:

$$n_{Fe} = 0.314 \text{ mol Fe} \times \frac{55.85 \text{ g Fe}}{1 \text{ mol Fe}} = 17.5 \text{ g Fe}$$

With practice, it will be easier to combine these three steps into a single calculation:

$$n_{Fe} = 25.0 \text{ g Fe}_2O_3 \times \frac{1 \text{ mol Fe}_2O_3}{159.70 \text{ g Fe}_2O_3} \times \frac{2 \text{ mol Fe}}{1 \text{ mol Fe}_2O_3} \times \frac{55.85 \text{ g Fe}}{1 \text{ mol Fe}} = 17.5 \text{ g Fe}$$

STUDY QUESTION 8.8
Sulfur dioxide can be removed from refinery stack gases by reaction with quicklime, CaO.

$$\textbf{SO}_2(g) + \textbf{CaO}(s) \rightarrow \textbf{CaSO}_3(s)$$

If 975 g of SO$_2$ is to be removed from stack gases by the above reaction, what mass of CaO is required to completely react with it?

Solution: First, be certain that the equation is correctly balanced, and it is. Then plan your road map.

$$\begin{array}{ccccccc} & & & \text{step 1} & & \text{step 2} & & \text{step 3} \\ \text{plan:} & & \text{g SO}_2 & \rightarrow & \text{mol SO}_2 & \rightarrow & \text{mol CaO} & \rightarrow & \text{g CaO} \end{array}$$

1. Convert the 975 g of SO$_2$ to moles using its molar mass.
2. Use the chemical equation to find the number of moles of CaO that reacts per mole of SO$_2$. From the balanced equation we can write:

$$1 \text{ mol CaO} \leftrightarrow 1 \text{ mol SO}_2$$

3. Convert moles of CaO to grams of CaO using the molar mass of CaO.

   Write out an equation which restates the problem:

$$? \text{ g CaO} = 9.75 \times 10^2 \text{ g SO}_2$$

String the conversion factors from the three steps one after the other.

$$? \text{ g CaO} = 9.75 \times 10^2 \text{ g SO}_2 \times \frac{1 \text{ mol SO}_2}{64.07 \text{ g SO}_2} \times \frac{1 \text{ mol CaO}}{1 \text{ mol SO}_2} \times \frac{56.08 \text{ g CaO}}{1 \text{ mol CaO}}$$

$$= 8.53 \times 10^2 \text{ g CaO}$$

PRACTICE QUESTION 8.8
How many grams of potassium hydroxide can be produced from the complete reaction of 19.55 g of potassium with water? How many grams of hydrogen will be produced?

PRACTICE QUESTION 8.9
Oxygen gas can be produced by the decomposition of mercury(II) oxide, HgO. How many grams of O_2 will be produced by the reaction of 24.2 g of the oxide?
$$2HgO(s) \rightarrow 2Hg(l) + O_2$$

PRACTICE QUESTION 8.10
The following reaction can be used to prepare hydrogen gas:
$$CaH_2 + 2H_2O \rightarrow Ca(OH)_2 + 2H_2$$
How many grams of H_2 will result from the reaction of 100 g of CaH_2 with excess H_2O?

PROBLEM TYPE 4: LIMITING REACTANTS.

A. Determining the Limiting Reactant.

Usually the reactants are not present in the exact ratio prescribed by the balanced chemical equation. A large excess of a less expensive reagent (reactant) might be used to ensure complete reaction of the more expensive reagent. In this situation, one reactant will be completely consumed before the other runs out. When this occurs, the reaction will stop and no more products will be made. The reactant that is consumed first is called the **limiting reactant** because it limits, or determines the amount of product formed. The reactant that is not completely consumed is called the **excess reactant**.

Considering the aluminum/iron oxide reaction,

$$Fe_2O_3 + 2Al \rightarrow 2Fe + Al_2O_3$$

Suppose that 1.2 moles of Al are reacted with 1 mole of Fe_2O_3. In any stoichiometry problem, it is important to determine which reactant limits formation of product (limiting reactant), to correctly predict the yield of products.

Step #	# reactant amounts		# product amounts	
	2 Al +	$Fe_2O_3 \rightarrow$	Al_2O_3 +	2 Fe
Initial	1.2 mol	1.0 mol	0 mol	0 mol
Change	−1.2 mol	−0.6 mol	+0.6 mol	+1.2 mol
Final	0 mol	0.4 mol	0.6 mol	1.2 mol

Explanation of the I.C.F table above:
- Initial amounts are the initial mole amounts of each reactant. If given grams, convert to moles to put into the table.
- Change amounts are how much reactant will be consumed and how much product will be formed. The balanced equation in this example shows that you have to have twice as much Al as Fe_2O_3. Since 1.2 mol Al isn't twice the 1.0 mol Fe_2O_3, Al is the limiting reactant. There will be 1/2 as much Fe_2O_3 formed based on the molar ratios in the equation. Note that the limiting reactant (1.2 mol Al) is not the reactant in lesser molar amount; 1.0 mol Fe_2O_3 is the lesser molar amount but is not the limiting reactant due to the molar ratios in the equation.

- **F**inal amount is what is left over after the reaction takes place.

If the problem then wants to know grams of product formed, or grams of excess reactant remaining, simply convert.

STUDY QUESTION 8.9
Phosphine (PH_3) burns in oxygen to produce phosphorus pentoxide (P_2O_5) and water.

$$2PH_3(g) + 4O_2(g) \rightarrow P_2O_5(s) + 3H_2O(l)$$

How many grams of P_2O_5 will be produced when 17.0 g of phosphine is mixed with 16.0 g of O_2 and reaction occurs?

Solution: When the amounts of **both reactants** are given, it is possible that one will be completely used up before the other. This is your indication that you are dealing with a limiting reactant problem. It is important that you look for this so you can plan your strategy. For limiting reactant problems, one tool is to construct an I.C.F. table, placing mole amounts into the initial line.

$$17.0 \text{ g PH}_3 \times \frac{1 \text{ mol PH}_3}{33.99 \text{ g PH}_3} = 0.500 \text{ mol PH}_3$$

$$16.0 \text{ g O}_2 \times \frac{1 \text{ mol O}_2}{32.00 \text{ g O}_2} = 0.500 \text{ mol O}_2$$

Step #	reactant amounts		product amounts	
	2 PH$_3$(g) + 4 O$_2$(g) →		P$_2$O$_5$(s) + 3 H$_2$O(l)	
initial	0.500 mol	0.500 mol	0.000 mol	0.000 mol
change	−0.250 mol	−0.500 mol	+ 0.125mol	+0.375 mol
finish	0.250 mol	0.000 mol	0.125 mol	0.375 mol

The key to the I.C.F. table is the change line. Looking at the mole ratio of PH_3 to O_2 we see that twice as many moles of O_2 are needed to react with the PH_3. There are not twice as many moles of O_2 as there are PH_3. Therefore, the limiting reactant is O_2. All of it will react. All of the other boxes in the change line are determined by the coefficients in the balanced equation. As shown in the textbook, this should be done using dimensional analysis. The mole–to–mole ratio of species in the balanced chemical reaction is used as the conversion factor to calculate the number of moles of PH_3 needed to react and the number of moles of P_2O_5 and H_2O produced from the limiting number of moles of O_2. We can see that there is 1/2 as much PH_3 in the balanced equation; (0.5 moles of O_2 × 2 moles of PH_3 / 4 moles of O_2), therefore a 0.250 is subtracted under PH_3. There is 1/4 as much P_2O_5 produced as O_2 used; therefore a 0.125 is added to the product side. And last, there is three times as much H_2O formed as P_2O_5; therefore, 3(0.125) = 0.375 is placed in the change box under H_2O. Finish filling out the table.

The problem is finished by solving for grams of P_2O_5 following the plan:

$$\text{mol P}_2O_5 \rightarrow \text{g P}_2O_5$$

$$0.125 \text{ mol P}_2O_5 \times \frac{141.74 \text{ g P}_2O_5}{1 \text{ mol P}_2O_5} = 17.7 \text{ g P}_2O_5$$

PRACTICE QUESTION 8.11
Given 5.00 mol of KOH and 2.00 mol of H_3PO_4, how many moles of K_3PO_4 can be prepared?

$$H_3PO_4 + 3KOH \rightarrow K_3PO_4 + 3H_2O$$

PRACTICE QUESTION 8.12
How many grams of excess reactant will be left–over in the reaction of 45.0 g of KCl and 145.0 g of Pb(NO$_3$)$_2$?

$$2KCl(aq) + Pb(NO_3)_2(aq) \rightarrow PbCl_2(s) + 2KNO_3(aq)$$

B. Reaction Yield.
The product amounts predicted from the balanced reaction are best case scenarios. In practice, the actual yield is usually less than that calculated, or "theoretical" yield. Sometimes it is even greater than 100% yield, which means that there is something besides the desired yield present. The theoretical yield is calculated based on the assumption that all the limiting reactant is consumed according to the balanced equation. The percent yield is based on the actual amount of product produced in the reaction. The percent yield is defined as

$$\% \text{ percent yield} = \frac{\text{actual yield}}{\text{theoretical yield}} \times 100\%$$

STUDY QUESTION 8.10
In the reaction of 4.0 moles of N$_2$ with 6.0 moles of H$_2$, a chemist obtained 1.6 moles of NH$_3$. What is the percent yield of NH$_3$?

$$3H_2 + N_2 \rightarrow 2NH_3$$

Solution: The actual yield is given (1.6 moles of NH$_3$); therefore we must calculate the theoretical yield of NH$_3$. That means we must calculate the number of moles NH$_3$ which should form. Notice that once again, this is a limiting reactant problem because data is provided for both reactants. This time let's see the textbook method (and not the ICF table method) for determining the theoretical yield.

Plan:

$$\text{mol } H_2 \rightarrow \text{mol } NH_3$$
$$\text{mol } N_2 \rightarrow \text{mol } NH_3$$

then choose the one that produces the least amount of NH$_3$. Note: since the actual yield is given in moles, we can stop at moles with the theoretical yield as well.

$$6.0 \text{ mol } H_2 \times \frac{2 \text{ mol } NH_3}{3 \text{ mol } H_2} = 4.0 \text{ mol } NH_3$$

$$4.0 \text{ mol } N_2 \times \frac{2 \text{ mol } NH_3}{1 \text{ mol } N_2} = 8.0 \text{ mol } NH_3$$

The limiting reactant is H$_2$. The maximum theoretical yield for NH$_3$ is 4.0 moles. The percent yield is found by dividing the actual yield by the theoretical yield and multiplying by 100 percent.

$$\% \text{ yield } NH_3 = \frac{\text{actual yield } NH_3}{\text{theoretical yield } NH_3} \times 100\%$$

$$= \frac{1.6 \text{ mol NH}_3}{4.0 \text{ mol NH}_3} \times 100\%$$

$$\% \text{ yield NH}_3 = 40\%$$

PRACTICE QUESTION 8.13
The reaction of iron ore with carbon follows the equation:
$$2Fe_2O_3 + 3C \rightarrow 4Fe + 3CO_2$$
    a.   **How many grams of Fe can be produced from a mixture of 200 g of Fe_2O_3 and 300 g of C?**
    b.   **If the actual yield of Fe is 110 g, what is the percent yield of iron?**

PRACTICE QUESTION 8.14
Hydrofluoric acid (HF) can be prepared according to the reaction:
$$CaF_2 + H_2SO_4 \rightarrow 2HF + CaSO_4$$
In one experiment 42.0 g of CaF_2 was treated with excess H_2SO_4 and a yield of 14.2 g of HF was obtained.

    a. **What is the theoretical yield of HF?**
    b. **Calculate the percent yield of HF.**

PRACTICE QUESTION 8.15
Consider the diagram below for the reaction $2 A + B_2 \rightarrow 2 AB_2$. Determine the limiting reactant and theoretical yield.

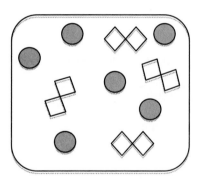

PROBLEM TYPE 5: PERIODIC TRENDS IN REACTIVITY OF THE MAIN GROUP ELEMENTS.

A. General Trends in Reactivity.
Although members of a group in the periodic table exhibit similar chemical and physical properties, the first member of each group tends to be significantly different from the other members. Hydrogen is essentially a group unto itself. The alkali metals (Group 1A) tend to be highly reactive toward oxygen, water, and acid. Group 2A metals are less reactive than Group 1A metals, but the heavier members all react with water to produce metal hydroxides and hydrogen gas. Groups that contain both metals and nonmetals (e.g., Groups 4A, 5A, and 6A) tend to show greater variability in their physical and chemical properties.

Another trend in the chemical reactivity of main group elements is the **diagonal relationship**. Diagonal relationships refer to similarities between pairs of elements in different groups and periods of the periodic table.

B. Reactions of the Active Metals.
The elements in Group 1A, with the exception of hydrogen, have their valence ns^1 electron well shielded from the nucleus and as a result it is easily lost. Low ionization energies make the alkali metals the most active (reactive) family of metals. These elements are found in nature as +1 ions in chemical combination with nonmetal ions and polyatomic ions. The densities of these elements are low, in part, because of their large radii. Li, Na, and K are even less dense than water.

Group 2A elements are called alkaline earth metals. The outermost electron configuration is ns^2. Although these two are not as easily lost as the ns^1 electron in alkali metal atoms, they are readily given up. Alkaline earth metals exist in nature as +2 ions in chemical combination with negative ions.

C. Reactions of Other Main Group Elements.
None of the Group 7A elements (halogens) are ever found free in nature. All have high electron affinities and so tend to acquire one electron to form −1 ions. Ionic compounds containing these anions are salts. Indeed, the name halogen means "salt–former." These nonmetals exist in the elemental form as diatomic molecules. At 25 °C, F_2 and Cl_2 are gases, Br_2 is a liquid, and I_2 is a solid.

Group 8A elements make up the noble gas elements. The term noble here means nonreactive. He, Ne, and Ar do not form any chemical compounds. Since 1960, a number of Kr and Xe compounds have been synthesized. The chemical inactivity of the noble gases is the result of the ns^2np^6 electron configuration of the outermost electrons in their atoms. The noble gas elements have negative electron affinities and high ionization energies.

STUDY QUESTION 8.11
Write the electron configurations for the atoms of all elements in Group 4A. Specify whether the element is a metal, a nonmetal, or a metalloid.

Solution: First, use a Periodic Table to identify the type of element. The outermost principle energy level of atoms in Group 4A contains four electrons. The general electron configuration for elements in this group is ns^2np^2. Their electron configurations are

C [He]$2s^22p^2$	nonmetal
Si [Ne]$3s^23p^2$	metalloid
Ge [Ar]$3d^{10}4s^24p^2$	metalloid
Sn [Kr]$4d^{10}5s^25p^2$	metal
Pb [Xe]$4f^{14}5d^{10}6s^26p^2$	metal

STUDY QUESTION 8.12
Compare the magnitudes of electron affinity and ionization energy for the alkali metal potassium with the halogen chlorine. Comment on their relative abilities to form ions.

Solution: From your textbook, the electron affinities of K and Cl are 48.4 and 349 kJ/mol, respectively; the ionization energies of K and Cl are 419 and 1256 kJ/mol, respectively. These values suggest that the metal atom K can lose an electron and form a K^+ ion much more readily than can the nonmetal Cl. Conversely, the nonmetal atom Cl can attract an electron to form a Cl^- much more readily than can the metallic K atom.

PRACTICE QUESTION 8.16
The first ionization energies of boron and silicon are 801 and 786 kJ/mol, respectively. The similarity of these values is an example of what relationship?

PRACTICE QUESTION 8.17
What charges do you expect for the ions of the Group 6A and the halogen elements? Why?

D. Comparison of Group 1A and Group 1B Elements.

The elements of Group 1B, copper, silver, and gold, are generally nonreactive and are called the **noble metals** or coinage metals. These metals are excellent conductors of heat and electricity. Atoms of the coinage metals have one electron in the outer s–orbital and 10 electrons in the underlying d-orbitals. The electron configurations of their outermost electrons are $ns^1 (n-1)d^{10}$. Note that this is an exception to the configuration expected from the Aufbau principle, which predicts $ns^2(n-1)d^9$. It is as if the $(n-1)d$-orbitals "borrowed" an electron from the higher energy ns orbital. For the coinage metal atoms the "borrowed" electron is used to complete an inner subshell. Apparently, the completed d–orbitals have an enhanced stability that corresponds to a lowering of the atom's energy. Thus, $ns^1(n-1)d^{10}$ is a lower energy configuration than $ns^2(n-1)d^9$.

In copper, silver, and gold this one outer electron is held much more tightly than the outermost electron in the alkali metal atoms. This can be most easily seen by comparing first ionization energies.

STUDY QUESTION 8.13
Write the electron configurations for the following atoms.

    a. Zn
    b. Cu

Solution:
a. Zn: $[Ar]4s^2 3d^{10}$
b. Reading the configuration directly from the periodic table we get Cu: $[Ar]4s^2 3d^9$. However, we see that the $3d$–orbitals are one away from being filled. We should move one electron from the $4s$ *to the* $3d$ to have lowest energy configuration. Cu: $[Ar]4s^1 3d^{10}$. Remember that this happens *only* with the $4s$–$3d$ and $5s$–$4d$ orbitals. We cannot move an electron from $2s$ to $2p$, for example.

PRACTICE QUIZ

1. Balance the following equations:
   a. $P_4O_{10} + H_2O \rightarrow H_3PO_4$
   b. $Ga + H_2SO_4 \rightarrow Ga_2(SO_4)_3 + H_2$
   c. $C_4H_{10} + O_2 \rightarrow CO_2 + H_2O$

2. Complete each line of the following table for the reaction:
   $2SO_2 + O_2 \rightarrow 2SO_3$

mol SO$_2$	grams O$_2$	mol SO$_3$	grams SO$_3$
1.50			
	20.0		
		5.21	

3. Sodium thiosulfate can be made by the reaction:

   $$Na_2CO_3 + 2Na_2S + 4\,SO_2 \rightarrow 3Na_2S_2O_3 + CO_2$$

   a. How many grams of Na_2CO_3 (sodium carbonate) are required to produce 321 g of sodium thiosulfate, $Na_2S_2O_3$?
   b. How many grams of Na_2S are required to react with 25.0 g of Na_2CO_3 if SO_2 is present in excess?

4. Hydrofluoric acid (HF) can be prepared according to the following equation:

   $$CaF_2 + H_2SO_4 \rightarrow 2HF + CaSO_4$$

   a. How many grams of HF can be prepared from 75.0 g of H_2SO_4 and 63.0 g of CaF_2?
   b. How many grams of the excess reagent will remain after the reaction ceases?
   c. If the actual yield of HF is 26.2 g, what is the percent yield?

5. Calculate the mass of antimony(III) sulfide, Sb_2S_3, produced from the reaction of 200. g of $SbCl_3$ and 200. g of $(NH_4)_2S$ if this reaction has an 86.7% yield.

   $$2SbCl_3 + 3(NH_4)_2S \rightarrow Sb_2S_3 + 6NH_4Cl$$

6. In the decomposition of 798 g of $Al(ClO_3)_3$, 128 g of O_2 are formed. What is the percent yield for the reaction?

   $$2Al(ClO_3)_3 \xrightarrow{\Delta} 3O_2 + 2AlCl_3$$

7. Identify the excess reactant and calculate the amount left over when 75.0 g of calcium react with 75.0 g of water.

   $$Ca + 2H_2O \rightarrow Ca(OH)_2 + H_2$$

8. Which two of the following would be most likely to have similar ionization energies?
   B, C, Si, Al, Ar

9. Of the following, which is the most reactive metal?
   V, Ge, Se, As, Zn

10. What charges do you expect for the ions of the alkali metals and the alkaline earth metals? Why?

Chapter 9
Chemical Reactions in Aqueous Solutions

PROBLEM-SOLVING STRATEGIES AND TUTORIAL SOLUTIONS

TYPES OF PROBLEMS

Problem Type 1:   General Properties of Aqueous Solutions.
        A.   Electrolytes and Nonelectrolytes.
        B.   Strong Electrolytes and Weak Electrolytes.

Problem Type 2:   Precipitation Reactions.
        A.   Solubility Guidelines for Ionic Compounds in Water.
        B.   Molecular Equations.
        C.   Ionic Equations.
        D.   Net Ionic Equations.

Problem Type 3:   Acid-Base Reactions.
        A.   Strong Acids and Bases.
        B.   Brønsted Acids and Bases.
        C.   Acid–Base Neutralization.

Problem Type 4:   Oxidation–Reduction Reactions.
        A.   Oxidation Numbers.

Problem Type 5:   Concentration of Solutions.
        A.   Molarity.
        B.   Dilution.
        C.   The pH Scale

Problem Type 6:   Aqueous Reactions and Chemical Analysis.
        A.   Gravimetric Analysis.
        B.   Acid–Base Titrations.

PROBLEM TYPE 1: GENERAL PROPERTIES OF AQUEOUS SOLUTIONS.

Many chemical reactions occur in aqueous solutions. Solutions are homogeneous mixtures of two or more substances. Solutions have two components, the solute, which is the substance present in the smaller amount, and the solvent, which is the substance present in the larger amount. A solution can be made with compounds in any of the states of matter. A special type of solution is an aqueous solution, in which water is the solvent.

A.   Electrolytes and Nonelectrolytes.
Many aqueous solutions of ionic compounds conduct electricity well, whereas pure water conducts very poorly. These solutions are called electrolytic solutions. An **electrolyte** is a substance that, when dissolved in water, produces ions that are capable of carrying an electrical current through the solution. For example, ionic compounds such as $KCl(s)$, readily separate into individual ions when dissolved in water. This process is referred to as **dissociation**, and it can be represented for $KCl(s)$ using the following equation:

$$KCl(s) \rightarrow K^+(aq) + Cl^-(aq)$$

Acids are a special kind of hydrogen–containing compound that can ionize in water. Hydrochloric acid, even though it is a molecule, will completely ionize in water. The ionization of $HCl(aq)$ can be represented by the following equation:

$$HCl(aq) \rightarrow H^+(aq) + Cl^-(aq)$$

Thus, a solution of HCl contains H^+ ions and Cl^- ions, but <u>no</u> HCl molecules! Molecular compounds that are able to dissolve in water and form solutions are classified as *electrolytes*. A **nonelectrolyte** is a substance that does not ionize or dissociate in aqueous solution. Such a solution does not conduct an electrical current. Sugar is an example of a nonelectrolyte. In solution, sugar exists as $C_6H_{12}O_6$ molecules that are neutral (no charge) and cannot conduct an electrical current. All nonelectrolytes exist as molecules in aqueous solution.

B. Strong Electrolytes and Weak Electrolytes.
Electrolytes can further be categorized as being *strong* electrolytes or *weak electrolytes*. A **strong electrolyte** dissociates or ionizes close to 100% into ions in solution. However, a **weak electrolyte**, dissociates or ionizes only partially in solution. This is one of the reasons for the difference in conductivity between a strong electrolyte and a weak electrolyte. An aqueous solution of a strong electrolyte will have a higher conductivity (i.e., ability to conduct electricity) than an aqueous solution of a weak electrolyte, as long as they have the same cation:anion ratio and the same initial number of moles of solute. For example, acetic acid (CH_3COOH) is a weak electrolyte, so that in solution it exists 99% as CH_3COOH molecules and 1% as the H^+ and CH_3COO^- ions as shown below:

$$CH_3COOH(aq) \rightleftharpoons CH_3COO^-(aq) + H^+(aq)$$

Because both hydrochloric acid and acetic acid have a 1:1 cation:anion ratio but HCl is a strong electrolyte, we can predict that an aqueous solution of HCl will have a higher conductivity than an aqueous solution of CH_3COOH if both solutions were prepared using the same number of moles of HCl or CH_3COOH, respectively.

A scheme to aid us in the classification of solutes based on their electrical conductivity can be found in Chapter 9 of your textbook.

STUDY QUESTION 9.1
Identify the following substances as strong, weak or nonelectrolytes:

    a.   **CH_3OH (methyl alcohol)**
    b.   **NH_3**
    c.   **Na_2CO_3**

Solution: Methyl alcohol is a molecular compound, therefore it is a nonelectrolyte. Since ammonia is a weak base it would be classified as a weak electrolyte. Sodium carbonate is an ionic compound and all ionic compounds dissociate completely in aqueous solution when they dissolve. Sodium carbonate is a strong electrolyte.

    PRACTICE QUESTION 9.1
    Which of the following are ionic compounds and therefore strong electrolytes?

        a.   **CaF_2**
        b.   **$C_6H_{12}O_6$**
        c.   **KBr**
        d.   **$LiNO_3$**
        e.   **$CuSO_4$**

PRACTICE QUESTION 9.2
Which of the following compounds are weak electrolytes?

    a. **KOH**
    b. **Hydrofluoric acid**
    c. **MgI_2**
    d. **Water**
    e. **HBr**
    f. **AgCl**

PRACTICE QUESTION 9.3
What species would exist in a solution of the following compounds and water?

    a. **CH_2CH_3OH**
    b. **$MgCl_2$**

PRACTICE QUESTION 9.4
Write a complete chemical equation that represents the dissociation of the following compounds in water.

    c. **$Mg(NO_3)_2$**
    d. **KOH**
    e. **CaF_2**

PROBLEM TYPE 2: PRECIPITATION REACTIONS.

A. Solubility Guidelines for Ionic Compounds in Water.

The solubility of a solute is the maximum amount that can be dissolved in a given quantity of solvent at a given temperature. For example, the solubility of $Pb(NO_3)_2$ is 56 g per 100 mL H_2O at 20 °C. The solubilities of ionic solids in water vary over a wide range. Compounds can be divided into two broad categories: *soluble* and *insoluble*. These terms are relative and classifying a substance as insoluble does not mean that no solute dissolves. An ionic compound can have a low solubility, but still be a strong electrolyte. We will use a set of empirical rules, summarized in Tables 9.2 and 9.3 in your textbook, to classify solutes as soluble in water or insoluble in water.

STUDY QUESTION 9.2
According to the solubility rules, which of the following compounds are soluble in water?

    a. $MgCO_3$
    b. $AgNO_3$
    c. $MgCl_2$
    d. Na_3PO_4
    e. KOH

Solution:

a. Compounds containing the CO_3^{2-} ion are water–insoluble. Mg^{2+} is not one of the soluble exceptions. $MgCO_3$ is insoluble.

b. Compounds containing the NO_3^- ion are soluble without exceptions, $AgNO_3$ is soluble.

c. Compounds containing the Cl^- ion are water soluble and Mg^{2+} is not an insoluble exception, $MgCl_2$ is soluble.

d. and e. Compounds containing alkali metal cations are water–soluble without exceptions, Na_3PO_4 and KOH are soluble.

PRACTICE QUESTION 9.5
Which of the following compounds are soluble in water?

    a.   **BaCl$_2$**
    b.   **PbSO$_4$**
    c.   **Ni(OH)$_2$**
    d.   **Ca$_3$(PO$_4$)$_2$**
    e.   **NH$_4$NO$_3$**

PRACTICE QUESTION 9.6
According to the solubility rules, which of the following compounds are insoluble in water?

    a.   **(NH$_4$)$_2$CO$_3$**
    b.   **AgBr**
    c.   **CaCO$_3$**
    d.   **FeCl$_3$**
    e.   **ZnS**

PRACTICE QUESTION 9.7
Would the addition of calcium nitrate to the following mixtures cause a precipitation reaction?

    a.   **NH$_4$ClO$_3$ and Na$_2$SO$_4$**
    b.   **RbNO$_3$ and Fe(C$_2$H$_3$O$_2$)$_3$**

PRACTICE QUESTION 9.8
True or False. A strong electrolyte is soluble in water?

B. Molecular Equations.

When certain solutions of electrolytes are mixed, the cation of one solute may combine with the anion of the other solute to form an insoluble compound. This leads to the formation of a precipitate that settles from the solution. For instance, when a solution of Pb(NO$_3$)$_2$ is mixed with a solution of KCl, a precipitate of insoluble PbCl$_2$(s) forms. In general, we can predict the products of any double–displacement reaction by exchanging the cations and anions of the two reactants. Let's take a look at the general molecular equation,

$$AB(aq) + CD(aq) \rightarrow ?$$

where AB and CD are electrolytes. A and C are cations, while B and D are anions. To form products as a result of the "double–exchange" or "double–displacement" of these ions, we just need to combine A with D and C with B, taking care of writing formulas of neutral compounds by careful examination of the charges of the ions. Our general molecular equation will now look like:

$$AB(aq) + CD(aq) \rightarrow AD(?) + CB(?)$$

To complete our molecular equation, we must use the solubility rules to assign the phase in which each product will occur.

C. Ionic Equations.

Chemical equations for precipitation reactions can be written in several ways. In the molecular equation, the formulas of the compounds are written as their usual chemical formulas. The precipitation reaction that occurs when aqueous solutions of lead(II) nitrate and potassium chloride would be written as:

$$Pb(NO_3)_2(aq) + 2KCl(aq) \rightarrow PbCl_2(s) + 2KNO_3(aq)$$

The precipitation reaction is more accurately represented by the ionic equation. The ionic equation shows the soluble compounds (except weak electrolytes) as being dissociated in solution.

$$Pb^{2+}(aq) + 2\,NO_3^-\,(aq) + 2K^+(aq) + 2Cl^-(aq) \rightarrow PbCl_2(s) + 2K^+(aq) + 2\,NO_3^-\,(aq)$$

D. Net Ionic Equations.

If you examine the ionic equation, you will notice that the K^+ ions and NO_3^- ions are not involved in the formation of the precipitate. We refer to the ions that are not involved in the reaction as spectator ions. Identical species on both sides of the chemical equation can be omitted from the equation in the same way that you would cancel equal variables on each side of an algebraic equation. This leads to one additional way of representing a precipitation reaction: a net ionic equation. The net ionic equation shows only the species that actually undergo a chemical change.

$$Pb^{2+}(aq) + 2Cl^-(aq) \rightarrow PbCl_2(s)$$

STUDY QUESTION 9.3
Write balanced molecular, ionic and net ionic equations for the reaction that occurs when aqueous solutions of $CaCl_2$ and Na_2CO_3 are mixed.

Solution: We were given the two reactants:

$$CaCl_2(aq) + Na_2CO_3(aq) \rightarrow ?$$

These two electrolytes will dissociate in solution yielding Ca^{2+} (A), Cl^-(B), Na^+(C), and CO_3^{2-} (D). To form new products we must combine Ca^{2+} with CO_3^{2-} (AD), and Na^+ with Cl^- (CB). Notice that the new combinations contain ions with charges having the same magnitude, but with opposite signs. This means that combining them in a 1:1 ratio will yield neutral compounds. We may now proceed and write:

$$CaCl_2(aq) + Na_2CO_3(aq) \rightarrow CaCO_3(?) + NaCl(?)$$

We need to balance the equation and assign the phases of the two products based on the solubility rules. Since carbonate–containing compounds are water–insoluble, calcium carbonate is insoluble. Compounds containing alkali metal cations are soluble without exceptions thus, sodium chloride is soluble. The molecular equation is:

$$CaCl_2(aq) + Na_2CO_3(aq) \rightarrow CaCO_3(s) + 2NaCl(aq)$$

The ionic equation will show all soluble compounds that are not weak electrolytes as dissociated. In this example, all of the soluble compounds are ionic compounds and therefore strong electrolytes. Thus, the ionic equation is:

$$Ca^{2+}(aq) + 2\,Cl^-(aq) + 2\,Na^+(aq) + CO_3^{2-}\,(aq) \rightarrow CaCO_3(s) + 2\,Na^+(aq) + 2\,Cl^-(aq)$$

The net ionic equation omits the **spectator ions** (Na^+ and Cl^- appear on both sides of the equation with the same stoichiometric coefficient). The net ionic equation is:

$$Ca^{2+}(aq) + CO_3^{2-}\,(aq) \rightarrow CaCO_3(s)$$

STUDY QUESTION 9.4
Will a precipitation reaction occur when aqueous solutions of NaCl and KOH are mixed?

Solution: The two reactants have been given:

$$NaCl(aq) + KOH(aq) \rightarrow ?$$

We just need to recognize that these two electrolytes will yield Na^+, Cl^-, K^+ and OH^- ions in solution. Combining them to form new products:

$$NaCl(aq) + KOH(aq) \rightarrow NaOH(?) + KCl(?)$$

We need to check that the equation is balanced and use the solubility rules to assign the phases of the products. By inspection, we can see that the equation is balanced. Compounds containing alkali metal cations are water–soluble without exceptions. Therefore, both sodium hydroxide and potassium chloride are soluble! No precipitate is formed. At the molecular level, this means that we had two solutions in which the solutes were dissociated into ions and as the two solutions were mixed, the ions remained in solution. No reaction took place! There was no chemical change! Therefore,

$$NaCl(aq) + KOH(aq) \rightarrow \text{no reaction}$$

PRACTICE QUESTION 9.9
Predict whether a precipitate will form when aqueous solutions of the following compounds are mixed. Write the formulas for any precipitates.

    a. $MgBr_2$ and $NaOH$
    b. NaI and $AgNO_3$
    c. NH_4Cl and $Mg(NO_3)_2$

PRACTICE QUESTION 9.10
Identify the spectator ions for the reactions that form precipitates in Practice Question 9.7.

PRACTICE QUESTION 9.11
Write net ionic equations for the reactions that occur when aqueous solutions of the following compounds are mixed.

    a. $MgBr_2$ and $Pb(NO_3)_2$
    b. $NaBr$ and $AgNO_3$
    c. K_2SO_4 and $Ba(NO_3)_2$

PRACTICE QUESTION 9.12
Describe on a molecular level, what the inside of a test tube would look like after mixing solutions of $CaNO_3$ and Li_2CO_3.

PROBLEM TYPE 3: ACID-BASE REACTIONS.

A. Strong Acids and Bases.
Svante Arrhenius proposed that an acid is a substance that produces hydrogen ions (H^+, also called protons) in aqueous solution, and a base is a substance that produces hydroxide ions (OH^-) in aqueous solution through their dissociation or ionization. Nitric acid is an example of an Arrhenius acid:

$$HNO_3(aq) \rightarrow H^+(aq) + NO_3^-(aq)$$

Calcium hydroxide is an example of an Arrhenius base:

$$Ca(OH)_2(aq) \rightarrow Ca^{2+}(aq) + 2\,OH^-(aq)$$

We can conclude that acids and bases are electrolytes undergoing dissociation in aqueous solution. Table 9.4 in your textbook lists the strong acids and bases.

B. Brønsted Acids and Bases.

The Arrhenius definition is somewhat restrictive because for an acid to produce an H^+ ion from its dissociation, that compound must possess one or more "ionizable hydrogen(s)" in its formula. Similarly, for a compound to be a base under the Arrhenius definition, the compound must contain an OH^- in its formula to be able to produce it as a result of its dissociation. There are many compounds that behave as acids or bases yet they do not contain H^+ or OH^- ions in their formulas.

According to Brønsted's theory, an acid is defined as a substance that is able to donate a proton, and a base is a substance that accepts a proton in a chemical reaction. For example, when HCl gas dissolves in water, it can be viewed as donating a proton to molecules of the solvent. This process is sometimes referred to as a proton–transfer reaction.

$$HCl(aq) + H_2O(l) \rightarrow H_3O^+(aq) + Cl^-(aq)$$
$$\text{acid} \qquad \text{base}$$

The water molecule is the proton acceptor, and is therefore a Brønsted base in this reaction.

Some acids are capable of donating more than one proton to a base. Acids are referred to as **monoprotic** (mono = 1; can donate one proton and has only one ionizable hydrogen in its formula), **diprotic** (di = 2; can donate two protons and has two ionizable hydrogens in its formula), and **triprotic** (tri = 3; can donate three protons and has three ionizable hydrogens in its formula). Sulfuric acid (H_2SO_4) and carbonic acid (H_2CO_3) are diprotic acids. Phosphoric acid (H_3PO_4) is a triprotic acid.

When ammonia dissolves in water, it accepts a proton and so is classified as a Brønsted base. In this reaction, water is the acid.

$$NH_3(aq) + H_2O(l) \rightarrow NH_4^+(aq) + OH^-(aq)$$
$$\text{base} \qquad \text{acid}$$

Ammonia is a weak electrolyte because only a small fraction of dissolved NH_3 molecules are actually ionized at any one time.

STUDY QUESTION 9.5
Write a molecular equation for the reactions:

    a. $HClO_4$ with water
    b. $(CH_3)_2NH$ with water

Solution:

a. $HClO_4$ is an acid since the formula begins with H. Therefore, according to the Brønsted definition, $HClO_4$ donates a proton to water. Water in this reaction behaves as a base by accepting the proton.

$$HClO_4(aq) + H_2O(l) \rightarrow H_3O^+(aq) + ClO_4^-(aq)$$

b. $(CH_3)_2NH$ is a Brønsted base. There are many nitrogen–containing compounds that behave as bases. Since $(CH_3)_2NH$ is a Brønsted base, it accepts a proton from water. Water is the acid in this reaction.

$$(CH_3)_2NH(aq) + H_2O(l) \leftrightarrow (CH_3)_2NH_2^+(aq) + OH^-(aq)$$

PRACTICE QUESTION 9.13
Write the molecular equations for the reaction of the following acids with water.

    a.   **HBr**
    b. $HC_2H_3O_2$
    c. H_2SO_4

C. Acid–Base Neutralization.

Acid–base neutralization reactions typically produce water and a salt and are another type of double–displacement reaction.

When an acid and a base react, the hydrogen ion from the acid and the hydroxide ion from the base combine to form water, a molecular substance. The metal ion from the base and the nonmetal ion from the acid constitute a solution of a salt. The neutralization reaction of hydrochloric acid and barium hydroxide, for example, is represented by the following molecular equation.

$$2HCl(aq) + Ba(OH)_2(aq) \rightarrow 2H_2O(l) + BaCl_2(aq)$$

acid	base	water	salt

Since this reaction involves a strong acid and a strong base, both are completely ionized in solution. We can write the ionic equation:

$$2H^+(aq) + 2Cl^-(aq) + Ba^{2+}(aq) + 2OH^-(aq) \rightarrow 2H_2O(l) + Ba^{2+}(aq) + 2Cl^-(aq)$$

Notice that Ba^{2+} and Cl^- ions are spectator ions. The net ionic equation is:

$$2H^+(aq) + 2OH^-(aq) \rightarrow 2H_2O(l)$$

The coefficients can be cancelled. Therefore, the net ionic equation for the reaction of any strong acid and strong base is:

$$H^+(aq) + OH^-(aq) \rightarrow H_2O(l)$$

PROBLEM TYPE 4: OXIDATION-REDUCTION REACTIONS.

A. Oxidation Numbers.

In order to be able to recognize oxidation–reduction (redox) reactions and to balance them, we must be able to assign oxidation numbers to all of the atoms participating in the reaction, both on the reactant side and the product side.

Electrons are transferred from one atom to another in a redox reaction. **Oxidation** is the process by which a chemical substance loses electrons (LEO = **L**oss of **E**lectrons is **O**xidation). This process must occur simultaneously with a reduction. **Reduction** is the chemical process by which a chemical substance gains electrons (GER = **G**ain of **E**lectrons is **R**eduction). In a redox reaction the substance oxidized is called the **reducing agent** because it supplies electrons which cause reduction in another substance. The substance being reduced is called the **oxidizing agent** because it acquires electrons from another substance.

The oxidation number (or **oxidation state**) of the reactive substances is used to help keep account of electrons in chemical reactions. The oxidation number is the charge an atom would have in a molecule or an ionic compound, if electrons were completely transferred. An element is oxidized in a reaction whenever its oxidation number *increases* as a result of the reaction. When the oxidation number of an element *decreases* in a reaction, that element is reduced. The rules employed to assign oxidation numbers are summarized in Table 9.6 in your textbook.

STUDY QUESTION 9.6
Assign oxidation numbers to all the atoms in the following.

a. Na_2SO_4
b. $CuCl$
c. SO_3^{2-}

Solution: When assigning oxidation numbers, there will always be the possibility of using the rules presented in your textbook for one or more of the atoms in the compound. Most of the time, a compound will have one atom with an unknown oxidation number, one that must be calculated. We could think of using the variable x as the unknown oxidation number for that particular atom.

a. In Na_2SO_4 we can assign oxidation numbers to Na and O because their oxidation numbers are almost always the same in all compounds (rule 4). Na is an alkali metal so its oxidation number is +1. Since this compound is not a peroxide, oxygen is –2. The oxidation number of sulfur varies from compound to compound. Since sodium sulfate is a neutral compound, the sum of the oxidation numbers of all atoms in the compound is zero.

(number of sodium atoms) × (oxidation number for sodium) + (number of sulfur atoms) × (oxidation number for sulfur) + (number of oxygen atoms) × (oxidation number for oxygen) = 0

Let's substitute based on the formula for sodium sulfate, using x for the oxidation number we don't know (the oxidation number of sulfur).

$$2 \times (+1) + 1(x) + 4(-2) = 0$$
$$x - 6 = 0$$
$$x = +6$$

Now we know that the oxidation numbers are:

$$Na = +1; S = +6; O = -2$$

b. In CuCl, the chlorine ion has an oxidation number of –1 according to rule 4. Although this example is easier to work out by inspection, let's apply the same general procedure:

(number of Cu atoms) × (oxidation number for Cu) + (number of Cl atoms) × (oxidation number for Cl) = 0

$$1(x) + 1(-1) = 0$$
$$x - 1 = 0$$
$$x = +1$$

The oxidation numbers are:

$$Cu = +1; \ Cl = -1$$

c. In SO_3^{2-}, we assume the oxidation number of oxygen is –2 because we are not dealing with a peroxide. We can apply rule 3 to calculate the oxidation number of sulfur.

(number of sulfur atoms) × (oxidation number of sulfur) + (number of oxygen atoms) × (oxidation number for oxygen) = –2

$$1(x) + 3(-2) = -2$$
$$x - 6 = -2$$
$$x = +4$$

The oxidation numbers are:

$$S = +4; \ O = -2$$

STUDY QUESTION 9.7
Determine whether iron in the reactant species is oxidized or reduced during the following reactions:

a. $2Fe + \dfrac{3}{2}O_2 + 2H_2O \rightarrow Fe_2O_3 \cdot 2H_2O$ (rust)

b. $FeO + CO \rightarrow Fe + CO_2$

Solution: By assigning oxidation numbers to iron in Fe and in Fe_2O_3 (ignore the waters of hydration), we see that the oxidation number changes from zero in Fe to +3 in Fe_2O_3. Since Fe becomes more positive, it must have lost electrons. Thus, iron is oxidized. Note that when a metal combines with oxygen, its electrons are always transferred to the oxygen atom. All elements except fluorine are oxidized when they combine with O_2, and oxygen is reduced in these reactions.

Using the oxidation number change as a guide, the oxidation number of iron changes from +2 in FeO to zero in Fe. Therefore, iron is gaining electrons in this reaction, and so it is reduced.

PRACTICE QUESTION 9.14
Assign oxidation numbers to the elements underlined in the following molecules or ions:

   a. $\underline{C}O_2$
   b. $K_2\underline{Cr}O_4$
   c. $H_2\underline{S}$
   d. $\underline{H}_2S$
   e. $Ca\underline{H}_2$
   f. $\underline{C}O_3^{2-}$
   g. $\underline{Fe}Br_3$
   h. $\underline{Cr}_2O_7^{2-}$
   i. $K\underline{Cl}O_4$

PRACTICE QUESTION 9.15
What is the oxidation number of Mn before and after the reaction?

$$5H_3AsO_3 + 2\,MnO_4^- + 6H^+ \rightarrow 2Mn^{2+} + 5H_3AsO_4 + 3H_2O$$

PRACTICE QUESTION 9.16
In the following reactions, identify the elements oxidized and the elements reduced.

   a. $3Ag + 4HNO_3 \rightarrow 3AgNO_3 + NO + 2H_2O$
   b. $6Fe^{2+} + Cr_2O_7^{2-} + 14H^+ \rightarrow 6Fe^{3+} + 2Cr^{3+} + 7H_2O$

PROBLEM TYPE 5: CONCENTRATION OF SOLUTIONS.

A. Molarity.

Solutions are characterized by their concentrations, that is, the amount of solute dissolved in a given amount of solvent. The most common unit of concentration used for aqueous solutions is molarity. The **molarity** of a solution is the number of moles of solute per liter of the solution.

$$molarity = \frac{moles\ of\ solute}{liters\ of\ solution}$$

STUDY QUESTION 9.8
How do we prepare 0.10 L of a 0.50 M solution of KCl?

Solution: The first thing we need to determine is the number of moles of KCl needed for this solution. For instance, 0.10 L of 0.50 M KCl will contain

$$\left(\frac{0.10\ \text{L soln}}{1}\right)\left(\frac{0.50\ \text{mol KCl}}{1\ \text{L soln}}\right) = 0.050\ \text{mol KCl}$$

Therefore, we need to measure out 0.050 mole of solid KCl, (3.70 g). The KCl is then added to a 0.10 L (100 mL) volumetric flask. This is a flask with a calibrated ring etched around the neck to mark the point at which the flask contains 0.10 liter of liquid. You should fill it about one–third full with water, add your solute and swirl to dissolve, then add water up to the mark. When water is added up to this mark, the solution will contain 0.50 mole of KCl.

PRACTICE QUESTION 9.17
What is the molarity of a solution consisting of 11.8 g of NaOH dissolved in enough water to make 300.0 mL of solution?

PRACTICE QUESTION 9.18
How many grams of HCl are contained in 250 mL of 0.500 M HCl?

PRACTICE QUESTION 9.19
Consider 125 mL of a 0.110 M magnesium chloride solution.
    a.   **How many moles of magnesium chloride are in contained in 125 mL of the 0.110 M solution?**
    b.   **What is the concentration of the magnesium ion?**
    c.   **What is the concentration of the chloride ion?**

B. Dilution.
Dilution is a procedure used to prepare a less concentrated solution from a more concentrated solution. The key to understanding dilution is to remember that adding more water to a given amount of an aqueous solution does not change the number of moles of solute present in solution.

$$\text{moles of solute before dilution} = \text{moles of solute after dilution.}$$

Since the moles of solute equals the solution volume (V) times the molarity (M) we have

$$\text{moles}_i = \text{moles}_f$$

$$M_i V_i = M_f V_f$$

where the subscripts i and f, stand for initial solution, and final solution respectively. The concentration after dilution is

$$M_f = M_i' \times \frac{V_i}{V_f}$$

Note that V_f is always larger than V_i.

STUDY QUESTION 9.9
What volume of a 6.0 M hydrochloric acid stock solution is required to prepare 725 mL of a 0.12 M hydrochloric acid solution?

Solution: Using the dilution formula:

$$M_i V_i = M_f V_f$$

Rearranging to solve for the initial volume, we get:

$$V_i = V_f \frac{M_f}{M_i}$$
$$V_i = ?$$

$$V_f = 725 \text{ mL}$$

$$M_f = 0.12 \, M$$

$$M_i = 6.0 \, M$$

$$V_i = 725 \text{ mL} \times \frac{0.120 \, M}{6.0 \, M} = 15 \text{ mL}$$

NOTE:
Because we are working with a ratio of molarities, there is no need to convert mL to L. The units in which the volume is given, as long as it is a multiple of the liter, will be the units of the volume you calculate using this method.

> *PRACTICE QUESTION 9.20*
> **When 147 mL of a 4.25 M ammonia aqueous solution is mixed with 353 mL of water, what is the concentration of the final solution?**

> *PRACTICE QUESTION 9.21*
> **How would you prepare 250.0 mL of 0.666 M potassium hydroxide solution, starting with a 5.0 M potassium hydroxide solution?**

C. **The pH Scale.**
The **pH** of a solution is defined as the negative logarithm of the hydrogen (hydronium) ion concentration.

$$pH = -\log [H^+]$$

Given the pH, how do we calculate the $[H^+]$? Rearrange the equation for pH:

$$pH = -\log [H^+]$$

$$\log [H^+] = -pH$$

taking the antilog of both sides:

$$\text{antilog} (\log [H^+]) = \text{antilog} (-pH)$$

gives:

$$[H^+] = 10^{-pH}$$

Any electronic calculator with a 10^x key will easily make the calculation of H^+ ion concentrations from pH values.

STUDY QUESTION 9.10
The pH of many cola–type soft drinks is about 3.0. How many times greater is the H^+ concentration in these drinks than in neutral water?

Solution: First, write out the H^+ ion concentrations in the cola drink and in neutral water.

$$[H^+]_{cola} = 1.0 \times 10^{-3} \ M, \text{ and } [H^+]_{neut} = 1.0 \times 10^{-7} \ M.$$

Then the ratio is

$$\frac{[H^+]_{cola}}{[H^+]_{neut}} = \frac{1.0 \times 10^{-3} \ M}{1.0 \times 10^{-7} \ M} = 10^4 = 10{,}000$$

NOTE:
Here is another approach. Since a change of 1.0 pH unit corresponds to a 10–fold change in H^+ concentration, then a change of 4.0 pH units corresponds to $10 \times 10 \times 10 \times 10 = 10^4$ or a 10,000–fold increase in H^+ concentration.

> **PRACTICE QUESTION 9.22**
> **The pH of a certain solution is 3.0. How many moles of $H_3O^+(aq)$ ions are there in 0.10 L of this solution?**

> **PRACTICE QUESTION 9.23**
> **Calculate the concentration of H^+ ions in an acid solution with a pH of 2.29.**

PROBLEM TYPE 6: AQUEOUS REACTIONS AND CHEMICAL ANALYSIS.

A. Gravimetric Analysis.
Quantitative analysis is the determination of the amount or concentration of a substance in a sample. Precipitation reactions form the basis of a type of chemical analysis called gravimetric analysis.
A **gravimetric analysis** experiment involves the formation of a precipitate, and measurement of its mass. By knowing the mass and chemical formula of the precipitate, we can calculate the mass (or concentration) of the cation or anion component in the original sample.

STUDY QUESTION 9.11
Calculate the number of moles of NaCl that must be added to 0.752 L of 0.150 M $AgNO_3$ in order to precipitate all the Ag^+ ions as AgCl.

Solution: We need to determine the chemical equation for the precipitation reaction. Recall that NaCl is a strong electrolyte and dissociates entirely into Na^+ and Cl^- ions in solution. $AgNO_3$ is also a strong electrolyte yielding Ag^+ and NO_3^- ions. At first, the mixed solution contains Ag^+, NO_3^-, Na^+, and Cl^- ions. The products formed are AgCl and $NaNO_3$. Since AgCl is insoluble, the net ionic reaction is:

$$Ag^+(aq) + Cl^-(aq) \rightarrow AgCl(s)$$

and the molecular equation is:

$$AgNO_3(aq) + NaCl(aq) \rightarrow AgCl(s) + NaNO_3(aq)$$

Plan: L $AgNO_3 \rightarrow$ mol $AgNO_3 \rightarrow$ mol NaCl

$$0.752 \ L \times \frac{0.150 \ \text{mol } AgNO_3}{1 \ L} \times \frac{1 \ \text{mol NaCl}}{1 \ \text{mol } AgNO_3} = 0.113 \ \text{mol NaCl}$$

PRACTICE QUESTION 9.24
Calculate the number of moles of NaBr that must be added to 450 mL of 0.250 M AgNO₃ solution in order to precipitate all the Ag⁺ ions as AgBr.

PRACTICE QUESTION 9.25
In order to precipitate AgCl, excess AgNO₃ was added to 10.0 mL of a solution containing Cl⁻ ion. If 0.339 g of AgCl was formed, what was the concentration of Cl⁻ in the original solution?

PRACTICE QUESTION 9.26
0.198 g sample of an ionic compound containing the Br⁻ ion was dissolved in water and treated with excess AgNO₃. If the mass of silver bromide precipitate that forms was 0.0964 g, what is the percent of Br⁻ by mass in the original solution?

B. Acid-Base Titrations.

A **titration** is a procedure for determining the concentration of an unknown acid (or base) solution using a known (**standardized**) concentration of a base (or acid) solution. In the titration of a measured volume of acid solution of unknown concentration, the volume of standardized base solution required to exactly neutralize the acid is carefully measured. Because the concentration (M) and the required volume (V) of the base are known, the number of moles of base needed to neutralize the acid can be calculated. The number of moles of acid that were neutralized is readily calculated using the mole ratios from the balanced neutralization reaction.

STUDY QUESTION 9.12
What volume of 0.900 M HCl is required to completely neutralize 25.0 g of Ca(OH)₂?

Solution: The first thing to do is write and balance the neutralization equation

$$2HCl + Ca(OH)_2 \rightarrow CaCl_2 + 2H_2O$$

Since the balanced equation relates moles of one reactant to moles of another reactant, first determine the number of moles of Ca(OH)₂ in 25.0 g.

Plan: g Ca(OH)₂ → mol Ca(OH)₂ → mol HCl → volume HCl

$$\left(\frac{25.0 \text{ g Ca(OH)}_2}{1}\right)\left(\frac{1 \text{ mol Ca(OH)}_2}{74.10 \text{ g Ca(OH)}_2}\right)\left(\frac{2 \text{ mol HCl}}{1 \text{ mol Ca(OH)}_2}\right)\left(\frac{1 \text{ L}}{0.900 \text{ mol HCl}}\right) = 0.750 \text{ L HCl}$$

STUDY QUESTION 9.13
A volume of 128 mL of 0.650 M barium hydroxide was required to completely neutralize 50.0 mL of nitric acid solution. What was the concentration of the acid solution?

Solution: First we need the balanced chemical equation to give us the important mole–ratio. Then we construct a roadmap:

$$Ba(OH)_2(aq) + 2HNO_3(aq) \rightarrow Ba(NO_3)_2(aq) + 2H_2O(\ell)$$

Plan: mL Ba(OH)₂ → L Ba(OH)₂ → mol Ba(OH)₂ → mol HNO₃ → M HNO₃

$$\left(\frac{128 \text{ mL}}{1}\right)\left(\frac{1 \text{ L}}{1000 \text{ mL}}\right)\left(\frac{0.650 \text{ mol Ba(OH)}_2}{1 \text{ L}}\right)\left(\frac{2 \text{ mol HNO}_3}{1 \text{ mol Ba(OH)}_2}\right) = 0.166 \text{ mol HNO}_3$$

$$\left(\frac{50.0 \text{ mL}}{1}\right)\left(\frac{1 \text{ L}}{1000 \text{ mL}}\right) = 0.0500 \text{ L}$$

$$\text{molarity} = \frac{0.166 \text{ mol HNO}_3}{0.0500 \text{ L}} = 3.32 \ M$$

PRACTICE QUESTION 9.27
How many milliliters of 1.00 M HI solution are required to neutralize 2.10 g KOH?

PRACTICE QUESTION 9.28
What volume of 0.210 *M* HNO$_3$ solution is needed to exactly neutralize 50.0 mL of 0.082 *M* Ca(OH)$_2$?

PRACTICE QUESTION 9.29
What volume of 0.0824 *M* NaOH solution is needed to neutralize completely 9.8 mL of 0.210 *M* H$_2$SO$_4$?

PRACTICE QUESTION 9.30
What is the molarity of an oxalic acid (H$_2$C$_2$O$_4$) solution if 22.50 mL of this solution requires 35.72 mL of 0.198 *M* NaOH for complete neutralization?

PRACTICE QUIZ

1. Identify each of the following substances as a strong electrolyte, a weak electrolyte or a nonelectrolyte.
    a. NaBr
    b. HCl
    c. CH_3COOH
    d. NH_3
    e. glucose

2. Write balanced *net ionic equations* for the reactions between:
    a. $HCl(aq)$ and $Mg(OH)_2(aq)$
    b. $CH_3COOH(aq)$ and $NaOH(aq)$
    c. $Ca(OH)_2(aq)$ and $H_2SO_4(aq)$

3. Write balanced net ionic equations for the reactions that occur when solutions of the following solutes are mixed:
    a. $Ba(NO_3)_2$ and Na_2CO_3
    b. RbCl and $AgNO_3$
    c. $Pb(NO_3)_2$ and K_2S

4. Characterize the following compounds as soluble or insoluble in water.
    a. $Mg(OH)_2$
    b. AgCl
    c. $BaSO_4$
    d. $CaCO_3$
    e. $Pb(NO_3)_2$
    f. Na_2CO_3

5. Write formulas for the acid and base whose reactions produce the following salts.
    a. $CuSO_4(aq)$
    b. $KBr(aq)$
    c. $Ca_3(PO_4)_2(s)$

6. Write ionic and net ionic equations for the reactions that occur when solutions of the following compounds are mixed.
    a. NaBr and $AgNO_3$
    b. $MgBr_2$ and $Pb(NO_3)_2$

7. Assign oxidation numbers to the underlined elements in the following molecules and ions:
    a. $N O_2^-$
    b. $\underline{N}H_3$
    c. $H\underline{Cl}O$
    d. $H\underline{Cl}O_3$
    e. $\underline{Cl}O_4^-$
    f. $H_2\underline{S}O_3$
    g. $K\underline{Mn}O_4$
    h. $\underline{C}H_4$
    i. $Na_2\underline{S}_2O_3$
    j. $\underline{Hg}_2Cl_2$
    k. $\underline{S}_4O_6^{2-}$

8. Identify which of the following reactions are oxidation-reduction reactions.
    a. $Mg(s) + HCl(aq) \rightarrow MgCl_2(aq) + H_2(g)$
    b. $HCl(aq) + NH_3(aq) \rightarrow NH_4Cl(aq)$

c. $Mg(s) + CO_2(g) \rightarrow MgO(s) + C(s)$
d. $Pb^{2+}(aq) + S^{2-} \rightarrow PbS(s)$

9. Identify the oxidizing agents and reducing agents in the following reactions:
   a. $S + O_2 \rightarrow SO_2$

   b. $BrO_3^- + 6I^- + 6H^+ \rightarrow 3I_2 + 3H_2O + Br^-$

   c. $As + H^+ + NO_3^- + H_2O \rightarrow H_3AsO_3 + NO$

10. What is the molarity of a solution consisting of 11.8 g of NaOH dissolved in enough water to make exactly 300 mL of solution?

11. How many grams of NaCl are present in 45.0 mL of 1.25 M NaCl?

12. How many liters of 0.50 M glucose, $C_6H_{12}O_6$, solution will contain exactly 100 g glucose?

13. If 30 mL of 0.80 M KCl is mixed with water to make a total volume of 0.400 L, what is the final concentration of KCl?

14. When aqueous solutions of $Pb(NO_3)_2$ and Na_2SO_4 are mixed a precipitate of $PbSO_4$ is formed. Calculate the mass of $PbSO_4$ formed when 655 mL of 0.150 M $Pb(NO_3)_2$ and 525 mL of 0.0751 M Na_2SO_4 are mixed.

15. Calculate the molar concentration of Pb^{2+} ions in 500.0 mL unknown aqueous solution if 1.07 g $PbSO_4$ is formed upon the addition of excess Na_2SO_4.

16. How many mL of 0.10 M H_2SO_4 would be required to neutralize 2.5 mL of 1.0 M NaOH?

17. In a titration experiment, a student finds that 23.6 mL of 0.755 M H_2SO_4 solution are required to completely neutralize 30.0 mL of NaOH solution. Determine the concentration of the NaOH solution.

18. If 10 mL of 1.0 M HCl are required to neutralize 50 mL of a NaOH solution, how many mL of 1.0 M H_2SO_4 will neutralize another 50 mL of the same NaOH solution?

19. Calculate the volume of 0.0300 M Ce^{4+} required to reach the equivalence point when titrating 41.0 mL of 0.0200 M $C_2O_4^{2-}$ (oxalate ion).

$$2Ce^{4+} + C_2O_4^{2-} \rightarrow 2Ce^{3+} + 2CO_2 \text{ (acidic solution)}$$

20. A sample of iron ore weighing 1.824 g is analyzed by converting the Fe to Fe^{2+} and then titrating with standard potassium dichromate.

$$6Fe^{2+} + C_2O_4^{2-} + 14H^+ \rightarrow 6Fe^{3+} + 2Cr^{3+} + 7H_2O$$

If 37.21 mL of 0.0213 M $K_2Cr_2O_7$ was required to reach the equivalence point, what was the mass percent of iron in the ore?

Chapter 10
Energy Changes in Chemical Reactions

PROBLEM-SOLVING STRATEGIES AND TUTORIAL SOLUTIONS

TYPES OF PROBLEMS
Problem Type 1:   Energy and Energy Changes.

Problem Type 2:   Introduction to Thermodynamics.
        A. States and State Functions.
        B. The First Law of Thermodynamics.
        C. Work and Heat.

Problem Type 3:   Enthalpy.
        A. Reactions Carried out at Constant Volume or at Constant Pressure.
        B. Enthalpy and Enthalpy Changes.
        C. Thermochemical Equations.

Problem Type 4:   Calorimetry.
        A. Specific Heat and Heat Capacity.
        B. Constant–Pressure Calorimetry.
        C. Constant–Volume Calorimetry.

Problem Type 5:   Hess's Law.

Problem Type 6:   Standard Enthalpies of Formation.

Problem Type 7: Bond Enthalpy and the Stability of Covalent Molecules.

Problem Type 8: Lattice Energy and the Stability of Ionic Solids.
        A. The Born–Haber Cycle.
        B. Comparison of Ionic and Covalent Compounds.

PROBLEM TYPE 1: ENERGY AND ENERGY CHANGES.

In order to study energy changes in chemical reactions we divide the universe into the **system** (specific part f the universe that is under study) and the **surroundings** (everything else). In an **open system**, mass and energy in the system are exchanged with the surroundings. In a **closed system**, only energy from the system is exchanged with the surroundings. In an **isolated system**, neither mass nor energy from the system is exchanged with the surroundings.

Heat is the thermal energy transferred between two systems at different temperatures. When a system undergoes change or reaction that produces heat, heat is released to the surroundings (heat is a product). The change or reaction is called **exothermic**. When a system undergoes change whereby heat is consumed or absorbed by the system from the surroundings (heat is a reactant), the change or reaction is called **endothermic**.

STUDY QUESTION 10.1
Indicate whether the following is endothermic or exothermic.

a.	**Vaporization**
b.	**Freezing**

c.   Deposition

Solution: First, we must determine whether each process requires energy (initial state is lower energy than the final state), or vice versa.

a. Vaporization is when a substance goes from liquid to gas. Gas has more energy than the liquid; hence, this process would require energy to occur. This is an **endothermic** process.

b. Freezing is when a substance goes from liquid to solid. Solid is a lower energy state than the liquid. Since our initial state or reactant is higher energy and the product is lower energy, energy is released during this process. It is **exothermic**.

c. Deposition is the opposite of sublimation. It is when a gas goes directly to solid. Gas is a higher energy state than the solid; hence this process is **exothermic**.

> **PRACTICE QUESTION 10.1**
> **Determine whether the following is endothermic or exothermic.**
>
> a.   **Melting**
> b.   **Condensation**
> c.   **Sublimation**

PROBLEM TYPE 2: INTRODUCTION TO THERMODYNAMICS.

A. States and State Functions.

While heat (q) and work (w) provide important information about the fate of the energy absorbed or released during a change, both q and w depend on the way the change is made; q and w are **"path" functions**. We prefer to simplify the quantitative treatment of energy changes by dealing with quantities that depend only on the initial and final states of the system, or **state functions**.

B. The First Law of Thermodynamics.

The first law of thermodynamics (The Law of Conservation of Energy) limits the energy absorbed or released by a reaction to two parts: work done on or by the system, w, and heat absorbed or released by the system, q. The First Law of Thermodynamics states that the energy of the Universe is constant, $\Delta U_{Universe} = 0$. (Remember that Δ means change.) Since the Universe is composed of system and surroundings, then:

$$\Delta U_{system} + \Delta U_{surroundings} = 0 \text{ or } \Delta U_{system} = -\Delta U_{surroundings}$$

where ΔU_{system} is the change in the internal energy of the system and $\Delta U_{surroundings}$ is the change in the internal energy of the surroundings.

C. Work and Heat.

We can divide the components of the internal energy of a system into the categories of work and heat.

$$\Delta U_{system} = q + w$$

We had defined heat as the thermal energy transferred between objects due to a difference in temperature. In chemistry we are mostly concerned with work that is done by expansion or compression of gases under a constant pressure. We can calculate the amount of work done by or on a system in which gases are present by taking the product of the pressure × change in volume. By definition, work done on the system is positive (work done by the system is negative) and heat absorbed by the system is positive (heat released from the system is negative).

$$w = -P \times \Delta V \text{ and } \Delta U = q - (P \times \Delta V)$$

STUDY QUESTION 10.2
Calculate the work done on the system when 6.0 L of a gas is compressed to 1.0 L by a constant external pressure of 2.0 atm.

Solution: The work done is:

$$w = -P \times \Delta V = -P(V_2 - V_1)$$

$$= -2.0 \text{ atm } (1.0 \text{ L} - 6.0 \text{ L}) = +10 \text{ L} \cdot \text{atm}$$

The answer can be converted to joules.

$$10 \text{ L} \cdot \text{atm} \times \frac{101.3 \text{ J}}{1 \text{ L} \cdot \text{atm}} = 1.0 \times 10^3 \text{ J}$$

NOTE:
Obtaining a positive value for work means that work is done *on* the system by the surroundings in a compression. A positive work value means the system gains energy.

STUDY QUESTION 10.3
A gas, initially at a pressure of 10.0 atm and having a volume of 5.0 L, is allowed to expand at constant temperature against a constant external pressure of 4.0 atm until the new volume is 12.5 L. Calculate the work done by the gas on the surroundings.

Solution: In this problem the system does work on the surroundings as it expands, and by convention the sign is negative:

$$w = -P\Delta V = -P(V_2 - V_1)$$

P is the pressure opposing the expansion, ΔV is the change in volume of the system.

$$w = -4.0 \text{ atm } (12.5 \text{ L} - 5.0 \text{ L}) = -30 \text{ L} \cdot \text{atm}$$

This quantity can be expressed in units of joules.

$$w = -30 \text{ L} \cdot \text{atm} \times \frac{101.3 \text{ J}}{1 \text{L} \cdot \text{atm}} = -3.0 \times 10^3 \text{ J}$$

STUDY QUESTION 10.4
A gas is allowed to expand at constant temperature from a volume of 10.0 L to 20.0 L against an external pressure of 1.0 atm. If the gas also absorbs 250 J of heat from the surroundings, what are the values of q, w, and ΔU?

Solution: The work done by the system is:

$$w = -P\Delta V = -P(V_2 - V_1)$$

$$= -1.0 \text{ atm } (20.0 \text{ L} - 10.0 \text{ L}) = -10 \text{ L} \cdot \text{atm}$$

$$w = -10 \text{ L} \cdot \text{atm} \times \frac{101.3 \text{ J}}{1 \text{L} \cdot \text{atm}} = -1.0 \times 10^3 \text{ J}$$

The amount of heat absorbed was 250 J, and so q = 250 J. (Heat is a positive value, since it is endothermic.) Substituting into the first law of thermodynamics gives the energy change, ΔU.

$$\Delta U = q + w$$

$$= 250\ J - 1000\ J$$

$$\Delta U = -750\ J$$

NOTE:
In this example, the system did more work than the energy absorbed as heat; therefore, the internal energy U decreased.

> *PRACTICE QUESTION 10.2*
> **Calculate the work done on a gas when 22.4 L of the gas is compressed to 2.24 L under a constant external pressure of 10.0 atm.**

> *PRACTICE QUESTION 10.3*
> **A system does 975 J of work on the surroundings. How much heat does the system absorb at the same time, if its energy change is –350 J?**

PROBLEM TYPE 3: ENTHALPY.

A. Reactions Carried Out at Constant Volume or at Constant Pressure.
The determination of the components of ΔU, heat and work, requires us to consider whether the change is taking place under constant volume or under constant pressure. q can be determined by following changes in temperature. However, work can be achieved through a change in volume under constant pressure, or a change in pressure under constant volume.

In a constant volume process, since $\Delta U = q - P\Delta V$ and there is no change in volume, $\Delta V = 0$ and the equation simplifies to $\Delta U = q_v$. The subscript V indicates that V is kept constant.

B. Enthalpy and Enthalpy Changes.
We can define a state function especially for reactions in which the pressure is constant. The state function is called the enthalpy, H, which is heat measured at constant pressure.

$$\Delta H = q_p = \Delta U + \Delta(PV)$$

Enthalpy, like energy, is a state function. We can represent it with the following equation.

$$\Delta H = H_{products} - H_{reactants}$$

C. Thermochemical Equations.
Most reactions are run at constant pressure, so the enthalpy is equal to the heat transferred by the reaction. When the enthalpy is included in a chemical reaction it is called a **thermochemical reaction**. For exothermic reactions, (heat is released by the system) heat is a product and the enthalpy change is negative:

$$CH_4(g) + 2O_2(g) \rightarrow CO_2(g) + 2H_2O(g) \quad \Delta H = -802.4\ kJ/mol$$

For endothermic reactions, (heat is absorbed by the system) heat can be considered a reactant and the enthalpy change is positive:

$$CaCO_3(s) \rightarrow CaO(s) + CO_2(g) \qquad \Delta H = 177.8\ kJ/mol$$

The balanced thermochemical equation can be used to convert between amount of reactant or product and the amount of heat. For example, let's consider the reaction:

$$CH_4(g) + 2O_2(g) \rightarrow CO_2(g) + 2H_2O(g) \qquad \Delta H = -802.4 \text{ kJ/mol}$$

The –802.4 kJ/mol means 802.4 kJ of heat are released per mole of reaction as it is balanced. This provides a series of conversion factors.

$$\frac{-802.4 \text{ kJ / mol}}{1 \text{ mol } CH_4} \quad \frac{-802.4 \text{ kJ / mol}}{2 \text{ mol } O_2} \quad \frac{-802.4 \text{ kJ / mol}}{1 \text{ mol } CO_2} \quad \frac{-802.4 \text{ kJ / mol}}{2 \text{ mol } H_2O}$$

To calculate the amount of heat produced when 80.0 g of CH_4 is burned, first convert grams to moles, then use the appropriate mol → kJ conversion factor.

$$80.0 \text{ g } CH_4 \times \frac{1 \text{ mol } CH_4}{16.032 \text{ g } CH_4} \times \frac{-802.4 \text{ kJ/mol}}{1 \text{ mol } CH_4} = -4.00 \times 10^3 \text{ kJ}$$

When a reaction is reversed, the sign of ΔH is reversed.

$$H_2O(s) \rightarrow H_2O(l) \qquad \Delta H = 6.01 \text{ kJ/mol}$$
$$H_2O(l) \rightarrow H_2O(s) \qquad \Delta H = -6.01 \text{ kJ/mol}$$

The enthalpy is an extensive property, its value depends on the amount of material in the system. Therefore, scaling the reaction (multiplying by a constant) also scales ΔH.

$$H_2O(s) \rightarrow H_2O(l) \qquad \Delta H = 6.01 \text{ kJ/mol}$$
$$2H_2O(s) \rightarrow 2H_2O(l) \qquad \Delta H = 12.0 \text{ kJ/mol}$$

The physical states of all reactants and products are crucial to enthalpy values.

$$CH_4(g) + 2O_2(g) \rightarrow CO_2(g) + 2H_2O(g) \qquad \Delta H = -802.4 \text{ kJ/mol}$$
$$CH_4(g) + 2O_2(g) \rightarrow CO_2(g) + 2H_2O(l) \qquad \Delta H = -890.4 \text{ kJ/mol}$$

STUDY QUESTION 10.5
The thermochemical equation for the combustion of propane is:
$$\mathbf{C_3H_8(g) + 5\,O_2(g) \rightarrow 3\,CO_2(g) + 4\,H_2O(l) \qquad \Delta H^o_{rxn} = -2220 \text{ kJ/mol}}$$

 a. **How many kJ of heat are released when 0.50 mole of propane reacts?**
 b. **How much heat is released when 88.2 g of propane reacts?**

Solution:
a. Let q = the heat absorbed or released by the reaction. The heat released by an exothermic reaction is an extensive property. This means that q depends on the amount of propane consumed. The equation indicates that 2220 kJ of heat is released per mole of propane burned.

$$q = 0.50 \text{ mol } C_3H_8 \times \frac{-2220 \text{ kJ}}{1 \text{ mol } C_3H_8}$$

$$q = -1.1 \times 10^3 \text{ kJ}$$

b. Since we know the heat of reaction per mole, we convert the number of grams of C_3H_8 to moles.
  Plan: g → mol → kJ

$$q = 88.2 \text{ g } C_3H_8 \times \frac{1 \text{ mol } C_3H_8}{44.1 \text{ g } C_3H_8} \times \frac{-2220 \text{ kJ}}{1 \text{ mol } C_3H_8}$$

$$q = -4440 \text{ kJ}$$

PRACTICE QUESTION 10.4
Given the thermochemical equation:

$$SO_2(g) + \frac{1}{2}O_2(g) \rightarrow SO_3(g) \qquad \Delta H^o_{rxn} = -99.0 \text{ kJ,}$$

how much heat is evolved when 75 g SO_2 undergoes combustion?

PRACTICE QUESTION 10.5
The reaction that occurs when a typical fat, glyceryl trioleate, is metabolized by the body is:
$$C_{57}H_{104}O_6(s) + 80O_2(g) \rightarrow 57CO_2(g) + 52H_2O(\ell)$$
$$\Delta H^o_{rxn} = -3.35 \times 10^4 \text{ kJ/mol.}$$
How much heat is evolved when 1 gram of this fat is completely oxidized?

PROBLEM TYPE 4: CALORIMETRY.

A. Specific Heat and Heat Capacity.

  A **calorimeter** is an apparatus used for measuring the heat changes associated with chemical reactions. All calorimetry experiments involve carrying out a reaction in a vessel immersed in a medium such as water and measuring the temperature change (ΔT) of the medium. The **heat capacity**, C, of a substance is the amount of heat, q, required to raise the temperature of a *given mass* of substance by 1 °C. The **heat capacity** of 100 g H_2O is 418.4 J/°C. This means that 418.4 J of heat will raise the temperature of 100 g water 1 °C. The heat capacity is a proportionality constant that relates the amount of heat absorbed or released by a material to its change in temperature:

$$q = C (T_{final} - T_{initial}) = C \, \Delta T.$$

  The **specific heat**, s, is a related term which is the amount of heat required to raise the temperature of one gram of a substance by 1 °C. In terms of the specific heat, the heat absorbed or released is:

$$q = sm\Delta T.$$

STUDY QUESTION 10.6
  a. **What is the heat capacity of a block of lead if the temperature of a 425 g block increases 2.31 °C when it absorbs 492 J of heat?**
  b. **What is the specific heat of lead?**

Solution:
a. The heat capacity of the block of lead is the heat absorbed divided by the temperature rise. Since q=CΔT, then:

$$C = \frac{q}{\Delta T} = \frac{492 \text{ J}}{(2.31 \text{ °C})} = 213 \text{ J/°C}$$

b. In terms of the specific heat (s), the amount of heat absorbed when an object of mass m is heated from T_i to T_f is:

$$q = ms\Delta T$$

Substituting into the equation the given quantities for q, m, and ΔT gives:

$$492 \text{ J} = (425 \text{ g})s(2.31 \text{ °C})$$

Rearranging to solve for the specific heats:

$$s = \frac{492 \text{ J}}{(425 \text{ g})(2.31 \text{ °C})} = 0.501 \text{ J/g·°C}$$

PRACTICE QUESTION 10.6
A piece of iron initially at a temperature of 25.2 °C absorbs 16.9 kJ of heat. If its mass is 82.0 g, calculate the final temperature. The specific heat of iron is 0.444 J/g·°C.

PRACTICE QUESTION 10.7
How much heat is absorbed by 52.0 g of iron when its temperature is raised from 25 °C to 275 °C? The specific heat of iron is 0.444 J/g·°C.

PRACTICE QUESTION 10.8
The specific heat of iron is 0.444 J/g·°C while that for titanium is 0.523 J/g·°C. If 10.0 g of both metals start out at 25°C, which will reach a higher temperature if they each absorb 85 kJ?

B. Constant-Pressure Calorimetry.
The heat evolved in non–combustion reactions is measured in a constant–pressure calorimeter. A coffee cup calorimeter is very inexpensive and works well. Typically, a reaction occurs in solution, absorbing or liberating heat that changes the temperature of the solution (mostly water), and the calorimeter:

$$q_{rxn} = -(q_{cal} + q_{water})$$

These calorimeters are constructed of low heat capacity materials, such as Styrofoam, so the heat absorbed by the calorimeter is minimal and sometimes can be neglected:

$$q_{rxn} = -q_{water}$$

Reactions in a coffee cup calorimeter occur under constant-pressure conditions, and so the heat released is equal to ΔH.

$$\Delta H = q_{rxn} = -s_{water}m\Delta T$$

STUDY QUESTION 10.7
A 3.80 g cube of an unknown metal at an initial temperature of 147.0 °C is dropped into a Styrofoam cup containing 65.0 mL of water at an initial temperature of 21.2 °C. When thermal equilibrium was achieved, the temperature of the water had risen to 22.1 °C. Use the following information to determine the identity of the metal.

Metal	Specific Heat (J/g·°C)
zinc	0.388
titanium	0.523
aluminum	0.900

Solution: The basic principle on which we will base the solution to this problem is on the total transfer of heat from the metal to the water. We will assume that the Styrofoam cup does not absorb any heat and that, if it does, the amount absorbed is negligible in comparison to the heat absorbed by the water.

Let's assume that the density of water is 1.00 g/mL and calculate the amount of heat absorbed by the water:

$$q_{water} = m_{water} s_{water} \Delta T_{water}$$

$$q_{water} = (65.0 \text{ mL})(\frac{1.00 \text{ g}}{1 \text{ mL}})(4.184 \text{ J/g°C})(22.1 \text{ °C} - 21.2 \text{ °C}) = 245 \text{ J}$$

The heat absorbed by the water is equal in magnitude to the heat released by the metal:

$$q_{metal} = -q_{water}$$

We will rearrange to solve for the specific heat of the metal, which is an intensive property that can be used to identify the unknown.

$$s_{metal} = \frac{q_{metal}}{m_{metal} \Delta T_{metal}} = \frac{-q_{water}}{m_{metal} \Delta T_{metal}}$$

Because the system and surroundings had reached thermal equilibrium, both the metal and the water had reached the same final temperature. Substitution of the values given leads to:

$$s_{metal} = \frac{-248 \text{ J}}{3.80 \text{ g}(22.1 \text{ °C} - 147.0 \text{ °C})} = 0.523 \text{ J/g} \cdot \text{°C}$$

The cube was made of titanium.

PRACTICE QUESTION 10.9
A 25.0 g piece of aluminum at 4.0 °C is dropped into a beaker of water. The temperature of water drops from 75.0 °C to 55.0 °C. What amount of heat did the aluminum absorb? The specific heat of Al is 0.902 J/g·°C.

C. Constant–Volume Calorimetry.

Constant–volume or bomb calorimeters are used to measure the heat evolved in combustion reactions. A high–pressure steel vessel, called a bomb, is loaded with a small amount of a combustible substance and O_2 at high pressure. The loaded bomb is immersed in a known amount of water. The heat evolved during combustion is absorbed by the water and the calorimeter:

$$q_{sys} = -q_{surr.}$$

The system is the reaction, the surroundings is made of the calorimeter parts as well as the water.

$$q_{sys} = -(q_{cal} + q_{water})$$

The change in the temperature of the water and calorimeter are related to the heat evolved by the reaction, q, and to the enthalpy, ΔH.

$$\Delta H \approx q_{rxn} = -(C_{cal}\Delta T + ms_{water}\Delta T)$$

Reactions in a bomb calorimeter occur under constant volume rather than constant pressure conditions, and so the heat released does not equal ΔH exactly. For most reactions, the difference is small and can be neglected. For instance, for the combustion of 1 mole of pentane, the difference is only 7 kJ out of 3500 kJ.

STUDY QUESTION 10.8
The combustion of benzoic acid is often used as a standard source of heat for calibrating combustion bomb calorimeters. The heat of combustion of benzoic acid has been accurately determined to be –26.42 kJ/g. When 0.8000 g of benzoic acid was burned in a calorimeter containing 950 g of water, a temperature rise of 4.08 °C was observed. What is the heat capacity of the bomb calorimeter (the calorimeter constant)?

Solution: The combustion of 0.8000 g of benzoic acid produces a known amount of heat.

$$q_{rxn} = 0.8000 \text{ g} \times \frac{-26.42 \text{ kJ}}{1 \text{ g}} = -21.14 \text{ kJ} = -2.114 \times 10^4 \text{ J}$$

And since ΔT and the amount of water are known, C_{cal}, the heat capacity of the calorimeter, can be calculated. All the heat from the combustion reaction is absorbed by the bomb calorimeter and water.

$$q_{rxn} = -q_{surr}$$

$$q_{rxn} = -(q_{cal} + q_{water})$$

$$q_{rxn} = -(C_{cal}\Delta T + s_{water}m\Delta T)$$

We know all of the variables in the above equation, except C_{cal}.

$$q_{rxn} = -2.114 \times 10^4 \text{ J}$$

$$m = 950 \text{ g water}$$

$$s_{water} = 4.184 \text{ J/g·°C}$$

$$\Delta T = 4.08 \text{ °C}$$

Solve for C_{cal} and plug in all of the other variables.

$$q_{rxn} = -(C_{cal}\Delta T + s_{water}m\,T)$$

$$-q_{rxn} - sm\Delta T = C_{cal}\Delta T$$

$$\frac{-q_{rxn} - ms\Delta T}{\Delta T} = C_{cal}$$

$$C_{cal} = \frac{-(-2.114 \times 10^4 \text{ J}) - (950 \text{ g})(4.184 \text{ J/g} \cdot {}^\circ\text{C})(4.08 \text{ }^\circ\text{C})}{4.08 \text{ }^\circ\text{C}}$$

$$C_{cal} = 1.21 \times 10^3 \text{ J/}^\circ\text{C}$$

STUDY QUESTION 10.9

The thermochemical equation for the combustion of pentane (C_5H_{12}) is:

$$C_5H_{12}(\ell) + 8O_2(g) \rightarrow 5CO_2(g) + 6H_2O(\ell) \quad \Delta H^\circ = -3509 \text{ kJ/mol}$$

From the following information calculate the heat of combustion, ΔH°, and compare your result to the value given above. 0.5521 g of C_5H_{12} was burned in the presence of excess O_2 in a bomb calorimeter. The heat capacity of the calorimeter was 1.800×10^3 J/°C, the temperature of the calorimeter and 1.000×10^3 g of water rose from 21.22 °C to 25.70 °C.

Solution: Again, as with all calorimetry problems:

$$q_{rxn} = -q_{surr}$$

The heat evolved by the combustion reaction is absorbed by the water and the calorimeter assembly (the surrounding).

$$q_{rxn} = -(q_{cal} + q_{water})$$

where

$$q_{rxn} = -(C_{cal}\Delta T + sm\Delta T) \text{ and } \Delta T = (25.70 \text{ }^\circ\text{C} - 21.22 \text{ }^\circ\text{C}) = 4.48 \text{ }^\circ\text{C}$$

Substituting:

$$q_{rxn} = -[(1.800 \times 10^3 \text{ J/}^\circ\text{C})(4.48 \text{ }^\circ\text{C}) + (1.000 \times 10^3 \text{ g})(4.184 \text{ J/g}^\circ\text{C})(4.48 \text{ }^\circ\text{C})]$$

$$= -2.68 \times 10^4 \text{ J} \quad \text{or} \quad -26.8 \text{ kJ}$$

ΔH° refers to the reaction of 1 mole of pentane. Therefore, if we can determine the amount of heat released for 1 mole of pentane, we will have the ΔH for the reaction as balanced.

- The 26,800 J was evolved by 0.5521 g C_5H_{12}.
- The molar mass of pentane is 72.15 g.

$$\Delta H^\circ = q_p$$

$$= 1 \text{ mol } C_5H_{12} \times \frac{72.15 \text{ g } C_5H_{12}}{1 \text{ mol } C_5H_{12}} \times \frac{26.8 \text{ kJ}}{0.5521 \text{ g}} = -3500 \text{ kJ}$$

$\Delta H^\circ = -3.50 \times 10^3$ kJ/mol (where mol represents the mol of reaction as balanced).

NOTE:

To three significant figures the result compares well with the value given in the thermochemical equation. Small differences will arise because the heat evolved at constant volume is not quite the same as the heat evolved at constant- pressure (which is equal to ΔU). Only the heat evolved at constant-pressure exactly equals ΔH, the enthalpy change.

PRACTICE QUESTION 10.10

Benzoic acid ($C_6H_5CO_2H$) was used to calibrate a bomb calorimeter. Its enthalpy of combustion is accurately known to be –3226.7 kJ/mol. When 1.0236 g of benzoic acid were burned in a bomb calorimeter, the temperature of the calorimeter and the 1.000 kg of water surrounding it rose from 20.66 °C to 24.47 °C. What is the heat capacity of the calorimeter?

PRACTICE QUESTION 10.11

To determine the heat capacity of a bomb calorimeter, a student added 150 g of water at 50.0 °C to the calorimeter. The calorimeter initially was at 20.0 °C. The final temperature of the water and calorimeter was 32.0 °C. What is the heat capacity of the calorimeter in J/°C?

PROBLEM TYPE 5: HESS'S LAW.

Some reactions don't proceed as written. For example, the reaction may proceed too slowly or generate side-products. In other cases, the enthalpy of formation (to be discussed in the next section) of one or more reactants or products may not be available. The alternative comes from Hess's law of heat summation. When a reaction is the sum of several reaction steps, then the enthalpy change for the reaction is equal to the sum of the enthalpy changes of the reaction steps. For example, it's not convenient to measure the enthalpy of the conversion of graphite to diamond. This reaction takes thousands of years at high pressure in the earth. However, it is easy to measure the enthalpies of combustion of graphite and diamond:

$$C_{graphite} + O_2 \rightarrow CO_2 \qquad \Delta H_{comb} = -393.51 \text{ kJ/mol}$$
$$C_{diamond} + O_2 \rightarrow CO_2 \qquad \Delta H_{comb} = -395.40 \text{ kJ/mol}$$

Notice that when we reverse the diamond thermochemical reaction, the sum of the combustion reactions is the graphite to diamond conversion:

$$C_{graphite} + O_2 \rightarrow CO_2 \qquad \Delta H_{comb} = -393.51 \text{ kJ/mol}$$
$$CO_2 \rightarrow C_{diamond} + O_2 \qquad \Delta H_{comb} = 395.40 \text{ kJ/mol}$$
$$\overline{C_{graphite} \rightarrow C_{diamond} \qquad \Delta H_{rxn} = 1.89 \text{ kJ/mol}}$$

STUDY QUESTION 10.10

The standard enthalpy change for the combustion of 1 mole of ethanol is

$$C_2H_5OH(\ell) + 3O_2(g) \rightarrow 2CO_2(g) + 3H_2O(\ell) \qquad \Delta H^\circ_{rxn} = -1367 \text{ kJ/mol}$$

and the heat of vaporization of water is

$$H_2O(l) \rightarrow H_2O(g) \qquad \Delta H_{vap} = 44 \text{ kJ/mol}$$

What is ΔH°_{rxn} for the following reaction in which H_2O is formed as a gas, rather than as a liquid?

$$C_2H_5OH(l) + 3O_2(g) \rightarrow 2CO_2(g) + 3H_2O(g)$$

Solution: We can imagine a two–step path for this reaction. In the first step, ethanol undergoes combustion to form liquid H_2O, followed by the second step in which $H_2O(l)$ is vaporized. The sum of the two steps gives the desired overall reaction, and according to Hess's law the sum of the two ΔH's gives the overall ΔH.

$$C_2H_5OH(l) + 3O_2(g) \rightarrow 2CO_2(g) + 3H_2O(l) \qquad \Delta H^\circ_{rxn} = -1367 \text{ kJ/mol}$$

$3H_2O(l) \rightarrow 3H_2O(g)$	$\Delta H^{\circ}_{rxn} = 132$ kJ/mol
$C_2H_5OH(l) + 3O_2(g) \rightarrow 2CO_2(g) + 3H_2O(g)$	$\Delta H^{\circ}_{rxn} = -1235$ kJ/mol

PRACTICE QUESTION 10.12

From the following thermochemical equations, calculate the enthalpy of formation of CH$_4$.

$CH_4(g) + 4F_2(g) \rightarrow CF_4(g) + 4HF(g)$ $\qquad \Delta H^{\circ}_{rxn} = -1942$ kJ/mol

$C(graphite) + F_2(g) \rightarrow CF_4(g)$ $\qquad \Delta H^{\circ}_{rxn} = -933$ kJ/mol

$H_2(g) + F_2(g) \rightarrow 2HF(g)$ $\qquad \Delta H^{\circ}_{rxn} = -542$ kJ/mol

PROBLEM TYPE 6: STANDARD ENTHALPIES OF FORMATION.

Since enthalpy is a state function, in principle, we should be able to calculate it if we knew the values for $H_{reactants}$ and $H_{products}$. However, there is no way to measure $H_{reactants}$ and $H_{products}$. Absolute energy values, such as $H_{reactants}$ and $H_{products}$, can't be measured because there is no way to define absolute zero energy. Relative energy values can be measured because the reference energy is arbitrary. The reference point for enthalpy values is the enthalpy of formation of the elements in their standard state, which is defined as zero at standard conditions (generally, 1 atm for gases and 1 M concentration of solutions) and 25 $^{\circ}$C. The standard enthalpy of allotropes and single atoms of diatomic elements (non-standard states) have non–zero values. For example,

$$\Delta H^{\circ}_f(O_2) = 0 \text{ kJ/mol} \qquad \Delta H^{\circ}_f(C_{graphite}) = 0 \text{ kJ/mol} \qquad \Delta H^{\circ}_f(H_2) = 0 \text{ kJ/mol}$$

$$\Delta H^{\circ}_f(O_3) = 142 \text{ kJ/mol} \qquad \Delta H^{\circ}_f(C_{diamond}) = 1.9 \text{ kJ/mol} \qquad \Delta H^{\circ}_f(H) = 218.2 \text{ kJ/mol}$$

We can calculate ΔH for any reaction using ΔH°_f of the reactants and products. The enthalpy change for any reaction is the difference of the sum of the standard enthalpies of formation of the products multiplied by their stoichiometric coefficients and the sum of the standard enthalpies of formation of the reactants multiplied by their stoichiometric coefficients. For example, the enthalpy of the decomposition of CaCO$_3$,

$$CaCO_3(s) \rightarrow CaO(s) + CO_2(g)$$

can be determined by

$$\Delta H^{\circ}_{rxn} = \sum nH^{\circ}_f(products) - \sum nH^{\circ}_f(reactants)$$

$$= H^{\circ}_f(CaO) + H^{\circ}_f(CO_2) - H^{\circ}_f(CaCO_3)$$

$$= 1(-635.6 \text{ kJ/mol}) + (1)(-393.5 \text{ kJ/mol}) - (1)(-1206.9 \text{ kJ/mol}) = 177.8 \text{ kJ/mol}$$

STUDY QUESTION 10.11

Using ΔH°_f values from Appendix 2 of your textbook, calculate the standard enthalpy change for the incomplete combustion of ethane (C$_2$H$_6$).

$$C_2H_6(g) + \frac{5}{2}O_2(g) \rightarrow 2CO(g) + 3H_2O(l)$$

Solution: The enthalpy change for this chemical reaction in terms of enthalpies of formation is:

$$\Delta H^{\circ}_{rxn} = \sum n \, \Delta H^{\circ}_{f} \, (\text{products}) - \sum m \, \Delta H^{\circ}_{f} \, (\text{reactants})$$

$$\Delta H^{\circ}_{rxn} = [2 \text{ mol} \times \Delta H^{\circ}_{f} \, (CO) + 3 \text{ mol} \times \Delta H^{\circ}_{f} \, (H_2O)] - [1 \text{ mol} \times \Delta H^{\circ}_{f} \, (C_2H_6) - \frac{5}{2} \text{mol} \times \Delta H^{\circ}_{f} \, (O_2)]$$

To avoid cumbersome notation, the physical states of the reactants and products were omitted from this equation. Be careful to obtain from Appendix 2 the appropriate value of ΔH°_{f}. Note that the coefficients from the chemical equation are equal to the number of moles of each substance and that <u>all</u> the terms involving ΔH°_{f} of the *reactants* are subtracted. Substituting the values from Appendix 2:

$$\Delta H^{\circ}_{rxn} = (2 \text{ mol})(-110.5 \text{ kJ/mol}) + (3 \text{ mol})(-285.8 \text{ kJ/mol}) - [(1 \text{ mol})(-84.68 \text{ kJ/mol}) + (5/2 \text{ mol})(0 \text{ kJ/mol})]$$

$$= (-221.0 \text{ kJ}) + (-857.4 \text{ kJ}) - (-84.68 \text{ kJ})$$

$$= -221.0 \text{ kJ} - 857.4 \text{ kJ} + 84.68 \text{ kJ}$$

$$\Delta H^{\circ}_{rxn} = -993.7 \text{ kJ}$$

Since the reaction is written with one mole of ethane, $\Delta H^{\circ}_{rxn} = -993.7 \text{ kJ/mol}$.

PRACTICE QUESTION 10.13
A reaction used for rocket engines is:
$$N_2H_4(l) + 2 \, H_2O_2(l) \rightarrow N_2(g) + 4 \, H_2O(l)$$
What is the enthalpy of reaction in kJ/mol given the following enthalpies of formation?

$\Delta H^{\circ}_{f} \, (N_2H_4) = \quad 95.1 \text{ kJ/mol}$

$\Delta H^{\circ}_{f} \, (H_2O_2) = -187.8 \text{ kJ/mol}$

$\Delta H^{\circ}_{f} \, (H_2O) = -285.8 \text{ kJ/mol}$

PROBLEM TYPE 7: BOND ENTHALPY AND THE STABILITY OF COVALENT MOLECULES.

Occasionally, the enthalpy change, ΔH, is needed for a reaction for which enthalpy of formation data does not exist. One approach that allows an estimate of ΔH° uses the concept of **bond enthalpy** (also called *bond dissociation energy*), the energy needed to break a covalent bond. Consider the gas phase atomization process:

$$CH_4(g) \rightarrow C(g) + 4H(g) \qquad \Delta H^{\circ}_{rxn} = 1664 \text{ kJ/mol}$$

In this reaction, four C—H bonds are broken; therefore, we can define the average C—H bond enthalpy as one–fourth of ΔH°_{rxn} for the reaction. Hence, the *average* C—H bond enthalpy in CH_4 is 416 kJ/mol. The actual bond enthalpies of the individual C—H bonds in CH_4 are not the same as the average value. Even so, the use of average bond enthalpies makes it possible to estimate the enthalpy changes of certain reactions.

We can estimate ΔH° for the combustion of methane as follows:

$$CH_4(g) + 2\ O_2(g) \rightarrow CO_2(g) + 2\ H_2O(g)$$

First, we break 4 C—H bonds and 2 O=O bonds. Then, let the atoms recombine to form the 2 C=O and 4 O—H bonds of the products. The enthalpy change for the reaction is taken as the sum of the bond enthalpies of the reactants (breaking a bond requires energy) minus the sum of the bond enthalpies of the products (forming a bond releases energy). Therefore,

$$\Delta H^\circ = \sum BE\ (\text{reactants}) - \sum BE\ (\text{products})$$

$$\Delta H^\circ = [4\text{mol} \times BE(C—H) + 2\ \text{mol} \times BE(O=O)] - [2\ \text{mol} \times BE(C=O) + 4\ \text{mol} \times BE(O—H)]$$

Inserting values from Table 10.4 in your textbook gives:

$$\Delta H^\circ = [4\ \text{mol}(414\ \text{kJ/mol}) + 2\ \text{mol}(499\ \text{kJ/mol})] - [2\ \text{mol}(802\ \text{kJ/mol}) + 4\ \text{mol}(460\ \text{kJ/mol})]$$

$$= 2654\ \text{kJ} - 3444\ \text{kJ}$$

$\Delta H^\circ = -790\ \text{kJ}$ or $-790\ \text{kJ/mol}$ since there is one mole of CH_4 reacting.

If we compare this value with the standard enthalpy of combustion of methane, we get

$$CH_4(g) + 2O_2(g) \rightarrow CO_2(g) + 2H_2O(\ell) \qquad \Delta H^\circ_{rxn} = -890\ \text{kJ/mol}$$

At first it seems there is a large discrepancy of 100 kJ. However, in our calculation, H_2O is present as a gas! But ΔH° refers to the standard state of H_2O as a liquid. Therefore, the 100 kJ difference is largely due to the ΔH of vaporization of two moles of H_2O, which is 81.4 kJ.

Two important conclusions can be drawn here:
(1) Bond enthalpies are used to estimate enthalpies of reaction of gas phase reactions.
(2) Enthalpies of reaction calculated from average bond enthalpies are only approximate values.

STUDY QUESTION 10.12
Estimate ΔH_{rxn} for the reaction:
$$Cl_2(g) + I_2(g) \rightarrow 2ICl(g)$$
using bond enthalpies given in Table 10.4 of your textbook, and given that BE (I—Cl) = 210 kJ/mol.

Solution: Recall that

$$\Delta H^\circ = \sum BE\ (\text{reactants}) - \sum BE\ (\text{products})$$

$$\sum BE\ (\text{reactants}) = BE(Cl—Cl) + BE(I—I) = 243\ \text{kJ/mol} + 151\text{kJ/mol} = 394\ \text{kJ, and}$$

$$\sum BE\ (\text{products}) = 2\ \text{mol} \times BE(I—Cl) = 2\ \text{mol}(210\ \text{kJ/mol}) = 420\ \text{kJ}$$

Subtraction yields:

$$\Delta H = 394\ \text{kJ} - 420\ \text{kJ}$$

$$= -26\ \text{kJ or } -26\ \text{kJ/mol of } Cl_2$$

PRACTICE QUESTION 10.14
Given the N≡N and H—H bond enthalpies in Table 10.4 in your textbook and the standard enthalpy for the reaction:

$$\frac{1}{2} N_2(g) + \frac{3}{2} H_2(g) \rightarrow NH_3(g) \qquad \Delta H^{\circ}_{rxn} = -46.3 \text{ kJ/mol},$$

calculate the average N—H bond enthalpy in ammonia.

PROBLEM TYPE 8: LATTICE ENERGY AND THE STABILITY OF IONIC SOLIDS.

A. The Born-Haber Cycle.

Lattice energies cannot be measured directly and must be calculated using Hess's law. The series of steps used to calculate the lattice energies of ionic solids is called the Born-Haber cycle. Each step is one you've seen previously in relation to the properties of atoms.

STUDY QUESTION 10.13

Given the following data calculate the lattice energy of potassium chloride:

Enthalpy of sublimation of potassium =	90.0 kJ/mol
Bond enthalpy (BE) of Cl—Cl =	242.7 kJ/mol
Ionization energy (IE) of K =	419 kJ/mol
Electron affinity (EA) of Cl =	+349 kJ/mol
ΔH°_f (KCl) =	−435.9 kJ/mol

Solution: First, let's identify the processes associated with the energy changes given:

(1) $\qquad K(s) \rightarrow K(g) \qquad \Delta H_{subl} = 90.0 \text{ kJ/mol}$

(2) $\qquad \frac{1}{2} Cl_2(g) \rightarrow Cl(g) \qquad \frac{1}{2} BE = 121.4 \text{ kJ/mol}$

(3) $\qquad K(g) \rightarrow K^+(g) + e^- \qquad IE = 419 \text{ kJ/mol}$

(4) $\qquad Cl(g) + e^- \rightarrow Cl^-(g) \qquad EA = 349 \text{ kJ/mol}$

(5) $\qquad K(s) + \frac{1}{2} Cl_2(g) \rightarrow KCl(s) \qquad \Delta H^{\circ}_f = -435.9 \text{ kJ/mol}$

The lattice energy for KCl is represented by the equation:

$$KCl(s) \rightarrow K^+(g) + Cl^-(g)$$

We need to apply Hess's Law in order to calculate the enthalpy change for this reaction. From inspection, we see that equation 5 would need to be reversed in order to place KCl(s) as a reactant. In order to get K$^+$(g) as a product, equations 1 and 3 need to be used as given. To get Cl$^-$ as a product, equations 2 and 4 should be used as given, remembering that the enthalpy change for equation 4 is equal to –EA.

$K(s) \rightarrow K(g)$	$\Delta H = 90.0 \text{ kJ/mol}$
$K(g) \rightarrow K^+(g) + e^-$	$\Delta H = IE = 419 \text{ kJ/mol}$
$\frac{1}{2} Cl_2(g) \rightarrow Cl(g)$	$\Delta H = \frac{1}{2} BE = 121.4 \text{ kJ/mol}$
$Cl(g) + e^- \rightarrow Cl^-(g)$	$\Delta H = EA = -349 \text{ kJ/mol}$
$KCl(s) \rightarrow K(s) + \frac{1}{2} Cl_2(g)$	$\Delta H = 435.9 \text{ kJ/mol}$

Adding up both the equations and the values for the enthalpy change of each step we get:

$$KCl(s) \rightarrow K^+(g) + Cl^-(g) \qquad\qquad \Delta H = 717 \text{ kJ/mol} = \text{lattice energy}$$

B. Comparison of Ionic and Covalent Compounds.

The covalent bond is an *intramolecular* force, the strength of which is measured by *bond enthalpy*. The *intermolecular forces* that exist between molecules are usually a much weaker force; hence, covalent compounds tend to be gases, liquids or low-melting solids.

The force that gives rise to the ionic bond is the electrical attraction existing between a positive ion and a negative ion. This force is usually very strong; hence ionic compounds are usually high-melting solids.

The **lattice energy** is the energy required to separate the ions in 1 mole of a solid ionic compound into gaseous ions. For example, the lattice energy of NaF(s) is 908 kJ/mol. This means 908 kJ are required to vaporize one mole of NaF(s) and form one mole of Na^+ ions and one mole of F^- ions in the gas phase.

$$NaF(s) \rightarrow Na^+(g) + F^-(g) \qquad \text{lattice energy} = 908 \text{ kJ}$$

Ionic solids with larger lattice energies are more stable. The lattice energies are reflected in the melting points of ionic crystals. During melting the ions gain enough kinetic energy to overcome the potential energy of attraction, and they move away from each other. Therefore, as the lattice energy increases, so does the melting point of the compound. In general, the smaller the ionic radius and the greater the ionic charge, the greater the lattice energy and higher the melting point.

STUDY QUESTION 10.14
Which member of the pair will have the higher melting point?

    a.   NaCl or CaO
    b.   NaCl or NaI.

Solution:

a. According to Coulomb's law, the doubly charged ions in CaO will attract each other more strongly than do the singly charged ions in NaCl. CaO will have a higher melting point.

b. According to Coulomb's law, the closer the centers of two ions can approach each other, the stronger the attraction between them will be. The sum of the ionic radii in NaCl is smaller than in NaI. NaCl will have the higher melting point.

PRACTICE QUESTION 10.15
Which member of the pair will have the higher melting point?

    a.   KCl or $CaCl_2$
    b.   RbI or NaI

PRACTICE QUESTION 10.16
Calculate the lattice energy of sodium bromide from the following information.

    ΔH_{subl} (Na) = 109 kJ/mol
    Ionization Energy (Na) = 496 kJ/mol
    Bond Enthalpy (Br—Br) = 192 kJ/mol
    Electron Affinity (Br) = 324 kJ/mol
    ΔH_f^o **(NaBr) = –359 kJ/mol**

PRACTICE QUIZ

1. A system does 975 J of work on its surroundings while at the same time it absorbs 625 J of heat. What is the change in energy, ΔE for the system?

2. During expansion of its volume from 1.0 L to 10.0 L against a constant external pressure of 2.0 atm, a gas absorbs 200 J of energy as heat. Calculate the change in internal energy of the gas.

3. How much work must be done on or by the system in a process in which the internal energy remains constant and 322 J of heat is transferred from the system?

4. How many grams of SO_2 must be burned to yield 251 kJ of heat? Given:

$$2SO_2(g) + O_2(g) \rightarrow 2SO_3(g) \qquad \Delta H^{\circ}_{rxn} = -198.0 \text{ kJ}$$

5. Octane, C_8H_{18}, a constituent of gasoline burns according to the equation:

$$C_8H_{18}(\ell) + \frac{25}{2}O_2(g) \rightarrow 8CO_2(g) + 9H_2O(\ell)$$

A 0.1111 g sample of C_8H_{18} was burned in the presence of excess O_2 in a bomb calorimeter. The heat capacity of the calorimeter was 1.726×10^3 J/°C. The temperature of the calorimeter and 1.200×10^3 g of water increased from 21.22 °C to 23.05 °C. Calculate the heat of combustion per gram of octane.

6. When a 1.36 g sample of butane, C_4H_{10}, was burned in a bomb calorimeter containing 2500 g water, a +3.06 °C temperature change was observed in the calorimeter. Calculate the enthalpy of combustion of butane. The heat capacity of the calorimeter is 11.6 kJ/°C.

7. Given the thermochemical equation

$$SO_2(g) + \frac{1}{2}O_2(g) \rightarrow SO_3(g) \qquad\qquad \Delta H^{\circ}_{rxn} = -99 \text{ kJ/mol},$$

how much heat is liberated when
a. 0.50 mole of SO_2 reacts
b. 2 moles of SO_2 reacts

8. Calculate ΔH°_{rxn} in kJ for the decomposition of HBr,

$$HBr(g) \rightarrow H(g) + Br(g)$$

given the following reactions and their associated enthalpy changes.

$$2H(g) \rightarrow H_2(g) \qquad\qquad \Delta H^{\circ}_{rxn} = -436 \text{ kJ/mol}$$
$$Br_2(g) \rightarrow 2\,Br(g) \qquad\qquad \Delta H^{\circ}_{rxn} = +224 \text{ kJ/mol}$$
$$H_2(g) + Br_2(g) \rightarrow 2HBr(g) \qquad\qquad \Delta H^{\circ}_{rxn} = -72 \text{ kJ/mol}$$

9. Nitrogen and oxygen react according to the following thermochemical equation:

$$N_2(g) + O_2(g) \rightarrow 2NO(g) \qquad\qquad \Delta H^{\circ}_{rxn} = 180 \text{ kJ/mol}$$

How many kJ of heat are absorbed when 50.0 g of N_2 reacts with excess O_2 to produce 107 g of NO?

10. How much energy is required to raise the temperature of 180 g of graphite from 25°C to 500 °C? The specific heat of graphite is 0.720 J/g·°C.

11. A 0.500 g sample of ethanol, $C_2H_5OH(l)$, was burned in a bomb calorimeter containing 2.000×10^3 g of water. The heat capacity of the bomb calorimeter was 950 J/°C, and the change in temperature was found to be 1.60 °C.
    a. Write a balanced equation for the combustion of ethanol.
    b. Calculate the amount of heat transferred to the calorimeter.
    c. Calculate ΔH for the reaction as written in part (a).

12. The enthalpy of combustion of sulfur is:

$$S(\text{rhombic}) + O_2(g) \rightarrow SO_2(g) \qquad \Delta H^\circ_{rxn} = -296 \text{ kJ/mol}$$

What is the enthalpy of formation of SO_2?

13. The combustion of methane occurs according to the equation

$$CH_4(g) + 2O_2(g) \rightarrow CO_2(g) + 2H_2O(l) \qquad \Delta H^\circ_{rxn} = -890 \text{ kJ/mol}$$

The standard enthalpies of formation for CO_2 is –393.5 kJ/mol and for H_2O is –285.8 kJ/mol. Determine the enthalpy of formation of methane.

14. Given:

$$2Al(s) + \frac{3}{2}O_2(g) \rightarrow Al_2O_3(s) \qquad \Delta H^\circ = -1670 \text{ kJ/mol}$$

What is ΔH° for the decomposition of Al_2O_3?

15. A 5.80 g piece of silver at 80.0°C is dropped into a Styrofoam cup containing 25.0 mL of water at 20.0°C. What is the final temperature of the water and silver? The specific heat of silver is 0.237 J/g°C.

16. Calculate the standard enthalpy for the decomposition of potassium chlorate using the standard enthalpies of formation given in Appendix 2 in your textbook.

$$\underline{\hspace{1cm}} KClO_3(s) \rightarrow \underline{\hspace{1cm}} KCl(s) + \underline{\hspace{1cm}} O_2(g) \text{ (unbalanced)}$$

a. Is the reaction endo– or exothermic?
b. How much energy would be produced/required when 4.0 kg of $KClO_3$ decomposes?

17. Calculate ΔH°_{rxn} for the reaction $3C_2H_2(g) \rightarrow C_6H_6(l)$ using the following data:

$$2C_2H_2(g) + 5O_2(g) \rightarrow 4CO_2(g) + 2H_2O(l) \qquad \Delta H^\circ_{rxn} = -1299.4 \text{ kJ/mol}$$
$$2C_6H_6(l) + 15O_2(g) \rightarrow 12CO_2(g) + 6H_2O(l) \qquad \Delta H^\circ_{rxn} = -3267.4 \text{ kJ/mol}$$

18. A microwave oven supplies 1000 joules of energy per second. What is the final temperature of 500.0 mL water initially at 25.3°C if you set the timer to 1.30 minutes? Assume only the water absorbs the heat.

19. Calculate the standard enthalpy for the dissolution of $NH_4Cl(s)$ using the ΔH_f° values from Appendix 2 in your textbook.

$$NH_4Cl(s) \xrightarrow{\;H_2O\;} NH_4^+(aq) + Cl^-(aq)$$

    a. Using this ΔH_{rxn}° determine the energy absorbed when 5.0 g of ammonium chloride dissolves in enough water to make 150.0 mL of an aqueous solution of ammonium chloride.

    b. By how much would the temperature change when the solid dissolves? Assume that the density of the solution is 1.03 g/mL and the specific heat of the solution is equal to that of water.

20. Why can't we determine the absolute value of the internal energy of a system, U_{sys}?

21. Calculate ΔH_f° for $H_2O_2(aq)$ from the following data:

$$2H_2O_2(aq) \rightarrow 2H_2O(l) + O_2(g) \qquad\qquad \Delta H_{rxn}^\circ = -189.3 \text{ kJ/mol}$$

    and $\Delta H_f^\circ = -285.8 \dfrac{\text{kJ}}{\text{mol}}$ for $H_2O(l)$.

22. Given the bond enthalpies in Table 10.4 in your textbook, calculate the enthalpy of formation for ammonia, NH_3.

$$\frac{1}{2}N_2(g) + \frac{3}{2}H_2(g) \rightarrow NH_3(g) \qquad\qquad \Delta H_f^\circ (NH_3) = ?$$

23. Write a chemical equation for the process that corresponds to the lattice energy of MgO.

24. Which member of the pair will have the higher lattice energy?
    a. NaCl or CaO   b. KI or KCl

Chapter 11
Gases

PROBLEM-SOLVING STRATEGIES AND TUTORIAL SOLUTIONS

TYPES OF PROBLEMS
Problem Type 1:   Kinetic Molecular Theory of Gases.
        A. Molecular Speed.

Problem Type 2:   Gas Pressure.
        A. Measurement of Pressure.

Problem Type 3:   The Gas Laws.
        A. Boyle's Law: The Pressure–Volume Relationship.
        B. Charle's and Gay-Lussac's Law: The Temperature –Volume Relationship.
        C. Avogadro's Law: The Amount–Volume Relationship.
        D. The Combined Gas Law: The Pressure–Temperature–Amount–Volume Relationship.

Problem Type 4:   The Ideal Gas Equation.
        A. Ideal Gas Equation.
        B. Applications of the Ideal Gas Equation.

Problem Type 5:   Gas Mixtures.
        A. Dalton's Law of Partial Pressures and Mole Fractions.

Problem Type 6:   Reactions with Gaseous Reactants and Products.

PROBLEM TYPE 1: KINETIC MOLECULAR THEORY OF GASES.

A. Molecular Speed.

Many gas laws let us predict gas behavior but don't explain it. The Kinetic Molecular Theory of Gases provides explanations of gas behavior at the molecular level. The principle hypotheses of the theory are:

1. Gas particles obey Newton's laws of motion.
2. Gas molecules are separated by distances much larger than their dimensions and can be considered as points (mass, but no volume). This hypothesis explains compressibility.
3. Gas molecules are in constant motion in random directions and often collide. Gas collisions are perfectly elastic, i.e., total amount of energy in the colliding molecules is the same after collision. Hypothesis 2 explains pressure–volume relationship.
4. There are no intermolecular forces (attraction or repulsion) between gas molecules. This hypothesis explains Dalton's Law of Partial Pressures.
5. The energy of gas molecules is primarily kinetic (internal E is small) and depends on mass and speed of molecules: $\overline{KE} = \frac{1}{2} m \overline{u}^2$, where m is the mass and $\overline{u^2}$ is the average of the squares of the

speeds of all the molecules. The kinetic energy is also proportional to the absolute temperature (temperature in Kelvin): $\overline{KE} = \dfrac{3}{2}\dfrac{RT}{N_A}$. This hypothesis explains Charles' Law.

The Kinetic Molecular Theory has several other applications. The molecules in a container of gas move randomly, but it is possible to determine the distribution of molecular speeds from kinetic molecular theory. As the temperature increases, the range of speed narrows. The average speed of the molecules in a gas can be calculated by combining the two features of H–4:

$$u_{rms} = \sqrt{\overline{u^2}} = \sqrt{\dfrac{3RT}{M}}$$

M is the molar mass of the substance. When using this equation, you must use the value for $R = 8.314$ J/K·mol and molar mass must be in units of kg/mol so units cancel properly leaving units of m/s for speed.

If two gases u_{rms} are compared under the same conditions of temperature, the following ratio can be derived:

$$\dfrac{u_{rms}(A)}{u_{rms}(B)} = \sqrt{\dfrac{M(B)}{M(A)}}$$

STUDY QUESTION 11.1
Calculate the root–mean–square speed of gaseous argon atoms at STP.

Solution: The root–mean-square speed is given by the equation

$$U_{rms} = \sqrt{\overline{u^2}} = \sqrt{\dfrac{3RT}{M}}$$

where $R = 8.314$ J/K·mol, T is the absolute temperature 273 K; and M is the molar mass of Ar in kilograms, 0.0399 kg/mol. Note that standard pressure is irrelevant here because the average kinetic energy depends only on the temperature. Before substituting into the equation, we recall that 1 J $= 1$ kg m^2/s^2.

$$U_{rms} = \sqrt{\dfrac{3(8.314 \text{ J/K} \cdot \text{mol})(273 \text{ K})}{(3.99 \times 10^{-2} \text{ kg/mol})} \times \dfrac{1 \text{ kg m}^2/\text{s}^2}{1 \text{ J}}}$$

$$U_{rms} = (1.71 \times 10^5 \text{ m}^2/\text{s}^2)^{1/2} = 413 \text{ m/s}$$

STUDY QUESTION 11.2
A sample of neon gas escapes through a pinhole at a rate of 6.5 molecules/s. Calculate the rate at which nitrogen gas molecules would escape under the same conditions.

Solution: Rates of effusion are inversely related to the square root of the molar mass of the gas. Therefore, when we compare two gases, their rates of effusion are related through the equation:

$$\dfrac{\text{rate}_{N_2}}{\text{rate}_{Ne}} = \sqrt{\dfrac{MM_{Ne}}{MM_{N_2}}}$$

Substitution of the molar masses for each gas and the rate of effusion for neon gives:

$$\text{rate}_{N_2} = (\text{rate}_{Ne})\sqrt{\frac{MM_{Ne}}{MM_{N_2}}}$$

$$\text{rate}_{N_2} = (6.5 \text{ molecules/s})\sqrt{\frac{20.18 \frac{g}{mol}}{28.02 \frac{g}{mol}}} = 5.5 \text{ molecules/s}$$

We would expect that particles with a larger mass will effuse more slowly and the result supports what we expected.

> **PRACTICE QUESTION 11.1**
> Calculate the root–mean–square speed of ozone molecules (O_3) in the stratosphere where the temperature is –83 °C.

> **PRACTICE QUESTION 11.2**
> If the root–mean–square speed of an O_2 molecule is 4.2×10^2 m/s at 25 °C, what is the average speed of a Cl_2 molecule at the same temperature?

> **PRACTICE QUESTION 11.3**
> Which diatomic gas would effuse at a rate that is 1.4 times faster than the rate at which $SO_2(g)$ effuses at any given temperature?

PROBLEM TYPE 2: GAS PRESSURE.

A. Measurement of Pressure.

Gas particles are in constant motion. Their collisions with the container exert pressure on the container. Pressure is force per unit area. Therefore, the SI unit for pressure (force/area) is the pascal ($Pa = N/m^2$). The pascal is not the most convenient pressure unit; the pressure of a small puff of air is > 100,000 Pa. The atmosphere is a more convenient unit:

$$1 \text{ atm} = 101,325 \text{ Pa}$$

The torr is based on using the barometer to measure atmospheric pressure. The simplest barometer consists of a tube of mercury inverted in a small pool of mercury. The height of the mercury column rises and falls in response to the pressure the atmosphere exerts on the mercury in the pool. Atmospheric pressure of 1 atm will support a mercury column that is 760 mm high.

STUDY QUESTION 11.3
The atmospheric pressure on Mars is about 0.22 inches of Hg as compared to about 30 inches of Hg on Earth. Express the Martian pressure in mmHg, atmospheres, and pascals.

Solution: Treat this like any other factor label problem. First, state the problem:

$$0.22 \text{ in Hg} = ? \text{ mmHg}$$

Recall that there are 2.54 cm/in, therefore there are 25.4 mm/in. The unit factor needed is

$$\frac{25.4 \text{ mm}}{1 \text{ in}}$$

Converting inches to millimeters:

$$? \text{ mmHg} = 0.22 \text{ in Hg} \times \frac{25.4 \text{ mm}}{1 \text{ in}} = 5.6 \text{ mmHg}$$

The number of atmospheres is:

$$? \text{ atm} = 5.6 \text{ mmHg} \times \frac{1 \text{ atm}}{760 \text{ mmHg}} = 7.4 \times 10^{-3} \text{ atm}$$

The pressure in pascals is:

$$? \text{ pascals} = 7.4 \times 10^{-3} \text{ atm} \times \frac{1.013 \times 10^5 \text{ Pa}}{1 \text{ atm}} = 750 \text{ Pa}$$

STUDY QUESTION 11.4
What is the pressure of the gas trapped in the J–tube shown below if the atmospheric pressure is 0.803 atm?

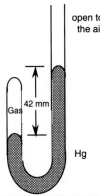

Solution: The pressure of the gas is sufficient to support a column of mercury 42 mm high and hold back the pressure of the atmosphere as well. The pressure from the mercury column is just the difference in heights of the two mercury surfaces, 42 mmHg, P_h. Therefore the pressure of the gas is:

$$P_{\text{gas}} = P_h + P_{\text{atm}}$$

Atmospheric pressure must be converted to the same units:

$$0.803 \text{ atm} \times \frac{760 \text{ mmHg}}{1 \text{ atm}} = 610 \text{ mmHg}$$

$$P_{\text{gas}} = 42 \text{ mmHg} + 610 \text{ mmHg}$$

$$P_{\text{gas}} = 652 \text{ mmHg}$$

PRACTICE QUESTION 11.4
Convert a pressure of 645 mmHg to

    a.   **atmospheres**
    b.   **kilopascals**

PRACTICE QUESTION 11.5
In each pairing, which pressure is greater?

    a.   540 mmHg or 98 kilopascals
    b.   1.45 atm or 910 torr

PRACTICE QUESTION 11.6
Do the following unit conversions.

    a.   125 mmHg to torr
    b.   725 mmHg to kilopascals

PROBLEM TYPE 3: THE GAS LAWS.

A. Boyle's Law: The Pressure–Volume Relationship

In the seventeenth century, Robert Boyle made many observations of the relationship between gas pressure and gas volume. He discovered that the volume occupied by molecules in a container decreases as the pressure increases at fixed temperature. This conclusion is summarized in Boyle's Law:

$$P = \frac{k_T}{V} \quad \text{or} \quad PV = k_T \quad \text{for constant amounts of gas at constant temperature.}$$

One of the most useful applications of Boyle's law is the prediction of pressure or volume changes of fixed amounts of gases at constant pressure, as might occur when gas is released from a container. The form of Boyle's law for this application is

$$P_{initial}V_{initial} = P_{final}V_{final} \quad \text{often abbreviated } P_1V_1 = P_2V_2.$$

STUDY QUESTION 11.5
At constant temperature, a sample of gas was compressed to 1/3 its original volume. How was the pressure changed?

Solution: Boyle's law is expressed by the equation $P_1V_1 = P_2V_2$. Pressure and volume are inversely proportional so if the volume is decreased by a factor of 3, the pressure must increase by a factor of 3. Therefore, the pressure is tripled.

To verify this, let's choose some actual numbers and plug them into the equation. $P_1 = 1$ atm; $V_1 = 3$ L; $V_2 = 1$ L (volume is compressed to 1/3 its original volume)

$$\frac{P_1V_1}{V_2} = P_2$$

$$\frac{1 \text{ atm}(3 \text{ L})}{1 \text{ L}} = 3 \text{ atm}$$

PRACTICE QUESTION 11.7
The pressure of a sample of gas doubles while the temperature is held constant. How was the volume changed?

B. Charle's and Gay-Lussac's Law: The Temperature–Volume Relationship.

In the nineteenth century Jacques Charles and Joseph Gay–Lussac observed that the volume occupied by molecules in a container increases as the temperature increases at fixed pressure. In other words, the volume occupied by a fixed amount of gas is proportional to the temperature at fixed pressure. Their observations are summarized by Charles' Law:

$$V = k_p T \quad \text{or} \quad \frac{V}{T} = k_p \quad \text{for constant amounts of gas at constant pressure.}$$

Applications of Charles' Law include the prediction of temperature changes with volume, such as the cooling that occurs when gases expand. The form of Charles' law for this application is

$$\frac{V_{initial}}{T_{initial}} = \frac{V_{final}}{T_{final}} \quad \text{or} \quad V_{initial} T_{final} = V_{final} T_{initial}$$

STUDY QUESTION 11.6

When 2.0 L of chlorine gas (Cl_2) is warmed at 0 °C constant pressure to 100 °C, what is the new volume?

Solution: The Boyle's volume of a fixed amount of a gas at constant pressure is proportional to the absolute temperature according to Charles's law, where $T_1 = 273$ K and $T_2 = (100 \,°C + 273 \,°C)$ K = 373 K.

$$\frac{V_1}{T_1} = \frac{V_2}{T_2}$$

Rearranging gives:

$$V_2 = V_1 \times \frac{T_2}{T_1}, \text{ where } V_1, \text{ the initial volume, is 2.0 L.}$$

Substituting, we get:

$$2.0 \text{ L} \times \frac{373 \text{ K}}{273 \text{ K}} = 2.7 \text{ L}$$

PRACTICE QUESTION 11.8
A sample of gas has a volume of 200 cm^3 at 25 °C and 700 mmHg. To which temperature should the gas sample be taken in order to reduce the volume to 100 cm^3 under constant pressure?

PRACTICE QUESTION 11.9
Two liters of oxygen gas at –15 °C are heated and the volume expands. At what temperature will the volume reach 2.31 L?

PRACTICE QUESTION 11.10
A 20.0 mL sample of a gas is enclosed in a gas–tight syringe at 50.0 °C. What will the volume of the gas be at the same pressure after the syringe has been immersed in ice water?

PRACTICE QUESTION 11.11
If 2.00 L of oxygen at –15 °C are allowed to warm to 25 °C at constant pressure, what is the new volume of oxygen gas?

C. Avogadro's Law: The Amount–Volume Relationship.

Around the same time, Amadeo Avogadro observed that equal volumes of different gases contained the same number of molecules. In other words, the volume occupied by molecules in a container increases as the number of gas molecules in the vessel increases at fixed pressure and temperature. These observations are summarized as Avogadro's Law:

$V=k_{TP}n$ at constant temperature and pressure.

One consequence of this relationship is the fact that, ideally, one mole quantities of different gases have the same volumes when the temperature and pressure are constant. The molar volume is the volume of one mole of gas at 0 °C and 1 atm (called standard temperature and pressure, STP): 22.41 L. This is a very important observation, because a reaction like

$$2H_2(g) + O_2(g) \rightarrow 2H_2O(g)$$

reduces the volume by a factor of 2/3. Avogadro's law was useful to scientists attempting to understand the stoichiometry of chemical reactions.

STUDY QUESTION 11.7
A container fitted with a movable piston is filled with 2.0 moles of Ar occupying a volume of 49.0 L. How many grams of Ar must be removed from the container under constant temperature and pressure in order to decrease the volume to 30.0L?

Solution: According to Avogadro's Law, a decrease in the amount of moles of gas under constant temperature and pressure is accompanied by a decrease in volume. In other words, volume and number of moles vary in a directly proportional way. Therefore,

$$\frac{V_1}{V_2} = \frac{n_1}{n_2}$$

Rearranging this equation for n_2 we get:

$$n_2 = \frac{n_1 V_2}{V_1}$$

Substituting the values given:

$$n_2 = \frac{(2.0 \text{ mol})(30.0 \text{ L})}{49.0 \text{ L}} = 1.2 \text{ mol}$$

The molar mass of Ar is 39.95 g/mol, thus;

$$1.2 \text{ mol} \times \frac{39.95 \text{ g}}{1 \text{ mol}} = 48 \text{ g of Ar}$$

> *PRACTICE QUESTION 11.12*
> **If during a hypothetical reaction the number of moles of species decreased as the reaction proceeds; as seen during a synthesis reaction, how would the volume of a cylinder with a movable piston be affected if the temperature and pressure remained constant?**

> *PRACTICE QUESTION 11.13*
> **A container fitted with a movable piston is initially filled with 6.5 grams of He and is found to occupy a volume of 22.0 L. If an additional 10 grams of He are added to the container, what is the expected volume?**

D. The Combined Gas Law: The Pressure–Temperature–Amount–Volume Relationship.
All of the primary gas laws discussed so far can be combined into a single equation that establishes the relative relationships between all of the variables of volume, temperature, amount and pressure. The

resulting equation is known as the combined gas law and it allows us to calculate any of the variables if we know the changes that the other three undergo.

$$\frac{P_1 V_1}{n_1 T_1} = \frac{P_2 V_2}{n_2 T_2}$$

PRACTICE QUESTION 11.14
A helium balloon contains 0.400 mol of the gas at a volume of 9.60 L. If 1.00 g of helium is added to the balloon under constant temperature and pressure, what will be its new volume?

PRACTICE QUESTION 11.15
A sample of gas occupies 155 mL at 21.5 °C, and at 305 mmHg. What is the pressure of the gas sample when it is placed in a 5.00 × 10² mL flask at a temperature of –10 °C?

PRACTICE QUESTION 11.16
Which sample contains more molecules:

> **a. 1.0 L of O₂ gas at 20 °C and 2.0 atm or**
> **b. 1.0 L of SF₄ gas at 20 °C and 2.0 atm? Which sample has more mass?**

PROBLEM TYPE 4: THE IDEAL GAS EQUATION.

A. Ideal Gas Equation.
All the gas relationships can be combined into a single equation called the Ideal Gas Equation:

$$PV = nRT$$

where n is the number of moles and R is the ideal gas constant, $R = 0.0821$ L·atm/K·mol. The dots in the units are to remind us that both L and atm are in the numerator, and K and mol are in the denominator. The units of R may also be written, $R = 0.0821$ L·atm·K⁻¹·mol⁻¹.

The molecules of an ideal gas
• have no intermolecular interactions (attraction or repulsion between molecules)
• take up no volume (compared to container volume)
The properties of many real gases can be described by the ideal gas equation at high temperatures (>0 °C) and low pressures (<10 atm).

What can you do with the ideal gas equation? Given any three of the four properties related by the ideal gas equation, the value of the fourth can be calculated. For example, a common quantity to know in order to work gas problems is the number of moles. Pressure, volume and temperature are all easily measured quantities. P, V and T can be plugged into the equation and you can solve for n.

STUDY QUESTION 11.8
What volume will be occupied by 0.833 mole of fluorine (F₂) at 645 mmHg and 15.0 °C?

Solution: The ideal gas law relates the gas volume to temperature, pressure and number of moles.

$$V = \frac{nRT}{P}$$

First, convert the pressure into atmospheres:

$$P = 645 \text{ mmHg} \times \frac{1 \text{ atm}}{760 \text{ mmHg}} = 0.849 \text{ atm}$$

$$= \frac{(0.833 \text{ mol})(0.0821 \text{ L} \cdot \text{atm/K} \cdot \text{mol})(288 \text{ K})}{0.849 \text{ atm}} = 23.2 \text{ L}$$

STUDY QUESTION 11.9

Given 10.0 L of neon gas at 5 °C and 630 mmHg, calculate the new volume at 400 °C and 2.5 atm.

Solution: First, note that both the pressure and the temperature of the gas are changed, but that the number of moles is constant. Write the ideal gas equation with all the constant terms on one side.

$$\frac{PV}{T} = nR = \text{a constant}$$

We see that PV/T is a constant. Therefore,

$$\frac{P_1 V_1}{T_1} = \frac{P_2 V_2}{T_2}$$

where the subscripts 2 refer to the final state, and the subscripts 1 refer to the initial state.

Rearranging to solve for V_2 gives

$$V_2 = V_1 \times \frac{P_1}{P_2} \times \frac{T_2}{T_1}$$

We need pressure in the same unit, so convert 630 mmHg to atm. (Equally correct would be to convert 2.5 atm to mmHg).

$$P_1 = 630 \text{ mmHg} \times \frac{1 \text{ atm}}{760 \text{ mmHg}} = 0.829 \text{ atm}$$

$$P_2 = 2.5 \text{ atm}$$

$$T_1 = (273 \text{ °C} + 5 \text{ °C}) \text{ K} = 278 \text{ K}$$

$$T_2 = (273 \text{ °C} + 400 \text{ °C}) \text{ K} = 673 \text{ K}$$

Substituting

$$V_2 = 10.0 \text{ L} \times \frac{0.829 \text{ atm}}{2.5 \text{ atm}} \times \frac{673 \text{ K}}{278 \text{ K}} = 8.0 \text{ L}$$

PRACTICE QUESTION 11.17
What is the pressure in atmospheres of 1.20×10^4 moles of methane, CH_4, when stored at 22 °C in a 3.00×10^3 L tank?

PRACTICE QUESTION 11.18
1.75 g sample of CO_2 is contained in a 7.50×10^2 mL flask at 35 °C. What is the pressure of the gas?

PRACTICE QUESTION 11.19
How many O_2 molecules occupy a 1.00 L flask at 75 °C and 777 mmHg?

B. Applications of the Ideal Gas Equation.

Another rearrangement relates gas properties to the gas density. Remembering that the number of moles is equal to the ratio of the sample mass to its molar mass, $n = \dfrac{m}{MM}$,

$$d = \frac{m}{V} = \frac{PMM}{RT}$$

One of the applications of this relationship is that gas density, which can be measured experimentally, can be combined with pressure and temperature to determine the molar mass.

$$MM = \frac{dRT}{P}$$

STUDY QUESTION 11.10

A gaseous compound has a density of 1.69 g/L at 25 °C and 714 torr. What is its molar mass?

Solution: The density of an ideal gas is directly proportional to its molecular mass. We use the equation

$$d = \frac{m}{V} = \frac{PMM}{RT}$$

Convert pressure into units of atmospheres.

$$P = 714 \text{ torr} \times \frac{1 \text{ atm}}{760 \text{ torr}} = 0.939 \text{ atm}$$

Rearranging and substituting, we get:

$$MM = \frac{dRT}{P} = \frac{(1.69 \text{ g/L})(0.0821 \text{ L} \cdot \text{atm/K} \cdot \text{mol})(298 \text{ K})}{0.939 \text{ atm}} = 44.0 \text{ g/mol}$$

PRACTICE QUESTION 11.20

What is the density of $H_2(g)$ at 35 °C and 650 torr?

PRACTICE QUESTION 11.21

When 2.96 g of mercuric chloride is vaporized in a 1.00 liter bulb at 680 K, the pressure is 450 mmHg. What is the molar mass and molecular formula of mercuric chloride?

PRACTICE QUESTION 11.22

Which one of the following gases will have the greatest density when they are all compared at the same temperature and pressure?

    a. **O_2**
    b. **CO_2**
    c. **NO_2**
    d. **CF_4**

PROBLEM TYPE 5: GAS MIXTURES.

A. Dalton's Law of Partial Pressures and Mole Fractions.

Another consequence of the ideal gas equation is that the pressure of a mixture of gases is the sum of the pressures of the individual components. The ideal gas equation reduces to $\frac{P}{n} = $ constant because the volume and temperature are the same for all the components. For example, the pressure of air is the sum of the pressures of several gases, mostly nitrogen, air and carbon dioxide:

$$P_{air} = P_{N_2} + P_{O_2} + P_{CO_2} + P_{\substack{trace \\ compounds}}$$

Another way to state this relationship is based on the mole fraction, X_i, the ratio of the number of moles of a component to the total number of moles of gas, $\frac{n_i}{n_T}$.

$$P_T = P_1 + P_2 + ... + P_C \qquad P_i = \frac{n_i}{n_T} P_T$$

STUDY QUESTION 11.11
When oxygen gas is collected over water and the total pressure is 645 mm Hg, what is the partial pressure of oxygen? Given is the vapor pressure of water at 30 °C is 31.8 mmHg.

Solution: Mixtures of gases obey Dalton's law of partial pressures which says that the total pressure is the sum of the partial pressures of oxygen and water vapor.

$$P_t = P_{O_2} + P_{H_2O}$$

$$P_{O_2} = P_t - P_{H_2O} = 645 \text{ mmHg} - 31.8 \text{ mmHg}$$

$$P_{O_2} = 613 \text{ mmHg}$$

STUDY QUESTION 11.12
What are the mole fractions of oxygen and water vapor in Study Question 11.11?

Solution: Recall that the partial pressures of O_2 and H_2O are related to their mole fractions,

$$P_{O_2} = X_{O_2} P_t \qquad P_{H_2O} = X_{H_2O} P_t$$

Therefore, on rearranging, we get:

$$X_{O_2} = \frac{P_{O_2}}{P_t} \qquad X_{H_2O} = \frac{P_{H_2O}}{P_t}$$

$$X_{O_2} = 613/645 = 0.950, \text{ and } X_{H_2O} = 31.8/645 = 0.0493$$

NOTE:
The sum of the mole fractions is 1.0, within the number of significant figures, given:

$$X_{O_2} + X_{H_2O} = 0.950 + 0.0493 = 0.999$$

of gases obey Dalton's law of partial pressures which says that the total pressure is the sum of the partial pressures of oxygen and water vapor.

$$P_t = P_{O_2} + P_{H_2O}$$

$$P_{O_2} = P_t - P_{H_2O} = 645 \text{ mmHg} - 31.8 \text{ mmHg}$$

$$P_{O_2} = 613 \text{ mmHg}$$

PRACTICE QUESTION 11.23
Hydrogen and helium are mixed in a 20.0 L flask at room temperature (20 °C). The partial pressure of hydrogen is 250 mmHg and that of helium is 75 mmHg. How many grams of H_2 and He are present?

PRACTICE QUESTION 11.24
A mixture contains two different types of gases. If the partial pressure of gas sample A is doubled and the partial pressure of gas sample B triples, what is the overall effect on the total pressure?

PRACTICE QUESTION 11.25
A 0.356 g sample of $XH_2(s)$ reacts with water according to the following equation:
$$XH_2(s) + 2H_2O(\ell) \rightarrow X(OH)_2(s) + 2H_2(g)$$
The hydrogen evolved is collected over water at 23 °C and occupies a volume of 431 mL at 746 mmHg total pressure. Find the number of moles of H_2 produced and the atomic mass of X. Vapor pressure of H_2O = 21 mmHg.

PROBLEM TYPE 6: REACTIONS WITH GASEOUS REACTANTS AND PRODUCTS.

Earlier the idea that balanced chemical reactions describe the relationships between moles (or molecules) of reactants and products was presented. The ideal gas equation relates the number of moles to gas volumes or pressures, making it possible to establish relationships between the amounts of reactants and products without converting volumes to moles first. Accordingly, you may treat the coefficients of a balanced equation as volume units instead of moles. Consider the reaction equation:

$$2H_2(g) + O_2(g) \rightarrow 2H_2O(g)$$

We already know that the equation states that **two moles** of hydrogen gas reacts with **one mole** of oxygen gas to produce **two moles** of water vapor. Due to the proportionality of V and n, we can now say that at constant temperature and pressure **two liters** of hydrogen gas reacts with **one liter** of oxygen gas to produce **two liters** of water vapor. For the above reaction, how many liters of oxygen would be needed to react with 4.30 L hydrogen at STP?

$$4.30 \text{ L H}_2 \times \frac{1 \text{ L O}_2}{2 \text{ L H}_2} = 2.15 \text{ L O}_2$$

If only one of the substances is a gas, then the ideal gas equation must be used to find number of moles and the rest of the calculation is approached as in any stoichiometric problem.

STUDY QUESTION 11.13
Calculate the volume of ammonia gas, measured at 645 torr and 21 °C, that is produced by the complete reaction of 25.0 g of quicklime, CaO, with excess ammonium chloride, NH₄Cl, solution.

Solution: First, write the balanced chemical equation.

$$CaO(s) + 2NH_4Cl(aq) \rightarrow 2NH_3(g) + CaCl_2(aq) + H_2O(\ell)$$

Here we must determine the number of moles of NH_3 formed in the reaction and convert this into the volume of an ideal gas. Three steps are required: (1) convert g CaO to moles CaO, (2) convert moles CaO to moles NH_3 produced, and (3) use the ideal gas equation to calculate the volume of NH_3.

Plan: grams CaO → moles CaO → moles NH_3 → volume NH_3

We already know how to perform the first two steps; therefore let's first find the number of moles of NH_3 formed. Stating the problem:

$$? \text{ mol } NH_3 = 25.0 \text{ g CaO}$$

$$? \text{ mol } NH_3 = 25.0 \text{ g CaO} \times \frac{1 \text{ mol CaO}}{56.1 \text{ g CaO}} \times \frac{2 \text{ mol } NH_3}{1 \text{ mol CaO}} = 0.891 \text{ mol } NH_3$$

For step 3, the volume of 0.891 mol NH_3 can be calculated from the ideal gas equation:

$$V = \frac{nRT}{P}$$

Remember that this conversion tool cannot be put in your chain of converters above. You must stop and use the ideal gas equation. Convert pressure into units of atm before substituting into the ideal gas equation.

$$P = 645 \text{ torr} \times \frac{1 \text{ atm}}{760 \text{ torr}} = 0.849 \text{ atm}$$

$$V = \frac{0.891 \text{ mol}(0.0821 \text{ L} \cdot \text{atm/K} \cdot \text{mol})(294 \text{ K})}{0.849 \text{ atm}} = 25.3 \text{ L}$$

PRACTICE QUESTION 11.26
The discovery of oxygen resulted from the decomposition of mercury(II) oxide.
$$2HgO \rightarrow 2Hg + O_2(g)$$
    a.   **What volume of oxygen will be produced by the decomposition of 25.2 grams of the oxide, if the gas is measured at STP?**
    b.   **How many grams of mercury(II) oxide must be decomposed to yield 10.8 L of O_2 gas at 1 atm and 298 K?**

PRACTICE QUESTION 11.27
Sodium azide decomposes according to the equation.
$$2NaN_3(s) \rightarrow 2Na(s) + 3N_2(g)$$
What volume of N_2 at 1.1 atm and 50.0 °C will be produced by the decomposition of 5.0 g NaN₃?

PRACTICE QUESTION 11.28

Consider the reaction of 20.0 g calcium oxide with carbon dioxide.

$$CaO(s) + CO_2(g) \rightarrow CaCO_3(s)$$

If you have 5.5 L of CO_2 at 7.50 atm and 22 °C, will you have enough carbon dioxide to react with all the CaO?

PRACTICE QUIZ

1. A barometer reads 695 mmHg. Calculate the pressure in units of:
   a. atm
   b. torr
   c. Pa

2. The pressure of H_2 gas in a 0.50 L cylinder is 1775 mmHg at 70 °F. What volume would the gas occupy at 1 atm and the same temperature?

3. If 30.0 L of oxygen is cooled at constant pressure from 200 °C to 0 °C, what is the new volume of oxygen?

4. A balloon filled with helium has a volume of 1500 L at 0.925 atm and 23 °C. At an altitude of 20 km the temperature is –50 °C and the pressure is 151.6 mmHg. What is the volume of this balloon at 20 km?

5. What is the pressure in mmHg of the gas trapped in the apparatus shown on the next figure when the atmospheric pressure is 0.950 atm?

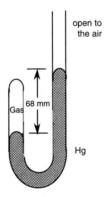

6. Given that the Martian atmosphere is mostly CO_2, at a pressure of 5.5 mmHg and a temperature of –31.4 °C. What is the density of the atmosphere?

7. How many grams of chlorine (Cl_2) occupy a 0.716 L cylinder when the pressure is 10.9 atm at 30 °C?

8. Calculate the volume occupied by 15.2 g of CO_2 at 0.74 atm and 24 °C.

9. What is the density of uranium hexafluoride gas, UF_6, at STP?

10. When 1.48 g of mercuric chloride is vaporized in a 1.00 L bulb at 680 K, the pressure is 225 mmHg. What is the molar mass and molecular formula of mercuric chloride vapor?

11. Determine the molar mass of chloroform gas if a sample weighing 0.495 g is collected as a vapor (gas) in a flask of volume 127 cm³ at 98 °C. The pressure of the chloroform vapor at this temperature in the flask was determined to be 754 mmHg.

12. A 150 mL sample of O_2 gas is collected over water at 20 °C and 758 torr. What volume will the same sample of oxygen occupy at STP when it is dry? The vapor pressure of water at 20 °C is 17.54 torr.

13. A sample of nitrogen gas was bubbled through liquid water at 25 °C and then collected in a volume of 750 cm³. The total pressure of the gas, which was saturated with water vapor, was found to be 740 mmHg at 25 °C. The vapor pressure of water at this temperature is 24 mmHg. How many moles of nitrogen were in the sample?

14. The volume of carbon monoxide gas (CO) collected over water at 25 °C was 680 cm^3 with a total pressure of 752 mmHg. The vapor pressure of water at 25 °C is 23.8 mmHg. Determine the partial pressure and mole fraction of CO in the container.

15. The partial pressures of N_2, O_2, and Ar in dry air are 570, 153, and 6 torr, respectively. What are the mole fractions of these three gases?

16. A mixture of 40.0 g of O_2 and 40.0 g of He has a total pressure of 0.900 atm. What are the partial pressures of O_2 and He in the mixture?

17. a. What volume of CO_2 at 1 atm and 225 °C would be produced by the reaction of 12.0 g $NaHCO_3$?

$$2NaHCO_3 + H_2SO_4 \rightarrow Na_2SO_4 + 2CO_2 + 2H_2O$$

   b. On cooling to 20 °C, what volume would the CO_2 occupy?

18. How many liters of ammonia at a constant pressure and pressure of 10 atm and 500 °C can be produced by the reaction of 6.0 g of hydrogen with excess N_2?

$$3H_2(g) + N_2(g) \rightarrow 2NH_3(g)$$

19. In the oxidation of ammonia

$$4NH_3(g) + 5O_2(g) \rightarrow 4NO(g) + 6H_2O(\ell)$$

how many liters of O_2, measured at 18 °C and 1.10 atm, must be used to produce 50 liters of NO at the same conditions?

Chapter 12
Liquids and Solids

PROBLEM-SOLVING STRATEGIES AND TUTORIAL SOLUTIONS

TYPES OF PROBLEMS
Problem Type 1:   Properties of Liquids.
                A. Surface Tension.
                B. Viscosity.
                C. Vapor Pressure.
                D. Boiling Point.

Problem Type 2:   Crystal Structure: Unit Cells and Packing Spheres.

Problem Type 3:   Types of Crystalline Solids.

Problem Type 4:   Phase Changes.
                A. Liquid–Vapor Phase Transition.
                B. Solid–Liquid Phase Transition.
                C. Solid–Vapor Phase Transition.

Problem Type 5: Phase Diagrams.

PROBLEM TYPE 1: PROPERTIES OF LIQUIDS.

A. Surface Tension.
    Surface tension is a force that tends to minimize the surface area of a drop of liquid. Energy is required to expand the surface of a liquid, and the surface tension is the amount of energy required to increase the surface area of a liquid by a unit area. Liquids in which strong intermolecular forces exist exhibit high surface tensions. **Cohesion** is the intermolecular attraction between like molecules in the drop. **Adhesion** is the intermolecular attractions between unlike molecules, such as between water and wax. Since there is very little attraction between water and wax (adhesion), the drop adopts a spherical shape because a sphere has the least surface area.

B. Viscosity.
    The unique pouring characteristics of each liquid are the result of its viscosity. **Viscosity** is a measure of a fluid's resistance to flow. Liquids whose molecules have strong intermolecular forces have greater viscosities than liquids that have weaker intermolecular forces.

STUDY QUESTION 12.1
The viscosity of ether is 0.000233 and of water is 0.00101 N s/m^2. Discuss their relative values in terms of the following molecular structures.

           water H—O—H   **diethyl ether** H_3C—CH_2—O—CH_2—CH_3

Solution: Molecules with strong intermolecular forces have greater viscosities than those that have weak intermolecular forces. Here, both molecules are polar: water because of the O—H bonds, and ether because of the polar C—O bond. However, water has a higher viscosity than ether because of its ability to form hydrogen bonds between molecules.

PRACTICE QUESTION 12.1
Which member of each pair should have the higher surface tension? Viscosity?

a. CO_2 or XeF_4
b. Cl_2 or $BrCl$
c. H_2O or H_2Se
d. NH_3 or PH_3

C. Vapor Pressure.

Ice, liquid water, and water vapor can exist together in a suitable container. As we previously learned, vaporization is the process in which liquids become gases. When a liquid substance is placed in a closed container, a portion evaporates and the gaseous molecules occupy the empty space above the liquid. The resulting gas molecules exert a pressure, called the **vapor pressure**. The term vapor is often applied to the gaseous state of a substance that is normally a liquid or solid at the temperature of interest.

Vaporization and condensation are constantly occurring, but at the same rates. When the rates of these opposing processes in the closed container become equal, the vapor pressure remains constant and is called the equilibrium vapor pressure, or just vapor pressure. Note that in the state of equilibrium, even though the amounts of vapor and liquid do not change, there is considerable activity on the molecular level. Such an equilibrium state is referred to as a dynamic equilibrium.

Vapor pressure is a function of the temperature. As temperature increases the rate of vaporization increases, but the rate of condensation remains the same as the molecules have too much kinetic energy to go back into the liquid. Therefore, more molecules exist in the vapor phase at higher temperatures than at lower temperatures.

D. Boiling Point.

The **boiling point** of a liquid is the temperature at which the vapor pressure is equal to the atmospheric pressure. Since the boiling point of a liquid depends on the atmospheric pressure, and the atmospheric pressure varies daily, the boiling point is not a constant. The **normal boiling point** is defined for purposes of comparing liquid substances. The normal boiling point is the boiling temperature when the external pressure is 1 atm.

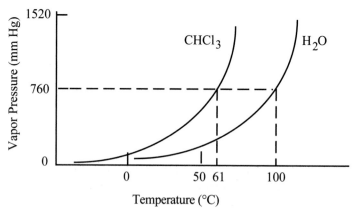

Figure 12.1. The effect of temperature on the vapor pressure of chloroform and water

The energy required to vaporize one mole of a liquid is called the molar heat of vaporization (ΔH_{vap}). The value of ΔH_{vap} is directly proportional to the strength of intermolecular forces. Liquids with relatively high heats of vaporization have low vapor pressures and high normal boiling points. The quantity ΔH_{vap} can be determined experimentally. The vapor pressure of a liquid is proportional to the temperature. The quantitative relationship between the vapor pressure (P) and the absolute temperature (T) is given by the Clausius–Clapeyron equation:

$$\ln P = -\frac{\Delta H_{vap}}{RT} + C$$

where–in P is the natural logarithm of P, R is the ideal gas constant (8.314 J/K·mol), and C is a constant. A useful form of this equation is

$$\ln\left(\frac{P_1}{P_2}\right) = \frac{\Delta H_{vap}}{R}\left(\frac{1}{T_2} - \frac{1}{T_1}\right)$$

The ΔH_{vap} can be calculated if the vapor pressures P_1 and P_2 are measured at two temperatures T_1 and T_2. Alternatively, if ΔH_{vap} is already known and one vapor pressure P_1 is known at temperature T_1, you can calculate the vapor pressure P_2 at some new temperature T_2.

STUDY QUESTION 12.2
The vapor pressure of water is 55.32 mmHg at 40.0 °C, and 92.51 mmHg at 50.0 °C. Calculate the molar heat of vaporization of water.

Solution: The Clausius–Clapeyron equation relates the vapor pressure of a liquid to its molar heat of vaporization.

$$\ln\left(\frac{P_1}{P_2}\right) = \frac{\Delta H_{vap}}{R}\left(\frac{1}{T_2} - \frac{1}{T_1}\right)$$

Substituting the values given above

$$\ln\left(\frac{55.32 \text{ mmHg}}{92.51 \text{ mmHg}}\right) = \frac{\Delta H_{vap}}{8.314 \text{ J/K}\cdot\text{mol}} \times \left(\frac{1}{323 \text{ K}} - \frac{1}{313 \text{ K}}\right)$$

$$\ln 0.5980 = \frac{\Delta H_{vap}}{8.314 \text{ J/K}\cdot\text{mol}}\left(-9.89 \times 10^{-5} \text{ K}^{-1}\right)$$

Taking the logarithm and rearranging yields:

$$\Delta H_{vap} = \frac{-0.5142(8.314 \text{ J/K}\cdot\text{mol})}{-9.89 \times 10^{-5} \text{ K}^{-1}}$$

$$\Delta H_{vap} = 43,000 \text{ J/mol}$$

NOTE:
Compare this result to the ΔH_{vap} listed in Table 12.6 of the text, 40,790 J/mol. The observed difference is real. The ΔH_{vap} is less at the boiling point than it is at a lower temperature. Over the entire temperature range of a liquid, ΔH_{vap}, is close to, but not really, a constant.

STUDY QUESTION 12.3
Using a ΔH_{vap} for water of 40.79 kJ/mol, calculate the boiling point of water in a pressure cooker at 2.00 atm pressure.

Solution: There are two pieces of information that we need to recall to solve this problem:
1. the Clausius–Clapeyron equation

$$\ln\left(\frac{P_1}{P_2}\right) = \frac{\Delta H_{vap}}{R}\left(\frac{1}{T_2} - \frac{1}{T_1}\right)$$

2. the normal boiling point of water (100 °C at a pressure of 1.00 atm)
   Before substituting any information into the Clausius–Clapeyron equation, we require the temperature to be expressed in Kelvin degrees and the enthalpy of vaporization expressed in J/mol.

$$\ln\left(\frac{2.0\ \text{atm}}{1.0\ \text{atm}}\right) = \frac{40790\ \text{J/mol}}{8.314\ \text{J/mol}\cdot\text{K}}\left(\frac{1}{373\ \text{K}} - \frac{1}{T_1}\right)$$

$$0.69 = (4906\ \text{K})\left(2.68 \times 10^{-3}\ \text{K}^{-1} - \frac{1}{T_1}\right)$$

$$\frac{1}{T_1} = 2.68 \times 10^{-3}\ \text{K}^{-1} - \left(\frac{0.69}{4906\ \text{K}}\right)$$

$$T_1 = 394\ \text{K}$$

It is very important to remember that temperatures obtained through the use of the Clausius–Clapeyron equation are in Kelvin. This corresponds to a temperature of 121 °C.

> *PRACTICE QUESTION 12.2*
> **Which substance in each pair has the higher vapor pressure at a given temperature?**
>   a. C_2H_5OH or CH_3OH
>   b. Cl_2 or Br_2
>   c. CH_3Br or CH_3Cl

> *PRACTICE QUESTION 12.3*
> **The vapor pressure of ethanol is 400 mmHg at 63.5 °C. Its vapor pressure at 34.9 °C is 100 mmHg. Calculate its molar heat of vaporization.**

> *PRACTICE QUESTION 12.4*
> **A liquid is placed in a sealed container at 25°C. What are the liquid's expected surface tension, viscosity, vapor pressure and boiling point at 60°C?**

PROBLEM TYPE 2: CRYSTAL STRUCTURE: UNIT CELLS AND PACKING SPHERES.

The atoms, molecules, or ions of a crystalline solid occupy specific positions in the solid, and possess long-range order. The **unit cell** is the smallest unit that when repeated over and over again, generates the entire crystal. The three types of unit cell are the simple cubic cell (scc), the body–centered cubic cell (bcc), and the face–centered cubic cell (fcc).

In a simple cubic cell, particles are located only at the corners of each unit cell. In a body–centered cubic cell, particles are located at the center of the cell as well as at the corners. In a face–centered cubic cell, particles are found at the center of each of the six faces of the cell as well as at the corners. In many calculations involving properties of a crystal, it is important to know how many atoms or ions are contained in each unit cell. Atoms at the corners of unit cells are shared by neighboring unit cells. For a cubic cell, each corner atom is shared by eight unit cells and is counted as 1/8 particle for each unit cell. A face–centered atom is shared by two unit cells and is counted as 1/2 of an atom for each unit cell. An atom located at the center belongs wholly to that unit cell. The **coordination number** is the number of particles surrounding each particle. It can be considered as the number of nearest neighbors.

STUDY QUESTION 12.4

If atoms of a solid occupy a face–centered cubic lattice, how many atoms are there per unit cell?

Solution: In a face–centered cubic cell there are atoms at each of the eight corners, and one in each of the six faces.

$$8 \text{ corners } (1/8 \text{ atom per corner}) + 6 \text{ faces } (1/2 \text{ atom per face})$$

$$= 4 \text{ atoms per unit cell}$$

STUDY QUESTION 12.5

Potassium crystallizes in a body–centered cubic lattice, and has a density of 0.856 g/cm^3 at 25 °C.

    a.   How many atoms are there per unit cell?
    b.   What is the length of the side of the cell?

Solution:

a. A body–centered cubic structure has one K atom in the center, and eight other K atoms, one at each corner of the cube. The corner atoms, however, are shared by 8 adjoining cells. Each corner atom contributes 1/8 of an atom to the unit cell. The total number of K atoms per unit cell is

$$1 + \frac{1}{8}(8) = 2 \text{ atoms}$$

b. Ideally, the crystal of potassium is made up of a large number of unit cells repeated over and over again. Thus, the density of the unit cell will be the same as the density of metallic K. Recall that density is an intensive property.

$$\text{density (unit cell)} = \frac{\text{mass of 2 K atoms}}{a^3}$$

where a is the length of the side of the unit cell.

The mass of a K atom is:

$$\frac{39.1 \text{ g}}{1 \text{ mol}} \times \frac{1 \text{ mol}}{6.02 \times 10^{23} \text{ atoms}} = 6.50 \times 10^{-23} \text{ g/atom}$$

Substituting into the density equation yields:

$$0.856 \text{ g/cm}^3 = \frac{2(6.50 \times 10^{-23} \text{ g/atom})}{a^3}$$

Rearranging yields:

$$a^3 = 1.52 \times 10^{-22} \text{ cm}^3$$

$$a = 5.34 \times 10^{-8} \text{ cm} = 534 \text{ pm}$$

PRACTICE QUESTION 12.5
Nickel crystallizes in a face–centered cubic lattice. Given that the density of Ni is 8.94 g/cm^3, calculate the edge length.

PROBLEM TYPE 4: TYPES OF CRYSTALLINE SOLIDS.

Each type of crystalline solid has characteristics determined in part by the types of interactions holding it together. In a solid, atoms, molecules, or ions occupy specific positions called **lattice points**. Crystals may be ionic, covalent, molecular, or metallic.

In **ionic crystals**, the lattice points are occupied by positive and negative ions. In all ionic compounds there are continuous three–dimensional networks of alternating positive and negative ions held together by strong electrostatic forces (ionic bonds). Most ionic crystals possess high lattice energies, high melting points, and high boiling points. In ionic solids there are no discrete molecules as are found in molecular or covalent substances.

In **covalent or networked crystals**, the lattice points are occupied by atoms that are held by a network of covalent bonds. That is, all lattice points (atoms) are covalently bonded to neighboring lattice points (atoms). Diamond, graphite, and quartz are well known examples. Materials of this type have high melting points and are extremely hard because of the large number of covalent bonds that have to be broken to melt or break up the crystal. The entire crystal can be thought of as one giant molecule.

Covalent compounds, like water and carbon dioxide, form crystals in which the lattice positions are occupied by discreet molecules. Such solids are called **molecular crystals**. Properties of molecular crystals range depending on the molecular polarity. Typically, they are soft and have relatively low melting points. These properties are the result of the fairly weak intermolecular forces that hold the molecules in the crystal. As a rule, polar compounds melt at higher temperatures than nonpolar compounds of comparable molecular mass.

Elemental metals form **metallic crystals.** The crystals are quite strong. Most transition metals have high melting points and densities. In metals, the array of lattice points is occupied by positive ions. The outer electrons of the metal atoms are loosely held and move freely from ion to ion throughout the metallic crystal. This mobility of electrons in metals accounts for one of the characteristic properties of metals, namely the ability to conduct electricity.

STUDY QUESTION 12.6
What type of force must be overcome in order to melt crystals of the following substances?

    a. **Mg**
    b. **Cl$_2$**
    c. **MgCl$_2$**
    d. **SO$_2$**
    e. **Si**

Solution:

a. Magnesium is a metal, which forms metallic crystals. Since the mobile electrons are shared between positive ions, the forces between Mg atoms could be viewed similar to covalent bonds, in this case called **metallic bonds**. The melting point is 1105 °C.

b. Cl_2, a nonpolar molecule, forms molecular crystals that are held together by dispersion forces. The melting point is –101 °C.

c. $MgCl_2$ is an ionic compound. When it melts, Mg^{2+} ions and Cl^- ions must overcome the electrostatic forces and break away from their lattice positions. To melt magnesium chloride, ionic bonds must be broken. The melting point is 1412 °C.

d. SO_2 is a polar molecule. Thus, dipole–dipole attractions and dispersion forces must be overcome. The melting point is –73 °C.

e. Si crystallizes in a diamond structure with each silicon atom bonded to four others by covalent bonds. It is a covalent crystal, and covalent bonds must be broken in order for it to melt. The melting point is 1410 °C.

> **PRACTICE QUESTION 12.6**
> **Arrange the following in order of increasing melting point. Diamond, dry ice, gold bracelet, table salt.**

> **PRACTICE QUESTION 12.7**
> **Could a salt with a formula of MX_2 have a cubic unit cell?**

PROBLEM TYPE 5: PHASE CHANGES.

A. Liquid–Vapor Phase Transition.

The boiling point and the heat of vaporization are convenient measures of the strength of intermolecular forces. In the liquid phase, molecules are very close together and are strongly influenced by intermolecular forces. In the gas phase, molecules are widely spaced, move rapidly, and have enough energy to overcome intermolecular forces. During vaporization, molecules are completely separated from one another. The heat of vaporization is the energy necessary to overcome intermolecular forces. The boiling point reflects the amount of kinetic energy that liquid molecules must achieve in order to overcome intermolecular forces and escape into the gas phase. When comparing molecules, as the strength of IMFs increase, the ΔH_{vap} and the normal boiling points increase.

When a liquid is heated in a closed container the vapor pressure increases, but boiling does not occur. The vapor cannot escape, and the vapor pressure continually rises. Eventually a temperature is reached at which the phase boundary between liquid and vapor disappears. This temperature is called the **critical temperature**. The critical temperature is the highest temperature at which the substance can exist as a distinguishable liquid. The **critical pressure** is the lowest pressure that will liquefy a gas at the critical temperature. As we saw previously, intermolecular forces are dominant in liquids and effectively determine many of their properties. The critical temperature and pressure are no exceptions. Substances with high critical temperatures have strong intermolecular forces of attraction.

STUDY QUESTION 12.7
How much heat is given off when 1.0 g of steam condenses at 100 °C?

Solution: The heat of vaporization of water at 100 °C is 40.8 kJ/mol. Condensation is the reverse of vaporization so it releases the same amount of heat.

$$H_2O(g) \rightleftharpoons H_2O(l) \quad \Delta H_{cond} = -40.8 \text{ kJ/mol}$$

Since ΔH_{cond} is in molar units, first convert 1.0 g H_2O into moles. The heat evolved when 1.0 g of water vapor condenses is:

$$q = 1.0 \text{ g} \times \frac{1 \text{ mol}}{18.0 \text{ g}} \times \frac{-40.8 \text{ kJ}}{1 \text{ mol}} = -2.27 \text{ kJ}$$

NOTE:
This is a lot of heat and shows us why burns from steam can be very serious.

> ### PRACTICE QUESTION 12.8
> **Arrange the following in order of increasing heat of vaporization: Diamond, dry ice, gold bracelet, and table salt**

B. Solid–Liquid Phase Transition.

The temperature at which the solid and liquid are in dynamic equilibrium at 1 atm pressure is called the **normal melting point** or **normal freezing point**. Melting is also called **fusion**. During melting the average distance between molecules is increased slightly as evidenced by the approximately 10% decrease in density for most substances.

$$\text{solid} \underset{\text{freezing}}{\overset{\text{fusion}}{\rightleftharpoons}} \text{liquid}$$

The energy required to melt one mole of a solid is called the **molar heat of fusion**, ΔH_{fus}. Upon freezing, the substance will release the same amount of energy. If this heat is removed from the system very quickly, the substance can be supercooled, that is cooled without solidifying. The ΔH_{fus} is always much less than ΔH_{vap} because during vaporization molecules are completely separated from each other, while melting only separates molecules to a small extent.

STUDY QUESTION 12.8
The heat of fusion of aluminum is 10.7 kJ/mol. How much energy is required to melt one ton of Al at the melting point, 660 °C?

Solution: The heat of fusion is the energy required to melt 1 mole of a substance. Therefore, we need to convert 2000 lb of Al to moles.

$$q = 2000 \text{ lb} \times \frac{454 \text{ g}}{1 \text{ lb}} \times \frac{1 \text{ mol Al}}{27.0 \text{ g Al}} \times \frac{10.7 \text{ kJ}}{1 \text{ mol Al}} = 3.60 \times 10^5 \text{ kJ (360 MJ)}$$

C. Solid–Vapor Phase Transition.

Sublimation is the transition of a solid directly into the vapor phase. Dry ice and iodine are substances that sublime readily. Ice also sublimes to some extent. As with liquids, the vapor pressure of a solid increases as the temperature increases. The direct conversion of solid to vapor is equivalent to melting the solid first and then vaporizing the liquid. From Hess's law we obtain:

$$\Delta H_{sub} = \Delta H_{fus} + \Delta H_{vap}$$

A heating curve is a convenient way to summarize the solid–liquid–gas transitions for a compound. The heating curve of water is shown in Figure 12.31 in your textbook. It is the result of an experiment in which a given amount of ice, 1 mole for instance, at some initial temperature below 0 °C is slowly heated at a constant rate.

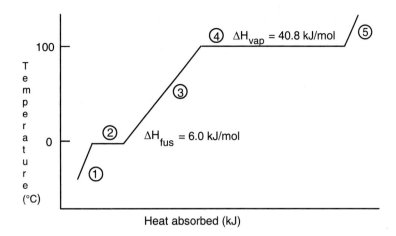

The curve shows that the temperature of ice increases on heating (line 1) until the melting point is reached. No rise in temperature occurs while ice is melting (line 2); as long as some ice remains, the temperature stays at 0 °C. The length of line 2 is a measure of the heat necessary to melt 1 mole of ice, which is ΔH_{fus}. Along line 3 the temperature of liquid water increases from 0 °C to 100 °C. The slope of the line Δt per Joule of heat depends inversely on the specific heat of liquid water. The greater slope for line 1 compared to line 3 means that less heat is required to raise the temperature of ice than of water. This is evidenced by the difference in specific heats of ice and liquid water. The specific heat of ice is 2.09 J/g· °C versus 4.18 J/g· °C for liquid H_2O.

At 100 °C the liquid begins to boil, and the heat added is used to bring about vaporization. The temperature does not rise until all the liquid has been transformed to gas. The length of line 4 is the heat required to vaporize 1 mole of liquid. Line 4 will always be longer than line 2 because $\Delta H_{vap} > \Delta H_{fus}$. Line 5 corresponds to the heating of steam. Again the slope of line 5 depends on the specific heat of steam, which is about 1.98 J/g· °C.

STUDY QUESTION 12.9
The heat of vaporization of iodine is 41.7 kJ/mol at the boiling point of iodine (456 K). How much heat is required to vaporize 20.0 g of I_2?

Solution: Since ΔH_{vap} is in molar units, first, convert g I_2 to moles. The heat required to vaporize 20.0 g of iodine is:

$$q = 20.0 \text{ g} \times \frac{1 \text{ mol}}{253.8 \text{ g}} \times \frac{41.7 \text{ kJ}}{1 \text{ mol}} = 3.29 \text{ kJ}$$

PRACTICE QUESTION 12.9
The heat of vaporization of ammonia (NH_3) is 23.2 kJ/mol at its boiling point. How many grams of ammonia can be vaporized with 12.5 kJ of heat?

PRACTICE QUESTION 12.10
Arrange the following in order of increasing heat needed per mole of substance.
Gas → liquid, solid → gas, liquid → solid, liquid → gas

PROBLEM TYPE 6: PHASE DIAGRAMS.

From the discussions in preceding sections we can see that the phase in which a substance exists depends on its temperature and pressure. In addition, two phases may exist in equilibrium at certain temperatures and pressures. Information about the stable phases for a specific compound is summarized by a phase diagram such as that shown below.

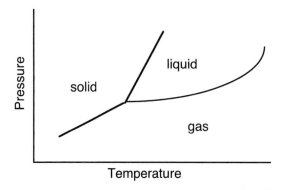

A **phase diagram** is a graph of pressure versus temperature. The lines represent P and T at which two phases can coexist. The regions between lines represent P and T at which only one phase can exist. The line between the solid and liquid regions is made up of P–T points at which the solid and liquid phases are in equilibrium. The line between the solid and gas regions is made up of P–T points at which solid and gas are in equilibrium. And the line between the liquid and gas regions gives temperatures and pressures at which liquid and gas are in equilibrium. All three lines intersect at the **triple point**. This one point describes the conditions under which gas, liquid, and solid all coexist in equilibrium.

The phase diagram shows at a glance several properties of a substance: melting point, boiling point, and triple point. If a P–T point describing a system falls in the solid region, the substance exists as a solid. If the point falls on a line such as that between liquid and gas regions, the substance exists as liquid and vapor in equilibrium.

STUDY QUESTION 12.10
Discuss the possibilities of liquefying oxygen and carbon dioxide at 25 °C by increasing the pressure. The critical temperature of O_2 is –119 °C and for CO_2 is 31 °C.

Solution: Above its critical temperature a substance cannot be liquefied by increasing the pressure. Substances having critical temperatures above 25 °C can be liquefied at 25 °C by application of sufficient pressure. With a critical temperature of 31 °C, CO_2 can be liquefied at 25 °C with application of enough pressure. The pressure required can be read off the phase diagram for CO_2 (Figure 12.34 in the text). Reading straight up from 25 °C you will cross the liquid–vapor equilibrium point at a pressure of 67 atm. This is the pressure required to liquefy CO_2 at 25 °C. Oxygen has a critical temperature of –119 °C; therefore, at any temperature above –119 °C oxygen cannot be liquefied, no matter how much pressure is applied. To liquefy O_2, its temperature must first be lowered below –119 °C and pressure applied.

PRACTICE QUIZ

1. What property of a liquid makes it possible to fill a glass of water to a level slightly above the rim?

2. Which of the two liquids has the higher viscosity at room temperature, C_2H_5OH or CH_3OCH_3?

3. The viscosities of liquids generally decrease with increasing temperature. Water has the following viscosities in units of $N \cdot s/m^2$; 0.0018 at 0 °C; 0.0010 at 20 °C; 0.0005 at 55 °C; and 0.0003 at 100 °C. Interpret this trend on a molecular basis.

4. The vapor pressure of liquid potassium is 1.00 mmHg at 341 °C and 10.0 mmHg at 443 °C. Calculate the molar heat of vaporization of potassium.

5. The vapor pressure of ethanol is 100 mmHg at 34.9 °C, and 400 mmHg at 63.5 °C. What is the molar heat of vaporization of ethanol?

6. The vapor pressure of liquid potassium is 10.0 mmHg at 443 °C. The heat of vaporization is 82.5 kJ/mol. Calculate the temperature in, °C, where the vapor pressure is 760 mmHg. This is the boiling point of liquid potassium.

7. The vapor pressure of ethanol is 400 mmHg at 63.5 °C. Its molar heat of vaporization is 41.7 kJ/mol. Calculate the vapor pressure of ethanol at 34.9 °C.

8. On top of Pikes Peak water boils at 86 °C. Why do liquids boil at lower temperatures in the mountains as compared to sea level?

9. Copper crystallizes in a cubic system and the edge of the unit cell is 361 pm. The density of copper is 8.96 g/cm³. How many atoms are contained in one unit cell? What type of cubic cell does Cu form?

10. Copper crystallizes in a face-centered cubic lattice. The density of copper is 8.96 g/cm³. What is the length of the edge of a unit cell in nanometers?

11. In a diffraction experiment, X rays of wavelength 0.154 nm were reflected from a gold crystal. The first–order reflection was at 22.2°. What is the distance between the planes of gold atoms, in pm?

12. The distance between the planes of gold atoms in a gold crystal is 204 pm. In a diffraction experiment the first–order reflection was observed at 22.2°. What was the wavelength of the X rays in nanometers?

13. The molar heat of vaporization of bromine (Br_2) is 30.04 kJ/mol. What is the heat of vaporization per gram of bromine?

14. The molar heats of fusion and vaporization of potassium are 2.4 and 79.1 kJ/mol, respectively. Estimate the molar heat of sublimation of potassium metal.

Chapter 13
Physical Properties of Solutions

PROBLEM-SOLVING STRATEGIES AND TUTORIAL SOLUTIONS

TYPES OF PROBLEMS

Problem Type 1:   A Molecular View of the Solution Process.
           A.   The Importance of Intermolecular Forces.

Problem Type 2:   Concentration Units.

Problem Type 3:   Factors that Affect Solubility.
           A.   Temperature.
           B.   Pressure.

Problem Type 4:   Colligative Properties.
           A.   Vapor–Pressure Lowering.
           B.   Boiling–Point Elevation.
           C.   Freezing–Point Depression.
           D.   Osmotic Pressure.
           E.   Electrolyte Solutions.

Problem Type 5:   Calculations Using Colligative Properties.

PROBLEM TYPE 1: A MOLECULAR VIEW OF THE SOLUTION PROCESS.

A.   The Importance of Intermolecular Forces.

Dissolving is a process that takes place at the molecular level and can be discussed in molecular terms. When one substance dissolves in another, the particles of the solute disperse uniformly throughout the solvent. The ease with which a solute particle may replace a solvent molecule depends on the relative strengths of three types of interactions:

• Solvent–solvent interaction
• Solute–solute interaction
• Solvent–solute interaction

Imagine that the solution process takes place in three steps as shown in the following figure. Step 1 is the separation of solvent molecules. Step 2 is the separation of solute molecules. These steps require input of energy to overcome attractive intermolecular forces. Step 3 is the mixing of solvent and solute molecules; it may be exothermic or endothermic. According to Hess's law, the heat of solution is given by the sum of the enthalpies of the three steps:

$$\Delta H_{soln} = \Delta H_1 + \Delta H_2 + \Delta H_3$$

The solute will be soluble in the solvent if the solute–solvent attraction is stronger than the solvent–solvent attraction and solute–solute attraction. Such a solution process is exothermic. Only a relatively small amount of the solute will be dissolved if the solute–solvent interaction is weaker than the solvent–solvent and solute–solute interaction; then the solution process will be endothermic. It turns out that all chemical processes are governed by two factors. The first of these is the energy factor. In other words, does the solution process absorb energy or release energy? Disorder or randomness is the other factor that must be considered. Processes that increase randomness or disorder are also favored. In the pure state, the solvent and solute possess a fair degree of order. Here, we mean the ordered arrangement of atoms, molecules, or ions in a three–dimensional crystal. Where order is high, randomness is low. The order in a crystal is lost when the solute dissolves and its molecules are dispersed in the solvent. The solution process is accompanied by an increase in disorder or randomness. It is the increase in disorder of the system that favors the solubility of any substance.

The most general solubility rule is that "like dissolves like." The term "like" refers to molecular polarity. "Like dissolves like" means that substances of like or similar polarity will mix to form solutions. Substances of different polarity will be insoluble or immiscible, or will tend only slightly to form solutions. The rule predicts that two polar substances will form a solution, and two nonpolar substances will form a solution, but a polar substance and a nonpolar substance will tend not to mix.

STUDY QUESTION 13.1
Which would be a better solvent for molecular $I_2(s)$: CCl_4 or H_2O?

Solution: Using the like–dissolves-like rule, first we identify I_2 as a nonpolar molecule. Therefore, it will be more soluble in the nonpolar solvent CCl_4 than in the polar solvent H_2O.

STUDY QUESTION 13.2
In which solvent will NaBr be more soluble, benzene or water?

Solution: Benzene is a nonpolar solvent and water is a polar solvent. Since NaBr is an ionic compound, it cannot dissolve in nonpolar benzene, but is very soluble in water.

> *PRACTICE QUESTION 13.1*
> **Isopropyl alcohol and water dissolve in each other regardless of the proportions of each.**
> **What term describes the solubilities of these liquids in each other?**

> *PRACTICE QUESTION 13.2*
> **Explain why hexane (C_6H_{14}) even though a liquid, is not miscible with water.**

PRACTICE QUESTION 13.3
Indicate whether each compound listed is soluble or insoluble in water.

    a. CH_3OH
    b.   LiBr
    c. C_8H_{18}
    d. CCl_4
    e. $BaCl_2$
    f. $HOCH_2CH_2OH$

PRACTICE QUESTION 13.4
Explain why ammonia gas, NH_3, is very soluble in water, but not in hexane, C_6H_{14}.

PRACTICE QUESTION 13.5
Predict which substance of the following pairs, will be more soluble in water.

    a. $NaCl(s)$ or $I_2(s)$
    b. $CH_4(g)$ or $NH_3(g)$
    c. $CH_3OH(l)$ or C_6H_6(benzene)

PROBLEM TYPE 2: CONCENTRATION UNITS.

The term concentration refers to how much of one component of a solution is present in a given amount of solution. You will recall that in an earlier chapter, we were introduced to the molarity concentration unit.

$$molarity = \frac{moles\ of\ solute}{liters\ of\ solution}$$

We will use three new concentration units in this chapter.

The percent by mass of solute

$$Percent\ by\ mass\ of\ solute = \frac{mass\ of\ solute}{mass\ solute + mass\ solvent} \times 100\%$$

$$= \frac{mass\ of\ solute}{mass\ of\ solution} \times 100\%$$

The mole fraction of a component of the solution

$$mole\ fraction\ of\ component\ A = X_A = \frac{moles\ of\ A}{sum\ of\ moles\ of\ all\ compounds}$$

The molality of a solute in a solution

$$molality = \frac{moles\ of\ solute}{mass\ of\ solvent\ (kg)}$$

Notice that molality is the only one of the mentioned concentration units which has the amount of *solvent* (and not *solution*) in the denominator. This is essential to keep in mind.

It will be very important to be able to convert between the different concentration units.

STUDY QUESTION 13.3
The dehydrated form of Epsom salts is magnesium sulfate.

    a.   **What is the percent MgSO₄ by mass in a solution made from 16.0 g MgSO₄ and 100 mL of H₂O at 25 °C? The density of water at 25 °C is 0.997 g/mL.**
    b.   **What is the mole fraction of each component?**
    c.   **Calculate the molality of the solution.**

Solution:

a. First, write the equation for percent by mass.

$$\text{percent MgSO}_4 = \frac{\text{mass MgSO}_4}{\text{mass MgSO}_4 + \text{mass water}} \times 100\%$$

The mas of water can be determined by using the density;

$$\text{mass H}_2\text{O} = 100 \text{ mL} \times \frac{0.997 \text{ g}}{1 \text{ mL}} = 99.7 \text{ g H}_2\text{O}$$

Therefore, the percent mass of MgSO₄,

$$\text{percent MgSO}_4 = \frac{16.0 \text{ g}}{16.0 \text{ g} + 99.7 \text{ g}} \times 100\% = \frac{16.0 \text{ g}}{115.7 \text{ g}} \times 100\%$$

$$\text{percent MgSO}_4 = 13.8\%$$

b. The following is the formula for the mole fraction of MgSO₄,

$$X_{\text{MgSO}_4} = \frac{\text{mol MgSO}_4}{\text{mol MgSO}_4 + \text{mol H}_2\text{O}}$$

Convert the masses in part (a) into moles to substitute into the equation.

$$16.0 \text{ g MgSO}_4 \times \frac{1 \text{ mol MgSO}_4}{120.4 \text{ g MgSO}_4} = 0.133 \text{ mol MgSO}_4$$

$$99.7 \text{ g H}_2\text{O} \times \frac{1 \text{ mol H}_2\text{O}}{18.0 \text{ g H}_2\text{O}} = 5.54 \text{ mol H}_2\text{O}$$

The mole fractions of MgSO₄ and H₂O are

$$X_{\text{MgSO}_4} = \frac{0.133 \text{ mol}}{0.133 \text{ mol} + 5.54 \text{ mol}} = 0.0235 \qquad X_{\text{H}_2\text{O}} = \frac{5.54 \text{ mol}}{5.67 \text{ mol}} = 0.977$$

Please notice that the sum of the two mole fractions is 1.000. The sum of the mole fractions of all solution components is **always** 1.00.

c. Molality is calculated using the following equation:

$$\text{molality} = \frac{\text{moles of MgSO}_4}{\text{kilograms of H}_2\text{O}}$$

By substituting our previously calculated values from parts (a and b) into this equation, we arrive at;

$$\text{molality} = \frac{0.133 \text{ mol MgSO}_4}{99.7 \text{ g}} \times \frac{10^3 \text{ g}}{1 \text{ kg}} = 1.33 \text{ m}$$

STUDY QUESTION 13.4

Concentrated hydrochloric acid is 36.5 percent HCl by mass. Its density is 1.18 g/mL. Calculate:

   a. **the molality of HCl.**
   b. **the molarity of HCl.**

Solution:

a. The formula for molality is;

$$\text{molality} = \frac{\text{mol HCl}}{\text{kg H}_2\text{O}}$$

Next, to find the number of moles of HCl per kilogram of H_2O, we take 100 g of solution, and determine how many moles of HCl and how many kilograms of the solvent it contains. A solution that is 36.5 percent HCl by mass corresponds to 36.5 g HCl/100 g solution. Since 100 g of solution contains 36.5 g HCl, the difference 100 g – 36.5 g must equal the mass of water which is 63.5 g. We have two ratios:

$$\frac{36.5 \text{ g HCl}}{100 \text{ g soln}} \quad \text{and} \quad \frac{36.5 \text{ g HCl}}{63.5 \text{ g H}_2\text{O}}$$

The moles of HCl is given by:

$$\text{moles HCl} = 36.5 \text{ g HCl} \times \frac{1 \text{ mol HCl}}{36.5 \text{ g HCl}} = 1.00 \text{ mol HCl}$$

The kilograms of water is given by:

$$\text{kg H}_2\text{O} = 63.5 \text{ g H}_2\text{O} \times \frac{1 \text{ kg}}{1 \times 10^3 \text{ g}} = 0.0635 \text{ kg H}_2\text{O}$$

We can now calculate the molality as:

$$\text{molality} = \frac{1.00 \text{ mol HCl}}{0.0635 \text{ kg H}_2\text{O}} = 15.7 \text{ m}$$

b. Molarity is calculated using the following equation:

$$\text{molarity} = \frac{\text{moles HCl}}{\text{liters soln}}$$

We must first find the number of moles of HCl and liters of solution are present in 100 g of solution

$$\frac{36.5 \text{ g HCl}}{100 \text{ g soln}}$$

Convert 100 g of solution to volume of solution using the density. (Note: 36.5 g HCl = 1.00 mol).

$$\text{Volume of soln} = (100 \text{ g soln}) \times \frac{1 \text{ mL}}{1.18 \text{ g soln}} \times \frac{10^{-3} \text{ L}}{1 \text{ mL}} = 0.0847 \text{ L}$$

Therefore, by substituting our calculated values in the molarity equation, we arrive at;

$$\text{molarity} = \frac{1.00 \text{ mol HCl}}{0.0847 \text{ L soln}} = 11.8 \ M$$

PRACTICE QUESTION 13.6
Calculate the percent by mass of the solute in the following aqueous solutions.

      a.   **6.50 g NaCl in 75.2 g of water.**
      b.   **27.2 g ethanol in 250 g of solution.**
      c.   **2.0 g I_2 in 125 g methanol.**

PRACTICE QUESTION 13.7
Calculate the molality of each of the following solutions.

      a.   **6.50 g NaCl in 75.2 g of water .**
      b.   **27.5 g glucose ($C_6H_{12}O_6$) in 425 g of water.**

PRACTICE QUESTION 13.8
Calculate the molarity of a 2.44 m NaCl solution given that its density is 1.089 g/mL.

PRACTICE QUESTION 13.9
Calculate the percent $AgNO_3$ by mass in a 0.650 m $AgNO_3$ solution.

PRACTICE QUESTION 13.10
Calculate the molality of a 5.5% $AgNO_3$ solution.

PROBLEM TYPE 3: FACTORS THAT AFFECT SOLUBILITY.

A. Temperature.

Temperature changes strongly affect the solubility of most solid solutes. The effect of temperature on the solubility of some common salts is shown in Figure 13.4 in your textbook. In most cases, the solubility of a solid in water increases with increasing temperature. However, this is not always true. Temperature effects must be determined experimentally.

For gases, the dependence of solubility on temperature is different. The solubility of gases in water decreases with increasing temperature. The solubility of an unreactive gaseous solute is due to intermolecular attractive forces between solute molecules and solvent molecules. As the temperature of such a solution is increased, more solute molecules attain sufficient kinetic energy to break away from these attractive forces and enter the gas phase, and so the solubility decreases.

B. Pressure.

The solubility of gases in liquids is directly proportional to the gas pressure. Henry's law relates gas concentration c, in moles per liter, to the gas pressure P in atmospheres

$$c = kP$$

where k is a constant for each gas and has units of mol/L·atm. The greater the value of k, the greater the solubility of the gas.

It is common, in a Henry's Law problem to be given a concentration for a corresponding pressure in order to find the Henry's law constant. Once k is determined, it can be used to convert between pressure of a gas and its concentration in solution.

STUDY QUESTION 13.5
What is the concentration of O_2 in air–saturated water at 25 °C and atmospheric pressure of 645 mmHg? Assume the mole fraction of oxygen in air is 0.209 and that the Henry's law constant for O_2 at 25 °C is 1.28×10^{-3} M·atm^{-1}.

Solution: Henry's law states that the concentration of dissolved O_2 (C_{O_2}) is proportional to its partial pressure (P_{O_2}) in atm.

$$C_{O_2} = k\,P_{O_2}$$

The partial pressure of O_2 in air is found by using Dalton's law of partial pressures.

$$P_{O_2} = X_{O_2}\,P_T = 0.209\,(645 \text{ mmHg}) \times \frac{1 \text{ atm}}{760 \text{ mmHg}} = 0.177 \text{ atm}$$

Finally, substitute the values into the formula:

$$C_{O_2} = k\,P_{O_2} = (1.28 \times 10^{-3} \text{ mol/L·atm})\,(0.177 \text{ atm})$$

$$= 2.27 \times 10^{-4} \text{ mol/L}$$

> *PRACTICE QUESTION 13.11*
> **The solubility of KNO_3 at 70 °C is 135 g per 100 g of water. At 10 °C the solubility is 21 g per 100 g of water. What mass of KNO_3 will crystallize out of solution if exactly 100 g of its saturated solution at 70 °C is cooled to 10 °C?**

> *PRACTICE QUESTION 13.12*
> **As temperature increases, the solubility of all gases in water (increases/decreases).**

> *PRACTICE QUESTION 13.13*
> **The Henry's law constant for argon is 1.5×10^{-3} mol/L·atm at 20 °C. Calculate the solubility or argon in water at 20 °C and 7.6 mmHg.**

PROBLEM TYPE 4: COLLIGATIVE PROPERTIES.

Properties of solutions that depend on the number of solute particles in solution are called colligative properties. The four colligative properties of interest here are vapor–pressure lowering, boiling–point elevation, freezing–point depression, and osmotic pressure.

A. Vapor–Pressure Lowering.

Vapor–pressure lowering, ΔP, is given by:

$$\Delta P = \chi_2\,P_1^{\circ}$$

where χ_2 is the mole fraction of the solute, and P_1° is the vapor pressure of the pure solvent. Recall that the vapor pressure of a liquid is the pressure exerted by a vapor in equilibrium, with its liquid phase. For a solution containing a nonvolatile solute, the vapor pressure due to the solvent is less than it is for the pure solvent. The amount of lowering of the vapor pressure can be seen to depend on χ_2, the mole fraction of the solute.

STUDY QUESTION 13.6
Calculate the vapor pressure of an aqueous solution at 30 °C made from 100 g of sucrose ($C_{12}H_{22}O_{11}$), and 100 g of water. The vapor pressure of water at 30 °C is 31.8 mmHg.

Solution: Sucrose is a nonvolatile solute, so the vapor pressure over the solution will be due to H_2O molecules. The problem can be worked in two ways.

The vapor–pressure lowering is proportional to the mole fraction of sucrose χ_2, and P_2^o, the vapor pressure of pure water at 30 °C.

$$\Delta P = \chi_2\, P_1^o$$

First, we calculate the mole fraction of sucrose, χ_2.

$$\chi_2 = \frac{n_2}{n_1 + n_2}$$

$$n_2 = 100\ \text{g} \times \frac{1\ \text{mol}}{342\ \text{g}} = 0.292\ \text{mol}$$

$$n_1 = 100\ \text{g} \times \frac{1\ \text{mol}}{18.0\ \text{g}} = 5.55\ \text{mol}$$

$$\chi_2 = \frac{0.292}{0.292 + 5.55} = \frac{0.292}{5.84}$$

$$\chi_2 = 0.0500$$

The vapor pressure lowering is:

$$\Delta P = (0.0500)\,(31.8\ \text{mmHg}) = 1.59\ \text{mmHg}$$

The vapor pressure, P_1, is $P_1^o - \Delta P$.

$$P_1 = 31.8 - 1.59$$
$$= 30.2\ \text{mmHg}$$

An alternative approach to solving this question, would involve using Raoult's law:

$$P_1 = \chi_1 P_1^o$$

where P_1^o and P_1 are the vapor pressures of pure solvent and of the solvent in solution, respectively. From above we calculated that, $\chi_2 = 0.0500$, therefore $\chi_1 = 1.00 - \chi_2 = 0.95$, and

$$P_1 = (0.95)(31.8\ \text{mmHg})$$
$$= 30\ \text{mmHg}$$

B. Boiling–Point Elevation.

The boiling point of a solution is higher than that of pure solvent because the vapor pressure of a solution is always less than the vapor pressure of pure solvent. Thus, a solution must be hotter than a pure solvent, if both vapor pressures are to be 1atm. The boiling–point elevation ΔT_b of a solution of molality m is given by:

$$\Delta T_b = K_b m$$

where K_b is a constant called the molal boiling–point elevation constant and has units of $^\circ C/m$. There is a K_b for each solvent, and Table 13.2 in the textbook lists K_b values for five solvents. The magnitude of ΔT_b, the increase in boiling point of the solution over the boiling point of the pure solvent, is proportional to the solute concentration m. The preceding equation is independent of the solute, but requires that the solute be nonvolatile and a nonelectrolyte.

STUDY QUESTION 13.7

What is the boiling point of an "antifreeze/coolant" solution made from a 50–50 mixture (by volume) of ethylene glycol, $C_2H_6O_2$ (density 1.11 g/mL), and water?

Solution: The boiling point depends on the molality of the 50–50 mixture. $\Delta T = K_b m$. For simplicity, assume 100 mL of the solution. Then, using the density of water, the mass of 50 mL (50 vol %) of H_2O is 50 g. The mass of 50 mL (50 vol %) of ethylene glycol, using the density given above is 55.5 g.

$$m = \frac{\text{mol } C_2H_6O_2}{\text{kg } H_2O}$$

$$\text{mol } C_2H_6O_2 = 55.5 \text{ g} \times \frac{1 \text{ mol}}{62.0 \text{ g}} = 0.895 \text{ mol}$$

$$m = \frac{0.895 \text{ mol } C_2H_6O_2}{0.050 \text{ kg}} = 17.9 \text{ mol/kg} = 17.9 \ m$$

$$\Delta T_b = K_b m = (0.52 \ ^\circ C/m) \times 17.9 \ m = 9.3 \ ^\circ C$$

Therefore the boiling point of the solution is 9.3 °C above the normal boiling point of water.

$$\text{Boiling point} = 109.3 \ ^\circ C.$$

C. Freezing–Point Depression.

In a similar way, the freezing point of a solution will always be lower than that of the pure solvent. The freezing–point depression, ΔT_f, is given by:

$$\Delta T_f = K_f m$$

where K_f is the molal freezing–point depression constant and has the units $^\circ C/m$. As is the case of K_b, K_f is different for each solvent. The equation applies to all nonelectrolyte solutes, even volatile ones, because freezing points are rather insensitive to vapor pressure.

STUDY QUESTION 13.8

Estimate the freezing point of a solution made by mixing 20.0 mL of isopropanol ($(CH_3)_2CHOH$; density = 0.786 g/mL) with 250 g of water. The K_f for water is 1.86°C/m and its normal freezing point is 0.0 °C.

Solution: First, we need to calculate the molality of the solute, isopropanol. The only information that needs to be calculated is the molar mass of the solute;

$$20.0 \text{ mL} \times \frac{0.786 \text{ g}}{1 \text{ mL}} \times \frac{1 \text{ mol}}{60.09 \text{ g}} = 0.262 \text{ mol}$$

Substituting into the equation that defines molality:

$$\text{molality} = \frac{\text{moles of solute}}{\text{kg of solvent}} = \frac{0.262 \text{ mol}}{0.250 \text{ kg}} = 1.05 \ m$$

Now, we can solve to find the change in the freezing point:

$$\Delta T_f = mK_f = (1.05m)(1.86 \ °C/m) = 1.95 \ °C$$

The freezing point of the solution must be lower than that of the pure solvent by 1.95°C. This solution will freeze at –2.0°C.

D. Osmotic Pressure.
   The osmotic pressure, π, is given by:

$$\pi = MRT$$

   where M is the molar concentration, R is the ideal gas constant (0.0821 L·atm/K·mol), and T is the absolute temperature.

To help memorize the equation, notice it's similarity to $PV = nRT$, solving for P gives $P = (n/V)RT$.

   Osmosis is the net flow of solvent molecules through a semipermeable membrane from a more dilute solution to a more concentrated solution. Note that in this process the solvent flows from a region of greater solvent concentration to a region of lesser solvent concentration. The osmotic pressure of a solution is the pressure required to stop osmosis.

STUDY QUESTION 13.9
A solution of glucose ($C_6H_{12}O_6$) is made by dissolving 15.0 grams of glucose in enough water to make 100.0 mL of solution. What is the osmotic pressure of the solution at 25 °C?

Solution: Osmotic pressure is determined according to the equation: $\pi = MRT$. Therefore we will first determine the molarity of the solution:

$$M = \frac{\text{mol of } C_6H_{12}O_6}{\text{L of solution}} = \frac{15.0 \text{ g } C_6H_{12}O_6 \times \dfrac{1 \text{ mol } C_6H_{12}O_6}{180.0 \text{ g } C_6H_{12}O_6}}{100.0 \text{ mL} \times \dfrac{1 \text{ L}}{1000 \text{ mL}}} = 0.833 \text{ mol/L}$$

Inserting the molarity into the osmotic pressure equation gives:

$$\pi = 0.833 \ \frac{\text{mol}}{\text{L}} \times 0.0821 \ \frac{\text{L} \cdot \text{atm}}{\text{mol} \cdot \text{K}} \times 298 \text{ K} = 20.4 \text{ atm}$$

E. Electrolyte Solutions.
   For the same solute concentrations, solutions of electrolytes always have more pronounced colligative effects than those of nonelectrolyte solutions. For a 0.1 m methanol solution, ΔT_f is the same as the calculated value. For 0.1 m NaCl, ΔT_f is 1.9 times greater than calculated. And for Na_2SO_4, ΔT_f is 2.7 times greater than the calculated value. Since colligative properties depend on the number of solute particles, the most likely explanation of this behavior is that NaCl and Na_2SO_4 dissociate into

ions and this produces a greater concentration of solute particles. Therefore, the effective concentration of solute particles in 0.10 *m* NaCl is essentially 2 × 0.10 *m*, and in Na_2SO_4 it is almost 3 × 0.10 *m*. The freezing–point depression depends on the total concentration of all solute particles. Each ion acts as an independent particle.

The ratio ΔT_f(obs)/ΔT_f(calc) is called the van't Hoff factor, *i*. For 1:1 electrolytes, such as NaCl, KCl, and $MgSO_4$, *i* ~ 2; for 2:1 electrolytes, such as $MgCl_2$ and K_2SO_4, *i* is a little less than 3. The deviation from a whole number results from the temporary formation of some ion pairs such as "NaCl" in the solution.

STUDY QUESTION 13.10
List the following aqueous solutions in the order of increasing boiling points: 0.10 *m* glucose ($C_6H_{12}O_6$); 0.10 *m* $Ca(NO_3)_2$; 0.10 *m* NaCl.

Solution: Ethanol is a nonelectrolyte, whereas NaCl and $Ca(NO_3)_2$ are electrolytes. NaCl dissociates into 2 ions per formula unit, and $Ca(NO_3)_2$ dissociates into three ions per formula unit. The effective molalities are approximately:

$$\text{glucose } 0.10 \ m$$
$$\text{NaCl} \cong 0.20 \ m$$
$$Ca(NO_3)_2 \cong 0.30 \ m$$

Therefore, the boiling–point elevation and the boiling point are greatest for a solution of $Ca(NO_3)_2$, second highest for NaCl, and lowest for glucose.

Please note that if you included the van't Hoff factors, the effective molalities would be affected slightly, but not enough to change the predicted results.

PRACTICE QUESTION 13.14
Calculate the vapor pressure at 30 °C above an aqueous solution which is 10.5% by mass glucose ($C_6H_{12}O_6$). The vapor pressure of pure water at 30 °C is 31.8 mmHg.

PRACTICE QUESTION 13.15
What is the freezing point of a solution made from 1.00 g of $C_6H_{12}O_6$ (glucose) and 100.0 g of H_2O?

PRACTICE QUESTION 13.16
Calculate the freezing point of an aqueous solution that boils at 100.5 °C.

PRACTICE QUESTION 13.17
Should pasta cook faster in a pot of boiling water or a pot of boiling water that contains 2 tablespoons of salt added to it?

8PRACTICE QUESTION 13.18
Calculate the approximate osmotic pressure at 25 °C of a solution that has a freezing point of –18 °C.

PRACTICE QUESTION 13.19
How many solute particles does each basic unit of the following compounds give in aqueous solution?

    a. $C_2H_6O_2$ (ethylene glycol)
    b. $(H_2N)_2CO$ (urea)
    c. HBr
    d. $(NH_4)_3PO_4$
    e. NaOH
    f. $Ca(OH)_2$

PRACTICE QUESTION 13.20
Would you expect the freezing point to be lower for a 1.0 *m* solution of NaI or a 0.75 m solution of $MgCl_2$?

PROBLEM TYPE 5: CALCULATIONS USING COLLIGATIVE PROPERTIES.

Colligative properties provide a means to determine the molar mass of the solute. Keep in mind that molar mass has units of grams per mole. If given the mass of the solute, the moles will need to be determined. (The mole term is embedded in molality – in the case of freezing point depression and boiling–point elevation, and is embedded in molarity in the case of osmotic pressure.) For instance, the freezing–point depression equation is:

$$\Delta T_f = K_f\, m = K_f \times \frac{\text{mol solute}}{\text{kg solvent}}$$

Since ΔT_f can be measured and K_f is known, then the molality, *m*, of the solution can be calculated. Recall that molality is

$$\text{molality} = \frac{\text{mol solute}}{\text{kg solvent}} = \frac{\text{mass solute/molar mass of solute}}{\text{kg solvent}}$$

If you already know the molality of a solution and a known mass of solute is dissolved in a known mass of solvent, this leaves the molar mass of the solute as the only unknown quantity.

Freezing–point depression is more sensitive to the number of moles of solute than the boiling–point elevation because for the same solvent, K_f is larger then K_b. The most sensitive method for the determination of the molar mass of a solute is to measure the osmotic pressure. The equation for osmotic pressure then becomes useful in the following form:

$$\pi = MRT = \frac{(\text{g solute/molar mass of solute})\, RT}{\text{L solution}}$$

STUDY QUESTION 13.11
Benzene has a normal freezing point of 5.51 °C. The addition of 1.25 g of an unknown compound to 85.0 g of benzene produces a solution with a freezing point of 4.52 °C. What is the molar mass of the unknown compound?

Solution: We first need to find the number of moles in 1.25 g of unknown compound X. The freezing–point depression is proportional to the number of moles of X per kilogram of solvent.

The freezing–point depression is:

$$\Delta T_f = 5.51\ °C - 4.52\ °C$$
$$= 0.99\ °C$$

Using the equation $\Delta T = K_f m$, we can calculate the molality of the solution because ΔT_f is given above. Substituting the K_f for benzene (Table 13.2 textbook).

$$m = \frac{0.99\ °C}{5.12\ °C/m} = 0.19\ m$$

This means that there are 0.19 mol of X per kg of benzene. The number of moles of X in 0.085 kg of the solvent benzene (the given amount) is:

$$0.085\ kg \times \frac{0.19\ mol\ X}{1\ kg\ benzene} = 1.6 \times 10^{-2}\ mol\ X$$

Therefore, 1.25 g X = 1.6×10^{-2} mol X, and

$$\text{molar mass of X} = \frac{1.25\ g}{1.6 \times 10^{-2}\ mol} = 78\ g/mol$$

PRACTICE QUESTION 13.21
Calculate the molar mass of naphthalene given that a solution of 2.11 g of naphthalene in 100 g of benzene has a freezing–point depression of 0.85 °C.

PRACTICE QUESTION 13.22
A solution contains 1.00 g of a compound (a nonelectrolyte) dissolved in 100.0 g of water. The freezing point of the solution is –0.103 °C. What is the molar mass of the compound?

PRACTICE QUIZ

1. For each of the following pairs predict which substance will be more soluble in water:
   a. $HCl(g)$ or $H_2(g)$
   b. CCl_4 or $AlCl_3$
   c. CH_3Cl or CCl_4
   d. CH_3OH or CH_3OCH_3

2. For each of the following pairs, predict which substance will be more soluble in $CCl_4(\ell)$.
   a. H_2O or oil
   b. C_6H_6 (benzene) or CH_3OH
   c. I_2 or $NaCl$

3. Calculate the molality of a solution prepared by dissolving 28.0 g of urea, $CO(NH_2)_2$, in 450 g of H_2O.

4. How many grams of KI must be dissolved in 300 g of water to make a 0.500 *m* solution?

5. What is the molarity of a 3.0% hydrogen peroxide (H_2O_2) aqueous solution? The density is essentially 1.0 g/cm^3.

6. Concentrated sulfuric acid is 96% H_2SO_4 by weight. Its density is 1.83 g/mL. Calculate the molarity of concentrated H_2SO_4.

7. What are the mole fractions of methanol and water in a solution that contains 40.0 g CH_3OH and 40.0 g H_2O?

8. An aqueous solution contains 167 g $CuSO_4$ in 820 mL of solution. The density of the solution is 1.195 g/mL. Calculate the following:
   a. Molarity
   b. Percent by mass
   c. Mole fraction
   d. Molality of the solution

9. The Henry's law constant for CO is 9.73×10^{-4} mol/L·atm at 25 °C. What is the concentration of dissolved CO in water if the partial pressure of CO in the air is 0.015 mmHg?

10. How many grams of isopropyl alcohol, C_3H_7OH, should be added to 1.0 L of water to give a solution that will not freeze above –16 °C?

11. When 48 g of glucose (a nonelectrolyte) is dissolved in 500 g of H_2O, the solution has a freezing point of –0.94 °C. What is the molar mass of glucose?

12. 7.85 g of a compound with an unknown formula is dissolved in 300 g of benzene. The freezing point of the solution is 2.10 °C below that of pure benzene. What is the molar mass of the compound?

13. The dart poison in root extracts used by the Peruvian Indians is called curare. It is a nonelectrolyte. The osmotic pressure at 20 °C of an aqueous solution containing 0.200 g of curare in 100 mL is 56.2 mmHg. Calculate the molar mass of curare.

14. What is the concentration of a NaCl solution with an osmotic pressure of 10 atm at 25 °C?

15. The walls of red blood cells are semipermeable membranes, and the solution of NaCl within those walls exerts an osmotic pressure of 7.82 atm at 37 °C. What concentration of NaCl must a surrounding solution have so that this pressure is balanced and cell rupture (hemolysis) is prevented?

16. Arrange the following aqueous solutions in order of increasing freezing points, lowest to highest.
    0.100 m ethanol
    0.050 m $Ca(NO_3)_2$
    0.100 m NaBr
    0.050 m HCl

17. Calculate the value of *i* for a 1:1 electrolyte, if a 1.0 *m* aqueous solution of the electrolyte freezes at −3.28 °C.

18. In the course of research a chemist isolated a new compound. An elemental analysis gave the following: C 50.7%, H 4.25%, O 45.1%. When 5.01 g of the compound was dissolved in exactly 100 g of water, it produced a solution with a freezing point of −0.65 °C. What is the molecular formula of the compound?

Chapter 14
Entropy and Free Energy

PROBLEM-SOLVING STRATEGIES AND TUTORIAL SOLUTIONS

TYPES OF PROBLEMS

Problem Type 1: Entropy Changes in a System.

        A. Calculating $\Delta S^{\circ}_{\text{system}}$.

        B. Standard Entropy, S°.

        C. Qualitatively Predicting the Sign of $\Delta S^{\circ}_{\text{system}}$.

Problem Type 2: Entropy Changes in the Universe.

        A. Calculating ΔS_{surr}.

        B. The Second Law of Thermodynamics.

        C. The Third Law of Thermodynamics.

Problem Type 3: Predicting Spontaneity.

        A. Gibbs Free–Energy Change, ΔG.

        B. Standard Free–Energy Changes, ΔG°.

        C. Relationship between ΔG and ΔG°.

PROBLEM TYPE 1: ENTROPY CHANGES IN A SYSTEM.

If we think about the number of molecules and the place they can be relative to their neighbors or the space they occupy, we are considering something called **microstates**. It turns out that there is a statistical calculation that can be performed for microstates. We won't get into a statistical treatment of microstates. However, we should realize that the more microstates for a system, the more disorder or randomness the system will exhibit. **Entropy** is a measure of this disorder or randomness of a system. Entropy (S) is a state function that increases in value as the disorder or randomness (microstates) of the system increases. Entropy has the units J/K·mol. Intuitively, we consider a system to be "ordered" if it is arranged according to some plan or pattern. The system is "disordered" when its parts are haphazard and their arrangement is random.

A. Calculating ΔS_{system}.

    As with any other state function, the calculation of the entropy change for a system can be performed by considering the final and initial states. A positive ΔS_{sys} represents a system that has become more disordered, whereas, a negative ΔS_{sys} corresponds to a system that has become more ordered.

$$\Delta S_{\text{system}} = \Delta S_{\text{final}} - \Delta S_{\text{initial}}$$

Equation 18.4 in the textbook also shows another relationship that can be employed to calculate entropy changes that occur during the expansion or compression of a gas.

$$\Delta S_{\text{system}} = nR\ln \frac{V_{\text{final}}}{V_{\text{initial}}}$$

STUDY QUESTION 14.1
Calculate the change in entropy for 0.875 g of oxygen gas undergoing an expansion from an initial volume of 1.4 L to a final volume of 3.6 L under a constant temperature.

Solution: In order to calculate the change in entropy, we must know the number of moles of oxygen gas undergoing expansion.

$$0.875 \text{ g} \times \frac{1 \text{ mol O}_2}{32.00 \text{ g O}_2} = 2.73 \times 10^{-2} \text{ mol O}_2$$

Now, we can substitute the values given on the equation:

$$\Delta S_{system} = nR\ln\frac{V_{final}}{V_{initial}} = 2.73 \times 10^{-2} \text{ mol} \times (8.314 \text{J/mol} \cdot \text{K}) \times \ln\frac{3.6 \text{ L}}{1.4 \text{ L}}$$

$$\Delta S_{system} = 0.21 \text{ J/K}$$

B.   Standard Entropy, $S°$.

In contrast to internal energy and enthalpy, the absolute value of the entropy of a system can be calculated. The **standard entropy**, $S°$, is the entropy of a substance at a pressure of 1 atm. These values have been tabulated and may be used to determine changes in entropy for reactions. Several general trends for the values of $S°$ can be pointed out:

1. $S°$ values increase with molecular complexity: the more atoms a compound has, the greater its $S°$. A good indicator for molecular complexity is the molar mass of the compound.
2. $S°$ increases in the order:

$$S°_{solid} < S°_{liquid} < S°_{gas}$$

The actual value of ΔS can also be calculated. The absolute value of the entropy of one mole of an element or compound can be determined by very careful experimentation. Appendix 2 of the textbook lists experimental values of the absolute entropy of a number of substances in their standard states at 1 atm and 25 °C. For a reaction

$$a\text{A} + b\text{B} \rightarrow c\text{C} + d\text{D},$$

the standard entropy change of the reaction is given by:

$$\Delta S°_{rxn} = \sum nS° \text{ (products)} - \sum mS° \text{ (reactants)}$$

where m and n are stoichiometric coefficients. When applied to the above reaction, we get:

$$\Delta S°_{rxn} = [cS°(\text{C}) + dS°(\text{D})] - [aS°(\text{A}) + bS°(\text{B})]$$

STUDY QUESTION 14.2

Use absolute entropies to calculate the standard entropy of reaction $\Delta S°_{rxn}$:

$$\text{H}_2(g) + \frac{1}{2}\text{O}_2(g) \rightarrow \text{H}_2\text{O}(\ell)$$

Solution: The standard entropy change is given by:

$$\Delta S°_{rxn} = S°(\text{H}_2\text{O}) - [S°(\text{H}_2) + \frac{1}{2}S°(\text{O}_2)]$$

Look up the standard entropy values in Appendix 2 of the text.

$$\Delta S^{\circ}_{rxn} = 1\ mol\left(\frac{69.9\ J}{K \cdot mol}\right) - \left[1\ mol\left(\frac{130.1\ J}{K \cdot mol}\right) + \frac{1}{2}mol\left(\frac{205.0\ J}{K \cdot mol}\right)\right]$$

$$\Delta S^{\circ}_{rxn} = 69.9\ J/K - 232.6\ J/K = -162.7\ J/K \cdot mol$$

C. Qualitatively Predicting the Sign of $\Delta S^{\circ}_{system}$.

We can estimate whether the change in entropy in a process is positive or negative. For chemical reactions in which solids or liquids are converted to gases, the entropy change, ΔS, is positive. It is negative for the condensation of a gas or the freezing of a liquid. When a crystal of a salt dissolves in water, the disorder increases due to the increase in freedom of motion of ions in the solution, compared to the highly ordered crystal. Heating also increases the entropy of a system. The higher the temperature, the more energetic the molecular motion and its accompanying disorder.

Hints for estimating the sign of ΔS:
1. If a reaction produces an increase in the number of moles of gaseous compounds, ΔS is positive.
2. If the total number of moles of gaseous compounds is decreased, ΔS is negative.
3. If there is no net change in the total number of gas molecules, then ΔS is either a small positive or a small negative number.

STUDY QUESTION 14.3
Predict the sign of ΔS_{sys} for each of the following reactions using just the qualitative ideas discussed in above.

a. $H_2O_2(\ell) \rightarrow H_2O(\ell) + \frac{1}{2}O_2(g)$

b. $H^+(aq) + OH^-(aq) \rightarrow H_2O(\ell)$

c. $CaO(s) + CO_2(g) \rightarrow CaCO_3(s)$

Solution:

a. The number of moles of gaseous compounds in the products is greater than in the reactant. Entropy increases during this reaction. The sign of ΔS is positive.

b. Two reactants combine into one product in this reaction. Order is increased and so entropy decreases. The sign of ΔS is negative.

c. The number of gas molecules is decreased. The sign of ΔS is negative.

PRACTICE QUESTION 14.1
Predict, using the intuitive ideas about entropy, whether ΔS_{rxn} will be positive, negative, or essentially zero for each of the following.

a. $CuSO_4(s) \rightarrow Cu^{2+}(aq) + SO_4^{2-}(aq)$

b. $SO_2(g) + \frac{1}{2}O_2(g) \rightarrow SO_3(g)$

c. $Ca(OH)_2(s) + CO_2(g) \rightarrow CaCO_3(s) + H_2O(g)$

d. $Ag^+(aq) + 2CN^-(aq) \rightarrow Ag(CN)_2^-(aq)$

PROBLEM TYPE 2: ENTROPY CHANGES IN THE UNIVERSE.

A. Calculating ΔS_{surr} .

The change in entropy for the surroundings can be calculated by knowing the change in enthalpy of the system and the temperature at which the process is taking place.

$$\Delta S_{surr} = -\frac{\Delta H_{sys}}{T}$$

B. The Second Law of Thermodynamics.

The **second law of thermodynamics** states that the entropy of the universe increases in a spontaneous change. The term **universe** used here refers to a system and all of its surroundings. For an isolated system:

$$\Delta S_{univ} = \Delta S_{sys} + \Delta S_{surr} \geq 0$$

where ΔS_{sys} stands for the entropy change of the system, and ΔS_{surr} is the entropy change of the surroundings. For any spontaneous change:

$$\Delta S_{sys} + \Delta S_{surr} > 0$$

and for a reaction at equilibrium (no net change):

$$\Delta S_{sys} + \Delta S_{surr} = 0$$

STUDY QUESTION 14.4

The solubility of silver chloride is so low that it precipitates spontaneously from many solutions. The entropy change of the system is negative for this process.

$$Ag^+(aq) + Cl^-(aq) \rightarrow AgCl(s) \quad \Delta H^\circ = -65 \text{ kJ/mol}$$

Since ΔS decreases in this spontaneous reaction, shouldn't the reaction be nonspontaneous?

Solution: For a spontaneous reaction, the second law states that $\Delta S_{univ} > 0$:

$$\Delta S_{sys} + \Delta S_{surr} > 0$$

In order to predict whether a reaction is spontaneous, both ΔS_{sys} and ΔS_{surr} must be considered, not just ΔS_{sys}. Since the entropy change of the system is negative, the only way for the inequality to be true is if ΔS_{surr} is positive and greater in amount than ΔS_{sys}. In this case, this is a reasonable assumption because the reaction is exothermic. This means that heat is released to the surroundings, which causes increased thermal motion and disorder of molecules in the surroundings. Therefore, the sum of ΔS_{sys} and ΔS_{surr} is greater than zero, even though ΔS_{sys} is negative.

C. The Third Law of Thermodynamics.

The Third Law of Thermodynamics states that the entropy of a perfect crystal at absolute zero temperature is zero. The importance of the Third Law lies in allowing the determination of absolute entropies for substances.

> #### PRACTICE QUESTION 14.2
>
> **When the environment is contaminated by a toxic chemical spill or an oil spill, the substance tends to disperse. How is this consistent with the second law of thermodynamics? In the same regard, which requires less work: cleaning the environment after a spill, or keeping the substance contained before a spill?**

PROBLEM TYPE 3: PREDICTING SPONTANEITY.

A. Gibbs Free–Energy Change, ΔG.

Quite often it is not possible to calculate ΔS_{surr}. This makes the second law difficult to apply when it is in the form given in the previous section. The Gibbs free–energy, expressed in terms of enthalpy and entropy, refers only to the system, yet can be used to predict spontaneity. The Gibbs free–energy change (ΔG) for a reaction carried out at constant temperature and pressure is given by:

$$\Delta G = \Delta H - T\Delta S$$

where both ΔH and ΔS refer to the system.

The **Gibbs free–energy** change is equal to the maximum possible work (w) that can be obtained from a process. Any process that occurs spontaneously can be utilized to perform useful work. The Gibbs free energy of the system will decrease ($\Delta G < 0$) in a spontaneous process, and will increase ($\Delta G > 0$) in a nonspontaneous process.

The free–energy criteria (at constant temperature and pressure) are summarized as follows:

1. If $\Delta G < 0$, the forward reaction is spontaneous.
2. If $\Delta G = 0$, the reaction is at equilibrium.
3. If $\Delta G > 0$, the forward reaction is nonspontaneous. The reverse reaction will have a negative ΔG and will be spontaneous.

The **standard free–energy** change of reaction ΔG°_{rxn} is the free–energy change for a reaction when it occurs under standard state conditions, when reactants in their standard states are converted to products in their standard states.

From the equation, $\Delta G = \Delta H - T\Delta S$, we can see that temperature will influence the spontaneity of reaction. If both ΔH and ΔS are positive then, as long as $\Delta H > T\Delta S$, at low temperature, ΔG is positive and the process will be nonspontaneous. However, as temperature increases, the $T\Delta S$ term increases and eventually $\Delta H = T\Delta S$. At this point, ΔG is zero. With further T increase, $T\Delta S > \Delta H$, making $\Delta G < 0$, and the reaction becomes spontaneous. Table 18.4 in your textbook summarizes the effect that temperature would have on the spontaneity of a reaction depending on the signs of ΔH and ΔS.

STUDY QUESTION 14.5

Hydrated lime Ca(OH)$_2$ can be reformed into quicklime CaO by heating.

$$Ca(OH)_2(s) \rightleftharpoons CaO(s) + H_2O(g)$$

At what temperatures is this reaction spontaneous under standard conditions (that is, where H$_2$O is formed at 1 atm pressure)? The following data on Ca(OH)$_2$ is not in the Appendix:

$$\Delta H^{\circ}_f \, [Ca(OH)_2] = -986.2 \text{ kJ/mol}$$

$$S^{\circ}[Ca(OH)_2] = 83.4 \text{ J/K·mol}$$

Solution: This reaction is nonspontaneous at room temperature. The temperature above which the reaction becomes spontaneous under standard conditions corresponds to $\Delta G^{\circ} = 0$, and is given by:

$$T = \frac{\Delta H^{\circ}}{\Delta S^{\circ}}$$

ΔH° and ΔS° must be calculated separately. From Appendix 3 and the given data:

$$\Delta H^{\circ} = [\, \Delta H^{\circ}_f (CaO) + \Delta H^{\circ}_f (H_2O)] - [\, \Delta H^{\circ}_f (Ca(OH)_2)]$$

$$\Delta H° = (-635.55 \text{ kJ /mol}) + (-241.83 \text{ kJ/mol}) - (-986.2 \text{ kJ/mol})$$

$$= 108.82 \text{ kJ/mol (or } 1.088 \times 10^5 \text{ J/mol)}$$

$$\Delta S° = S°(\text{CaO}) + S°(\text{H}_2\text{O}) - S°(\text{Ca(OH)}_2)$$

$$\Delta S° = (39.8 \text{ J/K·mol}) + (188.7 \text{ J/K·mol}) - (83.4 \text{ J/K·mol})$$

$$= + 145.1 \text{ J/K · mol}$$

The temperature at which $\Delta G°$ is equal to zero is:

$$T = \frac{\Delta H°}{\Delta S°} = \frac{1.088 \times 10^5 \text{ J}}{145.1 \text{ J/K}} = 750 \text{ K}$$

At temperatures above 750 K the reaction is spontaneous.

NOTE:
Recall that this is an approximate value because of the assumption that neither $\Delta H°$ nor $\Delta S°$ change appreciably from their values calculated at 25 °C.

> ### PRACTICE QUESTION 14.3
> **A reaction just becomes spontaneous at 786 K with $\Delta H° = -280.5$ kJ/mol. Calculate $\Delta S°$ of the reaction.**

> ### PRACTICE QUESTION 14.4
> **Determine the temperature conditions at which the following are spontaneous.**
> **a.   solid → liquid**
> **b.   gas → solid**

B.   Standard Free-Energy Changes $\Delta G°$.

The free energy change for a reaction can be calculated in two ways. When both ΔH and ΔS are known, then $\Delta G = \Delta H - T\Delta S$ will give the free energy change at the temperature T. $\Delta G°_{rxn}$ can also be calculated from standard free energies of formation in a manner analogous to the calculation of $\Delta H°$ by using enthalpies of formation of the reactants and products. The standard free energies of formation ($\Delta G°_f$) of selected compounds are tabulated in Appendix 2 of the text. Just as for the standard enthalpies of formation, the free energies of formation of elements in their standard states are equal to zero.

For a general reaction:

$$a\text{A} + b\text{B} \rightarrow c\text{C} + d\text{D}$$

The standard free energy change is given by:

$$\Delta G°_{rxn} = [c \, \Delta G°_f (\text{C}) + d \, \Delta G°_f (\text{D})] - [a \, \Delta G°_f (\text{A}) + b \, \Delta G°_f (\text{B})]$$

In general:

$$\Delta G°_{rxn} = \sum n \, \Delta G°_f (\text{products}) - \sum m \, \Delta G°_f (\text{reactants})$$

where n and m are stoichiometric coefficients.

STUDY QUESTION 14.6

Calculate ΔG°_{rxn} at 25 °C for the following reaction using Appendix 2 and given:

$$\Delta G^{\circ}_f \ (Fe_2O_3) = -741.0 \ kJ/mol$$
$$2Al(s) + Fe_2O_3(s) \rightarrow Al_2O_3(s) + 2Fe(s)$$

Solution:

$$\Delta G^{\circ}_{rxn} = [\ \Delta G^{\circ}_f \ (Al_2O_3) + 2 \ \Delta G^{\circ}_f \ (Fe)] - [2 \ \Delta G^{\circ}_f \ (Al) + \ \Delta G^{\circ}_f \ (Fe_2O_3)]$$

$$= [(-1576.41 \ kJ/mol) + 0] - [0 + (-741.0 \ kJ/mol)]$$

$$\Delta G^{\circ}_{rxn} = -1576.41 \ kJ + 741.0 \ kJ = -835.4 \ kJ/mol$$

For a phase transition, $\Delta G = 0$ when the two phases coexist in equilibrium. For example, at the boiling point (T_{bp}) the liquid and vapor phases are in equilibrium, and

$$\Delta H_{vap} - T_{bp}\Delta S_{vap} = 0$$

rearranging gives

$$S_{vap} = \frac{\Delta H_{vap}}{T_{bp}}$$

This equation allows the calculation of the entropy of vaporization from knowledge of the heat of vaporization and the boiling point.

STUDY QUESTION 14.7
The heat of fusion of water (ΔH_{fus}) at 0 °C is 6.02 kJ/mol. What is ΔS_{fus} for 1 mole of H_2O at the melting point?

Solution:

$$\Delta S_{fus} = \frac{\Delta H_{fus}}{T_{mp}} = \frac{6.02 \times 10^3 \ J/mol}{273 \ K} = +22.1 \ J/K \cdot mol$$

NOTE:
The increase in entropy upon melting of the solid corresponds to the higher degree of molecular disorder in the liquid state as compared to the solid state.

PRACTICE QUESTION 14.5

Calculate ΔG°_{rxn} for the following reaction: $3 \ NO_2(g) + H_2O(\ell) \rightarrow 2 \ HNO_3(\ell) + NO(g)$ given the following free energies of formation.

Substance	ΔG°_f (kJ/mol)
$H_2O(l)$	−237.2
$HNO_3(l)$	−79.9
$NO(g)$	86.7
$NO_2(g)$	51.8

C. Relationship between ΔG and $\Delta G°$.

Recall that $\Delta G°$ refers to the standard free–energy change. All the values we have calculated so far relate to processes in which the reactants are present in their standard states and are converted to products in their standard states. However, in many cases neither the reactants nor the products are present at standard concentration (1 M) and standard pressure (1 atm). Under nonstandard state conditions, we use the symbol ΔG.

The relationship between ΔG and $\Delta G°$ is:

$$\Delta G = \Delta G° + RT \ln Q$$

where R is the gas constant (8.314 J/K·mol), T is the absolute temperature, and Q is the reaction quotient. For a certain reaction at a given temperature the value of $\Delta G°$ is constant, but the value of Q depends on the composition of the reacting mixture; therefore, ΔG will depend on Q. To calculate ΔG, first find $\Delta G°$, then calculate Q from the given concentrations of reactants and products and substitute into the preceding equation.

STUDY QUESTION 14.8

The standard free energy change for the reaction,

$$\frac{1}{2}N_2(g) + \frac{3}{2}H_2(g) \rightleftharpoons NH_3(g)$$

is $\Delta G°_{rxn}$ = 26.9 kJ/mol at 700 K. Calculate ΔG at 700 K if the reaction mixture consists of 30.0 atm of H_2, 20.0 atm of N_2, and 0.500 atm of NH_3.

Solution: Under nonstandard conditions, ΔG is related to the reaction quotient Q by the equation:

$$\Delta G = \Delta G° + RT \ln Q_p$$

where,

$$Q_p = \frac{P_{NH_3}}{P_{N_2}^{1/2} P_{H_2}^{3/2}} = \frac{(0.500)}{(20.0)^{1/2}(30.0)^{3/2}} = 6.80 \times 10^{-4}$$

Since $\Delta G° = 26.9$ kJ/mol, substitution yields:

$$\Delta G = 26.9 \text{ kJ/mol} + (8.314 \text{ J/K·mol})(700 \text{ K}) \ln (6.80 \times 10^{-4})$$

$$= 26.9 \text{ kJ/mol} - (42,400 \text{ J/mol} \times \frac{1 \text{ kJ}}{10^3 \text{ J}}) = 26.9 \text{ kJ/mol} - 42.4 \text{ kJ/mol}$$

$$= -15.5 \text{ kJ/mol}$$

NOTE:

By making the partial pressures of N_2 and H_2 high and that of NH_3 low, the reaction is spontaneous in the forward direction. This condition corresponds to $Q_p < K_p$, and so the reaction proceeds in the forward direction until $Q_p = K_p$.

PRACTICE QUIZ

1. Which of the following processes are spontaneous?
   a. melting of ice at $-10\ ^\circ$C and 1 atm pressure
   b. evaporation of water at 30 $^\circ$C when the relative humidity is less than 100%
   c. Water + NaCl(s) $\rightarrow$ salt solution

2. Choose the substance that has the larger standard entropy at 25 $^\circ$C from each pair:
   a. $H_2O(\ell)$ or $H_2O(g)$
   b. $SiO_2(s)$ or $CO_2(g)$
   c. $Ag^+(g)$ or $Ag^+(aq)$
   d. $F_2(g)$ or $Cl_2(g)$
   e. $2Cl(g)$ or $Cl_2(g)$

3. Predict, using the intuitive ideas about entropy, whether ΔS_{sys} will be positive, negative, or essentially zero for each of the following:
   a. $Ca(OH)_2(s) + CO_2(g) \rightarrow CaCO_3(s) + H_2O(g)$
   b. $CuSO_4(s) \rightarrow Cu^{2+}\ (aq) + SO_4^{2-}\ (aq)$
   c. $2HCl(g) + Br_2(\ell) \rightarrow 2HBr(g) + Cl_2(g)$
   d. $3H_2(g) + N_2(g) \rightarrow 2NH_3(g)$
   e. $Cu^{2+}\ (aq) + 4NH_3(aq) \rightarrow Cu(NH_3)_4^{2+}\ (aq)$

4. At the boiling point, 35 $^\circ$C, the heat of vaporization of MoF_6 is 25 kJ/mol. Calculate ΔS for the vaporization of MoF_6.

5. Calculate ΔG°_{rxn} for the following reaction at 298 K:

   $$2H_2(g) + CO(g) \rightarrow CH_3OH(g)$$

   given that $\Delta H^\circ = -90.7$ kJ/mol and $\Delta S^\circ = -221.5$ J/K·mol for this process.

6. Calculate ΔS° at 298 K for the following reaction, given $\Delta H^\circ = -602$ kJ/mol and $\Delta G^\circ = -569$ kJ/mol.

   $$Mg(s) + \frac{1}{2}O_2(g) \rightarrow MgO(s)$$

7. Using Appendix 2 of the text, calculate ΔG° values for the following reactions:

   a. $3CaO(s) + 2Al(s) \rightarrow 3Ca(s) + Al_2O_3(s)$
   b. $ZnO(s) \rightarrow Zn(s) + \frac{1}{2}O_2(g)$

8. Which of the following three reactions will be least spontaneous at 25 $^\circ$C?
   a. $N_2 + O_2 \rightarrow 2\ NO$
   b. $N_2 + 2O_2 \rightarrow N_2O_4$
   c. $N_2 + \frac{1}{2}O_2 \rightarrow N_2O$
   Given at 25 $^\circ$C:

   $$\Delta G^{\circ}_f\ (NO) = +86.7\ \text{kJ/mol}$$

   $$\Delta G^{\circ}_f\ (N_2O_4) = +98.3\ \text{kJ/mol}$$

   $$\Delta G^{\circ}_f\ (N_2O) = +103.6\ \text{kJ/mol}$$

9. Calculate ΔG°_{rxn} at 25 °C for the following reaction:

$$NO(g) + \frac{1}{2}O_2(g) \rightarrow NO_2(g)$$

10. The synthesis of $O_2(g)$ can be carried out by the decomposition of $KClO_3$:

$$KClO_3(s) \rightarrow KCl(s) + \frac{1}{2}O_2(g)$$

for which $\Delta H^{\circ} = -44.7$ kJ/mol and $\Delta S^{\circ} = +247.2$ J/K·mol. Is this reaction spontaneous at 25 °C under standard conditions?

11. For the reaction $N_2 + O_2 \rightarrow 2$ NO, the following are given:
    $\Delta H^{\circ} = 180.7$ kJ/mol
    $\Delta S^{\circ} = 24.7$ J/K·mol
    a. Is this reaction spontaneous at 25 °C?
    b. Above what temperature will this reaction become spontaneous under standard conditions?

12. For the reaction,
$$2SO_2(g) + O_2(g) \rightarrow 2SO_3(g)$$

if we have the following partial pressures, what is ΔG?

$$P_{SO_2} = 1.2 \text{ atm} \qquad P_{O_2} = 0.5 \text{ atm} \qquad P_{SO_3} = 50 \text{ atm}$$

13. For the reaction
$$H_2(g) + Cl_2(g) \rightarrow 2HCl(g)$$
    a. Calculate ΔS° using the values of S° provided in Appendix 2.
    b. Calculate the value of ΔS°_{surr} at 550°C and use the Second Law of Thermodynamics to predict the spontaneity of the reaction at this temperature.

14. Use the values listed in Appendix 2 to calculate ΔG at 250°C for the reaction
$$2C_4H_{10}(g) + 13O_2(g) \rightarrow 8CO_2(g) + 10H_2O(g)$$
    when $P_{H_2O} = 0.150$ atm; $P_{CO_2} = 0.120$ atm; $P_{C_4H_{10}} = 0.680$ atm .

15. Calculate ΔS°_{sys} for the compression of 1000 g of CO_2 from an initial volume of 5.0 L to a final volume of 1.5 L.

16. Consider the reaction:
$$B_2O_3(s) + 3H_2O(l) \rightarrow 2H_3BO_3(aq)$$

Determine under which temperature conditions this reaction will be spontaneous.

Chapter 15
Chemical Equilibrium

PROBLEM-SOLVING STRATEGIES AND TUTORIAL SOLUTIONS

TYPES OF PROBLEMS

Problem Type 1:   Equilibrium Constant.
                A.   Calculating Equilibrium Constants.
                B.   Magnitude of the Equilibrium Constants.

Problem Type 2:   Equilibrium Expressions.
                A.   Heterogeneous Equilibria.
                B.   Manipulating Equilibrium Expressions.
                C.   Gaseous Equilibria.

Problem Type 3:   Chemical Equilibrium and Free Energy.
                A.   Using Q and K to Predict the Direction of Reaction.
                B.   Relationship between ΔG and $\Delta G°$.
                C.   Relationship between $\Delta G°$ and K.

Problem Type 4:   Calculating Equilibrium Concentrations.

Problem Type 4:   Le Châtelier's Principle: Factors that Affect Equilibrium.
                A.   Addition or Removal of a Substance.
                B.   Changes in Volume and Pressure.
                C.   Changes in Temperature.

PROBLEM TYPE 1: EQUILIBRIUM CONSTANT.

A. Calculating Equilibrium Constants.

A state of chemical equilibrium exists when the concentrations of reactants and products are observed to remain constant with time. When a mixture of SO_2 and O_2, for instance, is introduced into a closed reaction vessel at a temperature of 700 K, a reaction that produces SO_3 occurs:

$$2SO_2 + O_2 \; \rightleftharpoons \; 2SO_3$$

When a specific concentration of SO_3 is reached, the concentrations of SO_2, O_2, and SO_3 stay constant, and we say the system has reached chemical equilibrium.

The constant concentrations are the result of a reversible chemical reaction. When the rates of the forward reaction and reverse reactions are the same, no *net* chemical change occurs and a state of **chemical equilibrium** exists. The equilibrium state is referred to as dynamic, because of the continual conversion of reactants into products, and of products into reactants at the molecular level. A reversible reaction is represented by the double arrows in the chemical equation. At the point of equilibrium, all reactant and product species exist together in the reaction vessel. Equilibrium can be reached by the reverse reaction as well. That is, if only product is added to the reaction vessel it will form an equilibrium amount of reactants.

In general the, **equilibrium constant expression** (also called the mass action expression) has the form of a ratio of product concentrations over reactant concentrations at equilibrium. For the general equation

$$aA + bB \rightleftharpoons cC + dD$$

where *a*, *b*, *c*, and *d* are the stoichiometric coefficients for the balanced equation, the equilibrium constant expression is:

$$K_c = \frac{[C]^c [D]^d}{[A]^a [B]^b}$$

The numerator is found by multiplying together the equilibrium concentrations (illustrated as [ ] to represent moles per liter) of the products, each raised to an exponent equal to its stoichiometric coefficient. The denominator is found by multiplying together the equilibrium concentrations of the reactants, each raised to an exponent equal to its stoichiometric coefficient. The value of this expression is called the equilibrium constant.

STUDY QUESTION 15.1
Write the equilibrium constant expressions for the following reversible reaction:
$$4NH_3(g) + 5O_2(g) \rightleftharpoons 4NO(g) + 6H_2O(g).$$

Solution: Remember that the equilibrium constant expression has the concentrations of the products in the numerator and those of the reactants in the denominator. Raise each concentration to a power equal to the coefficient of that substance in the balanced equation.

$$K_c = \frac{[NO]^4 [H_2O]^6}{[NH_3]^4 [O_2]^5}$$

PRACTICE QUESTION 15.1
Write the equilibrium constant expressions (K_c) for the following equation:
$$2NO(g) + Br_2(g) \rightleftharpoons 2NOBr(g)$$

B. Magnitude of the Equilibrium Constants.
The magnitude of the equilibrium constant is related to the degree of conversion of reactants to products before chemical equilibrium is reached. Since K_c is proportional to the concentrations of products divided by the concentrations of reactants present at equilibrium, the value of K_c tells us the relative quantities of reactants and products present at equilibrium. When K_c is much larger than 1, more products are present than reactants at equilibrium. Conversely, when K_c is much smaller than 1, more reactants are present at equilibrium than products. In this case equilibrium is reached before appreciable concentrations of products build up. The larger the value of K_c, the greater the extent of reaction before equilibrium is reached. In general, we can say:

- When $K_c \gg 1$, the forward reaction will go nearly to completion. The equilibrium will lie to the right and favor the products.
- When $K_c \ll 1$, the reaction will not go forward to an appreciable extent. The equilibrium will lie to the left and favor the reactants.

STUDY QUESTION 15.2
Arrange the following reactions in order of their increasing tendency to proceed toward completion (least extent < greatest extent).

a. $CO + Cl_2 \rightleftharpoons COCl_2$	$K_c = 13.8$	
b. $N_2O_4 \rightleftharpoons 2NO_2$	$K_c = 2.1 \times 10^{-4}$	
c. $2NOCl \rightleftharpoons 2NO + Cl_2$	$K_c = 4.7 \times 10^{-4}$	

Solution: First, the larger the value of K_c, the more products there are at equilibrium compared to reactants, and the farther a reaction will proceed toward completion (the greater the extent of reaction). Therefore, b < c < a.

PROBLEM TYPE 2: EQUILIBRIUM EXPRESSIONS.

A. Heterogeneous Equilibria.

Whenever a reaction in a closed container involves reactants and products that exist in different phases, it is called a heterogeneous reaction, and a **heterogeneous equilibrium** will result. For example, when steam is brought into contact with charcoal the following equilibrium is established in a closed reaction vessel.

$$C(s) + H_2O(g) \rightleftharpoons H_2(g) + CO(g)$$

The equilibrium constant expression for this reaction will *not* be the same as for a homogeneous reaction. The usual expression for the constant would be

$$K_c = \frac{[H_2][CO]}{[C][H_2O]}$$

However, the concentration of a pure solid is itself a constant, and is not changed by the addition to, or removal of, some of that solid so long as some solid remains throughout the reaction. Remember that solids have their own volume, and do not fill their containers as gases do. The concentration (mol/L) of a solid such as charcoal depends only on its density.

This means that the equilibrium constant expression can be written:

$$K_c = K[C] = \frac{[H_2][CO]}{[H_2O]}$$

where the constant concentration of the pure solid [C] has been combined with the equilibrium constant. Thus, the equilibrium constant expression does not contain the solid. In general, *concentrations of solids and pure liquids do not appear in equilibrium constant expressions.* In this reaction, equilibrium will be maintained as long as some C(s) is present. The amount of solid carbon present does not affect the position of equilibrium.

STUDY QUESTION 15.3
Write the equilibrium constant expressions for the following reversible reaction:
$$BaO(s) + CO_2(g) \rightleftharpoons BaCO_3(s).$$

Solution: Remember that in equilibrium constant expressions, concentrations of products are written in the numerator, concentrations of reactants are written in the denominator and each is raised to the power equal to their stoichiometric coefficient in the balanced equation. Include only the concentration terms for gaseous components; leave out the concentrations of pure liquids and solids as they are not included in equilibrium constant expressions.

$$K_c = \frac{1}{[CO_2]}$$

> *PRACTICE QUESTION 15.2*
> **Write the equilibrium constant expressions for the following equations.**
>
> a. $2HgO(s) \rightleftharpoons 2Hg(\ell) + O_2(g)$
> b. $Ni(s) + 4CO(g) \rightleftharpoons Ni(CO)_4(g)$

B. Manipulating Equilibrium Expressions.

The equilibrium constant expression and its value depend on how the equation is balanced. An equation can be balanced with more than one set of coefficients, as shown below:

$$2SO_2 + O_2 \rightleftharpoons 2SO_3$$
$$SO_2 + \tfrac{1}{2}O_2 \rightleftharpoons SO_3$$

How does this affect the equilibrium constant? For the latter equation, the equilibrium constant expression is written:

$$K_c' = \frac{[SO_3]}{[SO_2][O_2]^{1/2}}$$

where the prime (') is just to help us keep track of the constant to which we are referring. Note that K_c' is the square root of K_c from the previous section.

$$K_c = \frac{[SO_3]^2}{[SO_2]^2[O_2]}$$

$$K_c' = \frac{[SO_3]}{[SO_2][O_2]^{1/2}}$$

Therefore, by inspection:

$$K_c' = \sqrt{K_c}$$

The equilibrium constant expression for the reaction written in the reverse direction

$$2SO_3 \rightleftharpoons 2SO_2 + O_2$$

is:

$$K_c'' = \frac{[SO_2]^2[O_2]}{[SO_3]^2}$$

Compare this to K_c for the original forward reaction:

$$K_c = \frac{[SO_3]^2}{[SO_2]^2[O_2]}$$

By inspection, you can tell that K_c'' is the reciprocal of K_c for the forward reaction.

$$K_c'' = \frac{[SO_2]^2[O_2]}{[SO_3]^2} = \frac{1}{K_c}$$

Always use the K_c expression and value that are consistent with the way in which the equation is written and balanced.

STUDY QUESTION 15.4
For the following reaction, $H_2(g) + I_2(g) \rightleftharpoons 2HI(g)$, $K_c = 64$ at 400 °C. What is the equilibrium constant value for the reverse reaction?

Solution: First, let's write the reverse reaction. $2HI(g) \rightleftharpoons H_2(g) + I_2(g)$

$$K_c' = \frac{[H_2][I_2]}{[HI]^2}$$

Therefore,

$$K_c' = \frac{1}{K_c} = \frac{1}{64} = 0.016.$$

PRACTICE QUESTION 15.3

For the reaction; $H_2(g) + Br_2(g) \rightleftharpoons 2HBr(g)$, $K_p = 7.1 \times 10^4$ at 700 K. What is the value of K_p for the following reactions at the same temperature?
a. $2HBr(g) \rightleftharpoons H_2(g) + Br_2(g)$
b. ½ $H_2(g)$ + ½ $Br_2(g) \rightleftharpoons HBr(g)$

When the product molecules of one equilibrium reaction become reactants in a second equilibrium process, we have an example of multiple equilibria. In such a case, the overall reaction is the sum of the two individual reactions.

For instance, consider the following reactions at 700 °C:

$$NO_2 \rightleftharpoons NO + \tfrac{1}{2}O_2 \quad K_1 = 0.012$$

followed by

$$SO_2 + \tfrac{1}{2}O_2 \rightleftharpoons SO_3 \quad K_2 = 20$$

The overall reaction is the sum of the two steps:

$$NO_2 + SO_2 \rightleftharpoons NO + SO_3 \quad K_c = ?$$

The equilibrium constant value for the overall reaction is given by *the product of the equilibrium constants of the individual reactions.*

$$K_c = K_1 K_2 = (0.012)(20) = 0.24$$

Guidelines for Writing Equilibrium Constant Expressions

- The concentrations of the reacting species in the solution phase are expressed in mol/L. In the gaseous phase, the concentrations can be expressed in mol/L or in atm. The constant K_c is related to K_p by a simple equation.
- The concentrations of pure solids, pure liquids, and solvents are constants, and do not appear in equilibrium constant expressions.
- The equilibrium constant (K_c or K_p) is a dimensionless quantity.
- In stating a value for the equilibrium constant, we must specify the balanced equation and the temperature.
- If a reaction can be expressed as the sum of two or more reactions, the equilibrium constant for the overall reaction is given by the product of the equilibrium constants of the individual reactions.

STUDY QUESTION 15.5
Consider the following equilibria:

$A(g) + 2B(g) \rightleftharpoons 2C(g)$	$K_c' = 10$	(equation 1)
$A(g) + B(g) \rightleftharpoons D(g)$	$K_c' = 0.10$	(equation 2)

What is the value of K_c for the reaction :
$$2D(g) \rightleftharpoons A(g) + 2C(g) \qquad \text{(equation 3)}$$
Solution: To create the reaction equation 3 from the first two reactions, equation 2 must be doubled and flipped. Then equation 1 and modified equation 2 can be added.

$$A(g) + 2B(g) \rightleftharpoons 2C(g) \qquad K_c' = 10$$

$$2D(g) \rightleftharpoons 2A(g) + 2B(g) \qquad K_c'' = (1/0.10)^2 \quad \text{(equation 4)}$$

$$2D(g) \rightleftharpoons A(g) + 2C(g) \qquad K_c = 10 \times 100 = 1000$$

Please note that doubling a reaction equation squares the value of K_c'' and when a reaction is flipped, take the reciprocal of K. This is demonstrated in equation 4. When equations are added to give equation 3, the K's are multiplied together.

> ### PRACTICE QUESTION 15.4
> **The following equilibrium constants were determined at 1123 K:**
> $$C(s) + CO_2(g) \rightleftharpoons 2CO(g) \qquad K_c = 1.4 \times 10^{12}$$
> $$CO(g) + Cl_2(g) \rightleftharpoons COCl_2(g) \qquad K_c = 5.5 \times 10^{-1}$$
> **Write the equilibrium constant expression K_c and calculate the equilibrium constant at 1123 K for the following reaction.**
> $$C(s) + CO_2(g) + 2Cl_2(g) \rightleftharpoons 2COCl_2(g)$$

C. Gaseous Equilibria.

For reactions involving gases, the equilibrium constant expression can be written in terms of the partial pressures, P_i, of each gaseous component. For the equation:

$$2SO_2 + O_2 \rightleftharpoons 2SO_3$$

$$K_p = \frac{P_{SO_3}^2}{P_{SO_2}^2 P_{O_2}}$$

The pressure, in atmospheres, of each gas at equilibrium is raised to a power corresponding to its coefficient in the balanced equation. The subscript p in K_p indicates that the equilibrium constant has a value that was calculated using equilibrium partial pressures in atmospheres, rather than units of moles per liter.

The values of K_p and K_c are related. Recall that for an ideal gas, $PV = nRT$, and so the pressure of an ideal gas is proportional to its concentration:

$$P = \frac{n}{V} RT$$

where $n/V =$ moles per liter. Substitution of $(n/V)RT$ for the pressure of each gas, gives the following equation relating K_p and K_c,

$$K_p = K_c (RT)^{\Delta n}$$

where R is the ideal gas constant (0.0821 L·atm/K·mol) and Δn is the change in the number of moles of gas when going from reactants to products. For the preceding reaction $\Delta n = -1$, so at 700 K, K_p is given by

$$K_p = K_c \, (RT)^{-1} = \frac{K_c}{RT} = \frac{4.3 \times 10^6}{(0.0821)(700)} = 7.5 \times 10^4$$

In general, $K_p \neq K_c$ except in the case when $\Delta n = 0$

STUDY QUESTION 15.6

What are the values of K_p and K_c at 1000 °C for the reaction, $CaCO_3(s) \rightleftharpoons CaO(s) + CO_2(g)$, if the pressure of CO_2 in equilibrium with $CaCO_3$ and CaO is 3.87 atm?

Solution: Enough information is given to find K_p. Writing the K_p expression for this heterogeneous reaction:

$$K_p = P_{CO_2} = 3.87$$

Then to get K_c, rearrange the equation,

$$K_p = K_c \, (RT)^{\Delta n}$$

where Δn, the change in the number of moles of gas in the reaction is +1.

$$K_c = [CO_2] = \frac{K_p}{(RT)^{\Delta n}} = \frac{3.87}{[(0.0821)(1273)]^1} = 0.0370$$

> ***PRACTICE QUESTION 15.5***
>
> **K_p for the decomposition of ammonium chloride at 427 °C is**
>
> $$NH_4Cl(s) \rightleftharpoons NH_3(g) + HCl(g) \quad K_p = 4.8$$
>
> **Calculate K_c for this reaction.**

PROBLEM TYPE 3: CHEMICAL EQUILIBRIUM AND FREE ENERGY.

A.   Using Q and K to Predict the Direction of Reaction.

Equilibrium constants provide useful information about chemical reaction systems. In this section, we will use them to predict the direction a reaction will proceed to establish equilibrium. We will also use them to determine the extent of reaction.

The reaction quotient, Q_c, is a useful aid in predicting whether or not a reaction system is at equilibrium. Take, for instance, the reaction,

$$2SO_2 + O_2 \rightleftharpoons 2SO_3$$

The reaction quotient is:

$$Q_c = \frac{[SO_3]_0^2}{[SO_2]_0^2[O_2]_0}$$

You will notice that Q has the same algebraic form of the concentrations terms as does K_c. The difference is that these concentrations substituted into this expression are not necessarily equilibrium concentrations. When a set of initial concentrations is substituted into the reaction quotient, Q_c takes on a certain value. In order to predict whether the system is at equilibrium at this point or not, the magnitude of Q_c must be compared with that of K_c.

- When $Q_c = K_c$, the reaction is at equilibrium, and no net reaction will occur.
- When $Q_c > K_c$, the system is not at equilibrium, and a net reaction will occur in the reverse direction until $Q_c = K_c$.
- When $Q_c < K_c$, the system is not at equilibrium, and a net reaction will occur in the forward direction until $Q_c = K_c$.

STUDY QUESTION 15.7
At a certain temperature the reaction, $CO(g) + Cl_2(g) \rightleftharpoons COCl_2(g)$, $K_c = 13.8$. Is the following mixture an equilibrium mixture? If not, in which direction (forward or reverse) will reaction occur to reach equilibrium? $[CO]_0 = 2.5\ M$; $[Cl_2]_0 = 1.2\ M$; and $[COCl_2]_0 = 5.0\ M$

Solution: By substitute the given concentrations into the reaction quotient for the reaction, and determine Q_c.

$$Q_c = \frac{[COCl_2]_0}{[CO]_0[Cl_2]_0} = \frac{(5.0)}{(2.5)(1.2)} = 1.7$$

Compare Q_c to K_c. Since $Q_c < K_c$ the reaction mixture is not an equilibrium mixture. The product concentrations are too low and a net forward reaction will occur until $Q_c = K_c$, and the reaction reaches equilibrium.

PRACTICE QUESTION 15.6
At 700 K, $K_c = 4.3 \times 10^6$ for the following the reaction,
$$2SO_2(g) + O_2(g) \rightleftharpoons 2\,O_3(g)$$
Is a mixture with the following concentrations at equilibrium?
$[SO_2] = 0.10\ M$; $[SO_3] = 10\ M$; $[O_2] = 0.10\ M$. If not at equilibrium, predict the direction in which a net reaction will occur to reach equilibrium.

B. **Relationship between ΔG and $\Delta G°$.**
Recall that $\Delta G°$ refers to the standard free–energy change. All the values we have calculated so far relate to processes in which the reactants are present in their standard states and are converted to products in their standard states. However, in many cases neither the reactants nor the products are present at standard concentration (1 M) and standard pressure (1 atm). Under nonstandard state conditions, we use the symbol ΔG.

The relationship between ΔG and $\Delta G°$ is:

$$\Delta G = \Delta G° + RT \ln Q$$

where R is the gas constant (8.314 J/K·mol), T is the absolute temperature, and Q is the reaction quotient. For a certain reaction at a given temperature the value of $\Delta G°$ is constant, but the value of Q depends on the composition of the reacting mixture; therefore, ΔG will depend on Q. To calculate ΔG, first find $\Delta G°$, then calculate Q from the given concentrations of reactants and products and substitute into the preceding equation.

STUDY QUESTION 15.8
The standard free energy change for the reaction,

$$\frac{1}{2} N_2(g) + \frac{3}{2} H_2(g) \rightleftharpoons NH_3(g)$$

is ΔG°_{rxn} = 26.9 kJ/mol at 700 K. Calculate ΔG at 700 K if the reaction mixture consists of 30.0 atm of H_2, 20.0 atm of N_2, and 0.500 atm of NH_3.

Solution: Under nonstandard conditions, ΔG is related to the reaction quotient Q by the equation:

$$\Delta G = \Delta G^\circ + RT \ln Q_p$$

where,

$$Q_p = \frac{P_{NH_3}}{P_{N_2}^{1/2} P_{H_2}^{3/2}} = \frac{(0.500)}{(20.0)^{1/2}(30.0)^{3/2}} = 6.80 \times 10^{-4}$$

Since ΔG° = 26.9 kJ/mol, substitution yields:

$$\Delta G = 26.9 \text{ kJ/mol} + (8.314 \text{ J/K·mol})(700 \text{ K}) \ln (6.80 \times 10^{-4})$$

$$= 26.9 \text{ kJ/mol} - (42,400 \text{ J/mol} \times \frac{1 \text{ kJ}}{10^3 \text{ J}}) = 26.9 \text{ kJ/mol} - 42.4 \text{ kJ/mol}$$

$$= -15.5 \text{ kJ/mol}$$

NOTE:
By making the partial pressures of N_2 and H_2 high and that of NH_3 low, the reaction is spontaneous in the forward direction. This condition corresponds to $Q_p < K_p$, and so the reaction proceeds in the forward direction until $Q_p = K_p$.

C. Relationship between ΔG° and K.
Under special conditions, equation 18.14 reduces to an extremely important relationship.
At equilibrium, $\Delta G = 0$ and therefore, $Q = K$. The equation then becomes:

$$0 = \Delta G^\circ + RT \ln K$$

or

$$\Delta G^\circ = - RT \ln K$$

This equation relates the equilibrium constant of a reaction to its standard free–energy change. Thus, if ΔG° can be calculated, K can be determined, and vice versa. In the equation K_p is used for gases and K_c for reactions in solution.
Three possible relationships exist between ΔG° and K, because ΔG° can be negative, positive, or zero.

1. When ΔG° is *negative*, ln K is positive and $K > 1$. The products are favored over reactants at equilibrium. The extent of reaction is large.
2. When ΔG° is *positive*, ln K is negative and $K < 1$. The reactants are favored over products at equilibrium. The extent of reaction is small.
3. When $\Delta G^\circ = 0$, ln K is zero and $K = 1$. The reactants and products are equally favored at equilibrium.

When calculating ΔG and ΔG° using the two equations listed on the previous page we need to express both free energy changes in units of kJ/mol. The "per mole" refers to "a mole of reaction" which is the reaction of exactly the molar amounts written in the balanced equation.

STUDY QUESTION 15.9
The standard free energy change for the reaction,

$$\frac{1}{2}N_2(g) + \frac{3}{2}H_2(g) \rightleftharpoons NH_3(g)$$

is ΔG_{rxn}° = 26.9 kJ/mol at 700 K. Calculate the equilibrium constant at this temperature.

Solution: The equilibrium constant is related to the standard free energy change by the equation:

$$\Delta G_{rxn}^{\circ} = -RT \ln K_p$$

Since the gas constant R has units involving joules and the free–energy change has units involving kilojoules, we must be careful to use consistent units. In terms of joules, we get

$$26.9 \times 10^3 \text{ J/mol} = -(8.314 \text{ J/mol·K})(700 \text{ K}) \ln K_p$$

$$-4.62 = \ln K_p$$

Taking the antilog of both sides:

$$K_p = e^{-4.62} = 9.8 \times 10^{-3}$$

PRACTICE QUESTION 15.7
The autoionization of water at 25 °C has the equilibrium constant,
$$2H_2O(\ell) \rightleftharpoons H_3O^+(aq) + OH^-(aq) \quad K_c = 1.0 \times 10^{-14}$$
Calculate the value of ΔG° for this reaction.

PROBLEM TYPE 4: CALCULATING EQUILIBRIUM CONCENTRATIONS.

The equilibrium constant for a reaction can be determined by measuring the concentrations of all components at equilibrium and substituting these values into the K_c expression. Two situations arise. In one, all of the equilibrium concentrations are given and a straightforward calculation yields the equilibrium constant. In the second, the initial concentrations of all reactants are given along with the equilibrium concentration of only one product or one reactant. You need to determine the other equilibrium concentrations first, and then K_c.

STUDY QUESTION 15.10
Ammonia is synthesized from hydrogen and nitrogen. An equilibrium mixture at a given temperature was analyzed and the following concentrations were found: 0.31 mol N_2/L; 0.90 mol H_2/L; and 1.4 mol NH_3/L. What is the equilibrium constant value?

Solution: First, we need to write the balanced chemical equation.

$$3H_2(g) + N_2(g) \rightleftharpoons 2NH_3(g)$$

When the concentrations of each component of a chemical system at equilibrium are known, the value of K_c can be determined readily by substituting these concentrations into the equilibrium constant expression.

$$K_c = \frac{[NH_3]^2}{[H_2]^3[N_2]}$$

$$K_c = \frac{(1.4)^2}{(0.90)^3(0.31)} = 8.7$$

STUDY QUESTION 15.11
At 400 °C, K_c = 64 for the reaction, $H_2(g) + I_2(g) \rightleftharpoons 2HI(g)$. If at equilibrium, the partial pressures of H_2 and I_2 in a container are 0.20 atm and 0.50 atm, respectively, what is the partial pressure of HI in the mixture?

Solution: Writing the equilibrium constant expression:

$$K_p = \frac{P_{HI}^2}{P_{H_2} P_{I_2}} \quad 64$$

and substituting the given pressures:

$$\frac{P_{HI}^2}{(0.20)(0.50)} = 64$$

the partial pressure of HI is

$$P_{HI} = \sqrt{(0.20)(0.50)(64)}$$

$$P_{HI} = 2.5 \text{ atm}$$

STUDY QUESTION 15.12
When 3.0 mol of I_2 and 4.0 mol of Br_2 are placed in a 2.0 L reaction chamber at 150 °C, the following reaction occurs until equilibrium is reached,
$$I_2(g) + Br_2(g) \rightleftharpoons 2\,IBr(g).$$
Chemical analysis then shows that the reactor contains 3.2 mol of IBr. What is the equilibrium constant K_c for the reaction?

Solution: In these types of problems, it will be very helpful to use the following approach.

---

1. Express the equilibrium concentrations of all species in terms of the initial concentrations and an unknown x, which represents *the change in concentration*, (ICE table in examples).
2. Substitute the equilibrium concentrations derived in part 1 into the equilibrium constant expression, and solve for x. The equilibrium concentration is given by:

   equilibrium concentration = initial concentration ± the change due to reaction equilibrium
   concentration = initial concentration ± x

   where the + sign is used for a product, and the – sign for a reactant.
3. Use x to calculate the equilibrium concentrations of all species.

For this question, we need to know the equilibrium concentrations of I_2, Br_2, and IBr, so that when they are substituted into the equilibrium expression K_c can be determined.

First, find the number of moles of I_2 and Br_2 at equilibrium. Initially, the amounts of each component in the reactor are 3.0 moles of I_2, 4.0 moles of Br_2, and zero moles of IBr. The number of moles of I_2 remaining at equilibrium is given by the initial number of moles of I_2 minus the moles of I_2 reacted. This is also true for Br_2. The information that 3.2 mol IBr is formed tells us that 1.6 mol I_2 and 1.6 mol Br_2 must have reacted. We know this because the balanced equation states that 2 mol IBr are formed whenever 1 mol I_2 and 1 mol of Br_2 react. This is best illustrated using an ICE table:

	I_2	+	Br_2	$\rightleftharpoons$	2 IBr
I	3.0 mol		4.0 mol		0 mol

C	−1.6 mol	−1.6 mol	+ 3.2 mol
E	1.4 mol	2.4 mol	3.2 mol

Substituting the equilibrium concentrations into the equilibrium constant expression gives the K_c value:

$$K_c = \frac{[IBr]^2}{[I_2][Br_2]} = \frac{\left(\dfrac{3.2\ \text{mol}}{2.0\ \text{L}}\right)^2}{\left(\dfrac{1.4\ \text{mol}}{2.0\ \text{L}}\right)\left(\dfrac{2.4\ \text{mol}}{2.0\ \text{L}}\right)}$$

$$K_c = 3.0$$

It is important to realize that the key to finding all the equilibrium concentrations was that the changes in I_2 and Br_2 could be related to the change in IBr through the balanced chemical equation. Keep in mind that the equilibrium concentration equals the initial concentration plus the change in concentration due to reaction to reach equilibrium. The change in concentration is positive for a product, and negative for a reactant.

STUDY QUESTION 15.13

At 400 °C, $K_c = 64$ for the equilibrium, $H_2(g) + I_2(g) \rightleftharpoons 2\,HI(g)$. If 1.00 mol H_2 and 2.00 mol I_2 are introduced into an empty 0.50 L reaction vessel, find the equilibrium concentrations of all components at 400 °C.

Solution: The equilibrium constant expression relates the concentrations H_2, I_2, and HI at equilibrium.

$$K_c = \frac{[HI]^2}{[H_2][I_2]} = 64$$

First we need expressions for the equilibrium concentrations of H_2, I_2, and HI. Begin by tabulating the initial concentrations.

Concentration	H_2 +	I_2 $\rightleftharpoons$	2 HI
Initial	1.00 mol/0.50 L	2.00 mol/0.50 L	0
Change	—	—	—
Equilibrium	—	—	—

Since the answer involves three unknowns, we will relate the concentrations to each other by introducing the variable x. Recall, that the equilibrium concentration = initial concentration ± change in concentration. Let x = the change in concentration of H_2. That is, let x = the number of moles of H_2 reacted per liter. From the coefficients of the balanced equation, we can tell that if the change in H_2 is $-x$, then the change in I_2 must also be $-x$, and the change in HI must be $+2x$.

The next step is to complete the table in units of molarity:

Concentration	H_2 +	I_2 $\rightleftharpoons$	2 HI
Initial (*M*)	2.0	4.0	0
Change (*M*)	$-x$	$-x$	$2x$
Equilibrium (*M*)	$(2.0 - x)$	$(4.0 - x)$	$2x$

Now, substitute the equilibrium concentrations from the table into the K_c expression,

$$K_c = \frac{(2x)^2}{(2.0 - x)(4.0 - x)} = 64$$

and solve for x.

$$\frac{(2x)^2}{x^2 - 6.0x + 8.0} = 64$$

Rearranging, we get:

$$4x^2 = 64x^2 - 384x + 512$$

and grouping yields a quadratic equation:

$$60x^2 - 384x + 512 = 0$$

We will use the general method of solving a quadratic equation of the form:

$$ax^2 + bx + c = 0$$

The root x is given by,

$$x = \frac{-b \pm \sqrt{b^2 - 4ac}}{2a}$$

In this case, $a = 60$, $b = -384$, and $c = 512$. Therefore,

$$x = \frac{-(-384) \pm \sqrt{(-384)^2 - 4(60)(512)}}{2(60)} = \frac{-(-384) \pm \sqrt{2.5 \times 10^4}}{120}$$

$$x = \frac{384 \pm 158}{120} = 1.9 \text{ and } 4.5 \text{ mol/L}$$

Recall that x = the number of moles of H_2 (or I_2) reacted per liter. Of the two answers (roots), only 1.9 is reasonable, because the value 4.5 M would mean that more H_2 (or I_2) reacted than was present at the start. This would result in a negative equilibrium concentration, which is physically meaningless. We therefore use the root $x = 1.9$ M to calculate the equilibrium concentrations:

$$[H_2] = 2.0 - x = 2.0\ M - 1.9\ M = 0.1\ M$$

$$[I_2] = 4.0 - x = 4.0\ M - 1.9\ M = 2.1\ M$$

$$[HI] = 2x = 2(1.9\ M) = 3.8\ M$$

The results can be checked by plugging these concentrations back into the K_c expression to see if $K_c = 64$:

$$K_c = \frac{[HI]^2}{[H_2][I_2]} = \frac{(3.8)^2}{(0.1)(2.1)} = 68$$

Thus, the concentrations we have calculated are correct (the difference between 64 and 68 results from rounding off to maintain the correct number of significant figures). Therefore, our result is correct only to the number of significant figures given in the problem.

PRACTICE QUESTION 15.8
A sample of nitrosyl bromide was heated to 100 °C in a 10.0 L container in order to partially decompose it.
$$2NOBr(g) \rightleftharpoons 2NO(g) + Br_2(g)$$
At equilibrium, the container was found to contain 0.0585 mol of NOBr, 0.105 mol of NO, and 0.0524 mol of Br_2. Calculate the value of K_c.

PRACTICE QUESTION 15.9
The decomposition of NOBr is represented by the equation:
$$2NOBr(g) \rightleftharpoons 2NO(g) + Br_2(g) \qquad K_c = 0.0169$$
At equilibrium the concentrations of NO and Br_2 are 1.05×10^{-2} M and 5.24×10^{-3} M, respectively. What is the concentration of NOBr?

PRACTICE QUESTION 15.10
The brown gas NO_2 and the colorless gas N_2O_4 exist in equilibrium.
$$N_2O_4(g) \rightleftharpoons 2 NO_2(g)$$
0.625 mol of N_2O_4 was introduced into a 5.00 L vessel and was allowed to decompose until it reached equilibrium with NO_2. The concentration of N_2O_4 at equilibrium was 0.0750 M. Calculate K_c for the reaction.

PRACTICE QUESTION 15.11
5.0 mol of ammonia were introduced into a 5.0 L reaction chamber in which it partially dissociated at high temperatures.
$$2NH_3(g) \rightleftharpoons 3H_2(g) + N_2(g)$$
At equilibrium, at a particular temperature, 80.0% of the ammonia had reacted. Calculate K_c for the reaction.

PRACTICE QUESTION 15.12
At 400 °C, the equilibrium constant is 64 for the reaction,
$$H_2(g) + I_2(g) \rightleftharpoons 2HI(g).$$
A mixture of 0.250 mol H_2 and 0.250 mol I_2 was introduced into an empty 0.75 L reaction vessel at 400 °C, find the equilibrium concentrations of all components.

PROBLEM TYPE 5: LE CHÂTELIER'S PRINCIPLE: FACTORS THAT AFFECT EQUILIBRIUM.

A. Addition or Removal of a Substance.
When a reaction reaches a state of chemical equilibrium under a particular set of conditions, then no further change in the concentrations of reactants and products occurs. If a change is made in the conditions under which the system is at equilibrium, chemical change will occur in such a way as to establish a new equilibrium. The chemical changes can qualitatively be predicted using Le Châtelier's principle: When a stress (or change) is applied to a system in a state of dynamic equilibrium, the system will, if possible, shift to a new position of equilibrium in which the stress is partially offset.

Let us apply Le Châtelier's principle to the following reaction, assuming that it is at equilibrium:

$$2NO_2(g) \rightleftharpoons N_2O_4(g)$$

Let's add more N_2O_4 as an illustration of a change in concentration. The concentration of N_2O_4 increases, and the equilibrium is disturbed. The system will respond by using up part of the additional N_2O_4 and forming NO_2. In this case, a net reverse reaction will partially offset the increased N_2O_4 concentration. The net reverse reaction brings the system to a new state of equilibrium. When equilibrium is reestablished, more NO_2 will be present than there was before the N_2O_4 was added. Thus, the position of equilibrium has shifted to the left.

Le Châtelier's principle will predict the direction of the net reaction or "shift in equilibrium" that brings the system to a new equilibrium. In this case, the stress of adding more N_2O_4 was partially offset by a net reverse reaction that consumed some of the additional N_2O_4. The key to the use of Le Châtelier's principle is to recognize which net reaction, forward or reverse, will partially offset the change in conditions.

B. Changes in Volume and Pressure.
The pressure of a system of gases in chemical equilibrium can be increased by decreasing the available volume. This change causes the concentration of all components to increase. The stress will be partially offset by a net reaction that will lower the total concentration of gas molecules. Consider our previous reaction:

$$2NO_2(g) \rightleftharpoons N_2O_4(g)$$

When the molecules of both gases are compressed into a smaller volume, their total concentration increases (this is the stress). A net forward reaction (shift to the right) will bring the system to a new state of equilibrium, in which the total concentration of molecules will be lowered somewhat. This partially offsets the initial stress on the system. Notice that when 2 moles of NO_2 react, only 1 mole of N_2O_4 is formed. When equilibrium is reestablished, more moles of N_2O_4 and fewer moles of NO_2 will be present than before the pressure increase occurred. In this case, the equilibrium has shifted to the right. In general, an increase in pressure by decreasing the volume will result in a net reaction that decreases the total concentration of gas molecules. A special case arises when the total numbers of moles of gaseous products and of gaseous reactants are equal in the balanced equation. In this case, no shift in equilibrium will occur.

The addition of an inert gas to a system at equilibrium will cause an increase in the total pressure within the reactor vessel. However, none of the partial pressures of reactants or products are changed, and so the equilibrium is not upset, and no shifting is needed to bring the system back to equilibrium

C. Changes in Temperature.
If the temperature of a system is changed, a change in the value of K_c occurs. An increase in temperature always shifts the equilibrium in the direction of the endothermic reaction, while a temperature decrease shifts the equilibrium in the direction of the exothermic reaction. Therefore, for endothermic reactions the value of K_c increases with increasing temperature, and for exothermic reactions the value of K_c decreases with increasing temperature. In the case of our example reaction, K_c will decrease as the temperature is increased, because the equilibrium will shift in the direction of the endothermic reaction, that is, in the reverse direction.

$$2NO_2(g) \rightleftharpoons N_2O_4(g) \qquad \Delta H^{\circ}_{rxn} = -58.0 \text{ kJ/mol}$$

We can explain this in terms of Le Châtelier's principle. As heat is added to the system, it represents a stress on the equilibrium. The equilibrium will shift in the direction that will consume some of the added heat. This partially offsets the stress. In this reaction, the equilibrium shifts to the left, and K_c decreases. Remember, of these three types of changes; concentration, pressure, and temperature, only changes in temperature will actually alter the K_c value.

STUDY QUESTION 15.14

For the reaction at equilibrium,

$$2NaHCO_3(s) \rightleftharpoons Na_2CO_3(s) + H_2O(g) + CO_2(g) \qquad \Delta H^\circ_{rxn} = 128 \text{ kJ/mol}$$

state the effects (increase, decrease, no change) of the following stresses on the number of moles of sodium carbonate, Na_2CO_3, at equilibrium in a closed container. Note that Na_2CO_3 is a solid (this is a heterogeneous equation); its concentration will remain constant, but its amount can change.

- **a.** **Removing $CO_2(g)$.**
- **b.** **Adding $H_2O(g)$.**
- **c.** **Raising the temperature.**
- **d.** **Adding $NaHCO_3(s)$.**

Solution: By applying Le Châtelier's principle, we can determine the effect of the stress on the equilibrium.

a. If the CO_2 concentration is lowered, the system will react in such a way as to offset the change. That is, a shift to the right will replace some of the missing CO_2. The number of moles of Na_2CO_3 increases.

b. The addition of $H_2O(g)$ exerts a stress on the equilibrium that is partially offset by a shift in the equilibrium to the left (net reverse reaction). This consumes Na_2CO_3 as well as some of the extra H_2O. The number of moles of Na_2CO_3 decreases.

c. An increase in temperature will increase the K_c value of an endothermic reaction. There is a shift to the right, and more Na_2CO_3 is formed.

d. The position of a heterogeneous equilibrium does not depend on the amounts of pure solids or liquids present. The same equilibrium is reached whether the system contains 1 g of $NaHCO_3(s)$ or 10 g of $NaHCO_3$. No shift in the equilibrium occurs. No change in the amount of Na_2CO_3 occurs.

PRACTICE QUESTION 15.13

Copper can be extracted from its ores by heating Cu_2S in air.

$$Cu_2S(s) + O_2(g) \rightleftharpoons 2Cu(s) + SO_2(g) \qquad \Delta H^\circ_{rxn} = -250 \text{ kJ/mol}$$

Predict the direction of the shift of the equilibrium position in response to each of the following changes in conditions. If no shift occurs, say so.

- **a.** **Adding more $O_2(g)$.**
- **b.** **Compressing the vessel volume in half.**
- **c.** **Raising the temperature.**
- **d.** **Adding more $SO_2(g)$.**
- **e.** **Adding more $Cu(s)$.**
- **f.** **Adding more catalyst.**

PRACTICE QUESTION 15.14

The following reaction is exothermic:

$$2SO_2(g) + O_2(g) \rightleftharpoons 2 SO_3(g)$$

Describe what will happen to the concentration of SO_2 when the temperature is increased.

PRACTICE QUESTION 15.15

Consider the following reaction at 400 °C:

$$H_2O(g) + CO(g) \rightleftharpoons H_2(g) + CO_2(g)$$

$H_2O(g)$, $CO(g)$, $H_2(g)$, and $CO_2(g)$ were put into a flask so that the composition corresponded to an equilibrium mixture. A lab technician added an iron catalyst to the mixture, but was surprised when no additional $H_2(g)$ and $CO_2(g)$ were formed, even after many days. Explain why the technician should not have been surprised.

PRACTICE QUIZ

1. Write the equilibrium constant expressions for the following reactions:
    a. $4NH_3(g) + 3O_2(g) \rightleftharpoons 2N_2(g) + 6H_2O(g)$
    b. $2N_2O(g) + 3O_2(g) \rightleftharpoons 2N_2O_4(g)$
    c. $2ClO_2(g) + F_2(g) \rightleftharpoons 2FClO_2(g)$
    d. $H_2(g) + Br_2(\ell) \rightleftharpoons 2HBr(g)$
    e. $C(s) + CO_2(g) \rightleftharpoons 2CO(g)$
    f. $CuO(s) + H_2(g) \rightleftharpoons Cu(s) + H_2O(g)$

2. The decomposition of $HI(g)$ is represented by the equation

    $$2HI(g) \rightleftharpoons H_2(g) + I_2(g) \quad K_c = 64$$

    If the equilibrium concentrations of H_2 and I_2 at 400 °C are found to be $[H_2] = 4.2 \times 10^{-4} M$ and $[I_2] = 1.9 \times 10^{-3} M$, what is the equilibrium concentration of HI?

3. For the reaction

    $$CH_4(g) + 2H_2S(g) \rightleftharpoons CS_2(g) + 4H_2(g) \qquad K_p = 2.05 \times 10^9 \text{ at } 25 \text{ °C.}$$

    Calculate K_p and K_c, at this temperature, for

    $$2H_2(g) + \frac{1}{2}CS_2(g) \rightleftharpoons H_2S(g) + \frac{1}{2}CH_4(g)$$

4. A 1.00 L vessel initially contains 0.777 moles of $SO_3(g)$ at 1100 K. What is the value of K_c for the following reaction, if 0.520 moles of SO_3 remain at equilibrium?

    $$2SO_3(g) \rightleftharpoons 2SO_2(g) + O_2(g)$$

5. Initially, a 1.0 L vessel contains 10.0 moles of NO and 6.0 moles of O_2 at a certain temperature. They react until equilibrium is established.

    $$2NO(g) + O_2(g) \rightleftharpoons 2NO_2(g)$$

    At equilibrium, the vessel contains 8.8 moles of NO_2. Determine the value of K_c at this temperature.

6. Given the reaction:

    $$N_2 + O_2 \rightleftharpoons 2NO \qquad K_c = 2.5 \times 10^{-3} \text{ at } 2130 \text{ °C}$$

    Decide whether the following mixture is at equilibrium, or if a net forward or reverse reaction will occur. $[NO] = 0.005; [O_2] = 0.25; [N_2] = 0.020$ mol/L.

7. Hydrogen iodide decomposes according to the equation:

    $$2HI(g) \rightleftharpoons H_2(g) + I_2(g) \quad K_c = 0.0156 \text{ at } 400 \text{ °C}$$

    A 0.55 mol sample of HI was injected into a 2.0 L vessel held at 400 °C. Calculate the concentration of HI at equilibrium.

8. For the reaction,

$$N_2(g) + O_2(g) \rightleftharpoons 2NO(g) \qquad K_p = 3.80 \times 10^{-4} \text{ at 2000 °C.}$$

What equilibrium pressures of N_2, O_2, and NO will result when a 10 L reactor vessel is filled with 2.00 atm of N_2 and 0.400 atm of O_2 and the reaction is allowed to come to equilibrium?

9. Arrange the following reactions in their increasing tendency to proceed toward completion:

   a. $2HF \rightleftharpoons H_2F_2$ $K_c = 1 \times 10^{-13}$

   b. $2H_2 + O_2 \rightleftharpoons 2H_2O$ $K_c = 3 \times 10^{81}$

   c. $2NOCl \rightleftharpoons 2NO + Cl_2$ $K_c = 4.7 \times 10^{-4}$

10. Which of the following three reactions will have the largest equilibrium constant?
    a. $N_2 + O_2 \rightleftharpoons 2\ NO$
    b. $N_2 + 2O_2 \rightleftharpoons N_2O_4$
    c. $N_2 + \frac{1}{2}O_2 \rightleftharpoons N_2O$

    Given: $\Delta G_f^{\circ} (NO) = +86.7 \text{ kJ/mol}$

    $\Delta G_f^{\circ} (N_2O_4) = +98.3 \text{ kJ/mol}$

    $\Delta G_f^{\circ} (N_2O) = +103.6 \text{ kJ/mol}$

11. Given the equilibrium constant at 400 °C for the reaction:

$$H_2(g) + I_2(g) \rightleftharpoons 2HI(g) \qquad K_p = 64$$

Calculate the value of ΔG_{rxn}° at this temperature.

12. Calculate ΔG_{rxn}° and K_p at 25 °C for the following reaction:

$$NO(g) + \frac{1}{2}O_2(g) \rightarrow NO_2(g)$$

13. For the reaction,

$$2SO_2(g) + O_2(g) \rightleftharpoons 2SO_3(g) \qquad K_p = 7.4 \times 10^4 \text{ at 700 K}$$

a. If, in a reaction vessel at 700 K, we have the following partial pressures, what is ΔG?

$$P_{SO_2} = 1.2 \text{ atm} \qquad P_{O_2} = 0.5 \text{ atm} \qquad P_{SO_3} = 50 \text{ atm}$$

b. Predict the direction of reaction.

14. Use the values listed in Appendix 2 to calculate ΔG at 250°C for the reaction

$$2C_4H_{10}(g) + 13O_2(g) \rightleftharpoons 8CO_2(g) + 10H_2O(g)$$

when $P_{H_2O} = 0.150$ atm; $P_{CO_2} = 0.120$ atm; $P_{C_4H_{10}} = 0.680$ atm.

15. The K_a for HF(aq) is 7.1×10^{-4}. Calculate ΔG for the acid ionization reaction of a 0.250 M aqueous solution of HF having a pH or 2.3 at 25°C.

16. What changes in the equilibrium composition of the reaction,

$$2SO_2(g) + O_2(g) \rightleftharpoons 2SO_3(g) \qquad \Delta H^{\circ}_{rxn} = -197 \text{ kJ/mol}$$

will occur if it experiences the following stresses?
a. The partial pressure of $SO_3(g)$ is increased.
b. Inert Ar gas is added.
c. The temperature of the system is decreased.
d. The total pressure of the system is increased by reducing the available volume.
e. The partial pressure of $O_2(g)$ is decreased.

17. For the chemical equilibrium,

$$PCl_5(g) \rightleftharpoons PCl_3(g) + Cl_2(g) \qquad \Delta H^{\circ}_{rxn} = 92.9 \text{ kJ/mol}$$

a. What is the effect on K_c of lowering the temperature?
b. What is the effect on the equilibrium concentration of PCl_3 of adding Cl_2?
c. What is the effect on the equilibrium concentrations of compressing the mixture to a smaller volume?
d. What is the effect on the equilibrium pressure of Cl_2 of removing PCl_3?

18. In a closed container at 1900 K nitrogen reacts with oxygen to yield NO:

$$N_2(g) + O_2(g) \rightleftharpoons 2NO(g) \qquad\qquad K_p = 2.3 \times 10^{-4}$$

However, $NO(g)$ quickly reacts with oxygen to produce $NO_2(g)$:

$$2NO(g) + O_2(g) \rightleftharpoons 2NO_2(g) \qquad K_p = 1.3 \times 10^{-4}$$

Calculate K_p and K_c at 1900 K for the net reaction.

$$N_2(g) + 2O_2(g) \rightleftharpoons 2NO_2(g)$$

19. A student prepared a saturated aqueous solution of KCl. Indicate whether there will be a change in the equilibrium when the following perturbations are applied:

$$KCl(s) \rightleftharpoons K^+(aq) + Cl^-(aq)$$

a. More water is added to the solution.
b. More solid is added to the solution.
c. The solution is heated.

20. Consider the reaction

$$2NH_3(g) \rightleftharpoons N_2(g) + 3H_2(g)$$

Initially, ammonia is introduced to a container at a pressure of 450 torr and kept at a constant temperature. When equilibrium was achieved, the partial pressure of H_2 in the system was 285 torr. Calculate K_p for this reaction.

Chapter 16
Acids, Bases and Salts

PROBLEM-SOLVING STRATEGIES AND TUTORIAL SOLUTIONS

TYPES OF PROBLEMS
Problem Type 1:   Brønsted Acids and Bases.

Problem Type 2:   Molecular Structure and Acid Strength.
    A.   Hydrohalic Acids.
    B.   Oxoacids.
    C.   Caboxylic Acids.

Problem Type 3:   The Acid–Base Properties of Water.

Problem Type 4:   The pH and pOH Scales.

Problem Type 5:   Strong Acids and Bases.
    A.   Strong Acids.
    B.   Strong Bases.

Problem Type 6:   Weak Acids and Acid Ionization Constants.
    A.   The Ionization Constant, K_a.
    B.   Percent Ionization and Calculating pH from K_a.
    C.   Using pH to Determine K_a.

Problem Type 7:   Weak Bases and Base Ionization Constants.
    A.   The Ionization Constant, K_b.
    B.   Calculating pH from K_b.
    C.   Using pH to Determine K_b.

Problem Type 8:   Conjugate Acid–Base Pairs.
    A.   The Strength of a Conjugate Acid or Base.
    B.   The Relationship Between K_a and K_b of a Conjugate Acid–Base Pair.

Problem Type 9:   Diprotic and Polyprotic Acids.

Problem Type 10: Acid–Base Properties of Salt Solutions.
    A.   Basic Salt Solutions.
    B.   Acidic Salt Solutions.
    C.   Neutral Salt Solutions.
    D.   Salts in Which Both the Cation and the Anion Hydrolyze.

Problem Type 11: Lewis Acids and Bases.

PROBLEM TYPE 1: BRØNSTED ACIDS AND BASES.

According to the Brønsted acid–base theory, an **acid** is defined as a substance that can donate a proton (H^+) to another substance. A **base** is a substance that can accept a proton from another substance. For example, when HBr(g) dissolves in water, it donates a proton to the solvent. Therefore, HBr is a Brønsted acid.

$$HBr(aq) + H_2O(\ell) \rightleftharpoons H_3O^+ (aq) + Br^- (aq)$$

acid   base

The water molecule is the proton acceptor. Therefore, water is a Brønsted base in this reaction. All acid–base reactions, according to Brønsted, involve the transfer of a proton. When a water molecule accepts a proton, a **hydronium ion** (H_3O^+) is formed.

A **conjugate acid–base pair** consists of two species that differ from each other by one H^+ unit. HBr and Br^- in the above example are a conjugate acid–base pair. Removing a proton (H^+) from the acid (HBr) gives its conjugate base (Br^-). H_3O^+ and H_2O are also a conjugate acid–base pair.

Substances that can behave as an acid in one proton transfer reaction and as a base in another reaction are called **amphoteric**. Water is an amphoteric substance. When ammonia dissolves in water, for example, a proton transfer reaction takes place. Only this time water is the acid and ammonia is the base.

$$NH_3(aq) + H_2O(\ell) \rightleftharpoons NH_4^+ (aq) + OH^- (aq)$$

The conjugate acid–base pairs in the above reaction are:

$$NH_4^+ / NH_3, \text{ and } H_2O/ OH^-$$

Such pairs are usually labeled as shown below in the following equation:

$$NH_3(aq) + H_2O(\ell) \rightleftharpoons NH_4^+ (aq) + OH^-(aq)$$

base$_1$ acid$_2$ acid$_1$ base$_2$

The subscript 1 designates one conjugate acid–base pair and the subscript 2 the other pair. The double arrow indicates that the reaction is reversible. In the reverse reaction OH^- acts as a base and NH_4^+ acts as an acid.

STUDY QUESTION 16.1
Write the formula of the conjugate base of H_2SO_4.

Solution: A conjugate base has one less proton than its acid. Hence, HSO_4^- is the conjugate base of H_2SO_4. Note that the charge on the conjugate base is one less positive (more negative) than the acid due to the loss of H^+.

STUDY QUESTION 16.2
For the reaction:

$$HSO_4^- (aq) + HCO_3^- (aq) \rightleftharpoons SO_4^{2-} (aq) + H_2CO_3(aq) ,$$

     a.   **Identify the acids and bases for the forward and reverse reactions.**
     b.   **Identify the conjugate acid–base pairs.**

Solution:

     a.   In the forward reaction, HSO_4^- ion is the proton donor, which makes it an acid. The HCO_3^- ion is the proton acceptor, and is a base. In the reverse reaction, H_2CO_3 is the proton donor (acid), and SO_4^{2-} is the proton acceptor (base).

b. $HSO_4^-(aq) + HCO_3^-(aq) \rightleftharpoons SO_4^{2-}(aq) + H_2CO_3(aq)$
   acid$_1$ base$_2$ base$_1$ acid$_2$
   The subscripts 1 and 2 designate the two conjugate acid–base pairs.

PRACTICE QUESTION 16.1
Identify the conjugate of HCO_3^- in the following acid-base reaction:

$$CO_3^{2-} + CH_3COOH \rightleftharpoons HCO_3^- + CH_3COO^-$$

PRACTICE QUESTION 16.2
Identify the Brønsted acid and base on the left side of the following equations, and identify the conjugate partner of each on the right side.

a. $NH_4^+(aq) + H_2O(\ell) \rightleftharpoons NH_3(aq) + H_3O^+(aq)$

b. $HNO_2(aq) + CN^-(aq) \rightleftharpoons NO_2^-(aq) + HCN(aq)$

PRACTICE QUESTION 16.3
For HPO_4^{2-}, identify its

a. **Conjugate acid, and**
b. **Conjugate base.**

PROBLEM TYPE 2: MOLECULAR STRUCTURE AND ACID STRENGTH.

Acid strength refers to the degree to which a substance ionizes in solution to produce H^+.

A. Hydrohalic Acids.
The acid strengths of nonmetal hydrides are related to two features of these molecules. These are the polarity of the H—X bond (X stands for a nonmetal atom), and the strength of the H—X bond. The polarity of an H—X bond increases with the electronegativity of the nonmetal atom.

When comparing acid strength of hydrides in a group of the periodic table versus those within a period, one must consider both the electronegativity of the nonmetal atom and the bond energy. Within a series of hydrides of elements in a group the acid strength increases with increasing atomic radius of the nonmetal. When reading across a row of the periodic table, the situation is reversed. The importance of bond polarity in determining acid strength outweighs that of atomic radius. Increasing acid strength parallels the increasing electronegativity on the nonmetal atom. The smaller decrease in atomic radius is insignificant in comparison to the larger change in electronegativity. The trend in acid strength results from significant increases in bond polarity when going across the period.

STUDY QUESTION 16.3
Which member of each of the following pairs is the stronger acid?

a. **HCl or HBr**
b. **HCl or H$_2$S**

Solution:
a. These acids are nonmetal hydrides of elements within Group 6A of the periodic table. Variations in atomic radius are more important within a group than electronegativity variations. The stronger acid is the one with the greater radius of its nonmetal atom. HBr is the stronger acid.

b. These acids are nonmetal hydrides of elements within the third row of the periodic table. The electronegativity changes more significantly going across a row than does the atomic radius. HCl with its more polar bond is the stronger acid.

B. Oxoacids.

For oxoacids with the same structure, but with different central atoms whose elements are in the same group, acid strength increases with increasing electronegativity of the central atom. Thus, acid strength increases in the following series where the central atom is a halogen element:

$$HOI < HOBr < HOCl$$

In each molecule, the O—H bond strength is approximately the same. In this series the ability of the halogen atom to withdraw electron density from the O—H bond increases with increasing electronegativity. As the O—H bond in a series of acids becomes more polar, the acid strength increases.

For oxoacids that have the same central atom, but differing numbers of attached oxygen atoms, the acid strength increases with increasing oxidation number of the central atom. For example, HNO_3 is a stronger acid than HNO_2.

The oxygen atoms draw electrons away from the nitrogen atom, making it more positive. The more positive the N atom, the more effective it is in withdrawing electrons from the O—H bond. This increases the polarity of the O—H bond, which, in turn, increases acid strength.

STUDY QUESTION 16.4

Which member of each of the following pairs is the stronger acid?

      a. $HClO_3$ or $HBrO_3$
      b. H_3PO_3 or H_3PO_4

Solution:

a. These two oxoacids have the same structure:

$$\underset{\displaystyle O-Cl-O-H}{\overset{\displaystyle O}{|}} \qquad \underset{\displaystyle O-Br-O-H}{\overset{\displaystyle O}{|}}$$

Within this group, the strengths of the acids increase with increasing electronegativity of the central atom. The electronegativity of Cl is greater than that of Br. $HClO_3$ is stronger.

b. H_3PO_3 and H_3PO_4 are oxoacids of the same element. The acid strength increases as the number of O atoms increases. H_3PO_4 is the stronger acid because of the greater number of oxygen atoms bonded to the phosphorus atom. In H_3PO_4, phosphorus has a higher oxidation number than in H_3PO_3.

C. Carboxylic Acids.

For organic acids (carboxylic acids), the strength of the acid depends on the polarity of the group attached to the carboxylic acid group. We can represent carboxylic acids with the short–hand notation R–COOH. If the R group contains electron–withdrawing (very electronegative) atoms, the acid becomes stronger than one in which the R group contains only carbon and hydrogen atoms.

PRACTICE QUESTION 16.4
Which member of each pair is the stronger acid?

  a. HNO_2 or HNO_3
  b. CH_4 or SiH_4
  c. $HOBr$ or HOI
  d. CH_4 or NH_3

PROBLEM TYPE 3: THE ACID-BASE PROPERTIES OF WATER.

Pure water is a very weak electrolyte and ionizes according to the equation:

$$H_2O(\ell) \rightleftharpoons H^+(aq) + OH^-(aq)$$

According to the Brønsted theory, the reaction is viewed as a proton transfer from one water molecule to another.

$$H_2O(\ell) + H_2O(\ell) \rightleftharpoons H_3O^+(aq) + OH^-(aq)$$
$$\text{acid}_1 \qquad \text{base}_2 \qquad\qquad \text{acid}_2 \qquad \text{base}_1$$

Since water can act as both an acid and a base, it is an *amphoteric* substance.

This reaction is reversible and H_2O, H_3O^+ and OH^- are in equilibrium. In pure water at 25 °C, $[H^+] = [OH^-] = 1.0 \times 10^{-7}$ *M*. These are low concentrations and tell us that very few H_2O molecules are ionized, and that the equilibrium lies to the left.

At equilibrium, the product of the hydrogen ion concentration and hydroxide ion concentration equals a constant, called the ion–product constant for water, K_w.

$$K_w = [H^+][OH^-] = (1.0 \times 10^{-7})(1.0 \times 10^{-7}) = 1.0 \times 10^{-14}$$

Like other equilibrium constants we treat K_w as unitless. The value of K_w applies to all aqueous solutions at 25 °C. When an acid is added to water, the $[H^+]$ increases. Therefore, the $[OH^-]$ must decrease in order for K_w to remain constant. In acidic solutions $[H^+] > [OH^-]$. Similarly, when a base is added to water, and $[OH^-]$ increases, then $[H^+]$ must decrease. In basic solutions $[OH^-] > [H^+]$. Acidic, basic, and neutral solutions are characterized by the following conditions:

$$\text{neutral} \quad [H^+] = [OH^-]$$
$$\text{acidic} \quad [H^+] > [OH^-]$$
$$\text{basic} \quad [H^+] < [OH^-]$$

The ion product provides a useful relationship for aqueous solutions. If the value of $[H^+]$ is known, then the concentration of OH^- can be calculated. Similarly, the $[H^+]$ ion concentration can be calculated, if the value of $[OH^-]$ is known. The product of the two concentrations in all aqueous solutions is 1.0×10^{-14}.

STUDY QUESTION 16.5
The H^+ ion concentration in a certain solution is 5.0×10^{-5} *M*. What is the OH^- ion concentration?

Solution: The ion–product constant of water is applicable to all aqueous solutions.

$$K_w = [H^+][OH^-] = 1.0 \times 10^{-14}$$

When $[H^+]$ is known, we can solve for $[OH^-]$.

$$[OH^-] = \frac{1.0 \times 10^{-14}}{[H^+]} = \frac{1.0 \times 10^{-14}}{5.0 \times 10^{-5}}$$

$$[OH^-] = 2.0 \times 10^{-10} \ M$$

PRACTICE QUESTION 16.5
Calculate the concentration of OH⁻ ions in an HNO$_3$ solution where [H⁺] = 0.0010 M.

PRACTICE QUESTION 16.6
What are the H⁺ ion and OH⁻ ion concentrations in a 0.0033 M NaOH solution? (Hint: NaOH is a strong electrolyte).

PRACTICE QUESTION 16.7
Arrange the following 0.10 M solutions in order of increasing concentration of OH⁻.
Ba(OH)$_2$ HBr   HNO$_3$ H$_2$SO$_4$ HF   KOH

PROBLEM TYPE 4: THE pH SCALE.

As we have seen in Chapter 9, the **pH** of a solution is defined as the negative logarithm of the hydrogen (hydronium) ion concentration.

$$pH = -\log [H^+]$$

Establishing the pH scale allows us to express concentrations of H⁺ and OH⁻ ions with numbers that normally fall between 0 and 14. First, let's find the pH of a neutral solution. In pure water at 25 °C; [H⁺] = 1 × 10⁻⁷ M. Using the definition of pH given above:

$$pH = -\log (1 \times 10^{-7})$$
$$pH = -(-7.0) = 7.0$$

The pH of a neutral solution is 7.0.

Likewise, for an acidic solution, where, for example, the H⁺ ion concentration is 1 × 10⁻⁵ M, the pH is 5.0.

$$pH = -\log (1 \times 10^{-5}) = -(-5.0) = 5.0$$

All acidic solutions have a pH < 7.0.

When [H⁺] is not an exact power of 10, the pH is not a round number. For example, if in a basic solution [H⁺] = 2.5 × 10⁻⁹ M, the pH is

$$pH = -\log (2.5 \times 10^{-9}) = -(-8.60)$$
$$pH = 8.60$$

Note that all basic solutions have a pH > 7.0. In terms of pH, solutions that are acidic, basic, and neutral are defined as follows:

neutral	pH = 7.0
acidic	pH < 7.0
basic	pH > 7.0

A scale just like the pH scale has been devised for the hydroxide ion concentration, where

$$pOH = -\log [OH^-]$$

Just as the H^+ ion and OH^- ion concentrations are related by the ion-product constant of water, K_w, the pH and pOH are also related.

$$K_w = [H^+][OH^-] = 1.0 \times 10^{-14}$$

pH + pOH = 14.00

The sum of the pH and pOH values of any solution is always 14 at 25 °C.

It is also important to notice that a change in pH of one unit corresponds to a 10–fold change in $[H^+]$. As H^+ drops from 10^{-8} to 10^{-9} M, for instance, the pH changes from 8 to 9. A change of 2.0 pH units corresponds to a 100–fold change in H^+ ion concentration. So, to say, "a pH of 2 is twice as acidic as a pH of 4" would be completely wrong as the pH is a logarithmic function. The solution with pH of 2 is really 100 times more acidic!

Also, recall that $[H^+]$ can be calculated from the pH, using the following equation;

$$[H^+] = 10^{-pH}$$

STUDY QUESTION 16.6
The OH^- ion concentration in a certain ammonia solution is 7.2×10^{-4} M. What is the pOH and pH?

Solution: The pOH is the negative logarithm of the OH^- ion concentration.

$$pOH = -\log [OH^-]$$

$$pOH = -\log (7.2 \times 10^{-4}) = -(-3.14) = 3.14$$

The pH and pOH are related by

$$pH + pOH = 14.00$$
$$pH = 14.00 - pOH = 14.00 - 3.14 = 10.86$$

STUDY QUESTION 16.7
What is the H^+ ion concentration in a solution with a pOH of 3.9?

Solution: Recall that pH + pOH = 14.00. Since the pOH given, the pH can be calculated, and then the H^+ ion concentration can be determined.

$$pH = 14.00 - pOH$$
$$pH = 14.00 - 3.90 = 10.10$$

Since pH is known, the hydrogen ion concentration is

$$[H^+] = 10^{-pH}$$
$$[H^+] = 10^{-10.10} = 7.9 \times 10^{-11} \ M$$

PRACTICE QUESTION 16.8
Consider the following at 25 $^\circ$C.

   0.0000050 M Mg(OH)$_2$
   Solution with pH = 7.3
   0.00000010 M HF
   pure water

Which contains the highest concentration of H$_3$O$^+$(aq) ions?

PROBLEM TYPE 5: STRONG ACIDS AND BASES.

A. Strong Acids.

The stronger the acid, the more completely it ionizes in water, producing H$_3$O$^+$(aq) and an anion. Strong acids are strong electrolytes and ionize completely in water. HBr, for example, is a strong acid.

$$HBr(aq) + H_2O(\ell) \rightarrow H_3O^+ (aq) + Br^-(aq)$$

A solution of hydrobromic acid consists of H$_3$O$^+$ (aq) and Br$^-$ (aq) ions. The concentration of HBr molecules is zero because all have ionized. The concentrations of H$_3$O$^+$ and Br$^-$ are equal to the initial concentration of the acid since the ionization occurs with a 1:1 molar ratio.

The pH of a strong acid solution depends on the hydrogen ion concentration. To calculate the pH of a 0.0052 M HBr solution, we start by determining whether it is a strong or weak acid. Knowing that HBr is a strong acid, when added to water it ionizes completely.

$$HBr(aq) + H_2O(\ell) \rightarrow H_3O^+(aq) + Br^-(aq)$$

For simplicity, we can write the reaction without the H$_2$O molecules.

$$HBr(aq) \rightarrow H^+(aq) + Br^-(aq)$$

Before any ionization occurs, the HBr concentration is 0.0052 M, but after ionization, its concentration is zero. Since each mole of HBr that reacts yields 1 mol of H$^+$ and 1 mol of Br$^-$, the concentrations of H$^+$ and Br$^-$ ions are both 0.0052 M. Or, 0.0052 mol HBr reacts to give 0.0052 mol H$^+$ ion and 0.0052 mol Br$^-$ ion per liter of solution.

The pH is given by:

$$pH = - \log (5.2 \times 10^{-3}) = 2.28$$

B. Strong Bases.

Like a strong acid, a strong base is one that ionizes completely in aqueous solution. KOH, a commonly used base, is an example. It is purchased as a white solid. It dissolves readily in water to give a solution of K$^+$ and OH$^-$ ions:

$$KOH(aq) \rightarrow K^+(aq) + OH^-(aq)$$

A solution of 0.10 M KOH is made by dissolving 0.10 mol of KOH in 1 L of solution. In this solution, [K$^+$] is 0.10 M, [OH$^-$] is 0.10 M, and [KOH] is essentially zero. Of the strong bases, only NaOH and KOH are commonly used in the laboratory.

STUDY QUESTION 16.8
What is the H$^+$ ion concentration in 0.0010 M NaOH?

Solution: Sodium hydroxide is a base and supplies practically all the OH$^-$ ions in the solution. The H$^+$ ions come from the autoionization of water.

$$H_2O(\ell) \rightleftharpoons H^+(aq) + OH^-(aq)$$

Sodium hydroxide is a strong base and therefore, dissociates 100% in aqueous solution.

$$NaOH(aq) \rightarrow Na^+(aq) + OH^-(aq)$$

This means that we can calculate the hydroxide ion concentration first, then we can plug its value in the ion–product constant of water (K_w) to solve for the H$^+$ concentration. When 0.0010 mol of NaOH per liter dissociates, 0.0010 mol Na$^+$ ions and 0.0010 mol OH$^-$ ions are formed per liter. Since the concentrations of H$^+$ and OH$^-$ ions are related to each other:

$$K_w = [H^+][OH^-] = 1.0 \times 10^{-14}$$

Then,

$$[H^+] = \frac{K_w}{[OH^-]} = \frac{1.0 \times 10^{-14}}{1.0 \times 10^{-3}}$$

$$[H^+] = 1.0 \times 10^{-11} \, M$$

PROBLEM TYPE 6: WEAK ACIDS AND ACID IONIZATION CONSTANTS.

Not all acids are strong proton donors in aqueous solutions. A weak acid is one that is only partially ionized in water. Hydrofluoric acid is a weak acid in water:

$$HF(aq) + H_2O(\ell) \rightleftharpoons H_3O^+(aq) + F^-(aq)$$

Notice that for a weak acid (or base) we write the ionization equation as equilibrium and not a complete reaction. Since HF is only partially ionized, the concentration of HF molecules is much greater than the concentrations of H$^+$(aq) and F$^-$(aq) ions. For example, a 1.0 M HF solution has the following concentrations: [H$^+$] = [F$^-$] = 0.026 M, and [HF] = 0.974 M. About 97% of the initial HF molecules are *not* ionized. For weak acids, the ionization equilibrium lies to the left, reactants are favored. The extent of reaction is small.

A. The Ionization Constant, K_a.
Since weak acids are ionized only to a small extent in aqueous solution, their ionization is reversible and can be mathematically described using an equilibrium constant expression. For instance, hydrocyanic acid ionizes in water as follows:

$$HCN(aq) \rightleftharpoons H^+(aq) + CN^-(aq)$$

The equilibrium constant for the ionization of a weak acid is called the acid ionization constant, K_a. The expression for this constant is:

$$K_a = \frac{[H^+][CN^-]}{[HCN]} = 4.9 \times 10^{-10}$$

The value of K_a is experimentally determined, as is done for any other equilibrium constant.

STUDY QUESTION 16.9
Calculate the pH of a 0.125 M aqueous solution of HCN.

Solution: The ionization of the weak acid HCN in aqueous solution is described by the following equilibrium:

$$HCN(aq) + H_2O(l) \rightleftharpoons H_3O^+(aq) + CN^-(aq) \qquad K_a = 4.9 \times 10^{-10}$$

In order to calculate the pH, determine the equilibrium concentration of H_3O^+ by using an ICE table.

	[HCN]	H₂O	[H₃O⁺]	[CN⁻]
Initial, M	0.125		0	0
Change, M	$-x$		$+x$	$+x$
Equilibrium, M	$0.125 - x$		x	x

The expression for the ionization equilibrium is:

$$K_a = \frac{[H_3O^+][CN^-]}{[HCN]} = \frac{x^2}{0.125 - x} = 4.9 \times 10^{-10}$$

This can be solved using a quadratic equation. However, since the value of the K_a is so small in comparison to the initial concentration of the acid (more than three orders of magnitude apart), we can make the approximation that a relatively small amount of the acid is ionized and the amount of the change will be negligible in comparison to the initial concentration.

$$\frac{x^2}{0.125 - x} \gg \frac{x^2}{0.125} = 4.9 \times 10^{-10}$$

$$x^2 = (4.9 \times 10^{-10})(0.125) = 6.1 \times 10^{-11}$$

$$x = \sqrt{6.1 \times 10^{-11}} = 7.8 \times 10^{-6}$$

Since $[H_3O^+] = x$, using the definition of pH ($pH = -\log [H_3O^+]$) we get:

$$pH = -\log (7.8 \times 10^{-6}) = 5.11$$

The K_a value is the quantitative measure of acid strength. The larger the K_a, the greater the extent of the ionization reaction and the stronger the acid. Therefore, acetic acid, with $K_a = 1.8 \times 10^{-5}$, is a stronger acid than hydrocyanic acid, with $K_a = 4.9 \times 10^{-10}$. Strong acids are essentially 100 percent ionized. It will help you to be able to recognize 5 or 6 strong acids by their names and formulas, so that you won't confuse them with weak acids.

STUDY QUESTION 16.10
To which side does the equilibrium lie in the following acid–base reaction?

$$HF(aq) + NO_2^-(aq) \rightleftharpoons HNO_2(aq) + F^-(aq)$$

Solution: First, we need to know which is the stronger acid, the one on the left (HF) or the one on the right (HNO_2) of the reaction. Second, we also need to know which is the stronger base and which is the weaker base. Look this information up in Table 16.7.

$$HF(aq) \ + \ NO_2^-(aq) \ \rightleftharpoons \ HNO_2(aq) + F^-(aq)$$

stronger	stronger	weaker	weaker
acid	base	acid	base

Since proton transfer reactions proceed from the stronger acid and stronger base toward the weaker acid and weaker base, the reaction favors HNO_2 and F^- ion. The equilibrium lies to the right.

B. Percent Ionization and Calculating pH from K_a.
In the previous section we learned about strong acids and bases. The main thing to remember about strong species is that they ionize (almost) to completion. That means that the percent ionization is near 100%. **Percent ionization** is calculated by

$$\text{percent ionization} = \frac{[H^+]}{[HA]_0} \times 100\%$$

For a weak acid, HA, the concentration of the acid that is ionized (A^-) is equal to the concentration of H^+ ions at equilibrium.

$$HA(aq) \rightleftharpoons H^+(aq) + A^-(aq)$$

STUDY QUESTION 16.11
For a 0.31 *M* HF solution,

    a. Calculate the concentrations of of H^+, F^-, and HF.
    b. What is the percent ionization?

Solution:
a. First, write the ionization reaction and the acid ionization constant expression for hydrofluoric acid. The value for K_a can be found in Table 16.7 in your textbook.

$$HF(aq) \rightleftharpoons H^+(aq) + F^-(aq)$$

$$K_a = \frac{[H^+][F^-]}{[HF]} = 7.1 \times 10^{-4}$$

The equilibrium concentrations of all species must be written in terms of a single unknown. We need to set up the ICE table and let x equal the moles of HF ionized per liter of solution. This means that x is the molarity of H^+ at equilibrium, because one H^+ is formed for each HF molecule that ionizes. Also, $[H^+] = [F^-] = x$.

Concentration	HF $\rightleftharpoons$	H^+ +	F^-
Initial (*M*)	0.31	0	0
Change (*M*)	$-x$	$+x$	$+x$
Equilibrium (*M*)	$(0.31-x)$	x	x

The value of x can be determined by substituting the equilibrium concentrations into the K_a expression.

$$K_a = \frac{(x)(x)}{(0.31 - x)} = 7.1 \times 10^{-4}$$

The small magnitude of K_a indicates that very little hydrofluoric acid actually ionizes. After all, it is a weak acid. Therefore, the value of x is much less than 0.31, and $0.31 - x \approx 0.31\ M$. This approximation simplifies the equation, and allows us to avoid solving a quadratic equation.

$$K_a = \frac{x^2}{0.31} = 7.1 \times 10^{-4}$$

Solving for x:

$$x = \sqrt{(0.31)(7.1 \times 10^{-4})} = 1.5 \times 10^{-2}\ M$$

Was our approximation valid? As a rule, if x is equal to or less than 5% of the initial concentration $[HF]_0$, then neglecting x in the calculation does not introduce an unacceptable amount of error into the answer.

$$[HF] = [HF]_0 - x \qquad \text{and} \qquad [HF] \approx [HF]_0$$

To test the validity of the approximation, divide x by the initial concentration of hydrofluoric acid $[HF]_0$.

$$\frac{x}{[HF]_0} \times 100\% = \frac{1.5 \times 10^{-2}\ M}{0.31\ M} \times 100\% = 4.8\%$$

Since x is less than 5 percent of the initial amount of HF, our approximate solution is close enough to the accurate result to be valid. Knowing x, we can utilize the bottom row in the table to answer part (a) of the problem.

$$[H^+] = [F^-] = 1.5 \times 10^{-2}\ M$$
$$[HF] = 0.31\ M - 0.015\ M = 0.30\ M$$

b. The percent ionization was already calculated.

$$\% \text{ ionization} = \frac{x}{[HF]_0} \times 100\% = 4.8\%$$

NOTE:
In this example we have followed three basic steps common to all problems of this type.
1. Write the acid dissociation equilibrium, and express the equilibrium concentration of all species in terms of an unknown (x), where x equals the change in the concentration of HA that is required to reach equilibrium.
2. Substitute these concentrations into the K_a expression, and look up the value of K_a in a table.
3. Solve for x, and calculate all equilibrium concentrations, the pH, and percent ionization.

C. Using pH to Determine K_a.
   Experimentally, the measurement of a pH of a solution yields the equilibrium concentration of the hydronium ion (and, indirectly, the equilibrium concentration of the hydroxide anion through their relationship from the autoionization of water). We can use the pH values to determine the value of the K_a for a weak acid.

STUDY QUESTION 16.12
The pH of a 0.160 M aqueous solution of propanoic acid, $HC_3H_5O_2$, is 2.85. What is the K_a of propanoic acid?

Solution: The equilibrium that describes the ionization of propanoic acid is:

$$HC_3H_5O_2(aq) + H_2O(l) \rightleftharpoons H_3O^+(aq) + C_3H_5O_2^-(aq) \qquad K_a = \frac{[H_3O^+][C_3H_5O_2^-]}{[HC_3H_5O_2]}$$

The equilibrium concentration of $H_3O^+(aq)$ is given by:

$$[H_3O^+] = 10^{-pH} = 10^{-2.85} = 1.4 \times 10^{-3}$$

Let's use an ICE table to complete our calculations. Based on the stoichiometry of the ionization reaction, the equilibrium concentrations of H_3O^+ and $C_3H_5O_2^-$ will be the same. The equilibrium concentration of H_3O^+ corresponds to the change in the concentrations of all the species that participate (x) and we can use it to determine the equilibrium concentration of the weak acid as well.

	$HC_3H_5O_2$	H_2O	H_3O^+	$C_3H_5O_2^-$
Initial, M	0.160		0	0
Change, M	$-x$		$+x$	$+x$
Equilibrium, M	$0.160 - 1.4 \times 10^{-3}$		1.4×10^{-3}	1.4×10^{-3}

Substituting into the expression for the ionization constant we get:

$$K_a = \frac{[H_3O^+][C_3H_5O_2^-]}{[HC_3H_5O_2]} = \frac{\left(1.4 \times 10^{-3}\right)^2}{0.160 - 1.4 \times 10^{-3}} = 1.2 \times 10^{-5}$$

PRACTICE QUESTION 16.9
What is the percent ionization of a weak acid in a 0.020 M HA solution given that its K_a is 2.3×10^{-5}?

PRACTICE QUESTION 16.10
What is the pH of a 0.050 M C_6H_5COOH (benzoic acid) solution?

PROBLEM TYPE 7: WEAK BASES AND BASE IONIZATION CONSTANTS.

Some bases are not ionized completely in aqueous solution. A weak base ionizes to a limited extent in water. Ammonia is an example of a weak base. In a 0.10 M solution of ammonia, only about 1% of the ammonia molecules are ionized. A dominant reverse reaction means that the equilibrium lies to the left.

$$NH_3(aq) + H_2O(\ell) \rightleftharpoons NH_4^+(aq) + OH^-(aq)$$
$$\text{base} \qquad\quad \text{acid}$$

Methylamine is an example of a weak base. It ionizes in water as shown in the following equation:

$$CH_3NH_2(aq) + H_2O(\ell) \rightleftharpoons CH_3NH_3^+(aq) + OH^-(aq)$$

The equilibrium constant for the ionization of a weak base is called the base ionization constant, K_b.

$$K_b = \frac{[CH_3NH_3^+][OH^-]}{[CH_3NH_2]} = 4.4 \times 10^{-4}$$

Table 16.8 in the text lists ionization constant values (K_b) for a number of weak bases. Note that the concentration of water is not shown in the K_b expression. The concentration of water is not measurably affected by the reaction and is essentially a constant.

STUDY QUESTION 16.13
What is the pH of a 0.010 M C$_5$H$_5$N (pyridine) solution?

Solution: Pyridine is a weak base. Write the equation for the reaction of pyridine in water and look up its K_b value in Table 16.8.

$$C_5H_5N + H_2O \rightleftharpoons C_5H_5NH^+ + OH^- \qquad K_b = 1.7 \times 10^{-9}$$

Let x equal the moles of C$_5$H$_5$N ionized per L to reach equilibrium. Note, when x mol/L of C$_5$H$_5$N are ionized, x mol/L of OH$^-$ and x mol/L of C$_5$H$_5$NH$^+$ are formed. These changes are listed in the following table:

Concentration	H$_2$O	+	C$_5$H$_5$N	$\rightleftharpoons$	C$_5$H$_5$NH$^+$	+	OH$^-$
Initial (*M*)			0.010		0		0
Change (*M*)			$-x$		$+x$		$+x$
Equilibrium (*M*)			$(0.010-x)$		x		x

Substitute the equilibrium concentrations into the base constant expression.

$$K_b = \frac{[C_5H_5NH^+][OH^-]}{[C_5H_5N]} = \frac{(x)(x)}{(0.010 - x)} = 1.7 \times 10^{-9}$$

Make the approximation, $[C_5H_5N] = 0.010 - x \approx 0.010\ M$. Substituting

$$\frac{x^2}{0.010} = 1.7 \times 10^{-9}$$

$$x = \sqrt{(0.010)(1.7 \times 10^{-9})}$$

$$x = 4.1 \times 10^{-6}\ M$$

Testing the assumption that the approximation is reasonable:

$$\frac{x}{[C_5H_5N]_0} \times 100\% = \frac{4.1 \times 10^{-6}\ M}{0.010\ M} \times 100\% = 0.041\%$$

Since x is less than 5% of the initial concentration of pyridine, then the assumption is valid. At equilibrium:

$$[OH^-] = x = 4.1 \times 10^{-6}\ M$$

To find the pH, first calculate the pOH, and then use the relationship pH + pOH = 14.

$$pOH = -\log(4.1 \times 10^{-6}) = 5.38$$

Since,

$$pH = 14.00 - pOH$$

Then,

$$pH = 14.00 - 5.38$$

$$pH = 8.62$$

STUDY QUESTION 16.14
A 0.145 M aqueous solution of ethylamine, $C_2H_5NH_2$, has a pH of 11.95. What is the K_b of ethylamine?

Solution: First, let's consider the equilibrium involved in the ionization of this weak base:

$$C_2H_5NH_2(aq) + H_2O(l) \rightleftharpoons C_2H_5NH_3^+(aq) + OH^-(aq)$$

$$K_b = \frac{[C_2H_5NH_3^+][OH^-]}{[C_2H_5NH_2]}$$

Since the hydroxide anion is the species involved in this equilibrium, from the pH and the pK_w we can calculate the pOH and this will give us the equilibrium concentration of hydroxide ions.

$$pOH = 14.00 - 11.80 = 2.05$$

$$[OH^-] = 10^{-pOH} = 8.9 \times 10^{-3}$$

Based on the stoichiometry, an equal amount of the conjugate acid, $C_2H_5NH_3^+$, will be formed. Let's summarize this using an ICE table:

	$C_2H_5NH_2$	H_2O	$C_2H_5NH_3^+$	OH
Initial, M	0.145		0	0
Charge, M	$-x$		$+x$	$+x$
Equilibrium, M	$0.145 - 8.9 \times 10^{-3}$		8.9×10^{-3}	8.9×10^{-3}

$$K_b = \frac{[C_2H_5NH_2][OH^-]}{[C_2H_5NH_2]} = \frac{\left(8.9 \times 10^{-3}\right)^2}{0.145 - 8.9 \times 10^{-3}} = 5.8 \times 10^{-4}$$

PRACTICE QUESTION 16.11
A 0.012 M solution of an unknown base has a pH of 10.10.

    a.   **What are the hydroxide ion and hydronium ion concentrations in the solution?**
    b.   **Is the base a weak base or a strong base?**
    c.   **If it is a weak base, what is the value of its K_b?**

PROBLEM TYPE 8: CONJUGATE ACID-BASE PAIRS.

A. The Strength of a Conjugate Acid or Base.

The strength of an acid is related to the strength of its conjugate base. Let's compare the ionization of HCl, a strong acid, and HF, a weak acid.

$$HCl(aq) + H_2O(\ell) \rightarrow H_3O^+(aq) + Cl^-(aq)$$

$$HF(aq) + H_2O(\ell) \rightleftharpoons H_3O^+(aq) + F^-(aq)$$

Chloride ion is the conjugate base of HCl. Here, we see that it must be a very weak base, because it has no tendency to accept a proton from H_3O^+ in the reverse reaction. In the ionization of the weak acid HF, the reverse reaction occurs to a much greater extent than for HCl. Fluoride ion accepts a proton from H_3O^+. The F^- ion is a much stronger base than Cl^-. The conjugate bases of strong acids have no measurable base strength.

When comparing the strength of different acids, we see that as acid strength increases, the strength of its conjugate base decreases.

STUDY QUESTION 16.15

The strengths of the following acids increase in the order; HCN < HF < HNO₃. Arrange the conjugate bases of these acids in order of increasing base strength.

Solution: We just saw that as the strength of acid increases, the strength of its conjugate base decreases. Increasing acid strength is given from left to right.

$$HCN < HF < HNO_3$$

Therefore, the strengths of their conjugate bases decrease from left to right.

$$CN^- > F^- > NO_3^-$$

decreasing base strength $\longrightarrow$

B. The Relationship Between K_a and K_b of a Conjugate Acid–Base Pair.

The K_a and K_b for *any* conjugate acid–base pair are always related by the equation:

$$K_a K_b = K_w$$

The K_b for a conjugate base of a weak acid is found by rearranging:

$$K_b = \frac{K_w}{K_a}$$

From this equation, you can see that as K_a decreases, K_b increases, an inverse relationship. The weaker the acid, the stronger its conjugate base – a relationship we have noted before.

STUDY QUESTION 16.16

Fluoride ion is a base. What is its K_b value?

Solution: $F^-(aq) + H_2O(\ell) \rightleftharpoons HF(aq) + OH^-(aq)$ $\qquad K_b = ?$

The base ionization constant K_b for F^- is related to the K_a for the conjugate acid HF by:

$$K_b = \frac{[HF][OH^-]}{[F^-]} = \frac{[H^+][OH^-]}{[H^+][F^-]/[HF]} = \frac{K_w}{K_a}$$

Substituting K_a for HF gives:

$$K_b = \frac{K_w}{7.1 \times 10^{-4}} = 1.4 \times 10^{-11}$$

STUDY QUESTION 16.17
Calculate:

a. K_b for NO_2^-.

b. K_a for NH_4^+.

Solution:

a. The K_a value of HNO_2 (the conjugate acid of NO_2^-) is 4.5×10^{-4}.

$$K_w = K_a K_b$$

therefore,

$$K_b = \frac{K_w}{K_a} = \frac{1.0 \times 10^{-14}}{4.5 \times 10^{-4}} = 2.2 \times 10^{-11}$$

b. The K_b value of NH_3 (the conjugate base of NH_4^+) is 1.8×10^{-5}.

$$K_a = \frac{K_w}{K_b} = \frac{1.0 \times 10^{-14}}{1.8 \times 10^{-5}} = 5.6 \times 10^{-10}$$

PRACTICE QUESTION 16.12
Which of the following acids has the strongest conjugate base?
HNO_2, HCN or H_3O^+

PRACTICE QUESTION 16.13
Given that K_a for formic acid (HCOOH) is 1.7×10^{-4}, find the K_b value for formate ion, $HCOO^-$.

PROBLEM TYPE 9: DIPROTIC AND POLYPROTIC ACIDS.

The acids listed in Table 16.7 of the text are monoprotic acids; that is, they produce one proton per acid molecule. Acetic acid (CH_3COOH) contains four hydrogen atoms per molecule, but it is still monoprotic, because only one of the hydrogen atoms is ionized in solution. The H atoms bonded to C do not ionize because the electronegativity difference between C and H is not great enough to produce sufficient positive charge on the hydrogen atoms for ionization.

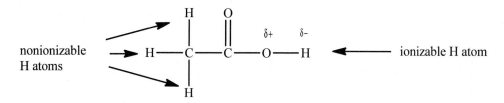

Acids that contain two ionizable H atoms are called diprotic acids. Those with three or more ionizable H atoms are called polyprotic acids. For example, sulfuric acid (H_2SO_4) is a diprotic acid, and phosphoric acid (H_3PO_4) with its three ionizable H atoms is a polyprotic acid. The ionization reactions occur stepwise; the conjugate base in the first step becomes the acid in the second step of ionization. For oxalic acid, $H_2C_2O_4$:

$$H_2C_2O_4 \rightleftharpoons H^+ + HC_2O_4^- \qquad\qquad K_{a1} = 6.5 \times 10^{-2}$$

$$HC_2O_4^- \rightleftharpoons H^+ + C_2O_4^{2-} \qquad\qquad K_{a2} = 6.1 \times 10^{-5}$$

This example is typical in that $K_{a1} \gg K_{a2}$. It is always more difficult to remove an H^+ ion from a negative ion such as $HC_2O_4^-$ than from a neutral molecule such as $H_2C_2O_4$. Table 16.9 of the text lists acid ionization constants for several polyprotic acids.

Here are some important observations concerning polyprotic acids:
1. The major product species in solution come from the first ionization step.
2. Since all reactants and products (ions and nonionized acid molecules) exist in the same solution, they are in equilibrium with each other. At equilibrium, each species can have only one concentration.
3. The concentration of all species in solution must be consistent with both K_{a1} and K_{a2}.

STUDY QUESTION 16.18
Calculate the concentrations of H_2XO_4, HXO_4^-, XO_4^{2-}, and H^+ in a 1.0 M H_2XO_4 solution.

Solution: First, write the equations for the stepwise ionization of H_2SO_4, given the K_a values shown.

$$H_2XO_4 \rightarrow H^+ + HXO_4^- \qquad\qquad K_{a1} = 1.3 \times 10^{-2}$$

$$HXO_4 \rightleftharpoons H^+ + XO_4^{2-} \qquad\qquad K_{a2} = 6.3 \times 10^{-8}$$

Since $K_{a1} \gg K_{a2}$, we assume that essentially all the H^+ comes from step 1. Let x be the concentration of H_2A ionized to reach equilibrium. We solve for x in the usual way.

Concentration	H_2A $\rightleftharpoons$	H^+ $+$	HA^-
Initial (M)	1.0	0	0
Change (M)	$-x$	$+x$	$+x$
Equilibrium (M)	$(1.0 - x)$	x	x

$$K_{a1} = \frac{[H^+][HA^-]}{[H_2A]} = \frac{(x)(x)}{1.0 - x} = 1.3 \times 10^{-2}$$

$$K_{a1} = \frac{x^2}{1.0 - x} = 1.3 \times 10^{-2}$$

For this acid, K_{a1} is quite large, and the percent ionization will be greater than 5%. Rearranging the preceding equation into quadratic form:

$$ax^2 + bx + c = 0$$

$$x^2 + (1.3 \times 10^{-2})x - (1.3 \times 10^{-2}) = 0$$

where,

$$a = 1, b = (1.3 \times 10^{-2}), \text{ and } c = (-1.3 \times 10^{-2}).$$

Substituting into the quadratic formula, and solving for x.

$$x = \frac{-b \pm \sqrt{b^2 - 4ac}}{2a}$$

$$x = \frac{-1.3 \times 10^{-2} \pm \sqrt{(1.3 \times 10^{-2})^2 - 4(1)(-1.3 \times 10^{-2})}}{2(1)}$$

$$x = 0.11 \ M$$

Therefore, at equilibrium,

$$[H^+] = [HA^-] = 0.11 \ M$$

and

$$[H_2A] = 1.00 - 0.11 = 0.89 \ M$$

which corresponds to 11% ionization.

To determine the concentration of A^{2-}, we must consider the second stage of ionization:

$$HA^- \rightleftharpoons H^+ + A^{2-} \qquad K_{a2} = 6.3 \times 10^{-8}$$

Let y be the equilibrium concentration of A^{2-}, and note that $x \gg y$.

Concentration	$HA^- \rightleftharpoons H^+$	$+$	A^{2-}
Initial (M)	0.11	0.11	0
Change (M)	$-y$	$+y$	$+y$
Equilibrium (M)	$(0.11-y)$	$(0.11+y)$	y

$$K_{a2} = \frac{[H^+][A^-]}{[HA]} = \frac{y(0.11 + y)}{(0.11 - y)} = 6.3 \times 10^{-8}$$

Since K_{a2} is very small, we can assume that less than 5% of the HA^- dissociates. Therefore,

$$0.11 \ M \pm y \approx 0.11 \ M$$

With this assumption:

$$K_{a2} = \frac{y(0.11)}{(0.11)}$$

and

$$y = K_{a2} = 6.3 \times 10^{-8}\ M$$

The concentration of A^{2-} is equal to K_{a2}.

$$[A^{2-}] = y = 6.3 \times 10^{-8}\ M$$

Checking the validity of the assumption that,

$$0.11\ M - (6.3 \times 10^{-8}\ M) \approx 0.11\ M$$

The percent ionization of step 2 is:

$$\text{percent ionization} = \frac{[H^+][A^-]}{[HA]} \times 100\%$$

$$\frac{6.3 \times 10^{-8}}{0.11} \times 100\% = 5.7 \times 10^{-5}\%$$

$$[H_2SO_4] = 0.89\ M$$
$$[HSO_4^-] = 0.11\ M$$
$$[SO_4^{2-}] = 6.3 \times 10^{-8}\ M$$
$$[H^+] = (0.11 + 6.3 \times 10^{-8})\ M = 0.11\ M$$

NOTE:
Because $K_{a1} \gg K_{a2}$, the species produced in step 1 will be present in much greater concentration than those from step 2. This can be seen by comparing the percent ionization of each step (11% vs. 0.000057%). Also, since H_2A is a weak acid, the concentration of undissociated H_2A should be greater than that of all ionic species. Therefore, we expect the following order of concentrations:

$$[H_2A] > [H^+] \approx [HA^-] > [A^{2-}]$$

PRACTICE QUESTION 16.14
Write chemical equations for the three stages of ionization of phosphoric acid, H_3PO_4.

PRACTICE QUESTION 16.15
Calculate the pH of a 0.50 M solution of phosphoric acid, H_3PO_4.

PROBLEM TYPE 10: ACID-BASE PROPERTIES OF SALT SOLUTIONS.

A **salt** is an ionic compound produced by the reaction of an acid and a base. In dealing with salts, it will be useful to recall that they are *strong* electrolytes and are completely dissociated in water. The term **salt**

hydrolysis refers to the reaction of a cation or an anion with water, after dissociation splits the salt into two parts. Salt hydrolysis usually affects the pH of a solution.

Hydrolysis can be cation hydrolysis if the cation of the salt reacts with water, and/or anion hydrolysis if the anion of the salt reacts with water.

A. Basic Salt Solutions.

Anions that cause hydrolysis are Brønsted bases because they can accept a proton from water.

$$A^- + H_2O \rightleftharpoons HA + OH^-$$

Not all anions cause hydrolysis. You must be able to recognize those that do and those that don't. Quite simply, those anions with the ability to undergo hydrolysis are conjugate bases of weak acids. The conjugate base of any weak acid is a weak base and will raise the pH of an aqueous solution. The extent of hydrolysis is proportional to the base ionization constant, K_b.

Keep in mind that the conjugate base of a *strong acid* has no tendency to accept protons from H_2O. Anions such as Cl^-, NO_3^-, and ClO_4^- do not cause hydrolysis. In contrast, the fluoride ion is the conjugate base of hydrofluoric acid (HF), a weak acid. As a weak base it can accept a proton from water. The resulting OH^- ions make the solution basic.

$$F^- + H_2O \rightleftharpoons HF + OH^- \qquad\qquad K_b = \frac{[HF][OH^-]}{[F^-]}$$

The force that makes this reaction possible is the tendency of the weak acid HF to remain undissociated in solution. Three areas of base strength can be distinguished:

1. The conjugate bases of strong acids ($K_a > 1.0$) are too weak to cause measurable hydrolysis. Thus, Cl^-, Br^-, and NO_3^- ions do not hydrolyze in aqueous solutions.

    $$Cl^- + H_2O \xleftarrow{\quad\longrightarrow\quad} HCl + OH^- \quad \text{(no reaction)}$$

2. The conjugate bases of acids with a K_a between 10^{-1} and 10^{-5} are weak bases. Thus, F^-, NO_2^-, $HCOO^-$ (formate), and CH_3COO^- (acetate) ions hydrolyze to a small extent and produce low concentrations of OH^- ions in solution.

    $$NO_2^- + H_2O \rightleftharpoons HNO_2 + OH^-$$

3. The conjugate bases of very weak acids ($K_a < 10^{-5}$) are moderately strong bases. Thus solutions containing CN^- and HS^- ions can be quite basic.

B. Acidic Salt Solutions.

It is important to know which metal ions cause hydrolysis and which do not. Those that cause hydrolysis are small, highly charged metal ions, such as Al^{3+}, Cr^{3+}, Fe^{3+}, Bi^{3+}, and Be^{2+}. These can split H_2O molecules, yielding acidic solutions. In the case of the hydrated Al^{3+} ion, the reaction is

$$Al(H_2O)_6^{3+} (aq) + H_2O(\ell) \rightleftharpoons Al(OH)(H_2O)_5^{2+} (aq) + H_3O^+(aq)$$

The singly charged metal ions associated with strong bases do not cause hydrolysis. The commonly encountered ions Na^+, K^+, Ca^{2+}, and Mg^{2+} do not have a great enough positive charge to attract an OH^- from an H_2O molecule and bond to it. Recall that NaOH and KOH are strong bases. That is, they ionize completely in solution. These ions, K^+ and OH^- for instance, have no tendency to stay together in solution.

$$KOH(aq) \xrightleftharpoons{} K^+(aq) + OH^-(aq)$$

The hydrolysis reaction that would bond an OH^- to a K^+ ion does not occur!

$$K^+(aq) + H_2O(\ell) \xrightleftharpoons{} KOH(aq) + H^+(aq) \quad \text{(no reaction)}$$

Another acidic cation that you will need to be familiar with is the ammonium ion (NH_4^+). Solutions of ammonium ion are acidic because NH_4^+ is a Brønsted acid (the conjugate acid of the weak base NH_3).

$$NH_4^+ + H_2O \xrightleftharpoons{} NH_3 + H_3O^+$$

or simply,

$$NH_4^+ \xrightleftharpoons{} NH_3 + H^+ \quad K_a = 5.6 \times 10^{-10}$$

C. Neutral Salt Solutions.

When a salt is dissolved in water the solution will have a pH consistent with the acid-base properties of the cation and anion making up the salt. In general, salts containing alkali metal ions or alkaline earth metal ions (except Be^{2+}) with the conjugate bases of strong acids do not undergo hydrolysis; their solutions are neutral. Consequently, a solution of KCl, for example, is neutral. A K^+ ion does not accept an OH^- ion from H_2O, and a Cl^- ion does not accept a H^+ ion from H_2O.

D. Salts in Which Both the Cation and the Anion Hydrolyze.

Salts derived from a weak acid and a weak base consist of cations and anions that will hydrolyze. Whether a solution containing such a salt is acidic, basic, or neutral depends on the relative strengths of the acidic cations and basic anions. For example, a solution of ammonium fluoride is *acidic* because NH_4^+ is a stronger acid than F^- is a base.

$$NH_4F(aq) \longrightarrow NH_4^+(aq) + F^-(aq) \quad \text{(strong electrolyte)}$$

$$NH_4^+ \xrightleftharpoons{} NH_3 + H^+ \qquad\qquad K_a = 5.6 \times 10^{-10}$$

$$F^- + H_2O \xrightleftharpoons{} HF + OH^- \qquad\qquad K_b = 1.4 \times 10^{-11}$$

NH_4F is a salt of a weak acid and a weak base. On the other hand, another such salt (NH_4CN) is a *basic* salt because CN^- ion is a stronger base than NH_4^+ is an acid.

STUDY QUESTION 16.19
Write net ionic equations to show which of the following ions hydrolyze in aqueous solution.

a. NO_3^-

b. NO_2^-

c. NH_4^+

Solution: Strong acids have extremely weak conjugate bases which have no affinity for H^+ ions. They have neutral properties. Weak acids have stronger conjugate bases with significant affinity for an H^+ ion. The conjugate bases of weak acids have measurable basic properties.

a. Nitrate ion is the conjugate base of a strong acid, HNO_3. Therefore, NO_3^- has no acidic or basic properties and has no tendency to react with water to form HNO_3.

$$NO_3^- + H_2O \rightarrow HNO_3 + OH^- \quad \text{(no reaction)}$$

b. Nitrite ion is the conjugate base of a weak acid, HNO_2. Therefore, NO_2^- is a stronger base than NO_3^- and will accept a proton from water to form HNO_2.

$$NO_2^- + H_2O \rightleftharpoons HNO_2 + OH^-$$

c. Ammonium ion is the conjugate acid of a weak base. Therefore, it is a weak acid:

$$NH_4^+ + H_2O \rightleftharpoons NH_3 + H_3O^+$$

STUDY QUESTION 16.20
Predict whether the following aqueous solutions will be acidic, basic, or neutral.

a. KI
b. NH_4I
c. CH_3COOK

Solution:

a. What we must decide is whether either of the ions of the salt undergoes hydrolysis. KI is a salt of a strong acid (HI) and a strong base (KOH). Neither K^+ nor I^- have acidic or basic properties. Thus, a solution containing KI remains neutral.

b. NH_4I is the salt of a strong acid (HI) and a weak base (NH_3). As explained in part (a) iodine does not hydrolyze. Since NH_4^+ is the conjugate acid of a weak base, it will donate protons to water.

$$NH_4^+ + H_2O \rightleftharpoons NH_3 + H_3O^+$$

Therefore, a solution containing NH_4I will be acidic.

c. CH_3COOK is a salt of a weak acid and a strong base. K^+ ion will not hydrolyze, but CH_3COO^- (acetate ion) is the conjugate base of a weak acid, acetic acid. Thus, acetate ions will accept protons from water and form acetic acid and OH^- ions.

$$CH_3COO^- + H_2O \rightleftharpoons CH_3COOH + OH^-$$

Solutions containing CH_3COOK will be basic.

STUDY QUESTION 16.21
Calculate the pH of 0.25 M $C_6H_7O_6Na$ (sodium ascorbate) solution. The K_a of ascorbic acid is 8.0×10^{-5}.

Solution: First, identify the ions that undergo hydrolysis. Then, use the ionization constant to calculate the equilibrium concentrations. Sodium ascorbate is a strong electrolyte.

$$C_6H_7O_6Na(aq) \rightleftharpoons C_6H_7O_6^-(aq) + Na^+(aq)$$

Ascorbate ion is a conjugate base of a weak acid and so it causes hydrolysis. Na^+ does not and is neutral.

$$C_6H_7O_6^-(aq) + H_2O(\ell) \qquad C_6H_8O_6(aq) + OH^-(aq)$$

It is important to use the correct equilibrium constant value. Here, we need the K_b for ascorbate ion. From the K_a for ascorbic acid, the conjugate acid of ascorbate ion, we can get this value.

$$K_b = \frac{K_w}{K_a} = \frac{1.0 \times 10^{-14}}{8.0 \times 10^{-5}} = 1.3 \times 10^{-10}$$

The OH^- ion concentration in a solution that is 0.25 M is obtained from the equilibrium constant expression:

$$K_b = \frac{[C_6H_8O_6][OH^-]}{[C_6H_7O_6^-]}$$

Let x = moles of ascorbate ion that undergo hydrolysis per L of solution.

Concentration	H_2O	+	$C_6H_7O_6^-$	$\rightleftharpoons$	$C_6H_8O_6$	+	OH^-
Initial (M)			0.25		0		0
Change (M)			$-x$		$+x$		$+x$
Equilibrium (M)			$(0.25-x)$		x		x

Substitute into the equilibrium constant expression.

$$K_b = 1.3 \times 10^{-10} = \frac{x^2}{(0.25 - x)}$$

Apply the approximation that $C_6H_7O_6^- = 0.25 - x \approx 0.25 \ M$

$$\frac{x^2}{0.25} = 1.3 \times 10^{-10}$$

$$x = 5.7 \times 10^{-6} \ M = [OH^-]$$

$$pOH = -\log [OH^-] = 5.24$$

$$pH = 14.00 - 5.24 = 8.76$$

PRACTICE QUESTION 16.16
Predict whether the pH is > 7, < 7, or = 7 for aqueous solutions containing the following salts.
   a. K_2CO_3
   b. $BaCl_2$
   c. $Fe(NO_3)_3$

PROBLEM TYPE 11: LEWIS ACIDS AND BASES.

G. N. Lewis, who also developed the use of electron dot structures, noticed that for an H^+ ion to be accepted by a base, the base must possess at least one unshared electron–pair. The electron–pair forms the bond to the proton. Consequently, a **Lewis base** is a molecule or ion that can donate a pair of electrons to form a bond. A **Lewis acid** is a molecule or ion that can accept a pair of electrons and form a bond.

$$H^+ \qquad + \qquad :NH_3 \quad \longrightarrow \quad H\!-\!NH_3^+ \quad (\text{or } NH_4^+)$$

Accepts	Donates	
electron pair	electron pair	

The ammonia molecule is the Lewis base because it donates a pair of electrons to H^+ to form a bond. By accepting the pair of electrons, the proton is a Lewis acid. Note that a new covalent bond is formed by the donation of the electron pair. Recall when both electrons of the shared pair are donated by the same atom, the bond is called a *coordinate covalent bond*. An acid–base reaction in the Lewis system is donation of an electron–pair to form a new covalent bond between the acid and the base, which allows for reactions to be categorized as acid–base reactions without having to have H^+ present.

Metal cations can be Lewis acids because they have lost their valence electrons and so have at least one vacant orbital in which to accept an electron-pair from a base. Consider the beryllium ion which can form four coordinate covalent bonds.

$$Be^{2+}(aq) + 4H_2O(\ell) \quad \longrightarrow \quad [Be(H_2O)_4]^{2+}(aq)$$

Water molecules with unshared electron-pairs on their oxygen atoms are Lewis bases and beryllium is a Lewis acid because it accepts electron-pairs to form $Be\!-\!OH_2$ bonds. Beryllium can form four $Be\!-\!OH_2$ bonds.

STUDY QUESTION 16.22
Identify the Lewis acids and bases in each of the following reactions.

a. $Ag^+(aq) + Cl^-(aq) \rightarrow AgCl(s)$

b. $Hg^{2+}(aq) + 4I^-(aq) \rightleftharpoons HgI_4^{2-}(aq)$

c. $BF_3(g) + NF_3(g) \rightarrow F_3N\!-\!BF_3(s)$

d. $SO_2(g) + H_2O(\ell) \rightleftharpoons H_2SO_3(aq)$

Solution: Lewis acids are species that accept electron–pairs. In each of the above reactions, the first reactant species is the Lewis acid, and the second reactant is the Lewis base (electron–pair donor). The Lewis acids in (a) and (b) are capable of accepting electron–pairs, because they are positive ions and have previously lost electrons. In (c), the boron atom has an incomplete octet. And in (d), the S atom in SO_2 is "oxidized," that is, its valence electrons are shifted toward the more electronegative oxygen atoms and away from the S atom. In all four reactions new coordinate covalent bonds are formed.

PRACTICE QUESTION 16.17
Why are nitrogen compounds and ions such as NH_3, CH_3NH_2, and NH_2^- good examples of Lewis bases?

PRACTICE QUIZ

1. The OH^- ion concentration in an ammonia solution is 7.5×10^{-3} M. What is the H^+ ion concentration?

2. HA is a weak acid. If a 0.020 M HA solution is 2.5% dissociated, what is the K_a value of the weak acid?

3. A 0.100 M solution of chloroacetic acid, $ClCH_2COOH$, has a pH of 1.95. Calculate the K_a for chloroacetic acid.

4. Determine accurately the percent ionization of formic acid, HCOOH, in a 0.0050 M solution.

5. Predict the direction in which the equilibrium will lie for the following reaction.

$$HNO_2(aq) + ClO_4^- \ (aq) \qquad NO_2^- \ (aq) + HClO_4(aq)$$

6. Which of the following is the strongest acid? HF, CH_3COOH, or C_6H_5COOH

7. Which of the following solutions has the lowest pH?
0.10 M HNO_2, 0.10 M CH_3COOH, or 0.10 M HCN

8. Which of the following statements, if any, are true with regard to a 0.10 M solution of a weak acid HA?
a. $[HA] > [H^+]$
b. the pH = 1.0
c. $[H^+] < [A^-]$
d. the pH > 1.0
e. $[OH^-] = [H^+]$

9. Calculate K_a for $C_5H_5NH^+$.

10. Calculate the pH of 0.021 M NaCN solution.

11. Why can a H^+ ion, Mg^{2+} ion, and Al^{3+} ion act as Lewis acids?

Chapter 17
Acid-Base Equilibria and Solubility Equilibria

PROBLEM-SOLVING STRATEGIES AND TUTORIAL SOLUTIONS

TYPES OF PROBLEMS

Problem Type 1:   The Common Ion Effect.

Problem Type 2:   Buffer Solutions.
- **A.** Calculating the pH of a Buffer.
- **B.** Preparing a Buffer Solution with a Specific pH.

Problem Type 3:   Acid–Base Titrations.
- **A.** Strong Acid–Strong Base Titrations.
- **B.** Weak Acid–Strong Base Titrations.
- **C.** Strong Acid–Weak Base Titrations.
- **D.** Acid–Base Indicators.

Problem Type 4:   Solubility Equilibria.
- **A.** Solubility Product Expression and K_{sp}.
- **B.** Calculations Involving K_{sp} and Solubility.
- **C.** Predicting Precipitation Reactions.

Problem Type 5:   Factors Affecting Solubility.
- **A.** The Common Ion Effect.
- **B.** pH.
- **C.** Complex Ion Formation.

PROBLEM TYPE 1: THE COMMON ION EFFECT.

In this chapter, we will examine the acid–base equilibria in buffer solutions and in titrations. Then, we will discuss two additional types of aqueous equilibria: those involving low solubility salts, and those involving the formation of metal complexes. We will start with the common ion effect.

First, consider the ionization of a weak acid such as HF for example. The species in equilibrium are HF molecules, H^+ ions, and F^- ions:

$$HF(aq) \rightleftharpoons H^+(aq) + F^-(aq)$$

The concentration of F^- ion can be increased by adding sodium fluoride, a strong electrolyte:

$$NaF(aq) \rightarrow Na^+(aq) + F^-(aq)$$

According to Le Chatelier's principle, the addition of F^- ions will shift the weak acid equilibrium to the left; this consumes some of the F^- ions and some $H^+(aq)$, and forms more HF. This shift lowers the percent ionization of HF. In effect, the percent ionization of a weak acid (HA) is suppressed by the addition to the solution of its conjugate base, A^- ion. The shift in equilibrium caused by the addition of an ion common to one of the products of a dissociation reaction is called the **common ion effect**.

STUDY QUESTION 17.1
Calculate the pH of a 0.125 M aqueous solution of formic acid, HCO_2H, that is 0.025 M in sodium formate, $NaCO_2H$. The K_a for formic acid is 1.8×10^{-4}.

Solution: The first thing we need to write is the acid ionization equilibrium for formic acid and the corresponding mathematical expression for its K_a.

$$HCO_2H(aq) + H_2O(g) \rightleftharpoons H_3O^+(aq) + CO_2H^-(aq) \qquad K_a = \frac{[H_3O^+][CO_2H^-]}{[HCO_2H]}$$

Now, we can calculate the equilibrium concentration of H_3O^+ using an ICE table:

	HCO_2H	H_2O	H_3O^+	CO_2H^-
Initial, M	0.125		0	0.025
Change, M	$-x$		$+x$	$+x$
Equilibrium, M	$0.125 - x$		x	$0.025 + x$

Substitution into the expression for K_a leads to:

$$K_a = \frac{x(0.025 + x)}{0.125 - x} = 1.8 \times 10^{-4}$$

As a first approximation, we may assume that the change in the amount of HCO_2H is small in comparison to its initial concentration, since the ratio of the initial concentration to the K_a is almost 1000–fold. Since according to Le Châtelier's principle, the presence of the formate anion should shift the equilibrium towards the reactants, we can assume that the change in concentration for CO_2H^- is also small. Therefore, our expression becomes,

$$\frac{x(0.025)}{0.125} = 1.8 \times 10^{-4}$$

$$0.025x = 2.3 \times 10^{-5}$$

$$x = 9.0 \times 10^{-4}$$

We need to check the validity of our assumption:

$$\frac{9.0 \times 10^{-4}}{0.125} \times 100 = 0.72\%$$

The amount of the change is less than 5% of the initial concentration; thus the approximation is valid. Since x is equal to the equilibrium concentration of H_3O^+:

$$pH = -\log[H_3O^+] = -\log(9.0 \times 10^{-4}) = 3.05$$

You should be able to check that in the absence of the formate anion the pH of the 0.125 M aqueous solution of formic acid would have been 2.32.

PRACTICE QUESTION 17.1
Calculate the pH of the following solutions:

     a.   **0.10 M CH$_3$COOH**
     b.   **0.020 M CH$_3$COONa and 0.10 M CH$_3$COOH**

Would a common ion effect exist in the following scenarios?

a. 0.10 *M* NaOH solution is added to a 0.10 *M* NH₄OH solution

b. 0.020 *M* NaCl solutions is added to a 0.10 *M* CH₃COOH

PROBLEM TYPE 2: BUFFER SOLUTIONS.

A. Calculating the pH of a Buffer

Any solution that resists significant changes in its pH, even when a strong acid or strong base is added, is called a buffered solution or just a **buffer.** A buffer must contain an acid to react with any OH^- ions that may be added, and a base to react with any added H^+ ions. A buffer solution may contain a weak acid and its salt or a weak base and its salt. That is, a buffer consists of a solution in which both the acid and its conjugate base (or the base and its conjugate acid) are present in solution with in a relatively large concentration.

A solution containing HF and NaF is a buffer solution. In a typical buffer, the weak acid and its salt are in approximately equal concentrations. The weak acid HF and its conjugate base F^- are called the buffer components. If you had a solution of just HF, the concentration of F^- ion would not be high enough to make it a buffer solution.

$$HF(aq) \rightleftharpoons H^+(aq) + F^-(aq)$$

The additional F^- ions needed in order to achieve approximately equal concentrations of the two components are supplied by the addition of NaF which is 100% dissociated in solution.

To calculate the pH of a buffer requires that the concentrations of the weak acid (such as HF) and a soluble salt of the weak acid (NaF) be substituted into the ionization constant expression for the weak acid.

$$HF(aq) \rightleftharpoons H^+(aq) + F^-(aq)$$

$$K_a = \frac{[H^+][F^-]}{[HF]}$$

$$[H^+] = \frac{[HF]}{[F^-]} K_a$$

The pH is found from pH = – log [H^+].

If we take the negative log of the equilibrium constant expression, we get

$$-\log K_a = -\log [H^+] + -\log \frac{[\text{conjugate base}]}{[\text{acid}]}$$

Applying the definition of the p–function, pX = –log [X], we get

$$pK_a = pH - \log \frac{[\text{conjugate base}]}{[\text{acid}]}$$

rearranging, we get

$$pH = pK_a + \log \frac{[\text{conjugate base}]}{[\text{acid}]}$$

This is known as the **Henderson–Hasselbach equation**, which is very useful for buffer calculations.

STUDY QUESTION 17.2
Calculate the pH of a buffer solution that is 0.25 *M* HF and 0.50 *M* NaF.

Solution: The pH of a HF/F⁻ buffer is given in terms of the Henderson–Hasselbach equation.

$$HF(aq) \rightleftharpoons H^+(aq) + F^-(aq)$$

$$pH = pK_a + \log \frac{[F^-]}{[HF]}$$

$$pK_a = -\log(7.1 \times 10^{-4}) = 3.15$$

The equilibrium concentrations of F⁻ ion and HF are determined as follows: the initial concentration of HF is 0.25 *M*, but at equilibrium, $[HF] = 0.25 - x$. We can assume that x is less than 5 percent of $[HF]_0$ because HF is a weak acid, and because the addition of F⁻ ion, a common ion, represses the dissociation of HF. We assume that $[HF] = [HF]_0 = 0.25$ *M*.

The F⁻ ion is contributed by two sources, sodium fluoride which is a strong electrolyte, and HF, a weak acid.

$$NaF(aq) \longrightarrow Na^+(aq) + F^-(aq)$$

Therefore, $[F^-] = 0.50$ *M* $+ x$. Since 0.50 *M* $\gg x$, then $[F^-] = 0.50$ *M*.

Substituting into the Henderson–Hasselbach equation gives:

$$pH = 3.15 + \log \frac{0.50}{0.25} = 3.15 + \log 2.0 = 3.15 + 0.30$$

$$= 3.45$$

STUDY QUESTION 17.3
Suppose 3.0 mL of 2.0 *M* HCl is added to exactly 100 mL of the buffer described in Example 17.2a, what is the new pH of the buffer after the HCl is neutralized.

Solution: The HCl is neutralized in this buffer by the following reaction, which goes to completion.

$$H^+ + F^- \rightarrow HF$$

This consumes some F⁻ ions and forms more HF. The number of moles of H⁺ added as HCl is

$$M \times V = 2.0 \text{ mol/L} \times 0.0030 \text{ L} = 0.0060 \text{ mol H}^+$$

The number of moles of HF originally present in 100 mL of buffer was

$$M \times V = 0.25 \text{ mol/L} \times 0.100 \text{ L} = 0.025 \text{ mol HF}$$

The number of moles of F⁻ originally present in 100 mL of buffer was

$$M \times V = 0.50 \text{ mol/L} \times 0.100 \text{ L} = 0.050 \text{ mol F}^-$$

After the added H^+ is neutralized by $H^+ + F^- \rightarrow HF$, 0.0060 H^+ reacts with 0.0060 mol of F^- ion to form 0.0060 mol HF.

The number of moles of HF is $0.025 + 0.006 = 0.031$ mol
The number of moles of F^- is $0.050 - 0.006 = 0.044$ mol

This is illustrated in the following ICE table:

	H^+ +	F^- $\rightarrow$	HF
I	0.0060 mol	0.050 mol	0.025 mol
C	–0.0060 mol	–0.0060 mol	+ 0.0060 mol
E	All neutralized	0.044 mol	0.031 mol

The new pH can be found as usual using the Henderson–Hasselbach equation, since the solution is still a buffer.

$$pH = pK_a + \log \frac{[F^-]}{[HF]}$$

$$= 3.15 + \log \frac{(0.044 \text{ mol}/0.103 \text{ L})}{(0.031 \text{ mol}/0.103 \text{ L})} = 3.15 + \log 1.42 = 3.15 + 0.15 = 3.30$$

Please note that there is only a 0.15 pH unit decrease from 3.45 (Study Question 17.2) due to the addition of 3.0 mL of 2.0 M HCl. Hence, the solution posed in this question is a buffer.

B. Preparing a Buffer Solution with a Specific pH

Each weak acid–conjugate base buffer has a characteristic pH range over which it is most effective. In terms of the Henderson–Hasselbach equation, the pH of a buffer depends on the pK_a of the weak acid, and the ratio of conjugate base concentration to weak acid concentration. A buffer has equal ability to neutralize either added acid or added base when the concentrations of its weak acid and conjugate base components are equal. When [conjugate base]/[weak acid] = 1, then,

$$\log \frac{[\text{conjugate base}]}{[\text{acid}]} = 0$$

and the Henderson–Hasselbach equation becomes $pH = pK_a$. In order to prepare a buffer of a certain pH, the weak acid must have a pK_a as close as possible to the desired pH.

Next, we substitute the desired pH and the pK_a of the chosen acid into the Henderson–Hasselbach equation in order to determine the ratio of [conjugate base] to [weak acid]. The ratio can then be converted into molar concentrations of the weak acid and its salt.

STUDY QUESTION 17.4
What ratio of $[F^-]$ to [HF] would you use to make a buffer of pH = 2.85?

Solution: First, we start with the Henderson–Hasselbach equation.

$$pH = pK_a + \log \frac{[\text{conjugate base}]}{[\text{acid}]}$$

Substitute in the desired pH and the pK_a value of HF.

$$2.85 = 3.15 + \log \frac{[F^-]}{[HF]}$$

$$\log \frac{[F^-]}{[HF]} = -0.30$$

$$\frac{[F^-]}{[HF]} = 10^{-0.30} = 0.50$$

The ratio of $[F^-]$ to $[HF]$ should be 0.50 to 1. Any amounts of F^- and HF that give a molar ratio of 0.5 will produce the desired pH.

PRACTICE QUESTION 17.3
A buffer solution prepared by dissolving 0.15 mol of benzoic acid (C_6H_5COOH) and 0.45 mol of sodium benzoate (C_6H_5COONa) in enough water to make 400 mL of solution.

   a. What is the pH?
   b. Does the volume of solution affect your answer?

PRACTICE QUESTION 17.4
How would the following additions affect the pH of the buffer discussed in Practice Question 17.3?

   a. The addition of NaOH.
   b. The addition of water.
   c. The addition of H_2SO_4.

PRACTICE QUESTION 17.5
Consider a buffer solution consisting of 0.165 M NH_3 and 0.120 M NH_4Cl with a final volume of 100mL.

   a. What is the change in pH when 40.0 mg of NaOH is added?
   b. Write the equation for the buffer equilibrium.
   c. Write the equation for the neutralization of NaOH by the buffer.

PRACTICE QUESTION 17.6
What mole ratio of benzoate ion to benzoic acid would you need to prepare a buffer solution with a pH of 5.00?

PRACTICE QUESTION 17.7
What is the optimum pH of a H_3PO_4/ $H_2PO_4^-$ buffer?

PROBLEM TYPE 3: ACID-BASE TITRATIONS.

A. Strong Acid–Strong Base Titrations
For a neutralization reaction, a graph of pH versus volume of **titrant**, acid or base of known concentration, added is called a **titration curve**. Initially, the pH is that of the unknown solution. As titrant is added, the pH becomes that of a partially neutralized solution of unknown plus titrant. The pH at the equivalence point refers to the H^+ ion concentration when just enough titrant has been added to completely neutralize the unknown. If more titrant is added after the equivalence point has been reached, the pH assumes a value consistent with the pH of excess titrant.

Figure 17.3 in the textbook shows a titration curve for the addition of a strong base to a strong acid. The main features of this curve can be stated briefly.

1. The pH starts out quite low because this is the pH of the original acid solution.
2. As base is slowly added, it is neutralized, and the pH is determined by the unreacted excess acid. Near the equivalence point, the pH begins to rise more rapidly.
3. At the equivalence point the pH changes sharply, increasing several pH units upon the addition of only two drops of base.
4. Beyond the equivalence point the pH is determined by the amount of excess base that is added and is higher than 7.

The pH at the equivalence point of an acid–base titration is the pH of the salt solution that is formed by neutralization. $NaCl$, $NaNO_3$, $NaBr$, KCl, and KI are examples of salts that can be formed in titrations of strong acids and strong bases. These salts yield ions that do not cause hydrolysis (Chapter 16). Therefore, the pH at the equivalence point in a titration of a strong acid with a strong base is 7.

STUDY QUESTION 17.5
Calculate the pH when 10.0 mL of 0.15 *M* HNO_3 is added to 50.0 mL of 0.10 *M* NaOH.

Solution: The net ionic reaction for the reaction between HNO_3 and NaOH is:

$$H^+ + OH^- \rightarrow H_2O$$

This is a limiting reactant problem and can be solved using the following table, plugging moles into the table. "I" = initial moles, "C" = change, "E" = final moles:

	H^+	+	OH^-	$\rightarrow$	H_2O
I	0.0015 mol		0.0050 mol		
C	−0.0015 mol		−0.0015 mol		
E	0		0.0035 mol		

The pH can be determined from the concentration of OH^-.

$$[OH^-] = 0.0035 \text{ mol} \div (0.0100 \text{ L} + 0.0500 \text{ L}) = 0.058 \text{ mol/L}$$

$$pOH = -\log(0.058 \text{ M}) = 1.23$$

$$pH = 14.00 - 1.23 = 12.77$$

B. **Weak Acid–Strong Base Titrations.**
    Figure 17.4 of the text shows a titration curve for the addition of a strong base to a weak acid. Because the acid is a weak acid, the initial pH is greater than in the titration of a strong acid. At the equivalence point, the pH is above 7.0 because of salt hydrolysis. The anion of the salt is the conjugate base of the weak acid used in the titration. In the titration of acetic acid with sodium hydroxide, the salt produced is CH_3COONa. The Na^+ ion does not hydrolyze, but CH_3COO^- does:

$$CH_3COO^- + H_2O \rightarrow CH_3COOH + OH^-$$

Hydrolysis of acetate ions makes the solution basic at the equivalence point.

C. **Strong Acid–Weak Base Titrations.**
    Figure 17.5 of the text shows a titration curve for the addition of a weak base to a strong acid. The first part of the curve is the same as in the strong acid versus strong base titration. However, the pH at the

equivalence point is below 7.0 because the cation of the salt is the conjugate acid of the weak base used in the titration. Hydrolysis of this salt yields an acidic solution. In the titration of hydrochloric acid with ammonia, the salt produced is NH_4Cl. The Cl^- ion does not hydrolyze, but NH_4^+ does:

$$NH_4^+ + H_2O \rightarrow NH_3 + H_3O^+$$

D. Acid–Base Indicators.

Indicators are used in the laboratory to reveal the equivalence point of a titration. The abrupt change in color of the indicator signals the *endpoint* of a titration which usually coincides very nearly with the equivalence point. Indicators are usually weak organic acids that have distinctly different colors in the nonionized (molecular) and ionized forms.

$$HIn(aq) \longrightarrow H^+(aq) + In^-(aq) \qquad K_a = \frac{[H^+][In^-]}{[HIn]}$$

$$\text{color 1} \qquad\qquad\qquad \text{color 2}$$

To determine the pH range in which an indicator will change color we can write the K_a expression in logarithmic form.

$$pH = pK_a + \log \frac{[In^-]}{[HIn]}$$

The color of an indicator depends on which form predominates. Typically, when $[In^-]/[HIn] \geq 10$ the solution will be color 2, and when $[In^-]/[HIn] \leq 0.1$, the solution will be color 1. Thus, the color change will occur between,

$$pH = pK_a + \log (10/1) = pK_a + 1.0$$

and

$$pH = pK_a + \log (1/10) = pK_a - 1.0$$

At the midpoint of the pH range over which the color changes, $[HIn] = [In^-]$, and $pH = pK_a$.

Like any weak acid, each HIn has a characteristic pK_a, and so each indicator changes color at a characteristic pH. Table 17.3 of the text lists a number of indicators used in acid–base titration and the pH ranges over which they change color. The choice of indicator for a particular titration depends on the expected pH at the equivalence point, as the following example shows.

STUDY QUESTION 17.6
Choose an indicator for the titration of 50 mL of a 0.10 *M* HI solution with 0.10 *M* NH₃.
$$HI(aq) + NH_3(aq) \rightarrow NH_4I(aq)$$

Solution: First, we need to know the pH at the equivalence point. This is the titration of a strong acid with a weak base. The net ionic equation is:

$$
\begin{array}{cccc}
H^+(aq) & + & NH_3(aq) & \longrightarrow & NH_4^+(aq) \\
5.0 \times 10^{-3} \text{ mol} & + & 5.0 \times 10^{-3} \text{ mol} & & 5.0 \times 10^{-3} \text{ mol}
\end{array}
$$

The product, NH_4^+ ion, is a weak acid and so we expect a slightly acidic solution at the equivalence point.

The concentration of NH_4^+ ion formed at the equivalence point is 5.0×10^{-3} mol/0.100 L $= 5.0 \times 10^{-2}$ *M*.

The [H$^+$] at the equivalence point is just the pH of a 0.050 M NH$_4^+$ solution. Consider the ionization of the weak acid, NH$_4^+$:

$$NH_4^+\,(aq) \rightarrow NH_3(aq) + H^+(aq) \qquad K_a = 5.6 \times 10^{-10}$$

	NH$_4^+$	$\rightarrow$	H$^+$	NH$_3$
I	0.050		0	0
C	$-x$		$+x$	$+x$
E	$(0.050 - x)$		x	x

$$K_a = \frac{[NH_3][H^+]}{[NH_4^+]} = \frac{x^2}{(0.050 - x)} = \frac{x^2}{0.050} = 5.6 \times 10^{-10}$$

Solving for x:

$$x = [NH_3] = [H^+] = 5.3 \times 10^{-6}\ M$$

and the pH is 5.28 at the equivalence point. According to Table 17.3 of the text, chlorophenol blue and methyl red are indicators that will change color in the vicinity of pH 5.28.

> **PRACTICE QUESTION 17.8**
> **A 25.0 mL sample of 0.222 M HBr was titrated with 0.111 M NaOH.**
> > **a. Write the overall balanced equation for the reaction.**
> > **b. After adding 30.0 mL of base solution, what is the pH?**

> **PRACTICE QUESTION 17.9**
> **Calculate the pH in the following titration after the addition of 12.0 mL of 0.100 M KOH to 20.0 mL of 0.200 M CH$_3$COOH?**

> **PRACTICE QUESTION 17.10**
> **Determine the pH at the equivalence point in the titration of 0.100 M NaOH with 0.100 M HNO$_3$?**

PROBLEM TYPE 4: SOLUBILITY EQUILIBRIA.

A. Solubility Product Expression and K_{sp}.
Compounds with low solubilities in water are described as slightly soluble or insoluble. In this chapter, solubility is treated quantitatively in terms of equilibrium. In a saturated solution of a slightly soluble compound such as silver bromide, for example, the solubility equilibrium is:

$$AgBr(s) \rightarrow Ag^+(aq) + Br^-(aq)$$

The equilibrium constant for the reaction in which a solid salt dissolves is called the **solubility product constant**, K_{sp}. For the above reaction:

$$K_{sp} = [Ag^+][Br^-]$$

The solubility product expression is always written in the form of the equilibrium constant expression for the solubility reaction. Two additional examples are $Ca(OH)_2$ and $AgCrO_4$:

$$Ca(OH)_2(s) \rightarrow Ca^{2+}(aq) + 2OH^-(aq)$$

$$K_{sp} = [Ca^{2+}][OH^-]^2$$

and

$$Ag_2CrO_4(s) \rightarrow 2Ag^+(aq) + CrO_4^{2-}(aq)$$

$$K_{sp} = [Ag^+]^2[CrO_4^{2-}]$$

B. Calculations Involving K_{sp} and Solubility.

The solubility product constant (also called a K_{sp} value) can be calculated by substituting the concentrations of the ions in a saturated solution into the solubility product expression. Take AgBr again as an example. Since AgBr is a strong electrolyte, all of the solid AgBr that dissolves is dissociated into Ag^+ and Br^- ions,

$$AgBr(s) \rightleftharpoons Ag^+(aq) + Br^-(aq)$$

The solubility of AgBr at 25 °C is 8.8×10^{-7} moles per liter of solution. Therefore, the ion concentrations in a saturated AgBr solution are:

$$[Ag^+] = [Br^-] = 8.8 \times 10^{-7} \ M$$

The solubility product constant is found by substituting these concentrations into the solubility product:

$$K_{sp} = [Ag^+][Br^-] = (8.8 \times 10^{-7})(8.8 \times 10^{-7})$$

$$K_{sp} = 7.7 \times 10^{-13}$$

This calculation points out the difference between the solubility and the K_{sp} value. These two quantities are *not* the same. Here, we see that the solubility is used to determine the ion concentrations, and that substitution of these into the solubility product expression gives the solubility product constant (K_{sp} value). Table 17.4 in the text lists solubility product constants for a number of slightly soluble salts, including silver bromide.

The K_{sp} value can be used to calculate the solubility of a compound in moles per liter (the molar solubility), or in grams per liter. For example, the K_{sp} value for $CaSO_4$ is 2.4×10^{-5}. What is the solubility of $CaSO_4$ in mol/L?

Start with the solubility equilibrium:

$$CaSO_4(s) \rightleftharpoons Ca^{2+}(aq) + SO_4^{2-}(aq) \qquad K_{sp} = [Ca^{2+}][SO_4^{2-}]$$

Let's use the solubility in mol/L. At equilibrium, both $[Ca^{2+}]$ and $[SO_4^{2-}]$ are equal to s because when s moles of $CaSO_4$ dissolve in 1 L, s mol/L of Ca^{2+} ions and s mol/L of SO_4^{2-} ions are produced. Substitute into the K_{sp} expression:

$$K_{sp} = [Ca^{2+}][SO_4^{2-}] = (s)(s) = s^2$$

The solubility product is given above: $K_{sp} = 2.4 \times 10^{-5}$

$$s^2 = 2.4 \times 10^{-5}$$

$$s = \sqrt{2.4 \times 10^{-5}} = 4.9 \times 10^{-3}\ M$$

It is important to notice that the molar solubility and K_{sp} are related, but are not the same quantity.

We can calculate s, the molar solubility of MgF_2, from its solubility of 7.3×10^{-3} g/100 mL solution. The molar mass of MgF_2 is 62.31 g/mol.

$$s = \frac{7.3 \times 10^{-3}\text{g}}{100\ \text{mL}} \times \frac{1\ \text{mL}}{10^{-3}\ \text{L}} \times \frac{1\ \text{mol}}{62.31\ \text{g}} = 1.17 \times 10^{-3}\ \text{mol/L}$$

The K_{sp} value can be calculated by substituting the ion concentrations in a saturated solution into the solubility product expression. When s mol of MgF_2 dissolves in 1 L of solution, the Mg^{2+} concentration is equal to s, and the F^- ion concentration is $2s$.

$$[Mg^{2+}] = s = 1.17 \times 10^{-3}\ M$$

$$[F^-] = 2s = 2(1.17 \times 10^{-3}\ M) = 2.34 \times 10^{-3}\ M$$

Now substitute the ion concentrations into the K_{sp} expression.

$$K_{sp} = (1.17 \times 10^{-3})(2.34 \times 10^{-3})^2$$

$$K_{sp} = 6.4 \times 10^{-9}$$

STUDY QUESTION 17.7
The K_{sp} value for Ag_2CrO_4 is 1.1×10^{-12}. Calculate the molar solubility of silver chromate.

Solution: First, we need to write the solubility equilibrium and the K_{sp} expression.

$$Ag_2CrO_4(s) \rightleftharpoons 2Ag^+(aq) + CrO_4^{2-}(aq)$$

$$K_{sp} = [Ag^+]^2[CrO_4^{2-}]$$

Let s = molar solubility of Ag_2CrO_4. Whenever s moles of Ag_2CrO_4 dissolve, $2s$ moles of Ag^+ and s moles of CrO_4^{2-} are produced.

$$[Ag^+] = 2s \quad [CrO_4^{2-}] = s$$

Summarize the changes in concentrations as follows.

	$Ag_2CrO_4(s) \rightleftharpoons$	$2Ag^+(aq)$ +	$CrO_4^{2-}(aq)$
I		0	0
C		$2s$	s
E		$2s$	s

Substitute these values into the K_{sp} expression.

$$K_{sp} = [2s]^2[s] = 1.1 \times 10^{-12}$$

and solve for s.

$$4s^2(s) = 4s^3 = 1.1 \times 10^{-12}$$

$$s = \sqrt[3]{\frac{1.1 \times 10^{-12}}{4}} = \sqrt[3]{2.75 \times 10^{-13}}$$

$$s = 6.5 \times 10^{-5}\ M$$

C. Predicting Precipitation Reactions.

When two solutions containing dissolved salts are mixed, formation of an insoluble compound is always a possibility. From knowledge of solubility rules and solubility products, you can predict whether a precipitate will form. For a dissolved salt $MX(aq)$, the ion product (Q) is $Q = [M^+]_0[X^-]_0$, where $[\ ]_0$ stands for the initial concentrations.

Any one of three conditions for the ion product Q may exist after two solutions are mixed.

1. $Q = K_{sp}$. For a saturated solution, the value of Q is equal to K_{sp}. No precipitate will form in this case, as no net reaction occurs in a system at equilibrium.
2. $Q < K_{sp}$. In an unsaturated solution, the value of Q is less than K_{sp}. No precipitate will form in this case.
3. $Q > K_{sp}$. In a supersaturated solution, more of the salt is dissolved than the solubility allows. In this case, an unstable situation exists, and some solute will precipitate from solution until a saturated solution is attained. At this point $Q = K_{sp}$ and equilibrium is reestablished.

Suppose 500 mL of a solution containing $2.0 \times 10^{-5}\ M\ Ag^+$ is mixed with 500 mL of solution containing $2.0 \times 10^{-4}\ M\ Br^-$. Will a precipitate form? First, determine the new concentrations of Ag^+ and Br^- in the mixture. The total volume is 1 L, and so accounting for dilution, the ion concentrations in the new solution are just half of what they were in the separate solutions. The initial concentrations before any precipitate forms are:

$$[Ag^+]_0 = 1.0 \times 10^{-5}\ M \quad [Br^-]_0 = 1.0 \times 10^{-4}\ M$$

The ion product for AgBr is; $Q = [Ag^+][Br^-]$.

To predict whether a precipitate of AgBr will form, calculate the ion product Q, and compare it to the K_{sp} value.

$$Q = [Ag^+][Br^-] = (1.0 \times 10^{-5})(1.0 \times 10^{-4}) = 1.0 \times 10^{-9}$$

$$K_{sp} = 7.7 \times 10^{-13}$$

Since $Q > K_{sp}$, the mixture corresponds to a supersaturated solution of AgBr, which means that some AgBr(s) will precipitate, until a new equilibrium is reached at which point $Q = K_{sp}$.

STUDY QUESTION 17.8
Predict whether or not a precipitate of PbI_2 will form when 200 mL of 0.015 M $Pb(NO_3)_2$ and 300 mL of 0.050 M NaI are mixed together.
Given: $K_{sp}(PbI_2) = 1.4 \times 10^{-8}$. If the answer is yes, what concentrations of $Pb^{2+}(aq)$ and $I^-(aq)$ will exist when equilibrium is reestablished.

Solution: Remember that $Pb(NO_3)_2$ and NaI are both strong electrolytes. When the solutions are mixed, will the following reaction occur?

$$\text{Pb}^{2+}(aq) + 2\text{NO}_3^{-}(aq) + 2\text{Na}^{+}(aq) + 2\text{I}^{-}(aq) \rightleftharpoons \text{PbI}_2(s) + 2\text{Na}^{+}(aq) + 2\text{NO}_3^{-}(aq)$$

Eliminating the spectator ions from both sides of the equation yields the net reaction:

$$\text{Pb}^{2+}(aq) + 2\text{I}^{-}(aq) \rightleftharpoons \text{PbI}_2(s)$$

A precipitate will form only if the ion product Q exceeds the K_{sp} value; $[\text{Pb}^{2+}][\text{I}^-]^2 > K_{sp}$. When the two solutions are mixed, 500 mL of new solution is formed. To calculate Q, we must calculate the initial concentration of each ion upon the mixing of the two solutions.

To determine the initial concentration of each ion, we must first determine the number of moles of each ion:

$$\text{moles of Pb}^{2+} = 0.2 \text{ L} \times 0.015 \ M = 0.0030 \text{ mol}$$

$$\text{moles of I}^{-} = 0.3 \text{ L} \times 0.050 \ M = 0.015 \text{ mol}$$

Immediately after mixing, the initial ion concentrations will be:

$$[\text{Pb}^{2+}] = \frac{\text{moles of Pb}^{2+}}{\text{liters of solution}} = \frac{0.0030 \text{ mol}}{0.500 \text{ L}} = 6.0 \times 10^{-3} \ M$$

$$[\text{I}^{-}] = \frac{\text{moles of I}^{-}}{\text{liters of solution}} = \frac{0.0150 \text{ mol}}{0.500 \text{ L}} = 3.0 \times 10^{-2} \ M$$

Substitution gives the value of the ion product.

$$Q = [\text{Pb}^{2+}][\text{I}^-]^2 = (6.0 \times 10^{-3})(3.0 \times 10^{-2})^2 = 5.4 \times 10^{-6}$$

Since $Q > K_{sp}$, then a precipitate of PbI_2 will form.

Therefore, the concentrations of Pb^{2+} and I^- remaining in solution depend on the number of moles initially and the number of moles that precipitate. The initial number of moles of Pb^{2+} was 0.0030 mol and for I^- was 0.015 mol. In order to determine how many moles of Pb^{2+} and I^- precipitated from solution, consider the net reaction. Look to see if there is a limiting reactant. Then, determine the amount of the excess reactant that remains after complete reaction of the limiting reactant.

We can easily determine that Pb^{2+} ion is the limiting reactant, and will react essentially completely. The 0.0030 mol of Pb^{2+} will react with 0.0060 mol of I^- (2 moles of I^- must react to every mole of Pb^{2+}), leaving 0.009 mol of excess I^- (0.015 – 0.0060) dissolved in solution.

Therefore,

$$[\text{I}^{-}] = 0.009 \text{ mol}/0.50 \text{ L} = 0.018 \ M$$

The Pb^{2+} concentration is controlled by the PbI_2 solubility equilibrium. In other words, some PbI_2 dissolves by the reverse reaction.

$$\text{PbI}_2(s) \rightleftharpoons \text{Pb}^{+2} + 2\text{I}^{-}$$

$$K_{sp} = [\text{Pb}^{2+}][\text{I}^-]^2 = 1.4 \times 10^{-8}$$

The Pb^{2+} ion concentration in equilibrium with 0.018 M I^- ion is:

$$[Pb^{2+}] = \frac{K_{sp}}{[I^-]^2} \quad \frac{1.4 \times 10^{-8}}{(0.018)^2} = 4.3 \times 10^{-5} \, M$$

Note, that we could also calculate the percentage of Pb^{2+} ion remaining unprecipitated as:

$$\frac{(4.3 \times 10^{-5} \, M)(0.500L)}{0.0030 \text{ mol}} \times 100\% = 0.72\%$$

This confirms our assumption that essentially all the Pb^{2+} ion precipitated.

PRACTICE QUESTION 17.11
Rank the following solutions from the most soluble to the least soluble.
- a. **Sample A has a $K_{sp} = 4.78 \times 10^{-14}$.**
- b. **Sample B has a $K_{sp} = 6.37 \times 10^{-9}$.**
- c. **Sample C has a $K_{sp} = 8.02 \times 10^{-10}$.**

PRACTICE QUESTION 17.12
The solubility of $PbBr_2$ is 0.392 g per 100 mL at 20 °C. What is the K_{sp} value for $PbBr_2$.

PRACTICE QUESTION 17.13
Solid barium fluoride was dissolved in pure water until a saturated solution was formed. If the fluoride concentration was 0.0150 M, determine the K_{sp} of BaF_2.

PRACTICE QUESTION 17.14
What is the solubility of CaF_2 in moles per liter given that $K_{sp} = 4.0 \times 10^{-11}$. Determine the F^- ion concentration in a saturated solution of CaF_2.

PRACTICE QUESTION 17.15
Will a precipitate form when 250 mL of 2.8×10^{-3} M $MgCl_2$ solution is added to 250 mL of 5.2×10^{-3} M Na_2CO_3? Identify the precipitate if any.

PROBLEM TYPE 5: FACTORS AFFECTING SOLUBILITY.

A. The Common Ion Effect.

The effect of adding an ion common to one already in equilibrium in a solubility reaction is to lower the solubility of the salt. In the case of AgBr solubility:

$$AgBr(s) \rightleftharpoons Ag^+(aq) + Br^-(aq)$$

The addition of either Ag^+ or Br^- ions will shift the equilibrium to the left, in accord with Le Chatelier's principle, thus decreasing the amount of AgBr dissolved. Ag^+ ions can be added by pouring in a solution of $AgNO_3$. Recall that $AgNO_3$ is very soluble, and is a strong electrolyte.

$$AgNO_3(s) \rightarrow Ag^+(aq) + NO_3^-(aq)$$

The nitrate ion will not interfere with AgBr solubility because it is not a common ion.

Additional Br⁻ ions could be supplied by adding NaBr, for instance. Sodium bromide is very soluble and is a strong electrolyte.

$$NaBr(s) \rightarrow Na^+(aq) + Br^-(aq)$$

The sodium ion will not affect the solubility of AgBr because it is not a common ion. Here, we see that the addition of an ion that is common to one already in the solubility equilibrium shifts the equilibrium to the left, which *decreases* the solubility. This is equivalent to the case of a weak acid, where the presence of a common ion decreases the percent ionization.

STUDY QUESTION 17.9
What is the solubility of $Pb(OH)_2$ in a buffer solution of pH 8.0 and in another of pH 9.0? Given: $K_{sp} = 1.2 \times 10^{-15}$.

Solution: Let's start off by writing the solubility equilibrium and the K_{sp} expression for $Pb(OH)_2$.

$$Pb(OH)_2(s) \rightleftharpoons Pb^{2+}(aq) + 2OH^-(aq)$$

$$K_{sp} = [Pb^{2+}][OH^-]^2 = 1.2 \times 10^{-15}$$

Let s equal the molar solubility of $Pb(OH)_2$. The concentration of Pb^{2+} will be equal to s. The concentration of OH^- ion will not be $2[Pb^{2+}]$ because in a buffer solution, the H^+ ion and OH^- ion concentrations are maintained constant by the buffer. At pH 8.0, the pOH is 6.0, and $[OH^-] = 1.0 \times 10^{-6}$ M. In the buffer solution, the $[OH^-]$ ion concentration is *maintained constant* at 1.0×10^{-6} M. The OH^- from dissolution of $Pb(OH)_2$ is neutralized by the buffer. Therefore, we can substitute as follows to obtain the solubility at pH 8.

$$[Pb^{2+}] = s$$

$$[Pb^{2+}] = \frac{K_{sp}}{[OH^-]^2}$$

$$s = \frac{1.2 \times 10^{-15}}{(1.0 \times 10^{-6})^2} = 1.2 \times 10^{-3}\ M$$

And at pH = 9, where $[OH^-] = 1.0 \times 10^{-5}$ M,

$$s = \frac{1.2 \times 10^{-15}}{(1.0 \times 10^{-6})^2} = 1.2 \times 10^{-5}\ M$$

The lower solubility of $Pb(OH)_2$ at pH 9.0, than at pH 8.0, is due to the common ion effect.

PRACTICE QUESTION 17.16
What is the molar solubility of Ag_3PO_4 in 0.20 M Na_3PO_4? $K_{sp} = 1.8 \times 10^{-18}$. It may also be helpful to be able to write the solubility equilibrium reaction for Ag_3PO_4.

PRACTICE QUESTION 17.17
Calculate the molar solubility AgCl in a solution containing 0.010 M $CaCl_2$.

PRACTICE QUESTION 17.18
What is the solubility of $Pb(OH)_2$ in a solution with a pH of 10.00?
Given: $K_{sp} = 1.43 \times 10^{-20}$.

B. pH.

The pH can affect the solubility of a solute in two ways. One of these is through the common ion effect. Consider the solubility equilibrium of an insoluble hydroxide such as $Mg(OH)_2$ or $Ca(OH)_2$.

$$Ca(OH)_2(s) \rightleftharpoons Ca^{2+}(aq) + 2OH^-(aq)$$

Upon the addition of NaOH, for instance, the pH of the solution will increase. The equilibrium position will shift to the left because of the added OH^- ion (a common ion); therefore, the solubility of $Ca(OH)_2$ will decrease proportionately.

The other case in which pH can affect solubility is when a salt contains a basic anion such as F^-, CH_3COO^-, or CN^-. Any basic anion will react with H^+ ions present in the solution and, thereby affect the solubility of the salt. Take, for instance, the effect of adding a strong acid on the solubility of silver acetate. To explain this, we need two reaction steps. In the first one, silver acetate dissolves, and in the second, the acetate ion combines with $H^+(aq)$ ions present in the system:

$$CH_3COOAg(s) \rightleftharpoons CH_3COO^-(aq) + Ag^+(aq)$$

$$H^+(aq) + CH_3COO^-(aq) \rightleftharpoons CH_3COOH(aq)$$

In the presence of H^+ ions, the concentration of acetate ion is lowered by the occurrence of the second reaction. The second reaction occurs essentially completely because it forms a weak acid. According to Le Châtelier's principle, the solubility equilibrium will shift to the right, causing more silver acetate to dissolve.

The solubility of salts containing anions that do *not* hydrolyze, such as Cl^-, Br^-, I^-, and NO_3^-, are not affected by pH. These anions are conjugate bases of strong acids.

STUDY QUESTION 17.10

Which of the following salts ($AgBr$, $Ba(OH)_2$, $MgCO_3$) will be more soluble at acidic pH than in pure water?

Solution: First, we need to identify the ions present in solutions of these compounds. Then we identify those with acid–base properties.

In silver bromide; $AgBr(s) \rightleftharpoons Ag^+ + Br^-$. Neither Ag^+ nor Br^- have acid–base properties and so the solubility of AgBr would not affected by pH.

Barium hydroxide dissociates following the following equation;

$$Ba(OH)_2 \rightleftharpoons Ba^{2+} + 2\ OH^-.$$

Comparing equilibria in water and in acidic solution, as $[H^+]$ increases, $[OH^-]$ decreases, and the solubility equilibrium will shift to the right, causing the solubility of $Ba(OH)_2$ to increase.

The carbonate ion is a base, and in acid solution it combines with $H^+(aq)$.

$$MgCO_3 \rightleftharpoons Mg^{2+} + CO_3^{2-}$$

$$H^+ + CO_3^{2-} \rightleftharpoons HCO_3^-$$

With the increase in $[H^+]$ in acid solution, $[CO_3^{2-}]$ decreases, and the solubility equilibrium shifts to the right, causing more $MgCO_3$ to dissolve.

PRACTICE QUESTION 17.19

Which of the following salts will be more soluble in an acidic solution than in a basic solution?

    a. AgI
    b. $BaCO_3$
    c. $Ca(CH_3COO)_2$
    d. $Zn(OH)_2$
    e. $CaCl_2$

PRACTICE QUESTION 17.20

What is the effect on the pH of the following solutions when:

    a. solid sodium acetate (CH_3COONa) is added to a dilute solution of acetic acid?
    b. solid $NaCl$ is added to a dilute solution of $NaOH$?
    c. solid KOH is added to a dilute solution of acetic acid?

PRACTICE QUESTION 17.21

Explain how changing the pH affects the solubility of $Zn(OH)_2$.

C. Complex Ion Formation.

When Ag^+ ion reacts with two CN^- ions:

$$Ag^+(aq) + 2CN^-(aq) \rightleftharpoons Ag(CN)_2^-(aq)$$

it forms a complex ion $Ag(CN)_2^-$. An ion made up of the metal ion bonded to one or more molecules or ions is called a **complex ion**. Some other examples are $Ag(NH_3)_2^+$, $Ni(H_2O)_6^{2+}$, and $Cu(CN)_4^{2-}$.

Complex ions are extremely stable and have an important effect on certain chemical species in solution. A measure of the tendency of a metal ion to form a certain complex ion is given by its formation constant, K_f. The **formation constant** is the equilibrium constant for the reaction that forms the complex ion. For example,

$$Ag^+ + 2CN^- \rightleftharpoons Ag(CN)_2^-$$

$$K_f = \frac{[Ag(CN)_2^-]}{[Ag^+][CN^-]^2} = 1.0 \times 10^{21}$$

Table 17.5 in the text lists K_f values for selected complex ions. The more stable the complex ion, the greater the extent of reaction, and the greater the value of K_f.

The formation of a complex ion has a strong effect on the solubility of a metal salt. For example, AgI, which has a very low solubility in water, will dissolve in an aqueous solution of $NaCN$. The stepwise process is:

$$
\begin{array}{lr}
AgI(s) \rightleftharpoons Ag^+(aq) + I^-(aq) & K_{sp} \\
Ag^+(aq) + 2CN^-(aq) \rightleftharpoons Ag(CN)_2^-(aq) & K_f \\
\hline
\text{Overall: } AgI(s) + 2CN^-(aq) \rightleftharpoons Ag(CN)_2^-(aq) + I^-(aq) & K
\end{array}
$$

The formation of the complex ion in the second step causes a decrease in $[Ag^+]$.

Therefore, the first equilibrium shifts to the right according to Le Châtelier's principle. This is why AgI is more soluble in CN^- solution than in pure water.

The overall reaction is the sum of the two steps. The equilibrium constant for the overall reaction is the product of the equilibrium constants of the two steps:

$$K = K_{sp} \times K_f$$

Tables 17.4 and 17.5 of the text give, respectively:

$$K_{sp} = 8.3 \times 10^{-17}, K_f = 1.0 \times 10^{21}.$$

Therefore,

$$K = 8.3 \times 10^4$$

The large value of K shows that AgI is very soluble in NaCN solution.

STUDY QUESTION 17.11
Calculate the concentration of free Ag^+ ions in a solution formed by adding 0.20 mol of $AgNO_3$ to 1 L of 1.0 M NaCN.

Solution: In this solution, Ag^+ ions will complex with CN^- ions, and the concentration of Ag^+ will be determined by the equation

$$Ag^+ + 2CN^- \rightleftharpoons Ag(CN)_2^- \qquad K_f = 1.0 \times 10^{21}$$

Since K_f is so large, we expect the Ag^+ to react essentially quantitatively to form 0.20 mol of $Ag(CN)_2^-$. We can find the concentration of free Ag^+ at equilibrium using the equilibrium constant expression:

$$K_f = \frac{[Ag(CN)_2^-]}{[Ag^+][CN^-]^2}.$$

Rearranging gives:

$$[Ag^+] = \frac{[Ag(CN)_2^-]}{K_f[CN^-]^2}$$

Substitute all known equilibrium concentrations into the above equation. At equilibrium; $[Ag(CN)_2^-] = 0.20$ M, and

$$[CN^-] = [CN^-]_0 - 2[Ag(CN)_2^-]$$

$$[CN^-] = 1.0 \ M - 0.40 \ M = 0.6 \ M$$

Therefore,

$$[Ag^+] = \frac{(0.20)}{(1.0 \times 10^{21})(0.6)^2}$$

$$[Ag^+] = 6 \times 10^{-22} \ M$$

STUDY QUESTION 17.12
Calculate the molar solubility of silver bromide in 6.0 M NH_3.

Solution: The solubility of AgBr is determined by two equilibria. The solubility and complex ion equilibria are

$$AgBr(s) \rightleftharpoons Ag^+ + Br^- \qquad K_{sp} = 7.7 \times 10^{-13}$$
$$Ag^+ + 2NH_3 \rightleftharpoons Ag(NH_3)_2^+ \qquad K_f = 1.5 \times 10^7$$
$$\text{Overall: } AgBr(s) + 2NH_3 \rightleftharpoons Ag(NH_3)_2^+ + Br^- \qquad K = K_{sp} \times K_f$$

The equilibrium constant of this net reaction controls the solubility of AgBr.

$$K = K_{sp} \times K_f = \frac{[Ag(NH_3)_2^+][Br^-]}{[NH_3]^2} = 1.2 \times 10^{-5}$$

Let s equal the solubility of AgBr in 6.0 M NH_3.

Concentration	AgBr(s) +	2NH$_3$	$\rightleftharpoons$ Ag(NH$_3$)$_2^+$	+ Br$^-$
Initial (M)		6.0	0	0
Change (M)		$-2s$	$+s$	$+s$
Equilibrium (M)		$(6.0 - 2s)$	s	s

Substitute into the K expression for the overall reaction:

$$K = \frac{(s)(s)}{(6.0 - 2s)^2} = 1.2 \times 10^{-5}$$

The left–hand side is a perfect square. Taking the square root of both sides gives:

$$\sqrt{\frac{s^2}{(6.0 - 2s)^2}} = \sqrt{1.2 \times 10^{-5}}$$

$$\frac{s}{6.0 - 2s} = 3.5 \times 10^{-3}$$

$$s = 2.1 \times 10^{-2} \ M$$

It is interesting to point out, that the solubility of AgBr in pure water is 8.8×10^{-7} M. The enhanced solubility in this example is due to the formation of the complex ion $Ag(NH_3)_2^+$.

PRACTICE QUESTION 17.22
Calculate the concentration of Cu^{2+} ions in a solution formed by adding 0.0500 mol of $CuSO_4$ to 0.500 L of 0.500 M NaCN?

PRACTICE QUESTION 17.23
Calculate the molar solubility of silver bromide in 1.0 M NH_3.

PRACTICE QUIZ

1. A buffer solution is prepared by mixing 500 mL of 0.60 M CH_3COOH with 500 mL of 1.00 M CH_3COONa solution. What is the pH of this solution?

2. A buffer solution is prepared by mixing 300 mL of 0.10 M HNO_2 with 200 mL of 0.40 M $NaNO_2$.
   a. Calculate the pH of the resulting solution.
   b. What is the new pH after 2.0 mL of 2.0 M HCl are added to this buffer?

3. What is the optimum pH of an HCN / CN^- buffer?

4. Which one of the following equimolar mixtures is suitable for making a buffer solution with an optimum pH of about 9.2?
   a. $NaC_2H_3O_2$ and $HC_2H_3O_2$
   b. NH_4Cl and NH_3
   c. HF and NaF
   d. HNO_2 and $NaNO_2$
   e. NaCl and HCl

5. A 20.0 mL portion of a solution of 0.0200 M HNO_3 is titrated with 0.0100 M KOH.
   a. How many mL of KOH are required to reach the equivalence point?
   b. What will the pH be after only 10.0 mL of KOH are added?
   c. What will the pH be after 45.0 mL of KOH are added?

6. Consider the titration of 50.0 mL of 0.10 M CH_3COOH with 0.10 M NaOH. What is the pH at the equivalence point?

7. Calculate the pH of a solution prepared by mixing 25.0 mL of 0.10 M HCl and 25.0 mL of 0.25 M CH_3COONa?

8. Which of the following is the *least* soluble in water?
   a. SrF_2 $K_{sp} = 2.8 \times 10^{-9}$
   b. $Zn(OH)_2$ $K_{sp} = 1.8 \times 10^{-14}$
   c. PbI_2 $K_{sp} = 1.4 \times 10^{-8}$
   d. BaF_2 $K_{sp} = 1.7 \times 10^{-6}$

9. At a certain temperature, the solubility of barium chromate ($BaCrO_4$) in water is 1.8×10^{-5} mol/L. What is the K_{sp} value at this temperature?

10. If the solubility of $Fe(OH)_2$ in water is 7.7×10^{-6} mol/L at a certain temperature, what is its K_{sp} value at that temperature?

Chapter 18
Electrochemistry

PROBLEM-SOLVING STRATEGIES AND TUTORIAL SOLUTIONS

TYPES OF PROBLEMS
Problem Type 1: Balancing Redox Reactions.

Problem Type 2: Galvanic Cells.

Problem Type 3: Standard Reduction Potentials.

Problem Type 4: Spontaneity of Redox Reactions under Standard–State Conditions.

Problem Type 5: Spontaneity of Redox Reactions under Conditions other than Standard State.
        A. The Nernst Equation.

Problem Type 6: Electrolysis.
        A. Electrolysis of an Aqueous Solution.
        B. Quantitative Applications of Electrolysis.

PROBLEM TYPE 1: BALANCING REDOX REACTIONS.

We learned in an earlier chapter that oxidation is a loss of electrons by a chemical species and reduction is a gain of electrons (**LEO** the lion says **GER**). Redox reactions are electron transfer reactions involving transfer of electrons from a **reducing agent**, the substance losing electrons, to an **oxidizing agent**, the substance gaining electrons.

For example, when copper metal is immersed in a solution containing Ag^+ ions, a reaction occurs in which electrons are transferred spontaneously from Cu atoms to the Ag^+ ions, forming Ag atoms that plate out on the copper surface. The newly formed Cu^{2+} ions go into solution:

$$Cu(s) + 2Ag^+(aq) \rightarrow Cu^{2+}(aq) + 2Ag(s)$$

Cu is oxidized to Cu^{2+} ion, and Ag^+ ion is reduced to Ag.

All redox reactions can be divided into half–reactions. The half–reactions are:

$$\begin{array}{ll} \text{Oxidation} & Cu \rightarrow Cu^{2+} + 2e^- \\ \text{Reduction} & 2e^- + 2Ag^+ \rightarrow 2Ag \end{array}$$

Copper metal is the reducing agent because it supplies electrons to the silver ions. The silver ions are the oxidizing agent because they accept electrons from copper. Two Ag^+ ions are required to accept the two electrons from one Cu atom.

Like all other reactions, redox reactions must be balanced. In order to do this, the overall reaction is divided into the two half–reactions, each is balanced according to mass and charge, and then the two half–reactions are added together to give the overall balanced equation. We will balance the equation on the next page to illustrate the steps involved.

This equation shows just the essential changes and is sometimes called a skeletal equation. To separate the equation into half–reactions, first identify which element is oxidized and which is reduced. Do this by writing oxidation numbers above each element.

$$\overset{+1\ -2\ \ +5\ -2}{H_2S\ +\ NO_3^-}\ \rightarrow\ \overset{0\ \ +2\ -2}{S\ +\ NO}$$

Note that S atoms and N atoms undergo oxidation number changes: S ($-2 \rightarrow 0$) and N ($+5 \rightarrow +2$). Sulfur is oxidized and nitrogen is reduced. Now we write the half–reactions.

$$\text{Oxidation} \qquad H_2S \rightarrow S$$
$$\text{Reduction} \qquad NO_3^- \rightarrow NO$$

Each half–reaction must be balanced by mass. In acidic solution, always add H^+ to balance H atoms, and add H_2O to balance O atoms. Here, we add H^+ to the oxidation half–reaction, and H_2O to the reduction half–reaction.

$$\text{Oxidation} \qquad H_2S \rightarrow S + 2H^+$$
$$\text{Reduction} \qquad NO_3^- \rightarrow NO + 2H_2O$$

The oxidation half–reaction is now balanced by mass. The reduction half–reaction needs $4H^+$ on the reactant side to achieve mass–balance.

$$\text{Reduction} \qquad 4H^+ + NO_3^- \rightarrow NO + 2H_2O$$

Next, the ionic charges must be balanced by adding electrons. In the oxidation half–reaction there is a net charge of $+2$ on the product side and zero on the reactant side. To balance the charge, 2 electrons are added to the product side.

$$\text{Oxidation} \qquad H_2S \rightarrow S + 2H^+ + 2e^-$$

Notice that both sides of the equation have the same charge, in this case, 0.

The reduction half–reaction can be charge balanced by adding three electrons to the reactant side.

$$\text{Reduction} \qquad 3e^- + 4H^+ + NO_3^- \rightarrow NO + 2H_2O$$

Note that electrons are added to the reactant side of a reduction half–reaction and to the product side of an oxidation half–reaction.

The balanced redox equation is obtained by adding the two half–reactions. But, first the number of electrons shown in the two half–reactions must be the same, since all the electrons lost during oxidation must be gained during reduction. Multiplying all the coefficients in the oxidation half–reaction by 3 and those in the reduction half–reaction by 2 will make the number of electrons in the two half–reactions equal. That is, 6 electrons are given up during oxidation and 6 are gained during reduction.

$$3 \times (H_2S \rightarrow S + 2H^+ + 2e^-) \quad \text{gives} \quad 3H_2S \rightarrow 3S + 6H^+ + 6e^-$$

$$2 \times (3e^- + 4H^+ + NO_3^- \rightarrow NO + 2H_2O) \quad \text{gives} \quad 6e^- + 8\,H^+ + 2\,NO_3^- \rightarrow 2NO + 4H_2O$$

The sum of the half–reactions is the overall balanced redox equation.

$$3H_2S \rightarrow 3S + 6H^+ + 6e^-$$

$$6e^- + 8H^+ + 2\,NO_3^- \rightarrow 2NO + 4H_2O$$

$$3H_2S + 8H^+ + 2\,NO_3^- \rightarrow 3S + 6H^+ + 2NO + 4H_2O$$

The $6H^+$ on the product side will cancel 6 of the $8H^+$ on the reactant side yielding:

$$3H_2S + 2H^+ + 2\,NO_3^- \rightarrow 3S + 2NO + 4H_2O$$

The following steps are useful in balancing redox equations.

1. Write the skeletal equation containing the oxidizing and reducing agents and the products in ionic form.
2. Separate the equation into two half–reactions.
3. Balance the atoms other than O and H in each half–reaction separately.
4a. For reactions in acidic medium, add H_2O to balance O atoms and H^+ to balance H atoms.
4b. For reactions in basic medium, first, balance the atoms as you would for an acidic solution. Then, for each H^+ ion, add an OH^- ion to both sides of the half–reaction. Whenever H^+ and OH^- appear on the same side, combine them to make H_2O.
5. Add electrons to one side of each half–reaction to equalize the charges. Electrons are added to the reactant side of a reduction half–reaction, and to the product side of an oxidation half–reaction. The number of electrons added to one side of a half–reaction should make the total charge of that side equal to the charge on the other side. This procedure is called a charge balance.
6. Add the two half–reactions together. Before this can be done, the number of electrons shown in both half–reactions must be the same. The number of electrons in the two half–reactions can be equalized by multiplying one or both half–reactions by appropriate coefficients. Now, add the two half–reactions.
7. Check the final equation by inspection. Recall that a properly balanced equation consists of a set of the smallest possible whole numbers.

STUDY QUESTION 18.1
Balance the following redox reaction by the ion–electron (half–reaction) method.
$$Cu + HNO_3 \rightarrow Cu^{2+} + NO_2 \quad \text{(in acidic solution)}$$

Solution: First, we need to write the skeletal equation and separate the given equation into half–reactions. Note that Cu is oxidized and nitrogen is reduced. However, it is not necessary to know which substance is reduced and which substance is oxidized. This will be evident as you work on balancing the half–reactions.

The half–reactions are:

Oxidation	$Cu \rightarrow Cu^{2+}$
Reduction	$HNO_3 \rightarrow NO_2$

We will continue with the steps of balancing redox reactions.

Step 3   The oxidation reaction's elements are already balanced. The nitrogen atoms in the reduction reaction are balanced.

Step 4a Balance the reduction half–reaction by adding one H_2O to the right–hand side, and one H^+ to the left–hand side. Step 4b is unnecessary, since the solution is acidic.

Oxidation	$Cu \rightarrow Cu^{2+}$
Reduction	$H^+ + HNO_3 \rightarrow NO_2 + H_2O$

Step 5   To balance according to charge, add electrons to balance the changes. For the oxidation reaction, add two electrons to the reactant side, to give it the same total neutral charge that exists on the reactant side. For the reduction reaction, one electron must be added to the reactant side to give it the same neutral charge that is on the product side. *At this point, it becomes evident which reaction is the oxidation reaction and which is the reduction reaction.* The loss of electron (copper reaction) is oxidation and the gain of electrons (nitrogen reaction) is reduction.

$$\text{oxidation} \qquad Cu \rightarrow Cu^{2+} + 2e^-$$
$$\text{reduction} \qquad e^- + H^+ + HNO_3 \rightarrow NO_2 + H_2O$$

Step 6   Next, equalize the number of electrons in the two half-reactions by multiplying the reduction half-reaction by 2.

$$2e^- + 2H^+ + 2HNO_3 \rightarrow 2NO_2 + 2H_2O$$

Now add the two half–reactions to obtain a balanced overall equation.

$$\text{oxidation} \qquad Cu \rightarrow Cu^{2+} + 2e^-$$
$$\underline{\text{reduction} \qquad 2e^- + 2H^+ + 2HNO_3 \rightarrow 2NO_2 + 2H_2O}$$
$$Cu + 2H^+ + 2HNO_3 \rightarrow Cu^{2+} + 2NO_2 + 2H_2O$$

Step 7   Check that the electrons on both sides cancel and that the equation is balanced in terms of atoms (mass) and charge.

PRACTICE QUESTION 18.1
Balance the following redox equations.

a. $Cr_2O_7^{2-}(aq) + Br^-(aq) \rightarrow Cr^{3+}(aq) + Br_2(g)$ (acidic solution)

b. $MnO_4^-(aq) + I^-(aq) \rightarrow MnO_2(s) + I_2(aq)$ (basic solution)

PROBLEM TYPE 2: GALVANIC CELLS.

Electrochemistry is the area of chemistry that deals with the interconversion of electrical energy and chemical energy. Electrochemical processes are redox reactions in which chemical energy is converted into electricity or in which electricity is used to cause a chemical reaction to occur.

A device which utilizes a spontaneous redox reaction to supply a constant flow of electrons is called a galvanic cell, or a voltaic cell. The design of a galvanic cell is such that the reactants are prevented from direct contact with each other. The oxidation half–reaction occurs at an electrode called the **anode**, and the reduction half–reaction occurs at an electrode called the **cathode**.

Figure 19.1 in your textbook shows a zinc–copper galvanic cell. The anode is a bar of zinc metal that is partially immersed in a solution of $ZnSO_4$. The cathode is a small bar of copper that is partially immersed in a solution of $CuSO_4$. The reducing agent, Zn metal, does not come into direct contact with the oxidizing agent, Cu^{2+} ion. Zn atoms lose two electrons and become zinc ions that enter the solution. Electrons travel through the outer circuit to the Cu electrode, where Cu^{2+} ions from the solution are reduced to copper atoms at the surface of the electrode. These atoms plate out on the electrode. As the reaction proceeds, the zinc electrode loses mass and the copper electrode gains mass.

While electrons travel through the outer circuit from Zn to Cu (from the anode to the cathode), negative ions move through a porous barrier or a "salt bridge" from the $CuSO_4$ solution into the $CuSO_4$ solution. This motion resupplies negative charge to the anode compartment and maintains electrical neutrality in the solutions surrounding the electrodes. An electric current will flow until either the Zn metal or the Cu^{2+} ions have completely reacted, provided that the flow of counter ions continues to keep the two compartments electrically neutral.

The fact that electrons flow from the anode to the cathode means that a difference in electrical potential energy exists between the two electrodes. Electrons flow from where they are at higher potential to where they are at a lower potential. The difference in electrical potential between the anode and cathode is measured in volts, and is called the **cell potential** or **cell emf**, which stands for *electromotive force*. The actual difference in electrical potential depends on the nature of the species involved in the half-reactions and their concentrations. The cell potential is represented by the symbol E_{cell}.

Rather than always representing a cell by a sketch, we can use a **cell diagram.** For instance, the zinc–copper cell discussed above and shown in Figure 19.1 is represented by:

$$Zn(s) \mid ZnSO_4(aq) \parallel CuSO_4(aq) \mid Cu(s)$$
$$\text{Anode} \qquad\qquad \text{Cathode}$$

In the diagram, the anode is on the left, and the cathode is on the right. The single vertical lines indicate phase boundaries, which, in this cell, are between the solid electrodes and the aqueous solutions. The sequence $Zn(s) \mid ZnSO_4(aq)$ represents oxidation, and the sequence $CuSO_4(aq) \mid Cu(s)$ represents reduction. The double vertical lines indicate a salt bridge or porous barrier between the two solutions.

STUDY QUESTION 18.2
Consider a galvanic cell constructed from the following half–cells, linked by a KNO$_3$ salt bridge. A Cu electrode immersed in 1.0 M Cu(NO$_3$)$_2$, and a Sn electrode in 1.0 M Sn(NO$_3$)$_2$. As the reaction proceeds, the Cu electrode gains mass and the Sn electrode loses mass.

    a.   **Which electrode is the anode and which is the cathode?**
    b.   **Write a balanced chemical equation for the overall cell reaction.**
    c.   **Sketch the half–cells and show the direction of flow of the electrons in the external circuit.**
    d.   **Which electrodes do the positive and the negative ions diffuse toward? Include a salt bridge in your sketch.**

Solution: Reduction occurs at the cathode and oxidation occurs at the anode. Since the Cu electrode gains mass, its half–reaction must involve the plating out of Cu^{2+} ions as Cu atoms. The copper electrode is the cathode.

$$Cu^{2+}(aq) + 2e^- \rightarrow Cu(s) \quad \text{reduction}$$

Since the Sn electrode loses mass, Sn metal must be reacting (oxidizing), and so Sn is the anode. The Sn^{2+} ions go into solution:

$$Sn(s) \rightarrow Sn^{2+}(aq) + 2e^- \quad \text{oxidation}$$

Adding the two half–reactions:

$$Sn(s) \rightarrow Sn^{2+}(aq) + 2e^-$$
$$\underline{Cu^{2+}(aq) + 2e^- \rightarrow Cu(s)}$$
$$\text{Overall} \quad Sn + Cu^{2+}(aq) \rightarrow Sn^{2+}(aq) + Cu$$

Electrons flow from the anode to the cathode. See the figure on the next page.

The positive ions (cations) diffuse toward the Cu electrode (the cathode) and the negative ions (anions) diffuse toward the Sn electrode (the anode).

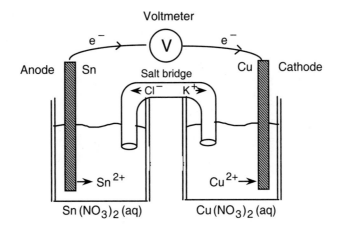

PROBLEM TYPE 3: STANDARD REDUCTION POTENTIALS.

The emf of a galvanic cell is dependent upon the nature of the half–reactions, the concentrations of the reactants and products, and the temperature. The standard cell emf is the voltage generated under standard state conditions. The standard cell emf has the symbol E°_{cell}. The superscript "°" denotes that all reactants and products are in their standard states. The standard state of all solute ions and molecules is a concentration of 1 M, and for all gases it is a partial pressure of 1 atm. *Unless stated otherwise, the temperature is assumed to be 25 °C.*

Rather than constructing a cell and measuring its voltage every time a cell emf is needed, a method has been devised that allows the use of a table of half–reactions and standard electrode potentials. The standard reduction potential, E°_{red}, is the voltage associated with a reduction reaction when all metals, solutes, and gases are in their standard states. This is the reduction potential when compared to a standard hydrogen electrode (the arbitrarily established reference point for reduction potential. A number of standard reduction potentials are listed in Table 18.1 of the textbook.

The following list summarizes the information contained in the table of standard reduction:

1. Standard electrode potentials are reduction potentials, and measure the ease of reduction under standard conditions. The E°_{red} values apply to the half–reactions when read in the forward direction. The more positive the E°_{red} value, the greater the tendency for the half–reaction to occur in the forward direction.
2. All half–cell reactions are reversible. Depending on the reducing strength of the other electrode that is chosen, any given electrode may act as an anode or as a cathode.
3. The species appearing on the reactant side of the half–reactions are oxidizing agents, and the E°_{red} value is related to the tendency of these oxidizing agents to accept electrons. The very strong oxidizing agents are at the top of the table. F_2 is the strongest oxidizing agent and Li^+ is the weakest.
4. The species shown on the product side of the half–reactions are reducing agents. For a reducing agent to supply electrons, the reaction must occur in the reverse direction. The strength of the reducing agents increases as you read down the table. Therefore, F^- ion is the weakest reducing agent and Li metal is the strongest reducing agent.
5. When predicting the direction of reaction under standard state conditions, you can use the diagonal rule. This rule states that any species on the reactant side of a given half–cell reaction (an oxidizing agent) will react spontaneously with a species that is shown on the product side of any half–reaction (a reducing agent) located below it in the table.
6. The standard reduction potential is not affected when stoichiometric coefficients of a half–cell reaction are changed. For example,

$$Ag^+ + e^- \rightarrow Ag(s) \qquad\qquad E^{\circ}_{red} = 0.80 \text{ V}$$

$$2Ag^+ + 2e^- \rightarrow 2Ag(s) \qquad\qquad E^{\circ}_{red} = 0.80 \text{ V}$$

Electrode potentials are intensive properties, and therefore do not depend on the size of the electrode or the manner in which the half–reaction is balanced.

Notice the 0.0 V potential assigned to hydrogen ion. A positive reduction potential means that the oxidizing agent (such as F_2) has a greater tendency to be reduced than the hydrogen ion, and is therefore a stronger oxidizing agent. A negative reduction potential for an oxidizing agent (such as Mg^{2+}) indicates that the hydrogen ion is more readily reduced than that substance.

Standard reduction potentials are used to calculate the potential of a galvanic cell. The standard cell emf (or standard cell potential), E°_{cell}, is the difference between the standard reduction potential of the cathode, and the standard reduction potential of the anode.

$$E^{\circ}_{cell} = E^{\circ}_{cathode} - E^{\circ}_{anode}$$

For example, the standard emf of the silver–copper cell in the cell diagram:

$$Cu(s) \mid CuSO_4(1 \ M) \parallel AgNO_3(1 \ M) \mid Ag(s)$$

can be calculated as follows. The overall reaction is:

$$Cu + 2Ag^+ \rightarrow Cu^{2+} + 2Ag$$

The half–reactions are:

$$
\begin{array}{ll}
\text{Oxidation} & Cu \rightarrow Cu^{2+} + 2e^- \\
\text{Reduction} & 2Ag^+ + 2e^- \rightarrow 2Ag
\end{array}
$$

Therefore, the standard cell emf is given by:

$$E^{\circ}_{cell} = E^{\circ}_{cathode} - E^{\circ}_{anode} = E^{\circ}_{Cu^{2+}/Cu} - E^{\circ}_{Ag^+/Ag}$$

The symbols Cu^{2+}/Cu and Ag/Ag^+ represent the reduction half-reactions that are related to the reduction potentials. The reduction potentials needed here are found in Table 19.1 of the text.

$$E^{\circ}_{cathode} = E^{\circ}_{Ag^+/Ag} = 0.80 \text{ V}$$

$$E^{\circ}_{anode} = E^{\circ}_{Cu^{2+}/Cu} = 0.34 \text{ V}$$

$$E^{\circ}_{cell} = 0.80 \text{ V} - 0.34 \text{ V}$$

$$E^{\circ}_{cell} = 0.46 \text{ V}$$

A positive E°_{cell} indicates that the reaction is spontaneous under standard conditions.

STUDY QUESTION 18.3
Calculate the standard cell emf (cell potential) for a galvanic cell having the overall cell reaction of tin metal and lead(II) ion and determine if the reaction is spontaneous under standard state conditions.

Solution: Let's begin by writing the overall cell reaction.

$$Sn(s) + Pb^{2+}(aq) \rightarrow Sn^{2+}(aq) + Pb(s)$$

The standard emf of the cell is $E^\circ_{cell} = E^\circ_{cathode} - E^\circ_{anode}$.

Separate the overall reaction into half-reactions, and identify the anode and cathode. We see that Pb^{2+} is reduced and tin is oxidized. Look up the necessary reduction potentials in Table 19.1 of the text.

$Pb^{2+}(aq) + 2e^- \rightarrow Pb(s)$	reduction/cathode
$Sn(s) \rightarrow Sn^{2+}(aq) + 2e^-$	oxidation/anode
$Sn(s) + Pb^{2+}(aq) \rightarrow Sn^{2+}(aq) + Pb(s)$	overall

$$E^\circ_{cell} = E^\circ_{cathode} - E^\circ_{anode} = E^\circ_{Pb^{2+}/Pb} - E^\circ_{Sn^{2+}/Sn}$$

$$E^\circ_{cell} = -0.13 \text{ V} - (-0.14 \text{ V}) = 0.01 \text{ V}$$

Since the standard cell potential is a positive value, the reaction is spontaneous under standard conditions.

> **PRACTICE QUESTION 18.2**
> **Predict whether Br^- ion can reduce I_2 under standard state conditions, in the following reaction:**
> $$2Br^-(aq) + I_2(s) \rightarrow Br_2(\ell) + 2I^-(aq).$$

> **PRACTICE QUESTION 18.3**
> **Calculate the standard emf (E°_{cell}), at 25°C, of cells that have the following overall reactions:**
> a. $2Co^{2+}(aq) + Zn^{2+}(aq) \rightarrow 2Co^{3+}(aq) + Zn(s)$
> b. $Cl_2(g) + 2Br^-(aq) \rightarrow 2Cl^-(aq) + Br_2(\ell)$

> **PRACTICE QUESTION 18.4**
> **Mn^{2+} will reduce which of the following ions?**
> F^- Ce^{4+} Ce^{3+} $Cr_2O_7^{2-}$ Hg_2^{2+} Cu^+ MnO_4^-

PROBLEM TYPE 4: SPONTANEITY OF REDOX REACTIONS UNDER STANDARD-STATE CONDITIONS.

For a process carried out at constant temperature and pressure, the Gibbs free energy change is equal to the maximum amount of work (w_{max}) that can be done by the process.

$$\Delta G = w_{max}$$

Any reaction that occurs spontaneously can be utilized to perform useful work. Both combustion of fuels and the oxidation of food components provide energy to do work. The difference in free energy between the reactants and the products is the energy available to do work.

Electrical work is equal to the product of the cell emf (E_{cell}°) and the total charge in coulombs carried through the circuit. The total charge in coulombs is nF, where n is the number of moles of electrons transferred from the reducing agent to the oxidizing agent, according to the balanced equation, and F is the Faraday constant. The Faraday constant equals the charge in coulombs carried by one mole of electrons which is 96,500 C/mol of electrons. In most thermodynamic calculations, F is expressed in Coulomb per mole of e^- units: 96,500 J/V·mol e^-.

The electrical work is given by:

$$w_{electric} = -nFE_{cell}$$

The maximum work from an electrochemical process is:

$$w_{max} = w_{electric} = -nFE_{cell}$$

The negative sign is in accord with the sign convention that when work is done on the surroundings, the system loses energy.

The Gibbs free energy change for a redox reaction is:

$$\Delta G = -nFE_{cell}$$

For a spontaneous redox reaction, ΔG will be negative, and E must be positive. For reactions in which reactants and products are in their standard states:

$$\Delta G = -nF\,E_{cell}^{\circ}$$

STUDY QUESTION 18.4
Calculate $\Delta G°$ at 25 °C for the reaction, given the standard cell potential:

$$2Br^-(aq) + I_2(s) \rightarrow Br_2(\ell) + 2I^-(aq) \qquad\qquad E_{cell}^{\circ} = -0.54 \text{ V.}$$

Solution: The standard Gibbs free energy change is related to the standard cell potential by the following equation in which is the Faraday constant and n is the number of moles of electrons transferred:

$$\Delta G = -nE°$$

Examination of the half–reactions clearly indicates that 2 mol of electrons are transferred in the balanced equation, $n = 2$.

The half–reactions are:

Reduction	$2e^- + I_2(s) \rightarrow 2I^-(aq)$
Oxidation	$2Br^-(aq) \rightarrow Br_2(\ell) + 2e^-$
Overall	$2Br^-(aq) + I_2(s) \rightarrow Br_2(\ell) + 2I^-(aq)$

Substituting into $\Delta G° = -n\underline{F}\,E_{cell}^{\circ}$

$$\Delta G° = -2(96,500 \text{ J/V·mol})(-0.54 \text{ V})$$

The units of the Faraday constant must be expressed in J/(V·mol) in order to cancel the units of volts from E_{cell}°. Continuing with our calculation:

$$\Delta G^{\circ} = 1.04 \times 10^5 \text{ J} = 104 \text{ kJ/mol}$$

The positive value of ΔG° indicates the reaction is not spontaneous under standard conditions.

PRACTICE QUESTION 18.5

Calculate the standard Gibbs free energy changes for the following overall reactions.

a. $2Co^{2+}(aq) + Zn^{2+}(aq) \rightarrow 2Co^{3+}(aq) + Zn(s)$
b. $Cl_2(g) + 2Br^-(aq) \rightarrow 2Cl^-(aq) + Br_2(\ell)$

PRACTICE QUESTION 18.6

What is the net ionic equation for the spontaneous reaction that will occur between two half-cells, one containing $FeSO_4$(aq) and an Fe electrode and the other $NiCl_2$(aq) and a Ni electrode.

PROBLEM TYPE 5: SPONTANEITY OF REDOX REACTIONS UNDER CONDITIONS OTHER THAN STANDARD STATE.

A.   The Nernst Equation.

We previously mentioned that the emf of a cell depends on the nature of the reactants and products and on their concentrations. Since the cell emf is a measure of the spontaneity of the cell reaction, we might reasonably expect the voltage to fall as reactants are consumed and products accumulate.

The equation that relates the cell emf to the concentrations of reactants and products is named after Walter Nernst. At 298 K, for a redox reaction of the type:

$$aA + bB \rightleftharpoons cC + dD$$

The Nernst equation is:

$$E = E^{\circ} - \frac{0.0257 \text{ V}}{n} \ln Q$$

where E° is the standard cell emf, E is the nonstandard cell emf, 0.0257 V is a constant at 298 K, and n is the number of moles of electrons transferred according to the balanced equation. The reaction quotient Q contains the concentrations of reactants and products.

$$Q = \frac{[C]^c [D]^d}{[A]^a [B]^b}$$

The Nernst equation predicts that E will decrease as reactant concentrations decrease, or as product concentrations increase. Both changes cause Q to increase. As Q increases, $\ln Q$ increases. Therefore, an increasingly larger number is subtracted from E°. As a result, E is not a constant, but decreases as a reaction proceeds. E° is a constant for a reaction and is characteristic of the reaction.

At equilibrium, this equation reduces to one we have seen before. When the reaction is at equilibrium, no net reaction occurs and no net transfer of electrons occurs, and $E = 0$. Also, at equilibrium, $Q = K$, and the Nernst equation becomes:

$$0 = E° - \frac{0.0257 \text{ V}}{n} \ln K$$

$$E° = \frac{0.0257 \text{ V}}{n} \ln K$$

Therefore, the standard cell emf is related to the equilibrium constant.

When both E_{cell} and $E°_{cell}$ are known, the Nernst equation can be used to calculate an unknown concentration. If any one of the three quantities $\Delta G°$, K, or $E°$ is known, both of the others can be calculated. Table 19.1 summarizes the criteria for spontaneous redox reactions.

Table 19.1 Criteria for Spontaneous Redox Reactions

$\Delta G°$	K	$E°_{cell}$	Reaction under Standard Conditions
Negative	> 1	Positive	Spontaneous
0	= 1	0	At equilibrium
Positive	< 1	Negative	Nonspontaneous -- (Reaction is spontaneous in the reverse direction)

STUDY QUESTION 18.5
Calculate the cell emf at 25 °C for a galvanic cell in which the following overall reaction occurs at the concentrations given:

$$\text{Zn}(s) + 2\text{H}^+(aq, 1.0 \times 10^{-4}\ M) \rightarrow \text{Zn}^{2+}(aq, 1.5\ M) + \text{H}_2(g, 1\text{ atm})$$

Solution: The cell emf can be calculated by use of the Nernst equation.

$$E = E° - \frac{0.0257 \text{ V}}{n} \ln \frac{[\text{Zn}^{2+}]P_{\text{H}_2}}{[\text{H}^+]^2}$$

Where,

$$E°_{cell} = E°_{cathode} - E°_{anode} = E°_{\text{H}^+/\text{H}_2} - E°_{\text{Zn}^{2+}/\text{Zn}} = 0.0 \text{ V} - (-0.76 \text{ V})$$

$$E°_{cell} = 0.76 \text{ V}$$

$n = 2$ (moles of electrons transferred according to the balanced equation)
$P_{\text{H}_2} = 1$, because hydrogen is in its standard state, 1 atm.

$$[\text{Zn}^{2+}] = 1.5\ M \text{ and } [\text{H}^+] = 1.0 \times 10^{-4}\ M$$

Substitution into the Nernst equation yields:

$$E = 0.76 \text{ V} - \frac{0.0257 \text{ V}}{2} \ln \frac{1.5}{(1.0 \times 10^{-4})^2}$$

$$= 0.76 \text{ V} - 0.24 \text{ V} = 0.52 \text{ V}$$

The low concentration of H^+ (1.0×10^{-4} M) compared to its standard state value (1 M) means that the driving force for reaction will be less than in the standard state, and as we see $E < E°$.

PRACTICE QUESTION 18.7

The cell emf, at 25°C, for the following reaction is 0.98 V.

$$2Ag^+(aq, 1\ M) + H_2(g, 1\ atm) \rightleftharpoons 2Ag(s) + 2H^+(aq,\ ?\ M)$$

Calculate the concentration of hydrogen ions.

PRACTICE QUESTION 18.8

Calculate E°_{cell}, E_{cell}, and ΔE°_{cell} for the following cell reaction at 25°C.

$$3Zn(s) + 2\ Cr^{3+}\ (0.0010\ M) \rightleftharpoons 3Zn^{2+}\ (0.010\ M) + 2Cr(s)$$

PROBLEM TYPE 6: ELECTROLYSIS.

A. Electrolysis of an Aqueous Solution.

Electrolysis is the process in which electrical energy is used to cause nonspontaneous redox reactions to occur. An electrolytic cell is the apparatus for carrying out electrolysis. A battery or other DC power supply serves as an "electron pump" that supplies electrons to the cathode, where chemical species are reduced. Electrons resulting from the oxidation of chemical species are withdrawn from the anode and return to the battery. In the electrolytic cell, the cathode is negative and the anode is positive. This is the opposite of a galvanic cell.

For example, molten calcium chloride can be decomposed into the elements; calcium and chlorine in an electrolytic cell. The reactions at the electrodes are:

Cathode/reduction	$Ca^{2+}(\ell) + 2e^- \rightarrow Ca(s)$
Anode/oxidation	$2\ Cl^-(\ell) \rightarrow Cl_2(g) + 2e^-$
Overall	$Ca^{2+}(\ell) + 2\ Cl^-(\ell) \rightarrow Ca(s) + Cl_2(g)$

The electrode in the electrolysis cell that is attached to the battery's negative terminal is the cathode. Electrons flow from the battery onto the cathode. Ca^{2+} ions are attracted to the cathode and are reduced to Ca metal. The anode of the electrolysis cell is attached to the positive terminal of the battery. Chloride ions drift toward the anode where they are oxidized. The electrons travel from the anode to the battery. For every electron leaving the battery, one must return. The minimum voltage required to bring about electrolysis can be *estimated* from the standard reduction potentials;

$$E^\circ_{cell} = E^\circ_{cathode} - E^\circ_{anode} = E^\circ_{Ca^{2+}/Ca} - E^\circ_{Cl_2/Cl^-}$$

$$E^\circ_{cell} = -2.87\ V - 1.36\ V = -4.23\ V$$

Note: the above values apply to species in aqueous solution. Here we are dealing with molten $CaCl_2$. The negative cell potential means that this is a nonspontaneous reaction. This reaction can be forced to occur by connecting the electrodes to an external source of electrical energy such as a battery or voltage supply. The electrolysis will occur when the external voltage is greater than the negative voltage of the nonspontaneous reaction.

A complicating factor in the electrolysis of aqueous solutions containing a dissolved salt is that water molecules can be reduced to H_2 or oxidized to O_2 in preference to the solute species.

$$\text{Reduction of water: } 2H_2O(\ell) + 2e^- \rightarrow H_2(g) + 2OH^- \qquad E^\circ_{red} = -0.83 \text{ V}$$

$$\text{Oxidation of water: } H_2O(\ell) \rightarrow \frac{1}{2}O_2(g) + 2H^+(aq) + 2e^- \qquad -E^\circ_{red} = -1.23 \text{ V}$$

Because oxidation is the reverse of reduction, the potentials measuring the ease of oxidation are $-E^\circ_{red}$.

We must consider whether water molecules are oxidized or reduced instead of ions of the salt. In the electrolysis of aqueous KI solution, for example, $H_2(g)$ is formed at the cathode, and I_2 at the anode. $H_2(g)$ is coming from the reduction of water. Water has a higher (more positive) standard reduction potential than K^+, which indicates that it is reduced more easily than potassium ion. K^+ ions drift to the cathode, but no K metal is formed.

$$2H_2O(\ell) + 2e^- \rightarrow H_2(g) + 2OH^-(aq) \qquad E^\circ_{red} = -0.83 \text{ V}$$

$$K^+(aq) + e^- \rightarrow K(s) \qquad E^\circ_{red} = -2.93 \text{ V}$$

Molecular iodine is formed at the anode during electrolysis of aqueous KI. This means the anode reaction is:

$$2I^-(aq) \rightarrow I_2(s) + 2e^-$$

rather than:

$$H_2O(\ell) \rightarrow \frac{1}{2}O_2(g) + 2H^+(aq) + 2e^-$$

This preference reflects that I^- is easier to oxidize than H_2O.

$$2I^- \rightarrow I_2(s) + 2e^- \qquad -E^\circ_{red} = -0.53 \text{ V}$$

$$H_2O(\ell) \rightarrow \frac{1}{2}O_2(g) + 2H^+(aq) + 2e^- \qquad -E^\circ_{red} = -1.23 \text{ V}$$

The higher (more positive) the value of $-E^\circ_{red}$, the more favorable is the oxidation. Iodide ions have a higher oxidation potential than H_2O, and so iodide ions are oxidized preferentially.

In summary, if there is more than one reducible species in solution, the species with the greater reduction potential will be preferentially reduced. If there is more than one oxidizable species in solution, the species with the higher $-E^\circ_{red}$ will be preferentially oxidized.

STUDY QUESTION 18.6
Predict the products of the electrolysis of aqueous MgCl₂ solution.

Solution: In aqueous solution the metal ion is not the only species that can be reduced. H_2O can be reduced as well. The two possible reduction reactions are:

$$Mg^{2+} + 2e^- \rightarrow Mg \qquad E^\circ_{red} = -2.37 \text{ V}$$

$$2H_2O + 2e^- \rightarrow H_2 + 2OH^- \qquad E^\circ_{red} = -0.83 \text{ V}$$

Of the two, water has a higher standard reduction potential, and therefore has a greater tendency to be reduced than Mg^{2+}. The two oxidizable species are H_2O and Cl^- ion. The possible oxidation reactions and their $-E^{\circ}_{red}$ potentials are:

$$H_2O \rightarrow \frac{1}{2} O_2 + 2H^+ + 2e^- \qquad -E^{\circ}_{red} = -1.23 \text{ V}$$

$$2Cl^- \rightarrow Cl_2 + 2e^- \qquad -E^{\circ}_{red} = -1.36 \text{ V}$$

The oxidation potentials indicate that H_2O is more readily oxidized than chloride ions. However, in this particular case, as discussed in Section 19.7 of the text, Cl^- ions are actually oxidized. The large overvoltage for O_2 formation prevents its production when Cl^- ion is there to compete. The overall reaction is:

$$2H_2O + 2Cl^- \rightarrow H_2 + 2OH^- + Cl_2$$

Please note that the only example of overvoltage that you will need to be concerned with is in the oxidation of H_2O. It lowers the $-E^{\circ}_{red}$ by about 0.6 V to -1.83 V.

B. Quantitative Applications of Electroylsis.

Our knowledge of the quantitative relationships in electrolysis is due mostly to the work of Michael Faraday. He observed that the quantity of a substance undergoing chemical change during electrolysis is proportional to the quantity of electrical charge that passes through the cell. The quantity of electrical charge is expressed in coulombs. The electrical charge of 1 mole of electrons is called 1 Faraday which is equal to 96,500 coulombs.

$$1 \text{ mole e}^- = 1 \, F = 96,500 \text{ C.}$$

The number of electrons shown in the half–reaction is the link between the number of moles of substance reacted and the number of faradays of electricity required. We obtain the following conversion factors for the two half-reactions shown below:

$$K^+ + e^- \rightarrow K \qquad\qquad Ca^{2+} + 2e^- \rightarrow Ca$$

$$\frac{1 \text{ mol K}}{1 \text{ mol e}^-} \qquad\qquad \frac{1 \text{ mol Ca}}{2 \text{ mol e}^-}$$

The number of coulombs passing through an electrolytic cell in a given period of time is related to the electrical current in amperes (A), where 1 ampere is equal to 1 coulomb per second.

$$1 \text{ A} = 1 \text{ C/s}$$

In an electrolysis experiment, the measured quantities are current and the time that the current flows. The number of coulombs passing through the cell is

$$\text{Total charge in coulombs (C)} = \text{current (A)} \times \text{time } (t)$$

$$\text{Charge in coulombs} = \frac{C}{s} \times s$$

When an electrical current of known amperes flows for a known time, the amount of chemical change can be calculated according to the following road map.

$$\text{current} \rightarrow \text{time} \rightarrow \text{coulombs} \rightarrow [\text{moles } e^- \text{ transferred}] \rightarrow \underset{\text{product}}{\text{moles}} \rightarrow \underset{\text{product}}{\text{grams}}$$

The number of moles of electrons is the key term. It links the experimental variables to the theoretical yield. Reversing the road map allows one to calculate the time required to produce a specific amount of chemical change with a given electrical current.

STUDY QUESTION 18.7
How many grams of copper metal will be deposited from a solution of CuSO₄ by the passage of 3.0 A of electrical current through an electrolytic cell for 2.0 hours?

Solution: Cu^{2+} is reduced according to the half–reaction:

$$Cu^{2+} + 2e^- \rightarrow Cu$$

This half–reaction equation tells us that 2 moles of electrons are required to produce 1 mol of Cu.

$$\frac{1 \text{ mol Cu}}{2 \text{ mol } e^-}$$

The number of coulombs passing through a cell is given by the current times the time.

$$3.0 \text{ A} \times 2.0 \text{ h} \times \frac{1 \text{ C/s}}{1 \text{ A}} \times \frac{3600 \text{ s}}{1 \text{ h}} = 2.16 \times 10^4 \text{ C}$$

From this number of coulombs we can find the number of moles of electrons passing through the cell. The road map will be:

$$\text{current} \times \text{time} \times \text{coulombs} \times [\text{moles } e^-] \times \text{mol Cu} \times \text{g Cu}$$

The number of moles of electrons is the link from the current measurements to the molar amount of Cu formed. The number of moles of electrons is:

$$2.16 \times 10^4 \text{ C} \times \frac{1 \text{ mol } e^-}{9.65 \times 10^4 \text{C}} = 0.224 \text{ mol } e^-$$

The number of moles of Cu is:

$$0.224 \text{ mol } e^- \times \frac{1 \text{ mol Cu}}{2 \text{ mol } e^-} = 0.112 \text{ mol Cu}$$

The number of grams of Cu is

$$0.112 \text{ mol Cu} \times \frac{63.5 \text{ g}}{1 \text{ mol Cu}} = 7.11 \text{ g Cu}$$

Of course, this calculation could be carried out using dimensional analysis.

$$3.0 \text{ A} \times \frac{1 \text{ C/s}}{1 \text{ A}} \times 2.0 \text{ h} \times \frac{3600 \text{ s}}{1 \text{ h}} \times \frac{1 \text{ mol } e^-}{9.65 \times 10^4 \text{C}} \times \frac{1 \text{ mol Cu}}{2 \text{ mol } e^-} \times \frac{63.5 \text{ g}}{1 \text{ mol Cu}} = 7.11 \text{ g Cu}$$

PRACTICE QUESTION 18.9

Write the half–reaction that occurs at the:

    a.   **anode during the electrolysis of aqueous lithium bromide**
    b.   **cathode during the electrolysis of aqueous lithium bromide**

PRACTICE QUESTION 18.10

How many minutes are required to electroplate 25.0 g of metallic Cu using a constant current of 20.0 A?

PRACTICE QUIZ

1. Draw the cell diagrams for each of the redox reactions given below. You may use platinum as an inert electrode.
   a. $2Al(s) + 3H_2SO_4(aq) \rightarrow Al_2(SO_4)_3(aq) + 3H_2(g)$
   b. $Fe_2(SO_4)_3(s) + 3Pb(s) \rightarrow 3PbSO_4(s) + 2Fe(s)$
   c. $CuSO_4(aq) + H_2(g) \rightarrow Cu(s) + H_2SO_4(aq)$
   d. $2\,NaBr(aq) + I_2(g) \rightarrow Br_2(\ell) + 2NaI(aq)$
   e. $2FeCl_2(aq) + SnCl_2(aq) \rightarrow 2FeCl_3(aq) + Sn(s)$
   f. $2FeCl_2(aq) + SnCl_4(aq) \rightarrow 2FeCl_3(aq) + SnCl_2(aq)$

2. Arrange the following species in order of increasing strength as oxidizing agents:
   $Ce^{4+}, O_2, H_2O_2, SO_4^{2-}$

3. a. Will $O_2(g)$ oxidize I^- ions in acid solution under standard conditions?
   b. Will $O_2(g)$ oxidize Br^-?
   c. Will $O_2(g)$ oxidize Cl_2?

4. Calculate $\Delta G°$ and the equilibrium constant (K) for the following reaction at 25 °C.

   $$Fe^{3+}(aq) + Ag(s) \rightarrow Fe^{2+}(aq) + Ag^+(aq)$$

5. At what ratio of $[Fe^{2+}]/[Fe^{3+}]$ would the reaction given in problem 6 become spontaneous if $[Ag^+] = 1.0\ M$?

6. Calculate the cell emf for the following reaction:

   $$2Ag^+(0.10\ M) + H_2(1\ atm) \rightarrow 2Ag(s) + 2H^+(pH = 8)$$

7. What is the minimum voltage required to bring about electrolysis of a solution of Cu^{2+} and Br^- at standard concentrations?

8. What change in the emf of a hydrogen-copper cell will occur when $NaOH(aq)$ is added to the solution in the hydrogen half–cell?

9. Consider a uranium–bromine cell in which U is oxidized and Br_2 is reduced. The half–reactions are:

   $$U^{3+}(aq) + 3e^- \rightarrow U(s) \qquad\qquad E°_{U^{3+}/U} = ?$$
   $$Br_2(\ell) + 2e^- \rightarrow 2Br^-(aq) \qquad\qquad E°_{Br_2/Br^-} = 1.07\ V$$

   If the standard cell emf is 2.91 V, what is the standard reduction potential for uranium?

10. When the concentration of Zn^{2+} ion is 0.15 M, the measured voltage of the Zn–Cu cell is 0.40 V. What is the Cu^{2+} ion concentration?

11. A hydrogen electrode is immersed in an acetic acid solution. This electrode is connected to another, consisting of an iron nail dipped into 0.10 M $FeCl_2$. If E_{cell} is found to be 0.24 V, what is the pH of the acetic acid solution?

12. How many moles of electrons are transferred in an electrolytic cell when a current of 2.0 amps flows for 6 hours?

13. How many moles of electrons are required to electroplate 6.0 g of chromium from a solution containing Cr^{3+}?

14. a. How many grams of nickel can be electroplated by passing a constant current of 5.2 A through a solution of $NiSO_4$ for 60.0 min?
    b. How many grams of cobalt can be electroplated by passing a constant current of 5.2 A through a solution of $CoCl_3$ for 60.0 min?

15. How long will it take to produce 54 kg of Al metal by the reduction of Al^{3+} in an electrolytic cell using a current of 500 amps?

16. Balance the following redox reactions, in acidic solution.
    a. $H_2O_2 + I^- \rightarrow I_2 + H_2O$
    b. $Cr_2O_7^{2-} + H_3AsO_3 \rightarrow Cr^{3+} + H_3AsO4$

17. Balance the following redox reaction, in basic solution.
    $Cl_2 \rightarrow ClO_4^- + Cl^-$

Chapter 19
Chemical Kinetics

PROBLEM-SOLVING STRATEGIES AND TUTORIAL SOLUTIONS

TYPES OF PROBLEMS

Problem Type 1: Measuring Reaction Progress and Expressing Reaction Rate.
    A. Average Reaction Rate.
    B. Instantaneous Rate.
    C. Stoichiometry and Reaction Rate.

Problem Type 2: Dependence of Reaction Rate on Reactant Concentration.
    A. The Rate Law.
    B. Experimental Determination of the Rate Law.

Problem Type 3: Dependence of Reactant Concentration on Time.
    A. First–Order Reactions.
    B. Second–Order Reactions.

Problem Type 4: Dependence of Reactant Concentration on Temperature: The Arrhenius Equation.

Problem Type 5: Reaction Mechanisms.
    A. Elementary Reactions.
    B. Rate–Determining Step.
    C. Experimental Support for Reaction Mechanisms.

Problem Type 6: Catalysis.
    A. Heterogeneous Catalysis.
    B. Homogeneous Catalysis.
    C. Enzymes: Biological Catalysts.

PROBLEM TYPE 1: MEASURING REACTION PROGRESS AND EXPRESSING REACTION RATE.

A. Average Reaction Rate.

As a chemical reaction proceeds, the concentrations of reactants and products change with time. For instance, as the reaction $A + B \rightarrow C$ progresses, the concentrations of A and B decrease while the concentration of C increases. The rate can be expressed as the change in the molar concentration of any of the species in the reaction per given amount of time, Δt.

$$\text{rate} = \frac{-[A]}{\Delta t} = \frac{-[B]}{\Delta t} = \frac{-[C]}{\Delta t}$$

For the generic reaction of $aA + bB \rightarrow cC$, the rate based on each species would be affected by the stoichiometry as follows.

$$\text{rate} = -\frac{1}{a}\frac{[A]}{\Delta t} = -\frac{1}{b}\frac{[B]}{\Delta t} = \frac{1}{c}\frac{[C]}{\Delta t}$$

Note that a negative sign is inserted before terms involving reactants. The change in A concentration, $\Delta[A]$, is negative because the concentration of A *decreases* with time. Inserting a negative sign in the expression makes the rate of reaction a positive quantity. No matter which reactant or product we use, the reaction rate will be positive and have the same value.

STUDY QUESTION 19.1
Write expressions for the rate of the following reaction in terms of each of the reactants and products.
$$2N_2O_5(g) \rightarrow 4NO_2(g) + O_2(g)$$

Solution: Recall that the rate is defined as the change in concentration of a reactant or product with time. Each "change–in–concentration" term is divided by the corresponding stoichiometric coefficient. Terms involving reactants are preceded by a minus sign. Therefore, the rate is expressed as:

$$\text{rate} = -\frac{\Delta[N_2O_5]}{2\Delta t} = \frac{\Delta[NO_2]}{4\Delta t} = \frac{\Delta[O_2]}{\Delta t}$$

STUDY QUESTION 19.2
Oxygen gas is formed by the decomposition of nitric oxide:

$$2NO(g) \rightarrow O_2(g) + N_2(g)$$

If the rate of formation of O_2 is 0.054 M/s, what is the rate of change of NO concentration?

Solution: From the stoichiometry of the reaction, 2 mol NO react for each mole of O_2 that forms.

$$\text{rate} = -\frac{\Delta[NO]}{2\Delta t} = \frac{\Delta[O_2]}{\Delta t}$$

$$\frac{\Delta[NO]}{2\Delta t} = -\frac{2\Delta[O_2]}{\Delta t} = -2(0.054 \text{ M/s})$$

$$= -0.11 \text{ } M\text{/s}$$

PRACTICE QUESTION 19.1
Write the rate expression with respect to each species in the following reaction.
$$N_2(g) + 3H_2(g) \rightarrow 2NH_3(g)$$

PRACTICE QUESTION 19.2
For the reaction in Practice Question 14.1, what is the rate of change for H_2 if the rate of appearance of NH_3 is 0.096 M/s?

B. Average Reaction Rate.
The average rate of reaction over any time interval is equal to the change in the concentration of a reactant $\Delta[A]$, or of a product $\Delta[C]$, divided by the time interval, Δt, during which the change occurred.

$$\text{average rate} = -\frac{\text{change in the concentration of A}}{\text{length of time interval}}$$

$$\text{average rate} = -\frac{\Delta[A]}{\Delta t} = -\frac{[A]_2 - [A]_1}{t_2 - t_1}$$

The concentration term $[A]_2$ is the concentration of A at time t_2, and $[A]_1$ is the concentration of A at time t_1.

STUDY QUESTION 19.3
Listed in the following table are experimental data for the hypothetical reaction

$$A \rightarrow 2B$$

Time (s)	[A] (mol/L)
0.00	1.000
10.0	0.891
20.0	0.794
30.0	0.707
40.0	0.630

a. Calculate the average rates of change of [A], and the average reaction rates for the two time intervals from 0 to 10 s and from 30 to 40 s.
b. Why does the rate decrease from one time interval to the next?

Solution:

a. The average rate of change of [A] is given by:

$$\frac{\Delta[A]}{\Delta t} = \frac{[A]_2 - [A]_1}{t_2 - t_1}$$

For the time interval 0 – 10 s, we get:

$$\frac{\Delta[A]}{\Delta t} = \frac{0.891 \text{ mol/L} - 1.00 \text{ mol/L}}{10.0 \text{ s} - 0 \text{ s}} = \frac{-0.109 \text{ mol/L}}{10.0 \text{ s}} = -0.0109 \text{ mol/L·s}$$

Since the average reaction rate $= -\dfrac{\Delta[A]}{\Delta t}$,

$$\text{average reaction rate} = -(-0.0109 \text{ mol/L·s}) = 0.0109 \text{ mol/L·s}$$

For the time interval 30 – 40 s, the rate of change of [A] is:

$$\frac{\Delta[A]}{\Delta t} = \frac{0.630 \text{ mol/L} - 0.707 \text{ mol/L}}{40.0 \text{ s} - 30.0 \text{ s}} = \frac{-0.077 \text{ mol/L}}{10.0 \text{ s}} = -0.0077 \text{ mol/L·s}$$

And the average reaction rate is:

$$\frac{\Delta[A]}{\Delta t} = \frac{0.630 \text{ mol/L} - 0.707 \text{ mol/L}}{40.0 \text{ s} - 30.0 \text{ s}} = \frac{-0.077 \text{ mol/L}}{10.0 \text{ s}} = -0.0077 \text{ mol/L·s}$$

$$\text{average rate} = -\frac{\Delta[A]}{\Delta t} = 0.0077 \text{ mol/L·s}$$

b. The reaction rate decreases as the total reaction time increases because the rate is proportional to the concentration of the reactant A, and the concentration of A decreases as the time of reaction increases.

PRACTICE QUESTION 19.3
Thiosulfate ion is oxidized by iodine in aqueous solution according to the equation

$$2S_2O_3^{2-}(aq) + I_2(aq) \rightarrow S_4O_6^{2-}(aq) + 2I^-(aq)$$

> **If 0.025 mol of $S_2O_3^{2-}$ is consumed in 0.50 L solution per minute:**
>
> > **a. Calculate the rate of removal of $S_2O_3^{2-}$ in M/s.**
> > **b. What is the rate of removal of I_2?**

PROBLEM TYPE 2: DEPENDENCE OF REACTION RATE ON REACTANT CONCENTRATION.

A.   The Rate Law.

The rate of a reaction is proportional to the reactant concentrations. For the reaction,

$NO + \dfrac{1}{2}O_2 \rightarrow NO_2$, the rate is proportional to the concentrations of NO and O_2. The **rate law** (or *rate equation*) for the reaction is:

$$\text{rate} = k[NO]^x[O_2]^y$$

The exponents x and y determine how strongly the concentrations affect the reaction rate. The exponent x is called the **order** with respect to NO, and y is the **order** with respect to O_2. The sum $(x + y)$ is the **overall order**.

The values of x and y must be determined from experiment, and cannot be derived by any other means.

The proportionality constant k is called the **rate constant**. The value of k depends on the reaction and the temperature. The values of x and y are often 1 or 2. However, other values (including fractions, zero, and negative values) are possible. The units on k are usually given as $M^{(1-\text{overall order})}s^{-1}$.

The NO reaction with O_2 experiment was measured experimentally at a particular temperature, and, accordingly, it was determined that $x = 2$ and $y = 1$. Therefore, the rate law for this reaction is:

$$\text{rate} = k[NO]^2[O_2]$$

This reaction is said to be second–order in nitric oxide and first–order in oxygen. It is third–order overall.

B.   Experimental Determination of the Rate Law.

From the previous rate law, the fact that the reaction is first order in O_2 means that the rate is directly proportional to the O_2 concentration. If $[O_2]$ doubles or triples, the rate will correspondingly double or triple. We can show this mathematically. Consider two experiments. In experiment 1 the concentration of O_2 is c. In experiment 2 the concentration of O_2 is doubled from c to $2c$. If the concentration of NO is the same in both experiments, it will have no effect on the rate. Use of the rate law allows us to write the ratio of the two rates:

$$\frac{\text{rate (expt 2)}}{\text{rate (expt 1)}} = \frac{k[NO]^2(2c)}{k[NO]^2(c)} = 2$$

If the concentration of O_2 is held constant in two experiments and the concentration of NO doubles (from c to $2c$), the rate law predicts that the rate will quadruple.

$$\frac{\text{rate (expt 2)}}{\text{rate (expt 1)}} = \frac{k[O_2]^2(2c)}{k[O_2]^2(c)} = 2^2 = 4$$

The fact that the reaction is second order in NO means that the rate is proportional to the square of the concentration of NO. Doubling or tripling of [NO] causes the rate to increase four– or nine–fold, respectively.

In general, if the concentration of one reactant is doubled while the other reactant concentration is unchanged, *and* the rate is:

1.   *unchanged*, the order with respect to the changing reactant is *zeroth–order*.
2.   *doubled*, the order with respect to the doubling the reactant concentration is *first–order*.
3.   *quadrupled*, the order with respect to the doubling the reactant concentration is *second–order*.

STUDY QUESTION 19.4

The reaction, A + 2B → products, was found to have the rate law: rate = $k[A][B]^3$. By what factor will the rate of reaction increase, if the concentration of B is increased from c to $3c$, while the concentration of A is held constant?

Solution: Write a ratio of the rate law expressions for the two different concentrations of B.

$$\frac{\text{rate (expt 2)}}{\text{rate (expt 1)}} = \frac{k[A][3c]^3}{k[A][c]^3} = \frac{3^3 c^3}{c^3} = 27$$

The rate will increase 27–fold when [B] is increased three–fold.

One procedure used to determine the rate law for a reaction involves the isolation method. In this method, the concentration of all but one reactant is fixed, and the rate of reaction is measured as a function of the concentration of the one reactant whose concentration is varied. Any variation in the rate is due to the variation of this reactant's concentration. In practice, the experimenter observes the dependence of the initial rate on the concentration of the reactant.

To determine the order with respect to A in the following chemical reaction,

$$2A + B \rightarrow C$$

the initial rate would be measured in several experiments in which the concentration of A is varied and the concentration of B is held constant. To determine the order with respect to B, the concentration of A must be held constant and the concentration of B is varied in several experiments.

STUDY QUESTION 19.5

The following rate data were collected for the reaction:
$$2NO + 2 H_2 \rightarrow N_2 + 2H_2O$$

Experiment	$[NO]_0$ (*M*)	$[H_2]_0$ (*M*)	$\Delta[N_2]/\Delta t$ (*M*/h)
1	0.60	0.15	0.076
2	0.60	0.30	0.15
3	0.60	0.60	0.30
4	1.20	0.60	1.21

a.   **Determine the rate law.**
b.   **Calculate the rate constant.**

Solution:

a. We want to determine the exponents in the equation

$$\text{rate} = k[NO]^x[H_2]^y$$

To determine the order with respect to H_2, first find two experiments in which [NO] is held constant. This can be done by comparing the data of experiments 1 and 2. When the concentration of H_2 is doubled, the reaction rate doubles. Thus, the reaction is first–order in H_2. When the NO concentration is doubled while H_2 concentration remains constant (experiments 3 and 4), the reaction rate quadruples. Therefore, the reaction is second–order in NO.

The rate law is:

$$\text{rate} = k[NO]^2[H_2]$$

Take a closer look at proving that $x = 2$. Write the ratio of rate laws for experiments 4 and 3.

$$\frac{\text{rate}_4}{\text{rate}_3} = \frac{k[NO]^x[H_2]}{k[NO]^x[H_2]}$$

$$\frac{1.21}{0.30} = \frac{k(1.20)^x(0.60)}{k(0.60)^x(0.60)} = \left(\frac{1.20}{0.60}\right)^x$$

$$4.0 = 2.0^x$$

Therefore, x = 2

b. Rearrange the rate law from part (a).

$$k = \frac{\text{rate}}{[NO]^2[H_2]}$$

Then, substitute data from any one of the experiments. Using experiment 1:

$$k = \frac{0.076 \; M/h}{(0.60 \; M)^2(0.15 \; M)} = 1.4/M^2h$$

NOTE:
You should get the same value of k from all four experiments. Note that the units of k are those of a third–order rate constant.

PRACTICE QUESTION 19.4
The rate law for the reaction $2A + B \rightarrow C$ was found to be rate $= k[A][B]^2$. If the concentration of B is tripled and the concentration of A is unchanged, by how many times will the reaction rate increase?

PRACTICE QUESTION 19.5
Use the following data to determine (a) the rate law and (b) the rate constant for the reaction
$2A + B \rightarrow C$

Expt	$[A]_0$	$[B]_0$	Rate (M/s)
1	0.25	0.10	0.012
2	0.25	0.20	0.048
3	0.50	0.10	0.024

PROBLEM TYPE 3: DEPENDENCE OF REACTANT CONCENTRATION ON TIME.

A. First–Order Reactions.

The first–order rate law is one of the most widely encountered kinetic forms, one where the exponent of [A] in the rate law is 1.

$$A \rightarrow products$$

$$rate = -\frac{\Delta[A]}{\Delta t} = k[A]$$

For a first–order reaction, the unit of the rate constant is reciprocal time, $1/t$. When we mathematically manipulate (integrate) the rate law, the resulting equation is in the form

$$\ln \frac{[A]_t}{[A]_0} = -kt$$

This is a very useful equation called the **integrated first–order equation.** Here $[A]_0$ is the initial concentration of A at time = 0, $[A]_t$ is the concentration of A at time = t, and k is the first–order rate constant. The concentration $[A]_t$ decreases as the time increases. This equation allows the calculation of the rate constant k when $[A]_0$ is known, and $[A]_t$ is measured at time t. Also, once k is known, $[A]_t$ can be calculated for any future time.

To determine whether a reaction is first order, we rearrange the first–order equation into the form:

$$\ln [A]_t = -kt + \ln [A]_0$$

corresponding to the linear equation $y = mx + b$. Here, m is the slope of the line and b is the intercept on the y axis. Comparing the last two equations, we can equate y and x to experimental quantities.

$$y = \ln [A]_t \quad and \quad x = t$$

A plot of ln [A] versus for a first–order reaction gives a straight line with a slope of $-k$ and y–intercept of ln $[A]_t$ as shown in the figure below. If a plot of ln [A] versus t yields a curved line, rather than a straight line, the reaction is not a first–order reaction. This graphical procedure is the method used by most chemists to determine whether or not a given reaction is first order.

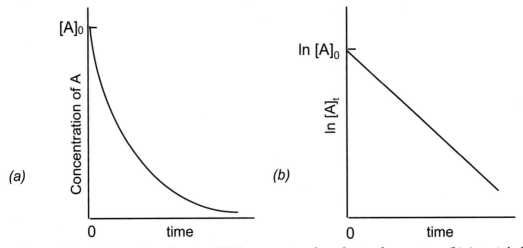

Figure 14.1 (a) Curved line for plot of [A] $_t$ versus time for a first-order reaction, (b) A straight line for ln [A] $_t$ versus time for a first-order reaction

Half-life, $t_{1/2}$, is the time required for the concentration of a reactant to drop to half of its initial value. For a first–order reaction, the half–life is:

$$t_{1/2} = \frac{0.693}{k}$$

where k is the rate constant. Knowledge of the half–life allows the calculation of the rate constant k. By definition, after one half–life, the ratio $[A]_t/[A]_0$ is equal to 0.5.

If the reaction continues, then $[A]_t$ will drop by 1/2 again during the second half–life as shown in Figure 14.13 in your textbook. After two half–life periods the fraction of the original concentration of A remaining, $[A]_t/[A]_0$, will be $0.5 \times 1/2 = 0.25$.

STUDY QUESTION 19.6
Methyl isocyanide undergoes a rearrangement to form methyl cyanide that follows first–order kinetics.

$$CH_3NC(g) \rightarrow CH_3CN(g)$$

The reaction was studied at 199 °C. The initial concentration of CH_3NC was 0.0258 mol/L, and after 11.4 min analysis showed the concentration of product was 1.30×10^{-3} mol/L.

    a. What is the first–order rate constant?
    b. What is the half–life of methyl isocyanide?
    c. How long will it take for 90 percent of the CH_3NC to react?

Solution:

a. This problem illustrates the use of the integrated first-order rate equation

$$\ln \frac{[CH_3NC]}{[CH_3NC]_0} = -kt$$

Where k is the rate constant and $[CH_3NC]$ is the reactant concentration at time t. The initial concentration is $[CH_3NC]_0 = 0.0258\ M$. After 11.4 min the product concentration is $1.30 \times 10^{-3}\ M$. This implies that the concentration of CH_3NC remaining unreacted is $0.0258 - 0.0013 = 0.0245\ M$. Substitution gives:

$$\ln \frac{0.0245\,M}{0.0258\,M} = -k\,(11.4\ \text{min})$$

$$\ln 0.950 = -k\,(11.4\ \text{min})$$

$$k = 4.54 \times 10^{-3}/\text{min}$$

b. For a first–order reaction the half–life equation is:

$$t_{1/2} = \frac{0.693}{k} = \frac{0.693}{4.54 \times 10^{-3}/\text{min}} = 153\ \text{min}$$

c. If 90 percent of the initial CH_3NC is consumed, then 10 percent remains. Therefore:

$$[CH_3NC] = 0.10[CH_3NC]_0$$

Substitution into the rate equation gives:

$$\ln \frac{0.10 \times [CH_3NC]_0}{[CH_3NC]_0} = -(4.54 \times 10^{-3}/min)t$$

Solving for *t*, we get:

$$t = -\frac{\ln 0.10}{4.54 \times 10^{-3}/min} = \frac{2.303}{4.54 \times 10^{-3}/min} = 507 \text{ min}$$

STUDY QUESTION 19.7
At 230 °C the rate constant for methyl isocyanide isomerization is 9.25×10^{-4}/s.
$$CH_3NC \rightarrow CH_3CN$$

a. **What fraction of the original isocyanide will remain after 60.0 min?**
b. **What is the half–life of methyl isocyanide at this temperature?**

Solution:
a. Again, applying the first–order equation:

$$\ln \frac{[CH_3NC]}{[CH_3NC]_0} = -kt$$

The fraction of methyl isocyanide remaining is given by the fraction $\frac{[CH_3NC]}{[CH_3NC]_0}$.

Plug in the rate constant and the time, and solve for the fraction remaining:

$$\ln \frac{[CH_3NC]}{[CH_3NC]_0} = -9.25 \times 10^{-4}/s \, (60 \text{ min}) \times \frac{60 \text{ s}}{1 \text{ min}} = -3.33$$

Taking the antilog of both sides gives:

$$\frac{[CH_3NC]}{[CH_3NC]_0} = e^{-3.33} = 0.0358$$

b. The half–life is

$$t_{1/2} = \frac{0.693}{9.25 \times 10^{-4}/s}$$

$$= 749 \text{ s or } 12.5 \text{ min}$$

B. Second–Order Reactions.
In a second–order reaction, the rate is proportional either (1) to the square of the concentration of one reactant,

$$A \rightarrow \text{product}$$

$$\text{rate} = -\frac{\Delta[A]}{\Delta t} = k[A]^2$$

or (2) to the product of the concentrations of two reactants, each raised to the first power:

$$A + B \rightarrow \text{product}$$

$$\text{rate} = -\frac{\Delta[A]}{\Delta t} = k[A][B]$$

This reaction is first order in A and first order in B, hence, it is second–order overall. We will only consider such reactions where [A] = [B]. The rate law of such reaction would then simplify to

$$\text{rate} = -\frac{\Delta[A]}{\Delta t} = k[A]^2$$

where the rate constant has units $1/M \cdot t$.

The **integrated second–order equation** is:

$$\frac{1}{[A]_t} = \frac{1}{[A]_0} + kt$$

As with the integrated first–order equation, this equation allows us to calculate the concentration of A at any time, t, after the reaction has begun. Alternatively, if $[A]_0$, $[A]_t$, and t are known, the rate constant can be calculated.

To determine whether a reaction is second order, we rearrange the second–order equation into the form:

$$\frac{1}{[A]_t} = kt + \frac{1}{[A]_0}$$

corresponding to the linear equation $y = mx + b$. We can equate y and x to experimental quantities.

$$y = \frac{1}{[A]_t} \quad \text{and} \quad x = t$$

A plot of $\dfrac{1}{[A]_t}$ versus t for a second-order reaction gives a straight line with a slope of k and

y–intercept of $\dfrac{1}{[A]_0}$. If a plot of $\dfrac{1}{[A]_t}$ versus t yields a curved line, rather than a straight line, the reaction is not a second–order reaction.

The half–life for a second–order reaction is given by:

$$t_{\frac{1}{2}} = \frac{1}{k[A]_0}$$

Here, we see that the half–life is inversely proportional to the initial concentration, $[A]_0$. This situation is different from that for a first–order reaction, $t_{\frac{1}{2}}$ is independent of $[A]_0$.

STUDY QUESTION 19.8
The rate law for

$$2AB \rightarrow A_2B_2$$

is rate = $k[AB]^2$. When $[AB]_0 = 7.50$ mM, it takes 2.4×10^{-7} s for the concentration to drop to 3.75 mM. How long will it take for the concentration of AB to drop to 0.34 mM when $[AB]_0 = 0.012$ M?

Solution: Knowing that the rate law for the reaction is second order overall, we can immediately look at the expression for the half–life of the reaction:

$$t_{1/2} = \frac{1}{k[AB]_0}$$

Since 3.75mM is half the initial concentration (7.50 mM), with the information provided we can determine the value of k. Re–arranging the equation for k we obtain:

$$k = \frac{1}{t_{1/2}[AB]_0} = \frac{1}{(2.42 \times 10^{-7}\ s)(7.5 \times 10^{-3}\ M)} = 5.51 \times 10^{8}\ M^{-1}s^{-1}$$

Now, we can use the value of k and the integrated second–order equation to make the calculation:

$$\frac{1}{[AB]_t} - \frac{1}{[AB]_0} = kt$$

$$\frac{1}{3.4 \times 10^{-4}\ M} - \frac{1}{0.012\ M} = (5.51 \times 10^{8}\ M^{-1}s^{-1})t$$

$$5.2 \times 10^{-6}\ s = t$$

Summary. The integrated rate equations and graphical methods allow us to distinguish between the various overall orders of reaction.

1. A reaction is first order when a plot of ln $[A]_t$ versus t is a straight line. A plot of $[A]_t$ versus time will be curved.
2. A reaction is second–order when a plot of $1/[A]_t$ versus t is a straight line, and ln $[A]_t$ versus t is curved.
3. The half–life equations also provide a way to distinguish between first– and second–order reactions. The half–life of a first–order reaction is independent of starting concentration, whereas the half–life of a second–order reaction is inversely proportional to the initial concentration.

> ***PRACTICE QUESTION 19.6***
> **A certain first–order reaction A → B is 40% complete (40% of the reactant is used up) in 75 s.**
>> a. **What is the rate constant?**
>> b. **What is the half-life of this reaction?**

PROBLEM TYPE 4: DEPENDENCE OF REACTANT CONCENTRATION ON TEMPERATURE.

A. The Arrhenius Equation.
Temperature plays a significant role on reaction kinetics. As a rule of thumb, reaction rates approximately double with a 10 °C increase in temperature. Looking at the rate law rate = $k[A]^x[B]^y$, since initial concentration of A and B are unaffected by temperature, we see that the rate constant is the only variable that changes with temperature. The Arrhenius equation,

$$k = Ae^{-E_a/RT}$$

shows the affect of temperature on k. In the equation, R is the ideal gas constant (in units of joules) and E_a is the **activation energy**. The logarithmic form of the

Arrhenius equation,

$$\ln k = -\frac{E_a}{RT} + \ln A$$

is useful, as it is in the form of $y = mx + b$. Both A and E_a are constants for a particular reaction.

The variable A is called the **frequency factor**. It is related to the frequency of molecular collisions, and the fraction of collisions that have the correct orientation. The **activation energy**, E_a, is related to the formation of the activated complex. The **activated complex** is the high–energy intermediate species that dissociates into the products. Figure 14.6 in the textbook shows that the **activation energy** is the difference in energy between the reactants and the activated complex. The factor $e^{-E_a/RT}$ that appears in the Arrhenius equation reflects the fraction of molecules with energies equal to or greater than the activation energy. This factor changes significantly with temperature. As temperature increases, a greater fraction of molecules have energy equal to or greater than the activation energy.

From the Arrhenius equation we see that as E_a increases, the negative exponent increases, and so k decreases. This leads to a slower reaction.

A convenient equation that can be used to calculate the activation energy is:

$$\ln\left(\frac{k_1}{k_2}\right) = \frac{E_a}{R}\left(\frac{T_1 - T_2}{T_1 T_2}\right)$$

where k_1 is the rate constant at temperature T_1, and k_2 is the rate constant at temperature T_2. Use of this equation requires that rate constants k_1 and k_2 be measured at two temperatures T_1 and T_2, respectively.

STUDY QUESTION 19.9
For the reaction, NO + O_3 → NO_2 + O_2, the following rate constants were determined.

Temperature, °C	k(1/M·s)
10.0	9.30×10^6
30.0	1.25×10^7

Calculate the activation energy for this reaction.

Solution: Here we are given two rate constants corresponding to two temperatures. The activation energy (E_a) can be calculated by substituting into the equation given earlier that shows the temperature dependence of the rate constant.

$$\ln\left(\frac{k_1}{k_2}\right) = \frac{E_a}{R}\left(\frac{T_1 - T_2}{T_1 T_2}\right)$$

Let

$$k_1 = 9.3 \times 10^6/M\text{·s} \quad \text{at} \quad T_1 = 273 + 10.0 = 283 \text{ K}$$
$$k_2 = 1.25 \times 10^7/M\text{·s} \quad \text{at} \quad T_2 = 273 + 30.0 = 303 \text{ K}$$

Recall $R = 8.314$ J/mol·K. Substitute into the preceding equation to solve for E_a:

$$\ln\left(\frac{9.3 \times 10^6 /M \cdot s}{1.25 \times 10^7 /M \cdot s}\right) = \frac{E_a}{8.314 \, \text{J/mol} \cdot \text{K}}\left(\frac{283 \text{ K} - 303 \text{ K}}{(283 \text{ K})(303 \text{ K})}\right)$$

Solving for E_a yields:

$$(\ln 0.744)(8.314 \text{ J/mol·K}) = E_a\left(\frac{-20 \text{ K}}{85749 \text{ K}^2}\right)$$

$$(-0.296)(8.314 \text{ J/mol K}) = E_a(-2.33 \times 10^{-4}/\text{K})$$

$$E_a = 1.06 \times 10^4 \text{ J/mol} \quad \text{(or 10.6 kJ/mol)}$$

STUDY QUESTION 19.10
Draw a reaction energy profile for the following endothermic reaction:
$$2HI(g) \rightarrow H_2(g) + I_2(g) \qquad \Delta H^o_{rxn} = 12.5 \text{ kJ}$$
Given the activation energy, $E_a = 185$ kJ/mol. Label the activation energy and the activated complex. What is the activation energy for the reverse reaction?

Solution: The reaction is endothermic so the products are at a higher level than the reactants.

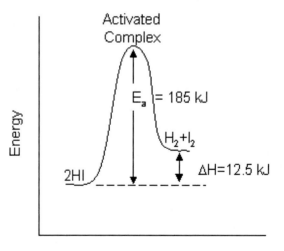

Using the diagram we can see that the activation energy for the reverse reaction is 185 kJ − 12.5 kJ = 173 kJ.

PRACTICE QUESTION 19.7
Calculate the activation energy for a reaction given that the rate constant is 4.60×10^{-4}/s at 350 °C and 1.87×10^{-4}/s at 320 °C.

PRACTICE QUESTION 19.8
The reaction $2NOCl \rightarrow 2NO + Cl_2$ has an E_a of 102 kJ/mol and a ΔH_{rxn} of 75.5 kJ/mol. Sketch the reaction energy diagram and determine the activation energy for the reverse reaction.

PROBLEM TYPE 5: REACTION MECHANISMS.

A. Elementary Reactions.

A **reaction mechanism** is the step–by–step process by which reactants are converted into products. It indicates which molecules must collide with each other and in what sequence. Each step is called an **elementary reaction.** A mechanism indicates the **intermediates** that are formed and consumed along the way, as well as which steps are fast and which are slow. This information is not supplied by the overall balanced chemical equation, because it is intended only to indicate the molar ratios of reactants and products in the reaction.

Molecularity of an elementary reaction pertains to the number of molecules that must collide in a single step. An elementary step that involves three molecules is called *termolecular.* A step that involves two molecules is *bimolecular*, and a step in which one molecule decomposes or rearranges is called *unimolecular.*

Thus, the equation:

$$A \rightarrow B$$

represents a unimolecular reaction in which a single molecule of A reacts to form a single molecule of B. Unimolecular reactions follow first–order rate laws.

$$rate = k[A]$$

The following equation represents a bimolecular reaction.

$$A + B \rightarrow C$$

The rate of a bimolecular step will depend on the number of molecules of A per unit volume, times the number of molecules of B per unit volume.

$$rate = k[A][B]$$

Therefore, bimolecular steps follow second–order kinetics.

This also can be understood by realizing that A and B must collide with each other if they are to react, and so the rate of a bimolecular process depends on the rate of collisions of A and B. Doubling the concentration of A will double the probability of collisions between A and B. Similarly, doubling the concentration of B will also double the rate of collisions of A and B. Reasoning further along these lines, we conclude that termolecular reactions are third order.

B. Rate–Determining Step.

It often turns out that one of the elementary steps in a mechanism is much slower than all the rest. This step determines the overall rate of reaction. The slowest step in a sequence of elementary steps is called the **rate–determining step.** This allows us to say that the overall law predicted by a mechanism will be the one corresponding to the rate–determining step.

C. Experimental Support for Reaction Mechanisms.

The sequence of events in a kinetic study of a reaction that leads to a proposed mechanism is as follows:

1. Determine the rate law experimentally.
2. Propose a mechanism for the reaction.
3. Test the mechanism.

To test a proposed mechanism, assume one step to be the rate–determining step, usually the one of highest molecularity. Then, establish the rate law for that step. This gives the rate law predicted by the mechanism. If the mechanism is adequate, when the predicted rate law is compared with the experimental rate law, the two will match. On the other hand, if the predicted and experimental rate laws do not match, the mechanism has been proved inadequate. Only those mechanisms consistent with all the data can be considered adequate.

STUDY QUESTION 19.11

The rate law for the following reaction has been experimentally determined to be third order:

$$2NO(g) + O_2(g) \rightarrow 2NO_2(g) \qquad \text{rate} = k[NO]^2[O_2]$$

Which of the two proposed mechanisms that follow is more satisfactory?

a.
$$NO + NO \xrightarrow{k_1} N_2O_2 \qquad \text{slow}$$

b.
$$NO + NO \underset{k_{-1}}{\overset{k_1}{\rightleftharpoons}} N_2O_2 \qquad \text{fast}$$

$$N_2O_2 + O_2 \xrightarrow{k_2} 2 NO_2 \qquad \text{fast} \qquad\qquad N_2O_2 + O_2 \xrightarrow{k_2} 2 NO_2 \qquad \text{slow}$$

Solution: Testing mechanism (a) first, we note that the first step is the slow or rate–determining step. The rate law for this bimolecular reaction is:

$$\text{rate} = k_1[NO]^2$$

This mechanism predicts a rate law that is second order in NO concentration and zeroth order in O_2. By comparison, the experimental rate law is second order in NO and first order in O_2. The predicted rate law and the experimental rate law do not match; therefore, *mechanism (a) is not satisfactory.*

Testing mechanism (b), we find that the rate–determining step is the second step. Since it is bimolecular, its rate law should be:

$$\text{rate} = k_2[N_2O_2][O_2]$$

It is not possible to compare this predicted rate law directly with the experimental rate law because of the unique term, which is the concentration of N_2O_2. In this mechanism, N_2O_2 is an **intermediate.** Because it is formed in step 1 and consumed in step 2, its concentration is always small and usually not measurable.

The way around this situation is to find a mathematical substitution for $[N_2O_2]$. Note that step 1 is reversible and equilibrium can be established. This means that the rates of the forward and reverse reactions are equal:

$$k_1[NO]^2 = k_{-1}[N_2O_2]$$

where k_1 is the rate constant for the forward reaction, and k_{-1} for the reverse. Rearranging terms gives us:

$$[N_2O_2] = \frac{k_1}{k_{-1}}[NO]^2$$

We now have an expression for $[N_2O_2]$, which we can substitute into the rate law, rate = $k_2[N_2O_2][O_2]$. This yields:

$$\text{rate} = \frac{k_2 k_1}{k_{-1}} [NO]^2[O_2]$$

Whenever a reaction step is reversible, we can use the equality of the forward and reverse rates to form an expression to substitute for the concentration of an intermediate. Now we compare this rate law with the experimental rate law, which is

$$\text{rate} = k[NO]^2[O_2]$$

We see that this mechanism predicts that the reaction will be second order in NO and first order in O_2, just as observed. Also, the collection of constants $k_2 k_1/k_{-1}$ will equal the rate constant.

$$k = \frac{k_2 k_1}{k_{-1}}$$

Therefore, the second mechanism predicts a rate law that matches the experimental order of reaction.

PRACTICE QUESTION 19.9
Carbon monoxide can be converted to carbon dioxide by the following overall reaction.
$$NO_2(g) + CO(g) \rightarrow NO(g) + CO_2(g)$$
The experimentally determined rate law is rate $= k[NO_2]^2$. The suggested mechanism involves two bimolecular elementary steps.
$$NO_2(g) + NO_2(g) \rightleftharpoons NO_3(g) + NO(g) \qquad \text{step 1}$$
$$NO_3(g) + CO(g) \rightarrow NO_2(g) + CO_2(g) \qquad \text{step 2}$$

a. **What is the rate law for each step?**
b. **Derive the predicted rate laws when step 1 is rate determining, and when step 2 is rate determining.**

PROBLEM TYPE 6: CATALYSIS.

A catalyst is a substance that increases the rate of a chemical reaction without being consumed in the reaction. For this reason a catalyst does not appear in the stoichiometric equation. Chromium(III) oxide, Cr_2O_3, is the catalyst present in catalytic converters in American automobiles. This compound accelerates the reaction of carbon monoxide with oxygen. By converting CO to CO_2 in the exhaust stream, CO emissions are reduced.

$$CO(g) + \frac{1}{2} O_2(g) \xrightarrow{Cr_2O_3} CO_2(g)$$

The lifetime of catalytic converters is quite long; the Cr_2O_3 and other catalysts do not need to be replaced. Catalysts accelerate reaction rates by providing a new reaction pathway (mechanism) which has a lower activation energy (see Figure 14.17 in your textbook). In the catalytic converter, both CO and O_2 are chemically adsorbed on the catalyst's surface. Adsorption of O_2 by Cr_2O_3 weakens the O—O bond enough so that oxygen atoms can react with adsorbed CO. The reaction path involving a weakened O—O bond has a significantly lower activation energy than that of the reaction occurring purely in the gas phase. Recall that as the activation energy is lowered, the rate constant for a reaction increases. After CO_2 is formed, it desorbs from the surface, leaving the Cr_2O_3 catalyst chemically unaltered. The catalyst is not consumed in the reaction. Catalysts are regenerated in one of the last steps of the mechanism. The textbook describes three types of catalysts, which we will review here.

Heterogeneous Catalysis
In heterogeneous catalysis, the reactants are in one phase and the catalyst is in another phase. The catalyst is usually a solid, and the reactants are gases or liquids. Ordinarily, the site of the reaction is the surface of the solid catalyst. Many industrially important reactions involve gases and are catalyzed by solid surfaces. For example, the Haber process for the synthesis of ammonia is catalyzed by iron plus a few percent of the oxides of potassium and aluminum. The famous Ziegler–Natta catalyst for polymerizing ethylene gas (C_2H_4) into a polyethylene polymer contains triethylaluminum and titanium or vanadium salts.

Homogeneous Catalysis
In homogeneous catalysis, the reactants, products, and catalysts are all in the same phase, which is usually the gas or the liquid phase. Many reactions are catalyzed by acids. The decomposition of formic acid (HCO_2H) is an example.

$$HCO_2H \rightarrow H_2O + CO$$

Formic acid is normally stable and lasts a long time on the shelf. However, when sulfuric acid is added, bubbles of carbon monoxide gas can be observed immediately. The hydrogen ions from the sulfuric acid initiate a new reaction path. The mechanism is:

$$HCO_2H + H^+ \rightarrow HCO_2H_2^+$$
$$HCO_2H_2^+ \rightarrow H_2O + HCO^+$$
$$HCO^+ \rightarrow CO + H^+$$
$$\overline{}$$
$$HCO_2H \rightarrow CO + H_2O$$

Note that the H^+ ion is consumed in the first step, but, as for all catalysts, H^+ is regenerated. The net reaction is the sum of the three steps. The intermediate $HCO_2H_2^+$ cancels out as does the catalyst.

Enzymes: Biological Catalysts
Nearly all chemical reactions that occur in living organisms require catalysts. Enzymes are biological catalysts. Most enzymes are protein molecules with molecular masses well over 10,000 amu. All enzyme molecules are highly specific with respect to the reactions they catalyze. That is, they can only affect the reaction rates of a few specific reactant molecules. Typically, an enzyme catalyzes a single reaction, or a group of closely related reactions. The molecule on which an enzyme acts is called a substrate. The simplest mechanism which explains enzyme activity, and is consistent with these trends is one in which the substrate S and the enzyme E form an enzyme–substrate complex, ES. This complex has a lower energy than the activated complex without the enzyme. The enzyme–substrate complex can either dissociate back into E and S, or break apart into the products P, and the regenerated enzyme E. The simplest mechanism has the two steps shown below.

$$E + S \rightleftharpoons ES$$
$$ES \rightarrow E + P$$

STUDY QUESTION 19.12
Given the following mechanism for the decomposition of ozone in the stratosphere, identify the intermediate and the catalyst:
$$O_3 + Cl \rightarrow O_2 + ClO$$
$$ClO + O \rightarrow Cl + O_2$$

Solution: Adding the two steps gives the overall reaction

$$O_3 + O \rightarrow 2\,O_2$$

An intermediate is formed in an early step and removed in a later step and does not appear in the overall equation; therefore ClO is the intermediate. A catalyst is consumed in an early step and is later regenerated. Atomic chlorine, Cl, is the catalyst.

PRACTICE QUIZ

1. The following experimental data were obtained for the reaction $2A + B \rightarrow$ products. What is the rate law for this reaction?

Exp.	$[A]_o$ (*M*)	$[B]_o$ (*M*)	Rate (*M*/s)
1	0.80	0.20	5.5×10^{-3}
2	0.40	0.20	5.6×10^{-3}
3	0.80	0.40	2.23×10^{-2}

2. For the reaction

    $$30CH_3OH + B_{10}H_{14} \rightarrow 10B(OCH_3)_3 + 22H_2$$

    express the rate in terms of the change in concentration with time for each of the reactants and each of the products.

3. The hydrolysis of sucrose ($C_{12}H_{22}O_{11}$) yields the simple sugars, glucose ($C_6H_{12}O_6$) and fructose ($C_6H_{12}O_6$), which just happen to be isomers.

    $$C_{12}H_{22}O_{11} + H_2O \rightarrow C_6H_{12}O_6 + C_6H_{12}O_6$$

    The rate law for the reaction is: rate $= k[C_{12}H_{22}O_{11}]$. At 27 °C the rate constant is 2.1×10^{-6} /s.
    a. Starting with a 0.10 *M* sucrose solution at 27 °C, what is the concentration of sucrose 24 hours later? (The solution is kept at 27 °C).
    b. What is the half–life of sucrose in seconds at 27 °C?

4. N_2O_5 decomposes according to first-order kinetics:

    $$2N_2O_5 \rightarrow 4NO_2 + O_2$$

    a. At a certain temperature 20.0% of the initial N_2O_5 decomposes in 2.10 h. Determine the rate constant.
    b. What fraction of the initial N_2O_5 will remain after 13.0 h?
    c. What is the half–life?
    d. If the initial concentration of N_2O_5 was 0.222 *M*, then what concentration remains after 24 h?

5. The following data were obtained on the rate of disintegration of a pesticide in soil at 30 °C.

Time elapsed (days)	Percent pesticide remaining
11	96
60	80
96	71

    a. What is the order of reaction with respect to pesticide concentration?
    b. What is the value of the rate constant?

6. It takes 30.0 min for the concentration of a reactant in a second–order reaction to drop from 0.40 *M* to 0.30 *M*. What is the value of the rate constant for this reaction? How long will it take for the concentration to drop from 0.40 *M* to 0.20 *M*?

7. The reaction $A + 2B \rightarrow C + D$ has for its experimental rate law rate $= k[A]^2[B]$. By what factor will the rate increase if the concentration of A is doubled and the concentration of B is tripled?

8. The rate of the reaction of hydrogen with iodine

$$H_2 + I_2 \rightarrow 2HI$$

has rate constants of 1.41×10^{-5} /M·s at 393 °C and 1.40×10^{-4} /M·s at 443 °C. Calculate the activation energy for this reaction.

9. At 300 K the rate constant is 1.5×10^{-5} /M·s for the reaction:

$$2NOCl \rightarrow 2NO + Cl_2$$

The activation energy is 90.2 kJ/mol. Calculate the value of the rate constant at 310 K. By what factor did the rate constant increase?

10. The activation energy for the reaction

$$CO + NO_2 \rightarrow CO_2 + NO$$

is 116 kJ/mol. How many times greater is the rate constant for this reaction at 250 °C than it is at 200 °C?

11. If the reaction $2HI \rightarrow H_2 + I_2$ has an activation energy of 190 kJ/mol and a $\Delta H = 10$ kJ, what is the activation energy for the reverse reaction?

$$H_2 + I_2 \rightarrow 2\,HI$$

12. The rate law for the net reaction

$$2\,NO + O_2 \rightarrow 2\,NO_2$$

is rate $= k[NO_2]^2[O_2]$. Could the mechanism of this reaction be a single termolecular process?

13. The reaction between the nitrite ion and oxygen gas, in aqueous solution is:

$$2\,NO_2^-\,(aq) + O_2(aq) \rightarrow 2\,NO_3^-\,(aq)$$

and proceeds at a rate that is first order in $[NO_2^-]$ and zero order in $[O_2]$. A mechanism has been proposed:

$$NO_2^- + O_2 \rightarrow NO_4^- \qquad \text{slow}$$
$$NO_4^- + NO_2^- \rightarrow 2\,NO_3^- \qquad \text{fast}$$

Show that this mechanism is consistent with the experimental rate law.

14. The rate law for the replacement of H_2O by NH_3 in the reaction of a complex ion:

$$Ni(H_2O)_6^{2+}\,(aq) + NH_3(aq) \rightarrow Ni(H_2O)_5(NH_3)^{2+}(aq) + H_2O(\ell)$$

is first order in $Ni(H_2O)_6^{2+}$, and zero order in NH_3.

rate $= k[\,Ni(H_2O)_6^{2+}\,]$

Show that the following mechanism is consistent with the experimental rate law.

$$Ni(H_2O)_6^{2+} \rightarrow Ni(H_2O)_5^{2+} + H_2O \qquad \text{slow}$$

$$Ni(H_2O)_6^{2+} + NH_3 \rightarrow Ni(H_2O)_5(NH_3)^{2+} \qquad \text{fast}$$

15. The rate law for the net reaction

$$H_2(g) + I_2(g) \rightarrow 2HI(g)$$

is rate $= k[H_2][I_2]$. A possible mechanism involves a bimolecular elementary step:

$$H_2 + I_2 \rightarrow 2\ HI$$

A second possibility has also been proposed:

$$I_2 \rightarrow 2\ I \qquad \text{fast}$$
$$2I + H_2 \rightarrow 2HI \qquad \text{slow}$$

Show that both mechanisms are consistent with the experimental rate law.

16. The following statements are sometimes made with reference to catalysts. Explain the fallacy in each one of them.
    a. A catalyst is a substance that accelerates the rate of a chemical reaction but does not take part in the reaction.
    b. A catalyst may increase the rate of a reaction going in one direction without increasing the rate of the reaction going in the reverse direction.

17. a. Identify the catalyst in the following mechanism.
    b. What is the overall reaction?
    c. Is this reaction an example of homogeneous or heterogeneous catalysis?

$$H_2O_2(aq) + Br_2(aq) \rightarrow 2H^+(aq) + O_2(aq) + 2Br^-(aq)$$
$$H_2O_2(aq) + 2H^+(aq) + 2Br^-(aq) \rightarrow Br_2(aq) + H_2O(\ell)$$

18. The following reaction, $A_2 \rightarrow 2A$, follows first-order kinetics. In an experiment, the initial concentration of A_2 was 0.100 M. What will be the concentration of A_2 after four half–lives?

19. The rate law for the reaction,

$$C + B_2 \rightarrow CB + B$$

is rate $= k[C][B_2]$. What will happen to the rate of the reaction when the following changes are made:
    a. the concentration of C is tripled while the concentration of B_2 is halved
    b. the concentration of C is reduced to ¼ of its initial value while the concentration of B_2 is halved
    c. the concentrations of C and B_2 are doubled

20. Use the following data to determine the activation energy for the reaction,

$$A_2 + B_2 \rightarrow \text{products}$$

Rate constant	Temperature (°C)
2.30×10^7	25.0
3.50×10^7	35.0
6.80×10^7	55.0
8.90×10^7	60.0

Chapter 20
Nuclear Chemistry

PROBLEM-SOLVING STRATEGIES AND TUTORIAL SOLUTIONS

TYPES OF PROBLEMS

Problem Type 1:   Nuclei and Nuclear Reactions.

Problem Type 2:   Nuclear Stability.
>   **A.**   Types of Nuclear Decay.
>   **B.**   Nuclear Binding Energy.

Problem Type 3:   Natural Radioactivity.
>   **A.**   Kinetics of Radioactive Decay.
>   **B.**   Dating Based on Radioactive Decay.

Problem Type 4:   Nuclear Transmutation.

Problem Type 5:   Nuclear Fission.

Problem Type 6:   Nuclear Fusion.

PROBLEM TYPE 1: NUCLEI AND NUCLEAR REACTIONS.

So far we have focused on the entire atom, its electronic structure and behavior. Now we turn our attention to inside the atom and focus on the nucleus and possible nuclear change.

We will consider two types of nuclear reactions: radioactive decay and nuclear transmutation.
Radioactive decay, or **radioactivity**, is described as the spontaneous emission of particles and/or radiation by unstable atomic nuclei. These processes often result in the formation of a new element. Table 20.1 lists types of radiation that can be emitted from radioactive decay.

A nucleus can also undergo change by **nuclear transmutation**. In this process, one nucleus reacts with another nucleus, an elementary particle, or a photon (gamma ray) to produce one or more new nuclei.

Table 20.1 Possible Radiation from Radioactive Decay

Radiation	Mass (amu)	Charge	Symbol
Alpha	4.0	+2	$_2^4\alpha$ or $_2^4\text{He}$
Beta	0.0005	−1	$_{-1}^{0}\beta$ or $_{-1}^{0}e$
Positron	0.0005	+1	$_{+1}^{0}\beta$ or $_{+1}^{0}e$
Gamma	0.0	0	$_0^0\gamma$

Radioactive decay and nuclear transmutation processes are described by nuclear equations. These equations use isotopic and elementary particle symbols to represent the reactants and products of nuclear reactions. For example, in the first nuclear transmutation ever observed (in 1919) alpha particles $_2^4\text{He}$ were used to bombard nitrogen–14 nuclei. The observed products were oxygen–17 nuclei and protons. The nuclear equation is:

$$_7^{14}\text{N} + _2^4\text{He} \rightarrow _8^{17}\text{O} + _1^1\text{H}$$

The balancing rules for nuclear equations are given below and are applied to all nuclear equations, using the above equation as an example.

1. The sum of the mass numbers of the reactants must equal the sum of the mass numbers of the products (conservation of mass number): $14 + 4 = 18 = 17 + 1$.
2. The sum of the nuclear charges of the reactants must equal the sum of the nuclear charges of the products (conservation of atomic number): $7 + 2 = 9 = 8 + 1$.

STUDY QUESTION 20.1
Complete the following nuclear equations. Label the nuclear reaction as radioactive decay or nuclear transmutation.

a. $^{14}_{7}\text{N} + ^{1}_{0}\text{n} \rightarrow ^{14}_{6}\text{C} + \underline{\quad}$

b. $^{226}\text{Ra} \rightarrow ^{4}_{2}\alpha + \underline{\quad}$

Solution:

a. According to rule #1, the sum of the mass numbers of the reactants must equal the sum of the mass numbers of the products: $14 + 1 = 15 = 14 + A$. Therefore, the unknown product will have a mass number of 1. According to rule 2, the sums of the nuclear charges on both sides of the equation must be the same: $7 + 0 = 7 = 6 + Z$. Therefore, the nuclear charge of the unknown product must be 1, making it a proton.

$$^{14}_{7}\text{N} + ^{1}_{0}\text{n} \rightarrow ^{14}_{6}\text{C} + ^{1}_{1}\text{p}$$

This reaction is a *nuclear transmutation*.

b. Note that the atomic number of radium is missing. The periodic table indicates that Ra is element number 88. Balancing the mass numbers first, we find that the unknown product must have a mass number of 222. Balancing the nuclear charges next, we find that the atomic number of the unknown must be 86. Element number 86 is radon.

$$^{226}_{88}\text{Ra} \rightarrow ^{4}_{2}\alpha + ^{222}_{86}\text{Rn}$$

This is *radioactive decay*.

As pointed out in Table 20.1 in your textbook, in chemical reactions the number of atoms of each element is conserved. Only changes in chemical bonding occur. However, in nuclear reactions the compositions of the atomic nuclei are altered, and so elements are converted from one to another.

PRACTICE QUESTION 20.1
How many protons (p), neutrons (n), and electrons (*e*) does an atom of the radioisotope phosphorus–32 contain?

PRACTICE QUESTION 20.2
Complete the following nuclear reactions.

a. $^{210}_{84}\text{Po} \rightarrow ^{4}_{2}\alpha + \underline{\quad}$

b. $^{4}_{2}\text{He} + ^{9}_{4}\text{Be} \rightarrow ^{12}_{6}\text{C} + \underline{\quad}$

c. $^{15}_{7}\text{N} \rightarrow \underline{\quad} + ^{0}_{-1}\beta$

PROBLEM TYPE 2: NUCLEAR STABILITY.

A. Types of Nuclear Decay.

If you were given the symbol of a radioisotope, without any experience, it would be impossible to tell its mode of decay. But with knowledge of the belt of stability, you can make accurate predictions of the expected mode of decay.

Isotopes with too many neutrons lie above the belt of stability. The nuclei of these isotopes decay in such a way as to lower their n:p ratio. For example, one neutron may decay into a proton and a beta particle.

$$\,_0^1n \rightarrow \,_1^1p + \,_{-1}^0\beta$$

The proton remains in the nucleus, and the beta particle is emitted from the atom. The loss of a neutron and the gain of a proton produces a new isotope with two important properties. It has a lower n : p ratio, and thus is more likely to be stable. Also, the daughter product has a nuclear charge that is one greater than the decaying isotope, due to the additional proton. Consider the decay of carbon–14, for example (Carbon–14 is continually produced in the upper atmosphere by the interaction of cosmic rays with nitrogen). Carbon–14 has a higher n : p ratio than either of carbon's stable isotopes (C–12 and C–13), and decays by beta decay.

$$\,_6^{14}C \rightarrow \,_7^{14}N + \,_{-1}^0\beta$$

Note that the product isotope $\,_7^{14}N$ has nuclear charge greater by one than carbon. Also, it is stable. Its n:p ratio is 1.0.

Isotopes with too many protons have low n:p ratios and lie below the belt of stability. These isotopes tend to decay by positron emission because this process produces a new isotope with a higher n:p ratio. During positron emission a proton emits a positron, $\,_{+1}^0\beta$, leaving behind a neutron.

$$\,_1^1p \rightarrow \,_0^1n + \,_{+1}^0\beta$$

The neutron remains in the nucleus, and the positron is emitted from the atom. Thus, the newly produced nucleus will contain one less proton and one more neutron than the parent nucleus. The n:p ratio increases due to positron decay.

Electron capture accomplishes the same end, that is, a higher n : p ratio. Some nuclei decay by capturing an atom's orbital electron.

$$\,_1^1p + \,_{-1}^0e \rightarrow \,_0^1n$$

Lanthanum–138, a naturally occurring isotope with an abundance of 0.089 percent, decays by electron capture.

$$\,_{57}^{138}La + \,_{-1}^0e \rightarrow \,_{56}^{138}Ba + h\nu$$

Electron capture is accompanied by X–ray emission, represented in the above reaction as $h\nu$ (the equation for the energy of a photon).

STUDY QUESTION 20.2
Using the band of stability, determine the mode of decay, if any, the following isotopes will undergo.

$$^{39}_{20}\text{Ca} \qquad ^{39}_{20}\text{Ca} \qquad ^{30}_{15}\text{P}$$

Solution: Calcium–39 has 20 protons and 19 neutrons, a n/p ratio of 0.95:1. It falls within the band of stability and will undergo positron emission. Calcium–40 has 20 protons and 20 neutrons, a n/p ratio of 1:1. This nucleus is stable. Phosphorus–30 should be the least stable because it has odd numbers of both protons (15) and neutrons (15), despite its 1:1 n/p ratio. Being on the lower edge of the band of stability, its nucleus will undergo positron emission.

STUDY QUESTION 20.3
The only stable isotope of sodium is sodium–23. What type of radioactivity would you expect from sodium–25?

Solution: Sodium–23 has 11 protons and 12 neutrons and is in the belt of stability. Sodium–25 has two more neutrons, and so has a higher n:p ratio than the stable isotope. Sodium–25 will decay by $^{0}_{-1}\beta$ emission.

B. Nuclear Binding Energy.

One of the important consequences of Einstein's theory of relativity was the discovery of the interconvertability of mass and energy. The total energy content (*E*) of a system of mass m is given by Einstein's theory:

$$E = mc^2$$

where *c* is the velocity of light (3.0×10^8 m/s). Therefore, the mass of a nucleus is a direct measure of its energy content. It was discovered in the 1930s that the measured mass of a nucleus is always smaller than the sum of the separate masses of its constituent **nucleons** (nuclear particles). This difference in mass is called the **mass defect**. When the mass defect is expressed as energy by applying Einstein's equation, it is called the binding energy of the nucleus. The **binding energy** is the energy required to break up a nucleus into its component nucleons. The binding energy provides a quantitative measure of nuclear stability. The greater its binding energy, the more stable a nucleus is toward decomposition.

In terms of the $^{17}_{8}\text{O}$ nucleus for instance, the binding energy (BE) is the energy required for the process:

$$^{17}_{8}\text{O} + \text{BE} \rightarrow 8\,^{1}_{1}\text{p} + 9\,^{1}_{0}\text{n}$$

The mass defect, Δm, is equal to the total mass of the products minus the total mass of the reactants.

$$\Delta m = [8(\text{proton mass}) + 9(\text{neutron mass})] - (^{17}_{8}\text{O nuclear mass})$$

Here, we can substitute atomic masses, which include the electrons for the nuclear masses:

$$\Delta m = [8(^{1}_{1}\text{H atomic mass}) + 9(\text{neutron mass})] - (^{17}_{8}\text{O atomic mass})$$

This works because the mass of 8 electrons in the 8 hydrogen atoms is canceled by the mass of the 8 electrons in the oxygen atom.

$$8(^{1}_{1}\text{H atomic mass}) = 8(\text{proton mass}) + 8(\text{electron mass})$$

$$(^{17}_{8}\text{O atom mass}) = (^{17}_{8}\text{O nuclear mass}) + 8(\text{electron mass})$$

Using the atomic masses given in the text:

$$\Delta m = [8(^1_1\text{H atomic mass}) + 9(\text{neutron mass})] - (^{17}_8\text{O atomic mass})$$

$$\Delta m = [8(1.007825 \text{ amu}) + 9(1.008665 \text{ amu})] - (16.999131 \text{ amu}) = 0.141454 \text{ amu}$$

The eight protons and nine neutrons have more mass than the oxygen–17 nucleus. The binding energy is:

$$\Delta E = (\Delta m)c^2$$

$$\Delta E = (0.141454 \text{ amu}) \times (1.6606 \times 10^{-27} \text{ kg/amu}) \times (3.0 \times 10^8 \text{ m/s})^2$$

$$\Delta E = 2.1141 \times 10^{-11} \text{J}$$

(Recall that $1 \text{ J} = 1 \text{ kg} \left(\dfrac{\text{m}^2}{\text{s}^2}\right)$)

In comparing the stabilities of two different nuclei, we must take into account the different numbers of nucleons per nucleus. A satisfactory comparison of nuclear stabilities can be made by using the binding energy per nucleon, that is, the binding energy of each nucleus divided by the total number of nucleons in the nucleus. This is one of the most important properties of a nucleus. When the BE per nucleon is plotted as a function of the nuclear charge, we get the curve of binding energy as shown in Figure 20.2 of the text. Note that at first, it rises rapidly with increasing mass, reaching a maximum at mass 56. Above mass 56, the binding energy drops slowly as mass increases.

STUDY QUESTION 20.4
Calculate the nuclear binding energy of the light isotope of helium, helium–3. The atomic mass of ^3_2He is 3.01603 amu.

Solution: The binding energy is the energy required for the process

$$^3_2\text{He} \rightarrow 2\ ^1_1\text{p} + \ ^1_0\text{n}$$

where,

$$\Delta m = [2(\text{proton mass}) + (\text{neutron mass})] - \ ^3_2\text{He (nuclear mass)}$$

The mass difference is calculated using atomic masses:

$$\Delta m = [2(1.007825 \text{ amu}) + 1.008665 \text{ amu}] - 3.01603 \text{ amu}$$

$$\Delta m = 8.29 \times 10^{-3} \text{ amu}$$

Using Einstein's equation:

$$\Delta E = (\Delta m)c^2 = (8.29 \times 10^{-3} \text{ amu}) \times (3.00 \times 10^8 \text{ m/s})^2$$

$$= 7.46 \times 10^{14} \text{ amu m}^2/\text{s}^2$$

$$= 7.46 \times 10^{14} \text{ amu m}^2/\text{s}^2 \ \times \ \frac{1.6606 \times 10^{-27} \text{ kg}}{1 \text{ amu}} \times \frac{1 \text{ J}}{1 \text{ kg m}^2/\text{s}^2}$$

The binding energy is:

$$\Delta E = 1.24 \times 10^{-12} \text{ J}$$

Each $^3_2$He atom contains three nucleons. The binding energy per nucleon is:

$$\text{BE per nucleon} = \frac{1.24 \times 10^{-12} \text{ J/atom}}{3 \text{ nucleons/atom}} = 4.13 \times 10^{-13} \text{ J/nucleon}$$

NOTE: By combining the above conversion factors into one constant, the number of steps in future calculations can be lessened.

$$? \text{ J/amu} = (3.00 \times 10^8 \text{ m/s})^2 \times \frac{1.6606 \times 10^{-27} \text{ kg}}{.1 \text{ amu}} \times \frac{1 \text{ J}}{1 \text{ kg m}^2/\text{s}^2}$$

$$= 1.49 \times 10^{-10} \text{J/amu}$$

This constant is a useful factor relating energy to mass in amu. Applying this to the mass defect (0.14145 amu) calculated earlier in the discussion for $^{17}_8$O , the binding energy is 2.10×10^{-11} J/atom.

PRACTICE QUESTION 20.3
Fluorine has only one stable isotope, fluorine–19.

    a.   **The nucleus of fluorine–18 lies below the belt of stability. Write an equation for the decay of fluorine–18.**

    b.   **The nucleus of fluorine–21 lies above the belt of stability. Write an equation for the decay of fluorine–21.**

PRACTICE QUESTION 20.4
For each pair of nuclei, predict which one is the more stable.

    a. $^3_2$He or $^4_2$He

    b. $^{26}_{13}$Al or $^{27}_{13}$Al

PRACTICE QUESTION 20.5
Which is expected to have higher binding energy, $^{29}_{13}$Al or $^{27}_{13}$Al?

PROBLEM TYPE 3: NATURAL RADIOACTIVITY.

A number of isotopes exist in nature that have a n : p ratio that places them outside the belt of stability. These isotopes, called **radioisotopes**, occur naturally on Earth and give rise to natural radioactivity. Uranium, thorium, radon, potassium–40, carbon–14, and tritium (hydrogen–3) are naturally occurring radioisotopes. For example, the radioactive decay of Uranium–238, which is fairly abundant in Earth's crust, begins a sequence of decay reactions that ultimately changes U–238 to a stable isotope of lead. The uranium series is shown in Figure 20.3 in your textbook.

A. Kinetics of Radioactive Decay.
    Radioactive decay rates obey first–order kinetics.

$$\text{decay rate} = \text{number of atoms disintegrating per unit time} = kN$$

where k is the first–order rate constant and N is the number of atoms of the particular radioisotope present in the sample being studied. Recall that the half–life is related to the rate constant by:

$$t_{1/2} = \frac{0.693}{k}$$

The integrated first-order equation is:

$$\ln \frac{N}{N_o} = -kt$$

where N is the number of atoms of the radioisotope present in the sample after time t has elapsed, and N_o is the number of atoms of the radioisotope present initially. If N_o, N, and t are known, we can calculate the rate constant, k.

B. Dating Based on Radioactive Decay.

The purpose of radioactive dating is to determine the age of geological and archaeological samples and specimens. The age (t in the calculation) of certain rocks, for instance, can be estimated from analysis of the number of atoms of a particular radioisotope present now (N), as compared to the number present when the rock was formed originally (N_o).

Rearranging the integrated first–order equation, the age t is given by:

$$t = -\frac{1}{k} \ln \frac{N}{N_o}$$

The value of the initial number of atoms N_o is the sum, $N + D$, where D is the number of daughter nuclei resulting from the decay of atoms of the radioisotope. The original number of atoms of a radioisotope present in a rock sample is equal to the number N remaining at time t, plus the number of daughter nuclei, (D).

STUDY QUESTION 20.5

The rubidium–87/strontium–87 method of dating rocks was used to analyze lunar samples. The half–life of Rb–87 is 4.9×10^{10} yr.

$$^{87}_{37}\text{Rb} \rightarrow ^{87}_{38}\text{Sr} + ^{0}_{-1}\beta$$

Estimate the age of moon rocks in which the mole ratio of Rb–87 to Sr–87 is 40.

Solution: The age of the rock can be calculated from first–order equation relating time to the number of nuclei remaining:

$$t = -\frac{1}{k} \ln \frac{N}{N_o}$$

Since the rate constant k is not given, we must use the equation relating k and $t_{1/2}$:

$$k = \frac{0.693}{t_{1/2}} = 1.41 \times 10^{-11}/\text{yr}$$

This can be substituted for k, yielding:

$$t = -\frac{1}{1.41 \times 10^{-11}/\text{yr}} \ln \frac{N^{Rb}}{N_0^{Rb}}$$

Since Rb decays to Sr, the initial number of Rb atoms is equal to the sum of the Rb atoms and Sr atoms present.

$$N_0^{Rb} = N^{Rb} + N^{Sr}$$

Given that: $N^{Rb}/N^{Sr} = 40$, then:

$$N^{Sr} = \frac{1}{40} N^{Rb}$$

Therefore, after substitution for N^{Sr}, we get:

$$N_0^{Rb} = N^{Rb} + \frac{1}{40} N^{Rb} = 1.025 N^{Rb}$$

Substituting into the rate equation, we get:

$$t = -\frac{1}{1.41 \times 10^{-11}/yr} \ln \frac{N^{Rb}}{1.025\, N_0^{Rb}} = -\frac{1}{1.41 \times 10^{-11}/yr} \ln \frac{1}{1.025}$$

$$t = 1.7 \times 10^9 \text{ yr}$$

PRACTICE QUESTION 20.6
The C–14 activity of some ancient corn was found to be 10 disintegrations per minute per gram of C. If present day plant life gives 15.3 dpm/g C, how old is the corn? The half–life of C–14 is 5730 y.

PRACTICE QUESTION 20.7
The uranium decay series starts with uranium–238 and ends with lead–206. Each step in the series involves the loss of either an alpha or a beta particle. In the entire series, how many alpha particles and how many beta particles are emitted?

PROBLEM TYPE 4: NUCLEAR TRANSMUTATION.

Nuclear transmutation is the process of converting one element into another. Artificial radioactivity results when an unstable nucleus is produced by transmutation. For example, when aluminum atoms are bombarded with alpha particles, the product is phosphorus-30, which is radioactive, equation (1). Phosphorus-30 decays by positron emission, equation (2), and has a half-life of 2.5 min. This isotope does not occur naturally in phosphorus compounds.

$$\begin{aligned}
{}^{27}_{13}\text{Al} + {}^{4}_{2}\text{He} &\rightarrow {}^{30}_{15}\text{P} + {}^{1}_{0}\text{n} &\qquad (1) \\
{}^{30}_{15}\text{P} &\rightarrow {}^{30}_{14}\text{Si} + {}^{0}_{+1}\beta &\qquad (2)
\end{aligned}$$

Using neutrons as bombarding particles is convenient because neutrons have no charge and therefore are not repelled by target nuclei. Neutrons readily produce artificial radioactivity because they are easily captured by stable nuclei, with the result that new nuclei with higher n : p ratios are formed. This leads to products that decay by beta decay. Neutron capture by chlorine–37 yields chlorine–38.

$$\begin{aligned}
{}^{1}_{0}\text{n} + {}^{37}_{17}\text{Cl} &\rightarrow {}^{38}_{17}\text{Cl} \\
{}^{38}_{17}\text{Cl} &\rightarrow {}^{38}_{18}\text{Ar} + {}^{0}_{-1}\beta
\end{aligned}$$

Chlorine–38 has a short half–life, and is not found naturally on Earth. Note that neutron capture followed by beta decay yields a new element (Ar) with one greater charge number than the original element. This procedure of neutron capture followed by beta decay of the product nucleus has been used to synthesize elements that were "missing" from the periodic table, such as Tc and Pm. All isotopes of these elements are radioactive.

When projectiles are positively charged, such as is the case for protons and alpha particles, they require considerable kinetic energy in order to overcome the electrostatic repulsion they feel upon approach and collision with the target nucleus.

STUDY QUESTION 20.6

The first transuranium element to be synthesized by scientists was neptunium, atomic number 93. Devise a means to produce neptunium starting with U–238 and neutrons.

Solution: Neutron capture by a nucleus followed by beta decay produces a new nucleus of one atomic number higher than the original. Neptunium is one atomic number beyond uranium. So start with uranium and bombard it with neutrons.

$$^{238}_{92}U + \,^{1}_{0}n \; \rightarrow \; ^{239}_{92}U$$

$$^{239}_{92}U \; \rightarrow \; ^{239}_{93}Np + \,^{0}_{-1}\beta$$

NOTE:

Uranium's other naturally occurring isotope, U–235, will not produce neptunium in the same way because it undergoes neutron–induced fission.

PRACTICE QUESTION 20.8

Predict the product (X) of the following bombardment reactions.

a. $^{27}_{13}Al + \,^{1}_{0}n \rightarrow X + \,^{1}_{1}H$

b. $^{11}_{5}B + \,^{4}_{2}He \rightarrow X + \,^{1}_{1}H$

PROBLEM TYPE 5: NUCLEAR FISSION.

During **fission**, a heavy nucleus (mass > 200 amu) with an odd number of neutrons reacts with a thermal neutron and splits into two lighter nuclei whose masses are usually between 80 and 160 amu. Since the two smaller nuclei are more stable than the larger nucleus, energy is released in the process.

The fission reaction is quite complex because the same two products are not formed by all fissioning nuclei. The two reactions below show just two out of many possibilities for the fission of uranium–235.

$$^{235}_{92}U + \,^{1}_{0}n$$
$$^{141}_{56}Ba + \,^{92}_{36}Kr + 3\,^{1}_{0}n$$
$$^{137}_{52}Te + \,^{96}_{40}Zr + 2\,^{1}_{0}n$$

The actual distribution of product yields is shown in Figure 20.7 of the text.

A significant feature of fission is that on the average 2 to 3 neutrons are released per fission event. Since neutrons are required to initiate fission, and because neutrons are also products of fission, a **nuclear chain reaction** is possible.

The energy released during nuclear fission depends somewhat on just what products are formed. The energy released from the fission of one mole of U–235 atoms can be calculated from the equation, $\Delta E = \Delta mc^2$. The calculation shows that about 2.0×10^{10} kJ are released per mole of uranium. This amount of energy is *70 million times* the amount released in the exothermic chemical reaction in which 1 mol of H_2 reacts with 1/2 mol of O_2 to form 1 mol of water!

> **PRACTICE QUESTION 20.9**
> **Complete the following fission reaction of plutonium.**
> $$^{239}_{94}Pu + {}^{1}_{0}n \rightarrow {}^{144}_{58}Ce + X + 3\,{}^{1}_{0}n$$

PROBLEM TYPE 6: NUCLEAR FUSION.

Nuclear fusion is the combining of small nuclei to form larger, more stable nuclei. These larger nuclei will have a higher average binding energy per nucleon, so fusion reactions will be exothermic. Because all nuclei are positively charged, they must collide with enormous force in order to combine (fuse). This means that the atoms undergoing fusion must be heated to millions of degrees. Fusion reactions are called thermonuclear reactions because they occur only at very high temperatures, such as those in the sun.

The reaction that accounts for the tremendous release of energy by the sun is believed to be the stepwise fusion of four hydrogen nuclei to produce one helium nucleus. The net process is:

$$4\,{}^{1}_{1}H \rightarrow {}^{4}_{2}He + 2\,{}^{0}_{1}e \quad \Delta E = -4.3 \times 10^{-12} \text{ J}$$

One gram of hydrogen, upon fusion, releases the energy equivalent to the combustion of 20 tons of coal. The fusion of four moles of H atoms by the preceding equation releases 2.6×10^9 kJ of energy.

STUDY QUESTION 20.7
Calculate the mass of hydrogen that must undergo nuclear fusion each day, in order to provide just the fraction of the daily energy output of the sun that reaches the earth, which is 1.5×10^{22} J.

Solution: Using the following equation:

$$4\,{}^{1}_{1}H \rightarrow {}^{4}_{2}He + 2\,{}^{0}_{+1}e \quad \Delta E = -4.3 \times 10^{-12} \text{ J}$$

we see that fusion of 4 H atoms yields 4.3×10^{-12} J. We can set up the calculation:

$$\text{g H atoms} = \frac{1.5 \times 10^{22}\text{J}}{\text{day}} \times \frac{4 \text{ H atoms}}{4.3 \times 10^{-12}\text{J}} \times \frac{1.66 \times 10^{-24}\text{g}}{\text{H atom}}$$

$$\text{g H atoms} = 2.3 \times 10^{10} \text{ g}$$

STUDY QUESTION 20.8
Compare fission and fusion with respect to the temperatures required and the nature of the by–products of these processes.

Solution: Neutron–induced fission occurs at ordinary temperatures, whereas nuclear fusion requires temperatures in the millions of degrees. Fission of heavy elements yields hundreds of isotopes of elements with intermediate atomic numbers. These isotopes, by and large, have an excess of neutrons and are beta emitters. On the other hand, fusion of "light" nuclei yields stable isotopes of low–to–medium atomic mass.

PRACTICE QUIZ

1. What similarities and differences exist between beta particles and positrons?

2. Complete and balance the following nuclear equations.

   a. $^{239}_{94}Pu \rightarrow \, ^{4}_{2}He + \underline{\quad}$

   b. $\underline{\quad} + \, ^{6}_{3}Li \rightarrow 2\,^{4}_{2}He$

   c. $^{90}_{38}Sr \rightarrow \underline{\quad} + \, ^{0}_{-1}\beta$

   d. $^{10}_{5}B + \, ^{4}_{2}He \rightarrow \underline{\quad} + \, ^{1}_{0}n$

   e. $^{56}_{26}Fe + \, ^{1}_{0}n \rightarrow \underline{\quad}$

3. Rank the following nuclides in order of increasing nuclear stability: $^{40}_{20}Ca \quad ^{39}_{20}Ca \quad ^{11}_{5}B$

4. With reference to the belt of stability, state the modes of decay you would expect for the following:

   a. $^{13}_{7}N$ b. $^{26}_{13}Al$ c. $^{28}_{13}Al$

5. a. Calculate the binding energy and the binding energy per nucleon of $^{27}_{13}Al$. The atomic mass of Al–27 is 26.98154 amu.

   b. Compare this result to the binding energy of $^{28}_{14}Si$, which has an even number of protons and neutrons. The atomic mass of Si–28 is 27.976928 amu.

6. Cobalt–60 is used in radiation therapy. It has a half–life of 5.26 years.
   a. Calculate the rate constant for radioactive decay.
   b. What fraction of a certain sample will remain after 12 years?

7. Radioactive decay follows first–order kinetics. If 20 percent of a certain radioisotope decays in 4.0 years, what is the half–life of this isotope?

8. Estimate the age of a bottled wine that has a tritium ($^{3}_{1}H$) content that is 3/4 that of environmental water obtained from the area where the grapes were grown. $t_{1/2} = 12.3$ yr.

9. Analysis of a sample of uranite ore shows that the ratio of U–238 atoms to Pb–206 atoms is 3.8. Assuming there was no Pb–206 present initially, how old is the rock? $t_{1/2}$ (U–238) $= 4.5 \times 10^{9}$ yr.

10. Consider the following fusion reactions:

    a. $^{1}_{1}H + \, ^{2}_{1}H \rightarrow \, ^{3}_{2}He$

    b. $^{3}_{2}He + \, ^{3}_{2}He \rightarrow \, ^{4}_{2}He + 2\,^{1}_{1}H$

    c. $^{2}_{1}H + \, ^{3}_{1}H \rightarrow \, ^{4}_{2}He + \, ^{1}_{0}n$

    Given the atomic masses: $^{3}_{2}He = 3.016029$ amu; $^{4}_{2}He = 4.002603$ amu; $^{3}_{1}H = 3.017005$ amu, which of the above has the largest change in energy as indicated by its change in mass Δm?

11. One atom of element 109 was prepared by bombardment of a target of bismuth–209 with accelerated nuclei of iron–58. Write a balanced nuclear equation to show the formation of the isotope of element 109 with a mass number of 266.

12. Write nuclear equations that show how Pu–239 is formed in a "breeder" reactor.

13. $^{137}_{55}Cs$ is a fission product of $^{235}_{92}U$. If it is formed along with three neutrons, what is the other isotope formed?

14. What is the mode of decay expected for "light" nuclei which are unstable because of low n : p ratio?

15. One natural radioactive series begins with $^{238}_{92}U$ and ends with $^{206}_{82}Pb$. All steps in the series are either alpha or beta decay. How many α particles and β particles are emitted?

16. The $^{0}_{-1}β$ particles emitted by carbon-14 atoms have a maximum energy of 2.5×10^{-14} J per particle. What is the dose in rads when 8.0×10^{10} carbon–14 atoms decay, and all the energy from their decay is absorbed by 2.0 kg of matter?

17. A sample of biological tissue absorbs a dose of 1.0 rad of alpha radiation. How many rems is this?

18. Radium–226 is an $^{4}_{2}α$ emitter with a half–life of 1600 years. If 2.0 g of radium were allowed to undergo decay for 10 years, and all of the $^{4}_{2}α$ particles were collected over that time as helium gas, what would be the mass and the volume of the He at STP?

Chapter 21
Metallurgy and the Chemistry of Metals

PROBLEM-SOLVING STRATEGIES AND TUTORIAL SOLUTIONS

TYPES OF PROBLEMS

Problem Type 1: Metallurgic Processes.
  A. Preparation of the Ore.
  B. Production of Metals.
  C. The Metallurgy of Iron.
  D. Steelmaking.
  E. Purification of Metals.

Problem Type 2: Band Theory of Conductivity.
  A. Conductors.
  B. Semiconductors.

Problem Type 3: The Alkali Metals.

Problem Type 4: The Alkaline Earth Metals.

Problem Type 5: Aluminum.

PROBLEM TYPE 1: METALLURGIC PROCESSES.

A. Preparation of the Ore.
The initial steps allow the separation of the metal from the waste material. Three processes, (1) flotation, (2) ferromagnetic separations, or (3) formation of **amalgams** (alloys with mercury) serve as separation techniques.

B. Production of Metals.
The process of producing a free metal is always one of reduction. Reduction of metal oxides is usually less complex than reduction of sulfides. **Roasting** is a preliminary operation used to convert a metal sulfide or carbonate into an oxide. Two types of reductions are ordinarily employed:

1. Chemical reduction involves the use of a reducing agent to prepare the elemental form of a metal from a compound. Calcium, magnesium, aluminum, hydrogen, and carbon are often used in chemical reductions.
2. Electrolytic reduction is necessary to produce the very electropositive metals such as sodium, magnesium and aluminum.

STUDY QUESTION 21.1
Chemical reduction involves the use of a reducing agent to prepare the elemental form of a metal from a compound. Look up the reduction potentials of calcium, magnesium, and hydrogen listed in the text and suggest five metals that calcium, and magnesium can reduce, but that hydrogen cannot reduce.

Solution: Ca and Mg are strong reducing agents. Their reduction potentials are quite low, which means that they have a strong tendency to transfer electrons to other metal cations.

$$Ca^{2+}(aq) + 2e^- \rightarrow Ca(s) \qquad\qquad E^{\circ}_{red} = -2.87$$
$$Mg^{2+}(aq) + 2e^- \rightarrow Mg(s) \qquad\qquad E^{\circ}_{red} = -2.37 \text{ V}$$
$$2H^+ + 2e^- \rightarrow H_2(g) \qquad\qquad E^{\circ}_{red} = 0 \text{ V}$$

H_2 is not as strong a reducing agent as the others.

$Ca(s)$ and $Mg(s)$ can be used to reduce ions of metals having higher reduction potentials (the diagonal rule) such as Zn^{2+}, Cr^{3+}, Co^{2+}, Ni^{2+}, and Sn^{2+}, but not Li^+ and K^+. H_2 cannot reduce any of these ions, but is used to reduce more noble metals such as Cu^{2+}.

C. The Metallurgy of Iron.

Iron is prepared by a chemical reduction process. The raw materials for making iron are (1) iron ore, either hematite Fe_2O_3 or magnetite Fe_3O_4, (2) coke, and (3) limestone. These three materials are mixed and fed into a huge blast furnace. A strong blast of air preheated to 1500 °C is blown in at the bottom, where oxygen in the air reacts with the coke. This reaction supplies heat for the furnace, and carbon monoxide which is the reducing agent in the reaction of the iron oxides.

$$2C(g) + O_2(g) \rightarrow 2CO(g)$$
$$Fe_2O_3(s) + 3CO(g) \rightarrow 2Fe(\ell) + 3CO_2(g)$$

A temperature gradient is set up in the furnace. The temperature at the bottom is 1500 °C, and it decreases with height; falling to about 250 °C at the top. Much of the carbon dioxide produced is reduced by reaction with excess coke as follows:

$$CO_2(g) + C(s) \rightarrow 2CO(g)$$

This reaction forms more of the principal reducing agent. The gas that escapes at the top of the furnace is mostly nitrogen and carbon monoxide. The molten iron runs to the bottom where it is withdrawn periodically.

Limestone serves as a flux in the removal of impurities such as silica (SiO_2) and alumina (Al_2O_3). The limestone decomposes at temperatures above 900 °C, forming calcium oxide and carbon dioxide.

$$CaCO_3(s) \rightarrow CaO(s) + CO_2(g)$$

Calcium oxide unites with silica and alumina to form a glassy, molten substance called slag which is composed mainly of $CaSiO_3$ and some calcium aluminate. Slag is less dense than iron, and so it collects as a pool on top of the metal. It is drawn off, leaving the molten iron. Iron prepared in this manner contains many impurities and is called pig iron, or cast iron. Cast iron is made into steel by further treatment in a basic oxygen furnace.

D. Steelmaking.

Steel is an alloy of iron containing between 0.03 and 1.4% carbon along with other elements. Its production involves an oxidation reaction (the "basic oxygen process"). The reaction with oxygen forms oxides with the common impurities which are silicon, phosphorus, and manganese. The properties of the various types of steel depend both on its chemical composition and the heat treatment it receives.

STUDY QUESTION 21.2
How are the impurities present in iron ore, such as silica and sulfur, removed during steel production?

Solution: Acidic impurities such as silica and sulfides combine with CaO, a base, to form a molten slag.

$$SiO_2 + CaO \rightarrow CaSiO_3$$
$$FeS + CaO \rightarrow CaS + FeO$$

The slag has a lower density than molten iron. Therefore, it collects as a pool on top of the metal. This slag is drawn off and used as a component in making cement.

E. Purification of Metals.

Once a metal is prepared it may need to be further purified. Distillation, electrolysis, and zone refining are three common procedures. Metals with low boiling points, such as mercury (357 °C), cadmium (767 °C), and zinc (907 °C), are referred to as volatile. They can be purified by fractional distillation. On heating, these metals vaporize, leaving behind any nonvolatile impurities. Condensation of the vapor yields the purified metal.

Metals can also be purified by electrolysis. The metals that plate out on a cathode can be controlled by the voltage at which the electrolysis is carried out. In the purification of copper, Cu^{2+} is reduced much more readily than iron or zinc ions, which are common impurities. At just the right voltage Cu is plated out and the metal impurities remain in solution.

Zone refining takes advantage of the fact that when liquids begin to freeze; the impurities tend to remain in the liquid phase. See Figure 21.8 of the textbook. Extremely pure metals can be obtained by repeating this process a number of times.

> **PRACTICE QUESTION 21.1**
> **Distinguish between an ore and a mineral.**
> _____

> **PRACTICE QUESTION 21.2**
> **Why must electrolytic reduction be used in the production of some metals such as lithium and sodium?**
> _____

> **PRACTICE QUESTION 21.3**
> **For what is zone refining used?**
> _____

> **PRACTICE QUESTION 21.4**
> **How is limestone used in iron production?**
> _____

> **PRACTICE QUESTION 21.5**
> **Write an equation for the chemical reduction of Cr_2O_3 with Al metal yielding chromium.**
> _____

PROBLEM TYPE 2: BAND THEORY OF CONDUCTIVITY.

A. Conductors.

The band theory is the result of the application of molecular orbital theory to metals. In metals, the atoms lie in a three–dimensional array and take part in bonding that spreads over the entire crystal. For example, consider an alkali metal atom that carries a single valence electron in an *s* orbital. You will recall that the atomic orbitals of two atoms will overlap when the atoms are close together, resulting in the formation of two molecular orbitals: a bonding orbital and an antibonding orbital. The total number of molecular orbitals produced always equals the number of atomic orbitals that overlap to produce them. The overlap of four atomic orbitals from four atoms, for instance, will form four molecular orbitals: two bonding and two antibonding orbitals. These bonding and antibonding orbitals are so closely spaced in terms of energy that they are called a "band."

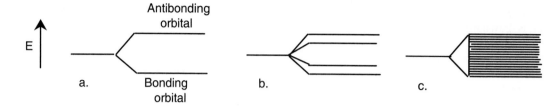

The band containing the valence shell electrons ($3s$ for Na) is called the valence band. Any band that is either vacant or partially filled is called a conduction band.

The band theory explains conduction in the alkali metals in the following way. When a voltage is applied across a piece of sodium metal, conduction occurs. The current is the result of electrons in the $3s$ band being free to jump from atom to atom. The free movement of electrons is possible for two reasons. In alkali metals the conduction band and the valence band are the same. The orbitals within the band are so similar in energy that an electron does not need to gain appreciable energy to reach the conduction band. And, as discussed above, the conduction band is only half filled to capacity which means that an electron is free to move through the entire metal.

In an insulator, such as glass or plastic, the valence band is filled. Thus, the next vacant higher–energy band becomes the conduction band. An energy gap exists between the valence band and the conduction band. This large separation prevents electrons in insulators from entering the conduction band.

B. Semiconductors.

A semiconductor, such as Si and Ge, has a filled valence band and an empty conduction band, but in contrast to an insulator, a relatively small gap exists between these bands. A relatively small amount of thermal energy will promote an electron into the conduction band. Thus as temperature increases, the conductivity of semiconductors increases correspondingly.

STUDY QUESTION 21.3
Show that the valence band is half–filled for sodium.

Solution: Sodium atoms have filled $1s$, $2s$, and $2p$ orbitals. Therefore, the corresponding bands in the solid are filled. The valence band for Na is the $3s$. In the metal, the $3s$ orbitals of N sodium atoms overlap to form a total of N molecular orbitals that make up the valence band. The capacity of this band is $2N$ electrons. Since each Na atom has one $3s$ electron, the N Na atoms have N electrons in the valence band. This means that the valence band is half–filled. The $3s$ band of Na is the conduction band. Remember, that the highest occupied band is called the valence band. This unfilled valence band is called the conduction band of sodium metal.

> **PRACTICE QUESTION 21.6**
> **Define the terms: valence band and conduction band.**

> **PRACTICE QUESTION 21.7**
> **Define the terms: conductor, insulator, and semiconductor.**

PROBLEM TYPE 3: THE ALKALI METALS.

As a group, the alkali metals are extremely reactive and never occur naturally in the elemental form. They are the least electronegative group of elements and exist as +1 cations, which combine with halides, sulfate, carbonate, and silicate ions. Compared with other metals, they are very soft, have low melting points and densities so low that lithium, sodium and potassium will float on water.

The preparation of Group 1A metals requires large amounts of energy because their positive ions are difficult to reduce. Lithium, being the most active metal, is prepared by electrolysis of molten LiCl.

Sodium and potassium occur in a wide variety of minerals, and are the sixth and seventh most abundant elements in Earth's crust. Sodium compounds are so abundant and widespread that it is difficult to find matter free of this element. Sodium chloride makes up about two–thirds of the solid matter dissolved in sea water. The minerals carnallite ($KMgCl_3 \cdot 6H_2O$), and sylvite (KCl) are found in ancient lake and seabeds and serve as commercial sources of potassium and its compounds. Metallic sodium is prepared commercially by the electrolysis of molten sodium chloride. Most of the sodium made in the United States is produced in the Downs cell, by the electrolysis of a mixture of sodium chloride and calcium chloride. This electrolyte mixture melts at 505 °C, whereas pure NaCl melts at 801 °C. The lower temperature reduces the cost of production.

A Downs cell consists of a carbon anode and an iron cathode. The chloride ions are oxidized at the anode, and the sodium ions are reduced at the cathode. The half–reactions are:

$$\text{Anode} \qquad 2Cl^- \rightarrow Cl_2(g) + 2e^-$$
$$\text{Cathode} \qquad 2Na^+ + 2e^- \rightarrow 2Na(\ell)$$

The liquid sodium is drawn off and kept from contact with chlorine and oxygen. Sodium can also be obtained by the electrolysis of molten sodium hydroxide.

Potassium is made by reaction of potassium chloride with sodium vapor in the absence of air at 900 °C.

$$KCl(\ell) + Na(g) \rightarrow K(g) + NaCl(\ell)$$

This reaction should occur to only a small extent. However, removal of potassium vapor as it is formed drives the reaction to completion.

Alkali metal reactivity increases with atomic mass. Sodium forms two oxides on reaction with oxygen: sodium oxide (Na_2O) and sodium peroxide (Na_2O_2).

$$2Na(s) + \frac{1}{2}O_2(g) \rightarrow Na_2O(s)$$
$$2Na(s) + O_2(g) \rightarrow Na_2O_2(s)$$

The reaction of potassium with oxygen forms the peroxide (K_2O_2) but, in addition, when K burns in air it forms potassium superoxide.

$$K(s) + O_2(g) \rightarrow KO_2(s)$$

Potassium superoxide is a source of O_2 in breathing equipment. KO_2 reacts with water:

$$2KO_2(s) + 2H_2O(\ell) \rightarrow 2KOH(aq) + O_2(g) + H_2O_2(aq)$$

Sodium chloride, sodium carbonate, sodium bicarbonate, sodium hydroxide, sodium nitrate, potassium hydroxide, and potassium nitrate are important compounds of these elements.

PRACTICE QUESTION 21.8
List four physical properties of the alkali metals.

PRACTICE QUESTION 21.9
List three chemical properties of the alkali metals.

PROBLEM TYPE 4: THE ALKALINE EARTH METALS.

The alkaline earth metals are somewhat less reactive than the alkali metals. Except for beryllium, the alkaline earth metals have similar chemical properties. The oxidation number of the Group 2A metals in almost all compounds is +2. The electronegativity of these elements is low enough that they form predominately ionic compounds with nonmetals. Table 21.5 in the text lists some common properties of these metals. As with the other groups of elements in the periodic table, metallic character increases as you move down the group.

Seawater is an important source of magnesium. It contains 0.13% Mg^{2+} by weight. Being an insoluble compound, $Mg(OH)_2$ is precipitated from seawater by adding $Ca(OH)_2$.

$$Mg^{2+}(aq) + Ca^{2+}(aq) + 2OH^-(aq) \rightarrow Mg(OH)_2(s) + Ca^{2+}(aq)$$

Magnesium hydroxide is then converted to magnesium chloride by neutralization with hydrochloric acid. Metallic magnesium is obtained by electrolysis of molten magnesium chloride.

$$\text{Electrolysis} \qquad Mg^{2+}(aq) + 2Cl^-(aq) \rightarrow Mg(\ell) + Cl_2(g)$$

Calcium is obtained from limestone. Heating limestone causes it to decompose into calcium oxide and carbon dioxide. Calcium metal is prepared by a thermal reduction process, rather than electrolysis. Calcium oxide is reduced with aluminum at a high temperature (1200 °C).

$$CaCO_3(s) \rightarrow CaO(s) + CO_2(g)$$
$$6CaO(s) + 2Al(\ell) \rightarrow 3Ca(\ell) + Ca_3Al_2O_6(s)$$

Magnesium is less reactive than calcium. Magnesium does not react with cold water, but it reacts slowly with steam. Calcium, on the other hand, reacts with cold water.

$$Mg(s) + H_2O(g) \rightarrow MgO(s) + H_2(g)$$
$$Ca(s) + 2H_2O(\ell) \rightarrow Ca(OH)_2(aq) + H_2(g)$$

Both MgO and CaO react with water to give hydroxides.

$$MgO(s) + H_2O(\ell) \rightarrow Mg(OH)_2(s) \quad \text{(reacts with steam)}$$
$$CaO(s) + H_2O(\ell) \rightarrow Ca(OH)_2(s) \quad \text{(reacts with cold water)}$$

When magnesium burns in air, considerable nitride is formed along with the oxide.

$$2Mg(s) + O_2(g) \rightarrow 2MgO(s)$$
$$3Mg(s) + N_2(g) \rightarrow Mg_3N_2(s)$$

Beryllium has excellent alloying qualities and is used to make alloys that are corrosion resistant. Beryllium is used as a "window" in X–ray tubes.

The major uses of magnesium are also in alloys. Because of its low atomic mass, it is a good lightweight structural metal. It is also used in batteries, and for cathodic protection of buried metal pipelines and storage tanks.

Metallic calcium finds use mainly in alloys. Calcium salts are used as dehydrating agents; anhydrous calcium chloride, for example, has a strong affinity for water. Quicklime (CaO) is used in steel production, and in the removal of SO_2 from coal–fired power plants. Slaked lime $(Ca(OH)_2)$ is used in water treatment.

Strontium nitrate and carbonate are used in fireworks and highway flares to provide their brilliant red color. Barium metal is the most active of the alkaline earth metals so it has very few uses.

STUDY QUESTION 21.4

Starting with Ca^{2+} ions in limestone, write equations to show how you would obtain pure calcium, using electrolysis, rather than thermal reduction.

Solution: Electrolysis of molten metal chlorides is often used to prepare pure metals. First, limestone, chalk, or sea shells, all of which contain $CaCO_3$, must be decomposed at 900 °C.

$$CaCO_3(s) \rightarrow CaO(s) + CO_2(g)$$

The $CaO(s)$ is slaked to yield calcium hydroxide.

$$CaO(s) + H_2O(\ell) \rightarrow Ca(OH)_2(s)$$

This base is then neutralized with hydrochloric acid to give the desired salt $CaCl_2$.

$$Ca(OH)_2(aq) + 2HCl(aq) \rightarrow CaCl_2(aq) + 2H_2O(\ell)$$

After drying, the $CaCl_2$ can be melted and then electrolyzed.

Electrolysis $$CaCl_2(\ell) \rightarrow Ca(\ell) + Cl_2(g)$$

PRACTICE QUESTION 21.10

Identify the element among the alkaline earth metals (excluding radium) that will have the following properties:

    a. **Most reactive oxide towards water.**
    b. **Lowest electronegativity.**
    c. **Smallest atomic radius.**
    d. **Lowest first ionization energy.**

PRACTICE QUESTION 21.11

What are the sources of magnesium and calcium?

PROBLEM TYPE 5: ALUMINUM.

Aluminum is the third most abundant element in the earth's crust, making up 7.5 percent by mass. It is too active chemically to occur free in nature, but is found in compounds in over 200 different minerals. The most important ore of Al is bauxite, which contains hydrated aluminum oxide $(Al_2O_3 \cdot 2H_2O)$, along with silica and hydrated iron oxide $(Fe_2O_3 \cdot 2H_2O)$. Another important mineral is cryolite (Na_3AlF_6), which is used in the metallurgy of aluminum. The Hall process for the production of aluminum metal is described in Figure 21.18 of the textbook.

Aluminum is obtained from bauxite ore $(Al_2O_3 \cdot 2H_2O)$. The first step in its preparation is to separate pure aluminum oxide from the silica and iron impurities. First, bauxite is pulverized and digested with sodium hydroxide solution. This converts the silica into soluble silicates, and aluminum oxide is converted to the aluminate ion AlO_2^- which remains in solution. However this digestion treatment has no effect on the iron which remains as insoluble $Fe_2O_3(s)$.

It is removed by filtration. Aluminum hydroxide is then precipitated by acidification (with carbonic acid) to about pH 6. The precipitate is heated strongly to produce pure Al_2O_3. The chemical changes involving aluminum are:

$$Al_2O_3(s) + 2OH^-(aq) \rightarrow 2\,AlO_2^-(aq) + H_2O(\ell)$$
$$AlO_2^-(aq) + H^+(aq) + H_2O(\ell) \rightarrow Al(OH)_3(s)$$
$$2Al(OH)_3(s) \rightarrow Al_2O_3(s) + 3H_2O(g)$$

The aluminum ions in Al_2O_3 can be reduced to metallic aluminum efficiently only by electrolysis. The melting point of Al_2O_3 is 2050 °C, which makes electrolysis of pure molten Al_2O_3 extremely expensive owing to the need to maintain the high temperature. In the Hall process, Al_2O_3 is dissolved in molten cryolite (Na_3AlF_6) which melts at 1000 °C. The use of cryolite makes it possible to lower the temperature of electrolysis by 1050 °C! The mixture is electrolyzed to produce aluminum and oxygen.

Cathode	$2Al^{3+} + 6e^- \rightarrow 2Al(\ell)$
Anode	$3O^{2-} \rightarrow \dfrac{3}{2}O_2 + 6e^-$

$$\text{Overall} \qquad 2Al^{3+} + 3O^{2-} \rightarrow 2Al(\ell) + \frac{3}{2}O_2$$

Pure aluminum is a silvery–white metal with low density (2.7 g/cm^3) and high tensile strength. It is malleable and can be rolled into thin foils. Its electrical conductivity is about 65 percent that of copper.

Aluminum is an amphoteric element, reacting with both acids and bases.

$$2Al(s) + 6HCl(aq) \rightarrow AlCl_3(aq) + 3H_2(g)$$
$$2Al(s) + 2NaOH(aq) + 2H_2O(\ell) \rightarrow 2NaAlO_2(aq) + 3H_2(g)$$

Aluminum has a strong affinity for oxygen, thus, it is usually covered by an oxide film.

$$4Al(s) + 3O_2(g) \rightarrow 2Al_2O_3(s)$$

This layer of Al_2O_3 forms a compact, adherent, protective, surface coating on the metal and is responsible for preventing further oxidation and corrosion. Because of this surface oxide layer, Al is practically insoluble in weak or dilute acids.

The large enthalpy of formation of aluminum oxide ($\Delta H_f^\circ = -1670$ kJ/mol) makes the metal an excellent reducing agent. Thus, a variety of metals can be produced in a series of similar reactions involving Al powder with the corresponding metal oxides. So much heat is liberated in these reactions, called aluminothermic reactions, that the metal is usually obtained in the molten state. The thermite reaction, used in welding steel and iron, is one example.

$$2Al(s) + Fe_2O_3(s) \rightarrow Al_2O_3(\ell) + 2Fe(\ell)$$

Aluminum hydroxide is an amphoteric hydroxide, dissolving in both acid and base.

$$Al(OH)_3(s) + 3H^+(aq) \rightarrow Al^{3+}(aq) + 3H_2O(\ell)$$
$$Al(OH)_3(s) + OH^-(aq) \rightarrow Al(OH)_4^-(aq)$$

Aluminum metal is used in high–voltage transmission lines because it is cheaper and lighter than copper. Its chief use is in aircraft construction. Aluminum is employed also as a solid propellant in rockets. This is another example of the great affinity that aluminum has for oxygen. Ammonium perchlorate, NH_4ClO_4, is the oxidizer.

The formation of aluminum hydroxide precipitate is used in water treatment plants. The process requires large amounts of aluminum sulfate.

$$Al_2(SO_4)_3(aq) + 3Ca(OH)_2(aq) \rightarrow 2Al(OH)_3(s) + 3CaSO_4(aq)$$

Aluminum hydroxide is a gelatinous substance. As it settles in treatment pools, it coprecipitates suspended matter such as bacteria and colloidal sized particles. This process clarifies drinking water.

Alums are compounds that have the general formula:

$$M^+M^{3+}(SO_4)_2 \cdot 12H_2O$$

where $M^+ = K^+$, Na^+, NH_4^+, and $M^{3+} = Al^{3+}$, Cr^{3+}, Fe^{3+}. They are used in the dying industry. The formation of the gelatinous aluminum hydroxide "fixes" the dye to the cloth.

STUDY QUESTION 21.5

Aluminum is a good reducing agent, and its reduction potential is quite negative, ($E^\circ_{red} = -1.66$ V) which means that aluminum should react with water and liberate hydrogen. But, we know that airplanes do not dissolve in thunderstorms. Explain.

Solution: Aluminum readily forms the oxide Al_2O_3 when exposed to air. The oxide forms a tenacious surface film that protects the aluminum metal from further corrosion due to water. Therefore, it is the presence of this oxide layer that makes aluminum practically insoluble even in dilute acids.

> *PRACTICE QUESTION 21.12*
> **What is the role of cryolite, Na_3AlF_6, in aluminum production?**

> *PRACTICE QUESTION 21.13*
> **The enthalpy of formation of Al_2O_3 is large and exothermic. Why doesn't aluminum metal just completely oxidize away?**

PRACTICE QUIZ

1. Distinguish between:
    a. Leaching and flotation.
    b. Roasting and reduction.

2. Can carbon be used to reduce Al_2O_3 to $Al(s)$? *Hint:* Is the following reaction spontaneous?

$$2Al_2O_3(s) + 2C(s) \rightarrow 4Al(s) + 3CO_2(g)$$

3. When a solution of NaOH is added, dropwise to a test tube containing an aqueous solution of Al^{3+}, a white gelatinous precipitate is formed. Upon continued addition of NaOH, the precipitate disappears. Write the chemical equations to explain this observation.

4. Distinguish between an alloy and an amalgam.

5. According to the band theory, why is copper a conductor?

6. Considering that electrons in metals have a random thermal motion like molecules of gas, explain why electrical conductivity of conductors decreases with increasing temperature.

7. Aluminum metal was first prepared in 1825 by the action of potassium metal on aluminum chloride. Write a balanced equation for this reaction.

8. What is the chemical form of aluminum in bauxite?

9. How is aluminum separated from Fe_2O_3?

10. What is the role of aluminum sulfate in water purification?

Chapter 22
Coordination Chemistry

PROBLEM-SOLVING STRATEGIES AND TUTORIAL SOLUTIONS

TYPES OF PROBLEMS
Problem Type 1: Coordination Compounds.
      A. Properties of Transition Metals.
      B. Ligands.
      C. Nomenclature of Coordination Compounds.

Problem Type 2: Structure of Coordination Compounds.

Problem Type 3: Bonding in Coordination Compounds: Crystal Field Theory.
      A. Crystal Field Splitting in Octahedral Complexes.
      B. Color.
      C. Magnetic Properties.

PROBLEM TYPE 1: COORDINATION COMPOUNDS.

A **complex ion** consists of a central metal cation to which several anions or molecules are bonded. The complex ion may be positively or negatively charged and is shown in brackets. A neutral species containing a complex ion is called a **coordination compound**. These compounds usually have complicated formulas. The **coordination number** is the number of donor atoms surrounding the central metal atom. As a result, the coordination number of Pt^{2+} in $[Pt(NH_3)_4]^{2+}$ is 4, and that of Co^{2+} in $[Co(NH_3)_6]^{2+}$ is 6.

A. Properties of Transition Metals.
For the representative elements, properties such as the atomic radius, ionization energy, and electronegativity vary markedly from element to element as the atomic number increases across any period. In contrast, the chemical and physical properties of the transition metal elements vary only slightly as we read across a period. Following is a summary of the transition metal characteristics.

1. *General Physical Properties*. Transition metals have relatively high densities, high melting and boiling points, and high heats of fusion and vaporization.

2. *Atomic Radius*. Atomic radii of representative elements decrease markedly as we read across a period of elements. In contrast, the atomic radii of transition metals decrease only slightly as we read across a series of these elements.

   The nuclear charge increases from scandium to copper, and electrons are being added to an inner $3d$ orbital. These $3d$ electrons shield the 4s electrons from the increasing nuclear charge for the most part. Consequently, the 4s electrons of the first-row transition elements feel only a slightly increasing effective nuclear charge as the atomic number increases, and the atomic radii makes only a gradual decrease.

3. *Ionization Energy*. The first ionization energies of the first transition metal series are remarkably similar, increasing very gradually from left to right. There is a slight increase over the first five elements, then the ionization energy barely changes from iron to copper.

4. *Variable Oxidation States.* The common oxidation states for the transition elements from Sc to Cu are +2 and +3. Some of the elements exhibit the +4 state, and Mn even shows +5, +6, and +7. Transition metals usually exhibit their highest oxidation states in compounds with oxygen, fluorine, or chlorine. $KMnO_4$ and $K_2Cr_2O_7$ are examples. The variability of oxidation states for transition metal ions results from the fact that the $4s$ and $3d$ orbitals are similar in energy. Therefore, an atom can form ions of roughly the same stability by losing different numbers of electrons. The +3 oxidation states are more stable at the beginning of the series, but toward the end, the +2 oxidation states are more stable.

STUDY QUESTION 22.1
What distinguishes a transition element from a representative element?

Solution: The representative elements are those in which all the inner subshells are filled and the last electron enters an outer s or p subshell. The transition elements are those in which an inner d subshell is partially filled. For example, the outer electron configuration of Sc is $4s^23d^1$ and for Co it is $4s^23d^7$.

STUDY QUESTION 22.2
Why do the transition metals have higher densities than the Group 1A and 2A metals?

Solution: Densities of solids are related to atomic and ionic radii. As the atoms get smaller, the solids tend to get denser. The effective nuclear charge experienced by the outermost electrons in transition metal atoms is greater than that experienced in Group 1A and 2A metal atoms. The atomic radii of the transition elements, in general, are smaller than those of the Group 1A and 2A elements of the same period.

> ### PRACTICE QUESTION 22.1
> **Using your periodic table, write the electron configurations of the following.**
>
> a. **Mn atom**
> b. **Mn^{2+} ion**
> c. **Mn^{3+} ion**

> ### PRACTICE QUESTION 22.2
> **Which of the species in Practice Question 22.1 is expected to form the strongest metal-ligand bond? Why?**

B. Ligands.
Ligand is the term used to refer to the neutral molecules or ions that form coordinate covalent bonds with the central metal atom. The ligands each have at least one unshared pair of electrons that can be donated to the electron-deficient metal ions. The donor atom is the atom in the ligand that is directly bonded to the metal. Common ligands are Cl^-, CN^-, NH_3, H_2O, and $H_2NCH_2CH_2NH_2$. Some ligands contain more than one donor atom. Ligands that coordinate through two bonds are called **bidentate ligands**. Those with more than two donor atoms are referred to as **polydentate ligands**. Ethylenediamine (en) and oxalate ion (ox) are bidentate ligands:

Ethylenediamine (en)   Oxalate ion (ox), $C_2O_4^{2-}$

Bidentate and polydentate ligands are often called **chelating ligands** (pronounced key-late-ing). The name derives from the Greek word *chele* meaning "claw." Chelate complexes are extra stable because two bonds must be broken to separate a metal from a ligand. $EDTA^{4-}$ (ethylenediaminetetraacetate) is an excellent chelating ligand because it has six donor atoms.

Each complex ion carries a net charge that is the sum of the charges on the central atom (or ion) and on each of the ligands. In $Fe(CN)_6^{3-}$, for example, each cyanide ion (CN^-) contributes a -1 charge, and the iron ion a $+3$ charge. The charge of the complex ion then is; $(+3) + 6(-1) = -3$. The oxidation number of iron is $+3$ which is the same as its ionic charge.

STUDY QUESTION 22.3
A certain coordination compound has the formula $[Co(NH_3)_4Cl_2]Cl$.

    a.   **Which atom is the central atom?**
    b.   **Name the ligands and point out the donor atoms.**
    c.   **What is the charge on the complex ion?**
    d.   **What is the oxidation number (O.N.) of the central atom?**
    e.   **What is the coordination number?**

Solution:
a.   Inside the brackets, the metal atom is written first, and this is the central atom to which all ligands are bonded. Cobalt is the central atom.
b.   The ligands are written next. These are ammonia and chloride ions. The donor atoms are N and Cl, respectively.
c.   Since only one chloride ion is needed to balance the charge of the complex ion, the complex ion must have a charge of $+1$.
d.   The charge of the central ion plus the sum of charges of ligands = the charge of the complex ion. Ammonia is a neutral ligand and will not affect the complex ion charge. Since two chloride ions net -2 charge, in order for the complex ion to have a $+1$ charge, then Co must be a $+3$ ion.
e.   Each of the ligands is monodentate. There is a total of six ligands. The coordination number is 6.

C. Nomenclature of Coordination Compounds.
Thousands of coordination compounds are known. The rules that have been developed for naming them are summarized below:

1.   The cation is named before the anion.
2.   Within the complex ion, the ligands are named first in alphabetical order followed by the metal ion.
3.   The names of anionic ligands end with the letter o, whereas neutral ligand names are usually the same as the names of the molecules. The exceptions are H_2O (aqua), CO (carbonyl), and NH_3 (ammine). Table 22.4 in the textbook lists the names of some common ligands.
4.   When several ligands of a particular kind are present, use the Greek prefixes di-, tri-, tetra-, penta-, and hexa- to indicate how many of each is present. Thus, the ligands in $[Co(NH_3)_4Cl_2]^+$ are "tetraamminedichloro." If the ligand itself contains a Greek prefix, use the prefixes; bis- (2), tris- (3), and tetrakis- (4) to indicate the number of ligands present. The ligand ethylenediamine already contains the term di; therefore, bis(ethylenediamine) is used to indicate two ethylenediamine ligands.
5.   The oxidation number of the metal is written in Roman numerals in parentheses directly following the name of the metal.
6.   If the complex is an anion, attach the ending -ate to the root name of the metal. $Fe(CN)_6^{3-}$ is named hexacyanoferrate (III) ion.

STUDY QUESTION 22.4
Write formulas for the following coordination compounds.

    a.   **diaquodicyanocopper(II)**
    b.   **potassium hexachloropalladate(IV)**
    c.   **dioxalatocuprate(II) ion**

Solution:
a.   diaquo refers to two water molecules and dicyano to two cyanide ions. Since the copper ion has a $+2$ charge, this coordination compound is neutral: $Cu(H_2O)_2(CN)_2$

b. The ending *-ate* indicates that the complex ion must be an anion: $K_2[Pd(Cl)_6]$.

c. The complex ion is an anion: $[Cu(C_2O_4)_2]^{2-}$

STUDY QUESTION 22.5
Give a systematic name for each of the following compounds.

    a. $[Ag(NH_3)_2]Br$
    b. $Ni(CO)_4$
    c. $K_2[Cd(CN)_4]$

Solution:

a. The two ammonia ligands are represented by the term "diammine." The oxidation state of the Ag ion is +1. The bromide ion is not complexed with silver. It is a counter ion. The systematic name is diamminesilver(I) bromide.

b. The term tetracarbonyl signifies that there are four carbon monoxide ligands. The oxidation state of the nickel atom is zero. Systematic name: tetracarbonylnickel(0).

c. In this case the complex ion is an anion; as a result, cadmium will be referred to as cadmate. Since K is +1 and each cyanide ligand is –1, Cd must have a +2 charge. The name is potassium tetracyanocadmate(II).

> *PRACTICE QUESTION 22.3*
> **Give the systematic name for $[CoCl_3Br_3]^{4-}$.**

> *PRACTICE QUESTION 22.4*
> **Write formulas for the following coordination compounds.**
>
>   a. **diammineoxolatocopper(II)**
>   b. **potassium amminetrichloroplatinate(II)**

> *PRACTICE QUESTION 22.5*
> **Other than the ligands, what is the main difference between $[CoCl_3Br_3]^{4-}$ and $[Co(NH_3)_3Br_3]$?**

PROBLEM TYPE 2: STRUCTURE OF COORDINATION COMPOUNDS.

The geometry of a coordination complex is defined by the arrangement of the donor atoms of the ligands around the central metal atom. The geometries are almost the same as what we learned from VSEPR theory.

When two or more compounds have the same composition but a different arrangement of atoms, the compounds are called **isomers**. Isomerism is a characteristic feature of coordination compounds. **Stereoisomers** are compounds that have the same types and numbers of atoms bonded together in the same sequence, but with different spatial arrangements. There are two types of stereoisomers: **geometric isomers** and **optical isomers**.

Geometric, or *cis-trans*, isomers are distinguished by the position of like ligands or groups. The isomer with the like-groups in adjacent positions is called the *cis* isomer, and the one with like-groups across from each other is called the *trans* isomer. *Cis* and *trans* isomers of coordination compounds generally have different physical and chemical properties.

Two forms of the complex $Pt(NH_3)_2Cl_2$ have been prepared. Both have square planar structures. The *trans* form has no dipole moment, whereas the *cis* form has an appreciable dipole moment.

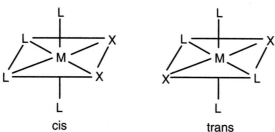

The simplest type of geometric isomerism of octahedral complexes occurs in cases where four of the six ligands of the complex are the same. The *cis* and *trans* forms correspond to those shown below. The two X ligands are closest to each other in the *cis* isomer. In the *trans* isomer they are across the complex ion from each other.

Optical isomers are nonsuperimposable mirror images of one another. They are also called **enantiomers**. Enantiomers have the same relationship to one another as do your right and left hands. If you place your left hand parallel to your right hand facing each other, you get the same effect that you would get if you placed one hand in front of a mirror. Your left hand is the mirror image of your right hand. However, your left hand is not superimposable upon your right hand. Your hands are enantiomers, that is, nonsuperimposable mirror images of one another.

Enantiomers of *cis*-dichlorobis (ethylenediamine) cobalt (III) ion exist. The mirror image of the *trans* isomer can be superimposed on the original after rotating it by 90°. Therefore, there is only one form of the *trans* isomer. The mirror image of the *cis* isomer cannot be superimposed on the original no matter how it is rotated. The *cis* isomer exists as enantiomers.

Not all objects are enantiomers. An ordinary chair, for example, looks the same in the mirror as when viewed directly. A chair is superimposable on its mirror image. An object which is not identical with its mirror image is said to be **chiral**, from the Greek word for hand.

Enantiomers are called optical isomers because they differ with respect to the direction in which they rotate the plane of polarized light, when light is passed through the substance. The isomer that rotates the plane of polarized light to the right is said to be dextrorotatory. Its mirror image will rotate the plane to the left, levorotatory. A **racemic mixture** is an equimolar mixture of the two optical isomers. Such a mixture produces no net rotation of plane polarized light.

STUDY QUESTION 22.6
For the complex $[Cr(en)_2Br_2]^+$,

    a. Sketch the geometric isomers.
    b. Indicate if either of the isomers exhibits optical isomerism, and, if so, if it is chiral.

Solution:

a. First, identify the ligands. Recall that "en" stands for ethylenediamine, a neutral bidentate ligand. Bromide ion is the other ligand. Then, the central ion is Cr^{3+} and its coordination number is 6 because it bonds to six donor atoms. The complex will be octahedral. N — N represents ethylenediamine.

For an octahedral complex, ligands at adjacent corners are referred to as *cis* to each other. The ligands at opposite corners are *trans* to each other. Two geometric isomers are possible. One has two Br⁻ ions in *cis* positions, and in the other isomer the Br⁻ ligands are *trans*.

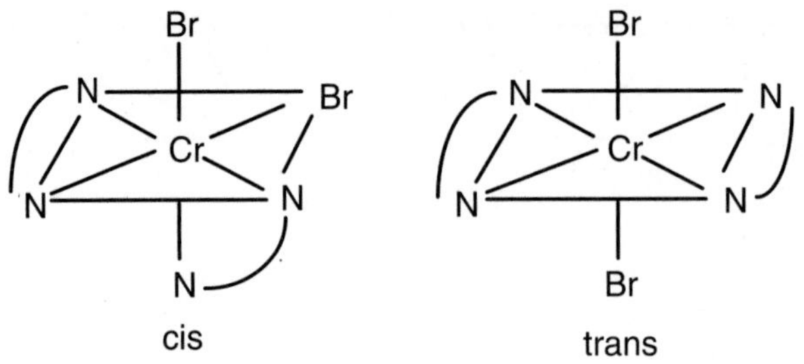

cis

trans

b. Sketch the mirror images of the *cis* and *trans* isomers, and look for nonsuperimposable mirror images, the enantiomers.

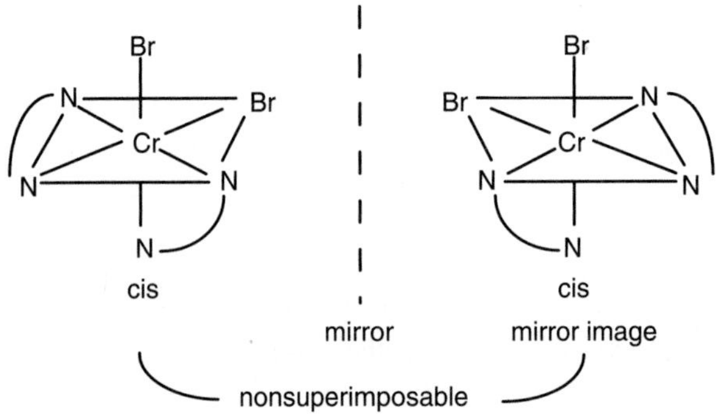

cis

mirror

cis

mirror image

nonsuperimposable

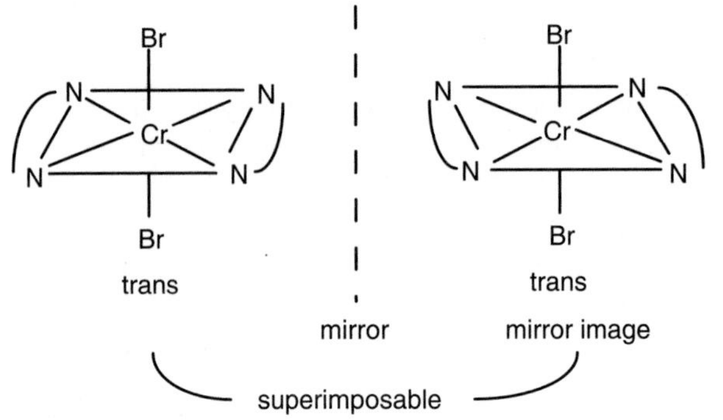

trans

mirror

trans

mirror image

superimposable

The mirror image of the *trans* isomer can be superimposed on the original. Therefore, there is only one form of the *trans* isomer. The mirror image of the *cis* isomer cannot be superimposed on the original no matter how it is rotated. The *cis* isomer exists in two forms (enantiomers). The *cis* isomer is chiral. The *trans* isomer is not.

PROBLEM TYPE 3: BONDING IN COORDINATION COMPOUNDS: CRYSTAL FIELD THEORY.

The crystal field model considers the ligand to metal bonding in a complex ion to be primarily electrostatic, rather than covalent. In an isolated transition metal ion all five *nd* orbitals have the same energy. The effect of the ligands is to change the relative energies of these orbitals through electrostatic interactions. There are two types of electrostatic interactions. One is the force that holds the ligands to the metal ion. This is the attraction between the positive charge of the metal ion and the lone electron pairs of the ligands. Second, there is the repulsion between the lone pairs on the ligand donor atoms and the electrons in the *nd* orbitals of the metal. It is this latter interaction that gives rise to **crystal field splitting**, and its effect on the color and magnetic properties of the complex ion.

A. Crystal-Field Splitting in Octahedral Complexes.

Consider the $Fe(CN)_6^{3-}$ complex ion, for example. An isolated Fe^{3+} ion has five *3d* electrons, all with the same energy:

$$Fe^{3+} \quad \boxed{\uparrow \,|\, \uparrow \,|\, \uparrow \,|\, \uparrow \,|\, \uparrow}$$

3d

When the six cyanide ligands become positioned in an octahedral arrangement, all the *3d* orbitals are raised in energy due to the repulsion just mentioned. The repulsion energy is not the same for all *3d* orbitals. Rather, those orbitals that are directed straight toward a ligand experience greater repulsion than those directed between ligands. Thus, in an octahedral complex, the five *3d* orbitals are split into two groups in terms of energy. One group, comprised of the d_{xy}, d_{yz} and d_{xz} orbitals, is not raised in energy as much as the other group consisting of the d_{z^2} and $d_{x^2-y^2}$ orbitals, as shown below. The energy difference between these two sets of d orbitals is called the **crystal field splitting energy** and is given the symbol Δ:

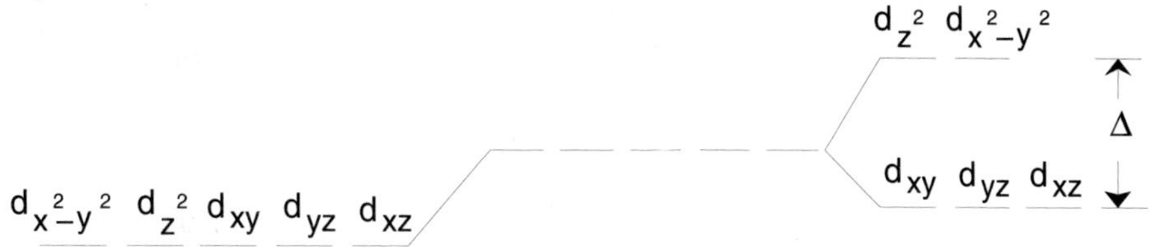

$$d_{z^2} \quad d_{x^2-y^2}$$

$$\Delta$$

$$d_{x^2-y^2} \quad d_{z^2} \quad d_{xy} \quad d_{yz} \quad d_{xz} \qquad\qquad d_{xy} \quad d_{yz} \quad d_{xz}$$

B. Color.

The color of a coordination compound results from electron transitions from the lower-energy set of orbitals to the higher-energy orbitals. Absorption of light occurs when the energy of an incoming photon is equal to the crystal field splitting energy, Δ, and an electron transition occurs. The requirement for light absorption is:

$$\Delta = E_{photon}$$

Recall from Chapter 7 that the energy of a photon is $E_{photon} = h\nu$, where h is Planck's constant, and ν is the frequency of the radiation. Therefore, $\Delta = h\nu$.

A substance has color because it absorbs light at certain wavelengths in the visible part of the electromagnetic spectrum (from 400 to 700 nm) and it reflects the other wavelengths. Light that is a combination of all visible wavelengths appears white. When white light strikes a coordination compound and light of a certain wavelength is absorbed, the reflected light is missing that component and no longer appears white to the eye.

The color that is seen depends on which color is absorbed. If the complex absorbs all colors except orange, the complex appears orange. The complex will also appear orange if it absorbs only light of the color blue. In a complementary manner, if the complex ion absorbed only orange, it would appear blue. Blue and orange are referred to as complementary colors.

The following "color wheel" allows you to estimate the color of reflected light when light of a given color (or wavelength) is absorbed. The colors that appear across from each other are complementary colors. For example, the hexacyanoferrate(II) ion absorbs light in the visible region of the spectrum at about 410 nm. The absorbed light is violet. The color of the complex is the color directly across the wheel from violet. The complex will appear yellow.

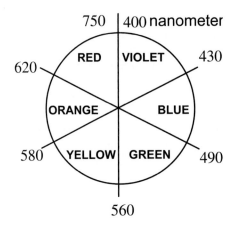

The extent of crystal field splitting, that is, the magnitude of Δ, determines the wavelength of light absorbed, and consequently, the color of the complex. Each ligand has a different strength of interaction with the d orbitals of the metal ion. Thus, each splits the orbital energies by a different amount. The ligands can be arranged according to increasing values of Δ. This establishes a **spectrochemical series:**

$$I^- < Br^- < Cl^- < OH^- < F^- < H_2O < NH_3 < en < CN^- < CO$$

CO and CN^- are called strong-field ligands because they produce a relatively large splitting, whereas I^- and Br^- are weak-field ligands because they split d orbital energies to a lesser extent.

STUDY QUESTION 22.7
$[Ti(H_2O)_6]^{3+}$ ion absorbs light at a wavelength of 498 nm.

    a. Calculate the crystal field splitting energy Δ.
    b. Determine the color of the complex.

Solution:
a. Absorption of light occurs when the energy of an incoming photon is equal to the difference in energy Δ between the lower energy 3d orbitals, and the higher energy 3d orbitals of the titanium ion. The requirement for light absorption is:

$$\Delta = E_{photon}$$

$$\Delta = h\nu = \frac{hc}{\lambda}$$

where h is Planck's constant, c is the velocity of light, and λ is the wavelength of light, 498 nm.

$$\Delta = \frac{(6.63 \times 10^{-34} \text{ J s})(3.00 \times 10^8 \text{ m/s})}{498 \text{ nm} \times \left(\dfrac{10^{-9} \text{ m}}{1 \text{ nm}}\right)}$$

$$\Delta = 3.99 \times 10^{-19} \text{ J}$$

b. The complex absorbs light with a wavelength of 498 nm. On the color wheel this is green light. A solution that absorbs green light will reflect all the colors except green. According to the color wheel the solution will be red.

C. Magnetic Properties.

Earlier in the text you learned that **paramagnetic** substances have at least one unpaired electron and **diamagnetic** substances have all their electrons paired. The Fe^{3+} ion has a $3d^5$ configuration. Two of its octahedral complexes, $[FeF_6]^{3-}$ and $[Fe(CN)_6]^{3-}$, are both paramagnetic, and yet they have different magnetic properties. $[FeF_6]^{3-}$ has five unpaired electrons, but $[Fe(CN)_6]^{3-}$ has only one. The former is called a high-spin complex, and the latter a low-spin complex.

The reason for this difference lies in the spectrochemical series and the value of Δ. Remember that according to Hund's rule, electrons in orbitals with similar energies prefer to be unpaired. Energy is required to make electrons pair up. Therefore, if Δ is small, as in the case of the weak-field ligand F^-, the electrons enter all five orbitals one at a time without pairing, in accordance with Hund's rule. For small Δ values, pairing of electrons would not occur until the sixth electron is added.

When Δ is large enough, Hund's rule no longer applies to the entire set of *d*- orbitals. The lowest energy configuration corresponds to completely filling the three lower-energy $3d$ orbitals first, before the two higher-energy orbitals begin to fill. This is the case for $[Fe(CN)_6]^{3-}$, because CN^- is a strong-field ligand and results in a large Δ. The following figure shows the energy level diagrams for hexafluoroferrate(III) ion and hexacyanoferrate(III) ion:

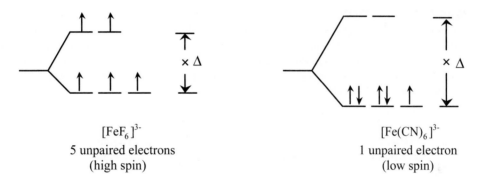

$[FeF_6]^{3-}$	$[Fe(CN)_6]^{3-}$
5 unpaired electrons	1 unpaired electron
(high spin)	(low spin)

So far, we have focused on octahedral complexes. Crystal field splitting occurs for tetrahedral and square planar complexes as well. See Figures 22.19 and 22.20 in the text for the splitting diagrams of these complexes.

STUDY QUESTION 22.8

The complex $[CoF_6]^{4-}$ is a high-spin paramagnetic complex, and $[Co(en)_3]^{2+}$ is a low-spin paramagnetic complex. Draw the crystal field splitting diagrams for these two complex ions and show the proper relationship of the two Δ values.

Solution: The crystal field splitting diagram for a Co^{2+} ion ($3d^7$) must show 7 electrons in the $3d$ orbitals. In a high-spin complex, Δ is small enough so that electrons can enter $3d$ orbitals one at a time, according to Hund's rule. Pairing of electrons will not occur until after the fifth electron has been added. In the low-spin complex, Δ is much larger and the electrons enter the lower-energy orbitals first. Only after 6 electrons are positioned in the lower-energy set of orbitals, can electrons then be placed in the upper two orbitals.

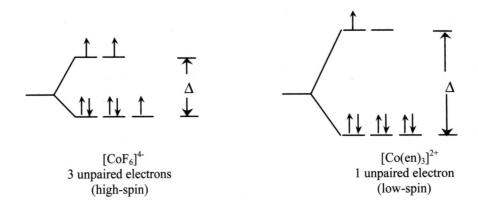

$[CoF_6]^{4-}$
3 unpaired electrons
(high-spin)

$[Co(en)_3]^{2+}$
1 unpaired electron
(low-spin)

PRACTICE QUESTION 22.6

In the complex ion, ML_6^{n+}, M^{n+} has six *d*-electrons and L is a weak field ligand. How many unpaired electrons are there in this complex?

PRACTICE QUESTION 22.7

The color of oxyhemoglobin is due to complex ions involving Fe(II) ions. Oxyhemoglobin (with O_2 bound to iron) is a low-spin, Fe(II) complex. Five of the donor atoms are N atoms and do not change. The sixth donor atom is O in the O_2 molecule. Deoxyhemoglobin (without the O_2) has a water ligand in place of O_2 and is a high-spin complex. Both complexes are octahedral Fe^{2+} complexes.

    a. How many unpaired electrons occupy the 3d orbitals of the iron(II) ion in each case?

    b. Oxyhemoglobin is red and deoxyhemoglobin is blue. Explain in terms of the crystal field theory why the two forms of hemoglobin have different colors.

PRACTICE QUESTION 22.8

In the complex ion, ML_6^{n+}, M^{n+} has six *d*-electrons and L is a strong field ligand. What is the magnetic property of this complex?

PRACTICE QUIZ

1. Using a periodic table, write the electron configuration of an Fe atom; an Fe^{2+} ion; an Fe^{3+} ion.

2. Write the formula of the coordination compound containing a Pt^{4+} central atom, three chloride ligands, three ammonia ligands, and one uncomplexed nitrate ion.

3. Find the oxidation number of the central atom in $Na_2[Co(H_2O)_2I_4]$.

4. For the coordination compound $[Co(NH_3)_6]Cl_3$, give the oxidation state and coordination number of the central atom.

5. For the coordination compound $[Ni(en)_2(NH_3)_2]Cl_2$, give the oxidation state and coordination number of the central atom.

6. Name the following coordination compounds and complex ions:
   a. $[Pt(NH_3)_4Cl_2]Cl_2$
   b. $[Ni(NH_3)_6]^{2+}$
   c. $[Cr(OH)_4]^-$
   d. $[Co(CN)_6]^{3-}$
   e. $K_2[Cu(CN)_4]$

7. Write the formula for each of the following coordination compounds and ions:
   a. tetrahydroxoaluminate(III) ion
   b. tetraiodomercurate(II) ion
   c. potassium dichlorobis(oxalato)cobaltate(III)
   d. tris(ethylenediamine)nickel(II) sulfate

8. Sketch two geometrical isomers for $[Ni(NH_3)_2(CN)_4]^{2-}$.

9. Sketch the structures of the square planar complex *trans*-$[Pt(Cl)_2(NH_3)_2]$ and its mirror image. Are the two structures enantiomers? Explain.

10. Why are chiral substances said to be optically active?

11. Illustrate each of the following by giving formulas, geometry, or energy diagrams:
    a. Octahedral complex of $NiCl_6^{4-}$
    b. *Cis* and *trans* isomers of $Ni(CO)_2(CN)_2$
    c. $[Fe(H_2O)_6]^{3+}$ is a high-spin complex
    d. A bidentate ligand
    e. A coordination number of 2

12. If green light is absorbed by a complex in a solution, what is the color of the solution?

13. Complexes of zinc are never paramagnetic. Explain.

14. $[Ni(H_2O)_6]^{2+}$ is green and $[Ni(en)_3]^{2+}$ is violet. Which has the larger value of Δ?

Chapter 23
Organic Chemistry

PROBLEM-SOLVING STRATEGIES AND TUTORIAL SOLUTIONS

TYPES OF PROBLEMS
Problem Type 1: Classes of Organic Compounds.
        A. Basic Nomenclature.
        B. Molecules with Substituents.
        C. Molecules with Specific Functional Groups.

Problem Type 2: Representing Organic Molecules.
        A. Condensed Structural Formulas.
        B. Kekulé Structures.
        C. Skeletal Structures.
        D. Resonance.

Problem Type 3: Isomerism.
        A. Constitutional Isomerism.
        B. Stereoisomerism.

Problem Type 4: Organic Reactions.
        A. Addition Reactions.
        B. Substitution Reactions.
        C. Other Types of Organic Reactions.

PROBLEM TYPE 1: CLASSES OF ORGANIC COMPOUNDS

A. Basic Nomenclature.
The simplest hydrocarbons (molecules made up of carbon and hydrogen) are the alkanes, alkenes, alkynes, and cycloalkanes. The general formula for an **alkane** is C_nH_{2n+2}, where n is the number of carbon atoms in the molecule, n = 1, 2, 3···. The name of a **normal alkane**, one that consists of a single chain of carbon atoms, always ends in the suffix *–ane*. The first part of each name represents the number of C atoms in the molecule and it corresponds to the roots given on page 964 of your textbook.

An **alkene** contains at least one $C = C$ double bond and has the general formula C_nH_{2n}. Alkenes are named with the root word indicating the number of carbon atoms and end in "*-ene*"

C_2H_4	$CH_2 = CH_2$	ethene (ethylene)
C_3H_6	$CH_3CH = CH_2$	propene
C_4H_8	$CH_3CH_2CH = CH_2$	1-butene

An **alkyne** contains at least one $C \equiv C$ triple bond and has the general formula C_nH_{2n-2}. The names of alkynes end with "*-yne*."

C_2H_2	$CH \equiv CH$	ethyne (acetylene)
C_3H_4	$CH_3C \equiv CH$	propyne
C_4H_6	$CH_3CH_2C \equiv CH$	1-butyne

There is also a type of alkane that has atoms bonded into ring configurations. These are the **cycloalkanes**. They have the same general formula as the alkenes, C_nH_{2n}, but do not have double bonds. Two cycloalkanes are shown below.

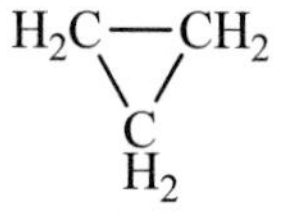

$$H_2C—CH_2$$
$$\backslash \ /$$
$$C$$
$$H_2$$

cyclopropane

$$H_2C—CH_2$$
$$| \quad |$$
$$H_2C—CH_2$$

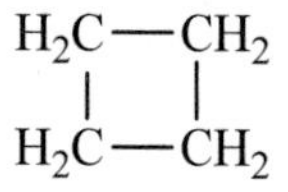

cyclobutane

B. Molecules with Substituents.

As we mentioned, a normal alkane consists of a single chain of carbon atoms. However, we also find compounds in which "branches" of carbon-hydrogen chains may be bonded to some of the carbon atoms within the main skeleton.

Straight chain hydrocarbons are called *normal* and use the symbol *n*. The straight chain form of C_4H_{10} has the name *n*-butane. The following guidelines are useful when naming branched hydrocarbons.

1. The systematic name of an alkane is based on the number of carbon atoms in the longest carbon chain. For alkanes, the longest carbon chain is given a name corresponding to the alkane with the same number of C atoms.
2. Groups attached to the main chain are called substituent groups. Substituent groups that contain only hydrogen and carbon atoms are called alkyl groups. When an H atom is removed from an alkane the fragment is called an alkyl group. This group can be attached to the longest chain. Alkyl groups are named by dropping the ending *-ane* and adding *-yl* to the alkane name. Therefore, CH_3- is a methyl group, C_2H_5- is an ethyl group, and C_3H_7- is a propyl group. Table 24.1 in the text lists the names of eleven common alkyl groups.
3. The locations of alkyl groups that are attached to the main chain must also be included in the name. Number the carbon atoms in the longest chain. The numbering should start at the end of the chain such that the side groups will have the smallest numbers.
4. Prefixes such as *di*, *tri*, and *tetra* are used when more than one substituent of the same kind is present. Combine the names of substituent groups, their locations, and the parent chain into the hydrocarbon name. Arrange the substituent groups alphabetically, followed by the parent name of the main chain.
5. Alkanes can have many different types of substituents besides alkyl groups. Halogenated hydrocarbons contain fluorine, chlorine, bromine and iodine atoms which are named fluoro, chloro, bromo, and iodo, respectively. NH_2-, NO_2-, and $CH_2\!=\!CH-$ are called amino, nitro, and vinyl, respectively.
6. When naming alkenes we indicate the positions of the carbon-carbon double bonds; in alkynes the position of the triple bond is numbered. For alkenes and alkynes, the parent name is derived from the longest chain that contains the double or triple bond. The parent name will end in *-ene* for alkenes and *-yne* for alkynes. Number the C atoms in the main chain by starting at the end nearer to the double or triple bond and use the lower number of the carbon atom in the $C\!=\!C$ bond.

STUDY QUESTION 23.1
Name the molecule that has the following structure:

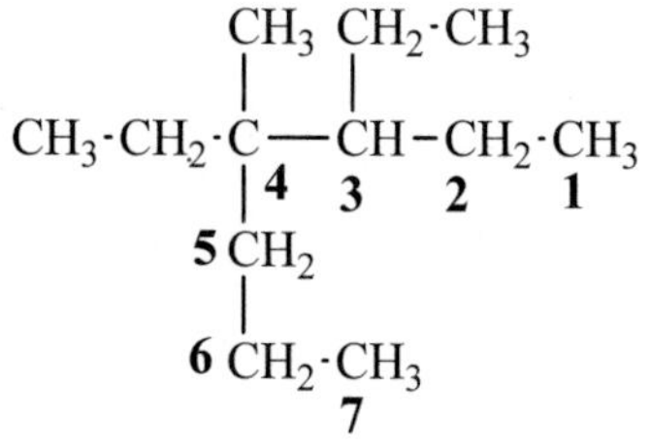

$$CH_3 \quad CH_2 \cdot CH_3$$
$$| \qquad |$$
$$CH_3 \cdot CH_2 \cdot C—CH−CH_2 \cdot CH_3$$
$$|4 \quad 3 \quad 2 \quad 1$$
$$5\,CH_2$$
$$|$$
$$6\,CH_2 \cdot CH_3$$
$$7$$

Solution: First identify the longest continuous carbon chain. Two equivalent chains containing seven carbon atoms are noticeable. Number the carbon atoms as shown. There are two ethyl groups, one at carbon 3 and one at carbon 4, and one methyl group also bonded to carbon 4. The parent name is heptane. Placing the names of the substituent groups (alphabetically, ethyl is before methyl) and their position numbers in front of the parent alkane name we get: 3,4-diethyl-4-methylheptane.

NOTE:
If the chain had been numbered in reverse order the name would be 4,5-diethyl-5-methylheptane. Since 3,4 is smaller than 4,5 the first name is the correct one.

STUDY QUESTION 23.2
Name the following hydrocarbon:

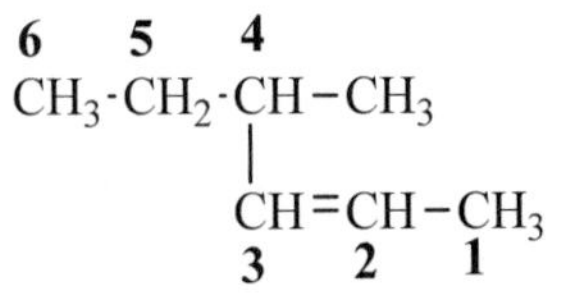

$$
\begin{array}{cccc}
6 & 5 & 4 & \\
CH_3\text{-}CH_2\text{-}CH & - & CH_3 \\
& & | & \\
& & CH = CH - CH_3 \\
& & 3 \quad\; 2 \quad\; 1
\end{array}
$$

Solution: Locate the longest chain that contains the double bond. This chain contains six C atoms so the root name will be hexene. Number the chain so that the double bond has the smallest number, as shown. Name the methyl group on carbon 4. The name is 4-methyl-2-hexene.

NOTE:
The 2 indicates the position of the $C = C$ double bond, which connects carbons numbered 2 and 3.

PRACTICE QUESTION 23.1
Write the general formulas of alkanes, alkenes, and alkynes.

PRACTICE QUESTION 23.2
Give the IUPAC name for the compound with the following structure.

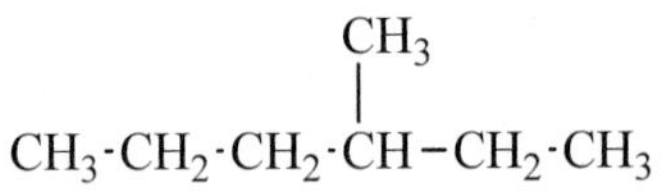

$$
\begin{array}{c}
CH_3 \\
| \\
CH_3\text{-}CH_2\text{-}CH_2\text{-}CH - CH_2\text{-}CH_3
\end{array}
$$

C. **Molecules with Specific Functional Groups.**
   Certain groups of atoms (other than C and H) within organic compounds have similar chemical and physical properties. The group of atoms that is largely responsible for the chemical behavior of a molecule is called a **functional group**. Various functional groups are listed in Table 24.2 of your textbook.

   Alcohols are molecules that have a hydroxyl group, OH, in the place of one of the hydrogen atoms. The molecule is named according to its hydrocarbon name with the final –e replaced with –ol.

methanol ethanol 1-propanol 2-propanol

The **carboxylic acid** contains the carboxyl functional group, $-\overset{\overset{\displaystyle O}{\|}}{C}-OH$ abbreviated as —COOH. These molecules are weak acids.

$$CH_3COOH(aq) \rightleftharpoons CH_3COO^-(aq) + H^+(aq)$$

Organic acids are named as the hydrocarbon with the final –e replaced by –oic acid.

HCOOH	methanoic acid (common name of formic acid)
CH_3COOH	ethanoic acid (common name of acetic acid)
CH_3CH_2COOH	propanoic acid
$CH_3CH_2CH_2COOH$	butanoic acid

The **ester** functional group is $-\overset{\overset{\displaystyle O}{\|}}{C}-OR$ or —COOR, where R stands for any alkyl group. This functional group resembles the carboxylic acid group except that the R group replaces the H atom.

The functional group in **aldehydes** and **ketones** is the carbonyl group, $C=O$. In aldehydes, the carbonyl group is at the end of the alkane chain. In ketones it is not a terminal group.

$CH_3 \cdot CH_2 \cdot CH_2 \cdot \overset{\overset{\displaystyle O}{\|}}{CH}$ or $CH_3CH_2CH_2CHO$ $CH_3 \cdot CH_2 \cdot \overset{\overset{\displaystyle O}{\|}}{C} - CH_3$ or $CH_3CH_2COCH_3$

butanal (an aldehyde)   2-butanone (a ketone)

Aldehydes and ketones are named according to the name of the hydrocarbon. The final –e is replaced with –al for aldehydes and –one for ketones.

Organic bases contain the —NH$_2$ functional group, which is called an **amino group**. Molecules containing an amino group are called **amines**. The amines may be considered as derivatives of ammonia (NH_3). Amines have the general formula, R_3N, where R may be H, an alkyl group, or an aromatic group. Molecules of the type RNH_2 are called **primary amines**. One with R_2NH is a **secondary amine**. R_3N is a **tertiary amine**.

methylamine   dimethylamine   trimethylamine
a primary amine   a secondary amine   a tertiary amine

STUDY QUESTION 23.3
Indicate the functional group for each.

   a. **C₅H₁₁OH**
   b. **CH₃CHO**
   c. **C₃H₇COOH**
   d. **CH₃COC₂H₅**
   e. **CH₃COOCH₃**

Solution: Learning to recognize functional groups requires memorization of their structural formulas. Table 24.2 (textbook) shows a number of the important functional groups.

   a. $C_5H_{11}OH$ contains a hydroxyl group (OH). It is an *alcohol*.
   b. CH_3CHO is a way to represent a molecule on a single line of type.

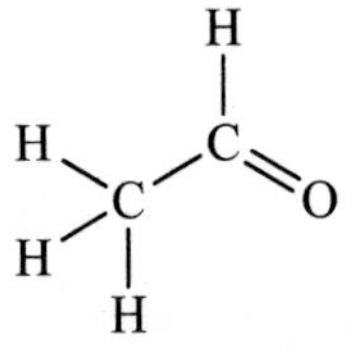

C＝O is a carbonyl group. Because a hydrogen is connected to the carbon atom of the carbonyl (it is a terminal carbonyl), CH_3CHO is an *aldehyde*.

   c. C_3H_7COOH contains the carboxyl group —COOH and is a *carboxylic acid*.
   d. $CH_3COC_2H_5$ is a way to represent a *ketone* on a single line.

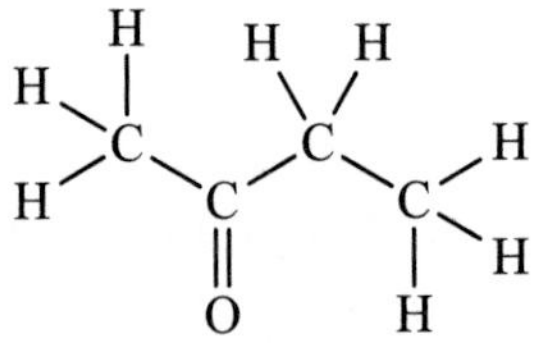

C＝O is a carbonyl group.

   e. CH_3COOCH_3 is a condensed structural formula for:

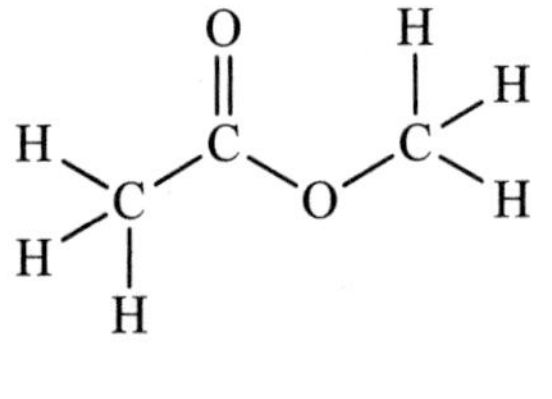

CH_3COOCH_3 is an *ester*.

PRACTICE QUESTION 23.3
Draw structures for the following functional groups: aldehyde, amine, carboxylic acid, ester.

PRACTICE QUESTION 23.4
Identify the functional group or type of molecule in the following.

   a. **CH₃OH**
   b. **CH₃CHO**
   c. **CH₃COCH₃**
   d. **CH₃CH=CH₂**
   e. **CH₃CH₂CH₃**

PROBLEM TYPE 2: REPRESENTING ORGANIC MOLECULES.

A. Condensed Structural Formulas.

We could consider a condensed structural formula as a combination of a molecular formula and a structural formula. The sequence of "units" or "pieces" provide clues about the ways the atoms are connected. For example, consider the condensed structural formula $CH_3(CH_2)_2CO(CH_2)_2CH_3$. The way the formula is written tells us that there is a terminal propyl group bonded to carbonyl group to which another propyl group is attached as shown in the Lewis structure below.

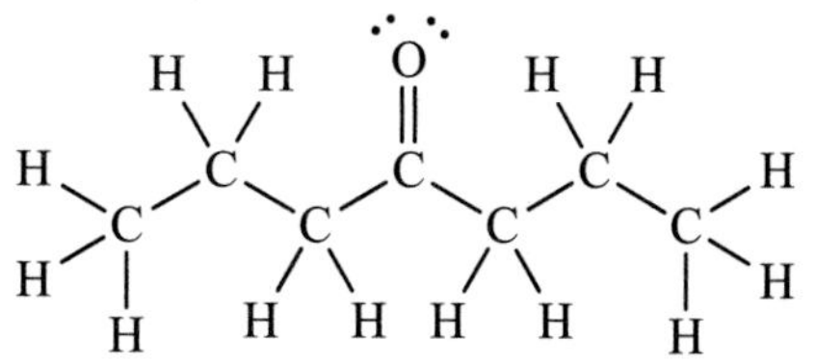

B. Kekulé Structures

A Kekulé structure is a Lewis structure in which lone pairs are not shown. Thus, the Lewis structure shown above would be represented as the following Kekulé structure.

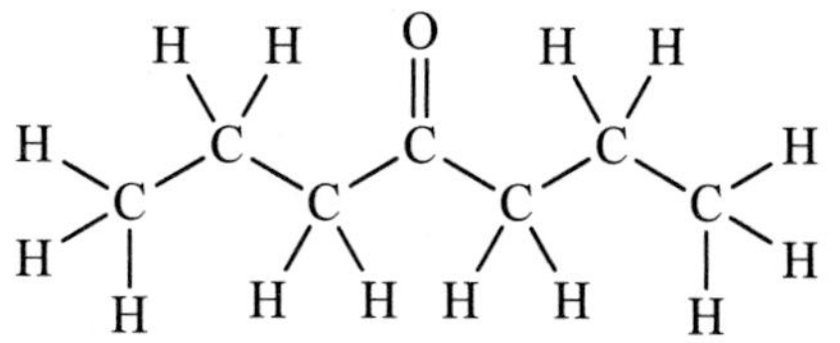

C. Skeletal Structures

Skeletal structures are sometimes referred to as line structures. In a skeletal structure, lines are drawn to represent the covalent bonds. A carbon atom is assumed to lie at the intersection of any pair of lines and at the end of any line. Carbon atoms and the hydrogen atoms attached to them are not written in a skeletal structure. All other atoms (heteroatoms) are explicitly written, as well as the hydrogen atoms attached to them. The line structure for the molecule shown above is:

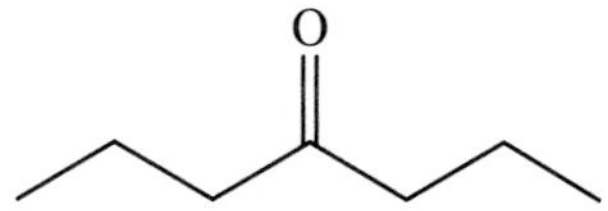

STUDY QUESTION 23.4

Write a condensed structural formula and a Kekulé structure for the following organic compound:

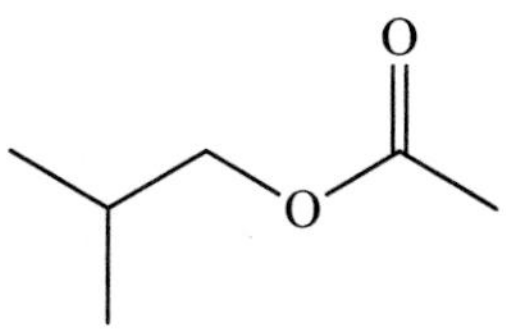

Solution: Looking at the left hand side of the structure, there are three lines intersecting, which means there is a –CH– group (to complete the octet a bond to a hydrogen atom is needed). The ends of the lines represent terminal methyl groups, $-CH_3$. To the right of the three-way intersection there is a $-CH_2-$ group followed by an ester group.

The condensed structural formula is $CH_3COOCH_2CH(CH_3)_2$. Note that you cannot represent the ester group as $-CH_2OOC-$, since this does not clearly indicate to which carbon the second oxygen is bonded.

The Kekulé structure is:

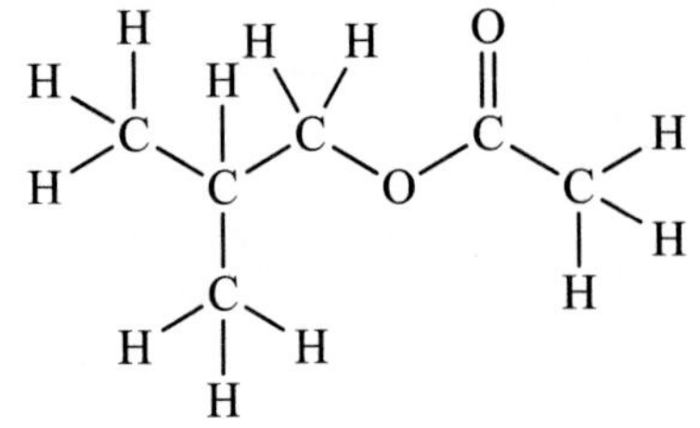

PRACTICE QUESTION 23.5
Draw the Kekulé and skeletal structures for:

    a. $CH_3(CH)_2(CH_2)_2CH_3$

    b. $CH_2CHCH_2N(CH_3)_2$

D. Resonance

Resonance occurs in a molecule when electrons can be delocalized over more than two atoms. The "skeleton" of the Lewis structures is the same but the positions of the electrons are different.

A common example of resonance structures can be seen with the carbonate anion CO_3^{2-}.

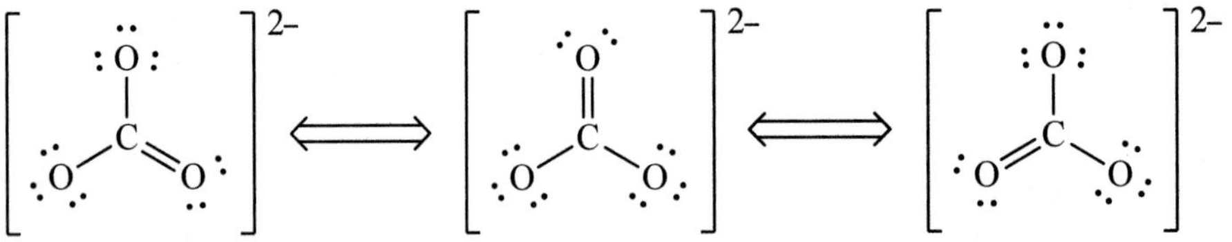

STUDY QUESTION 23.5

Draw all the possible resonance forms for the hydrogen phosphate anion HPO_4^{2-}. There is no P – H bond.

Solution: This compound has a total of 32 valence electrons. One possible Lewis structure is:

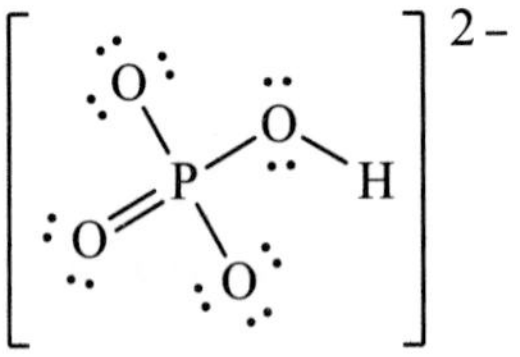

Because the formal charge on the oxygen atom to which the hydrogen atom is bonded is zero, we do not want to delocalize those lone pairs on that particular oxygen atom. We can therefore draw only two additional Lewis structures to show the resonance forms for this anion.

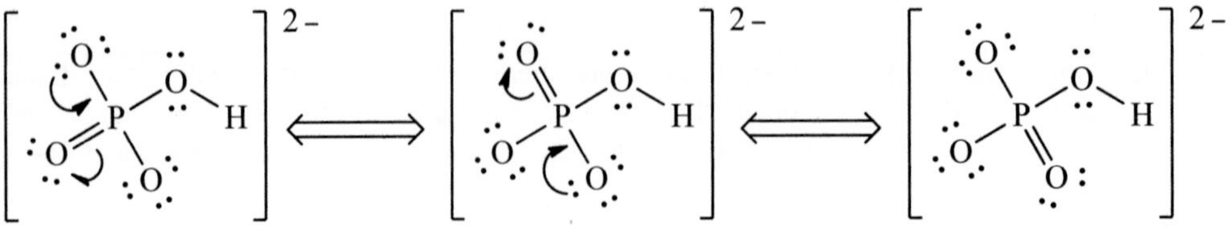

PRACTICE QUESTION 23.6

Draw all the possible resonance forms for the thiocyanate anion NCS⁻ :

PROBLEM TYPE 3: ISOMERISM.

A. Constitutional Isomerism.

Isomers are compounds that share a common molecular formula but have different structural formulas. **Constitutional isomers** have the same molecular formula but their atoms are bonded following a different sequence. We have seen that alcohols and ethers may fall into this category.

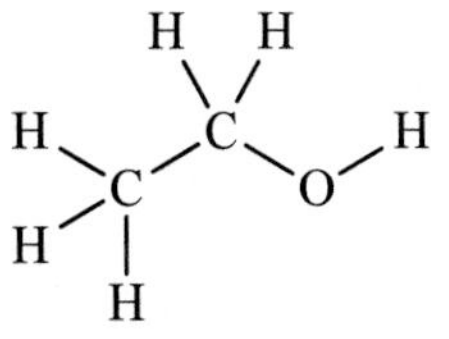

ethanol 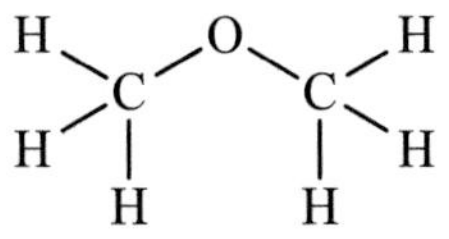   dimethylether

B. Stereoisomerism

Stereoisomers are compounds that have the same molecular formula, their atoms are bonded following the same sequence, but their orientation in space is different. Two types of stereoisomers were discussed in your textbook:

    1.   Geometrical isomers
    2.   Optical isomers

Geometrical isomers are found among compounds that contain carbon – carbon double bonds. One common type of geometrical isomerism is *cis – trans* isomerism. These isomers differ in the relative position of the substituents with respect to the double bond: if the substituents are on the "same side" with respect to the double bond it is a *cis* isomer; if they are on "opposite sides" it is the *trans* isomer.

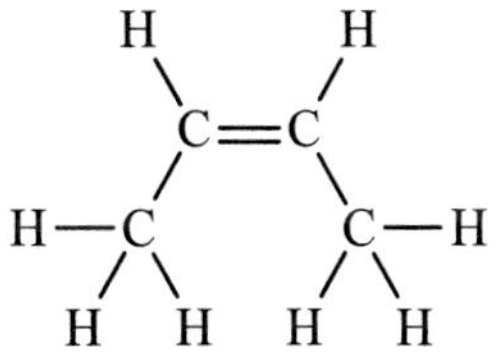

cis-2-butene 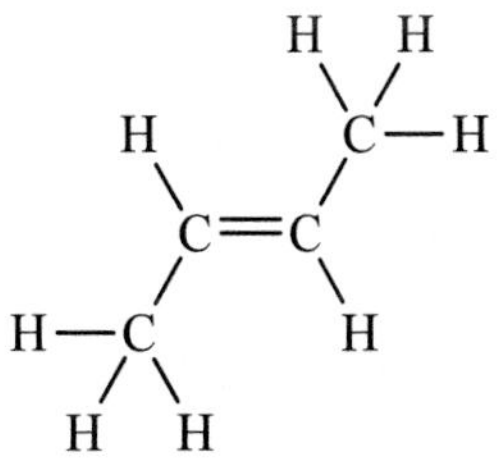   *trans*-2-butene

To recognize **optical isomers**, we must be able to visualize the molecules in three-dimensions. Optical isomerism results when molecules of a compound cannot be superimposed on each other by rotation or "flipping" it horizontally or vertically. These molecules possess at least one carbon atom in which all the four substituents are different. These molecules are said to be **chiral** and a pair of these nonsuperimposable molecules is referred to as **enantiomers**. We find common examples among the amino acids. Figure 24.3 in your textbook illustrates an example of enantiomers.

STUDY QUESTION 23.6

Give two constitutional isomers for C₃H₈O.

Solution: Constitutional isomers have atoms in different sequence. Limiting our option to alcohols, the -OH group can be bonded either to a terminal C or an internal C atom. The resulting constitutional isomers are $CH_3CH(OH)CH_3$ and $CH_3CH_2CH_2OH$.

> **PRACTICE QUESTION 23.7**
>
> **Draw the structure of another constitutional isomers of C₃H₈O.**

PROBLEM TYPE 4: ORGANIC REACTIONS.

In organic chemistry, some of the most common reactions take place as a result of movement of electrons between compounds with low electron densities and compounds with high electron densities. When a compound possesses an atom or region bearing a partial positive charge, the compound behaves as an electrophile ("electron loving"). On the same token, if a compound possesses an atom or region bearing a partial negative charge then it behaves as a nucleophile ("proton loving"). We can understand this designation by keeping in mind that opposite charges attract each other.

A. Addition Reactions

Two compounds "add" together as a result of either an electrophilic or nucleophilic attack. The particular type of addition reaction will depend on the nature of the species that initiates the reaction.

STUDY QUESTION 23.7

Predict the product(s) for the following reaction: $CH_3(CH)_2CH_3 + HCl \longrightarrow$

Solution: This is an electrophilic addition. It is initiated by the attack of the electrophile (HCl) on the double bond.

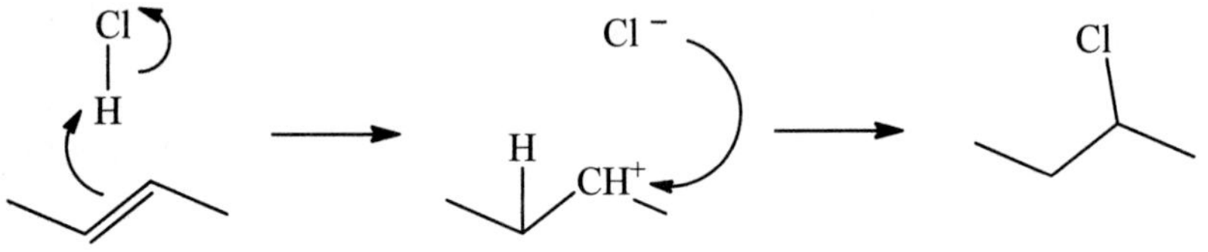

STUDY QUESTION 23.8

Use curved arrows to show the movement of electrons in the following acid-catalyzed nucleophilic addition of water to dipropylketone.

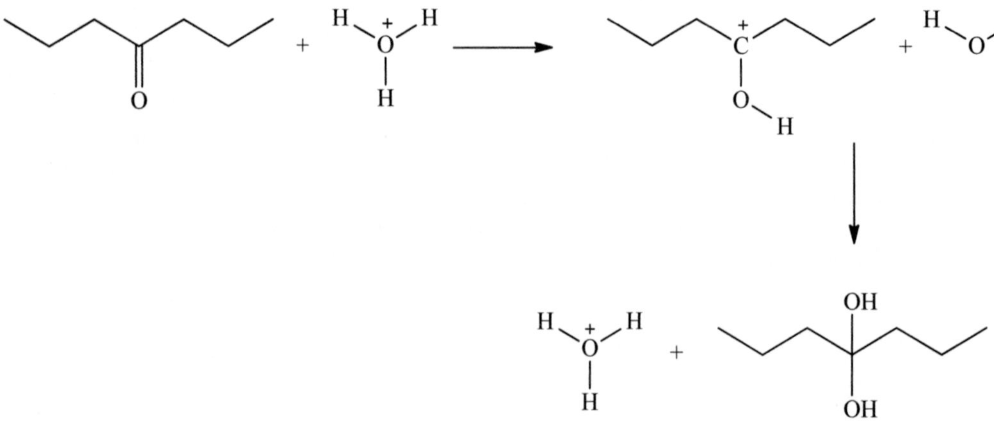

346

Solution: Water is a rather weak nucleophile. In order to have a nucleophilic addition to a ketone, it must be protonated in order to generate a carbocation that would become a stronger electrophile and provide the driving force for this reaction.

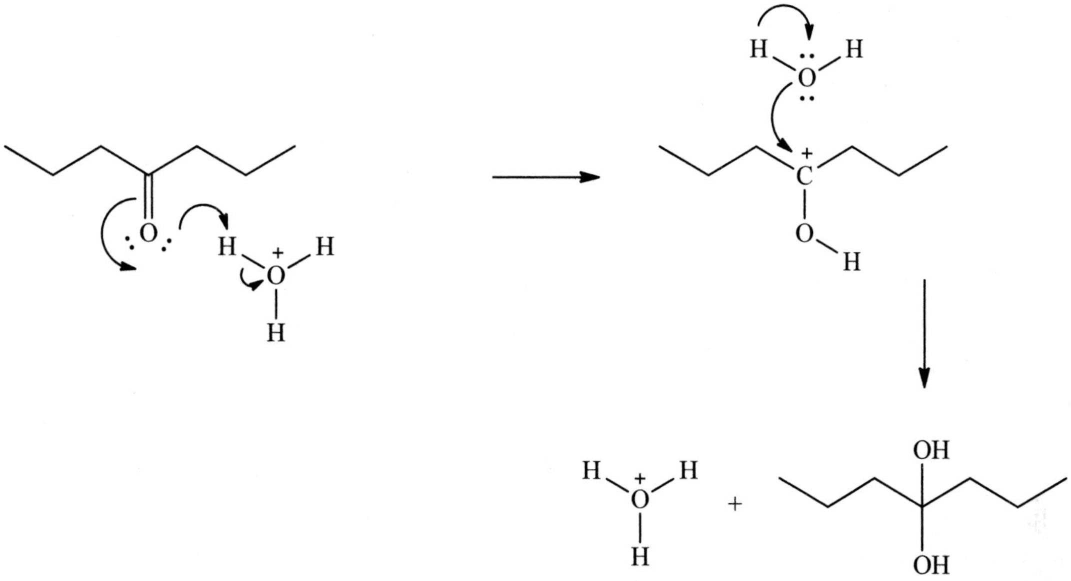

PRACTICE QUESTION 23.8
What is the product of the following addition reaction: $Cl_2 + CH_3CH = CHCH_3$?

PRACTICE QUESTION 23.9
Draw the two possible products for the following reaction:

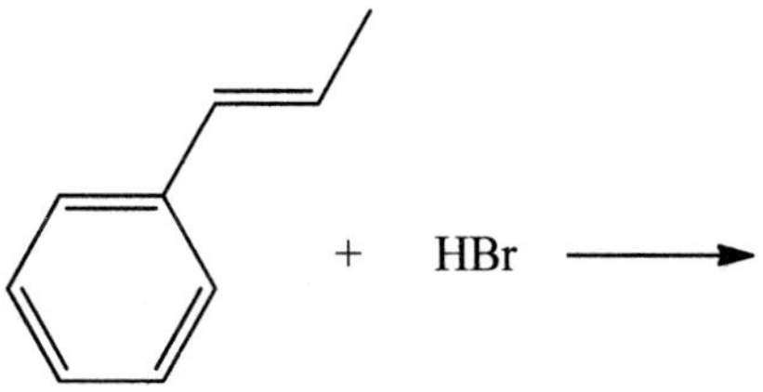

B. Substitution Reactions

In a substitution reaction, an atom or group of atoms in one compound is replaced or substituted by another atom or group of atoms from a different compound. Just as the addition reactions, we can have electrophilic substitution or nucleophilic substitution reactions depending on the nature of the species that initiates the reaction.

STUDY QUESTION 23.9

Predict the product(s) for the following reaction: $CH_3(CH)_2CH_3 + HCl \longrightarrow$

Solution: This is an electrophilic addition. It is initiated by the attack of the electrophile (HCl) on the double bond.

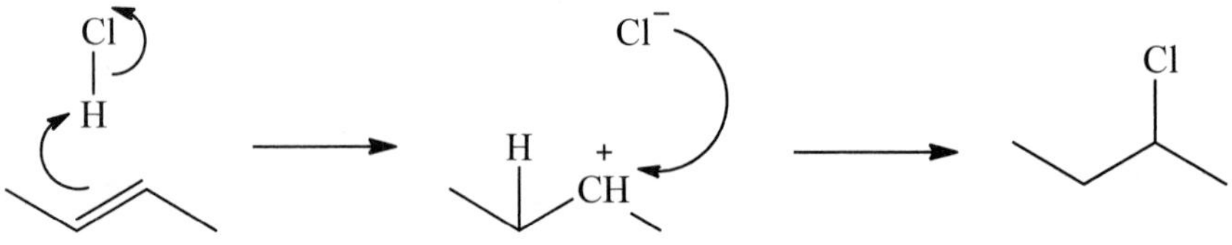

STUDY QUESTION 23.10

Using curved arrows to show electron movement, provide a mechanism for the following nucleophilic substitution:

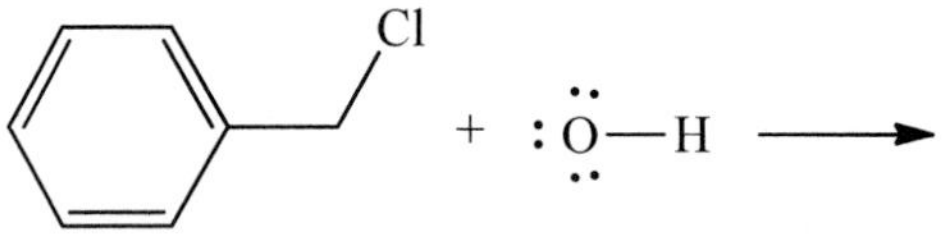

Solution: The hydroxide anion acts as a nucleophile attacking the carbon atoms to which the chlorine atom is bonded since it bears a partial positive charge.

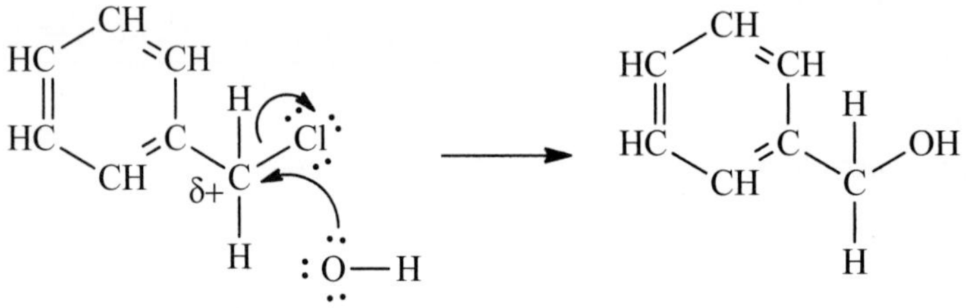

PRACTICE QUESTION 23.10

The following reactants participate in a nucleophilic substitution reaction. Identify the nucleophile and the electrophile and predict the major product.

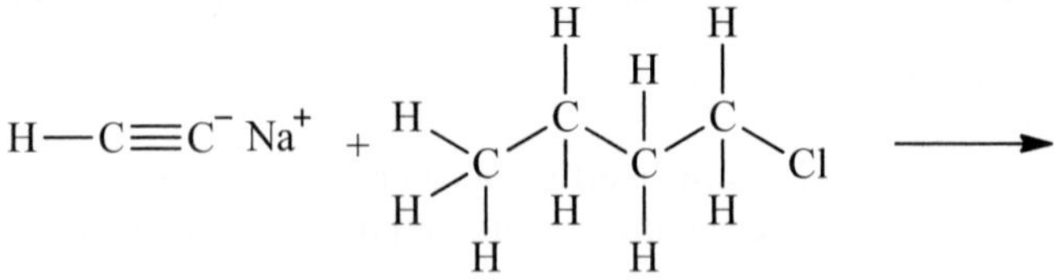

C. Other Types of Organic Reactions

Oxidation and reduction are two other common reactions for organic compounds. They are easy to identify by following these guidelines:

1. When a molecule gains an O atom or loses an H atom, it is oxidized.
2. When a molecule loses an O atom or gains an H atom, it is reduced.

Isomerization is a reaction in which a molecule is transformed into a constitutional, geometrical, or optical isomer.

STUDY QUESTION 23.11
Predict the major product for the following reactions.

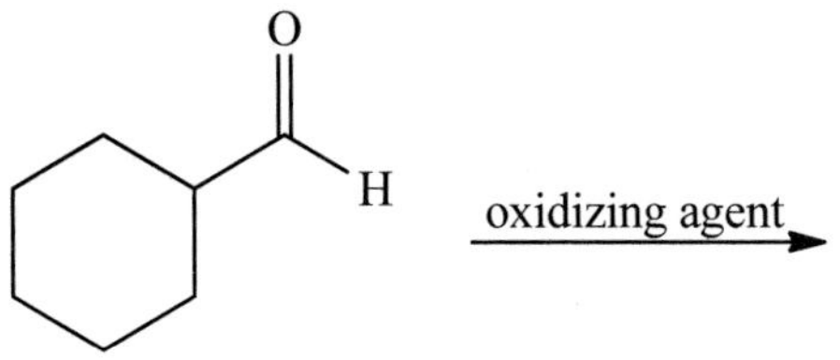

a.

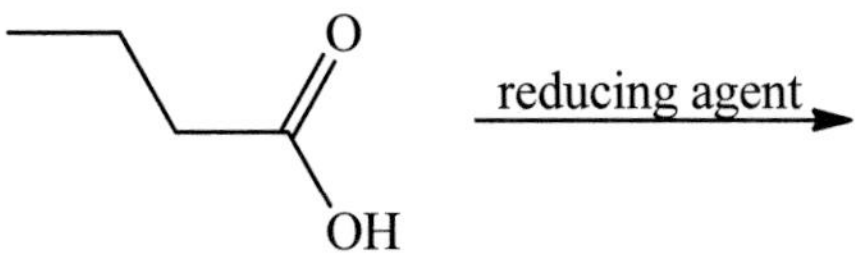

b.

Solution:

a. In an oxidation, either one or more oxygen atoms are added or one or more hydrogen atoms are lost. The oxidation of an aldehyde generates a carboxylic acid.

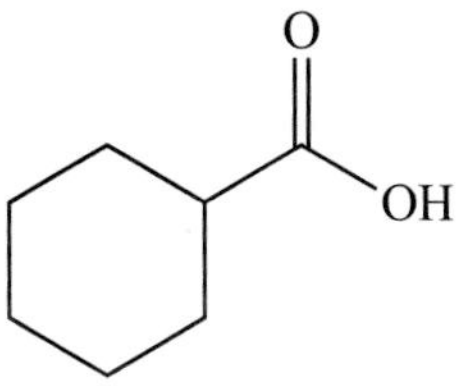

b. In a reduction, either one or more oxygen atoms are lost or one or more hydrogen atoms are gained. The reduction of a carboxylic acid generates an alcohol.

PRACTICE QUESTION 23.11
Identify each of the following reactions as an oxidation, a reduction, or an isomerization.

a.

b.

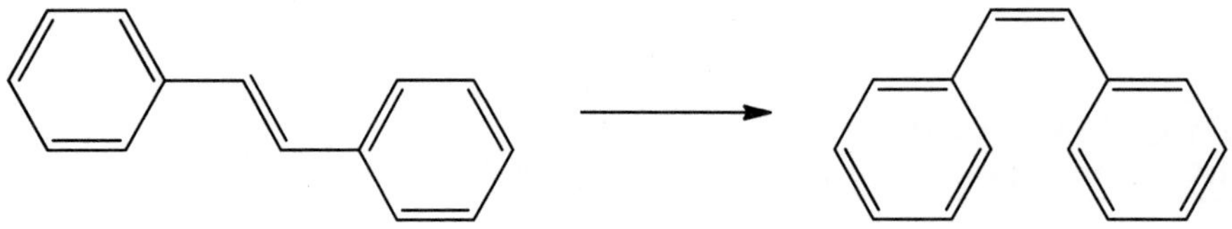

PRACTICE QUIZ

1. Give the systematic name for the following structural formulas:

   a. $CH_3(CH_2)_7CH_3$

   b.

$$
\begin{array}{c}
CH_3 \\
| \\
CH_3 \quad CH_2 \\
CH \\
| \\
CH \\
CH_2
\end{array}
$$

2. Give the systematic name for the compounds represented by the following structural formulas:

   a.

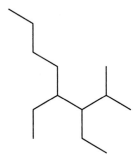

   b.

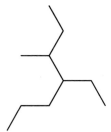

3. Give the systematic name for the following compound:

4. Give the systematic name for the following compound:

5. Give the systematic name for the following:

$$CH_3$$
$$H_2C$$
$$CH$$
$$HC$$
$$CH_3$$

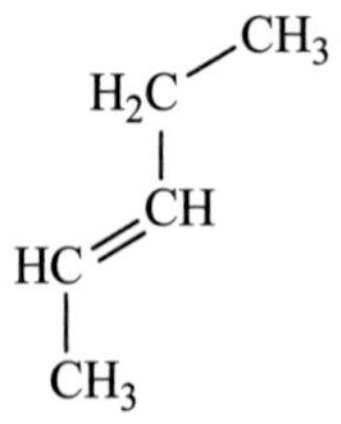

6. Give the systematic name for the following:

$$CH_3 \quad CH_2$$
$$CH \quad CH_3$$
$$OH$$

a. 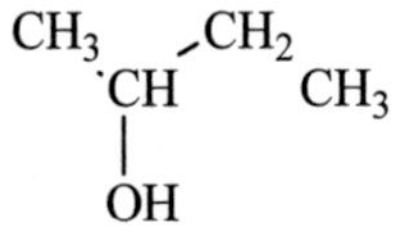
b. CH_3—OH

7. Draw Kekulé structures for:

a. 3-ethyl-4-methyl-4-isopropylheptane
b. 3-bromo-2,5-dimethyl-*trans*-3-hexene
c. 3-methyl-3-hexanol
d. trichloroacetic acid

8. Draw structural formulas of all isomers of C_5H_{10}. Include both structural and geometric isomers.

9. Name the functional groups that are shown in the following molecule:

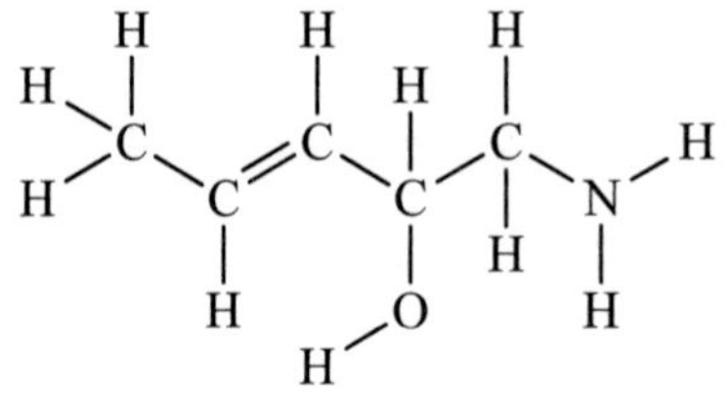

10. Name the functional groups that are shown in the following molecule:

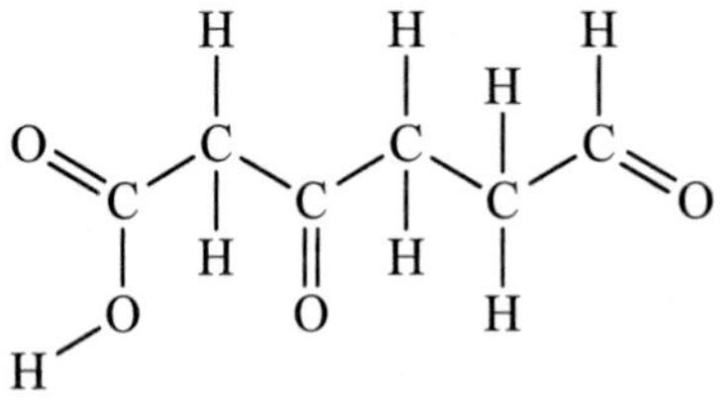

11. Give the structure of the product of each of the following organic reactions:

a.

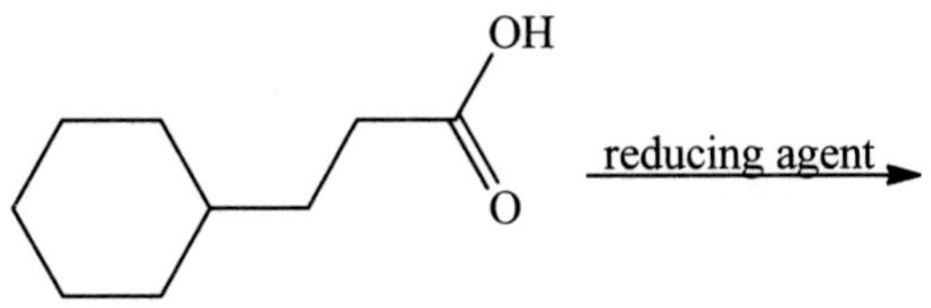

$\xrightarrow{\text{reducing agent}}$

b.

$\xrightarrow{\text{oxidizing agent}}$

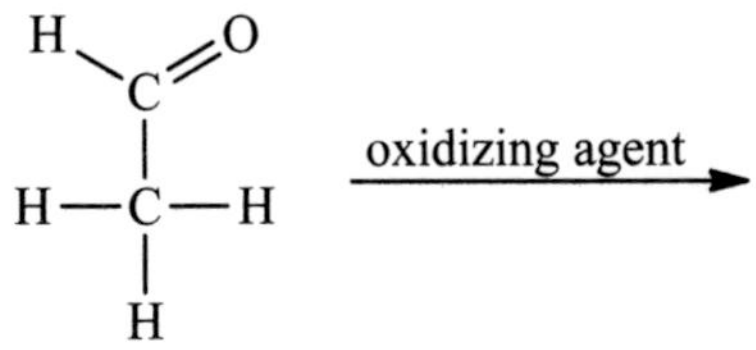

12. Give the structure of the product of each of the following organic reactions:

a. $\quad CH_3C\equiv CH \;+\; Cl_2 \longrightarrow$

b. $\;+\; H_3C-\overset{-}{O} \longrightarrow$

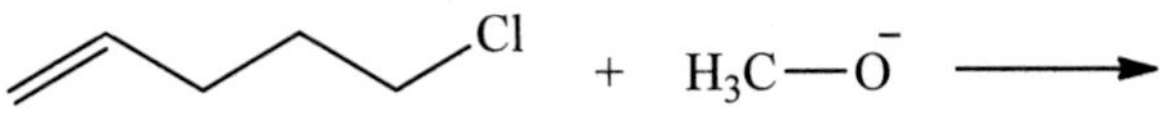

13. Give the structure of the product of each of the following organic reactions:

a.

$\;+\; H_2O \xrightarrow{\text{acid}}$

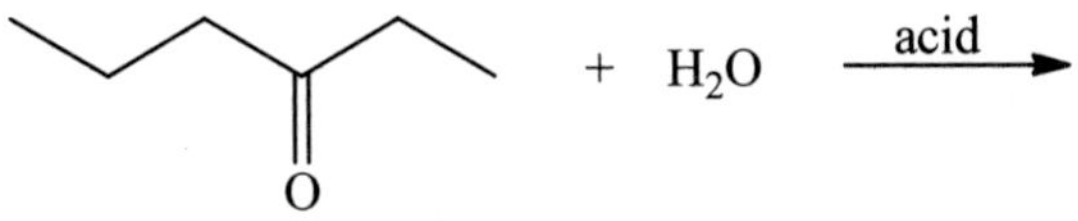

b.

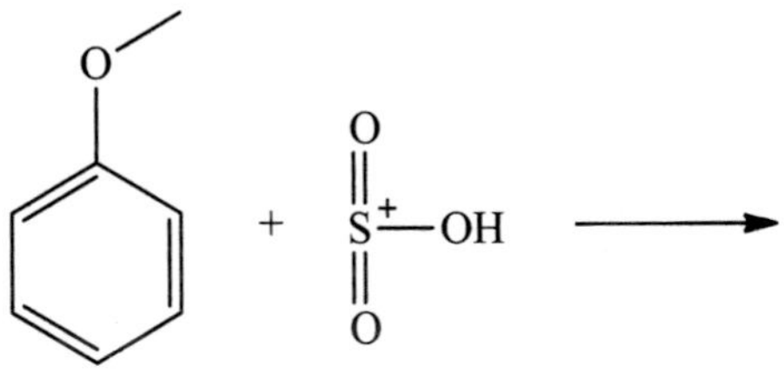

14. Define the following terms:
    a. alkene
    b. homologous series
    c. geometrical isomer
    d. functional group

15. Draw all the possible constitutional isomers for $C_4H_{10}O$.

16. Identify which of the following species are nucleophiles:

    a. NH_3
    b. CH_3Cl
    c. I^-
    d. $O=\overset{+}{N}=O$
    e. SO_3

17. Which monomer was the precursor for the following addition polymer?

18. Show all possible resonance forms for

19. Draw all the possible geometrical isomers for 3,5-octadiene.

20. Draw skeletal structures for:
    a. methyl butanoate
    b. propylamine

Chapter 24
Modern Materials

PROBLEM-SOLVING STRATEGIES AND TUTORIAL SOLUTIONS

TYPES OF PROBLEMS
Problem Type 1:   Polymers.
      A. Addition Polymers.
      B. Condensation Polymers.

Problem Type 2:   Ceramics and Composite Materials.

Problem Type 3:   Liquid Crystals.

Problem Type 4:   Biomedical Materials.

Problem Type 5:   Nanotechnology.

Problem Type 6:   Semiconductors.

PROBLEM TYPE 1: POLYMERS.

The word polymer means "many parts." A **polymer** is a compound with an unusually high molecular mass, consisting of a large number of small molecular units that are linked together. The small unit that is repeated many times is called a **monomer**. A typical polymer molecule contains a chain of monomers several thousand units long. Polymers are often called macromolecules.

A. **Addition Polymers.**
Addition polymers are made by adding monomer to monomer until a long chain is produced. Ethylene and its derivatives are excellent monomers for addition polymers. In an addition reaction, the polymerization process is initiated by a radical. When ethylene is heated to 250 °C under high pressure (1000–3000 atm) in the presence of a little oxygen or benzoyl peroxide (the initiator), addition polymers with molecular masses of about 30,000 amu are obtained. This reaction is represented by:

ethylene   a segment of polyethylene

The general equation for addition polymerization is:

monomer   repeating unit

Polyethylene is an example of a homopolymer, which is a polymer made up of only one type of monomer. Substitution of one or more hydrogen atoms in ethylene with Cl atoms, phenyl groups, acetate, cyano groups, and F atoms provides a wide selection of monomers from which to make various homopolymers. For example, substitution of a chlorine atom, Cl, for a hydrogen atom, H, in ethylene gives the monomer called vinyl chloride, $CH_2{=}CHCl$. Polymerization of vinyl chloride yields the polymer, polyvinyl chloride.

vinyl chloride
monomer

polyvinyl chloride
repeating unit

The addition polymerization process occurs by several steps which are summarized below. The steps are: initiation, chain growth process, and a termination reaction.

1. *Initiation.* Polymerization is initiated by a radical. A radical (also called a free radical) is a species which has an unpaired electron. The symbol for a radical contains the usual letters for the elements followed by a dot to represent the unpaired electron. $CH_3\cdot$ is the symbol for a methyl radical. The letter R is the general symbol for an alkyl group and R· for an alkyl radical.

   In a polymerization process, first the initiator molecule is dissociated by heating to yield two radicals. Then, the radical adds to an ethylene molecule, which generates a new radical.

   $$R{-}R \longrightarrow 2\,R\cdot$$

   $$R\cdot\ +\ H_2C{=}CH_2 \longrightarrow R{-}CH_2{-}CH_2\cdot$$

2. *Chain Growth Process.* Next, the free radical formed above can add to another molecule of ethylene.

   $$R{-}CH_2{-}CH_2\cdot\ +\ H_2C{=}CH_2 \longrightarrow R{-}CH_2{-}CH_2{-}CH_2{-}CH_2\cdot$$

   The length of the carbon chain grows rapidly as the last free radical formed reacts with yet another ethylene molecule, and so on.

   $$R{-}(CH_2{-}CH_2)_n{-}CH_2{-}CH_2\cdot$$

3. *Termination.* The polymerization process continues until a termination reaction occurs. When the radical ends of two chains meet, they may combine. When this happens, there is no new radical formed and the chain lengthening process ceases. Of course, the result of this termination is the formation of a molecule of polyethylene which may contain up to 800 carbon atoms.

   $$R{-}(CH_2{-}CH_2)_n{-}CH_2{-}CH_2\cdot\ +\ \cdot CH_2{-}CH_2{-}(CH_2{-}CH_2)_n{-}R \longrightarrow$$

   $$R{-}(CH_2{-}CH_2)_n{-}CH_2{-}CH_2{-}CH_2{-}CH_2{-}(CH_2{-}CH_2)_n{-}R$$

B. **Condensation Polymers.**

Copolymers are polymers that contain two or more different monomers. When monomers A and B are linked by condensation reactions, a uniform copolymer—ABABAB—can be formed. Nylon and polyesters, such as the well-known Dacron, are copolymers.

Polyester is made by an esterification reaction. When one monomer is an alcohol and the other is a carboxylic acid, they can be joined by an esterification reaction. The alcohol and the acid both must contain two functional groups. The monomers in polyester are the dicarboxylic acid called phthalic acid and the dialcohol, ethylene glycol.

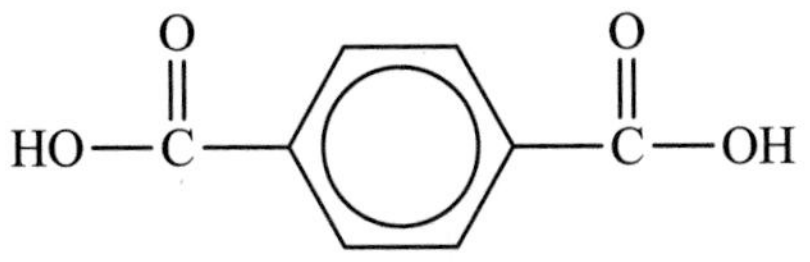

phthalic acid   ethylene glycol

Condensation reactions differ from addition reactions in that the former always results in the formation of a small molecule such as water. Polyesters are produced from the esterification reaction between an alcohol and an acid. When phthalic acid and ethylene glycol react to form an ester, the first products are:

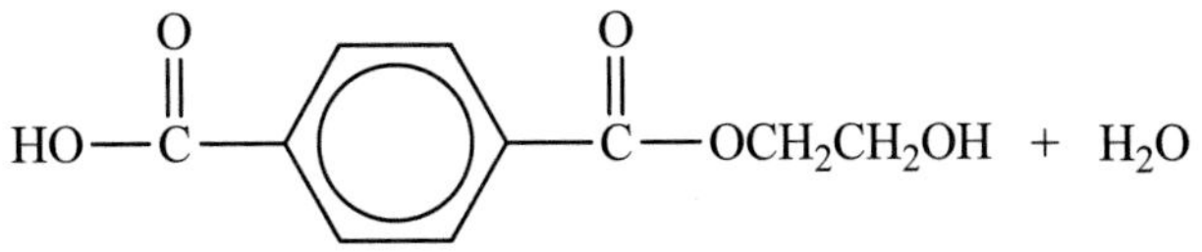

This molecule can react further from both ends. When this product reacts with another molecule of the diacid, the polymer chain grows longer.

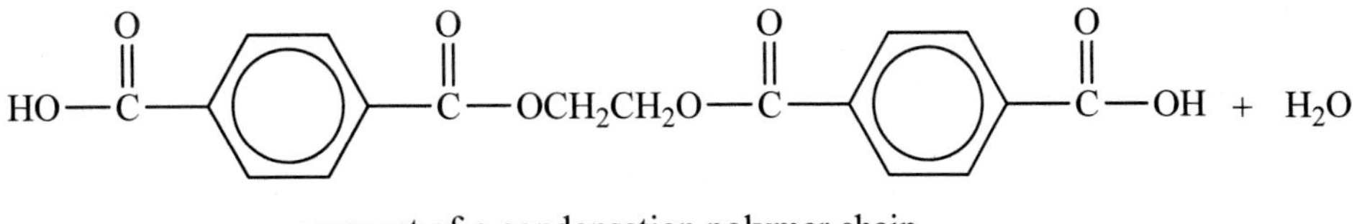

segment of a condensation polymer chain

The general formula for this polyester is:

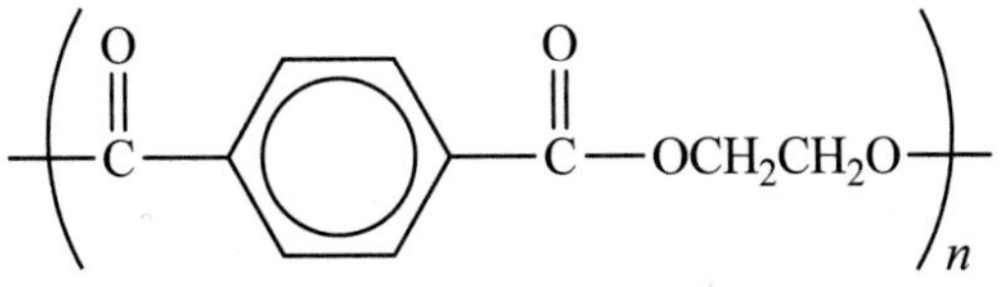

STUDY QUESTION 24.1
Write the formulas of the monomers used to prepare the following polymers:

   a.   **Teflon**
   b.   **Polystyrene**
   c.   **PVC**

Solution: Teflon is an addition polymer with the formula $-(\,CF_2-CF_2\,)-_n$. It is prepared from the monomer tetrafluoroethylene ($CF_2 = CF_2$).

The monomer used to prepare polystyrene is styrene.

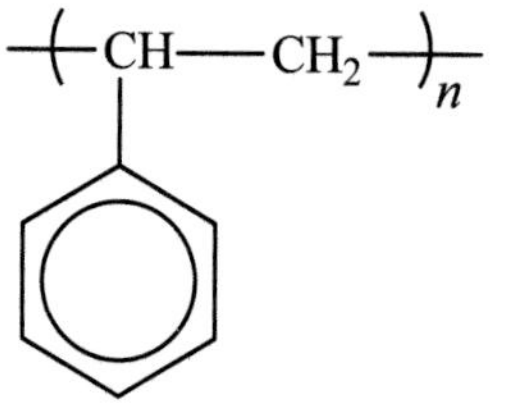

            polystyrene 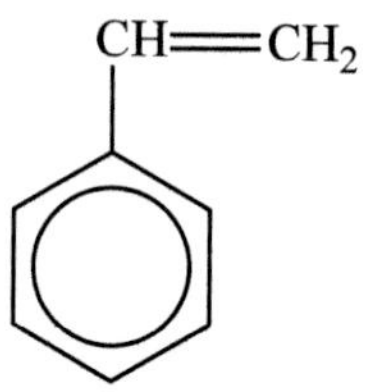   styrene

Polyvinylchloride is prepared by the successive addition of vinyl chloride molecules, $CH_2 = CHCl$.

> **PRACTICE QUESTION 24.1**
> **List three steps occurring during addition polymerization.**

PROBLEM TYPE 2: CERAMICS AND COMPOSITE MATERIALS.

Ceramics are polymeric inorganic compounds that share the properties of hardness, strength, and high melting points. Ceramics are usually formed by melting and then solidifying clay, sand, and similar inorganic material.

Ceramics are often electrical insulators at room temperature. At very cold temperatures some ceramic mixtures conduct electricity exceptionally well. Be sure to read about *superconductors* later. Ceramics are also good heat insulators, which is why the tiles protecting the Space Shuttle are made of ceramics. They can withstand the very hot temperatures of re-entry into the atmosphere (over a thousand degrees centigrade). Making ceramics can be done by mixing inorganic substances with water then heating the mixture to a very high temperature under high pressure in a process, called **sintering.** An example of sintering can be seen making yttrium/oxygen ceramic.

$$Y + 3\ CH_3CH_2OH \longrightarrow Y(OCH_2CH_3)_3 + 3\ H^+$$

                                yttrium alkoxide

The metal hydroxide, once formed, undergoes condensation polymerization to form a chain with bridging oxygen atoms between the metal atoms:

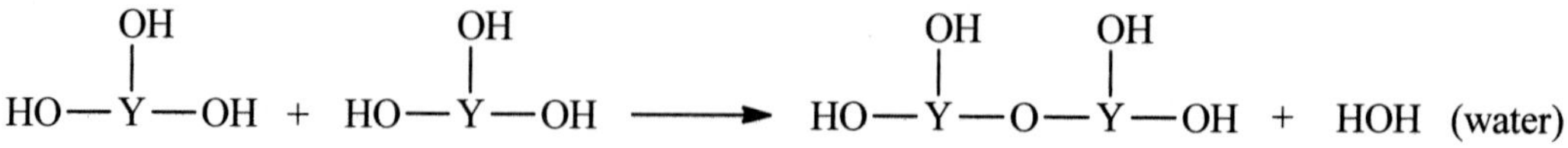

Just as in condensation polymerization between organic molecules, a small molecule (water, in this case) is produced by the combination of atoms from the two monomers. Repeated condensation produces an insoluble chain.

A **composite material** is made from two or more ingredients which both remain in the finished product. Think of the ancients stomping mud, grass, and water to make bricks. In modern times fiberglass might replace the straw and plastics the mud. Fiberglass and Kevlar are such composites. Biological composites are found throughout your body in substances like bone or tooth enamel.

STUDY QUESTION 24.2
A ceramic is made from mixing a slurry of silicon dioxide and additives to help bind the mixture, then it is baked into a form to fit an electrical motor. This type of ceramic could correctly be called a _____ .

Solution: Since the ceramic derives its name from the fundamental ingredient, this would be properly called a silicate.

> *PRACTICE QUESTION 24.2*
> **A ceramic is made from carbon oxides and additives to help it hold its form. This substance could be correctly termed a _____ .**

PROBLEM TYPE 3: LIQUID CRYSTALS.

Liquid crystals show some of the characteristics of liquids (like taking the shape of the container which holds them) and crystals (order between the atoms). As a liquid they are *isotropic* (their bulk properties don't depend on the observer's point of view). As a crystal, liquid crystals are *anisotropic* (other properties can change as the observer moves about the sample). An example of liquid crystal is cholesteryl benzoate. Depending upon the temperature, the samples can be found to show order. The higher temperature liquid form can be poured and the lower temperature form is found to line up (like crayons in a box).

Liquid crystals can be used to trap, pass or polarize (align) light. Clever engineering can use liquid crystals to form areas of dark or light in a display window. While a simple layer of liquid crystals might be used to form the lens of Polaroid sunglasses, more complex arrangements in layers can be controlled to form numbers in the small, thin displays of a watch or a calculator. The watch display is called an LCD (liquid crystal display). Areas of darkness can be created with two layers of liquid crystal aligned at 90 degrees. You can try this yourself with two pairs of polarized sunglasses. Hold the lenses over each other and rotate one with respect to the other.

PROBLEM TYPE 4: BIOMEDICAL MATERIALS.

Biomedical materials find use in the construction of replacement knees, dental implants, and artificial organs. Polymers, composites, and ceramics all might be used in a successful implant.

Dental fillings have been made out of lead, tin, platinum, gold, and various mixtures of metal (called an amalgam). Modern fillings now use resins, and crowns even make use of porcelain and stainless steel.

Artificial skin can be made from polymers like Dacron, and "thread" to stitch wounds shut may be made by the same material (but they must be removed). "Dissolving stitches" are likely made of biological material like leather or "gut." The Dacron has carbonyl and hydroxyl groups, which likely help the material be so readily accepted by the body (as those functional groups are so prevalent in the body).

Dacron:

glycolic acid   lactic acid

Joint replacements, such as artificial knees, are often made from stainless steel, Teflon, and other materials that wear out slowly and are easily accepted by the body.

STUDY QUESTION 24.3
A polymer is created by reacting glycolic acid and lactic acid by condensation. This material is thin, readily accepted by the body, but not durable. What application for this biomaterial would be best and why?

Solution: Because the material is readily accepted by the body but not strong or durable, it would make an excellent artificial skin.

PROBLEM TYPE 5: NANOTECHNOLOGY.

Atoms and the bonds between them are sized and spaced at distances on the order of a nanometer (a billionth of a meter). Nanotechnology is the newfound science which strives to make use of objects on this scale. To examine objects of this size, scanning electron microscopes have been developed to take advantage of the force exerted on a tiny stylus or the refractions of lasers of nanometer-wavelength.

Carbon atoms have recently been shown to arrange themselves into groups of about 60 atoms in the shape of a soccer ball. When arranged in a ball, the carbon spheres (called "buckyballs") can be made to carry ions or molecules in research focused on targeted drug delivery. If rolled into tubes, these nanometer-size "straws" can be used to separate or deliver compounds to selected destinations.

PRACTICE QUESTION 24.3
A polymer is made from condensation of glycolic acid and lactic acid. It forms a thin sheet of material readily accepted by the body. It is easily manipulated but not durable. It would/would not be a good material for heart value replacement. Why or why not?

PROBLEM TYPE 6: SEMICONDUCTORS.

If you line atoms "in a row" you can imagine their orbitals lining up like the hats of soldiers in formation. The row of hats (orbitals) can be thought of as a "band" from the final valance electrons (the valance band). If one then imagined gaps above the hats and wires over the formation you would be seeing an empty area (a gap between bands) and the wires above as a place where electricity could flow if the electrons were to reach them (the conduction band).

If the space (**band gap**) is large, the substance will be an insulator. If the gap is negligible, the substance will conduct. However, if the gap is of medium separation, the new classification is a "semiconductor." Silicon, germanium, and carbon are all likely to form semiconductors. Gallium, phosphorus, zinc, and cadmium can be used to maximize a semiconducting mixture to a desired end. Adding trace amounts (**doping**) of phosphorus will add an "extra" electron and increase conductivity (these mixtures are called **n-type**). Adding a trace of gallium (for example) will create an electron deficiency, lowering the conductivity (**p-type** semiconductor). Clever combinations of conductors, insulators, n-type, p-type semiconductors and light can create electrical circuits of amazing complexity (forming diodes and our modern electronic devices).

STUDY QUESTION 24.4
A sample of silicon is "doped" with trace amounts of gallium. This is a _____ type semiconductor.

Solution: This is a p-type semiconductor. Addition of an atom with fewer valance electrons than the substrate creates a p-type semiconductor while addition of a "contaminant" with a higher number of valance electrons creates a n-type semiconductor.

STUDY QUESTION 24.5
Which of the following would be the correct choice for a semiconductor? Copper has a resistivity of 3.0×10^{-5}, germanium has a resistivity of 1.0×10^{-2}, or rubber has a resistivity of 1.0×10^{12}?

Solution: Resistors have resistivities well into the ten-thousands and higher. Rubber certainly falls into this category. Conductors have resistivities approaching zero, copper clearly falls into this category. Germanium has resistivities in the tenths of an ohm making it a semiconductor.

PRACTICE QUESTION 24.4
 What could be added to silicon to make a p-type semiconductor?

PRACTICE QUESTION 24.5
 What class of conduction material would tungsten fall into (resistivity = 5.6×10^{-8})?

PRACTICE QUIZ

1. Sketch the structures of the monomers from which these polymers are formed.

    a.

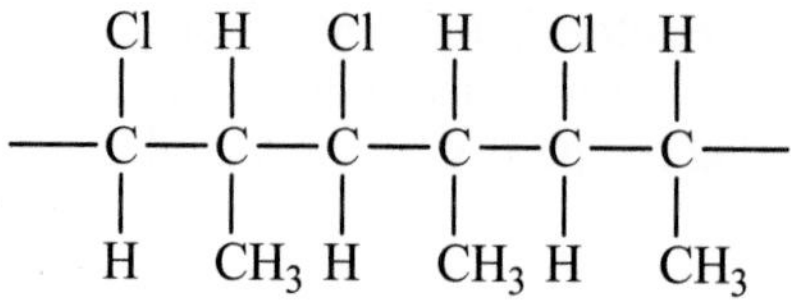

    $$\begin{array}{cccccc}
    Cl & H & Cl & H & Cl & H \\
    | & | & | & | & | & | \\
    -C- & C- & C- & C- & C- & C- \\
    | & | & | & | & | & | \\
    H & CH_3 & H & CH_3 & H & CH_3
    \end{array}$$

    b.

    $$-CH_2-CCl_2-CH_2-CCl_2-CH_2-$$

2. Draw structures for the monomers used to make the following polyester.

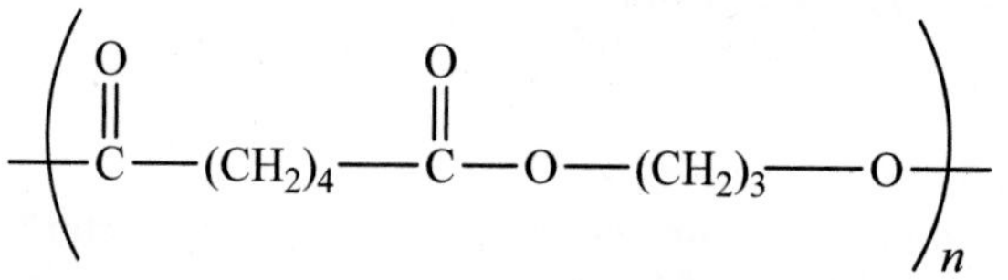

    $$\left(\begin{array}{c}
    \overset{O}{\overset{\|}{C}}-(CH_2)_4-\overset{O}{\overset{\|}{C}}-O-(CH_2)_3-O
    \end{array}\right)_n$$

3. A ceramic is made by mixing various nitrogen oxides with additives to bind the mixture, then baked into a part for an MP3 player. This ceramic could be correctly called a _____.

4. A silicon lattice is "doped" with arsenic. This material would be a _____ type semiconductor.

Solutions Manual

Chapter 1
Chemistry: The Science of Change

Key Skills

1.1   d   1.2   e   1.3   a   1.4   b

Problems

1.5   a. **Law** – Newton's Second Law of Motion

       b. **Theory** – Big Bang Theory

       c. **Hypothesis** – It may be possible but we need data to support this statement.

1.9   a. The sea is a heterogeneous mixture of seawater and biological matter, but seawater with the biomass filtered out is a **homogeneous mixture**.

       b. **pure substance**

       c. **pure substance**

       d. **homogeneous mixture**

       e. **homogeneous mixture**

       f. **pure substance**

       g. **heterogeneous mixture**

       h. **pure substance**

       i. **pure substance**

1.15   a. **Quantitative**. This statement involves a measurable distance.

       b. **Qualitative**. This is a value judgment. There is no numerical scale of measurement for artistic excellence.

       c. **Qualitative**. If the numerical values for the densities of ice and water were given, it would be a quantitative statement.

       d. **Qualitative**. The statement is a value judgment.

       e. **Qualitative**. Even though numbers are involved, they are not the result of measurement.

1.17   a. **Physical change.** The material is helium regardless of whether it is located inside or outside the balloon.

       b. **Chemical change** in the battery.

       c. **Physical change.** The orange juice concentrate can be regenerated by the addition of water.

d. **Chemical change.** Photosynthesis changes water, carbon dioxide, etc., into complex organic matter.

e. **Physical change.** The sugar can be recovered unchanged by evaporation.

1.23 **Strategy:** Use the density equation:

$$d = \frac{m}{V}$$

Solution:

$$d = \frac{m}{V} = \frac{586 \text{ g}}{188 \text{ mL}} = \textbf{3.12 g/mL}$$

1.25 **Strategy:** Find the appropriate equations for converting Fahrenheit to Celsius and Celsius to Fahrenheit given in Section 1.4 of the text. Substitute the temperature values given in the problem into the appropriate equation.

Setup: Conversion from °F to °C:

$$°C = (°F - 32°F) \times \frac{5°C}{9°F}$$

Conversion from °C to °F:

$$°F = \left(°C \times \frac{9°F}{5°C}\right) + 32°F$$

Solution:

a. $°C = (95°F - 32°F) \times \dfrac{5°C}{9°F} = \textbf{35°C}$

b. $°C = (12°F - 32°F) \times \dfrac{5°C}{9°F} = \textbf{--11°C}$

c. $°C = (103°F - 32°F) \times \dfrac{5°C}{9°F} = \textbf{39°C}$

d. $°C = (1852°F - 32°F) \times \dfrac{5°C}{9°F} = \textbf{1011°C}$

e. $°F = \left(-273.15°C \times \dfrac{9°F}{5°C}\right) + 32°F = \textbf{--459.67°F}$

1.27 **Strategy:** Use the density equation.

Solution:

$$\text{Volume of water} = V = \frac{m}{d} = \frac{2.50 \text{ g}}{0.992 \text{ g/mL}} = \textbf{2.52 mL}$$

1.29 **Strategy:** Use the equation for converting °C to K.

Setup: Conversion from °C to K:

$$K = °C + 273.15$$

Solution:
a. $K = 113°C + 273 = \textbf{386 K}$

b. $K = 37°C + 273 = \textbf{310 K}$

c. $K = 357°C + 273 = \textbf{630 K}$

Note that when there are no digits to the right of the decimal point in the original temperature, we use 273 instead of 273.15.

1.31   Mass is extensive and additive: 37.2 + 62.7 = **99.9 g**
Temperature is intensive: **20°C**
Density is intensive: **11.35 g/cm^3**

1.37   **Strategy:**   To convert an exponential number $N \times 10^n$ to a decimal number, move the decimal n places to the left if $n < 0$, or move it n places to the right if $n > 0$. While shifting the decimal, add place-holding zeros as needed.

Solution: a. $1.52 \times 10^{-2} = \textbf{0.0152}$

b. $7.78 \times 10^{-8} = \textbf{0.0000000778}$

c. $3.29 \times 10^{-6} = \textbf{0.00000329}$

d. $8.41 \times 10^{-1} = \textbf{0.841}$

1.39   a. Addition using scientific notation

Strategy: A measurement is in *scientific notation* when it is written in the form $N \times 10^n$, where $0 \leq N < 10$ and n is an integer. When adding measurements that are written in scientific notation, rewrite the quantities so that they share a common exponent. To get the "N part" of the result, we simply add the "N parts" of the rewritten numbers. To get the exponent of the result, we simply set it equal to the common exponent. Finally, if need be, we rewrite the result so that its value of N satisfies $0 \leq N < 10$.

Solution: Rewrite the quantities so that they have a common exponent. In this case, choose the common exponent $n = -3$.
$$0.0095 = 9.5 \times 10^{-3}$$

Add the "N parts" of the rewritten numbers and set the exponent of the result equal to the common exponent.
$$9.5 \times 10^{-3}$$
$$+ \ 8.5 \times 10^{-3}$$
$$\overline{18.0 \times 10^{-3}}$$

Rewrite the number so that it is in scientific notation (so that $0 \leq N < 10$).
$$18.0 \times 10^{-3} = \mathbf{1.8 \times 10^{-2}}$$

b. Division using scientific notation

Strategy: When dividing two numbers using scientific notation, divide the "N parts" of the numbers in the usual way. To find the exponent of the result, *subtract* the exponent of the divisor from that of the dividend.

Solution: Make sure that all numbers are expressed in scientific notation.

$$653 = 6.53 \times 10^2$$

Divide the "N parts" of the numbers in the usual way.

$$6.53 \div 5.75 = 1.14$$

Subtract the exponents.

$$1.14 \times 10^{+2 - (-8)} = 1.14 \times 10^{+2 + 8} = \mathbf{1.14 \times 10^{10}}$$

c. Subtraction using scientific notation

Strategy: When subtracting two measurements that are written in scientific notation, rewrite the quantities so that they share a common exponent. To get the "N part" of the result, we simply subtract the "N parts" of the rewritten numbers. To get the exponent of the result, we simply set it equal to the common exponent. Finally, if need be, we rewrite the result so that its value of N satisfies $0 \le N < 10$.

Solution: Rewrite the quantities so that they have a common exponent. Rewrite 850,000 in such a way that $n = 5$.

$$850{,}000 = 8.5 \times 10^5$$

Subtract the "N parts" of the numbers and set the exponent of the result equal to the common exponent.

$$
\begin{array}{r}
8.5 \times 10^5 \\
- \ 9.0 \times 10^5 \\
\hline
- \ 0.5 \times 10^5
\end{array}
$$

Rewrite the number so that $0 \le N < 10$ (ignore the sign of N when it is negative).

$$-0.5 \times 10^5 = \mathbf{-5 \times 10^4}$$

d. Multiplication using scientific notation

Strategy: When multiplying two numbers using scientific notation, multiply the "N parts" of the numbers in the usual way. To find the exponent of the result, *add* the exponents of the two measurements.

Solution: Multiply the "N parts" of the numbers in the usual way.
$$3.6 \times 3.6 = 12.96$$

Add the exponents.

$$12.96 \times 10^{-4 + (+6)} = 12.96 \times 10^2$$

Rewrite the number so that it is in scientific notation (so that $0 \le N < 10$). Round the final result to two significant figures.

$$\mathbf{13 \times 10^2 = 1.3 \times 10^3}$$

1.41   a. **one**   d. **four**   g. **one or two**

      b. **three**   e. **three**   h. **three**

      c. **three**   f. **one**

1.43   a. Division

> **Strategy:**   The number of significant figures in the answer is determined by the original number having the smallest number of significant figures.
>
> **Solution:**
> $$\frac{7.310 \text{ km}}{5.70 \text{ km}} = 1.28\mathbf{2}$$
>
> The **2** (bolded) is a nonsignificant digit because the original number 5.70 only has three significant digits. Therefore, the answer has only three significant digits.
>
> The correct answer rounded off to the correct number of significant figures is:
>
> 1.28

b. Subtraction

> **Strategy:**   The number of significant figures to the right of the decimal point in the answer is determined by the lowest number of digits to the right of the decimal point in any of the original numbers.
>
> **Solution:**   Writing both numbers in the decimal notation, we have
>
> $$\begin{array}{r} 0.00326 \quad \text{mg} \\ -\ 0.0000788 \text{ mg} \\ \hline 0.003181\mathbf{2}\text{ mg} \end{array}$$
>
> The bold numbers are nonsignificant digits because the number 0.00326 has five digits to the right of the decimal point. Therefore, we carry five digits to the right of the decimal point in our answer.
>
> The correct answer rounded off to the correct number of significant figures is:
>
> 0.00318 mg = 3.18 × 10^{−3} mg

c. Addition

> **Strategy:**   The number of significant figures to the right of the decimal point in the answer is determined by the lowest number of digits to the right of the decimal point in any of the original numbers.
>
> **Solution:**   Writing both numbers with exponents = +7, we have
>
> $$(0.402 \times 10^7 \text{ dm}) + (7.74 \times 10^7 \text{ dm}) = \mathbf{8.14 \times 10^7 \text{ dm}}$$
>
> Since 7.74×10^7 has only two digits to the right of the decimal point, two digits are carried to the right of the decimal point in the final answer.

1.45   **Carpenter Z's measurements are the most accurate. Carpenter Y's measurements are the least accurate. Carpenter X's measurements are the most precise. Carpenter Y's measurements are the least precise.**

1.47 **Upper thermometer: 41.85°**

Lower thermometer: 41.85°

Since both thermometers are marked to tenths of a degree, the number of significant figures is the same, regardless of whether or not the tenths of a degree are labeled.

1.49 **Strategy:** The difference between the masses of the empty and filled bulbs is the mass of the gas in the bulb. The density is then determined using the density equation, $d = \dfrac{m}{V}$.

Solution: $$243.22 \text{ g} - 243.07 \text{ g} = 0.15 \text{ g gas}$$

$$d = \frac{m}{V} = \frac{0.15 \text{ g}}{111.5 \text{ mL}} = 1.3 \times 10^{-3} \text{ g/mL}$$

$$\frac{1.3 \times 10^{-3} \text{ g}}{1 \text{ mL}} \times \frac{1 \times 10^{3} \text{ mL}}{\text{L}} = \mathbf{1.3\, g\,/\,L}$$

Because the density of gases is generally low, the density is typically expressed in g/L.

1.51 a. **Strategy:** The solution requires a two-step dimensional analysis because we must first convert pounds to grams and then grams to milligrams.

Setup: The necessary conversion factors as derived from the equalities: 1 g = 1000 mg and 1 lb = 453.6 g.

$$\frac{453.6 \text{ g}}{1 \text{ lb}} \text{ and } \frac{1 \text{ mg}}{1 \times 10^{-3} \text{ g}}$$

Solution: $$242 \text{ lb} \times \frac{453.6 \text{ g}}{1 \text{ lb}} \times \frac{1 \text{ mg}}{1 \times 10^{-3} \text{ g}} = \mathbf{1.10 \times 10^{8} \ mg}$$

b. **Strategy:** We need to convert from cubic centimeters to cubic meters.

Setup: 1 m = 100 cm. When a unit is raised to a power, the corresponding conversion factor must also be raised to that power in order for the units to cancel.

Solution: $$68.3 \text{ cm}^3 \times \left(\frac{1 \text{ m}}{100 \text{ cm}} \right)^3 = \mathbf{6.83 \times 10^{-5} \ m^3}$$

c. **Strategy:** In Chapter 1 of the text, a conversion is given between liters and cm³ (1 L = 1000 cm³). If we can convert m³ to cm³, we can then convert to liters. Recall that 1 cm = 1×10^{-2} m. We need to set up two conversion factors to convert from m³ to L. Arrange the appropriate conversion factors so that m³ and cm³ cancel, and the unit liters is obtained in your answer.

Setup: The sequence of conversions is m³ → cm³ → L. Use the following conversion factors:

$$\left(\frac{1 \text{ cm}}{1 \times 10^{-2} \text{ m}} \right)^3 \text{ and } \frac{1 \text{ L}}{1000 \text{ cm}^3}$$

Solution:

$$7.2 \text{ m}^3 \times \left(\frac{1 \text{ cm}}{1 \times 10^{-2} \text{ m}} \right)^3 \times \frac{1 \text{ L}}{1000 \text{ cm}^3} = \textbf{7.2} \times \textbf{10}^3 \text{ L}$$

d. Strategy: A relationship between pounds and grams is given on the end sheet of the text (1 lb = 453.6 g). This relationship will allow conversion from grams to pounds. If we can convert from micrograms to grams, we can then convert from grams to pounds. Recall that $1 \ \mu g = 1 \times 10^{-6}$ g. Arrange the appropriate conversion factors so that micrograms and grams cancel, and the unit pounds is obtained in your answer.

Setup: The sequence of conversions is $\mu g \rightarrow g \rightarrow$ lb. Use the following conversion factors:

$$\frac{1 \times 10^{-6} \text{ g}}{1 \ \mu g} \text{ and } \frac{1 \text{ lb}}{453.6 \text{ g}}$$

Solution:

$$28.3 \ \mu g \times \frac{1 \times 10^{-6} \text{ g}}{1 \ \mu g} \times \frac{1 \text{ lb}}{453.6 \text{ g}} = \textbf{6.24} \times \textbf{10}^{-8} \textbf{ lb}$$

1.53 **a. Strategy:** The given unit is grams and the desired unit is amu. Use a conversion factor to convert $g \rightarrow$ amu.

Setup: Use the conversion factor:

$$\frac{1 \text{ amu}}{1.6605378 \times 10^{-24} \text{ g}}$$

Solution:

$$1.1 \times 10^{-22} \text{ g} \times \frac{1 \text{ amu}}{1.6605378 \times 10^{-24} \text{ g}} = \textbf{66 amu}$$

b. Strategy: The given unit is kilograms and the desired unit is amu. Use a conversion factor to convert $kg \rightarrow$ amu.

Setup: Use the conversion factor:

$$\frac{1 \text{ amu}}{1.6605378 \times 10^{-27} \text{ kg}}$$

Solution:

$$1.08 \times 10^{-29} \text{ kg} \times \frac{1 \text{ amu}}{1.6605378 \times 10^{-27} \text{ kg}} = \textbf{6.50} \times \textbf{10}^{-3} \textbf{ amu}$$

c. Strategy: The given unit is meters and the desired unit is Å. Use a conversion factor to convert $m \rightarrow$ Å.

Setup: Use the conversion factor:

$$\frac{1 \text{ Å}}{1 \times 10^{-10} \text{ m}}$$

Solution:

$$8.3 \times 10^{-9} \text{ m} \times \frac{1 \text{ Å}}{1 \times 10^{-10} \text{ m}} = \textbf{83 Å}$$

d. Strategy: The given unit is picometers and the desired unit is angstroms. Use conversion factors to convert pm $\rightarrow$ m $\rightarrow$ Å.

Setup: Use the conversion factors:

$$\frac{1\times10^{-12}\ \text{m}}{\text{pm}} \text{ and } \frac{1\ \text{Å}}{1\times10^{-10}\ \text{m}}$$

Solution:

$$132\ \text{pm}\times\frac{1\times10^{-12}\ \text{m}}{\text{pm}}\times\frac{1\ \text{Å}}{1\times10^{-10}\ \text{m}} = \mathbf{1.32\ Å}$$

1.55 **Strategy:** You should know conversion factors that will allow you to convert between days and hours, hours and minutes, and minutes and seconds. Make sure to arrange the conversion factors so that days, hours, and minutes cancel, leaving units of seconds for the answer.

Setup: The sequence of conversions is d → h → min → s. Use the following conversion factors:

$$\frac{24\ \text{h}}{1\ \text{d}},\ \frac{60\ \text{min}}{1\ \text{h}},\text{ and }\frac{60\ \text{s}}{1\ \text{min}}$$

Solution:

$$365.24\ \text{d}\times\frac{24\ \text{h}}{1\ \text{d}}\times\frac{60\ \text{min}}{1\ \text{h}}\times\frac{60\ \text{s}}{1\ \text{min}} = \mathbf{3.1557\times10^{7}\ s}$$

1.57 a. **Strategy:** The measurement is given in mi/min. We are asked to convert this rate to in/s. Use conversion factors to convert mi → ft → in and to convert min → s.

Setup: Use the conversion factors:

$$\frac{5280\ \text{ft}}{1\ \text{mi}},\ \frac{12\ \text{in}}{1\ \text{ft}},\text{ and }\frac{1\ \text{min}}{60\ \text{s}}$$

Be sure to set the conversion factors up so that the appropriate units cancel.

Solution:

$$\frac{1\ \text{mi}}{13\ \text{min}}\times\frac{5280\ \text{ft}}{1\ \text{mi}}\times\frac{12\ \text{in}}{1\ \text{ft}}\times\frac{1\ \text{min}}{60\ \text{s}} = \mathbf{81\ in\,/\,s}$$

b. **Strategy:** The measurement is given in mi/min. We are asked to convert this rate to m/min. Use a conversion factor convert mi → m.

Setup: Use the conversion factor:

$$\frac{1609\ \text{m}}{1\ \text{mi}}$$

Solution:

$$\frac{1\ \text{mi}}{13\ \text{min}}\times\frac{1609\ \text{m}}{1\ \text{mi}} = \mathbf{1.2\times10^{2}\ m\,/\,min}$$

c. **Strategy:** The measurement is given in mi/min. We are asked to convert this rate to km/h. Use conversion factors to convert mi → m → km and convert min → h.

Setup: Use the conversion factors:

$$\frac{1609\ \text{m}}{1\ \text{mi}},\ \frac{1\ \text{km}}{1000\ \text{m}},\text{ and }\frac{60\ \text{min}}{1\ \text{h}}$$

Solution:

$$\frac{1\ \text{mi}}{13\ \text{min}}\times\frac{1609\ \text{m}}{1\ \text{mi}}\times\frac{1\ \text{km}}{1000\ \text{m}}\times\frac{60\ \text{min}}{1\ \text{h}} = \mathbf{7.4\ km\,/\,h}$$

1.59   **Strategy:** The rate is given in the units mi/h. The desired units are km/h. Use conversion factors to convert mi → m → km.

      Setup: Use the conversion factors:

$$\frac{1609 \text{ m}}{1 \text{ mi}} \text{ and } \frac{1 \text{ km}}{1000 \text{ m}}$$

      Solution:

$$\frac{65 \text{ mi}}{1 \text{ h}} \times \frac{1609 \text{ m}}{1 \text{ mi}} \times \frac{1 \text{ km}}{1000 \text{ m}} = \mathbf{1.0 \times 10^2 \text{ km/h}}$$

1.61   **Strategy:** We seek to calculate the mass of Pb in a 6.0×10^3 g sample of blood. Lead is present in the blood at the rate of $0.62 \text{ ppm} = \dfrac{0.62 \text{ g Pb}}{1 \times 10^6 \text{ g blood}}$. Use the rate to convert g blood → g Pb.

      Setup: Be sure to set the conversion factor up so that g blood cancels.

      Solution:

$$6.0 \times 10^3 \text{ g blood} \times \frac{0.62 \text{ g Pb}}{1 \times 10^6 \text{ g blood}} = \mathbf{3.7 \times 10^{-3} \text{ g Pb}}$$

1.63   a. **Strategy:** The given unit is nm and the desired unit is m. Use a conversion factor to convert nm → m.

      Setup: Use the conversion factor:

$$\frac{1 \times 10^{-9} \text{ m}}{1 \text{ nm}}$$

      Solution:

$$185 \text{ nm} \times \frac{1 \times 10^{-9} \text{ m}}{1 \text{ nm}} = \mathbf{1.85 \times 10^{-7} \text{ m}}$$

      b. **Strategy:** The given unit is yr and the desired unit is s. Use conversion factors to convert yr → d → h → s.

      Setup: Use the conversion factors:

$$\frac{365 \text{ d}}{1 \text{ yr}}, \frac{24 \text{ h}}{1 \text{ d}}, \text{ and } \frac{3600 \text{ s}}{1 \text{ h}}$$

      Solution:

$$4.5 \times 10^9 \text{ yr} \times \frac{365 \text{ d}}{1 \text{ yr}} \times \frac{24 \text{ h}}{1 \text{ d}} \times \frac{3600 \text{ s}}{1 \text{ h}} = \mathbf{1.4 \times 10^{17} \text{ s}}$$

      c. **Strategy:** The given unit is cm^3 and the desired unit is m^3. Use a conversion factor to convert $cm^3 \rightarrow m^3$.

      Setup: Use the conversion factor:

$$\left(\frac{0.01 \text{ m}}{1 \text{ cm}}\right)^3$$

      Solution:

$$71.2 \text{ cm}^3 \times \left(\frac{0.01 \text{ m}}{1 \text{ cm}}\right)^3 = \mathbf{7.12 \times 10^{-5} \text{ m}^3}$$

      d. **Strategy:** The given unit is m^3 and the desired unit is L. Use conversion factors to convert $m^3 \rightarrow cm^3 \rightarrow L$.

Setup: Use the conversion factors:

$$\left(\frac{1\ cm}{1 \times 10^{-2}\ m}\right)^3 \quad \text{and} \quad \frac{1\ L}{1000\ cm^3}$$

Solution:

$$88.6\ m^3 \times \left(\frac{1\ cm}{1 \times 10^{-2}\ m}\right)^3 \times \frac{1\ L}{1000\ cm^3} = \mathbf{8.86 \times 10^4\ L}$$

1.65 **Strategy:** The given unit s g/L and the desired unit is g/cm^3. Use a conversion factor to convert L → cm^3.

Setup: Use the conversion factor:

$$\frac{1\ L}{1000\ cm^3}$$

Solution:

$$\frac{0.625\ g}{1\ L} \times \frac{1\ L}{1000\ cm^3} = \mathbf{6.25 \times 10^{-4}\ g/cm^3}$$

1.67 a. **Upper ruler: 2.5 cm**   b. **Lower ruler: 2.55 cm**

1.69 a. **chemical**   b. **chemical**   c. **physical**   d. **physical**   e. **chemical**

1.71 $(95.0 \times 10^9\ lb$ sulfuric acid$) \times \dfrac{1\ ton}{2.0 \times 10^3\ lb} = \mathbf{4.75 \times 10^7\ tons\ of\ sulfuric\ acid}$

1.73 a. **Strategy:** Calculate the volume of the sphere using:

$$V = \frac{4}{3}\pi r^3$$

Then use the density equation, $d = \dfrac{m}{V}$, to find the mass.

Setup: Solve the density equation for m to get $m = dV$. Find the volume and substitute it into the equation for m.

Solution:

$$V = \left(\frac{4}{3}\right)(3.14159)(10.0\ cm)^3 = 4189\ cm^3$$

$$m = dV = \frac{19.3\ g}{1\ cm^3} \times 4189\ cm^3 = \mathbf{8.08 \times 10^4\ g}$$

b. **Strategy:** Compute the volume of the cube using:

$$V = s^3$$

Then, find the mass using the density equation:

$$d = \frac{m}{V}$$

Setup: Solve the density equation for m to get $m = dV$. Find the volume and substitute it into the equation for m.

Solution: The edge of the cube is $s = 0.040\ mm = 0.0040\ cm$, and $V = (0.0040\ cm)^3 = 6.4 \times 10^{-8}\ cm^3$.

$$m = dV = \frac{21.4\ g}{1\ cm^3} \times \left(6.4 \times 10^{-8}\ cm^3\right) = \mathbf{1.4 \times 10^{-6}\ g}$$

c. **Strategy:** Use the density equation:

$$d = \frac{m}{V}$$

Setup: Solve the density equation for m to get $m = dV$.

Solution:

$$50.0 \text{ mL} \times \frac{0.798 \text{ g}}{1 \text{ mL}} = \textbf{39.9 g}$$

1.75 **Strategy:** The difference between the masses of the empty and filled flasks is the mass of the water in the flask. The volume of the water (and the flask) can be found using the density equation.

Setup: Solve the density equation for V:

$$V = \frac{m}{d}$$

Solution: $87.39 - 56.12 = 31.27$ g water

$$V = \frac{m}{d} = \frac{31.27 \text{ g}}{0.9976 \text{ g/cm}^3} = \textbf{31.35 cm}^3$$

1.77 **Strategy:** The volume of the piece of silver is the same as the volume of water it displaces. Once the volume is found, use the density equation to compute the density.

Setup: $V = 260.5 - 242.0 = 18.5$ mL

Solution:

$$d = \left(\frac{194.3 \text{ g}}{18.5 \text{ mL}} \right) = \textbf{10.5 g / mL}$$

The density of a solid is generally reported in g/cm^3 (1 mL = 1 cm^3). Therefore, the density is reported as **10.5 g/cm^3**.

1.79 **Strategy:** Use the density equation.

Setup:

$$d = \frac{m}{V}$$

Solution:

$$d = \frac{m}{V} = \frac{1.20 \times 10^4 \text{ g}}{1.05 \times 10^3 \text{ cm}^3} = \textbf{11.4 g / cm}^3$$

1.81 **Strategy:** For the Fahrenheit thermometer, we must convert the possible error of 0.1°F to °C. For each thermometer, use the percent error equation to find the percent error for the measurement.

Setup:

$$0.1°F \times \frac{5°C}{9°F} = 0.056°C \ .$$

For the Fahrenheit thermometer, we expect $\left| \text{true value} - \text{experimental value} \right| = 0.056°C$.

For the Celsius thermometer, we expect $\left| \text{true value} - \text{experimental value} \right| = 0.1°C$.

$$\text{Percent error} = \frac{\left| \text{true value} - \text{experimental value} \right|}{\text{true value}} \times 100\%$$

Solution: For the Fahrenheit thermometer,

$$\textbf{Percent error} = \frac{0.056°C}{38.9°C} \times 100\% = \textbf{0.1\%}$$

For the Celsius thermometer,

$$\textbf{Percent error} = \frac{0.1°C}{38.9°C} \times 100\% = \textbf{0.3\%}$$

1.83 There are 78.3 + 117.3 = 195.6 Celsius degrees between 0°S and 100°S. We can write this as a conversion factor.

$$\frac{195.6 \ C}{100 \ S}$$

Set up the equation like a Celsius to Fahrenheit conversion. We need to subtract 117.3°C, because the zero point on the new scale is 117.3°C lower than the zero point on the Celsius scale.

$$? °C = \left(\frac{195.6°C}{100°S}\right)(? °S) - 117.3°C$$

Solving for ? °S gives:

$$? °S = (? °C + 117.3°C)\left(\frac{100°S}{195.6°C}\right)$$

For 25°C we have:

$$? °S = (25°C + 117.3°C)\left(\frac{100°S}{195.6°C}\right) = \textbf{72.8°S}$$

1.85 **Strategy:** The key to solving this problem is to realize that all the oxygen needed must come from the 4% difference (20% − 16%) between inhaled and exhaled air. The 240 mL of pure oxygen/min requirement comes from the 4% of inhaled air that is oxygen.

Setup: 240 mL of pure oxygen/min = (0.04)(volume of inhaled air/min)

Solution:

$$\text{Volume of inhaled air/min} = \frac{240 \text{ mL of oxygen/min}}{0.04} = 6000 \text{ mL of inhaled air/min}$$

Since there are 12 breaths per min,

$$\text{Volume of air / breath} = \frac{6000 \text{ mL of inhaled air}}{1 \text{ min}} \times \frac{1 \text{ min}}{12 \text{ breaths}} = \textbf{5} \times \textbf{10}^\textbf{2} \textbf{ mL / breath}$$

1.87 **Strategy:** The volume of seawater is given. The strategy is to use the given conversion factors to convert L seawater → g seawater → g NaCl. This result can then be converted to kg NaCl and to tons NaCl. Note that 3.1% NaCl by weight means 100 g seawater = 3.1 g NaCl.

Setup: Use the conversion factors:

$$\frac{1000 \text{ mL seawater}}{1 \text{ L seawater}}, \ \frac{1.03 \text{ g seawater}}{1 \text{ mL seawater}}, \text{ and } \frac{3.1 \text{ g NaCl}}{100 \text{ g seawater}}$$

Solution:

$$1.5 \times 10^{21} \text{ L seawater} \times \frac{1000 \text{ mL seawater}}{1 \text{ L seawater}} \times \frac{1.03 \text{ g seawater}}{1 \text{ mL seawater}} \times \frac{3.1 \text{ g NaCl}}{100 \text{ g seawater}} = 4.8 \times 10^{22} \text{ g NaCl}$$

$$\text{Mass of NaCl (kg)} = 4.8 \times 10^{22} \text{ g NaCl} \times \frac{1 \text{ kg}}{1000 \text{ g}} = \textbf{4.8} \times \textbf{10}^{\textbf{19}} \textbf{ kg NaCl}$$

$$\text{Mass of NaCl (tons)} = 4.8 \times 10^{22} \text{ g NaCl} \times \frac{1 \text{ lb}}{453.6 \text{ g}} \times \frac{1 \text{ ton}}{2000 \text{ lb}} = \textbf{5.3} \times \textbf{10}^{\textbf{16}} \textbf{ tons NaCl}$$

1.89   a. **Strategy:**   Use the given conversion factor to convert troy oz → g.

       Setup: Conversion factor:

$$\frac{31.103 \text{ g Au}}{1 \text{ troy oz Au}}$$

       Solution:

$$2.41 \text{ troy oz Au} \times \frac{31.103 \text{ g Au}}{1 \text{ troy oz Au}} = \textbf{75.0 g Au}$$

   b. **Strategy:**   Use the given conversion factors to convert 1 troy oz → g → lb → oz.

       Setup: Conversion factors:

$$\frac{31.103 \text{ g}}{1 \text{ troy oz}}, \frac{1 \text{ lb}}{453.6 \text{ g}}, \text{ and } \frac{16 \text{ oz}}{1 \text{ lb}}$$

       Solution:

$$1 \text{ troy oz} \times \frac{31.103 \text{ g}}{1 \text{ troy oz}} \times \frac{1 \text{ lb}}{453.6 \text{ g}} \times \frac{16 \text{ oz}}{1 \text{ lb}} = 1.097 \text{ oz}$$

$$1 \text{ troy oz} = 1.097 \text{ oz}$$

A troy ounce is heavier than an ounce.

1.91   a. **Strategy:**   Use the percent error equation.

       Setup: The percent error of a measurement is given by:

$$\frac{|\text{true value} - \text{experimental value}|}{\text{true value}} \times 100\%$$

       Solution:

$$\frac{|0.798 \text{ g/mL} - 0.802 \text{ g/mL}|}{0.798 \text{ g/mL}} \times 100\% = \textbf{0.5\%}$$

   b. **Strategy:**   Use the percent error equation.

       Setup: The percent error of a measurement is given by:

$$\frac{|\text{true value} - \text{experimental value}|}{\text{true value}} \times 100\%$$

       Solution:

$$\frac{|0.864 \text{ g} - 0.837 \text{ g}|}{0.864 \text{ g}} \times 100\% = \textbf{3.1\%}$$

1.93   **Strategy:**   Use the percent composition measurement to convert kg ore → g Cu. Note that 34.63% Cu by mass means 100 g ore = 34.63 g Cu.

Setup: Use the conversion factors:

$$\frac{34.63 \text{ g Cu}}{100 \text{ g ore}} \text{ and } \frac{1000 \text{ g}}{1 \text{ kg}}$$

Solution:

$$(5.11 \times 10^3 \text{ kg ore}) \times \frac{34.63 \text{ g Cu}}{100 \text{ g ore}} \times \frac{1000 \text{ g}}{1 \text{ kg}} = \mathbf{1.77 \times 10^6 \text{ g Cu}}$$

1.95 **Strategy:** Use the given rates to convert cars → kg CO_2.

Setup: Conversion factors:

$$\frac{5000 \text{ mi}}{1 \text{ car}}, \frac{1 \text{ gal gas}}{20 \text{ mi}}, \text{ and } \frac{9.5 \text{ kg } CO_2}{1 \text{ gal gas}}$$

Solution:

$$(40 \times 10^6 \text{ cars}) \times \frac{5000 \text{ mi}}{1 \text{ car}} \times \frac{1 \text{ gal gas}}{20 \text{ mi}} \times \frac{9.5 \text{ kg } CO_2}{1 \text{ gal gas}} = \mathbf{9.5 \times 10^{10} \text{ kg } CO_2}$$

1.97 **Strategy:** Use the given rate to convert J → yr.

Setup: Conversion factor:

$$\frac{1 \text{ yr}}{1.8 \times 10^{20} \text{ J}}$$

Solution:

$$(2.0 \times 10^{22} \text{ J}) \times \frac{1 \text{ yr}}{1.8 \times 10^{20} \text{ J}} = \mathbf{1.1 \times 10^2 \text{ yr}}$$

1.99 **Strategy:** To calculate the density of the pheromone, you need the mass of the pheromone and the volume that it occupies. The mass is given in the problem.

Setup: volume of a cylinder = area × height = $\pi r^2 \times h$

Converting the radius and height to cm gives:

$$0.50 \text{ mi} \times \frac{1609 \text{ m}}{1 \text{ mi}} \times \frac{1 \text{ cm}}{0.01 \text{ m}} = 8.05 \times 10^4 \text{ cm}$$

$$40 \text{ ft} \times \frac{12 \text{ in}}{1 \text{ ft}} \times \frac{2.54 \text{ cm}}{1 \text{ in}} = 1.22 \times 10^3 \text{ cm}$$

Solution:

$$\text{volume} = \pi (8.05 \times 10^4 \text{ cm})^2 \times (1.22 \times 10^3 \text{ cm}) = 2.48 \times 10^{13} \text{ cm}^3$$

Density of gases is usually expressed in g/L. Let's convert the volume to liters.

$$(2.48 \times 10^{13} \text{ cm}^3) \times \frac{1 \text{ mL}}{1 \text{ cm}^3} \times \frac{1 \text{ L}}{1000 \text{ mL}} = 2.48 \times 10^{10} \text{ L}$$

$$\mathbf{density} = \frac{\text{mass}}{\text{volume}} = \frac{1.0 \times 10^{-8} \text{ g}}{2.48 \times 10^{10} \text{ L}} = \mathbf{4.0 \times 10^{-19} \text{ g / L}}$$

1.101   **Strategy:**   We wish to calculate the density and radius of the ball bearing. For both calculations, we need the volume of the ball bearing. The data from the first experiment can be used to calculate the density of the mineral oil. In the second experiment, the density of the mineral oil can then be used to determine what part of the 40.00 mL volume is due to the mineral oil and what part is due to the ball bearing. Once the volume of the ball bearing is determined, we can calculate its density and radius.

   Solution: From the first experiment:

$$\text{Mass of oil} = 159.446 \text{ g} - 124.966 \text{ g} = 34.480 \text{ g}$$

$$\text{Density of oil} = \frac{34.480 \text{ g}}{40.00 \text{ mL}} = 0.8620 \text{ g/mL}$$

   From the second experiment:

$$\text{Mass of oil} = 50.952 \text{ g} - 18.713 \text{ g} = 32.239 \text{ g}$$

$$\text{Volume of oil} = 32.239 \text{ g} \times \frac{1 \text{ mL}}{0.8620 \text{ g}} = 37.40 \text{ mL}$$

   The volume of the ball bearing is obtained by difference.

$$\text{Volume of ball bearing} = 40.00 \text{ mL} - 37.40 \text{ mL} = 2.60 \text{ mL} = 2.60 \text{ cm}^3$$

   Now that we have the volume of the ball bearing, we can calculate its density and radius.

$$\text{Density of ball bearing} = \frac{18.713 \text{ g}}{2.60 \text{ cm}^3} = \textbf{7.20 g / cm}^3$$

   Using the formula for the volume of a sphere, we can solve for the radius of the ball bearing.

$$V = \frac{4}{3}\pi r^3$$

$$2.60 \text{ cm}^3 = \frac{4}{3}\pi r^3$$

$$r^3 = 0.621 \text{ cm}^3$$

$$\textbf{\textit{r} = 0.853 cm}$$

1.103   **It would be more difficult to prove that the unknown substance is a pure substance. Most mixtures can be separated into two or more substances.** For example, upon distillation a mixture of salt and water can be separated.

1.105   **Gently heat the liquid to see if any solid remains after the liquid evaporates. Also, collect the vapor and then compare the densities of the condensed liquid with the original liquid. The composition of a mixed liquid frequently changes with evaporation along with its density.**

1.107   **Strategy:**   As water freezes, it expands. First, calculate the mass of the water at 20°C. Then, determine the volume that this mass of water would occupy at −5°C.

Solution:

$$\text{Mass of water} = 242 \text{ mL} \times \frac{0.998 \text{ g}}{1 \text{ mL}} = 241.5 \text{ g}$$

$$\text{Volume of ice at } -5°C = 241.5 \text{ g} \times \frac{1 \text{ mL}}{0.916 \text{ g}} = 264 \text{ mL}$$

The volume occupied by the ice is larger than the volume of the glass bottle. The glass bottle would break.

1.109   a.   **Strategy:**   The threshold level is given in units of micrograms per deciliter and the given units are grams per liter. Use conversion factors to convert g $\rightarrow \mu$g and L $\rightarrow$ dL.

   Setup: Use the conversion factors:

$$\frac{1 \times 10^6 \ \mu\text{g}}{1 \text{ g}} \text{ and } \frac{1 \text{ L}}{10 \text{ dL}}$$

   Solution:

$$\frac{3.0 \times 10^{-4} \text{ g}}{1 \text{ L}} \times \frac{1 \times 10^6 \ \mu\text{g}}{1 \text{ g}} \times \frac{1 \text{ L}}{10 \text{ dL}} = 30 \ \mu\text{g} / \text{dL}$$

   Yes, 30 μg/dL would exceed the CDC's threshold level of 10 μg/dL.

   b.   **Strategy:**   The threshold level is given in units of micrograms per deciliter and the given units are milligrams per milliliter. Use conversion factors to convert mg $\rightarrow \mu$g and mL $\rightarrow$ dL.

   Setup: Use the conversion factors:

$$\frac{1 \times 10^3 \ \mu\text{g}}{1 \text{ mg}} \text{ and } \frac{1 \times 10^2 \text{ mL}}{1 \text{ dL}}$$

   Solution:

$$\frac{2.0 \times 10^{-5} \text{ mg}}{\text{mL}} \times \frac{1 \times 10^3 \ \mu\text{g}}{1 \text{ mg}} \times \frac{1 \times 10^2 \text{ mL}}{1 \text{ dL}} = 2.0 \ \mu\text{g} / \text{dL}$$

   No, 2.0 μg/dL would not exceed the CDC's threshold level of 10 μg/dL.

   c.   **Strategy:**   The threshold level is given in units of micrograms per deciliter. The given units are grams per cubic centimeter. Since 1 g/cm^3 = 1 g/mL, use conversion factors to convert g $\rightarrow \mu$g and mL $\rightarrow$ dL.

   Setup: Use the conversion factors:

$$\frac{1 \times 10^6 \ \mu\text{g}}{1 \text{ g}} \text{ and } \frac{1 \times 10^2 \text{ mL}}{1 \text{ dL}}$$

   Solution:

$$6.5 \times 10^{-8} \text{ g/cm}^3 = 6.5 \times 10^{-8} \text{ g/mL}$$

$$\frac{6.5 \times 10^{-8} \text{ mg}}{\text{mL}} \times \frac{1 \times 10^6 \ \mu\text{g}}{1 \text{ g}} \times \frac{1 \times 10^2 \text{ mL}}{1 \text{ dL}} = 6.5 \ \mu\text{g} / \text{dL}$$

   No, 6.5 μg/dL would not exceed the CDC's threshold level of 10 μg/dL.

1.111   a. **Strategy:**   Substitute the temperature values given in the problem into the appropriate equation.

Setup: Conversion from °F to °C:

$$°C = (°F - 32°F) \times \frac{5°C}{9°F}$$

Conversion from °C to K:

$$K = °C + 273.15$$

Solution:

$$°C = (980°F - 32°F) \times \frac{5°C}{9°F} = \mathbf{527°C}$$

$$°C = (2240°F - 32°F) \times \frac{5°C}{9°F} = \mathbf{1227°C}$$

$$K = 527°C + 273 = \mathbf{800\ K}$$

$$K = 1227°C + 273 = \mathbf{1500\ K}$$

The range of 700°C is equal to a range of 700 K.

Note that when there are no digits to the right of the decimal point in the original temperature, we use 273 instead of 273.15

b. **Strategy:** We know the temperature change takes place over the course of 6 h. To determine the rate of temperature change per second, we need to use conversion factors to convert h → min → s. The rate is then equal to the value of the temperature change divided by the number of seconds.

Setup:

$$\text{Rate of temperature change per second} = \frac{\text{change in temperature}}{\text{number of seconds}}$$

Use the conversion factors:

$$\frac{1\ h}{60\ min} \quad \text{and} \quad \frac{1\ min}{60\ s}$$

Solution: Fahrenheit:

$$2240°F - 980°F = 1260°F$$

$$\frac{1260°F}{6\ h} \times \frac{1\ h}{60\ min} \times \frac{1\ min}{60\ s} = \mathbf{5.833 \times 10^{-2}°F/s}$$

Celsius:

$$1227°C - 527°C = 700°C$$

$$\frac{700°C}{6\ h} \times \frac{1\ h}{60\ min} \times \frac{1\ min}{60\ s} = \mathbf{3.241 \times 10^{-2}°C/s}$$

Kelvin:

$$1500\ K - 800\ K = 700\ K$$

$$\frac{700°K}{6\,h} \times \frac{1\,hr}{60\,min} \times \frac{1\,min}{60\,s} = \mathbf{3.241 \times 10^{-2}°K/s}$$

1.113 **a. Strategy:** We are asked to determine when °C = °F. Since both °C and °F are used in the conversion factor, we can replace both °C and °F with a common variable, such as °C. Solving the algebraic equation for °C will yield the temperature at which the values are numerically equal.

Setup: Conversion from °C to °F:

$$°F = \left(°C \times \frac{9°F}{5°C}\right) + 32°F$$

Solution: °C = °F

Replacing °F in the equation with °C yields:

$$°C = \left(°C \times \frac{9}{5}\right) + 32 \quad \text{or} \quad °C = \frac{9}{5}°C + 32$$

Combine like terms to yield:

$$-\frac{4}{5}°C = 32$$

Solving for °C gives **–40°**.

b. Strategy: We are asked to determine when °F = K. If °F = K, we can set the conversion factors equal to one another. Then, we solve for °C which is the variable common to both equations. This value, converted to both °F and K, will yield the value at which both scales are numerically equivalent.

Setup: Conversion from °C to °F:

$$°F = \left(°C \times \frac{9°F}{5°C}\right) + 32°F$$

Conversion from °C to K:

$$K = °C + 273.15$$

Solution: If we set °F = K, then:

$$\left(°C \times \frac{9}{5}\right) + 32 = °C + 273.15$$

Solve the equation for °C:

$$1.8\,°C + 32 = °C + 273.15$$

$$0.8\,°C = 241.15$$

$$°C = 301.44$$

Use this value of °C to solve for K or °F:

$$K = 301.44°C + 273.15 = \textbf{574.59 K}$$

$$°F = \left(301.44°C \times \frac{9°F}{5°C}\right) + 32°F = \textbf{574.59°F}$$

c. **Strategy:** Use the equation which shows the conversion between °C and K.

Setup: Conversion from °C to K:
$$K = °C + 273.15$$

Solution: **No. Since the value of Kelvin is always equal to a value that is 273.15 greater than the Celsius value, there is no temperature at which the values can be numerically equal.**

1.115 a. **Strategy:** Substitute the temperature values given in the problem into the equation for converting °C to °F.

Setup: Conversion from °F to °C:
$$°C = (°F - 32°F) \times \frac{5°C}{9°F}$$

Solution:
$$°C = (68°F - 32°F) \times \frac{5°C}{9°F} = 20°C$$

$$°C = (77°F - 32°F) \times \frac{5°C}{9°F} = 25°C$$

Storage temperature range: **20°C to 25°C**

b. **Strategy:** Use the density equation,
$$d = \frac{m}{V}$$

Solution:
$$d = \frac{m}{V} = \frac{5.89 \text{ g}}{5.00 \text{ mL}} = \textbf{1.18 g / mL}$$

c. **Strategy:** The density was determined in g/mL. We are asked to convert this value to g/L and kg/m^3. For g/L, use conversion factors to convert mL $\rightarrow$ L. For kg/m^3, use conversion factors to convert g/mL $\rightarrow$ kg/m^3.

Setup: For mL $\rightarrow$ L, use the conversion factor:
$$\frac{1 \times 10^3 \text{ mL}}{1 \text{ L}}$$

For g/mL $\rightarrow$ kg/m^3:
$$1 \text{ g/mL} = 1000 \text{ kg/m}^3$$

Solution:

$$\frac{1.18 \text{ g}}{\text{mL}} \times \frac{1 \times 10^3 \text{ mL}}{\text{L}} = \mathbf{1180 \text{ g} / \text{L}}$$

$$1.18 \text{ g/mL} \times \frac{1000 \text{ kg/m}^3}{1 \text{ g/mL}} = \mathbf{1180 \text{ kg} / \text{m}^3}$$

d. **Strategy:** The dosage is given in mg per kg of body weight. The given body weight is in pounds so conversion factors should be used to convert lb → g → kg.

Setup: Use the conversion factors:

$$\frac{453.6 \text{ g}}{1 \text{ lb}} \quad \text{and} \quad \frac{1 \text{ kg}}{1 \times 10^3 \text{ g}}$$

Solution:

$$185 \text{ lb} \times \frac{453.6 \text{ g}}{1 \text{ lb}} \times \frac{1 \text{ kg}}{1 \times 10^3 \text{ g}} = 83.92 \text{ kg}$$

$$\frac{5 \text{ mg}}{\text{kg}} \times 83.92 \text{ kg} = \mathbf{420 \text{ mg}}$$

Chapter 2
Atoms and the Periodic Table

Key Skills

2.1   c   2.2   d   2.3   e   2.4   d

Problems

2.9   Note that you are given information to set up the conversion factor relating meters and miles.

$$r_{\textbf{atom}} = 10^4\, r_{\text{nucleus}} = 10^4 \times 1.0 \text{ cm} \times \frac{1 \text{ m}}{100 \text{ cm}} = \textbf{1.0} \times \textbf{10}^2 \text{ m}$$

$$1.0 \times 10^2 \text{ m} \times \frac{1 \text{ mi}}{1609 \text{ m}} = \textbf{6.2} \times \textbf{10}^{-2} \text{ mi}$$

2.15   **Strategy:** The 243 in Pu-243 is the mass number. The mass number (A) is the total number of neutrons and protons present in the nucleus of an atom of an element. The number of protons in the nucleus of an atom is the atomic number (Z). The atomic number (Z) of plutonium is 94 (see inside front cover of the text).

   Setup: mass number (A) = number of protons (Z) + number of neutrons

   Therefore,
              number of neutrons = mass number (A) – number of protons (Z)

   Solution:
              number of neutrons = 243 – 94 = **149**

2.17   **Strategy:** The superscript denotes the mass number (A) and the subscript denotes the atomic number (Z). Since all the atoms are neutral, the number of electrons is equal to the number of protons.

   Setup: Number of protons = Z. Number of neutrons = $A - Z$. Number of electrons = number of protons.

   Solution: $^{17}_{8}\text{O}$: The atomic number is 8, so there are **8 protons**. The mass number is 17, so the number of **neutrons** is 17 – 8 = **9**. The number of electrons equals the number of protons, so there are **8 electrons**.

   $^{29}_{14}\text{Si}$: The atomic number is 14, so there are **14 protons**. The mass number is 29, so the number of **neutrons** is 29 – 14 = **15**. The number of electrons equals the number of protons, so there are **14 electrons**.

   $^{58}_{28}\text{Ni}$: The atomic number is 28, so there are **28 protons**. The mass number is 58, so the number of **neutrons** is 58 – 28 = **30**. The number of electrons equals the number of protons, so there are **28 electrons**.

$^{89}_{39}Y$: The atomic number is 39, so there are **39 protons**. The mass number is 89, so the number of **neutrons** is 89 – 39 = **50**. The number of electrons equals the number of protons, so there are **39 electrons**.

$^{180}_{73}Ta$: The atomic number is 73, so there are **73 protons**. The mass number is 180, so the number of **neutrons** is 180 – 73 = **107**. The number of electrons equals the number of protons, so there are **73 electrons**.

$^{203}_{81}Tl$: The atomic number is 81, so there are **81 protons**. The mass number is 203, so the number of **neutrons** is 203 – 81 = **122**. The number of electrons equals the number of protons, so there are **81 electrons**.

2.19   The superscript denotes the mass number (*A*) and the subscript denotes the atomic number (*Z*).

a. $^{187}_{75}Re$ b. $^{209}_{83}Bi$ c. $^{75}_{33}As$ d. $^{236}_{93}Np$

2.21   **Strategy:**   The mass number (*A*) is the total number of neutrons and protons present in the nucleus of an atom of an element. The number of protons in the nucleus of an atom is the atomic number (*Z*). The atomic number (*Z*) can be found on the periodic table.

Setup: mass number (*A*) = number of protons (*Z*) + number of neutrons

Solution: a. The atomic number of chlorine (Cl) is 17, so there are 17 protons. The mass number is 17 + 18 = **35**.

b. The atomic number of phosphorus (P) is 15, so there are 15 protons. The mass number is 15 + 17 = **32**.

c. The atomic number of antimony (Sb) is 51, so there are 51 protons. The mass number is 51 + 70 = **121**.

d. The atomic number of palladium (Pd) is 46, so there are 46 protons. The mass number is 46 + 59 = **105**.

2.27   The principal factor for determining the stability of a nucleus is the *neutron-to-proton ratio* (n/p). For stable elements of low atomic number, the n/p ratio is close to 1. As the atomic number increases, the n/p ratios of stable nuclei become greater than 1. The following rules are useful in predicting nuclear stability.

1) Nuclei that contain 2, 8, 20, 50, 82, or 126 protons or neutrons are generally more stable than nuclei that do not possess these numbers of particles. These numbers are called *magic numbers*.

2) Nuclei with even numbers of both protons and neutrons are generally more stable than those with odd numbers of these particles (see Table 2.2 of the text).

a. **Lithium-9** should be less stable. The neutron-to-proton (n/p) ratio is too high. For the smaller atoms, the n/p ratio will be close to 1:1.

b. **Sodium-25** is less stable. Its neutron-to-proton ratio is too high.

c. **Scandium-48** is less stable because of odd numbers of protons and neutrons. Calcium-48 has a magic number of both protons (20) and neutrons (28), so we would expect it to be more stable, even though its n/p ratio is higher.

2.29   a. **Neon-17** should be radioactive. It falls below the belt of stability (low n/p ratio).

    b. **Calcium-45** should be stable, because it falls on the belt of stability (n/p ratio 1.25), but its atomic number is located where the slope of the belt of stability changes. Actually this isotope is radioactive.

    c. **Technetium-92.** (All technetium isotopes are radioactive.)

2.35 $(203.973020 \text{ amu})(0.014) + (205.974440 \text{ amu})(0.241)$

$$+ (206.975872 \text{ amu})(0.221) + (207.976627 \text{ amu})(0.524) = \textbf{207.2 amu}$$

2.37   **Strategy:**   Each isotope contributes to the average atomic mass based on its relative abundance. Multiplying the mass of an isotope by its fractional abundance (not percent) will give the contribution to the average atomic mass of that particular isotope.

    It would seem that there are two unknowns in this problem, the fractional abundance of ^{6}Li and the fractional abundance of ^{7}Li. However, these two quantities are not independent of each other; they are related by the fact that they must sum to 1. Start by letting x be the fractional abundance of ^{6}Li. Since the sum of the two fractional abundances must be 1, we can write:

$$(6.0151 \text{ amu})(x) + (7.0160 \text{ amu})(1 - x) = 6.941 \text{ amu}$$

    Solution: Solving for x gives 0.075, which corresponds to the fractional abundance of ^{6}**Li = 7.5%.** The expression $(1 - x)$ has the value 0.925, which corresponds to the fractional abundance of ^{7}**Li = 92.5%.**

2.39   **Strategy:**   Each isotope contributes to the average atomic mass based on its relative abundance. Multiplying the mass of an isotope by its fractional abundance (percent value divided by 100) will give the contribution to the average atomic mass of that particular isotope.

    We are asked to solve for the atomic mass of ^{24}Mg, x, given the contribution to the average atomic mass of ^{25}Mg and ^{26}Mg, and the average atomic mass of magnesium.

    Setup: Each percent abundance must be converted to a fractional abundance: 10.00% to 10.00/100 or 0.1000, 11.01% to 11.01/100 or 0.1101, and 78.99% to 78.99/100 or 0.7899.

    We can then write:

$$(0.1000)(24.9858374 \text{ amu}) + (0.1101)(25.9825937 \text{ amu}) + (0.7899)(x) = 24.3050 \text{ amu}$$

    Solution: Solving for x gives:

$$(0.1000)(24.9858374 \text{ amu}) + (0.1101)(25.9825937 \text{ amu}) + (0.7899)(x) = 24.3050 \text{ amu}$$

$$2.4986 + 2.8607 + 0.7899x = 24.3050$$

$$0.7899x = 18.9457$$

$$x = \textbf{23.98 amu}$$

2.45   a. **Metallic character increases as you progress down a group of the periodic table.** For example, moving down Group 4A, the nonmetal carbon is at the top, and the metal lead is at the bottom of the group.

    b. **Metallic character decreases from the left side of the table** (where the metals are located) **to the right side of the table** (where the nonmetals are located).

2.47   **Na and K** are both Group 1A elements; they should have similar chemical properties. **N and P** are both Group 5A elements; they should have similar chemical properties. **F and Cl** are Group 7A elements; they should have similar chemical properties.

2.49

1A																	8A
	2A											3A	4A	5A	6A	7A	
Na	Mg	3B	4B	5B	6B	7B	—	8B	—	1B	2B			P	S		
							Fe										
															I		

Atomic number 26, iron, Fe (present in hemoglobin for transporting oxygen)
Atomic number 53, iodine, I (present in the thyroid gland)
Atomic number 11, sodium, Na (present in intra- and extracellular fluids)
Atomic number 15, phosphorus, P (present in bones and teeth)
Atomic number 16, sulfur, S (present in proteins)
Atomic number 12, magnesium, Mg (present in chlorophyll molecules)

2.53   **Strategy:**   Determine the diameter of the atoms in micrometers. Then divide the diameter of a human hair by the diameter of the atoms.

           Setup: Use the conversion factor (see Table 1.2):

$$\frac{1 \times 10^{-6} \ \mu m}{1 \ pm}$$

          Solution: First, convert picometers to micrometers:

$$121 \ pm \times \frac{1 \times 10^{-6} \ \mu m}{1 \ pm} = 1.21 \times 10^{-4} \ \mu m$$

            Then, divide the diameter of the human hair by the diameter of the atoms:

$$25.4 \ \mu m \times \frac{1 \ atom}{1.21 \times 10^{-4} \ \mu m} = \mathbf{2.10 \times 10^5 \ atoms}$$

2.55 $(5.00 \times 10^9 \ \text{Ni atoms}) \times \dfrac{1 \ \text{mol Ni}}{6.022 \times 10^{23} \ \text{Ni atoms}} = \mathbf{8.30 \times 10^{-15} \ mol \ Ni}$

2.57 **Strategy:** We are given moles of platinum and asked to solve for grams of platinum. What conversion factor do we need to convert between moles and grams? Arrange the appropriate conversion factor so that moles cancel, and the unit grams is obtained for the answer.

Setup: The conversion factor needed to convert between moles and grams is the molar mass. In the periodic table, we see that the molar mass of Pt is 195.1 g. This can be expressed as:

$$1 \text{ mol Pt} = 195.1 \text{ g Pt}$$

From this equality, we can write two conversion factors:

$$\frac{1 \text{ mol Pt}}{195.1 \text{ g Pt}} \text{ and } \frac{195.1 \text{ g Pt}}{1 \text{ mol Pt}}$$

The conversion factor on the right is the correct one. Moles will cancel, leaving the unit grams for the answer.

Solution: We write:

$$? \text{ g Pt} = 26.4 \text{ mol Pt} \times \frac{195.1 \text{ g Pt}}{1 \text{ mol Pt}} = \mathbf{5.15 \times 10^3 \text{ g Pt}}$$

2.59 a. **Strategy:** We can look up the molar mass of antimony (Sb) on the periodic table (121.8 g/mol). We want to find the mass of a single atom of antimony (unit of g/atom). Therefore, we need to convert from the unit mole in the denominator to the unit atom in the denominator. What conversion factor is needed to convert between moles and atoms? Arrange the appropriate conversion factor so that mole in the denominator cancels, and the unit atom is obtained in the denominator.

Setup: The conversion factor needed is Avogadro's number. We have:

$$1 \text{ mol} = 6.022 \times 10^{23} \text{ particles (atoms)}$$

From this equality, we can write two conversion factors:

$$\frac{1 \text{ mol Sb}}{6.022 \times 10^{23} \text{ Sb atoms}} \text{ and } \frac{6.022 \times 10^{23} \text{ Sb atoms}}{1 \text{ mol Sb}}$$

The conversion factor on the left is the correct one. Moles will cancel, leaving the unit atoms in the denominator of the answer.

Solution: We write:

$$? \text{ g / Sb atom} = \frac{121.8 \text{ g Sb}}{1 \text{ mol Sb}} \times \frac{1 \text{ mol Sb}}{6.022 \times 10^{23} \text{ Sb atoms}} = \mathbf{2.023 \times 10^{-22} \text{ g / Sb atom}}$$

b. Follow the same method as part (a).

$$? \text{ g / Pd atom} = \frac{106.4 \text{ g Pd}}{1 \text{ mol Pd}} \times \frac{1 \text{ mol Pd}}{6.022 \times 10^{23} \text{ Pd atoms}} = \mathbf{1.767 \times 10^{-22} \text{ g / Pd atom}}$$

2.61 **Strategy:** The question asks for atoms of Sc. We cannot convert directly from grams to atoms of scandium. What unit do we need to convert grams of Sc to moles of Sc in order to convert to atoms? What does Avogadro's number represent?

Setup: To calculate the number of Sc atoms, we must first convert grams of Sc to moles of Sc. We use the molar mass of Sc as a conversion factor. Once moles of Sc are obtained, we can use Avogadro's number to convert from moles of scandium to atoms of scandium.

$$1 \text{ mol Sc} = 44.96 \text{ g Sc}$$

The conversion factor needed is:

$$\frac{1 \text{ mol Sc}}{44.96 \text{ g Sc}}$$

Avogadro's number is the key to the second conversion. We have:

$$1 \text{ mol} = 6.022 \times 10^{23} \text{ particles (atoms)}$$

From this equality, we can write two conversion factors.

$$\frac{1 \text{ mol Sc}}{6.022 \times 10^{23} \text{ Sc atoms}} \quad \text{and} \quad \frac{6.022 \times 10^{23} \text{ Sc atoms}}{1 \text{ mol Sc}}$$

The conversion factor on the right is the one we need because it has number of Sc atoms in the numerator, which is the unit we want for the answer.

Solution: Let's complete the two conversions in one step.

$$\text{grams of Sc} \rightarrow \text{moles of Sc} \rightarrow \text{number of Sc atoms}$$

$$? \text{ atoms of Sc} = 4.09 \text{ g Sc} \times \frac{1 \text{ mol Sc}}{44.96 \text{ g Sc}} \times \frac{6.022 \times 10^{23} \text{ Sc atoms}}{1 \text{ mol Sc}} = \mathbf{5.48 \times 10^{22} \text{ Sc atoms}}$$

2.63

$$173 \text{ Au atoms} \times \frac{1 \text{ mol Au}}{6.022 \times 10^{23} \text{ Au atoms}} \times \frac{197.0 \text{ g Au}}{1 \text{ mol Au}} = 5.66 \times 10^{-20} \text{ g Au}$$

$$7.5 \times 10^{-22} \text{ mol Ag} \times \frac{107.9 \text{ g Ag}}{1 \text{ mol Ag}} = 8.1 \times 10^{-20} \text{ g Ag}$$

7.5×10^{-22} mole of silver has a greater mass than 173 atoms of gold.

2.65 **Strategy:** Molar mass of an element is numerically equal to its average atomic mass. Use the molar mass of francium to convert from mass to moles. Then, use Avogadro's constant to convert from moles to atoms.

Setup: The molar mass of francium is 223 g/mol. Once the number of moles is known, we multiply by Avogadro's constant to convert to atoms.

Solution: $$\text{atoms of Fr} = 30 \text{ g Fr} \times \frac{1 \text{ mol Fr}}{223 \text{ g Fr}} \times \frac{6.022 \times 10^{23} \text{ Fr atoms}}{1 \text{ mol Fr}} = \mathbf{8.1 \times 10^{22} \text{ Fr atoms}}$$

2.67   **Strategy:**   The superscript denotes the mass number (A), and the subscript denotes the atomic number (Z). The atomic number (Z) is the number of protons in the nucleus. For atoms that are neutral, the number of electrons is equal to the number of protons.

   Setup: mass number (A) = number of protons (Z) + number of neutrons

   Solution: The atom has 54 electrons. Since the number of electrons equals the number of protons, there are 54 protons. The element with an atomic number (Z) of 54 is xenon (Xe).

   The mass number (A) is 131.

   Therefore **the symbol for this atom is:**

$$^{131}_{54}\text{Xe}$$

2.69   **Strategy:**   The superscript denotes the mass number (A), and the subscript denotes the atomic number (Z). The mass number (A) is the total number of neutrons and protons present in the nucleus of an atom of an element. The atomic number (Z) is the number of protons in the nucleus. The atomic number (Z) can be found on the periodic table. For atoms that are neutral, the number of electrons is equal to the number of protons.

   Setup: mass number (A) = number of protons (Z) + number of neutrons

   Solution: Atom A:

   The atom has 10 electrons. Since the number of electrons equals the number of protons, there are 10 protons.

   The element with an atomic number (Z) of 10 is neon (Ne).

$$\text{mass number } (A) = 10 + 12 = 22$$

   The symbol for **Atom A** is:

$$^{22}_{10}\text{Ne}$$

   Atom B:

   The element with an atomic number (Z) of 75 is rhenium (Re).

$$\text{mass number } (A) = 75 + 110 = 185$$

   The symbol for **Atom B** is:

$$^{185}_{75}\text{Re}$$

   Atom C:

   The element with an atomic number (Z) of 21 is scandium (Sc).

$$\text{mass number } (A) = 21 + 21 = 42$$

The symbol for **Atom C is:**

$$^{42}_{21}Sc$$

Atom D:

Without the number of neutrons, the mass number (A) cannot be determined. **To write a correct symbol for Atom D, the number of neutrons would need to be known.**

2.71　All masses are relative, which means that the mass of every object is compared to the mass of a standard object (such as the piece of metal in Paris called the "standard kilogram"). The mass of the standard object is determined by an international committee, and that mass is an arbitrary number to which everyone in the scientific community agrees.

Atoms are so small that it is hard to compare their masses to the standard kilogram. Instead, we compare atomic masses to the mass of one specific atom. In the nineteenth century, the atom was ^1H, and for a good part of the twentieth century it was ^{16}O. Now it is ^{12}C, which **establishes a standard mass unit that permits the measurement of masses of all other isotopes relative to ^{12}C.**

2.73
a. Isotope	^4_2He	$^{20}_{10}Ne$	$^{40}_{18}Ar$	$^{84}_{36}Kr$	$^{132}_{54}Xe$
No. Protons	2	10	18	36	54
No. Neutrons	2	10	22	48	78
b. **neutron/proton ratio**	1.00	1.00	1.22	1.33	1.44

The neutron/proton ratio increases with increasing atomic number.

2.75　**Strategy:**　Each isotope contributes to the average atomic mass based on its relative abundance. Multiplying the mass of each isotope by its fractional abundance (percent value divided by 100) will give its contribution to the average atomic mass.

Setup:　Each percent abundance must be converted to a fractional abundance: 37.3% to 37.3/100 or 0.373 and 62.7% to 62.7/100 or 0.627. Once we find the contribution to the average atomic mass for each isotope, we can then add the contributions together to obtain the average atomic mass.

Solution:　$(0.373)(190.960584 \text{ amu}) + (0.627)(192.962917 \text{ amu}) = $ **192 amu**

This is the atomic mass that appears in the periodic table.

2.77　a. **iodine**　　　b. **radon**　　　c. **selenium**　　　d. **sodium**　　　e. **lead**

2.79　The mass number (A) is given. The atomic number (Z) is found in the periodic table. The problem is to find the number of neutrons, which is $A - Z$.

a.　^{40}Mg: $40 - 12 = $ **28** neutrons

c.　^{48}Ca: $48 - 20 = $ **28** neutrons

b.　^{44}Si: $44 - 14 = $ **30** neutrons

d.　^{43}Al: $43 - 13 = $ **30** neutrons

2.81 The mass number (A) is the total number of neutrons and protons present in the nucleus of an atom of an element. The number of protons in the nucleus of an atom is the atomic number (Z). The atomic number (Z) can be found on the periodic table. The superscript denotes the mass number (A) and the subscript denotes the atomic number (Z).

$$\text{mass number } (A) = \text{number of protons } (Z) + \text{number of neutrons}$$

Symbol	^{101}Ru	^{181}Ta	^{150}Sm
Protons	**44**	**73**	62
Neutrons	**57**	108	88
Electrons	**44**	73	**62**

2.83 a. The following strategy can be used to convert from the volume of the Pt cube to the number of Pt atoms.

$$\text{cm}^3 \rightarrow \text{grams} \rightarrow \text{atoms}$$

$$1.0 \text{ cm}^3 \times \frac{21.45 \text{ g Pt}}{1 \text{ cm}^3} \times \frac{1 \text{ atom Pt}}{3.240 \times 10^{-22} \text{ g Pt}} = \mathbf{6.6 \times 10^{22} \text{ Pt atoms}}$$

b. Since 74% of the available space is taken up by Pt atoms, 6.6×10^{22} atoms occupy the following volume:

$$0.74 \times 1.0 \text{ cm}^3 = 0.74 \text{ cm}^3$$

We are trying to calculate the radius of a single Pt atom, so we need the volume occupied by a single Pt atom.

$$\text{Volume Pt atom} = \frac{0.74 \text{ cm}^3}{6.6 \times 10^{22} \text{ Pt atoms}} = 1.1 \times 10^{-23} \text{ cm}^3/\text{Pt atom}$$

The volume of a sphere is $\frac{4}{3}\pi r^3$. Solving for the radius:

$$V = 1.1 \times 10^{-23} \text{ cm}^3 = \frac{4}{3}\pi r^3$$
$$r^3 = 2.6 \times 10^{-24} \text{ cm}^3$$
$$r = 1.4 \times 10^{-8} \text{ cm}$$

Converting to picometers:

$$\text{radius Pt atom} = 1.4 \times 10^{-8} \text{ cm} \times \frac{0.01 \text{ m}}{1 \text{ cm}} \times \frac{1 \text{ pm}}{1 \times 10^{-12} \text{ m}} = \mathbf{1.4 \times 10^2 \text{ pm}}$$

Chapter 3
Quantum Theory and the Electronic Structure of Atoms

Visualizing Chemistry

VC 3.1 b   VC 3.2 a   VC 3.3 c   VC 3.4 b

Key Skills

3.1   b   3.2   d   3.3   c   3.4   d

Problems

3.5   **Strategy:** Use Equation 3.1 $\left(E_k = \frac{1}{2}mu^2 \right)$ to calculate the kinetic energy. Note that for units to cancel properly, giving E_k in *joules*, the mass must be expressed in *kilograms* and the velocity in *meters per second*.

a.   **Solution:**

$$E_k = \frac{1}{2}mu^2$$

$$E_k = \frac{1}{2}(7.5 \text{ kg})(7.9 \text{ m/s})^2$$

$$E_k = 2.3 \times 10^2 \text{ kg·m}^2/\text{s}^2 = \mathbf{2.3 \times 10^2 \text{ J}}$$

b.   **Setup:** The mass of the car is 3250 lb. Its mass in kilograms is:

$$3250 \text{ lb} \times \frac{453.6 \text{ g}}{1 \text{ lb}} \times \frac{1 \text{ kg}}{1 \times 10^3 \text{ g}} = 1474 \text{ kg}$$

The velocity of the car is 55 mph. Its velocity in meters per second is:

$$\frac{55 \text{ mi}}{1 \text{ h}} \times \frac{1.61 \text{ km}}{1 \text{ mi}} \times \frac{1 \times 10^3 \text{ m}}{1 \text{ km}} \times \frac{1 \text{ h}}{60 \text{ min}} \times \frac{1 \text{ min}}{60 \text{ s}} = 24.6 \text{ m/s}$$

Solution:

$$E_k = \frac{1}{2}mu^2$$

$$E_k = \frac{1}{2}(1474 \text{ kg})(24.6 \text{ m/s})^2$$

$$E_k = 4.5 \times 10^5 \text{ kg·m}^2/\text{s}^2 = \mathbf{4.5 \times 10^5 \text{ J}}$$

c.　　**Setup:**　The mass of an electron is 9.10938×10^{-28} g (Table 2.1). The mass in kilograms is:

$$9.10938 \times 10^{-28} \text{ g} \times \frac{1 \text{ kg}}{1 \times 10^3 \text{ g}} = 9.10938 \times 10^{-31} \text{ kg}$$

　　　Solution:

$$E_k = \frac{1}{2} mu^2$$

$$E_k = \frac{1}{2}(9.10938 \times 10^{-31} \text{ kg})(315 \text{ m/s})^2$$

$$E_k = 4.52 \times 10^{-26} \text{ kg·m}^2/\text{s}^2 = \textbf{4.52} \times \textbf{10}^{-26} \textbf{ J}$$

d.　　**Setup:**　The mass of a helium atom is 4.003 amu. The factor for conversion of amu → g is 1.661×10^{-24} g/1 amu (Section 1.4). Therefore, the mass of a helium atom in kilograms is:

$$4.003 \text{ amu} \times \frac{1.661 \times 10^{-24} \text{g}}{1 \text{ amu}} \times \frac{1 \text{ kg}}{1 \times 10^3 \text{g}} = 6.649 \times 10^{-27} \text{ kg}$$

　　　Solution:

$$E_k = \frac{1}{2} mu^2$$

$$E_k = \frac{1}{2}(6.649 \times 10^{-27} \text{ kg})(275 \text{ m/s})^2$$

$$E_k = 2.51 \times 10^{-22} \text{ kg·m}^2/\text{s}^2 = \textbf{2.51} \times \textbf{10}^{-22} \textbf{ J}$$

3.7　　a. **Strategy:** Use Equation 3.1 $\left(E_k = \frac{1}{2} mu^2 \right)$ to solve for velocity. Note that for units to cancel properly, giving velocity in *meters per second*, the mass must be expressed in *kilograms*.

　　Setup:　Solving Equation 3.1 for velocity gives:

$$u = \sqrt{\frac{2E_k}{m}}$$

The mass of a Ne atom is 20.18 amu. The factor for conversion of amu → g is 1.661×10^{-24} g/1 amu (Section 1.4). Therefore, the mass of a Ne atom in kilograms is:

$$20.18 \text{ amu} \times \frac{1.661 \times 10^{-24} \text{ g}}{1 \text{ amu}} \times \frac{1 \text{ kg}}{1 \times 10^3 \text{ g}} = 3.352 \times 10^{-26} \text{ kg}$$

Note that 1.12×10^{-20} J $= 1.12 \times 10^{-20}$ kg·m²/s² for the purpose of making the unit cancellation obvious.

　　Solution:

$$u = \sqrt{\frac{2E_k}{m}}$$

$$u = \sqrt{\frac{2 \times (1.12 \times 10^{-20} \text{ kg·m}^2/\text{s}^2)}{3.352 \times 10^{-26} \text{ kg}}}$$

$$u = \textbf{817 m/s}$$

b. Strategy: Use Equation 3.1 $\left(E_k = \dfrac{1}{2}mu^2 \right)$ to solve for velocity. Note that for units to cancel properly, giving velocity in *meters per second*, the mass must be expressed in *kilograms*.

Setup: Solving Equation 3.1 for velocity gives:

$$u = \sqrt{\dfrac{2E_k}{m}}$$

The mass of a Kr atom is 83.80 amu. The factor for conversion of amu $\to$ g is 1.661×10^{-24} g/1 amu (Section 1.4). Therefore, the mass of a Kr atom in kilograms is:

$$83.80 \text{ amu} \times \dfrac{1.661 \times 10^{-24}\text{g}}{1 \text{ amu}} \times \dfrac{1 \text{ kg}}{1 \times 10^3 \text{g}} = 1.392 \times 10^{-25} \text{ kg}$$

Note that 1.12×10^{-20} J $= 1.12 \times 10^{-20}$ kg·m²/s² for the purpose of making the unit cancellation obvious.

Solution:

$$u = \sqrt{\dfrac{2E_k}{m}}$$

$$u = \sqrt{\dfrac{2 \times (1.12 \times 10^{-20} \text{ kg} \cdot \text{m}^2/\text{s}^2)}{1.392 \times 10^{-25} \text{ kg}}}$$

$$u = \mathbf{401 \ m/s}$$

c. Strategy: Use Equation 3.1 $\left(E_k = \dfrac{1}{2}mu^2 \right)$ to solve for mass. The mass of the atom will be expressed in kilograms. In order to determine the identity of the atom, the mass must be converted to amu.

Setup: Solving Equation 3.1 for mass gives:

$$m = \dfrac{2E_k}{u^2}$$

Note that 3.704×10^{-21} J $= 3.704 \times 10^{-21}$ kg·m²/s² for the purpose of making the unit cancellation obvious.

To convert kg $\to$ amu, use the conversion factors:

$$\dfrac{1 \times 10^3 \text{ g}}{1 \text{ kg}} \quad \text{and} \quad \dfrac{1 \text{ amu}}{1.661 \times 10^{-24} \text{ g}}$$

Solution:

$$m = \dfrac{2E_k}{u^2}$$

$$m = \dfrac{2 \times (3.704 \times 10^{-21} \text{ kg·m}^2/\text{s}^2)}{(315 \text{ m/s})^2}$$

$$m = 7.466 \times 10^{-26} \text{ kg}$$

$$7.466 \times 10^{-26} \text{ kg} \times \frac{1 \times 10^3 \text{ g}}{1 \text{ kg}} \times \frac{1 \text{ amu}}{1.661 \times 10^{-24} \text{ g}} = \textbf{44.9 amu}$$

The atom with a mass of 44.9 amu is **scandium.**

3.9 a. **Strategy:** Use Equation 3.2 $\left(E_{el} \propto \dfrac{Q_1 Q_2}{d} \right)$ to compare the magnitudes of the two E_{el} values. Because

the distance between the charges is the same in both cases, we can solve for the ratio of E_{el} values without actually knowing the distance. Both the distance and the proportionality constant cancel in the solution.

Setup:

$$E_{el(+3,-3)} = c \frac{Q_1 Q_2}{d}, \quad Q_1 = +3 \text{ and } Q_2 = -3$$

$$E_{el(+2,-3)} = c \frac{Q_1 Q_2}{d}, \quad Q_1 = +2 \text{ and } Q_2 = -3$$

Solution:

$$\frac{c \times \left(\dfrac{3 \times (-3)}{d} \right)}{c \times \left(\dfrac{2 \times (-3)}{d} \right)} = 1.5$$

The electrostatic energy between charges of +3 and –3 is **1.5 times as large** as the electrostatic energy between charges of +2 and –3.

b. **Strategy:** Use Equation 3.2 $\left(E_{el} \propto \dfrac{Q_1 Q_2}{d} \right)$ to compare the magnitudes of the two d values. Because

the E_{el} value is the same in both the cases, we can solve for the ratio of d values without actually knowing the E_{el} or d by equating the two equations. The proportionality constant cancels in the solution.

Setup: Charges +3 and –3 are separated by a distance of d and charges +2 and –3 are separated by an unknown distance, d_x.

For charges +3 and –3:

$$E_{el(+3,-3)} = c \frac{Q_1 Q_2}{d}, \quad Q_1 = +3 \text{ and } Q_2 = -3$$

For charges +2 and –3:

$$E_{el(+2,-3)} = c \frac{Q_1 Q_2}{d_x}, \quad Q_1 = +2 \text{ and } Q_2 = -3$$

Since $E_{el(+3,-3)} = E_{el(+2,-3)}$:

Solution:

$$c\frac{Q_1 Q_2}{d} = c\frac{Q_1 Q_2}{d_x}$$

$$c\left(\frac{3\times(-3)}{d}\right) = c\left(\frac{2\times(-3)}{d_x}\right)$$

$$\frac{-9}{d} = \frac{-6}{d_x}$$

$$-9d_x = -6d$$

$$d_x = \frac{-6}{-9}d = \frac{2}{3}d$$

Charges of +2 and –3 separated by a distance of $\frac{2}{3}d$ would have the same electrostatic energy as charges of +3 and –3 separated by a distance of d.

3.15 **a. Setup:** We are given the frequency of an electromagnetic wave and asked to calculate the wavelength. Rearranging Equation 3.3 of the text to solve for wavelength gives:

$$\lambda = \frac{c}{\nu}$$

Solution: Substituting the frequency and the speed of light (3.00×10^8 m/s) into the above equation, the wavelength is:

$$\lambda = \frac{c}{\nu} = \frac{3.00\times10^8 \text{ m/s}}{8.6\times10^{13}\text{ /s}} = 3.5\times10^{-6}\text{ m} = \mathbf{3.5 \times 10^3 \text{ nm}}$$

b. Setup: We are given the wavelength of an electromagnetic wave and asked to calculate the frequency. Rearranging Equation 3.3 of the text to solve for frequency gives:

$$\nu = \frac{c}{\lambda}$$

Solution: Because the speed of light is given in meters per second, it is convenient to first convert wavelength to units of meters. Recall that $1 \text{ nm} = 1 \times 10^{-9}$ m (see Table 1.3 of the text). We write:

$$566 \text{ nm}\times\frac{1\times10^{-9}\text{ m}}{1 \text{ nm}} = 566\times10^{-9}\text{ m or } 5.66\times10^{-7}\text{ m}$$

Substituting in the wavelength and the speed of light (3.00×10^8 m/s), the frequency is:

$$\nu = \frac{c}{\lambda} = \frac{3.00\times10^8 \text{ m/s}}{566\times10^{-9}\text{ m}} = 5.30\times10^{14}\text{ /s} = \mathbf{5.30\times10^{14} \text{ Hz}}$$

3.17

$$\lambda = \frac{c}{\nu} = \frac{3.00\times10^8 \text{ m/s}}{9,192,631,770 \text{ s}^{-1}} = 3.26\times10^{-2}\text{ m} = \mathbf{3.26\times10^7 \text{ nm}}$$

This radiation falls in the **microwave region** of the spectrum. (See Figure 3.1 of the text.)

3.19   A radio wave is an electromagnetic wave that travels at the speed of light. The speed of light is in units of m/s, so convert distance from units of miles to meters. $\left(28 \text{ million mi} = 2.8 \times 10^7 \text{ mi}\right)$

$$? \text{ distance (m)} = \left(2.8 \times 10^7 \text{ mi}\right) \times \frac{1.61 \text{ km}}{1 \text{ mi}} \times \frac{1000 \text{ m}}{1 \text{ km}} = 4.5 \times 10^{10} \text{ m}$$

Now, we can use the speed of light as a conversion factor to convert from meters to minutes $(c = 3.00 \times 10^8 \text{ m/s})$.

$$? \min = \left(4.5 \times 10^{10} \text{ m}\right) \times \frac{1 \text{ s}}{3.00 \times 10^8 \text{ m}} = 1.5 \times 10^2 \text{ s} = \textbf{2.5 min}$$

3.25   **Setup:**   We are given the wavelength of photon and asked to calculate the energy. Rearranging Equation 3.3 of the text to solve for frequency gives:

$$v = \frac{c}{\lambda}$$

Solution: Substituting this into Equation 3.4, the energy is:

$$E = hv = \frac{hc}{\lambda} = \frac{\left(6.63 \times 10^{-34} \text{ J} \cdot \text{s}\right)\left(3.00 \times 10^8 \text{ m/s}\right)}{7.05 \times 10^{-7} \text{ m}} = \textbf{2.82} \times \textbf{10}^{-19} \textbf{ J}$$

3.27   a. Rearranging Equation 3.3 to solve for wavelength:

$$\lambda = \frac{c}{v} = \frac{3.00 \times 10^8 \text{ m/s}}{6.5 \times 10^9 \text{ s}^{-1}} = 0.046 \text{ m} = \textbf{4.6} \times \textbf{10}^7 \textbf{ nm}$$

No, the radiation does not fall in the visible region; it is microwave radiation (see Figure 3.1 of the text).

b. Using Equation 3.4, the energy of each photon is:

$$E = hv = \left(6.63 \times 10^{-34} \text{ J} \cdot \text{s}\right)\left(6.5 \times 10^9 \text{ s}^{-1}\right) = \textbf{4.3} \times \textbf{10}^{-24} \textbf{ J} \text{ per photon}$$

c. Using Avogadro's number to convert to J/mol:

$$E = \frac{4.3 \times 10^{-24} \text{ J}}{1 \text{ photon}} \times \frac{6.022 \times 10^{23} \text{ photons}}{1 \text{ mol}} = \textbf{2.6 J/mol}$$

3.29   **Strategy:**   We are given two wavelengths of light. Use Equation 3.3 ($c = \lambda v$) to convert wavelength to frequency; then use Equation 3.4 ($E = hv$) to determine the energy of the photon for each wavelength. The difference in energy between the photons is the absolute value of the difference between the energy of the red photon and the energy of the blue photon.

Setup: The wavelengths must be converted from nanometers to meters:

$$680 \text{ nm} \times \frac{1 \times 10^{-9} \text{ m}}{1 \text{ nm}} = 6.80 \times 10^{-7} \text{ m}$$

$$442 \text{ nm} \times \frac{1 \times 10^{-9} \text{ m}}{1 \text{ nm}} = 4.42 \times 10^{-7} \text{ m}$$

Planck's constant, h, is 6.63×10^{-34} J·s.

Solution: For $\lambda = 6.80 \times 10^{-7}$ m (red):

$$v = \frac{c}{\lambda} = \frac{3.00 \times 10^8 \text{ m/s}}{6.80 \times 10^{-7} \text{ m}} = 4.412 \times 10^{14} \text{ s}^{-1}$$

$$E = hv = \left(6.63 \times 10^{-34} \text{ J} \cdot \text{s}\right)\left(4.412 \times 10^{14} \text{ s}^{-1}\right) = \mathbf{2.93 \times 10^{-19} \text{ J}}$$

For $\lambda = 4.42 \times 10^{-7}$ m (blue):

$$v = \frac{c}{\lambda} = \frac{3.00 \times 10^8 \text{ m/s}}{4.42 \times 10^{-7} \text{ m}} = 6.787 \times 10^{14} \text{ s}^{-1}$$

$$E = hv = \left(6.63 \times 10^{-34} \text{ J} \cdot \text{s}\right)\left(6.787 \times 10^{14} \text{ s}^{-1}\right) = \mathbf{4.50 \times 10^{-19} \text{ J}}$$

The energy difference, ΔE, is:

$$\Delta E = \left| E_{\text{(red)}} - E_{\text{(blue)}} \right| = \left| \left(2.93 \times 10^{-19} \text{ J}\right) - \left(4.50 \times 10^{-19} \text{J}\right) \right| = \mathbf{1.57 \times 10^{-19} \text{ J}}$$

3.31   We can rearrange Equation 3.3 to solve for frequency,

$$v = \frac{c}{\lambda}$$

and substitute the result into Equation 3.4 to solve for the energy of a photon.

$$E = hv = \frac{hc}{\lambda} = \frac{\left(6.63 \times 10^{-34} \text{ J} \cdot \text{s}\right)\left(3.00 \times 10^8 \text{ m/s}\right)}{\left(0.154 \times 10^{-9} \text{ m}\right)} = \mathbf{1.29 \times 10^{-15} \text{ J}}$$

3.33   **Infrared photons have insufficient energy to cause the chemical changes.**

3.35   We are given the energy of a photon (gamma particle) in J/mol. We begin by using Avogadro's number to convert to J/photon:

$$E = \frac{1.29 \times 10^{11} \text{ J}}{1 \text{ mol}} \times \frac{1 \text{ mol}}{6.022 \times 10^{23} \text{ photons}} = 2.14 \times 10^{-13} \text{ J / photon}$$

Rearranging Equation 3.4 to solve for frequency, we can convert J/photon to frequency:

$$v = \frac{E}{h} = \frac{2.14 \times 10^{-13} \text{ J}}{6.63 \times 10^{-34} \text{ J} \cdot \text{s}} = \mathbf{3.23 \times 10^{20} \text{ s}^{-1}}$$

Finally, using Equation 3.3, we can solve for wavelength:

$$\lambda = \frac{c}{v} = \frac{3.00\times10^8 \text{ m/s}}{3.23\times10^{20}\text{ s}^{-1}} = 9.29\times10^{-13}\text{ m}$$

3.37 **Strategy:** Use Equation 3.5 ($hv = \text{KE} + W$) to determine the kinetic energy of the ejected electron. The binding energy, W, is 5.86×10^{-19} J, according to Problem 3.36.

Setup: Solving Equation 3.5 for KE gives:
$$\text{KE} = hv - W$$
Planck's constant, h, is 6.63×10^{-34} J·s

Solution:
$$\text{KE} = \left(6.63\times10^{-34}\text{ J·s}\right)\left(2.00\times10^{15}\text{ s}^{-1}\right) - 5.86\times10^{-19}\text{ J} = \mathbf{7.40\times10^{-19}\text{ J}}$$

3.39 A "blue" photon (shorter wavelength) has higher energy than a "yellow" photon. For the same amount of energy delivered to the metal surface, there must be fewer "blue" photons than "yellow" photons. Thus, the **yellow light would eject more electrons** since there are more "yellow" photons. Since the "blue" photons are of higher energy, **blue light will eject electrons with greater kinetic energy**.

3.45 Note that we use more significant figures than we usually do for the values of h and c for this problem.

$$E = \frac{hc}{\lambda} = \frac{\left(6.6256\times10^{-34}\text{ J·s}\right)\left(2.998\times10^8\text{ m/s}\right)}{656.3\times10^{-9}\text{ m}} = \mathbf{3.027\times10^{-19}\text{ J}}$$

3.47 **Strategy:** We are given the initial and final states in the emission process. We can calculate the energy of the emitted photon using Equation 3.8 of the text. Then, from this energy, we can solve for the frequency of the photon, and from the frequency, we can solve for the wavelength.

Solution: From Equation 3.8 we write:

$$\Delta E = -2.18\times10^{-18}\text{ J}\left(\frac{1}{n_f^2} - \frac{1}{n_i^2}\right) = -2.18\times10^{-18}\text{ J}\left(\frac{1}{3^2} - \frac{1}{4^2}\right) = -1.06\times10^{-19}\text{ J}$$

The negative sign for ΔE indicates that this is energy associated with an emission process. To calculate the frequency, we will omit the minus sign for ΔE because the frequency of the photon must be positive. We know that:

$$\Delta E = hv$$

Rearranging the equation and substituting in the known values:

$$v = \frac{\Delta E}{h} = \frac{1.06\times10^{-19}\text{ J}}{6.63\times10^{-34}\text{ J·s}} = 1.60\times10^{14}\text{ s}^{-1} \text{ or } \mathbf{1.60\times10^{14}\text{ Hz}}$$

We also know that $\lambda = \dfrac{c}{v}$. Substituting the frequency calculated above into this equation gives:

$$\lambda = \frac{3.00\times10^8\text{ m/s}}{1.60\times10^{14}\text{ s}^{-1}} = 1.88\times10^{-6}\text{ m} = \mathbf{1.88\times10^3\text{ nm}}$$

This wavelength is in the infrared region of the electromagnetic spectrum (see Figure 3.1 of the text).

3.49

$$\Delta E = -2.18 \times 10^{-18} \text{ J} \left(\frac{1}{2^2} - \frac{1}{n_i^2} \right)$$

n_f is given in the problem, but we need to calculate ΔE. The photon energy is:

$$E = \frac{hc}{\lambda} = \frac{\left(6.63 \times 10^{-34} \text{ J} \cdot \text{s} \right) \left(3.00 \times 10^8 \text{ m/s} \right)}{434 \times 10^{-9} \text{ m}} = 4.58 \times 10^{-19} \text{ J}$$

Since this is an emission process, the energy change ΔE must be negative or -4.58×10^{-19} J. Substitute ΔE into the following equation, and solve for n_i.

$$-4.58 \times 10^{-19} \text{ J} = -2.18 \times 10^{-18} \text{ J} \left(\frac{1}{2^2} - \frac{1}{n_i^2} \right)$$

$$\frac{1}{n_i^2} = \left(\frac{-4.58 \times 10^{-19} \text{ J}}{2.18 \times 10^{-18} \text{ J}} \right) + \frac{1}{2^2} = -0.210 + 0.250 = 0.040$$

$$n_i = \frac{1}{\sqrt{0.040}} = \mathbf{5}$$

3.51 **Analyze the emitted light by passing it through a slit and a prism, or a spectroscope. The number of emission lines in the spectrum is the number of wavelengths in the light.**

3.53 **Excited atoms of the chemical elements emit the same characteristic frequencies or lines in a terrestrial laboratory, in the sun, or in a star many light-years distant from earth.** Therefore, if the same spectrum is seen in the laboratory and a distant star, then the same element is in the laboratory and the star.

3.57 **Strategy:** We are given the mass and the speed of the neutron and asked to calculate the wavelength. We need the de Broglie equation, which is Equation 3.11 of the text. Note that because the unit of Planck's constant is J·s, m must be in kg and u in m/s (1 J = 1 kg·m^2/s^2).

Solution: Using Equation 3.11 we write:

$$\lambda = \frac{h}{mu}$$

$$\lambda = \frac{h}{mu} = \frac{6.63 \times 10^{-34} \text{ J} \cdot \text{s}}{\left(1.675 \times 10^{-27} \text{ kg} \right) \left(7.00 \times 10^2 \text{ m/s} \right)} = 5.65 \times 10^{-10} \text{ m} = \mathbf{0.565 \text{ nm}}$$

3.59 **Strategy:** We are given the mass and the speed of the honeybee and asked to calculate the de Broglie wavelength. We need the de Broglie equation, which is Equation 3.11 of the text. Note that because the unit of Planck's constant is J·s, m must be in kg and u in m/s(1 J = 1 kg·m^2/s^2).

Solution: Because mass in this problem is given in g and speed is given in mph, we must first convert these to kg and m/s, respectively.

$$m = 8.45 \text{ g} \times \frac{1 \text{ kg}}{1000 \text{ g}} = 0.00845 \text{ kg}$$

$$u = \frac{6.28 \text{ mi}}{1 \text{ h}} \times \frac{1.61 \text{ km}}{1 \text{ mi}} \times \frac{1000 \text{ m}}{1 \text{ km}} \times \frac{1 \text{ h}}{3600 \text{ s}} = 2.81 \text{ m/s}$$

Using these values in Equation 3.11 we write:

$$\lambda = \frac{h}{mu} = \frac{6.63 \times 10^{-34} \text{ J} \cdot \text{s}}{(0.00845 \text{ kg})(2.81 \text{ m/s})} = 2.79 \times 10^{-32} \text{ m}$$

Convert the final answer to cm:

$$2.79 \times 10^{-32} \text{ m} \times \frac{100 \text{ cm}}{1 \text{ m}} = \mathbf{2.79 \times 10^{-30} \text{ cm}}$$

3.61   **Strategy:** Use Equation 3.11 to calculate velocity from the de Broglie wavelength:

$$\lambda = \frac{h}{mu}$$

Setup: Solving Equation 3.11 for velocity gives:

$$u = \frac{h}{\lambda m}$$

Planck's constant, h, is 6.63×10^{-34} J·s or 6.63×10^{-34} kg·m^2/s. For the purpose of making the unit cancellation obvious, the mass must be in kilograms and the wavelength in meters.

The mass of a neutron is 1.67493×10^{-24} g (Table 2.1). The mass in kilograms is:

$$1.67493 \times 10^{-24} \text{ g} \times \frac{1 \text{ kg}}{1 \times 10^3 \text{ g}} = 1.67493 \times 10^{-27} \text{ kg}$$

The wavelength is 10.5Å. The wavelength in meters is:

$$10.5 \text{ Å} \times \frac{1 \times 10^{-10} \text{ m}}{1 \text{ Å}} = 1.05 \times 10^{-9} \text{ m}$$

Solution:
$$u = \frac{h}{\lambda m}$$

$$u = \frac{6.63 \times 10^{-34} \text{ kg} \cdot \text{m}^2/\text{s}}{(1.05 \times 10^{-9} \text{ m})(1.67493 \times 10^{-27} \text{ kg})} = \mathbf{377 \text{ m/s}}$$

3.69   Use Equation 3.13 to calculate Δu.

$$\Delta x \cdot m \Delta u \geq \frac{h}{4\pi}$$

The uncertainty in position, Δx = diameter of sac
$$= 5.0 \times 10^{-5} \text{ m}.$$

Mass of oxygen molecule, $m = 5.3 \times 10^{-26}$ kg.

Setup: Rearranging to solve for Δu, we have

$$\Delta u = \frac{h}{4\pi \, m \, \Delta x}$$

$$\Delta u = \frac{6.63 \times 10^{-34} \text{ kg} \cdot \text{m}^2/\text{s}}{4\pi \left(5.3 \times 10^{-26} \text{ kg}\right)\left(5.0 \times 10^{-5}\text{m}\right)}$$

$$= \mathbf{2.0 \times 10^{-5} \text{ m/s}}$$

3.71   Rearranging Equation 3.13 to solve for the uncertainty in velocity, Δu, we write:

$$\Delta u \geq \frac{h}{4\pi m \Delta x} = \frac{6.63 \times 10^{-34} \text{J} \cdot \text{s}}{\left(4\pi\right)\left(2.80 \times 10^{-3}\text{kg}\right)\left(4.30 \times 10^{-7}\text{m}\right)} = \mathbf{4.38 \times 10^{-26} \text{ m / s}}$$

This uncertainty is far smaller than what can be measured. Therefore, we are able to determine the speed of a macroscopic object with great certainty using a visible wavelength of light.

3.75   **Strategy:**   What are the relationships among n, ℓ, and m_ℓ?

           Solution: The angular momentum quantum number ℓ can have integral (that is, whole number) values from 0 to $n - 1$. In this case $n = 2$, so the allowed values of the angular momentum quantum number, $\boldsymbol{\ell}$, are **0**, corresponding to an s orbital; or **1**, corresponding to a p orbital.

                Each allowed value of the angular momentum quantum number labels a subshell. Within a given subshell (label ℓ), there are $2\ell + 1$ allowed energy states (orbitals) each labeled by a different value of the magnetic quantum number. The allowed values run from $-\ell$ through 0 to $+\ell$ (whole numbers only). For the subshell labeled by the angular momentum quantum number $\boldsymbol{\ell = 1}$, the allowed values of the magnetic quantum number, m_ℓ, are **–1, 0, or 1**. For the other subshell in this problem labeled by the angular momentum quantum number $\boldsymbol{\ell = 0}$, the allowed value of the magnetic quantum number is **0**.

3.77   **For $n = 4$, the allowed values of ℓ are 0, 1, 2, and 3 [$\ell = 0$ to $(n - 1)$, integer values].** These ℓ values correspond to the **4s, 4p, 4d, and 4f** subshells. **These subshells each have one, three, five, and seven orbitals**, respectively (number of orbitals = $2\ell + 1$).

3.83   a. **2p: $n = 2$, $\ell = 1$, $m_\ell = -1$, 0, or 1**

         b. **3s: $n = 3$, $\ell = 0$, $m_\ell = 0$** (only allowed value)

         c. **5d: $n = 5$, $\ell = 2$, $m_\ell = -2, -1, 0, 1$, or 2**

An orbital in a subshell can have any of the allowed values of the magnetic quantum number for that subshell. All the orbitals in a subshell have exactly the same energy.

3.85   **A 2s orbital is larger than a 1s orbital and exhibits a node. Both have the same spherical shape. The 1s orbital is lower in energy than the 2s.**

3.87   **In H, energy depends only on *n*, but for all other atoms, energy depends on *n* and *ℓ*.** This is because in the many-electron atom, the 3*p* orbital electrons are more effectively shielded by the inner electrons of the atom (that is, the 1*s*, 2*s*, and 2*p* electrons) than the 3*s* electrons. The 3*s* orbital is said to be more "penetrating" than the 3*p* and 3*d* orbitals. In the hydrogen atom there is only one electron (no shielding), so the 3*s*, 3*p*, and 3*d* orbitals have the same energy.

3.89   Equation 3.7 of the text gives the orbital energy in terms of the principal quantum number, *n*, alone (for the hydrogen atom). The energy does not depend on any of the other quantum numbers. If two orbitals in the hydrogen atom have the same value of *n*, they have equal energy.

   a. **2*s* > 1*s***

   b. **3*p* > 2*p***

   c. **equal**

   d. **equal**

   e. **5*s* > 4*f***

3.91   a. Based on the relative size of the orbitals, **orbital (b)** has the greatest value of *n*. The principal quantum number, *n*, designates the size of the orbital. As *n* increases, the average distance of an electron in the orbital from the nucleus increases. Therefore, the orbital is larger.

   b. The angular momentum quantum number, *ℓ*, describes the shape of the atomic orbital. A value of *ℓ* = 1 corresponds to a *p* orbital. Figure 3.19 shows that each *p* orbital can be thought of as two lobes on opposite sides of the nucleus. **Orbitals (a) and (d)** represent *p* orbitals with *ℓ* =1.

   c. **None.** The shape of orbital (b) is that of a *d* orbital, specifically the d_{z^2} orbital (see Figure 3.20). For each value of $n \geq 3$, there is only one d_{z^2} orbital. Therefore, there can be no other orbitals with the same general shape as (b) with the same value of *n*.

3.95   The electron configurations for the elements are:

   a. N: $1s^2 2s^2 2p^3$ There are **three** *p*-type electrons.

   b. Si: $1s^2 2s^2 2p^6 3s^2 3p^2$ There are **six** *s*-type electrons.

   c. S: $1s^2 2s^2 2p^6 3s^2 3p^4$ There are **none** *d*-type electrons.

3.97

	n value	Orbital sum	Total number of electrons
a.	1	1	**2**
b.	2	$1 + 3 = 4$	**8**
c.	3	$1 + 3 + 5 = 9$	**18**
d.	4	$1 + 3 + 5 + 7 = 16$	**32**

3.99   For aluminum, there are **two 2*p* electrons missing and too many 3*p* electrons**. The electron configuration should be $\mathbf{1s^2 2s^2 2p^6 3s^2 3p^1}$.

   For boron, there are **too many 2*p* electrons**. The electron configuration should be $\mathbf{1s^2 2s^2 2p^1}$.

For fluorine, there are **too many 2*p* electrons**. The correct electron configuration is $1s^2 2s^2 2p^5$.
(The configuration shown is that of the F⁻ ion.)

3.101   To determine the number of unpaired electrons, we must look at the electron configuration of each of the
        elements. Since a *p* shell has three orbitals, according to Hund's rule once all of the orbitals are singly
        occupied, additional electrons will have to pair with those already in orbitals (see Section 3.9 of text).

B: $1s^2 2s^2 2p^1$ There is **one unpaired electron.**

C: $1s^2 2s^2 2p^2$ There are **two unpaired electrons.**

N: $1s^2 2s^2 2p^3$ There are **three unpaired electrons.**

O: $1s^2 2s^2 2p^4$ There are **two unpaired electrons.**

F: $1s^2 2s^2 2p^5$ There is **one unpaired electron.**

In order of increasing number of unpaired electrons, we have **B = F < C = O < N.**

3.103   To determine the number of unpaired electrons, we must look at the electron configuration of each of the
        elements. Those with all paired electrons are diamagnetic; and those with one or more unpaired electrons are
        paramagnetic.

a. **Rb:** $[Kr]5s^1$ There is **one unpaired electron; paramagnetic**

b. **As:** $[Ar]4s^2 3d^{10} 4p^3$ There are **three unpaired electrons; paramagnetic**

c. **I:** $[Kr]5s^2 4d^{10} 5p^5$ There is **one unpaired electron; paramagnetic**

d. **Cr:** $[Ar]4s^1 3d^5$ There are **six unpaired electrons; paramagnetic**

e. **Zn:** $[Ar]4s^2 3d^{10}$ There are **no unpaired electrons; diamagnetic**

3.105   a. **is wrong** because the magnetic quantum number m_ℓ **can have only whole number values.**

        b. **is wrong** because the magnetic quantum number m_ℓ **can have only the value 0 when** the angular
           momentum quantum number ℓ **is 0.**

        c. **is wrong** because the magnetic quantum number m_ℓ **can have only the value 0 when** the angular
           momentum quantum number ℓ **is 0.**

        d. **is acceptable.**

        e. **is wrong** because the electron spin quantum number m_s **can have only half-integral values.**

3.115   The ground state electron configuration of Ir is:**$[Xe]6s^2 4f^{14} 5d^7$**

3.117   **Strategy:**   How many electrons are in the Ge atom ($Z = 32$)? We start with $n = 1$ and proceed to fill orbitals
                    in the order shown in Figure 3.23 of the text. Remember that any given orbital can hold at most
                    two electrons. However, do not forget about degenerate orbitals. Starting with $n = 2$, there are
                    three *p* orbitals of equal energy, corresponding to $m_\ell = -1, 0, 1$. Starting with $n = 3$, there are five
                    d orbitals of equal energy, corresponding to $m_\ell = -2, -1, 0, 1, 2$. We can place electrons in the
                    orbitals according to the Pauli exclusion principle and Hund's rule. The task is simplified if we
                    use the noble gas core preceding Ge for the inner electrons.

Solution: Germanium has 32 electrons. The noble gas core in this case is [Ar]. (Ar is the noble gas in the period preceding germanium.) [Ar] represents $1s^2 2s^2 2p^6 3s^2 3p^6$. This core accounts for 18 electrons, which leaves 14 electrons to place.

See Figure 3.23 of the text to check the order of filling subshells past the Ar noble gas core. You should find that the order of filling is 4s, 3d, 4p. There are 14 remaining electrons to distribute among these orbitals. The 4s orbital can hold two electrons. Each of the five 3d orbitals can hold two electrons for a total of 10 electrons. This leaves two electrons to place in the 4p orbitals.

The electron configuration for **Ge** is:

$$[Ar]4s^2 3d^{10} 4p^2$$

You should follow the same reasoning for the remaining atoms.

Fe: $[Ar]4s^2 3d^6$	**Zn:** $[Ar]4s^2 3d^{10}$	**Ni:** $[Ar]4s^2 3d^8$
W: $[Xe]6s^2 4f^{14} 5d^4$	**Tl:** $[Xe]6s^2 4f^{14} 5d^{10} 6p^1$	

3.119   a. **Ga**   b. **Y**   c. **Cr**   d. **Ac**

3.121   **Part (b) is correct in the view of contemporary quantum theory. Bohr's explanation of emission and absorption line spectra appear to have universal validity. Parts (a) and (c) are artifacts of Bohr's early planetary model of the hydrogen atom and are *not* considered to be valid today.**

3.123   a. With $n = 2$, there are n^2 orbitals $= 2^2 = 4$. $m_s = +\frac{1}{2}$, specifies one electron per orbital, for a total of **4 electrons** (one electron in the 2s and 2p orbitals).

      b. $n = 4$ and $m_\ell = +1$, specifies one orbital in each subshell with $\ell = 1, 2$, or 3 (that is, a 4p, 4d, and 4f orbital). Each of the three orbitals holds two electrons for a total of **6 electrons**.

      c. If $n = 3$ and $\ell = 2$, m_ℓ has the values 2, 1, 0, –1, or –2. Each of the five orbitals can hold two electrons for a total of **10 electrons** (two electrons in each of the five 3d orbitals).

      d. If $n = 2$ and $\ell = 0$, then m_ℓ can only be 0. $m_s = -\frac{1}{2}$ specifies one electron in this orbital for a total of **1 electron** (one electron in the 2s orbital).

      e. $n = 4$, $\ell = 3$, and $m_\ell = -2$, specifies one 4f orbital. This orbital can hold **two electrons**.

3.125   The energy given in the problem is the energy of 1 mole of gamma rays. We need to convert this to the energy of one gamma ray, then we can calculate the wavelength and frequency of this gamma ray.

$$\frac{3.14 \times 10^{11} \text{ J}}{1 \text{ mol}} \times \frac{1 \text{ mol}}{6.022 \times 10^{23} \text{ gamma particles}} = 5.214 \times 10^{-13} \text{ J/gamma particle}$$

Now, we can calculate the wavelength and frequency from this energy.

$$E = \frac{hc}{\lambda}$$

$$\lambda = \frac{hc}{E} = \frac{\left(6.63 \times 10^{-34} \text{ J} \cdot \text{s}\right)\left(3.00 \times 10^8 \text{ m/s}\right)}{5.214 \times 10^{-13} \text{ J}} = 3.81 \times 10^{-13} \text{ m} = \textbf{0.381 pm}$$

and

$$E = h\nu$$

$$\nu = \frac{E}{h} = \frac{5.214 \times 10^{-13} \text{ J}}{6.63 \times 10^{-34} \text{ J} \cdot \text{s}} = \mathbf{7.86 \times 10^{20} \, s^{-1}}$$

3.127   a. First, we can calculate the energy of a single photon with a wavelength of 633 nm.

$$E = \frac{hc}{\lambda} = \frac{\left(6.63 \times 10^{-34} \text{ J} \cdot \text{s}\right)\left(3.00 \times 10^8 \text{ m/s}\right)}{633 \times 10^{-9} \text{ m}} = 3.14 \times 10^{-19} \text{ J}$$

The number of photons produced in a 0.376 J pulse is:

$$0.376 \text{ J} \times \frac{1 \text{ photon}}{3.14 \times 10^{-19} \text{ J}} = \mathbf{1.20 \times 10^{18} \text{ photons}}$$

b. Since 1 W = 1 J/s, the power delivered per 1.00×10^{-9} s pulse is:

$$\frac{0.376 \text{ J}}{1.00 \times 10^{-9} \text{ s}} = 3.76 \times 10^8 \text{ J/s} = \mathbf{3.76 \times 10^8 \text{ W}}$$

Compare this with the power delivered by a 100-W light bulb!

3.129   **Strategy:** Use Equation 3.11 to calculate velocity from the de Broglie wavelength.

$$\lambda = \frac{h}{mu}$$

Setup: Solving Equation 3.11 for mass gives:

$$m = \frac{h}{\lambda u}$$

We are told the velocity, u, is equal to 15% of the speed of light. The value for u is:

$$u = \frac{15}{100}\left(3.00 \times 10^8 \text{ m/s}\right) = 4.50 \times 10^7 \text{ m/s}$$

Next, since mass will be determined in kilograms, we need to convert to g → amu (to determine the element) using the conversion factors:

$$\frac{1 \times 10^3 \text{ g}}{1 \text{ kg}} \quad \text{and} \quad \frac{1 \text{ amu}}{1.661 \times 10^{-24} \text{ g}}$$

Planck's constant, h, is 6.63×10^{-34} J·s or 6.63×10^{-34} kg·m^2/s.

Solution:

$$m = \frac{h}{\lambda u}$$

$$m = \frac{6.63 \times 10^{-34} \text{ kg·m}^2/\text{s}}{\left(1.06 \times 10^{-16} \text{ m}\right)\left(4.50 \times 10^7 \text{ m/s}\right)} = 1.39 \times 10^{-25} \text{ kg}$$

$$1.39 \times 10^{-25} \text{ kg} \times \frac{1 \times 10^3 \text{ g}}{1 \text{ kg}} \times \frac{1 \text{ amu}}{1.661 \times 10^{-24} \text{ g}} = 83.7 \text{ amu}$$

The element with a mass closest to 83.7 amu is **krypton**.

3.131   Since 1 W = 1 J/s, the energy output of the light bulb in 1 s is 75 J. The actual energy converted to visible light is 15 percent of this value or 11 J.

First, we need to calculate the energy of one 550 nm photon. Then, we can determine how many photons are needed to provide 11 J of energy. The energy of one 550 nm photon is:

$$E = \frac{hc}{\lambda} = \frac{\left(6.63 \times 10^{-34} \text{ J·s}\right)\left(3.00 \times 10^8 \text{ m/s}\right)}{550 \times 10^{-9} \text{ m}} = 3.62 \times 10^{-19} \text{ J/photon}$$

The number of photons needed to produce 11 J of energy in 1 s is:

$$11 \text{ J} \times \frac{1 \text{ photon}}{3.62 \times 10^{-19} \text{ J}} = \mathbf{3.0 \times 10^{19}} \text{ \textbf{photons} per second}$$

3.133   The Balmer series corresponds to transitions to the $n = 2$ level. Rearranging Equation 3.6 to solve for wavelength we write:

$$\lambda = \frac{1}{R_\infty \left(\dfrac{1}{n_1^2} - \dfrac{1}{n_2^2} \right)}$$

The Rydberg constant for He^+ is $4.39 \times 10^7 \text{ m}^{-1}$. Therefore, for the transition $n = 3 \longrightarrow 2$:

$$\lambda = \frac{1}{4.39 \times 10^7 \text{ m}^{-1} \left(\dfrac{1}{3^2} - \dfrac{1}{2^2} \right)} = -1.64 \times 10^{-7} \text{ m}$$

Note that the negative sign indicates the *emission* of light. Wavelengths are always positive quantities. For **He^+**:

For the transition $n = 3 \longrightarrow 2$: $\lambda = \mathbf{164}$ **nm**

For the transition $n = 4 \longrightarrow 2$: $\lambda = \mathbf{121}$ **nm**

For the transition $n = 5 \longrightarrow 2$: $\lambda = 108$ nm

For the transition $n = 6 \longrightarrow 2$: $\lambda = 103$ nm

For **H**, the calculations are identical to those above, except the Rydberg constant for H is 1.097×10^7 m^{-1}.

For the transition $n = 3 \longrightarrow 2$: $\lambda = 656$ nm

For the transition $n = 4 \longrightarrow 2$: $\lambda = 486$ nm

For the transition $n = 5 \longrightarrow 2$: $\lambda = 434$ nm

For the transition $n = 6 \longrightarrow 2$: $\lambda = 410$ nm

All the Balmer transitions for He$^+$ are in the ultraviolet region, whereas the transitions for H are all in the visible region.

3.135   Since the energy corresponding to a photon of wavelength λ_1 equals the energy of photon of wavelength λ_2 plus the energy of photon of wavelength λ_3, then the equation must relate the wavelength to energy.

Energy of photon 1 = (energy of photon 2 + energy of photon 3)

Since $E = \dfrac{hc}{\lambda}$, then:

$$\frac{hc}{\lambda_1} = \frac{hc}{\lambda_2} + \frac{hc}{\lambda_3}$$

Dividing by *hc*:

$$\frac{1}{\lambda_1} = \frac{1}{\lambda_2} + \frac{1}{\lambda_3}$$

3.137   The excited atoms are still neutral, so the total number of electrons is the same as the atomic number of the element.

a.   **He** (2 electrons); $1s^2$

b.   **N** (7 electrons); $1s^2 2s^2 2p^3$

c.   **Na** (11 electrons); $1s^2 2s^2 2p^6 3s^1$

d.   **As** (33 electrons); **[Ar]$4s^2 3d^{10} 4p^3$**

e.   **Cl** (17 electrons); **[Ne]$3s^2 3p^5$**

3.139   Applying the Pauli exclusion principle and Hund's rule:

a.   ⇅   ⇅   ⇅ ⇅ ↑

$1s^2$ $2s^2$ $2p^5$

b. **[Ne]**   ⇅   ↑ ↑ ↑

$3s^2$ $3p^3$

c. **[Ar]**   ⇅   ⇅ ⇅ ↑ ↑ ↑
           $4s^2$ $3d^7$

3.141 $n_i = 236$, $n_f = 235$

$$\Delta E = (-2.18 \times 10^{-18} \text{ J})\left(\frac{1}{235^2} - \frac{1}{236^2}\right) = -3.34 \times 10^{-25} \text{ J}$$

$$\lambda = \frac{hc}{\Delta E} = \frac{(6.63 \times 10^{-34} \text{ J} \cdot \text{s})(3.00 \times 10^8 \text{ m/s})}{3.34 \times 10^{-25} \text{ J}} = \textbf{0.596 m}$$

This wavelength is in the ***microwave/radio*** region. (See Figure 3.1 of the text.)

3.143   a. **False.** $n = 2$ is the first excited state.

b. **False.** In the $n = 4$ state, the electron is (on average) farther from the nucleus and hence easier to remove.

c. **True.**

d. **False.** The $n = 4$ to $n = 1$ transition is a higher energy transition, which corresponds to a *shorter* wavelength.

e. **True.**

3.145   **Strategy:**   We are given the wavelengths for blue, green, and red light. Use Equation 3.3 ($c = \lambda \nu$) to convert wavelength to frequency; then use Equation 3.4 ($E = h\nu$) to determine the energy of one photon at each of the given wavelengths. Next, determine how many photons are needed to provide 2.0×10^{-17} J of energy.

Setup: The wavelengths must be converted from nanometers to meters:

$$419 \text{ nm} \times \frac{1 \times 10^{-9} \text{ m}}{1 \text{ nm}} = 4.19 \times 10^{-7} \text{ m}$$

$$531 \text{ nm} \times \frac{1 \times 10^{-9} \text{ m}}{1 \text{ nm}} = 5.31 \times 10^{-7} \text{ m}$$

$$558 \text{ nm} \times \frac{1 \times 10^{-9} \text{ m}}{1 \text{ nm}} = 5.58 \times 10^{-7} \text{ m}$$

Solution: For a photon of blue light, $\lambda = 4.19 \times 10^{-7}$ m:

$$\nu = \frac{c}{\lambda} = \frac{3.00 \times 10^8 \text{ m/s}}{4.19 \times 10^{-7} \text{ m}} = 7.16 \times 10^{14} \text{ s}^{-1}$$

$$E = h\nu = \left(6.63 \times 10^{-34} \text{ J} \cdot \text{s}\right)\left(7.16 \times 10^{14} \text{ s}^{-1}\right) = 4.75 \times 10^{-19} \text{ J}$$

For a photon of green light, $\lambda = 5.31 \times 10^{-7}$ m:

$$\nu = \frac{c}{\lambda} = \frac{3.00 \times 10^8 \text{ m/s}}{5.31 \times 10^{-7} \text{ m}} = 5.65 \times 10^{14} \text{ s}^{-1}$$

$$E = h\nu = \left(6.63 \times 10^{-34} \text{ J} \cdot \text{s}\right)\left(5.65 \times 10^{14} \text{ s}^{-1}\right) = 3.75 \times 10^{-19} \text{ J}$$

For a photon of red light, $\lambda = 5.58 \times 10^{-7}$ m:

$$\nu = \frac{c}{\lambda} = \frac{3.00 \times 10^8 \text{ m/s}}{5.58 \times 10^{-7} \text{ m}} = 5.38 \times 10^{14} \text{ s}^{-1}$$

$$E = h\nu = \left(6.63 \times 10^{-34} \text{ J} \cdot \text{s}\right)\left(5.38 \times 10^{14} \text{ s}^{-1}\right) = 3.56 \times 10^{-19} \text{ J}$$

Use the calculated energy values per photon of each wavelength to determine the number of photons needed to produce 2.0×10^{-17} J of energy.

blue light: $2.0 \times 10^{-17} \text{ J} \times \dfrac{1 \text{ photon}}{4.75 \times 10^{-19} \text{ J}} = \textbf{42 photons}$

green light: $2.0 \times 10^{-17} \text{ J} \times \dfrac{1 \text{ photon}}{3.75 \times 10^{-19} \text{ J}} = \textbf{53 photons}$

red light: $2.0 \times 10^{-17} \text{ J} \times \dfrac{1 \text{ photon}}{3.56 \times 10^{-19} \text{ J}} = \textbf{56 photons}$

3.147   Energy of a photon at 360 nm:
$$E = h\nu = \frac{hc}{\lambda} = \frac{\left(6.63 \times 10^{-34} \text{ J} \cdot \text{s}\right)\left(3.00 \times 10^8 \text{ m/s}\right)}{360 \times 10^{-9} \text{ m}} = 5.53 \times 10^{-19} \text{ J}$$

Area of exposed body in cm^2:

$$0.45 \text{ m}^2 \times \left(\frac{1 \text{ cm}}{1 \times 10^{-2} \text{ m}}\right)^2 = 4.5 \times 10^3 \text{ cm}^2$$

The number of photons absorbed by the body in 2.5 h is:

$$0.5 \times \frac{2.0 \times 10^{16} \text{ photons}}{\text{cm}^2 \cdot \text{s}} \times \left(4.5 \times 10^3 \text{ cm}^2\right) \times \frac{3600 \text{ s}}{1 \text{ h}} \times 2.5 \text{ h} = 4.1 \times 10^{23} \text{ photons}$$

The factor of 0.5 is used above because only 50% of the radiation is absorbed.

4.1×10^{23} photons with a wavelength of 360 nm correspond to an energy of:

$$4.1 \times 10^{23} \text{ photons} \times \frac{5.53 \times 10^{-19} \text{ J}}{1 \text{ photon}} = \textbf{2.2} \times \textbf{10}^5 \textbf{ J}$$

3.149 Based on the *selection rule*, which states that $\Delta\ell = \pm 1$, **only (b) and (d) are allowed transitions. Any of the transitions in Figure 3.11 is possible as long as ℓ for the final state differs from ℓ of the initial state by 1.**

3.151 Rearranging Equation 3.13 to solve for the uncertainty in velocity, Δu, we write:

$$\Delta u \geq \frac{h}{4\pi m \Delta x} = \frac{6.63 \times 10^{-34}\ \text{J} \cdot \text{s}}{(4\pi)(2.80 \times 10^{-3}\ \text{kg})(6.75 \times 10^{-7}\ \text{m})} = \mathbf{2.79 \times 10^{-26}\ m/s}$$

This uncertainty is far smaller than can be measured. Therefore, we are able to determine the speed of a macroscopic object with great certainty using a visible wavelength of light.

3.153 a. First, we need to calculate the moving mass of the proton, and then we can calculate its wavelength using the de Broglie equation.

$$m_{\text{moving}} = \frac{m_{\text{rest}}}{\sqrt{1 - \left(\dfrac{u}{c}\right)^2}} = \frac{1.67 \times 10^{-27}\ \text{kg}}{\sqrt{1 - \left[\dfrac{(0.50)(3.00 \times 10^8\ \text{m/s})}{3.00 \times 10^8\ \text{m/s}}\right]^2}} = 1.93 \times 10^{-27}\ \text{kg}$$

$$\lambda = \frac{h}{mu} = \frac{6.63 \times 10^{-34}\ \text{J} \cdot \text{s}}{\left(1.93 \times 10^{-27}\ \text{kg}\right)\left[(0.500)(3.00 \times 10^8\ \text{m/s})\right]}$$

$$\lambda = 2.29 \times 10^{-15}\ \text{m} = \mathbf{2.29 \times 10^{-6}\ nm}$$

b.

$$m_{\text{moving}} = \frac{m_{\text{rest}}}{\sqrt{1 - \left(\dfrac{u}{c}\right)^2}} = \frac{6.0 \times 10^{-2}\ \text{kg}}{\sqrt{1 - \left(\dfrac{63\ \text{m/s}}{3.00 \times 10^8\ \text{m/s}}\right)^2}} \approx \mathbf{6.0 \times 10^{-2}\ kg}$$

The equation is important only for speeds close to that of light. Note that photons have a rest mass of zero; otherwise, their moving mass would be infinite!

Chapter 4
Periodic Trends of the Elements

Key Skills

4.1   e   4.2   d   4.3   e   4.4   d

Problems

4.15   **Strategy:**   To write the electron configuration, we refer to the information in Section 3.9 of the text. We start with principal quantum number $n = 1$ and continue upward in energy until all electrons are accounted for.

   Solution: We know that for $n = 1$, we have a $1s$ orbital (two electrons). For $n = 2$, we have a $2s$ orbital (two electrons) and three $2p$ orbitals (six electrons). This constitutes the neon noble gas core. For $n = 3$, we have a $3s$ orbital (two electrons). The number of electrons left to place is $16 - 12 = 4$. These four electrons are placed in the $3p$ orbitals. A neutral atom with 16 electrons (and 16 protons) is sulfur, **S. The electron configuration is $1s^2 2s^2 2p^6 3s^2 3p^4$ or [Ne]$3s^2 3p^4$**.

4.17   Elements that have the same number of valence electrons will have similarities in chemical behavior. Looking at the periodic table, elements with the same number of valence electrons are in the same group. Therefore, the pairs that would represent similar chemical properties of their atoms are **(a) and (d), (b) and (e), (c) and (f)**.

4.19   a. **Group 1A or 1**   b. **Group 5A or 15**   c. **Group 5A or 15**   d. **Group 8B or 8**

4.21   Refer to Figure 3.26, which provides the outermost ground-state electron configurations for the known elements.

   a. The fifth period element that is designated by the electron configuration $ns^2(n-1)d^{10}np^2$ is **Sn** ([Kr]$5s^2 4d^{10} 5p^2$).

   b. The fourth period element that is designated by the electron configuration $ns^2(n-1)d^3$ is **V** ([Ar]$4s^2 3d^3$).

   c. The third period element that is designated by the electron configuration $ns^2 np^5$ is **Cl** ([Ne]$3s^2 3p^5$).

   d. The sixth period element that is designated by the electron configuration ns^2 is **Ba** ([Xe]$6s^2$).

4.25   a. Carbon has a nuclear charge of +6. If each core electron were totally effective in screening the valence electrons from the nucleus, and the valence electrons did not shield one another, the shielding constant (σ) would be 2 and the nuclear charge (Z_{eff}) would be +4 for both the $2s$ and the $2p$ electrons. Therefore, **$\sigma = 2$ and $Z_{eff} = +4$.**

   b. In reality, **for $2s$ electrons, $Z_{eff} = +6 - 2.78 = +3.22$; for $2p$ electrons, $Z_{eff} = +6 - 2.86 = +3.14$.**

   These values are lower than those in part (a) because the $2s$ and $2p$ electrons actually do shield one another to some extent.

4.39   The atomic number of mercury is 80. We carry an extra significant figure throughout this calculation to avoid rounding errors.

$$\Delta E = \left(2.18 \times 10^{-18} \text{ J}\right)\left(80^2\right)\left(\frac{1}{1^2} - \frac{1}{\infty^2}\right) = 1.395 \times 10^{-14} \text{ J/ion}$$

$$\boldsymbol{\Delta E} = \frac{1.395 \times 10^{-14} \text{ J}}{1 \text{ ion}} \times \frac{6.022 \times 10^{23} \text{ ions}}{1 \text{ mol}} \times \frac{1 \text{ kJ}}{1000 \text{ J}} = \boldsymbol{8.40 \times 10^6 \text{ kJ/mol}}$$

4.41   According to Coulomb's law (Equation 4.2), the force between two ions is directly proportional to the product of the two charges and *inversely* proportional to the distance (*d*) between the objects squared.

$$F \propto \frac{Q_1 \times Q_2}{d^2}$$

In this type of comparison, it does not matter what units we use. We are not trying to calculate a particular attractive force, only the change in attractive forces.

a. Since the force between the ions is directly proportional to the product of the two charges and *inversely* proportional to the distance squared, doubling both charges will make the force **four times greater.**

$$\frac{2 \times 2}{1^2} = 4$$

b. Since the force between the ions is directly proportional to the product of the two charges and *inversely* proportional to the distance squared, doubling the positive charge and tripling the negative charge will make the force **six times greater**.

$$\frac{2 \times 3}{1^2} = 6$$

c. Since the force between the ions is directly proportional to the product of the two charges and *inversely* proportional to the distance squared, doubling the negative charge and the distance will **reduce the force by one-half.**

$$\frac{1 \times 2}{2^2} = \frac{1}{2}$$

d. Since the force between the ions is directly proportional to the product of the two charges and *inversely* proportional to the distance squared, doubling the positive charge and halving the distance will make the force **eight times greater.**

$$\frac{2 \times 1}{0.5^2} = 8$$

e. Since the force between the ions is directly proportional to the product of the two charges and *inversely* proportional to the distance squared, halving both charges and the distance will result in **no change.**

$$\frac{0.5 \times 0.5}{0.5^2} = 1$$

4.43   **Strategy:**   First, calculate the attractive force between each of the particles. Then arrange them in order of increasing magnitude.

According to Coulomb's law (Equation 4.2), the force between two ions is directly proportional to the product of the two charges and *inversely* proportional to the distance (*d*) between the objects squared.

$$F \propto \frac{Q_1 \times Q_2}{d^2}$$

The product of a positive number and a negative number is always a negative number. When comparing the magnitudes of attractive forces, it is unnecessary to include the sign.

Solution:

i: $F \propto \dfrac{(2) \times (-3)}{(2d)^2} = \dfrac{6}{4d^2} = \dfrac{1.5}{d^2}$

ii: $F \propto \dfrac{(2) \times (-4)}{(3d)^2} = \dfrac{8}{9d^2} = \dfrac{0.9}{d^2}$

iii: $F \propto \dfrac{(1) \times (-4)}{(2d)^2} = \dfrac{4}{4d^2} = \dfrac{1}{d^2}$

iv: $F \propto \dfrac{(4) \times (-3)}{(3d)^2} = \dfrac{12}{9d^2} = \dfrac{1.3}{d^2}$

In order of increasing magnitude of attractive force: **ii < iii < iv < i**

4.45 **Strategy:** Recall that the general periodic trends in atomic size are:

(1) Moving from left to right across a row (period) of the periodic table, the atomic radius *decreases* due to an increase in effective nuclear charge.

(2) Moving down a column (group) of the periodic table, the atomic radius ***increases*** since the orbital size increases with increasing principal quantum number.

Solution: The atoms that we are considering are all in the same period of the periodic table. Hence, the atom furthest to the left in the row will have the largest atomic radius, and the atom farthest to the right in the row will have the smallest atomic radius. Arranged in order of increasing atomic radius, we have

Cl < P < Si < Al < Mg

4.47 **Oxygen** is the smallest atom in Group 6A. Atomic radius increases moving down a group since the orbital size increases with increasing principal quantum number, *n*.

4.49 **Left to right: S, Se, Ca, and K**.

4.51 The general periodic trend for first ionization energy is that it increases across a period (row) of the periodic table and decreases down a group (column). Of the choices, K will have the smallest ionization energy. Ca, just to the right of K, will have a higher first ionization energy. Moving to the right across the periodic table, the ionization energies will continue to increase as we move to P. Continuing across to Cl and moving up the halogen group, F will have higher ionization energy than P. Finally, Ne is to the right of F in Period 2, thus it will have higher ionization energy. The correct order of increasing first ionization energy is:

K < Ca < P < F < Ne

You can check the above answer by looking up the first ionization energies for these elements in Figure 4.8 of the text.

4.53   **The Group 3A elements (such as Al) all have a single electron in the outermost *p* subshell, which is well shielded from the nuclear charge by the inner electrons and the *ns*² electrons. Therefore, less energy is needed to remove a single *p* electron than to remove a paired *s* electron from the same principal energy level (such as for Mg).**

4.55   **Strategy:**   Removal of the outermost electron requires less energy if it is shielded by a filled inner shell.

   Solution: The lone electron in the 3*s* orbital will be much easier to remove. This lone electron is shielded from the nuclear charge by the filled inner shell. Therefore, the ionization energy of **496 kJ/mol is paired with** the electron configuration $1s^2 2s^2 2p^6 3s^1$.

   A noble gas electron configuration such as $1s^2 2s^2 2p^6$ is a very stable configuration, making it extremely difficult to remove an electron. The 2*p* electron is not as effectively shielded by electrons in the same energy level. The high ionization energy of **2080 kJ/mol is paired with** $1s^2 2s^2 2p^6$, **a very stable noble gas electron configuration**.

4.57   **Strategy:**   The energy required to remove core electrons is much greater than the energy required to remove valence electrons.

   Solution: **A:** Element A exhibits the largest increase in ionization energy in going from the third to the fourth ionization. This energy increase is due to the removal of core electrons in the fourth ionization. The first three ionizations remove three valence electrons. Elements in **Group 3A or 3** have three valence electrons.

   B: Element B exhibits the largest increase in ionization energy in going from the second to the third ionization. This energy increase is due to the removal of core electrons in the third and fourth ionizations. The first two ionizations remove two valence electrons. Elements in **Group 2A or 2** have two valence electrons.

   C: Element C exhibits the largest increase in ionization energy in going from the first to the second ionization. This energy increase is due to the removal of core electrons in the second, third, and fourth ionizations. The first ionization removes one valence electron. Elements in **Group 1A or 1** have one valence electron.

4.59   **Strategy:**   What are the trends in electron affinity in a periodic group and in a particular period? Which of the above elements are in the same group and which are in the same period?

   Solution: One of the general periodic trends for electron affinity is that the tendency to accept electrons increases (that is, electron affinity values become more positive) as we move from left to right across a period. However, this trend does not include the noble gases. We know that noble gases are extremely stable, and they do not want to gain or lose electrons.

   Based on the above periodic trend, **Cl** would be expected to have the highest electron affinity. Addition of an electron to Cl forms Cl⁻, which has a stable noble gas electron configuration.

4.61   **Alkali metals have a valence electron configuration of *ns*¹, so they can accept another electron in the *ns* orbital. On the other hand, alkaline earth metals have a valence electron configuration of *ns*², so they have little tendency to accept another electron unless it goes into a higher energy *p* orbital.**

4.63   **Strategy:**   Use Coulomb's law (Equation 4.2) to calculate a number to which the attractive force will be proportional in each case.

$$F \propto \frac{Q_1 \times Q_2}{d^2}$$

Setup: From Figure 4.6, the atomic radii of sodium and magnesium are 186 pm and 160 pm, respectively. The effective nuclear charges for sodium and magnesium are 2.51 and 3.31, respectively. The charge on the valence electron in each case is –1.

Solution: For Na:

$$F \propto \frac{(2.51) \times (-1)}{(186 \text{ pm})^2} = -7.255 \times 10^{-5}$$

For Mg:

$$F \propto \frac{(3.31) \times (-1)}{(160 \text{ pm})^2} = -1.293 \times 10^{-4}$$

Note that in this type of comparison, it doesn't matter what units we use for the distance between the charges. We are not trying to calculate a particular attractive force, only to compare the magnitudes of these two attractive forces.

The calculated number for magnesium is about 78.2 percent larger than that for sodium.

4.69 Determine the number of electrons, and then use Figure 3.23 of the text to determine the order for filling energy sublevels.

a. **[Kr]**

b. **[Kr]**

c. **[Kr]$5s^2 4d^{10}$**

d. **[Kr]$5s^2 4d^{10} 5p^6$**

e. **[Xe]**

f. **[Kr]$5d^{10}$**

g. **[Xe]$6s^2 4f^{14} 5d^{10}$**

h. **[Xe]$4f^{14} 5d^{10}$**

4.71 **Strategy:** First write electron configurations for the atoms. Then add electrons (for anions) or remove electrons (for cations) to account for the charge. The electrons removed from a *d*-block element must come first from the outermost *s* subshell, not the partially filled *d* subshell.

Setup: a. Fe^{2+} forms when Fe ($1s^2 2s^2 2p^6 3s^2 3p^6 4s^2 3d^6$ or [Ar]$4s^2 3d^6$) loses two electrons.

b. Cu^{2+} forms when Cu ($1s^2 2s^2 2p^6 3s^2 3p^6 4s^1 3d^{10}$ or [Ar]$4s^1 3d^{10}$) loses two electrons, one from the 4*s* subshell and another from the 3*d* subshell.

c. Co^{2+} forms when Co ($1s^2 2s^2 2p^6 3s^2 3p^6 4s^2 3d^7$ or [Ar]$4s^2 3d^7$) loses two electrons.

d. Mn^{2+} forms when Mn ($1s^2 2s^2 2p^6 3s^2 3p^6 4s^2 3d^5$ or [Ar]$4s^2 3d^5$) loses two electrons.

Solution: a. **[Ar]$3d^6$**

b. **[Ar]$3d^9$**

      c. **[Ar]3d^7**

      d. **[Ar]3d^5**

      Be sure to *add* electrons to form an anion and *remove* electrons to form a cation. Also, double-check to make sure that the electrons removed from a *d*-block element come first from the *ns* subshell and then, if necessary, from the $(n - 1)d$ subshell.

4.73   This exercise simply depends on determining the total number of electrons and using Figure 3.23 of the text to determine the order of filling the energy sublevels.

    a. **[Ar]3d^8**   b. **[Ar]3d^{10}**   c. **[Kr]4d^{10}**   d. **[Xe]4f^{14}5d^{10}**   e. **[Xe]4f^{14}5d^8**

4.75   a. **Cr^{3+}**   b. **Sc^{3+}**   c. **Rh^{3+}**   d. **Ir^{3+}**

4.77   Isoelectronic means that the species have the same number of electrons and the same electron configuration.

      Be^{2+} and He ($2e^-$) N^{3-} and F^- ($10e^-$) Fe^{2+} and Co^{3+} ($24e^-$) S^{2-} and Ar ($18e^-$)

4.79   **Strategy:**   To determine the number of unpaired electrons, we must know the electron configurations of the ions. Begin by writing the electron configurations for the atoms. Then remove electrons to account for the positive charge. The electrons removed from a *d*-block element must come first from the outermost *s* subshell, not the partially filled *d* subshell.

                        To determine the number of unpaired electrons, recall from Section 3.8 that there are five *d* orbitals of equal energy.

    Setup: Ti^{3+} forms when Ti ($1s^2 2s^2 2p^6 3s^2 3p^6 4s^2 3d^2$ or [Ar]$4s^2 3d^2$) loses three electrons.

                        Fe^{2+} forms when Fe ($1s^2 2s^2 2p^6 3s^2 3p^6 4s^2 3d^6$ or [Ar]$4s^2 3d^6$) loses two electrons.

                        V^{3+} forms when V ($1s^2 2s^2 2p^6 3s^2 3p^6 4s^2 3d^3$ or [Ar]$4s^2 3d^3$) loses three electrons.

                        Cu^+ forms when Cu ($1s^2 2s^2 2p^6 3s^2 3p^6 4s^1 3d^{10}$ or [Ar]$4s^1 3d^{10}$) loses one electron.

                        Mn^{4+} forms when Mn ($1s^2 2s^2 2p^6 3s^2 3p^6 4s^2 3d^5$ or [Ar]$4s^2 3d^5$) loses four electrons.

    Solution: Ti^{3+}: [Ar]3d^1 one unpaired electron

                        Fe^{2+}: [Ar]3d^6 four unpaired electrons

                        V^{3+}: [Ar]3d^2 two unpaired electrons

                        Cu^+: [Ar]3d^{10} zero unpaired electrons

                        Mn^{4+}: [Ar]3d^3 three unpaired electrons

                        In order of increasing number of unpaired electrons: **$Cu^+ < Ti^{3+} < V^{3+} < Mn^{4+} < Fe^{2+}$**

4.81   An anion is an ion whose net charge is *negative* due to an *increase* in the number of electrons. An anion with a net –2 charge has gained two electrons. Since this anion has a total of 36 electrons, the neutral element from which the anion is derived must have 34 electrons (34 + 2 = 36 electrons). The element with 34 electrons is **selenium, Se**.

4.83 The metal ion in question is a transition metal ion because it has five electrons in the $3d$ subshell. Remember that in a transition metal ion, the $(n-1)d$ orbitals are more stable than the ns orbital. Hence, when a cation is formed from an atom of a transition metal, electrons are *always* removed first from the ns orbital and then from the $(n-1)d$ orbitals if necessary. Since the metal ion has a +3 charge, three electrons have been removed. Since the $4s$ subshell is less stable than the $3d$, two electrons would have been lost from the $4s$ and one electron from the $3d$. Therefore, the electron configuration of the neutral atom is $[Ar]4s^2 3d^6$. This is the electron configuration of iron. Thus, the metal is **iron, Fe**.

4.85 **Strategy:** The transition metal ion electron configurations shown here have empty s orbitals. Recall that the electrons removed from a d-block element must come first from the outermost s subshell, not the partially filled d subshell.

Setup: Once electrons have been removed from the outermost s subshell, they may be removed from the partially filled d subshell. Refer to Figure 4.12 to see the common monatomic ions of the transition metal elements to determine the number of d electrons that could have been removed.

Solution: a. The ion with an electron configuration of $[Ar]3d^6$ could have lost two electrons from the s subshell. The electron configuration $[Ar]4s^2 3d^6$ corresponds to iron, Fe. The electron configuration of this ion could represent **Fe^{2+}**.

The ion with an electron configuration of $[Ar]3d^6$ could have also lost one electron from the d subshell in addition to the two from the s subshell. This would give an electron configuration of $[Ar]4s^2 3d^7$. Therefore, it is also possible that the electron configuration of this ion represents **Co^{3+}**.

Since Ni does not form Ni^{4+}, Fe^{2+} and Co^{3+} are the only two ions represented by this electron configuration.

b. The ion with an electron configuration of $[Kr]4d^{10}$ could have lost as many as two electrons from the s subshell. Silver, Ag, has an electron configuration of $[Kr]5s^1 4d^{10}$. Losing one electron would leave an empty s subshell and the electron configuration $[Kr]4d^{10}$. The electron configuration of this ion could represent **Ag^+**.

Losing two electrons would give the electron configuration $[Kr]5s^2 4d^{10}$, which corresponds to cadmium, Cd. The electron configuration of this ion could also represent **Cd^{2+}**.

Since the remaining d subshell is filled, Cd^{2+} and Ag^+ are the only two ions represented by this electron configuration.

c. The ion with an electron configuration of $[Ar]3d^9$ could have lost two electrons from the s subshell. However, the electron configuration $[Ar]4s^2 3d^9$ does not exist. Copper is an anomaly with an electron configuration of $[Ar]4s^1 3d^{10}$ due to the greater stability associated with a filled $3d^{10}$ subshell. Losing one electron from the s subshell and one from the d subshell would give $[Ar]3d^9$. Therefore, the electron configuration of this ion could represent **Cu^{2+}**.

Since Zn does not form Zn^{3+}, Cu^{2+} is the only ion represented by this electron configuration.

d. The ion with an electron configuration of $[Ar]3d^{10}$ could have lost as many as two electrons from the s subshell. Losing one electron would give the electron configuration $[Ar]4s^1 3d^{10}$, which corresponds to copper, Cu. The electron configuration of this ion could represent **Cu^+**.

Losing two electrons would give the electron configuration $[Ar]4s^2 3d^{10}$, which corresponds to zinc, Zn. The electron configuration of this ion could represent **Zn^{2+}**.

Since Zn is the last of the transition metals in this period, Cu^+ and Zn^{2+} are the only two ions represented by this electron configuration.

e. The ion with an electron configuration of $[Ar]3d^3$ could have lost two electrons from the s subshell. The electron configuration $[Ar]4s^23d^3$ corresponds to vanadium, V. The electron configuration of this ion could represent $\mathbf{V^{2+}}$.

The ion with an electron configuration of $[Ar]\,3d^3$ could have also lost one electron from the d subshell in addition to the two from the s subshell. This would give an electron configuration of $[Ar]4s^13d^5$ (recall $[Ar]4s^23d^4$ does not exist). Therefore, it is also possible that the electron configuration of this ion represents $\mathbf{Cr^{3+}}$.

Since Mn does not form Mn^{4+}, V^{2+}, and Cr^{3+} are the only two ions represented by this electron configuration.

f. The ion with an electron configuration of $[Ar]3d^7$ could have lost two electrons from the s subshell. $[Ar]4s^23d^7$ corresponds to cobalt, Co. The electron configuration of this ion could represent $\mathbf{Co^{2+}}$.

The ion with an electron configuration of $[Ar]3d^7$ could have also lost one electron from the d subshell in addition to the two from the s subshell. This would give an electron configuration of $[Ar]4s^23d^8$. Therefore, it is also possible that the electron configuration of this ion represents $\mathbf{Ni^{3+}}$.

Since Cu does not form Cu^{4+}, Co^{2+} and Ni^{3+} are the only two ions represented by this electron configuration.

4.89 **Strategy:** In comparing ionic radii, it is useful to classify the ions into three categories: (1) isoelectronic ions (see Section 4.6 of the text), (2) ions that carry the same charges and are generated from atoms of the same group, and (3) ions that carry different charges but are generated from the same atom. In case (1), ions with fewer protons (smaller atomic numbers) are always larger; in case (2), ions from atoms having a greater atomic number are always larger; in case (3), ions that have a smaller positive charge (or a larger negative charge) are always larger.

Solution: a. **Cl** is smaller than Cl^-. An atom gets bigger when more electrons are added.

b. $\mathbf{Na^+}$ is smaller than Na. An atom gets smaller when electrons are removed.

c. $\mathbf{O^{2-}}$ is smaller than S^{2-}. Both elements belong to the same group, and the ionic radius increases going down a group.

d. $\mathbf{Al^{3+}}$ is smaller than Mg^{2+}. The two ions are isoelectronic and in such cases the radius gets smaller as the charge becomes more positive.

e. $\mathbf{Au^{3+}}$ is smaller than Au^+. The ion gets smaller as more electrons are removed.

4.91 **The Cu^+ ion is larger than the Cu^{2+} ion because it has one more electron.**

4.93 Elements in Group 6A form ions with a net charge of –2. The formation of an anion with a –2 net charge requires that the atom gains two electrons. When an atom becomes an anion, its radius increases due to increased electron–electron repulsions.

Therefore, **(c)** is the most realistic representation of an atom from Group 6A becoming an ion. The atom gains two electrons to become an anion with a –2 net charge. Additionally, the anion is larger in radius than the neutral atom.

420

4.95   Having the same valence electron configuration results in similar chemical properties. To determine which ion would be most suitable to replace Ga^{3+}, we need to look at the electron configurations of the ions to predict which ion would have chemical properties similar to Ga^{3+}. The electron configurations are:

Ga^{3+}: $[Ar]3d^{10}$
Fe^{3+}: $[Ar]3d^{5}$
Ca^{2+}: $[Ar]$
Ti^{3+}: $[Ar]3d^{1}$
In^{3+}: $[Kr]4d^{10}$
Tl^{+}: $[Xe]6s^{2}4f^{14}5d^{10}$

Both **Ga^{3+}** and **In^{3+}** have the same valence electron configurations, with completely filled d subshells and the same ionic charges. Although the radius of In^{3+} is larger than that of Ga^{3+}, its chemical properties should be similar.

4.97   a. **bromine, Br**   b. **nitrogen, N**   c. **rubidium, Rb**   d. **magnesium, Mg**

4.99   **Strategy:**   The ionization energy is the amount of energy required to remove an electron from an atom or ion. For isoelectronic ions, the ionization energy increases as the charge becomes more positive, since the effective nuclear charge increases.

             Setup: The ion with the largest net *negative* charge will have the smallest ionization energy. The ion with the largest net *positive* charge will have the highest ionization energy.

             Solution: $O^{2-} < F^{-} < Na^{+} < Mg^{2+}$

4.101   **O^{+} and N**   **S^{2-} and Ar**   **N^{3-} and Ne**   **As^{3+} and Zn**   **Cs^{+} and Xe**

4.103   **(a)** and **(d)**

4.105   **The binding of a cation to an anion results from electrostatic attraction. As the +2 cation gets smaller (from Ba^{2+} to Mg^{2+}), the distance between the opposite charges decreases and the electrostatic attraction increases.**

4.107   **Strategy:**   The atomic radius *decreases* as we move from left to right across a period and *increases* from top to bottom as we move down within a group.

                         The smaller the atomic radius, the larger the ionization energy (due to a larger effective nuclear charge). The ionization energy (in general) has a periodic trend opposite to that of the atomic radius.

             Setup: The higher the principal quantum number (n), the higher the period number of the element.

                         Looking at the principal quantum number of the valence orbital diagrams, we know that element A is located in Period 3, element B in Period 2, element C in Period 5, and element D is also in Period 3. Since element D has two additional electrons in the p subshell, it will be located to the right of element A in Period 3.

            Solution: a. In order of increasing atomic size: **B < D < A < C**

                         b. In order of increasing ionization energy: **C < A < D < B**

4.109 **This anomaly is a result of the lanthanide contraction. The lanthanide contraction is a term used to describe a decrease in atomic radius of the lanthanide elements. The lanthanide contraction also decreases the radius of the elements following the lanthanides in the periodic table, including Hf. The lanthanide contraction is caused by poor shielding of the nuclear charge from the partially filled 4f electron orbitals.**

4.111 H⁻ is the larger species. Both H⁻ and He are isoelectronic species with two electrons. **Since H⁻ has only one proton compared to two protons for He, the nucleus of H⁻ will attract the two electrons less strongly compared to He. Therefore, H⁻ is larger.**

4.113 Replacing Z in the equation given in Problem 4.38 with $(Z - \sigma)$ gives:

$$E_n = \left(2.18 \times 10^{-18} \text{ J}\right)\left(Z - \sigma\right)^2 \left(\frac{1}{n^2}\right)$$

For helium, the atomic number (Z) is 2, and in the ground state its two electrons are in the first energy level, so $n = 1$. Substitute Z, n, and the first ionization energy into the above equation to solve for σ.

$$E_1 = 3.94 \times 10^{-18} \text{ J} = \left(2.18 \times 10^{-18} \text{ J}\right)\left(2 - \sigma\right)^2 \left(\frac{1}{1^2}\right)$$

$$(2 - \sigma)^2 = \frac{3.94 \times 10^{-18} \text{ J}}{2.18 \times 10^{-18} \text{ J}}$$

$$2 - \sigma = \sqrt{1.81}$$

$$\sigma = 2 - 1.34 = \mathbf{0.66}$$

4.115 The change in the effective nuclear charge as we move from the top of a group to the bottom is generally less significant than the change as we move across a period. Although each step down a group represents a large increase in the nuclear charge, there is also an additional shell of core electrons to shield the valence electrons from the nucleus. Consequently, **the effective nuclear charge changes less than the nuclear charge as we move down a column of the periodic table. Gallium has an additional ten *core* electrons in its 3d shell to shield valence electrons. Therefore, the effective nuclear charge is not as large as might be expected.**

4.117 The volume of a sphere is $\frac{4}{3}\pi r^3$. The percentage of volume occupied by K⁺ compared to K is:

$$\frac{\text{volume of K}^+ \text{ ion}}{\text{volume of K atom}} \times 100\% = \frac{\frac{4}{3}\pi \left(138 \text{ pm}\right)^3}{\frac{4}{3}\pi \left(227 \text{ pm}\right)^3} \times 100\% = 22.5\%$$

Therefore, there is a decrease in atomic volume of $(100 - 22.5)\% = \mathbf{77.5\%}$ when K⁺ is formed from K.

4.119 Rearrange the given equation to solve for ionization energy.

$$IE = h\nu - \frac{1}{2}mu^2$$

or

$$IE = \frac{hc}{\lambda} - KE$$

The kinetic energy of the ejected electron is given in the problem. Substitute h, c, and λ into the above equation to solve for the ionization energy.

$$IE = \frac{\left(6.63 \times 10^{-34} \text{ J} \cdot \text{s}\right)\left(3.00 \times 10^{8} \text{ m/s}\right)}{162 \times 10^{-9} \text{ m}} - \left(5.34 \times 10^{-19} \text{ J}\right)$$

$$\mathbf{IE = 6.94 \times 10^{-19} \text{ J/electron}}$$

If there are no other electrons with lower kinetic energy, then this is the electron from the valence shell. To ensure that the ejected electron is the valence electron, UV light of the *longest* wavelength (lowest energy) that can still eject electrons should be used.

4.121 **Strategy:** The atomic radius *decreases* as we move from left to right across a period and *increases* from top to bottom as we move down within a group.

Additionally, when an atom loses an electron and becomes a cation, its radius decreases due in part to a reduction in electron-electron repulsions in the valence shell. When an atom gains electrons and becomes an anion, its radius increases due to increased electron-electron repulsions.

Setup: Fluorine is located in Period 2, Group 7A. It has the lowest principal quantum number, n, and therefore the smallest radius. F^- will be larger than F due to the anion's increase in size.

Mg is located in Period 3, Group 2A. Mg will be larger than F but smaller than Rb.

Rb is located in Period 5, Group 1A. Rb will be the largest of these elements. Rb^+ will be smaller than Rb due to the cation's decrease in size.

It is difficult to compare Mg and Rb^+, since they have different charges and are in different groups. You would expect them to be similar in size. Rb^+ would be slightly smaller, since the valence electrons are in the $4s$ and $4p$ subshells (the electron in the $5s$ subshell has been lost to make the cation) versus the $3s$ subshell for Mg, yet the number of protons in Rb is 25 more than in Mg.

Solution: Based on their relative sizes, the spheres can be identified as follows:

A: Rb

B: F

C: Mg

D: Rb$^+$

E: F$^-$

4.123   The plot is:

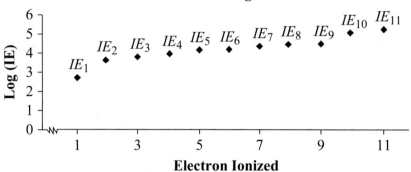

a. IE_1 corresponds to the electron in $3s^1$ IE_7 corresponds to the electron in $2p^1$

   IE_2 corresponds to the first electron in $2p^6$ IE_8 corresponds to the first electron in $2s^2$

   IE_3 corresponds to the first electron in $2p^5$ IE_9 corresponds to the electron in $2s^1$

   IE_4 corresponds to the first electron in $2p^4$ IE_{10} corresponds to the first electron in $1s^2$

   IE_5 corresponds to the first electron in $2p^3$ IE_{11} corresponds to the electron in $1s^1$

   IE_6 corresponds to the first electron in $2p^2$

b. Each break ($IE_1 \longrightarrow IE_2$ and $IE_9 \longrightarrow IE_{10}$) represents the transition to another shell ($n = 3 \longrightarrow 2$ and $n = 2 \longrightarrow 1$).

4.125   **Considering electron configurations,** $\mathbf{Fe^{2+}[Ar]3d^6 \longrightarrow Fe^{3+}[Ar]3d^5}$

$$\mathbf{Mn^{2+}[Ar]3d^5 \longrightarrow Mn^{3+}[Ar]3d^4,}$$

a half-filled shell has extra stability. In oxidizing Fe^{2+}, the product is a half-filled d^5 shell. In oxidizing Mn^{2+}, a half-filled d^5 shell electron is being lost, which requires more energy.

4.127   **Z_{eff} increases from left to right across the table, so electrons are held more tightly. (This explains the electron affinity values of C and O.) Nitrogen has a zero value of electron affinity because of the stability of the half-filled $2p$ subshell (that is, N has little tendency to accept another electron).**

4.129   **Once an atom gains an electron forming a negative ion, adding additional electrons is typically an unfavorable process due to electron-electron repulsions. Second and third electron affinities do not occur spontaneously and are therefore difficult to measure.**

4.131   a. **It was determined that the periodic table was based on atomic number, not atomic mass.**

   b. Argon:
$$(0.00337 \times 35.9675 \text{ amu}) + (0.00063 \times 37.9627 \text{ amu}) + (0.9960 \times 39.9624 \text{ amu}) = \mathbf{39.95 \text{ amu}}$$

   Potassium:
$$(0.93258 \times 38.9637 \text{ amu}) + (0.000117 \times 39.9640 \text{ amu}) + (0.0673 \times 40.9618 \text{ amu}) = \mathbf{39.10 \text{ amu}}$$

4.133   **Element 119**

Electron configuration: [Rn]$7s^2 5f^{14} 6d^{10} 7p^6 8s^1$

4.135   **There is a large jump from the second to the third ionization energy, indicating a change in the principal quantum number, n.** In other words, the third electron removed is an inner, noble gas core electron, which is difficult to remove. Therefore, the element is in **Group 2A**.

4.137   **Strategy:**   The energy required to remove core electrons is much greater than the energy required to remove valence electrons.

Recall from Sections 3.2 and 3.3 that wavelength is *inversely* proportional to energy. From Equations 3.3 and 3.4, we have $c = \lambda v$ and $E = hv$; therefore, $E = \dfrac{hc}{\lambda}$.

Setup: Calculating the ionization energy for each wavelength, we have:

218 nm:

$$E = \frac{hc}{\lambda} = \frac{\left(6.63 \times 10^{-34} \ \text{J} \cdot \text{s}\right)\left(3.00 \times 10^8 \ \text{m/s}\right)}{2.18 \times 10^{-7} \ \text{m}} = 9.12 \times 10^{-19} \ \text{J}$$

113 nm:

$$E = \frac{hc}{\lambda} = \frac{\left(6.63 \times 10^{-34} \ \text{J} \cdot \text{s}\right)\left(3.00 \times 10^8 \ \text{m/s}\right)}{1.13 \times 10^{-7} \ \text{m}} = 1.76 \times 10^{-18} \ \text{J}$$

29 nm:

$$E = \frac{hc}{\lambda} = \frac{\left(6.63 \times 10^{-34} \ \text{J} \cdot \text{s}\right)\left(3.00 \times 10^8 \ \text{m/s}\right)}{2.9 \times 10^{-8} \ \text{m}} = 6.86 \times 10^{-18} \ \text{J}$$

22 nm:

$$E = \frac{hc}{\lambda} = \frac{\left(6.63 \times 10^{-34} \ \text{J} \cdot \text{s}\right)\left(3.00 \times 10^8 \ \text{m/s}\right)}{2.2 \times 10^{-8} \ \text{m}} = 9.04 \times 10^{-18} \ \text{J}$$

17 nm:

$$E = \frac{hc}{\lambda} = \frac{\left(6.63 \times 10^{-34} \ \text{J} \cdot \text{s}\right)\left(3.00 \times 10^8 \ \text{m/s}\right)}{1.7 \times 10^{-8} \ \text{m}} = 1.17 \times 10^{-17} \ \text{J}$$

14 nm:

$$E = \frac{hc}{\lambda} = \frac{\left(6.63 \times 10^{-34} \ \text{J} \cdot \text{s}\right)\left(3.00 \times 10^8 \ \text{m/s}\right)}{1.4 \times 10^{-8} \ \text{m}} = 1.42 \times 10^{-17} \ \text{J}$$

Solution: Compare the difference in ionization energy between successive ionizations:

$$IE_1 \longrightarrow IE_2: \ 8.48 \times 10^{-19} \, \text{J}$$

$$IE_2 \longrightarrow IE_3: \ 5.1 \times 10^{-18} \, \text{J}$$

$$IE_3 \longrightarrow IE_4: \ 2.18 \times 10^{-18} \, \text{J}$$

$$IE_4 \longrightarrow IE_5: \ 2.66 \times 10^{-18} \, \text{J}$$

$$IE_5 \longrightarrow IE_6: \ 2.5 \times 10^{-18} \, \text{J}$$

The largest increase in ionization energy in going from the second to the third ionization. This energy increase is due to the removal of core electrons in the third ionization. The first two ionizations remove two valence electrons. Elements in **Group 2A or 2** have two valence electrons.

4.139   Having the same valence electron configuration results in similar chemical properties. To determine which ion might successfully replace Zn^{2+} while maintaining the activity of the metalloenzyme, we need to look at the electron configurations of the ions to predict which ion would have chemical properties similar to Zn^{2+}. The electron configurations are:

Zn^{2+}: $[Ar]3d^{10}$

Cu^{2+}: $[Ar]3d^{9}$

Sn^{4+}: $[Kr]4d^{10}$

Sc^{3+}: $[Ar]$

Pb^{2+}: $[Xe]6s^{2}4f^{14}5d^{10}$

Co^{2+}: $[Ar]3d^{7}$

Since Zn^{2+} and Sn^{4+} both have completely filled d subshells, they have similar chemical properties. Thus, **Sn^{4+}** might successfully replace Zn^{2+} in the metalloenzyme.

Chapter 5
Ionic and Covalent Compounds

Key Skills

5.1   d   5.2   c   5.3   a   5.4   c

Problems

5.7   **Strategy:**   A Lewis dot symbol consists of the element's symbol surrounded by dots, where each dot represents a valence electron. For the main group elements, the number of dots in the Lewis dot symbol is the same as the group number.

Starting with the Lewis dot symbol for each element, add dots (for anions) or remove dots (for cations) as needed to achieve the correct charge on each ion. Don't forget to include the appropriate charge on the Lewis dot symbol.

Solution: a. Br is located in Group 7A or 17. The Lewis dot symbol for Br is:

$$:\overset{\displaystyle ..}{\underset{\displaystyle ..}{Br}}\cdot$$

b. N is located in Group 5A or 15. The Lewis dot symbol for N is:

$$\cdot\overset{\displaystyle ..}{\underset{\displaystyle \cdot}{N}}\cdot$$

c. I is located in Group 7A or 17. The Lewis dot symbol for I is:

$$:\overset{\displaystyle ..}{\underset{\displaystyle ..}{I}}\cdot$$

d. As is located in Group 5A or 15. The Lewis dot symbol for As is:

$$\cdot\overset{\displaystyle ..}{\underset{\displaystyle \cdot}{As}}\cdot$$

e. F is located in Group 7A or 17. The Lewis dot symbol for F is:

$$:\overset{\displaystyle ..}{\underset{\displaystyle ..}{F}}\cdot$$

f. P is located in Group 5A or 15. The Lewis dot symbol for P is:

$$\cdot\overset{\displaystyle ..}{\underset{\displaystyle \cdot}{P}}\cdot$$

Adding three dots to account for the –3 charge gives:

$$\left[:\overset{\displaystyle ..}{\underset{\displaystyle ..}{P}}:\right]^{3-}$$

g. Na is located in Group 1A or 1. The Lewis dot symbol for Na is:

$$\cdot Na$$

Removing one dot to account for the +1 charge gives:

$$\mathbf{Na^+}$$

h. Mg is located in Group 2A or 2. The Lewis dot symbol for Mg is:

$$\cdot Mg \cdot$$

Removing two dots to account for the +2 charge gives:

$$\mathbf{Mg^{2+}}$$

i. As is located in Group 5A or 15. The Lewis dot symbol for As is:

$$\cdot \overset{\cdot\cdot}{\underset{\cdot}{As}} \cdot$$

Removing three dots to account for the +3 charge gives:

$$\left[\cdot As \cdot \right]^{3+}$$

j. Pb is located in Group 4A or 14. The Lewis dot symbol for Pb is:

$$\cdot \overset{\cdot}{\underset{\cdot}{Pb}} \cdot$$

Removing two dots to account for the +2 charge gives:

$$\left[\cdot Pb \cdot \right]^{2+}$$

5.9 **Strategy:** For the main group elements, the number of dots in the Lewis dot symbol is the same as the group number. Dots are added for anions and removed for cations to achieve the correct charge on each ion.

Solution: a. Element X has four valence electrons in its Lewis dot symbol. Element X is located in **Group 4A or 14**.

b. Element X has six valence electrons in its Lewis dot symbol. Element X is located in **Group 6A or 16**.

c. Element X has a charge of −1, which means that one dot is added to its Lewis dot symbol. Since the anion with a net charge of −1 has eight electrons, the neutral element must have seven valence electrons. Element X is located in **Group 7A or 17**.

5.15   We use Coulomb's law to answer this question: $E \propto \dfrac{Q_1 \times Q_2}{d}$ where Q_1 is the charge on the cation, A^+, and Q_2 is the charge on the anion, B^-.

a. Doubling the radius of the cation would increase the distance, d, between the centers of the ions. A larger value of d results in a smaller energy, E, of the ionic bond. Doubling the radius of the cation **decreases the ionic bond energy**.

b. Tripling the charge on A^+ **triples the ionic bond energy**, since the energy of the bond is directly proportional to the charge on the cation, Q_1.

c. Doubling the charge on both the cation and anion **increases the bond energy by a factor of 4**.

d. Decreasing the radius of both the cation and the anion to half of their original values is the same as halving the distance, d, between the centers of the ions. Decreasing the radii of A^+ and B^- to half their original **values increases the bond energy by a factor of 2**.

5.17   Lewis representations for the ionic reactions are as follows.

a. $\dot{B}a \; + \; \cdot \ddot{O} \cdot \;\; \longrightarrow \;\; Ba^{2+} \left[: \ddot{O} : \right]^{2-}$

b. $\dot{A}\dot{l} \cdot \; + \; \cdot \ddot{N} \cdot \;\; \longrightarrow \;\; Al^{3+} \left[: \ddot{N} : \right]^{3-}$

5.23   **Strategy:**   To determine the formula of the compounds formed from each pair of ions, first determine their ratios of combination using the charges on the cation and anion. The sum of the charges on the cation and anion in the formula must be zero.

Combine the names for each cation and anion, eliminating the word *ion* to name the compounds.

Solution: a. K^+ and F^- combine in a 1:1 ratio to form **KF**. KF contains the potassium ion and the fluoride ion.

Combining the cation and anion names, and eliminating the word *ion* from each of the individual ions' names, we get **potassium fluoride** as the name for KF.

b. Rb^+ and O^{2-} combine in a 2:1 ratio to form **Rb_2O**. Rb_2O contains the rubidium ion and the oxide ion.

Rb_2O is **rubidium oxide**.

c. Ba^{2+} and P^{3-} combine in a 3:2 ratio to form **Ba_3P_2**. Ba_3P_2 contains the barium ion and the phosphide ion.

Ba_3P_2 is **barium phosphide**.

d. Ga^{3+} and Se^{2-} combine in a 2:3 ratio to form **Ga_2Se_3**. Ga_2Se_3 contains the gallium ion and the selenide ion.

Ga_2Se_3 is **gallium selenide**.

5.25   **Strategy:**   Identify the ions in each compound and determine their ratios of combination using the charges on the cation and anion in each.

  Setup: a. Sodium oxide is a combination of Na^+ and O^{2-}. In order to prepare a neutral compound, these two ions must combine in a 2:1 ratio.

     b. Iron sulfide is a combination of Fe^{2+} and S^{2-}. These ions combine in a 1:1 ratio.

     c. Cobalt telluride is a combination of Co^{3+} and Te^{2-}. These ions combine in a 2:3 ratio.

     d. Barium fluoride is a combination of Ba^{2+} and F^-. These ions combine in a 1:2 ratio.

  Solution: The formulas are:

     a. **Na_2O**   b. **FeS**   c. **Co_2Te_3**   d. **BaF_2**

5.27   **Strategy:**   Begin by identifying the cation and anion in each compound, and then combine the names for each, eliminating the word *ion*.

  Setup: a. NaH contains Na^+ and H^-, the sodium ion and the hydride ion.

     b. Li_3N contains Li^+ and N^{3-}, the lithium ion and the nitride ion.

     c. Na_2O contains Na^+ and O^{2-}, the sodium ion and the oxide ion.

     d. Na_2O_2 contains Na^+ and O_2^{2-}, the sodium ion and the peroxide ion. We know that oxygen is O_2^{2-} because it combines with Na^+ in a 2:1 ratio to give a neutral formula.

  Solution: a. Combining the cation and anion name, and eliminating the word *ion* from each of the individual ions' names, we get **sodium hydride** as the name of NaH.

     b. Li_3N is **lithium nitride**.

     c. Na_2O is **sodium oxide**.

     d. Na_2O_2 is **sodium peroxide**.

5.29   **Strategy:**   Green spheres represent cations, and red spheres represent anions.

  Setup: BaS contains Ba^{2+} and S^{2-} combined in a 1:1 ratio. The compound contains one cation and one anion.

     CaF_2 contains Ca^{2+} and F^- combined in a 1:2 ratio. The compound contains one cation and two anions.

     Mg_3N_2 contains Mg^{2+} and N^{3-} combined in a 3:2 ratio. The compound contains three cations and two anions.

     K_2O contains K^+ and O^{2-} combined in a 2:1 ratio. The compound contains two cations and one anion.

Solution: a.   G = Green R = Red

b.

c.

d.

5.41   a. This is a polyatomic molecule that is an **element**. It is not a compound.

      b. This is a polyatomic molecule that is a **compound**.

      c. This is a diatomic molecule that is a **compound**.

5.43   **Elements:** N_2, S_8, H_2

      Compounds: NH_3, NO, CO, CO_2, SO_2

5.45 $\dfrac{\text{g blue: 1.00 g red (right)}}{\text{g blue: 1.00 g red (left)}} = \dfrac{2/3}{1/1} \approx 0.667 \approx$ **2:3**

5.47   a. Dividing all subscripts by 2, the simplest whole number ratio of the atoms in C_2N_2 is **CN**.

      b. Dividing all subscripts by 6, the simplest whole number ratio of the atoms in C_6H_6 is **CH.**

      c. The molecular formula as written, **C_9H_{20}**, contains the simplest whole number ratio of the atoms present. In this case, the molecular formula and the empirical formula are the same.

      d. Dividing all subscripts by 2, the simplest whole number ratio of the atoms in P_4O_{10} is **P_2O_5**.

      e. Dividing all subscripts by 2, the simplest whole number ratio of the atoms in B_2H_6 is **BH_3**.

5.49   **$C_3H_7NO_2$**

5.55   a. **nitrogen trichloride**

      b. **iodine heptafluoride**

      c. **tetraphosphorus hexoxide**

      d. **disulfur dichloride**

5.57   All of these are molecular compounds. We use prefixes to express the number of each atom in the molecule. The molecular formulas and names are:

a. **NF$_3$: nitrogen trifluoride**

b. **PBr$_5$: phosphorus pentabromide**

c. **SCl$_2$: sulfur dichloride**

5.59   **Strategy:**   A functional group is a group of atoms that determines the chemical properties of an organic compound. Table 5.9 contains a list of common organic functional groups.

        Solution: The molecule contains an –OH group and a –CHO group. These are the **alcohol and aldehyde functional groups**.

5.61   **Strategy:**   Begin by identifying the cation and the anion in each compound, and then combine the names for each, eliminating the word *ion*.

        Setup: a. KClO contains K^+ and ClO^-, the potassium ion and the hypochlorite ion.

b. Ag_2CO_3 contains Ag^+ and CO_3^{2-}, the silver ion and the carbonate ion.

c. HNO_2 contains H^+ and NO_2^-, the hydrogen ion and the nitrite ion.

d. $KMnO_4$ contains K^+ and MnO_4^-, the potassium ion and the permanganate ion.

e. $CsClO_3$ contains Cs^+ and ClO_3^-, the cesium ion and the chlorate ion.

f. KNH_4SO_4 contains K^+, NH_4^+, and SO_4^{2-}, the potassium ion, the ammonium ion, and the sulfate ion.

g. $Fe(BrO_4)_2$ contains Fe^{2+} and BrO_4^-, the iron(II) ion and the perbromate ion. We know that the iron in $Fe(BrO_4)_2$ is iron(II), because it is combined with the perbromate ion in a 1:2 ratio.

h. K_2HPO_4 contains K^+ and HPO_4^{2-}, the potassium ion and the hydrogen phosphate ion.

        Solution: a. Combining the cation and anion names, and eliminating the word *ion* from each of the individual ions' names, we get **potassium hypochlorite** as the name of KClO.

b. Ag_2CO_3 is **silver carbonate**.

c. HNO_2 is **hydrogen nitrite**.

d. $KMnO_4$ is **potassium permanganate**.

e. $CsClO_3$ is **cesium chlorate**.

f. KNH_4SO_4 is **potassium ammonium sulfate**.

g. $Fe(BrO_4)_2$ is **iron(II) perbromate**.

h. K_2HPO_4 is **potassium hydrogen phosphate**.

5.63   **Strategy:**   Identify the ions in each compound, and determine their ratios of combination using the charges on the cation and anion in each. The sum of the charges on the cation and anion in the formula must be zero.

For binary acids, remove the *-gen* ending from hydrogen (leaving *hydro-*), change the *-ide* ending on the second element to *-ic*, combine the two words, and add the word *acid*.

Setup: a. Copper(I) cyanide is a combination of Cu^+ and CN^-. In order to prepare a neutral compound, these two ions must combine in 1:1 ratio.

b. Strontium chlorite is a combination of Sr^{2+} and ClO_2^-. These ions combine in a 1:2 ratio.

c. Perbromic acid is a combination of H^+ and BrO_4^-. These ions combine in a 1:1 ratio.

d. Hydroiodic acid is a combination H^+ and I^-. These ions combine in a 1:1 ratio.

e. Disodium ammonium phosphate is a combination of Na^+, NH_4^+, and PO_4^{3-}. These ions combine in a 2:1:1 ratio.

f. Lead(II) carbonate is a combination of Pb^{2+} and CO_3^{2-}. These ions combine in a 1:1 ratio.

g. Tin(II) sulfite is a combination of Sn^{2+} and SO_3^{2-}. These ions combine in a 1:1 ratio.

h. Cadmium thiocyanate is a combination of Cd^{2+} and SCN^-. These ions combine in a 1:2 ratio.

Solution: The formulas are

a. **CuCN**

b. **Sr(ClO₂)₂**

c. **HBrO₄**

d. **HI**

e. **Na₂NH₄PO₄**

f. **PbCO₃**

g. **SnSO₃**

h. **Cd(SCN)₂**

5.67 **Strategy:** Add the masses of all atoms in the formula. Remember that the absence of a subscript means that there is one atom of that element present.

Solution: a. CH_3Cl $1(12.01 \text{ amu}) + 3(1.008 \text{ amu}) + 1(35.45 \text{ amu}) = \textbf{50.48 amu}$

b. N_2O_4 $2(14.01 \text{ amu}) + 4(16.00 \text{ amu}) = \textbf{92.02 amu}$

c. SO_2 $1(32.07 \text{ amu}) + 2(16.00 \text{ amu}) = \textbf{64.07 amu}$

d. C_6H_{12} $6(12.01 \text{ amu}) + 12(1.008 \text{ amu}) = \textbf{84.16 amu}$

e. H_2O_2 $2(1.008 \text{ amu}) + 2(16.00 \text{ amu}) = \textbf{34.02 amu}$

f. $C_{12}H_{22}O_{11}$ $12(12.01 \text{ amu}) + 22(1.008 \text{ amu}) + 11(16.00 \text{ amu}) = \textbf{342.3 amu}$

g. NH_3 $1(14.01 \text{ amu}) + 3(1.008 \text{ amu}) = \textbf{17.03 amu}$

5.69 Using the appropriate atomic masses,

a. CH_4 $1(12.01 \text{ amu}) + 4(1.008 \text{ amu}) = \textbf{16.04 amu}$

b. NO_2 $1(14.01 \text{ amu}) + 2(16.00 \text{ amu}) = \textbf{46.01 amu}$

c. SO_3 $1(32.07 \text{ amu}) + 3(16.00 \text{ amu}) = \textbf{80.07 amu}$

d. C_6H_6 $6(12.01 \text{ amu}) + 6(1.008 \text{ amu}) = \textbf{78.11 amu}$

e. NaI $1(22.99 \text{ amu}) + 1(126.9 \text{ amu}) = \textbf{149.9 amu}$

f. K_2SO_4 $2(39.10 \text{ amu}) + 1(32.07 \text{ amu}) + 4(16.00 \text{ amu}) = \textbf{174.27 amu}$

g. $Ca_3(PO_4)_2$ $3(40.08 \text{ amu}) + 2(30.97 \text{ amu}) + 8(16.00 \text{ amu}) = \textbf{310.2 amu}$

5.71 **Strategy:** First, write the formula of each of ionic substance based their lattice diagrams. Recall that the structure of an ionic compound consists of a vast array of interspersed cations and anions called a lattice. The lattice diagram shows a three-dimensional network of alternating cations and anions.

Then, calculate the formula mass for each compound by summing all the atomic masses.

Setup: a. The cation is Rb^+ and the anion is Cl^-. In order to prepare a neutral compound, these ions combine in a 1:1 ratio to give RbCl.

b. The cation is Ba^{2+} and the anion is Se^{2-}. These ions combine in a 1:1 ratio to give BaSe.

c. The cation is Cd^{2+} and the anion is O^{2-}. These ions combine in a 1:1 ratio to give CdO.

Solution: For each compound, sum the atomic mass of each element.

a. The formula mass of RbCl is $85.47 \text{ amu} + 35.45 \text{ amu} = \textbf{120.92 amu}$.

b. The formula mass of BaSe is $137.3 \text{ amu} + 78.96 = \textbf{216.3 amu}$.

c. The formula mass of CdO is $112.4 \text{ amu} + 16.00 \text{ amu} = \textbf{128.4 amu}$.

5.75 **Strategy:** Recall the procedure for calculating a percentage. Assume that we have 1 mole of SnO_2. The percent by mass of each element (Sn and O) is given by the mass of that element in 1 mole of SnO_2 divided by the molar mass of SnO_2, then multiplied by 100 to convert from a fractional number to a percentage.

Solution: Molar mass of $SnO_2 = (118.7 \text{ g/mol}) + 2(16.00 \text{ g/mol}) = 150.7 \text{ g/mol}$

$$\%Sn = \frac{118.7 \text{ g/mol}}{150.7 \text{ g/mol}} \times 100\% = \mathbf{78.77\%}$$

$$\%O = \frac{(2)(16.00 \text{ g/mol})}{150.7 \text{ g/mol}} \times 100\% = \mathbf{21.23\%}$$

5.77 **Strategy:** In a formula this complicated, it is easy to count the atoms incorrectly, so rewrite the formula by gathering atoms of the same kind together: $Ca_5(PO_4)_3(OH) = Ca_5(P_3O_{12})(OH) = Ca_5P_3O_{13}H$

Formula mass of $Ca_5P_3O_{13}H = 5(40.08 \text{ amu}) + 3(30.97 \text{ amu}) + 13(16.00 \text{ amu}) + (1.008 \text{ amu}) = 502.32 \text{ amu}$

Solution:

$$\% \text{ Ca} = \frac{5 \times 40.08 \text{ amu}}{502.32 \text{ amu}} \times 100\% = \mathbf{39.89\%}$$

$$\% \text{ P} = \frac{3 \times 30.97 \text{ amu}}{502.32 \text{ amu}} \times 100\% = \mathbf{18.50\%}$$

$$\% \text{ O} = \frac{13 \times 16.00 \text{ amu}}{502.32 \text{ amu}} \times 100\% = \mathbf{41.41\%}$$

$$\% \text{ H} = \frac{1 \times 1.008 \text{ amu}}{502.32 \text{ amu}} \times 100\% = \mathbf{0.2007\%}$$

Check: $39.89 + 18.50 + 41.41 + 0.2007 = 100.00$

Note that in problems such as this, rounding error may result in the final sum of percentages not being exactly equal to 100.

5.79

	Compound	Molar mass (g/mol)	N% by mass
a.	$(NH_2)_2CO$	60.06	$\dfrac{2(14.01 \text{ g/mol})}{60.06 \text{ g/mol}} \times 100\% = 46.65\%$
b.	NH_4NO_3	80.05	$\dfrac{2(14.01 \text{ g/mol})}{80.05 \text{ g/mol}} \times 100\% = 35.00\%$
c.	$HNC(NH_2)_2$	59.08	$\dfrac{3(14.01 \text{ g/mol})}{59.08 \text{ g/mol}} \times 100\% = 71.14\%$

d. NH_3 17.03 $\dfrac{14.01 \text{ g/mol}}{17.03 \text{ g/mol}} \times 100\% = 82.27\%$

Ammonia, NH_3, is the richest source of nitrogen on a mass percentage basis.

5.83   To find the molar mass (g/mol), we simply divide the mass (in g) by the number of moles.

$$\frac{152 \text{ g}}{0.372 \text{ mol}} = \textbf{409 g / mol}$$

5.85   **Strategy:**   We are given grams of glucose and asked to solve for atoms of C, H, and O. We cannot convert directly from grams glucose to atoms of each element. What conversions must we do to convert from grams of a compound to atoms of its constituent elements? How should Avogadro's number be used here?

Solution: To calculate number of atoms, we first must convert grams of glucose to moles of glucose. We use the molar mass of glucose as a conversion factor. Once moles of glucose are obtained, we can use Avogadro's number to convert from moles of glucose to molecules of glucose. From molecules of glucose we can determine atoms of individual elements using the molecular formula of glucose.

molar mass of $C_6H_{12}O_6$ = 6(12.01 g/mol) + 12(1.008 g/mol) + 6(16.00 g/mol) = 180.16 g/mol

The conversion factor needed is:
$$\frac{1 \text{ mol } C_6H_{12}O_6}{180.16 \text{ g } C_6H_{12}O_6}$$

Avogadro's number is the key to the second conversion. We have:

$$1 \text{ mol} = 6.022 \times 10^{23} \text{ particles (molecules)}$$

From this equality, we can write the conversion factor:

$$\frac{6.022 \times 10^{23} \text{ molecules } C_6H_{12}O_6}{1 \text{ mol } C_6H_{12}O_6}$$

The subscript for each element in the molecular formula is the key to the third conversion.

Let's complete these three conversions in one step to determine the number of C atoms.

grams of glucose → moles of glucose → number of glucose molecules
→ number of carbon atoms

$$1.50 \text{ g glucose} \times \frac{1 \text{ mol glucose}}{180.16 \text{ g glucose}} \times \frac{6.022 \times 10^{23} \text{ molecules glucose}}{1 \text{ mol glucose}} \times \frac{6 \text{ C atoms}}{1 \text{ molecule glucose}}$$

$$= \textbf{3.01} \times \textbf{10}^{\textbf{22}} \textbf{ C atoms}$$

The ratio of H atoms to C atoms in glucose is 2:1. Therefore, there are twice as many H atoms in glucose as C atoms, so the number of H atoms = 2(3.01 × 10^{22} atoms) = **6.02 × 10^{22} H atoms**. The ratio of O atoms to C atoms in glucose is 1:1. Therefore, there are the same number of O atoms in glucose as C atoms, so the number of O atoms = **3.01 × 10^{22} O atoms**.

5.87   The molar mass of $C_{19}H_{38}O$ is 282.5 g.

$$1.0 \times 10^{-12} \text{ g} \times \frac{1 \text{ mol}}{282.5 \text{ g}} \times \frac{6.022 \times 10^{23} \text{ molecules}}{1 \text{ mol}} = \textbf{2.1} \times \textbf{10}^9 \textbf{ molecules}$$

Notice that even though 1.0×10^{-12} g is an extremely small mass, it still contains over a billion pheromone molecules!

5.89   The molar mass of cinnamic alcohol is 134.17 g/mol.

a.
$$\%C = \frac{(9)(12.01 \text{ g/mol})}{134.17 \text{ g/mol}} \times 100\% = \textbf{80.56\%}$$

$$\%H = \frac{(10)(1.008 \text{ g/mol})}{134.17 \text{ g/mol}} \times 100\% = \textbf{7.51\%}$$

$$\%O = \frac{16.00 \text{ g/mol}}{134.17 \text{ g/mol}} \times 100\% = \textbf{11.93\%}$$

b.
$$1.028 \text{ g } C_9H_{10}O \times \frac{1 \text{ mol } C_9H_{10}O}{134.17 \text{ g } C_9H_{10}O} \times \frac{6.022 \times 10^{23} \text{ molecules } C_9H_{10}O}{1 \text{ mol } C_9H_{10}O}$$

$$= \textbf{4.614} \times \textbf{10}^{21} \textbf{ molecules } C_9H_{10}O$$

5.91   **Strategy:**   Tin(II) fluoride is composed of Sn and F. The mass due to F is based on its percentage by mass in the compound. How do we calculate mass percent of an element?

Solution: First, we must find the mass % of fluorine in SnF_2. Then, we convert this percentage to a fraction and multiply by the mass of the compound (7.10 g) to find the mass of fluorine in 7.10 g of SnF_2.

The percent by mass of fluorine in tin(II) fluoride is calculated as follows:

$$\text{mass } \% \text{ F} = \frac{\text{mass of F in 1 mol } SnF_2}{\text{molar mass of } SnF_2} \times 100\%$$

$$= \frac{2(19.00 \text{ g})}{156.7 \text{ g}} \times 100\% = 24.25\% \text{ F}$$

Converting this percentage to a fraction, we obtain $24.25/100 = 0.2425$.

Next, multiply the fraction by the total mass of the compound.

$$? \text{ g F in 7.10 g } SnF_2 = (0.2425)(7.10 \text{ g}) = \textbf{1.72 g F}$$

5.93   First, calculate the number of grams of arsenic(III) oxide corresponding to the LD_{50}, noting that the units of LD_{50} are g/kg (grams of poison per kg of body weight). The formula for arsenic(III) oxide is As_2O_3.

$$? \text{ g As}_2O_3 = 212 \text{ lb} \times \frac{1 \text{ kg}}{2.20 \text{ lb}} \times \frac{0.018 \text{ g As}_2O_3}{1 \text{ kg}} = 1.735 \text{ g As}_2O_3$$

Next, convert grams to number of molecules using the molar mass of As_2O_3 and Avogadro's number.

$$? \text{ As}_2O_3 = 1.735 \text{ g} \times \frac{\text{mol As}_2O_3}{197.84 \text{ g}} \times \frac{6.022\times10^{23} \text{ AsO}_3 \text{ molecules}}{\text{mol AsO}_3} = \mathbf{5.28 \times 10^{21} \text{ molecules}}$$

5.95   **Strategy:**   Assume a 100-g sample so that the mass percentages of carbon, hydrogen, and oxygen, given in the problem statement correspond to the masses of C, H, and O in the compound. Then, using the appropriate molar masses, convert the grams of each element to moles. Use the resulting numbers as subscripts in the empirical formula, reducing them to the lowest possible whole numbers. To calculate the molecular formula, first divide the molar mass given in the problem statement by the empirical formula mass. Then, multiply the subscripts in the empirical formula by the resulting number to obtain the subscripts in the molecular formula.

Setup: The molar masses of C, H, and O are 12.01 g/mol, 1.008 g/mol, and 16.00 g/mol, respectively. One hundred grams of a compound that is 80.56 percent C, 7.51 percent H, and 11.92 percent O by mass contains 80.56 g C, 7.51 g H, and 11.92 g O.

Solution:
$$80.56 \text{ g C} \times \frac{1 \text{ mol C}}{12.01 \text{ g C}} = 6.708 \text{ mol C}$$

$$7.51 \text{ g H} \times \frac{1 \text{ mol H}}{1.008 \text{ g H}} = 7.45 \text{ mol H}$$

$$11.92 \text{ g O} \times \frac{1 \text{ mol O}}{16.00 \text{ g O}} = 0.745 \text{ mol O}$$

This gives a formula of $C_{6.708}H_{7.45}O_{0.745}$. Dividing the subscripts by the smallest value to get the smallest possible whole numbers ($6.708/0.745 \approx 9$, $7.45/0.745 = 10$, $0.745/0.745 = 1$) gives an empirical formula of $C_9H_{10}O$.

Finally, dividing the approximate molar mass (268 g/mol) by the empirical formula mass [9(12.01 g/mol) + 10(1.008 g/mol) + 16.00 g/mol = 134.2 g/mol] gives $268/134.2 \approx 2$. Then, multiplying the subscripts in the empirical formula by 2 gives the molecular formula, $\mathbf{C_{18}H_{20}O_2}$.

5.97   **Strategy:**   Refer to the labels on the atoms to determine the molecular formula. Then, calculate the molar mass and use the molar mass to convert to grams of acid.

Setup: There are two C atoms, seven H atoms, two O atoms, and one As atom, so the subscript on C will be 2, the subscript on H will be 7, the subscript on O will be 2, and there will be no subscript on As. The molecular formula of cacodylic acid is $C_2H_7O_2As$.

The molar mass of $C_2H_7O_2As$ is 2(12.01 g/mol) + 7(1.008 g/mol) + 2(16.00 g/mol) + 74.92 g/mol = 138.00 g/mol.

Solution: The number of grams is:

$$3.57 \text{ mol } C_2H_7O_2As \times \frac{138.00 \text{ g } C_2H_7O_2As}{1 \text{ mol } C_2H_7O_2As} = \textbf{493 g } C_2H_7O_2As$$

5.99   Sodium is an alkali metal. The charge of the metal ion is +1. Since the charge on oxygen is –2, sodium and oxygen must combine in a 2:1 ratio to give a neutral compound: **Na_2O**

Strontium is an alkaline earth metal. You should expect the charge of the metal to be the same (+2) as calcium: **SrO**.

5.101   **NaCl is an ionic compound; it does not consist of molecules**.

5.103   The species and their identification are as follows:

a. SO_2 **molecule and compound**

b. S_8 **element and molecule**

c. Cs   **element**

d. N_2O_5 **molecule and compound**

e. O   **element**

f. O_2 **element and molecule**

g. O_3 **element and molecule**

h. CH_4 **molecule and compound**

i. KBr   **compound, not molecule**

j. S   **element**

k. P_4 **element and molecule**

l. LiF   **compound, not molecule**

5.105   a.

	Ethane	Acetylene
	2.65 g C	4.56 g C
	0.665 g H	0.383 g H

Let's compare the ratio of the hydrogen masses in the two compounds. To do this, we need to start with the same mass of carbon. If we were to start with 4.56 g of C in ethane, how much hydrogen would combine with 4.56 g of carbon?

$$0.665 \text{ g H} \times \frac{4.56 \text{ g C}}{2.65 \text{ g C}} = 1.14 \text{ g H}$$

We can calculate the ratio of H in the two compounds.

$$\frac{1.14 \text{ g}}{0.383 \text{ g}} \approx 3$$

Yes; This is consistent with the Law of Multiple Proportions which states that if two elements combine to form more than one compound, the masses of one element that combine with a fixed mass of the other element are in ratios of small whole numbers. In this case, the ratio of the masses of hydrogen in the two compounds is 3:1.

b. For a given amount of carbon, there is three times the amount of hydrogen in ethane compared to acetylene. A reasonable formula for **ethane** would be **any formula with C:H = 1:3 (CH_3, C_2H_6, etc.);** a reasonable formula for **acetylene** would be **any formula with C:H = 1:1 (CH_3, C_2H_6, etc.):**

Ethane	Acetylene
CH_3	CH
C_2H_6	C_2H_2

5.107   Compounds of metals with nonmetals are usually ionic. Nonmetal-nonmetal compounds are usually molecular.

Ionic: **LiF, $BaCl_2$, KCl**

Molecular: **$SiCl_4$, B_2H_6, C_2H_4**

5.109   Recall that you can classify bonds as ionic or covalent based on electronegativity difference.

The melting points (°C) are shown in parentheses following the formulas.

Ionic: **NaF** (993)   **MgF_2** (1261)   **AlF_3** (1291)

Covalent: **SiF_4** (−90.2)   **PF_5** (−83)   **SF_6** (−50)   **ClF_3** (−83)

5.111

Cation	Anion	Formula	Name
Mg^{2+}	HCO_3^-	$Mg(HCO_3)_2$	Magnesium bicarbonate
Sr^{2+}	Cl^-	$SrCl_2$	**Strontium chloride**
Fe^{3+}	NO_2^-	$Fe(NO_2)_3$	**Iron(III) nitrite**
Mn^{2+}	ClO_3^-	$Mn(ClO_3)_2$	Manganese(II) chlorate
Sn^{4+}	Br^-	$SnBr_4$	**Tin(IV) bromide**

5.113   **Strategy:**   Use Equation 5.1 to determine the percent by mass contributed by carbon in each of these compounds.

        Setup: a. There are two atoms of carbon in C_2H_2. The molecular mass of C_2H_2 is 2(12.01 amu) + 2(1.008 amu) = 26.04 amu.

                         b. There are six atoms of carbon in C_6H_6. The molecular mass of C_6H_6 is 6(12.01 amu) + 6(1.008 amu) = 78.11 amu.

c. There are two atoms of fluorine in C_2H_6. The molecular mass of C_2H_6 is
2(12.01 amu) + 6(1.008 amu) = 30.07 amu.

d. There are three atoms of carbon in C_3H_8. The molecular mass of C_3H_8 is
3(12.01 amu) + 8(1.008 amu) = 44.09 amu.

Solution: Multiply the number of atoms of carbon by its atomic mass, divide by the molecular mass, and multiply by 100 percent.

a.
$$\% \, C = \frac{2 \times 12.01 \text{ amu C}}{26.04 \text{ amu C}_2\text{H}_2} \times 100\% = \textbf{92.24\%}$$

b.
$$\% \, C = \frac{6 \times 12.01 \text{ amu C}}{78.11 \text{ amu C}_6\text{H}_6} \times 100\% = \textbf{92.25\%}$$

c.
$$\% \, C = \frac{2 \times 12.01 \text{ amu C}}{30.07 \text{ amu C}_2\text{H}_6} \times 100\% = \textbf{79.88\%}$$

d.
$$\% \, C = \frac{3 \times 12.01 \text{ amu C}}{44.09 \text{ amu C}_3\text{H}_8} \times 100\% = \textbf{81.72\%}$$

Therefore, $\textbf{C}_2\textbf{H}_6 < \textbf{C}_3\textbf{H}_8 < \textbf{C}_2\textbf{H}_2 = \textbf{C}_6\textbf{H}_6$

5.115   a. B and F should form a **covalent** compound; both elements are nonmetals. One possibility would be **BF_3, boron trifluoride**.

b. K and Br will form an **ionic** compound; K is a metal while Br is a nonmetal. The substance will be **KBr, potassium bromide**.

5.117   **Strategy:** Refer to the labels on the atoms to determine the molecular formula. Then, calculate the molar mass of isoflurane and use the molar mass to convert to moles of isoflurane.

Solution: There are three carbon atoms, two hydrogen atoms, one oxygen atom, five fluorine atoms, and one chlorine atom, so the subscript on C will be 3, the subscript on H will be 2, the subscript on F will be 5, and there will be no subscript on O or Cl. The **molecular formula of isoflurane is $C_3H_2OF_5Cl$**.

The **molar mass** of $C_3H_2OF_5Cl$ is
3(12.01 g/mol) + 2(1.008 g/mol) + 16.00 g/mol + 5(19.00 g/mol) + 35.45 g/mol = **184.50 g/mol**.

The number of moles is:

$$3.82 \text{ g C}_3\text{H}_2\text{OF}_5\text{Cl} \times \frac{1 \text{ mol C}_3\text{H}_2\text{OF}_5\text{Cl}}{184.50 \text{ g C}_3\text{H}_2\text{OF}_5\text{Cl}} = \textbf{0.0207 mol C}_3\textbf{H}_2\textbf{OF}_5\textbf{Cl}$$

5.119   The symbol "O" refers to moles of oxygen atoms, not oxygen molecule (O_2). Look at the molecular formulas given in parts (a) and (b). What do they tell you about the relative amounts of carbon and oxygen?

a. $0.212 \text{ mol C} \times \dfrac{1 \text{ mol O}}{1 \text{ mol C}} = \textbf{0.212 mol O}$

b. $0.212 \text{ mol C} \times \dfrac{2 \text{ mol O}}{1 \text{ mol C}} = \textbf{0.424 mol O}$

5.121 Typically, **ionic** compounds are composed of a metal cation and a nonmetal anion. **RbCl and KO$_2$** are ionic compounds.

Typically, **covalent** compounds are composed of nonmetals. **PF$_5$, BrF$_3$, and CI$_4$** are covalent compounds.

5.123 **Strategy:** Identify the ions in the compound, and determine the charges on the cation and anion by using their ratio of combination. The subscript for the cation is numerically equal to the charge on the anion and the subscript on the anion is numerically equal to the charge on the cation.

Setup: Al$_2$(MoO$_4$)$_3$ combines the two ions in a 2:3 ratio. Al$_2$(MoO$_4$)$_3$ contains Al^{3+} and MoO$_4^{2-}$.

Mg is located in Group 2A or 2 and has a charge of +2. Mg^{2+} will combine with the MoO$_4^{2-}$ ion in a 1:1 ratio.

Solution: The formula of magnesium molybdate is **MgMoO$_4$**.

5.125 The molar mass of chlorophyll is 893.48 g/mol. Finding the mass of a 0.0011-mol sample:

$$0.0011 \text{ mol chlorophyll} \times \frac{893.48 \text{ g chlorophyll}}{1 \text{ mol chlorophyll}} = 0.98 \text{ g chlorophyll}$$

This is greater than 0.72, so the **0.011 mol chlorophyll** sample has the greater mass.

5.127 **Strategy:** First, use Avogadro's number and the molar mass of hydrogen to convert from number of atoms to moles of hydrogen. Using the appropriate molar masses, convert grams of Cl and O to moles. Use the resulting numbers as subscripts in the empirical formula, reducing them to the lowest possible whole numbers for the final answer.

Setup: Starting with the number of atoms of hydrogen (6.02×10^{23}), we use Avogadro's number to convert to the number of moles of hydrogen.

The molar masses of H, Cl, and O are 1.008 g/mol, 35.45 g/mol, and 16.00 g/mol, respectively.

Solution: Convert number of atoms of hydrogen to moles:

$$6.02 \times 10^{23} \text{ H atoms} \times \frac{1 \text{ mol H}}{6.022 \times 10^{23} \text{H atoms}} = 1.00 \text{ mol H}$$

Convert the grams of each remaining element to moles:

$$35.45 \text{ g Cl} \times \frac{1 \text{ mol Cl}}{35.45 \text{ g Cl}} = 1.000 \text{ mol Cl}$$

$$64.00 \text{ g O} \times \frac{1 \text{ mol O}}{16.00 \text{ g O}} = 4.000 \text{ mol O}$$

This gives a formula of $H_{1.00}Cl_{1.000}O_{4.000}$. Dividing the subscripts by the smallest value to get the smallest possible whole numbers (1.00/1.00 = 1, 1.000/1.00 = 1, 4.000/1.00 = 4) gives an empirical formula of **$HClO_4$**.

5.129   Molar mass of $C_4H_8Cl_2S$ = 4(12.01 g/mol) + 8(1.008 g/mol) + 2(35.45 g/mol) + 32.07 g/mol = 159.07 g/mol

$$\%C = \frac{4(12.01 \text{ g/mol})}{159.07 \text{ g/mol}} \times 100\% = \textbf{30.20\%}$$

$$\%H = \frac{8(1.008 \text{ g/mol})}{159.07 \text{ g/mol}} \times 100\% = \textbf{5.069\%}$$

$$\%Cl = \frac{2(35.45 \text{ g/mol})}{159.07 \text{ g/mol}} \times 100\% = \textbf{44.57\%}$$

$$\%S = \frac{32.07 \text{ g/mol}}{159.07 \text{ g/mol}} \times 100\% = \textbf{20.16\%}$$

5.131   a. The molar mass of hemoglobin is:

$$2952(12.01 \text{ g/mol}) + 4664(1.008 \text{ g/mol}) + 812(14.01 \text{ g/mol}) + 832(16.00 \text{ g/mol})$$
$$+ 8(32.07 \text{ g/mol}) + 4(55.85 \text{ g/mol})$$

$$= \textbf{6.532} \times \textbf{10}^\textbf{4} \textbf{ g/mol}$$

b. To solve this problem, the following conversions need to be completed:

L → mL → red blood cells → hemoglobin molecules → mol hemoglobin → mass hemoglobin

We will use the following abbreviations: RBC = red blood cells, HG = hemoglobin

$$5.0 \text{ L} \times \frac{1 \text{ mL}}{0.001 \text{ L}} \times \frac{5.0 \times 10^9 \text{ RBC}}{1 \text{ mL}} \times \frac{2.8 \times 10^8 \text{ HG molecules}}{1 \text{ RBC}} = 7 \times 10^{18} \text{ HG molecules}$$

$$7 \times 10^{18} \text{ HG molecules} \times \frac{1 \text{ mol HG}}{6.022 \times 10^{23} \text{ molecules HG}} \times \frac{6.532 \times 10^4 \text{ g HG}}{1 \text{ mol HG}} = \textbf{7.6} \times \textbf{10}^\textbf{-2} \textbf{ g HG}$$

5.133   a. $$8.38 \text{ g KBr} \times \frac{1 \text{ mol KBr}}{119.0 \text{ g KBr}} \times \frac{6.022 \times 10^{23} \text{ KBr}}{1 \text{ mol KBr}} \times \frac{1 \text{ K}^+ \text{ ion}}{1 \text{ KBr}} = \textbf{4.24} \times \textbf{10}^\textbf{22} \textbf{ K}^\textbf{+} \textbf{ ions}$$

Since there is one Br^- for every one K^+, the number of Br^- ions = **4.24×10^{22} Br$^-$ ions**

b. Since there are two Na^+ for every one SO_4^{2-}, the number of SO_4^{2-} ions = **2.29×10^{22} SO$_4^{2-}$ ions**

c.

$$7.45 \text{ g Ca}_3(\text{PO}_4)_2 \times \frac{1 \text{ mol Ca}_3(\text{PO}_4)_2}{310.18 \text{ g Ca}_3(\text{PO}_4)_2} \times \frac{6.022 \times 10^{23} \text{ Ca}_3(\text{PO}_4)_2}{1 \text{ mol Ca}_3(\text{PO}_4)_2} \times \frac{3 \text{ Ca}^{2+} \text{ ions}}{1 \text{ Ca}_3(\text{PO}_4)_2}$$

$$= \mathbf{4.34 \times 10^{22} \text{ Ca}^{2+} \text{ ions}}$$

Since there are three Ca^{2+} for every two PO_4^{3-}, the number of PO_4^{3-} ions is:

$$4.34 \times 10^{22} \text{ Ca}^{2+} \text{ ions} \times \frac{2 \text{ PO}_4^{3-} \text{ ions}}{3 \text{ Ca}^{2+} \text{ ions}} = \mathbf{2.89 \times 10^{22} \text{ PO}_4^{3-} \text{ ions}}$$

5.135   a. The mass of chlorine is 5.0 g.

b. From the percent by mass of Cl, we can calculate the mass of chlorine in 60.0 g of $NaClO_3$.

$$\text{mass \% Cl} = \frac{35.45 \text{ g Cl}}{106.44 \text{ g compound}} \times 100\% = 33.31\% \text{ Cl}$$

$$\text{mass Cl} = 60.0 \text{ g} \times 0.3331 = 20.0 \text{ g Cl}$$

c. 0.10 mol of KCl contains 0.10 mol of Cl.

$$0.10 \text{ mol Cl} \times \frac{35.45 \text{ g Cl}}{1 \text{ mol Cl}} = 3.5 \text{ g Cl}$$

d. From the percent by mass of Cl, we can calculate the mass of chlorine in 30.0 g of $MgCl_2$.

$$\text{mass \% Cl} = \frac{(2)(35.45 \text{ g Cl})}{95.21 \text{ g compound}} \times 100\% = 74.47\% \text{ Cl}$$

$$\text{mass Cl} = 30.0 \text{ g} \times 0.7447 = 22.3 \text{ g Cl}$$

e. The mass of Cl can be calculated from the molar mass of Cl_2.

$$0.50 \text{ mol Cl}_2 \times \frac{70.90 \text{ g Cl}}{1 \text{ mol Cl}_2} = 35 \text{ g Cl}$$

Thus, **(e) 0.50 mol Cl$_2$** contains the greatest mass of chlorine.

5.137   The molar mass of air can be calculated by multiplying the mass of each component by its abundance and adding them together. Recall that nitrogen gas and oxygen gas are diatomic.

molar mass of air = (0.7808)(28.02 g/mol) + (0.2095)(32.00 g/mol) + (0.0097)(39.95 g/mol) = **28.97 g/mol**

Chapter 6
Representing Molecules

Key Skills

6.1   a, c

6.2   b, e

6.3   c

6.4   a

Problems

6.9   a. **covalent; BF$_3$; boron trifluoride**   b. **ionic; KBr; potassium bromide**

6.13   **Strategy:** Use Equation 6.1 to solve for the dipole moment, μ, of AlF.

      Setup: The partial charges on Al and F are +0.019 and –0.019, respectively. In units of electronic charge, this is 0.019 e^-.
The charge units of electronic charge must be converted to coulombs and the distance between the partial charges converted to meters. The calculated dipole moments should be expressed as debyes (1 D = 3.336×10^{-30} C $\cdot$ m).

$$0.019\ e^- \times \frac{1.6022 \times 10^{-19}\ \text{C}}{1\ e^-} = 3.044 \times 10^{-21}\ \text{C}$$

$$165\ \text{pm} \times \frac{1.0 \times 10^{-12}\ \text{m}}{1\ \text{pm}} = 1.65 \times 10^{-10}\ \text{m}$$

      Solution:

$$\mu = Q \times r$$

$$\mu = \left(3.044 \times 10^{-21}\ \text{C}\right) \times \left(1.65 \times 10^{-10}\ \text{m}\right)$$

$$\mu = 5.023 \times 10^{-31}\ \text{C} \cdot \text{m}$$

Converting to debyes gives

$$\mu = 5.023 \times 10^{-31}\ \text{C} \cdot \text{m} \times \frac{1\ \text{D}}{3.336 \times 10^{-30}\ \text{C} \cdot \text{m}} = \mathbf{0.151\ D}$$

6.15   **Strategy:** Use Equation 6.1 to calculate the dipole moment of BaO assuming that the charges on Ba and O are +2 and –2, respectively. Then, rearrange Equation 6.2 to solve for the measured (observed) dipole moment. The magnitude of the charges must be expressed in coulombs (1 e^- = 1.6022 × 10^{-19} C; therefore 2 e^- = 2 × 1.6022 × 10^{-19} C, or 3.2044 × 10^{-19} C). The bond length (r) must be expressed in meters (1 pm = 1 × 10^{-12} m); and the measured dipole moments should be expressed in debyes (1 D = 3.336 × 10^{-30} C·m).

Setup: The bond length is 194 pm (1.94×10^{-10} m) and the percent ionic character of the Ba–O bond is 85.3% or 85.3/100 = 0.853.

Solution: The expected dipole moment if the magnitude of charges were 3.2044×10^{-19} C is:

$$\mu = Q \times r$$

$$\mu = (3.2044 \times 10^{-19} \text{ C}) \times (1.94 \times 10^{-10} \text{ m})$$

$$\mu = 6.217 \times 10^{-29} \text{ C} \cdot \text{m}$$

Converting to debyes gives:

$$6.217 \times 10^{-29} \text{ C} \cdot \text{m} \times \frac{1 \text{ D}}{3.336 \times 10^{-30} \text{ C} \cdot \text{m}} = 18.636 \text{ D}$$

The measured dipole moment of the Ba–O bond is:

$$\frac{\mu_{measured}}{18.636 \text{ D}} = 0.853$$

$$\mu_{measured} = (18.636 \text{ D}) \times (0.853)$$

$$\mu_{measured} = 15.9 \text{ D}$$

The measured dipole moment of the Ba–O bond is **15.9 D.**

6.17 **Strategy:** We can look up electronegativity values in Figure 6.4 of the text. The amount of ionic character is based on the electronegativity difference between the two atoms. The larger the electronegativity difference, the greater the ionic character.

Solution: Let ΔEN = electronegativity difference. The bonds arranged in order of increasing ionic character are:

C–H (ΔEN = 0.4) < Br–H (ΔEN = 0.7) < F–H (ΔEN = 1.9) < Li–Cl (ΔEN = 2.0) < Na–Cl (ΔEN = 2.1) < K–F (ΔEN = 3.2)

6.19 a. The bond is **covalent. The two silicon atoms are the same.**

b. The bond is **polar covalent. The electronegativity difference between Cl and Si is 3.0 – 1.8 = 1.2.**

c. The bond is **ionic. The electronegativity difference between F and Ca is 4.0 – 1.0 = 3.0.**

d. The bond is **polar covalent. The electronegativity difference between N and H is 3.0 – 2.1 = 0.9.**

6.21 The order of increasing ionic character is:

Cl–Cl (zero difference in electronegativity) **< Br–Cl** (difference 0.2) **< Si–C** (difference 0.7) **< Cs–F** (difference 3.3)

6.23 The Lewis structures are

a. 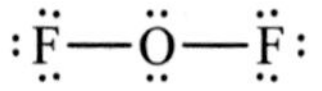

b. 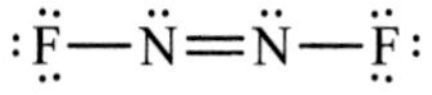

c. 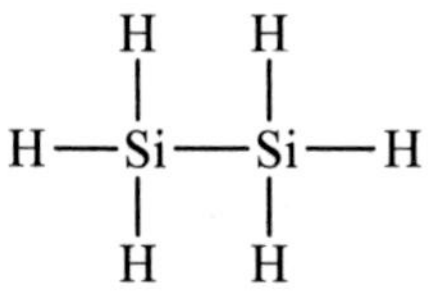

d. 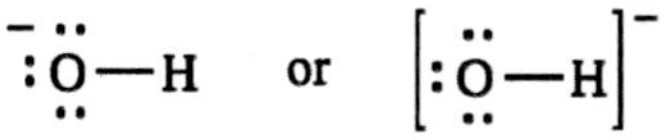

e. 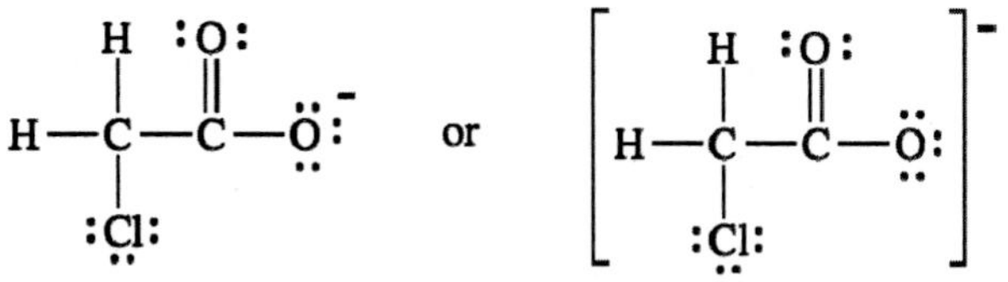

f. 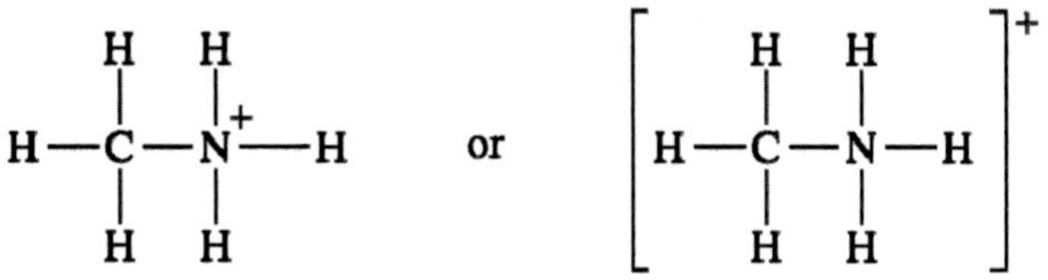

6.25 The Lewis structures are:

a. 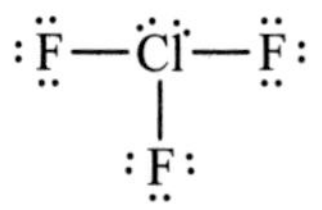

b. 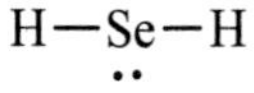

c. 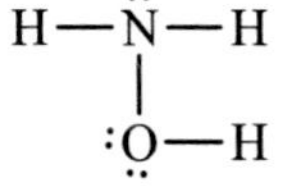

d. 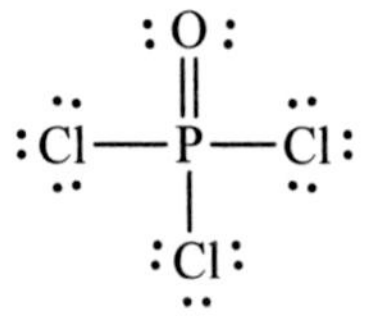

e.
$$
\begin{array}{ccc}
& H & H \\
& | & | \\
H & - C - & C - \ddot{B}r\!: \\
& | & | \\
& H & H
\end{array}
$$

f.
$$
:\ddot{C}l - \ddot{N} - \ddot{C}l: \\
\quad\quad | \\
\quad :\ddot{C}l:
$$

g.
$$
\begin{array}{ccc}
& H & H \\
& | & | \\
H & - C - & N\!: \\
& | & | \\
& H & H
\end{array}
$$

6.29   **Strategy:**   We follow the procedure for drawing Lewis structures outlined in Section 6.3 of the text.

Solution: a. *Step 1:*   The atoms in the Lewis structure are: O N O

Step 2:   The outer-shell electron configuration of O is $2s^2 2p^4$. Each O has six valence electrons. The outer-shell electron configuration of N is $2s^2 2p^3$. N has five valence electrons. Also, we must subtract an electron to account for the positive charge. Thus, there are:

$$(2 \times 6) + 5 - 1 = 16 \text{ valence electrons}$$

Step 3:   We draw single covalent bonds between the atoms and then attempt to complete the octets—first for the O atoms.

$$:\ddot{O} - N - \ddot{O}:$$

Because this uses all the available electrons, Step 5 outlined in the text is not required, but Step 6 is required to move electron pairs to satisfy the octet for nitrogen.

$$\ddot{O} = N = \ddot{O}$$

We determine the formal charges on the atoms using the method outlined in Section 6.4 and Equation 6.2. For each oxygen atom, formal charge = 6 − 6 = 0. For the nitrogen atom, formal charge = 5 − 4 = +1.

$$\ddot{O} = \overset{+}{N} = \ddot{O} \quad \text{or} \quad \left[\ddot{O} = N = \ddot{O}\right]^{+}$$

Check: As a final check, we verify that there are 16 valence electrons in the Lewis structure of NO_2^{+}.

Follow the same procedure as part (a) for parts (b), (c), and (d).

b. $:\overset{-}{\ddot{S}} - C \equiv N:$ or $\left[:\ddot{S} - C \equiv N:\right]^{-}$

c. :S̈—S̈:⁻ or [:S̈—S̈:]²⁻

d. :F̈—C̈l⁺—F̈: or [:F̈—C̈l—F̈:]⁺

6.31   a. **Neither oxygen atom has a complete octet, and the leftmost hydrogen atom shows two bonds (four electrons). Hydrogen can hold only two electrons in its valence shell.**

b. **The correct structure is:**

6.35   **Strategy:** We follow the procedure for drawing Lewis structures outlined in Section 6.3 of the text. After we complete the Lewis structure, we draw the resonance structures.

Solution: Following the procedure in Sections 6.3 and 6.5 of the text, we come up with the following resonance structures:

a.

b.

6.37   The structures of the most important resonance forms are:

6.39   Three reasonable resonance structures for OCN⁻ are:

6.41   Resonance structures differ only in the positions of their *electrons,* not in the positions of their atoms. Structures **(b) and (e)** are resonance structures of $COCl_2$. The structures differ only in the positions of their electrons. Structures (a) and (c) are not resonance structures because they violate the octet rule for C (carbon cannot have more than four bonds). Structure (d) is not a resonance structure because C and O do not obey the octet rule in this structure.

6.43 **Strategy:** Draw a Lewis structure for the guanine molecule in which the atoms are arranged in the same way but the electrons are arranged differently.

Setup: A correct structural formula for guanine is

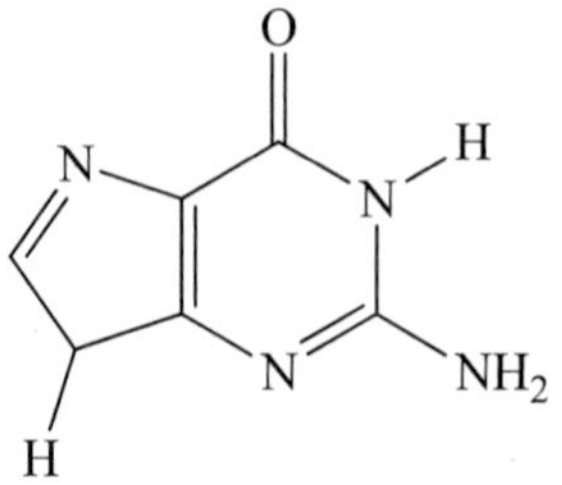

But the double bonds can be put in different positions in the molecule.

Solution:

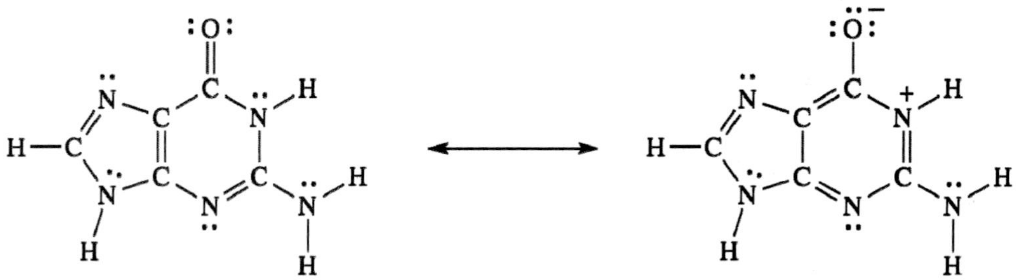

In order for all the atoms in the second structure to obey the octet rule, there are two atoms with nonzero formal charges. When there are nonzero formal charges, they should be consistent with the electronegativities of the atoms in the molecule. A positive formal charge on oxygen, for example, is inconsistent with oxygen's high electronegativity.

6.49 **Strategy:** We follow the procedure outlined in Section 6.3 of the text for drawing Lewis structures. We assign formal charges as discussed in Section 6.4 of the text.

Solution: Drawing the structure with single bonds between Be and each of the Cl atoms, we see that Be is "electron deficient." **No, the octet rule is not followed for Be.** The Lewis structure is:

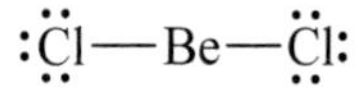

$$:\ddot{\text{C}}\text{l}\!-\!\text{Be}\!-\!\ddot{\text{C}}\text{l}:$$

An octet of electrons on Be can only be formed by making two double bonds as shown below:

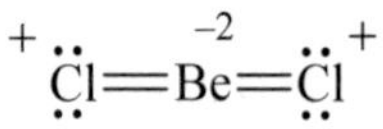

$$\overset{+}{\ddot{\text{C}}\text{l}}\!=\!\overset{-2}{\text{Be}}\!=\!\overset{+}{\ddot{\text{C}}\text{l}}$$

It is not a plausible Lewis structure because this places a high negative formal charge on Be and positive formal charges on the Cl atoms. This structure distributes the formal charges counter to the electronegativities of the elements.

6.51 The reaction can be represented as:

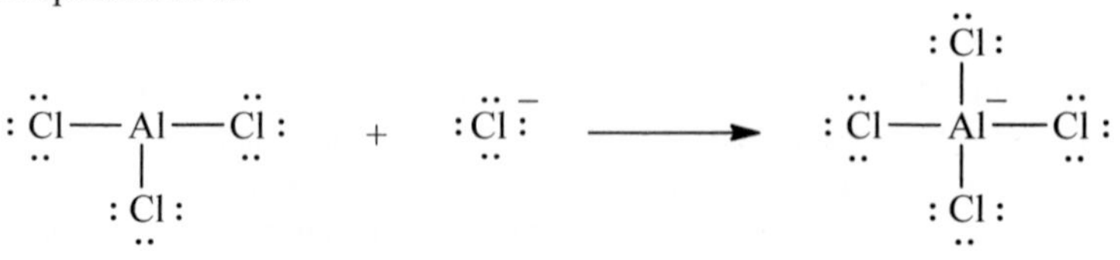

The new bond formed is called a **coordinate covalent bond**.

6.53   The outer electron configuration of antimony is $5s^2 5p^3$. The Lewis structure is shown below. All five valence electrons are shared in the five covalent bonds. **No, the octet rule is not obeyed**.

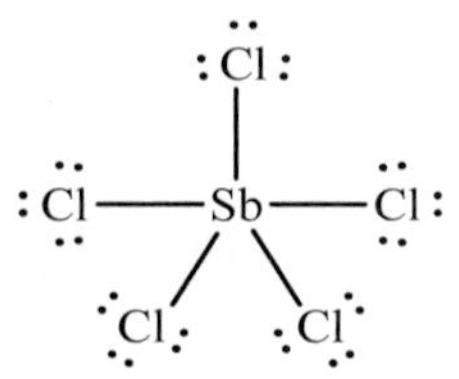

$$H—O—O$$

(structure: $:\overset{..}{C}l:$ central Sb bonded to five Cl atoms)

6.55   **Strategy:**   For HO_2, the structural formula is:

$$H—O—O$$

Because hydrogen has only one electron, it can form only one covalent bond.

For ClO, the structural formula is:

$$Cl—O$$

Setup: **HO_2:** There are a total of 13 valence electrons (one from the H atom and six from each of the two O atoms). We subtract four electrons to account for the two bonds in the skeleton, leaving us with nine electrons to distribute as follows: three lone pairs on the terminal O atom, one lone pair on the central O atom, and the last remaining electron also on the central O atom.

ClO: There are a total of 13 valence electrons (seven from the Cl and six from the O atom). We subtract two electrons to account for the bond in the skeleton, leaving us with 11 electrons to distribute at follows: three lone pairs on the O atom, two lone pairs on the Cl atom, and the last remaining electron also on the Cl atom.

Solution: For HO_2, the Lewis structure is:

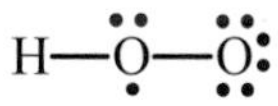

$$H—\overset{..}{\underset{.}{O}}—\overset{..}{\underset{..}{O}}:$$

For ClO, the Lewis structure is

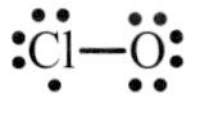

$$:\overset{..}{\underset{..}{C}l}—\overset{..}{\underset{.}{O}}:$$

6.57   **Completed octet on S:** $\left[:\overset{..}{O}—\overset{..}{\underset{:\overset{..}{O}:}{S}}—\overset{..}{O}: \right]^{2-}$; **zero formal charge on S:** $\left[:\overset{..}{\underset{..}{O}}—\overset{:\overset{..}{O}:}{\underset{\parallel}{S}}—\overset{..}{\underset{..}{O}}: \right]^{2-}$

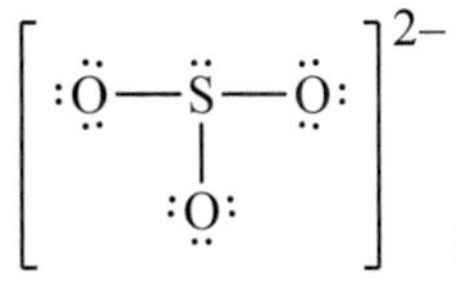

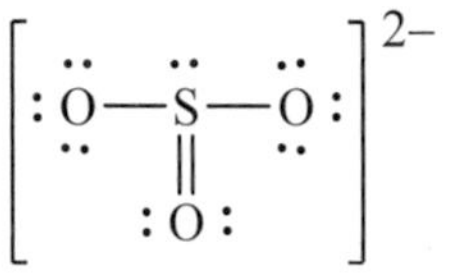

6.59   The resonance structures are:

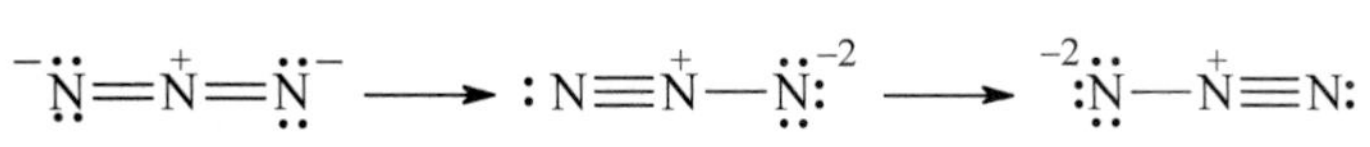

$$\overset{-}{\overset{..}{\underset{..}{N}}}=\overset{+}{N}=\overset{-}{\overset{..}{\underset{..}{N}}} \longrightarrow :N\equiv\overset{+}{N}—\overset{..}{\underset{..}{N}}:^{-2} \longrightarrow {}^{-2}\overset{..}{\underset{..}{N}}—\overset{+}{N}\equiv N:$$

6.61   a. **An example of an aluminum species that satisfies the octet rule is the anion** $AlCl_4^-$. The Lewis dot structure is shown below.

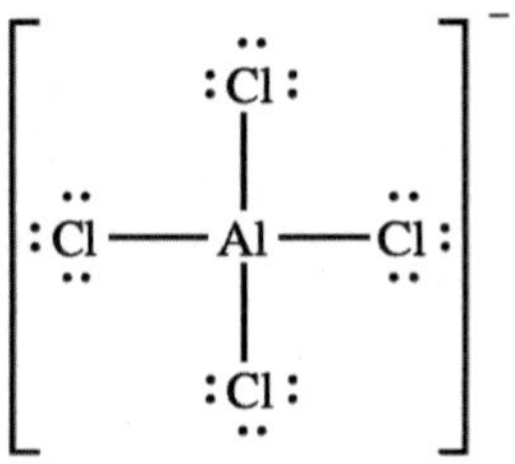

b. **An example of an aluminum species containing an expanded octet is anion** AlF_6^{3-}. The Lewis dot structure is shown below.

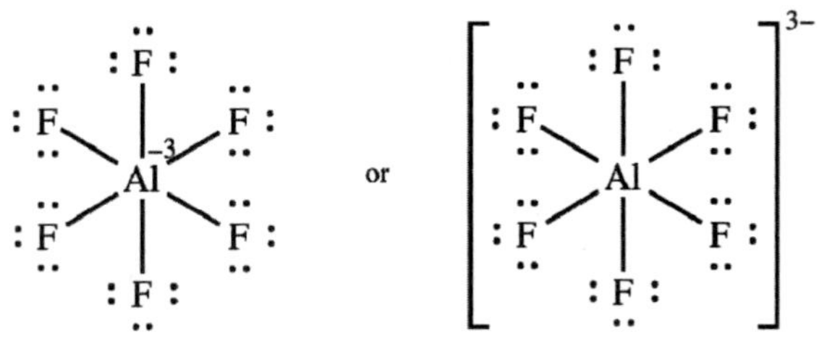

c. **An aluminum species that has an incomplete octet is the compound** $AlCl_3$. The dot structure is shown below.

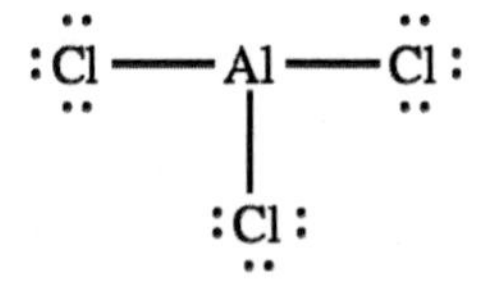

6.63   **CF_2 would be very unstable because carbon does not have an octet.**

LiO_2 would not be stable because the lattice energy between Li^+ and superoxide O_2^- would be too low to stabilize the solid. A charge of +2 would be impossible on Li anyway, because it has only one valence electron.

$CsCl_2$ requires a Cs^{2+} cation. The second ionization energy is too large to be compensated by the increase in lattice energy.

PI_5 appears to be a reasonable species. However, the iodine atoms are too large to have five of them "fit" around a single P atom.

6.65   a. **False**   b. **False**   c. **False**

6.67   **Only N_2 has a triple bond. Therefore, it has the shortest bond length.**

6.69   To be isoelectronic, molecules must have the same number and arrangement of valence electrons. **CH_4 and NH_4^+** are isoelectronic (8 valence electrons), as are **N_2 and CO** (10 valence electrons), as are **C_6H_6 and $B_3N_3H_6$** (30 valence electrons). Draw Lewis structures to convince yourself that the electron arrangements are the same in each isoelectronic pair.

452

6.71   The complete Lewis structure is:

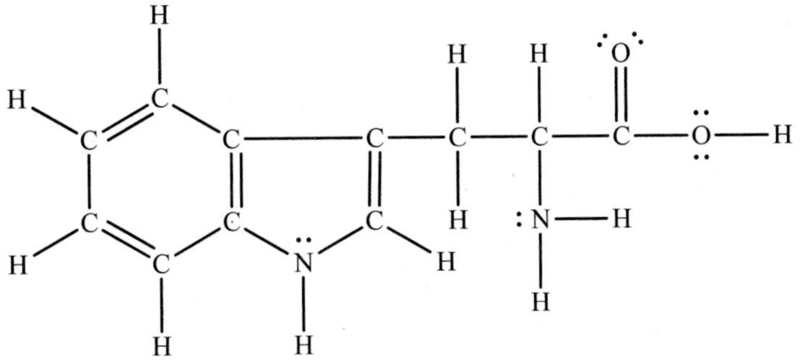

6.73   **Strategy:**   For SO_2^-, the structural formula is:

$$O—S—O$$

This puts the unique atom, S, in the center and puts the more electronegative O atoms in terminal positions.

For SO_3^-, the structural formula is:

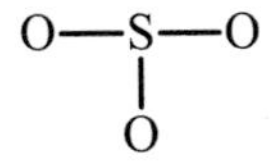

$$O—S—O$$
$$|$$
$$O$$

This puts the unique atom, S, in the center and puts the more electronegative O atoms in terminal positions.

Setup: SO_2^-: There are a total of 18 valence electrons (six from the S and six from each of the two O atoms). We add one electron to account for the –1 charge, which gives us a total of 19 electrons. We subtract four electrons to account for the two bonds in the skeleton, leaving us with 15 electrons. Distribute the electrons as three lone pairs on each O atom and one lone pair as well as the last remaining electron on the S atom. Since the S atom has less than an octet, use one lone pair from one of the O atoms to make a double bond to the S atom.

SO_3^-: There are a total of 24 valence electrons (six from the S and six from each of the three O atoms). We add one electron to account for the –1 charge which gives us a total of 25 electrons. We subtract six electrons to account for the three bonds in the skeleton, leaving us with 18 electrons. Distribute the electrons as three lone pairs on each O atom and the last remaining electron on the S atom. Since the S atom has less than an octet, use one lone pair from two of the O atoms to make two double bonds to the S atom.

Solution: Lewis structure of SO_2^-

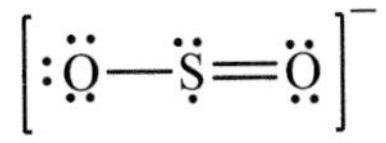

Lewis structure of SO_3^-

$$\left[:\ddot{O}\!-\!\overset{\textstyle .}{S}\!=\!\ddot{O}: \right]^{-}$$
$$\underset{\textstyle \ddot{O}}{\overset{\textstyle \|}{}}$$

6.75   **The central iodine atom in I_3^- has *ten* electrons surrounding it: two bonding pairs and three lone pairs. The central iodine has an expanded octet. Elements in the second period such as fluorine cannot have an expanded octet, as would be required for F_3^-.**

6.77   The atoms in methyl isocyanate are

$$\text{H}$$
$$\text{H} \quad \text{C} \quad \text{N} \quad \text{C} \quad \text{O}$$
$$\text{H}$$

The number of valence electrons is: $(1 \times 3) + (2 \times 4) + 5 + 6 = 22$ valence electrons

We can draw two resonance structures for methyl isocyanate.

$$\text{H}\!-\!\underset{\overset{\textstyle |}{\text{H}}}{\overset{\overset{\textstyle \text{H}}{\textstyle |}}{\text{C}}}\!-\!\ddot{\text{N}}\!=\!\text{C}\!=\!\ddot{\text{O}} \longleftrightarrow \text{H}\!-\!\underset{\overset{\textstyle |}{\text{H}}}{\overset{\overset{\textstyle \text{H}}{\textstyle |}}{\text{C}}}\!-\!\overset{+}{\text{N}}\!\equiv\!\text{C}\!-\!\ddot{\ddot{\text{O}}}:^{-}$$

6.79   **Form (a) is the most important resonance structure with no formal charges and all satisfied octets. Form (b) is likely not as important as (a) because of the positive formal charge on O. Forms (c) and (d) do not satisfy the octet rule for all atoms and are likely not important.**

6.81   a.

$$:\ddot{\text{Cl}}:$$
$$|$$
$$:\ddot{\text{F}}\!-\!\text{C}\!-\!\ddot{\text{Cl}}:$$
$$|$$
$$:\ddot{\text{Cl}}:$$

b.

$$:\ddot{\text{F}}:$$
$$|$$
$$:\ddot{\text{F}}\!-\!\text{C}\!-\!\ddot{\text{Cl}}:$$
$$|$$
$$:\ddot{\text{Cl}}:$$

c.

$$:\ddot{\text{F}}:$$
$$|$$
$$\text{H}\!-\!\text{C}\!-\!\ddot{\text{Cl}}:$$
$$|$$
$$:\ddot{\text{F}}:$$

d.

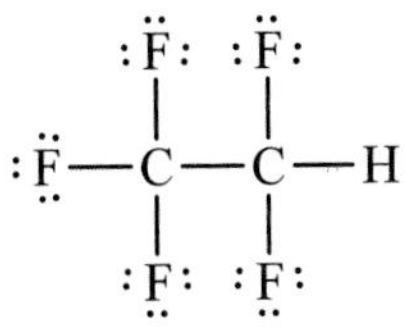

$$:\ddot{F}:\quad:\ddot{F}:$$
$$\quad\;|\qquad\;|$$
$$:\ddot{F}-C-C-H$$
$$\quad\;|\qquad\;|$$
$$:\ddot{F}:\quad:\ddot{F}:$$

6.83   The Lewis structures are:

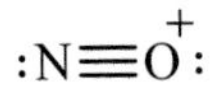

a. 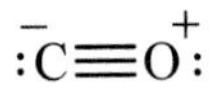 $:\bar{C}\!\equiv\!\overset{+}{O}:$

b. $:N\!\equiv\!\overset{+}{O}:$

c. $:\bar{C}\!\equiv\!N:$

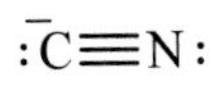

d. 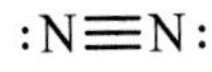 $:N\!\equiv\!N:$

6.85   **The noble gases violate the octet rule. This is because each noble gas atom already has completely filled *ns* and *np* subshells.**

6.87   a.

$$:\dot{N}\!=\!\ddot{O}:\;\longleftrightarrow\;:\bar{\ddot{N}}\!=\!\dot{\overset{+}{O}}:$$

The first structure is the most important. Both N and O have formal charges of zero. In the second structure, the more electronegative oxygen atom has a formal charge of +1. Having a positive formal charge on a highly electronegative atom is not favorable. In addition, both structures leave one atom with an incomplete octet. This cannot be avoided due to the odd number of electrons.

b.   **No,** it is not possible to draw a structure with a triple bond between N and O.

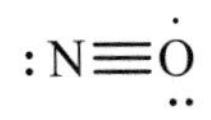

$$:N\!\equiv\!\dot{\ddot{O}}$$

Any structure drawn with a triple bond will lead to an expanded octet. Elements in the second row of the periodic table cannot exceed the octet rule.

6.89

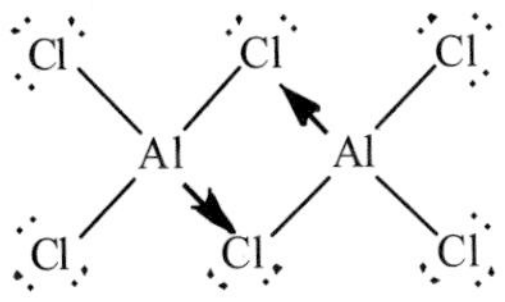

The arrows indicate coordinate covalent bonds. This dimer does not possess a dipole moment.

6.91   **Strategy:**   Use Equation 6.1 to calculate the dipole moments in RbF and RbI assuming that the charges on Rb, F, and I are +1, –1, and –1, respectively. Use Equation 6.2 to calculate percent ionic character. The magnitude of the charges must be expressed in coulombs ($1\,e^{-} = 1.6022 \times 10^{-19}$ C); the bond length (r) must be expressed in meters (1 pm $= 1 \times 10^{-12}$ m); and the calculated dipole moments should be expressed in debyes (1 D $= 3.336 \times 10^{-30}$ C·m).

Setup: The dipole moments of RbF and RbI are 8.55 D and 11.48 D, respectively. Using Figure 4.13, the bond distance of RbF is 285 pm (152 pm (Rb^{+}) + 133 pm (F^{-}) = 285 pm) and the bond distance of RbI is 372 pm (152 pm (Rb^{+}) + 220 pm (I^{-}) = 372 pm). In meters, this gives 2.85×10^{-10} m and 3.72×10^{-10} m, respectively.

Solution: **RbF:**

For a magnitude of charges of 1.6022×10^{-19} C is, the expected dipole moment is:

$$\mu = Q \times r$$

$$\mu = (1.6022 \times 10^{-19} \text{ C}) \times (2.85 \times 10^{-10} \text{ m})$$

$$\mu = 4.566 \times 10^{-29} \text{ C·m}$$

Converting to debyes gives:

$$4.566 \times 10^{-29} \text{ C} \cdot \text{m} \times \frac{1 \text{ D}}{3.336 \times 10^{-30} \text{ C} \cdot \text{m}} = 13.687 \text{ D}$$

The percent ionic character of the Rb–F bond is:

$$\frac{8.55 \text{ D}}{13.687 \text{ D}} \times 100 \% = \mathbf{62.5 \%}$$

RbI:

For a magnitude of charges of 1.6022×10^{-19} C, the expected dipole moment is:

$$\mu = Q \times r$$

$$\mu = (1.6022 \times 10^{-19} \text{ C}) \times (3.72 \times 10^{-10} \text{ m})$$

$$\mu = 5.960 \times 10^{-29} \text{ C} \cdot \text{m}$$

Converting to debyes gives

$$5.960 \times 10^{-29} \text{ C} \cdot \text{m} \times \frac{1 \text{ D}}{3.336 \times 10^{-30} \text{ C} \cdot \text{m}} = 17.866 \text{ D}$$

The percent ionic character of the Rb–I bond is

$$\frac{11.48 \text{ D}}{17.866 \text{ D}} \times 100 \% = \mathbf{64.25\%}$$

We would expect the percent ionic character to be higher for RbF than it is for RbI, given that fluorine is more electronegative.

6.93   **Strategy:** Use Equation 6.1 to solve for the dipole moments, μ. The electrostatic potential map shows regions where electrons spend a lot of time in red and regions where electrons spend very little time in blue. (Regions where electrons spend a moderate amount of time are in green.) The electrons in covalent bonds, with low dipole moments, spend roughly the same amount of time in the vicinity of each atom. The electrostatic potential map will be symmetrical, with the electrons spending time in the vicinity of each atom. The higher the dipole moment, the more time an electron will spend in the vicinity of one atom over the other. In this case, the electrostatic potential map will show the electrons spending most of their time near one atom (red) and not much time near the other (blue).

Setup: First calculate the dipole moments:

Partial charges of +0.19 and –0.19:

In units of electronic charge, this is $0.19 \ e^-$.

The charge units of electronic charge must be converted to coulombs and the distance between the partial charges converted to meters.

$$0.19 \ e^- \times \frac{1.6022 \times 10^{-19} \ C}{1 \ e^-} = 3.04 \times 10^{-20} \ C$$

$$213 \ \text{pm} \times \frac{1 \times 10^{-12} \ m}{1 \ \text{pm}} = 2.13 \times 10^{-10} \ m$$

The dipole moment is:

$$\mu = Q \times r$$

$$\mu = (3.04 \times 10^{-20} \ C) \times (2.13 \times 10^{-10} \ m)$$

$$\mu = 6.475 \times 10^{-30} \ C \cdot m$$

Converting to debyes gives:

$$6.475 \times 10^{-30} \ C \cdot m \times \frac{1 \ D}{3.336 \times 10^{-30} \ C \cdot m} = \mathbf{1.94 \ D}$$

Partial charges of +0.051 and –0.051:

In units of electronic charge, this is $0.051 \ e^-$.

$$0.051 \ e^- \times \frac{1.6022 \times 10^{-19} \ C}{1 \ e^-} = 8.171 \times 10^{-21} \ C$$

$$214 \ \text{pm} \times \frac{1 \times 10^{-12} \ m}{1 \ \text{pm}} = 2.14 \times 10^{-10} \ m$$

The dipole moment is:

$$\mu = Q \times r$$

$$\mu = (8.171 \times 10^{-21} \ C) \times (2.14 \times 10^{-10} \ m)$$

$$\mu = 1.749 \times 10^{-30} \ C \cdot m$$

Converting to debyes gives:

$$1.749 \times 10^{-30} \ C \cdot m \times \frac{1 \ D}{3.336 \times 10^{-30} \ C \cdot m} = \mathbf{0.524 \ D}$$

Partial charges of +0.68 and –0.68: In units of electronic charge, this is 0.68 e^-.

$$0.68 \ e^- \times \frac{1.6022 \times 10^{-19} \ C}{1 \ e^-} = 1.09 \times 10^{-19} \ C$$

$$315 \ pm \times \frac{1 \times 10^{-12} \ m}{1 \ pm} = 3.15 \times 10^{-10} \ m$$

The dipole moment is:

$$\mu = Q \times r$$

$$\mu = (1.09 \times 10^{-19} \ C) \times (3.15 \times 10^{-10} \ m)$$

$$\mu = 3.433 \times 10^{-29} \ C \cdot m$$

Converting to debyes gives

$$3.433 \times 10^{-29} \ C \cdot m \times \frac{1 \ D}{3.336 \times 10^{-30} \ C \cdot m} = \textbf{10.3 D}$$

Solution:	Compound	Partial charges	Distance between charges (pm)	Dipole moment
	B	±0.19	213	**1.94 D**
	C	±0.051	214	**0.524 D**
	A	±0.68	315	**10.3 D**

6.95   Pyridine has a structure similar to benzene (see Section 6.5), but one of the carbon atoms is replaced by a nitrogen atom. The two resonance structures of pyridine are:

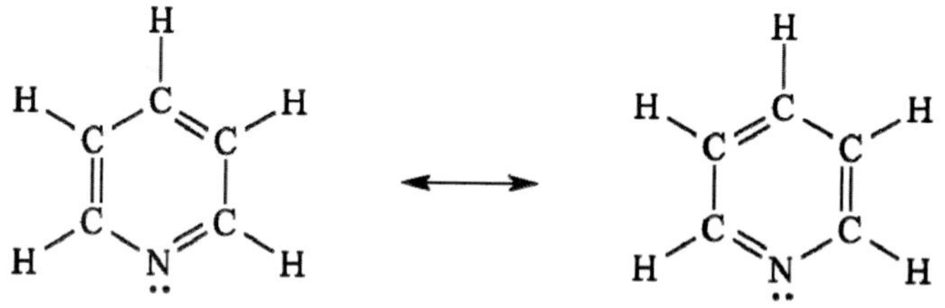

In each structure, there is one single C–N bond and one double C–N bond. Therefore, **the C–N bond length will be shorter than that of a single bond (143 pm) and longer than that of a double bond (138 pm), approximately 141 pm.**

6.97   :N≡Ṅ⁺—Ṅ̈⁻=Ṅ⁺=Ṅ̈⁻ ⟷ ⁻Ṅ̈=Ṅ⁺=Ṅ̈=Ṅ⁺≡N: ⟷ :N≡Ṅ⁺—Ṅ̈⁻—Ṅ⁺≡N:

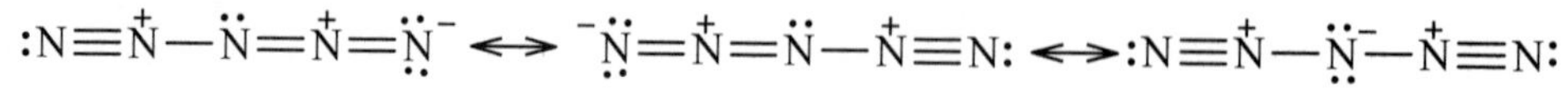

6.99 **Using Mulliken's definition,**

$$EN(O) = \frac{1314 + 141}{2} = \textbf{727.5} \qquad\qquad EN(F) = \frac{1681 + 328}{2} = \textbf{1004.5}$$

$$EN(Cl) = \frac{1256 + 349}{2} = \textbf{802.5}$$

When converted to the Pauling scale,

$$\textbf{EN(O)} = \frac{727.5}{230} = \textbf{3.16} \text{ or } \textbf{3.2}, \quad \textbf{EN(F)} = \frac{1004.5}{230} = \textbf{4.37} \text{ or } \textbf{4.4}, \quad \textbf{EN(Cl)} = \frac{802.5}{230} = \textbf{3.49} \text{ or } \textbf{3.5}$$

These values compare to the Pauling values for oxygen of 3.5, fluorine of 4.0, and chlorine of 3.0.

Chapter 7
Molecular Geometry, Intermolecular Forces, and Bonding Theories

Visualizing Chemistry

VC 7.1 c   VC 7.2 b   VC 7.3 b   VC 7.4 b

Key Skills

7.1   c   7.2   d   7.3   b   7.4   a, e

Problems

7.7

	Lewis Structure	Electron-domain geometry	Molecular geometry
(a)	$\ddot{Br}$ $\ddot{Br}$—C—$\ddot{Br}$ $\ddot{Br}$	tetrahedral	**tetrahedral**
(b)	$\ddot{Cl}$—B—$\ddot{Cl}$ $\ddot{Cl}$	trigonal planar	**trigonal planar**
(c)	$\ddot{F}$—N—$\ddot{F}$ $\ddot{F}$	tetrahedral	**trigonal pyramidal**
(d)	H—$\ddot{Se}$—H	tetrahedral	**bent**
(e)	$\ddot{O}$=N—$\ddot{O}$$^{-}$	trigonal planar	**bent**

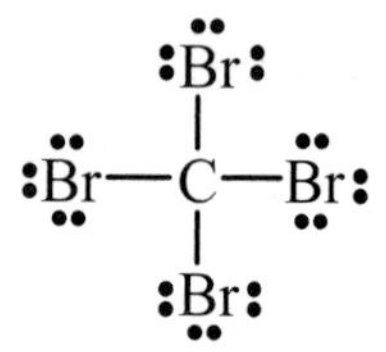

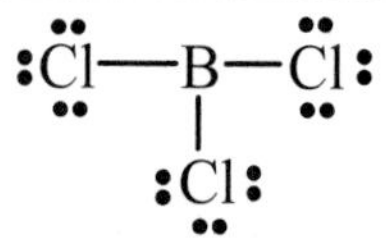

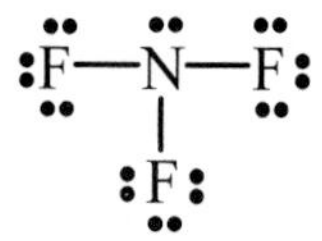

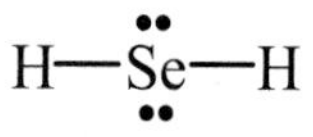

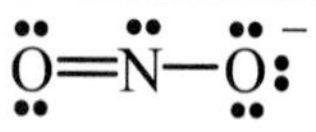

7.9 **Strategy:** The sequence of steps in determining molecular geometry is as follows:

draw Lewis $\longrightarrow$ count electron- $\longrightarrow$ find electron- $\longrightarrow$ determine molecular
structure   domains around   domain geometry   geometry based on
                             the central atom   position of atoms

Solution: a. The Lewis structure of PCl_3 is shown below. In the VSEPR method, it is the number of bonding pairs and lone pairs of electrons around the *central atom* (phosphorus, in this case) that is important in determining the structure; the lone pairs of electrons around the chlorine atoms have been omitted for simplicity. There are four electron domains around the central atom. The electron-domain geometry is tetrahedral. There is one lone pair around the central atom, phosphorus, which makes this an AB_3 case with one lone pair. The information in Table 7.2 shows that the structure is **trigonal pyramidal** like ammonia.

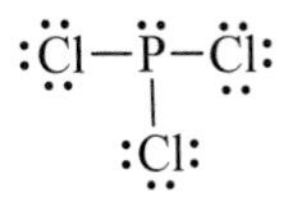

$$:\ddot{Cl}-\overset{..}{\underset{|}{P}}-\ddot{Cl}:$$
$$:\underset{..}{\overset{|}{Cl}}:$$

b. The Lewis structure of $CHCl_3$ is shown below. There are four electron domains, all of which are bonds, around carbon, which makes this an AB_4 case with no lone pairs. The molecule should be **tetrahedral** like the AB_4 example shown in Figure 7.2.

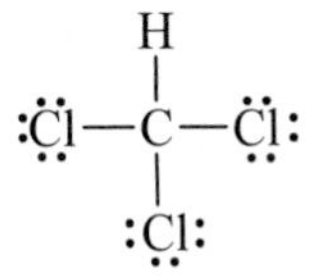

$$\overset{\displaystyle H}{\underset{\displaystyle :\underset{..}{\overset{|}{Cl}}:}{:\ddot{Cl}-\overset{|}{\underset{|}{C}}-\ddot{Cl}:}}$$

c. The Lewis structure of SiH_4 is shown below. Like part (b), it is a **tetrahedral** AB_4 molecule.

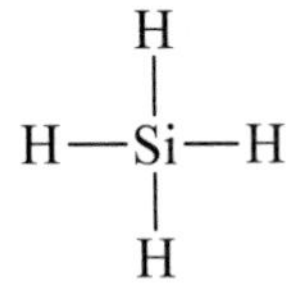

$$\overset{\displaystyle H}{\underset{\displaystyle H}{H-\overset{|}{\underset{|}{Si}}-H}}$$

d. The Lewis structure of $TeCl_4$ is shown below. There are five electron domains, four of which are bonds, which make this an AB_4 case with one lone pair. Consulting Table 7.2 shows that the structure should be **seesaw-shaped** like SF_4.

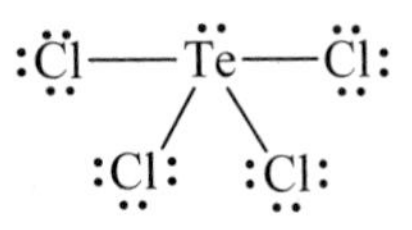

$$:\ddot{Cl}-\overset{..}{Te}-\ddot{Cl}:$$
$$:\ddot{Cl}: \quad :\ddot{Cl}:$$

7.11   a. AB_2 (no lone pairs on the central atom)   The geometry of $HgBr_2$ is **linear**.

b. AB_2 (no lone pairs on the central atom)   The geometry of N_2O is **linear**.

c. AB_2 (no lone pairs on the central atom)   The geometry of SCN^- is **linear**.

7.13   The Lewis structure is:

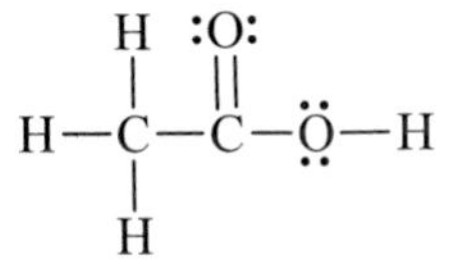

$$\overset{\displaystyle H \quad :\overset{..}{O}:}{\underset{\displaystyle H}{H-\overset{|}{\underset{|}{C}}-\overset{||}{C}-\ddot{O}-H}}$$

carbon at the center of H_3C—

electron-domain geometry = tetrahedral, molecular geometry = tetrahedral

$$:\overset{..}{O}:$$
carbon at center of $-\overset{||}{C}-\ddot{O}-H$

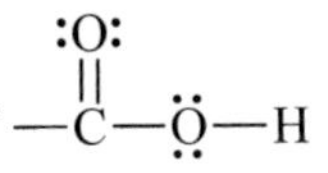

electron-domain geometry = trigonal planar, molecular geometry = trigonal planar

oxygen in $-\ddot{O}-H$

electron-domain geometry = tetrahedral, molecular geometry = bent

7.15   A lone pair takes up more space than the bonding pairs. As a result, the **bond angles may deviate slightly from the ideal angle.**

T-shaped (a): A T-shaped molecular geometry indicates five electron domains (three single bonds and two lone pairs) on the central atom. The lone pair will repel the other bonds more strongly than the bonds repel one another. It therefore "squeezes" them closer together than the ideal angle of 90°. Therefore, **angles of <90° are possible.**

Seesaw-shaped (b): A seesaw-shaped molecular geometry indicates five electron domains (four single bonds and one lone pair) on the central atom. The lone pair will repel the other bonds more strongly than the bonds repel one another. The bonds will be squeezed closer together than the ideal angle of 120°. Therefore, **it is not possible for this angle to be >120°.**

Trigonal pyramidal (c): A trigonal pyramidal molecular geometry indicates four electron domains (three single bonds and one lone pair) on the central atom. The lone pair will repel the other bonds more strongly than the bonds repel one another. It therefore squeezes them closer together and **angles of <109.5° are possible.**

Square pyramidal (d): A square pyramidal molecular geometry indicates six electron domains (five single bonds and one lone pair) on the central atom. The lone pair will repel the other bonds more strongly than the bonds repel one another. The bonds will be squeezed closer together than the ideal angle of 90°. Therefore, **it is not possible for this angle to be >90°.**

7.19   **Strategy:**   For each molecule, draw the Lewis structure, use the VSEPR model to determine its molecular geometry, and then determine whether the individual bond dipoles cancel.

   Setup: a. The Lewis structure of BrF_5 is:

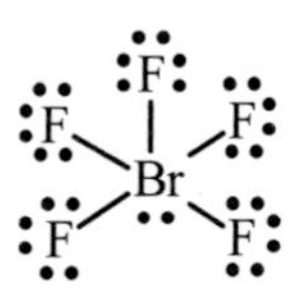

There are six electron domains around the central atom: five single bonds and one lone pair of electrons. The electron-domain geometry is octahedral. Because one of the domains is a lone pair, the molecular geometry is square pyramidal. The equatorial bond dipoles will cancel each other, but the axial dipoles will not sum to zero.

   b. The Lewis structure of ClF_3 is:

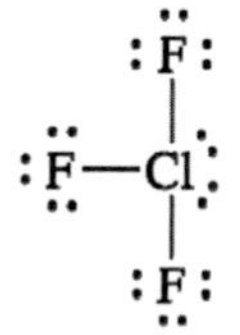

There are five electron domains around the central atom: three single bonds and two lone pairs. The electron-domain geometry is trigonal bipyramidal. Because two of the domains are lone pairs, the molecular geometry is T-shaped. The axial bond dipoles will cancel each other, but the equatorial dipoles will not sum to zero.

   c. The Lewis structure of BCl_3 is:

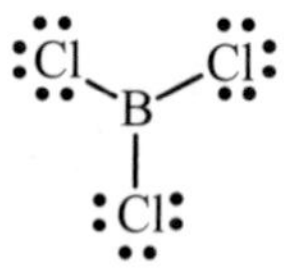

There are three electron domains around the central atom: three single bonds. The electron-domain geometry is trigonal planar. Because there are no lone pairs, the molecular geometry is also trigonal planar. The bond dipoles are symmetrically distributed around the B atom and therefore cancel each other.

Solution: a. BrF_5 is **polar**.

b. ClF_3 is **polar**.

c. BCl_3 is **nonpolar**.

7.21 **Only (c) is polar.** The others are symmetrical distributions of identical bonds, with the bond dipoles summing to zero.

7.29 **Strategy:** Classify the species into three categories: ionic, polar (possessing a dipole moment), and nonpolar. Keep in mind that dispersion forces exist between *all* species.

Solution: The three molecules are essentially nonpolar. There is little difference in electronegativity between carbon and hydrogen. Thus, the only type of intermolecular attraction in these molecules is dispersion forces. Other factors being equal, the molecule with the greater number of electrons will exert greater intermolecular attractions. By looking at the molecular formulas you can predict that the order of increasing boiling points will be $CH_4 < C_3H_8 < C_4H_{10}$.

Butane would be a liquid in winter (boiling point −44.5°C), and on the coldest days even propane would become a liquid (boiling point −0.5°C). Only methane would remain gaseous (boiling point −161.6°C).

7.31 a. Benzene (C_6H_6) molecules are nonpolar. Only **dispersion** forces will be present.

b. Chloroform (CH_3Cl) molecules are polar. **Dispersion and dipole-dipole** forces will be present.

c. Phosphorus trifluoride (PF_3) molecules are polar. **Dispersion and dipole-dipole** forces will be present.

d. Sodium chloride (NaCl) is a lattice of alternating ions, so single molecules do not exist. **Dispersion and ionic** forces will be present, but the dispersion force is insignificant.

e. Carbon disulfide (CS_2) molecules are nonpolar. Only **dispersion** forces will be present.

7.33 In this problem, you must identify the species capable of hydrogen bonding among themselves, not with water. In order for a molecule to be capable of hydrogen bonding with another molecule like itself, it must have at least one hydrogen atom bonded to N, O, or F. Of the choices, only **(e) CH₃COOH (acetic acid)** shows this structural feature. The others cannot form hydrogen bonds among themselves.

7.35 **Strategy:** The molecule with the stronger intermolecular forces will have the higher boiling point. If a molecule contains an N–H, O–H, or F–H bond it can form intermolecular hydrogen bonds. A hydrogen bond is a particularly strong dipole-dipole intermolecular attraction.

Solution: **1-Butanol has greater intermolecular forces because it can form hydrogen bonds.** (It contains an O–H bond.) **Therefore, it has the higher boiling point.** Diethyl ether molecules do contain both oxygen atoms and hydrogen atoms. However, all the hydrogen atoms are bonded to carbon, not oxygen. There is no hydrogen bonding in diethyl ether, because carbon is not electronegative enough.

7.37   a. **Xe: it is larger and therefore has stronger dispersion forces.**

b. **CS$_2$: it is larger (both the molecules are nonpolar) and therefore has stronger dispersion forces.**

c. **Cl$_2$: it is larger (both the molecules are nonpolar) and therefore has stronger dispersion forces.**

d. **LiF: it is an ionic compound, and the ion-ion attractions are much stronger than the dispersion forces between F$_2$ molecules.**

e. **NH$_3$: it can form hydrogen bonds and PH$_3$ cannot.**

7.39   **Strategy:**   Classify the species into three categories: ionic, polar (possessing a dipole moment), and nonpolar. Also look for molecules that contain an N–H, O–H, or F–H bond, which are capable of forming intermolecular hydrogen bonds. Keep in mind that dispersion forces exist between *all* species.

Solution: a. Water has O–H bonds. Therefore, water molecules can form hydrogen bonds. The attractive forces that must be overcome are **dispersion and dipole-dipole, including hydrogen bonding.**

b. Bromine (Br$_2$) molecules are nonpolar. The forces that must be overcome are **dispersion only**.

c. Iodine (I$_2$) molecules are nonpolar. The forces that must be overcome are **dispersion only**.

d. In this case, the F–F bond must be broken. This is an *intra*molecular force between two F atoms, not an *inter*molecular force between F$_2$ molecules. The attractive forces that must be overcome are **covalent bonds**.

7.41   The Lewis structures for PH$_3$ and SiH$_4$ are:

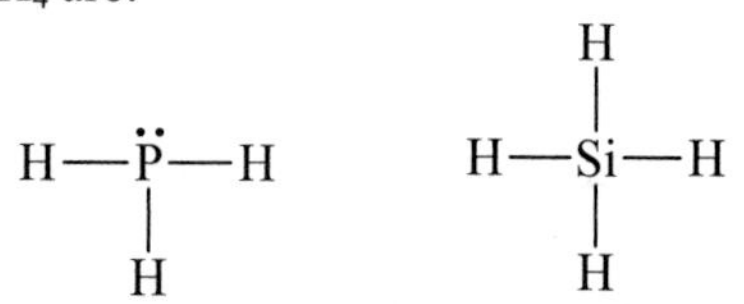

The strength of intermolecular forces determines the physical state of a substance under a given set of conditions. Stronger intermolecular attractions will draw and hold the molecules together, causing the gas to condense to a liquid.

For molecules of comparable molar mass, the magnitude of intermolecular attractions generally increases with increasing polarity. In the case of SiH$_4$, the molecule is tetrahedral (four electron domains around the central atom) and the bond dipoles are symmetrically distributed around the Si atom. Therefore, the bond dipoles sum to zero and the molecule is nonpolar.

In the case of PH$_3$, the molecule is trigonal pyramidal (three single bonds and one lone pair around the central atom). Because of the lone pair of electrons, the bond dipoles do not sum to zero. Therefore, the molecule is polar.

Therefore, **PH$_3$ will condense to a liquid at a higher temperature because it is polar and has stronger intermolecular forces.**

7.43   **The compound with –NO$_2$ and –OH groups on adjacent carbons can form hydrogen bonds with itself (*intra*molecular hydrogen bonds). Such bonds do not contribute to *inter*molecular attraction and do not help raise the melting point of the compound. The other compound with the –NO$_2$ and –OH groups on the opposite sides of the ring can form only *inter*molecular hydrogen bonds; therefore, it will take a higher temperature to escape into the gas phase.**

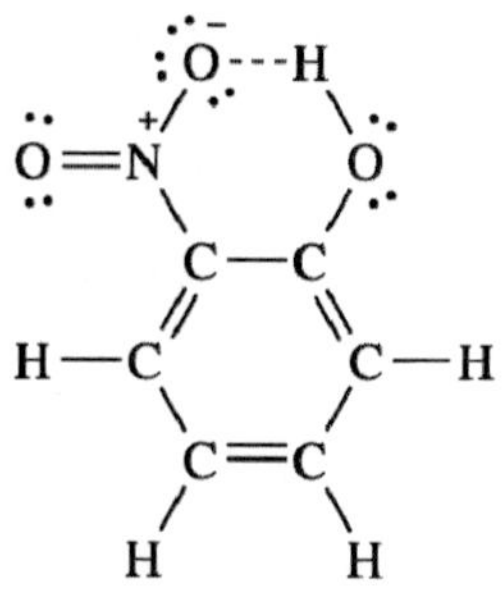

7.49 **Valence bond theory fails to explain the bonding in molecules in which the central atom in its ground state does not have enough unpaired electrons to form the observed number of bonds.**

For N_2, the N atoms have a ground state configuration of $[He]2s^2 2p^3$. There are three unpaired electrons and N forms a triple bond with N. There are enough unpaired electrons to form the observed number of bonds. Valence bond theory explains the bonding in N_2.

For BF_3, the central B atom has a ground state electron configuration of $[He]2s^2 2p^1$. There is one unpaired electron although B forms three bonds with F. There are not enough unpaired electrons to form the observed number of bonds. Valence bond theory cannot explain the bonding in **BF_3.**

For HI, the ground state configuration of I is $[Kr]4d^{10}5s^2 5p^5$ and for H it is $1s^1$. There is one unpaired electron on both atoms, and they form a single bond. There are enough unpaired electrons to form the observed number of bonds. Valence bond theory explains the bonding in HI.

7.55 **Strategy:** The steps for determining the hybridization of the central atom in a molecule are:

draw Lewis structure of the molecule $\longrightarrow$ count the number of electron-domains on the central atom $\longrightarrow$ use Table 7.6 of the text to determine the hybridization state of the central atom

Solution: Draw the Lewis structure of the molecule **SiH_4**:

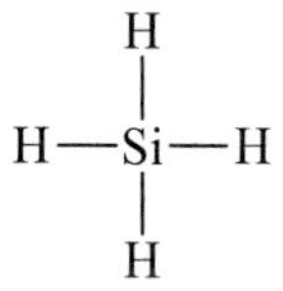

$$H - \overset{\overset{\displaystyle H}{|}}{\underset{\underset{\displaystyle H}{|}}{Si}} - H$$

Count the number of electron domains around the central atom. Since there are four electron domains around Si, the electron-domain geometry is tetrahedral and we conclude that Si is *sp*³ **hybridized**.

Draw the Lewis structure of the molecule **H_3Si–SiH_3**:

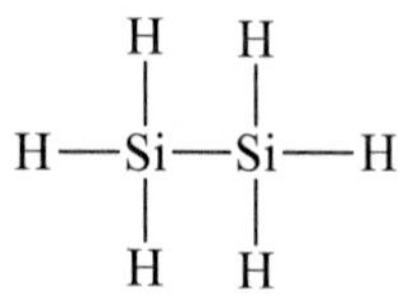

$$H - \overset{\overset{\displaystyle H}{|}}{\underset{\underset{\displaystyle H}{|}}{Si}} - \overset{\overset{\displaystyle H}{|}}{\underset{\underset{\displaystyle H}{|}}{Si}} - H$$

Count the number of electron domains around the "central atoms." Since there are four electron domains around each Si, the electron-domain geometry for each Si is tetrahedral, and we conclude that each Si is *sp*³ **hybridized**.

7.57 **Strategy:** The steps for determining the hybridization of the central atom in a molecule are:

draw Lewis structure of the molecule $\longrightarrow$ count the number of electron-domains on the central atom $\longrightarrow$ use Table 7.6 of the text to determine the hybridization state of the central atom

Solution: Draw the Lewis structure of the molecule.

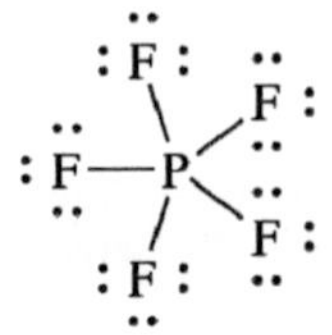

Count the number of electron domains around the central atom. **Since there are five electron domains around P, the electron-domain geometry is trigonal bipyramidal (AB_5 with no lone pairs), and we conclude that P is sp^3d hybridized.**

7.59 The Lewis structures of BF_3, NH_3, and F_3B—NH_3 are shown below:

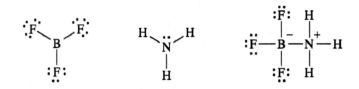

Before the reaction, **boron is sp^2 hybridized in BF_3** (three electron domains around the central atom, trigonal planar electron-domain geometry) **and nitrogen is sp^3 hybridized in NH_3** (four electron domains around the central atom, tetrahedral electron-domain geometry). **After the reaction, boron and nitrogen are both sp^3 hybridized** (tetrahedral electron-domain geometry).

7.63 a. *sp*   b. *sp*   c. *sp*

7.65 **Strategy:** The steps for determining the hybridization of the central atom in a molecule are:

draw Lewis structure of the molecule $\longrightarrow$ count the number of electron-domains on the central atom $\longrightarrow$ use Table 7.6 of the text to determine the hybridization state

Solution: Draw the Lewis structure of the molecule. Several resonance forms with formal charges are shown.

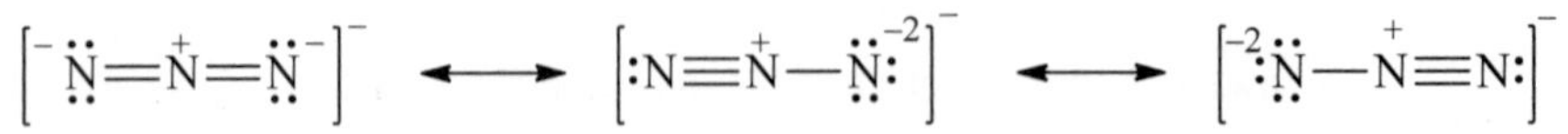

$$\left[\ddot{\underset{..}{N}} = \overset{+}{N} = \ddot{\underset{..}{N}} \right]^{-} \longleftrightarrow \left[:N \equiv \overset{+}{N} - \ddot{\underset{..}{N}}:^{-2} \right]^{-} \longleftrightarrow \left[{}^{-2}:\ddot{\underset{..}{N}} - \overset{+}{N} \equiv N: \right]^{-}$$

Count the number of electron domains around the central atom. Since there are two electron domains around N, the electron-domain geometry is linear (AB_2 with no lone pairs), and we conclude that N is *sp* **hybridized**.

Remember, a multiple bond is just *one* electron domain.

7.67 A single bond is usually a σ bond, a double bond is usually a σ bond and a π bond, and a triple bond is always a σ bond and two π bonds. Therefore, there are **nine σ bonds** and **nine π bonds** in the molecule.

7.69   The benzo[*a*]pyrene molecule contains **36 σ bonds** (some hydrogen atoms are omitted from the figure) and **10 π bonds**.

7.75   **In order for the two hydrogen atoms to combine to form an H_2 molecule, the electrons must have opposite spins. If two H atoms collide and their electron spins are parallel, no bond will form.**

7.77   The energy level diagrams are shown below.

	Li_2	Li_2^+	Li_2^-
σ_{2s}^*	___	___	↑
σ_{2s}	↑↓	↑	↑↓

The bond order of Li_2 is 1, that of Li_2^+ is $\dfrac{1}{2}$, and that of Li_2^- is $\dfrac{1}{2}$.

Order of increasing stability: $Li_2^- = Li_2^+ < Li_2$

In reality, Li_2^+ is more stable than Li_2^- because there is less electrostatic repulsion in Li_2^+ .

7.79   See Figure 7.23 of the text. Removing an electron from B_2 gives B_2^+ , which has a bond order of 1/2.

Therefore, **B_2^+ , with a bond order of 1/2, has the longer B—B bond. (B_2 has a bond order of 1.)**

7.81   **The Lewis diagram has all electrons paired (incorrect) and a double bond (correct). The MO diagram has two unpaired electrons (correct), and a bond order of 2 (correct).**

7.83   We refer to Figure 7.23 of the text for O_2, and then add or subtract electrons for the other MO diagrams.

O_2 has a bond order of 2 and is paramagnetic (two unpaired electrons).

O_2^+ has a bond order of 2.5 and is paramagnetic (one unpaired electron).

O_2^- has a bond order of 1.5 and is paramagnetic (one unpaired electron).

O_2^{2-} has a bond order of 1 and is diamagnetic.

7.85   As shown in the text (see Figure 7.23), **the two shared electrons that make up the single bond in B_2 both lie in pi *bonding* MOs and constitute a pi bond. The four shared electrons that make up the double bond in C_2 all lie in pi molecular orbitals and constitute two pi bonds.**

7.87   A diamagnetic species is one in which there are no unpaired electrons. Since BeO is diamagnetic, its molecular orbital diagram will not have any unpaired electrons.

The molecular orbital diagrams for Be_2 and O_2 are shown in Figure 7.23 (we assume the ordering of molecular orbitals for Be_2 to be like that in Li_2).

Be has an electron configuration of $[He]2s^2$ and O has an electron configuration of $[He]2s^22p^4$. Therefore, BeO has eight valence electrons that must be placed in molecular orbitals.

If the eight electrons fill the orbitals ordered as they are for Be_2, there will be no unpaired electrons. If the eight electrons fill the orbitals ordered as they are for O_2, there will be two unpaired electrons. Since BeO is diamagnetic, the orbitals must be arranged as they are in Be_2.

	Be_2				O_2	
$\sigma^*_{2p_x}$	___			$\sigma^*_{2p_x}$	___	
$\pi^*_{2p_y}, \pi^*_{2p_z}$	___ ___			$\pi^*_{2p_y}, \pi^*_{2p_z}$	___ ___	
σ_{2p_x}	___			π_{2p_y}, π_{2p_z}	↑ ↑	
π_{2p_y}, π_{2p_z}	↑↓ ↑↓			σ_{2p_x}	↑↓	
σ^*_{2s}	↑↓			σ^*_{2s}	↑↓	
σ_{2s}	↑↓			σ_{2s}	↑↓	

The molecular orbital diagram of BeO will be ordered **like that of Be₂.**

7.91 **The left symbol shows three delocalized double bonds (correct). The right symbol shows three localized double bonds and three single bonds (incorrect).**

7.93   a. Two Lewis resonance forms are shown below. (Formal charges different than zero are indicated.)

$$ \overset{..}{\underset{..}{O}} = \underset{+}{N} - \overset{..}{\underset{..}{O}}:^{-} \quad \longleftrightarrow \quad {}^{-}:\overset{..}{\underset{..}{O}} - \underset{+}{N} = \overset{..}{\underset{..}{O}} $$

(with $:\overset{..}{\underset{..}{F}}:$ attached above N in each structure)

   b. There are no lone pairs on the nitrogen atom; it should have a trigonal planar electron-domain geometry and therefore use *sp²* **hybrid orbitals**.

   c. In the bonding of FNO₂, **sigma bonds join the nitrogen atom to the fluorine and oxygen atoms.** In addition, **there is a pi molecular orbital delocalized over the N and O atoms.**

7.95   The Lewis structures of ozone are:

$$ ^{-}:\overset{..}{O} \diagdown \overset{\overset{+}{\overset{..}{O}}}{} \diagup \overset{..}{O}: \quad \longleftrightarrow \quad :\overset{..}{O} \diagup \overset{\overset{+}{\overset{..}{O}}}{} \diagdown \overset{..}{O}:^{-} $$

The central oxygen atom is *sp²* hybridized (AB₂ with one lone pair). **The unhybridized 2p_z orbital on the central oxygen overlaps with the 2p_z orbitals on the two terminal atoms.**

7.97 **Strategy:**   The sequence of steps in determining molecular geometry is as follows:

draw Lewis ⟶ count electron- ⟶ find electron- ⟶ determine molecular
structure   domains around   domain geometry   geometry based on
                 the central atom   position of atoms

Solution: Write the Lewis structure of the molecule.

$$:\overset{..}{\underset{..}{Br}} - Hg - \overset{..}{\underset{..}{Br}}: $$

Count the number of electron domains around the central atom. There are two electron domains around Hg.

Since there are two electron domains around Hg, the electron-domain geometry is **linear.**

In addition, since there are no lone pairs around the central atom, the geometry is also **linear** (AB_2).

You could establish the geometry of $HgBr_2$ by measuring its dipole moment. If mercury(II) bromide were bent, it would have a measurable dipole moment. Experimentally, it has no dipole moment and therefore must be linear.

7.99   Geometry: **bent;** sp^3 **hybridization**

7.101   a. $C_6H_8O_6$

b. The hybridizations of the **five central O atoms** are sp^3; the **hybridization of the sixth (double bonded) peripheral O atom is** sp^2. The **three C atoms** bonded with three other atoms are sp^2; the other **three C atoms** bonded with four other atoms are sp^3.

c. The geometry about the **five central O atoms** is **bent.** The geometry about the **three** sp^2 **C atoms** is **trigonal planar;** the geometry about the **three** sp^3 **C atoms** is **tetrahedral.**

7.103   To predict the bond angles for the molecules, you would have to draw the Lewis structure and determine the geometry using the VSEPR model. From the geometry, you can predict the bond angles.

a. $BeCl_2$: AB_2, **180°** (linear)

b. BCl_3: AB_3, **120°** (trigonal planar)

c. CCl_4: AB_4, **109.5°** (tetrahedral)

d. CH_3Cl: AB_4, **109.5°** (tetrahedral with a possible slight distortion resulting from the different sizes of the chlorine and hydrogen atoms).

e. Hg_2Cl_2:   Each mercury atom is of the AB_2 type. The entire molecule is linear, **180°** bond angles.

f. $SnCl_2$: AB_2 w/one lone pair, roughly **120°** (bent).

g. H_2O_2:   The atom arrangement is HOOH. Each oxygen atom is of the AB_2 with two lone pairs and the H–O–O angles will be roughly **109.5°**.

h. SnH_4: AB_4, **109.5°** (tetrahedral)

7.105   a. a.   The Lewis structure is:

The shape will be trigonal **planar** (AB_3).

b. The Lewis structure is (lone pairs on O atoms not shown):

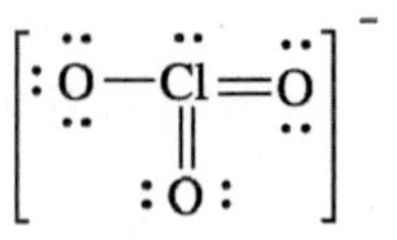

$$\left[:\overset{..}{O}-\overset{..}{Cl}=\overset{..}{O}: \right]^{-}$$
$$\underset{:O:}{\overset{\|}{}}$$

The molecule will be a trigonal pyramid (**nonplanar**).

c. The Lewis structure is:

$$H-C\equiv N:$$

The molecule is **polar**.

d. The Lewis structure is (lone pairs on F atoms not shown):

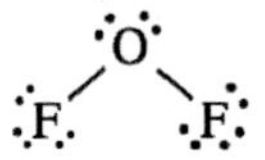

The molecule is bent and therefore **polar**.

e. The Lewis structure is:

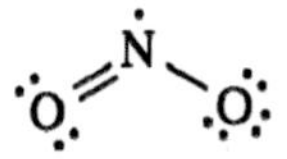

The nitrogen atom is an AB_2 with one lone, unshared electron rather than the usual *pair*. As a result, the repulsion will not be as great and the O–N–O angle will be **greater than 120°** expected for AB_2 (with one lone pair) geometry. Experiment shows the angle to be around 135°.

7.107 **CCl₄**. Generally, the larger the molecule greater its polarizability. Recall that polarizability is the ease with which the electron distribution in an atom or molecule can be distorted.

7.109 **Only ICl_2^- and CdBr₂ will be linear**. The rest are bent.

7.111 a. **Strategy:** The steps for determining the hybridization of the central atom in a molecule are:

draw Lewis Structure of the molecule $\longrightarrow$ count the number of electron-domains on the central atom state of the central $\longrightarrow$ use Table 7.6 of the text to determine the hybridization atom

Solution: The geometry around each nitrogen is identical. To complete an octet of electrons around N, you must add a lone pair of electrons. Count the number of electron domains around N. There are three electron domains around each N.

Since there are three electron domains around N, and none of them is a lone pair, the electron-domain geometry is trigonal planar, and we conclude that each N is *sp²* **hybridized**.

b. **Strategy:** Keep in mind that the polarity of a molecule depends on both the difference in electronegativities of the elements present and its geometry. A molecule can have polar bonds (if the bonded atoms have different electronegativities), but it may not be polar if it has a highly symmetrical geometry.

Solution: An N–F bond is polar because F is more electronegative than N. **The structure on the right is polar** because the two N–F bond moments do not cancel each other out, and so the molecule has a net dipole moment. On the other hand, the two N–F bond moments in the left-hand structure cancel. The sum of bond dipoles will be *zero*. Therefore, the **molecule on the left will be nonpolar.**

7.113 **In 1,2-dichloroethane, the two C atoms are joined by a sigma bond. Rotation about a sigma bond does not destroy the bond and the bond is therefore free (or relatively free) to rotate. Thus, all angles are permitted and the molecule is nonpolar because the C–Cl bond moments cancel each other because of the averaging effect brought about by rotation. In *cis*-dichloroethylene, the two C–Cl bonds are locked in position. The π bond between the C atoms prevents rotation (in order to rotate, the π bond must be broken, using an energy source such as light or heat). Therefore, there is no rotation about the C=C in *cis*-dichloroethylene, and the molecule is polar.**

7.115 a. **K$_2$S: Ionic forces are much stronger than the dipole-dipole forces in (CH$_3$)$_3$N.**

   b. **Br$_2$: Both the molecules are nonpolar, but Br$_2$ has a larger molar mass.** (The boiling point of Br$_2$ is 50°C and that of C$_4$H$_{10}$ is −0.5°C.)

7.117 The Lewis structure is:

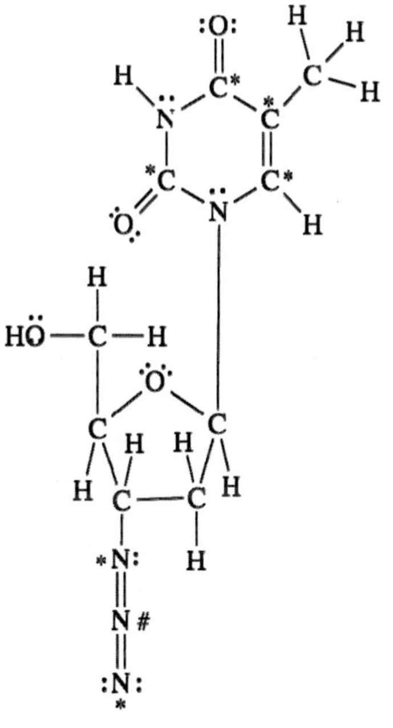

The carbon atoms and nitrogen atoms marked with an asterisk (C* and N*) are *sp^2* hybridized; unmarked carbon atoms and nitrogen atoms are *sp^3* hybridized; and the nitrogen atom marked with # is *sp* hybridized.

7.119 **Strategy:** Looking at the Lewis structure, we can count the number of electron domains surrounding each atom. This is the number of hybrid orbitals necessary to account for the atom's geometry. (This is also the number of atomic orbitals that must undergo hybridization.) Table 7.6 shows how the number of electron domains on an atom corresponds to a set of hybrid orbitals.

Solution: H atoms, having only one electron in 1*s* orbitals, do not undergo hybridization.
The leftmost O atom has four electron domains (two single bonds and two lone pairs). Four electron domains require four hybrid orbitals, and four hybrid orbitals require the hybridization of four atomic orbitals: one *s* and three *p*. This corresponds to *sp^3* hybridization.

The rightmost O has three electron domains (one double bond and two lone pairs). Three electron domains require three hybrid orbitals, and three hybrid orbitals require the hybridization of three atomic orbitals: one s and two p. This corresponds to sp^2 hybridization.

The leftmost C atom has three electron domains (two single bonds and one double bond). Three electron domains require three hybrid orbitals, and three hybrid orbitals require the hybridization of three atomic orbitals: one s and two p. This corresponds to sp^2 hybridization.

The central C atom has four electron domains (four single bonds). Four electron domains require four hybrid orbitals, and four hybrid orbitals require the hybridization of four atomic orbitals: one s and three p. This corresponds to sp^3 hybridization.

The rightmost C atom has four electron domains (four single bonds). Four electron domains require four hybrid orbitals, and four hybrid orbitals require the hybridization of four atomic orbitals: one s and three p. This corresponds to sp^3 hybridization.

The N atom has four electron domains (three single bonds and one lone pair). Four electron domains require four hybrid orbitals, and four hybrid orbitals require the hybridization of four atomic orbitals: one s and three p. This corresponds to sp^3 hybridization.

The Se atom has four electron domains (two single bonds and two lone pairs). Four electron domains require four hybrid orbitals, and four hybrid orbitals require the hybridization of four atomic orbitals: one s and three p. This corresponds to sp^3 hybridization.

Figure (a) shows a sp hybrid orbital. There are **no sp hybrid orbitals** in this molecule.

Figure (b) shows a sp^2 hybrid orbital. There are **two sp^2 hybrid orbitals** in this molecule.

Figure (c) shows a sp^3 hybrid orbital. There are **five sp^3 hybrid orbitals** in this molecule.

Figure (d) shows a sp^3d hybrid orbital. There are **no sp^3d hybrid orbitals** in this molecule.

7.121 **The carbons are all sp^2 hybridized. The nitrogen double bonded to the carbon in the ring is sp^2 hybridized. The other nitrogens are sp^3 hybridized.**

7.123 **The molecules are all polar. The F atoms can form H-bonds with water and other –OH and –NH groups in the membrane, so water solubility plus easy attachment to the membrane would allow these molecules to pass the blood-brain barrier.**

7.125 O_3, CO, CO_2, NO_2, N_2O, CH_4, and $CFCl_3$ are greenhouse gases. **All** are greenhouse gases **except** the **homonuclear diatomic molecules N_2 and O_2.**

7.127 **The S—S bond is a single covalent bond that links proteins together by their sulfur atoms. Each S is sp^3 hybridized, so the X—S—S angle is about 109°.**

7.129 a. A paramagnetic species is one in which there are unpaired electrons. Since BO^+ is paramagnetic, its molecular orbital diagram will have unpaired electrons.

The molecular orbital diagrams for B_2 and O_2 are shown in Figure 7.23.

B has an electron configuration of $[He]2s^2 2p^1$ and O has an electron configuration of $[He]2s^2 2p^4$. We must remove one electron to account for the positive ion. Therefore, BO^+ has eight valence electrons that must be placed in molecular orbitals.

If the eight electrons fill the orbitals ordered as they are for B_2, there will be no unpaired electrons. If the eight electrons fill the orbitals ordered as they are for O_2, there will be two unpaired electrons. Since BO^+ is paramagnetic, the orbitals must be arranged as they are in O_2.

B_2 O_2

	B_2			O_2	
$\sigma^*_{2p_x}$	—		$\sigma^*_{2p_x}$	—	
$\pi^*_{2p_y}, \pi^*_{2p_z}$	— —		$\pi^*_{2p_y}, \pi^*_{2p_z}$	— —	
σ_{2p_x}	—		π_{2p_y}, π_{2p_z}	↑ ↑	
π_{2p_y}, π_{2p_z}	↑↓ ↑↓		σ_{2p_x}	↑↓	
σ^*_{2s}	↑↓		σ^*_{2s}	↑↓	
σ_{2s}	↑↓		σ_{2s}	↑↓	

The molecular orbital diagram of BO^+ will be ordered **like that of O_2.**

b. Equation 7.1 states that bond order =

$$\frac{\text{number of electrons in bonding molecular orbitals} - \text{number of electrons in antibonding orbitals}}{2}$$

The **bond order** for BO^+ is $[(6 - 2)/2] = 2$

c. According to the molecular orbital diagram in part (a), there are **two unpaired electrons** in the BO^+ ion.

7.131 **Strategy:** Use Lewis structures and the VSEPR model to determine first the electron-domain geometry and then the molecular geometry (shape).

Setup: The Lewis structures for SO_3 and SO_3^{2-} are:

:Ö—S—Ö: [:Ö—S—Ö:]²⁻
      ‖ |
      :Ö: :Ö:

There are three electron domains for SO_3: two single bonds and one double bond. SO_3^{2-} has four electron domains: three single bonds and one lone pair.

Solution: According to VSEPR model, three electron domains will be arranged in a trigonal plane. Since there are no lone pairs on the central atom in SO_3, the molecular geometry is the same as the electron-domain geometry, trigonal planar.

Four electron domains will be arranged in a tetrahedron. Since there is one lone pair on the central atom in SO_3^{2-}, the molecular geometry is trigonal pyramidal.

The change in molecular geometry in going from SO_3 (trigonal planar) to SO_3^{2-} (trigonal pyramidal) is illustrated by **figure (c)**.

7.133 **The second ("asymmetric stretching mode") and third ("bending mode")** vibrational motions are responsible for CO_2 behaving as a greenhouse gas. CO_2 is a nonpolar molecule. The second and third vibrational motions create a changing dipole moment. The first vibration, a symmetric stretch, does *not* create a dipole moment. Since CO, NO_2, and N_2O are all polar molecules, they will also act as greenhouse gases.

7.135 The complete structure of progesterone is shown below.

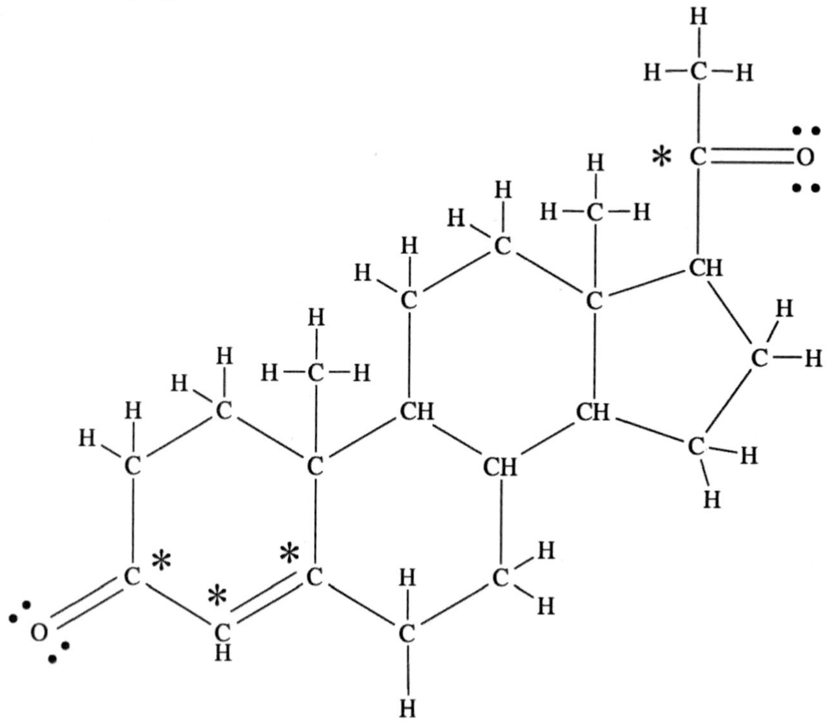

The four carbons marked with an asterisk are *sp²* hybridized. The remaining carbons are *sp³* hybridized.

7.137 a. Looking at the electronic configuration for N_2 shown in Figure 7.23 of the text, we write **the electronic configuration for P_2:**

$$[Ne_2](\sigma_{3s})^2(\sigma_{3s}^*)^2(\pi_{3p_y})^2(\pi_{3p_z})^2(\sigma_{3p_x})^2$$

b. Past the Ne_2 core configuration, there are eight bonding electrons and two antibonding electrons. The bond order is:

bond order = ½(8 − 2) = 3

c. All the electrons in the electronic configuration are paired. P_2 is **diamagnetic**.

7.139 **The Lewis structure of O_2 shows four pairs of electrons on the two oxygen atoms. From Figure 7.23 of the text, we see that these eight valence electrons are placed in the σ_{2p_x}, π_{2p_y}, π_{2p_z}, $\pi_{2p_y}^*$, and $\pi_{2p_z}^*$ orbitals. For all the electrons to be paired, energy is needed to flip the spin in one of the antibonding molecular orbitals ($\pi_{2p_y}^*$ or $\pi_{2p_z}^*$). According to Hund's rule, this arrangement is less stable than the ground-state configuration shown in Figure 7.23, and hence the Lewis structure shown actually corresponds to an excited state of the oxygen molecule.**

7.141   a.   **Although the O atoms are *sp*³ hybridized, they are locked in a planar structure by the benzene rings. The molecule is symmetrical and therefore is not polar.**

         b.   **6 pi bonds and 20 sigma bonds.**

7.143   a.

$$:\overset{-}{C}\!\!\equiv\!\!\overset{+}{O}:$$

         This is the only reasonable Lewis structure for CO. **The electronegativity difference between O and C suggests that electron density should concentrate on the O atom, but assigning formal charges places a negative charge on the C atom. Therefore, we expect CO to have a small dipole moment.**

         b.   **CO is isoelectronic with N_2 and has a bond order of 3. This agrees with the triple bond in the Lewis structure.**

         c.   **Since C has a negative formal charge, it is more likely to form bonds with Fe^{2+} (OC—Fe^{2+} rather than CO—Fe^{2+}).** More elaborate analysis of the orbitals involved shows that this is indeed the case.

7.145   There are eight valence electrons (four from tin and one each from four hydrogens), and the dot structure indicates four electron domains, all bonding. The electron-domain and molecular geometries are both **tetrahedral**.

7.147   The CF bond shortens when CF loses an electron to form CF^+. The bond order indicates the stability of the molecule or the strength of the bond. As the bond order increases, the bond will become more stable, which is indicative of a shorter bond. When CF loses an electron, its bond order increases; therefore, the bond becomes more stable and the bond length decreases.

         We can assume that the electron is removed from an antibonding orbital. This will increase the ratio of bonding to antibonding electrons and increase the bond order.

Chapter 8
Chemical Reactions

Visualizing Chemistry

VC 8.1 a   VC 8.2 b   VC 8.3 c   VC 8.4 b

Key Skills

8.1   b   8.2   c   8.3   e   8.4   a

Problems

8.5   **Strategy:**   (1) translate each compound name into the correct chemical formula;
(2) translate "and" into "+"; this applies to compound phrases separated by commas;
(3) translate "react to form" into a reaction arrow " $\longrightarrow$ ".

Translation of a compound name into the correct chemical formula is the most difficult part of the solution. If the compound name is a **systematic name**, it will have Greek prefixes such as "di-" (2), which tell you what the subscript of a particular element is in the formula. However, in most cases the name used is a **common name** and you must simply look up (or learn and remember) the name/formula translation.

a. $KOH + H_3PO_4 \longrightarrow K_3PO_4 + H_2O$

b. $Zn + AgCl \longrightarrow ZnCl_2 + Ag$

c. $NaHCO_3 \longrightarrow Na_2CO_3 + H_2O + CO_2$

d. $NH_4NO_2 \longrightarrow N_2 + H_2O$

e. $CO_2 + KOH \longrightarrow K_2CO_3 + H_2O$

8.7   **Strategy:**   (1) translate each chemical formula into the correct compound name;
(2) translate "+" into "and";
(3) translate the reaction arrow " $\longrightarrow$ " into the phrase "react to form."

Translation of a chemical formula into the correct compound name is the most difficult part of the solution. Watch your spelling! In many cases you must simply look up (or learn and remember) the formula/name translation. In general, these formulas and their corresponding names are in your text.

a. **Potassium and water react to form potassium hydroxide and hydrogen.**

b. **Barium hydroxide and hydrochloric acid react to form barium chloride and water.**

c. **Copper and nitric acid react to form copper nitrate, nitrogen monoxide, and water.**

d. **Aluminum and sulfuric acid react to form aluminum sulfate and hydrogen.**

e. **Hydrogen iodide reacts to form hydrogen and iodine.**

8.9 **Strategy:** The goal is to have the same number of atoms of a given element on both sides of the equation. You do this by adjusting the **stoichiometric coefficients** in front of each formula (remember that if no coefficient is written, it is understood to be "1"). Remember also that you **cannot** change the **subscripts** in the formulas!

Everybody balances chemical equations the same way: we **guess** what the coefficients must be. Sometimes there are clues about what the numbers must be (see equation (a) below). Don't be afraid to guess wrong—just erase (or better yet, cross out) the wrong answer and try again. Each time you guess, count all the atoms again very carefully.

There is no "one way" or "best way" to balance chemical equations, but with practice you will learn how to do it and make it look as easy as your instructor does. Remember, your instructor is guessing too, but is also probably doing a lot of mental arithmetic along the way!

a.
$$N_2O_5 \longrightarrow N_2O_4 + O_2$$

First, count **carefully** the atoms of each element on both sides of the reaction arrow:

$$N_2O_5 \longrightarrow N_2O_4 + O_2$$
$$2 - N - 2$$
$$5 - O - 6$$

In order to maintain the nitrogen atom balance, the coefficients of N_2O_5 and N_2O_4 must be the same (we will leave them at 1 and 1 for now). By using only half of the oxygen molecule we can write a balanced equation:

$$N_2O_5 \longrightarrow N_2O_4 + \frac{1}{2}O_2$$
$$2 - N - 2$$
$$5 - O - 5$$

Although this equation is balanced, very often we are asked to balance chemical equations with the smallest whole number stoichiometric coefficients. To clear the fraction, multiply all three coefficients by 2 (the denominator of the fraction):

$$2N_2O_5 \longrightarrow 2N_2O_4 + O_2$$
$$4 - N - 4$$
$$10 - O - 10$$

b. $2KNO_3 \longrightarrow 2KNO_2 + O_2$

c. $NH_4NO_3 \longrightarrow N_2O + 2H_2O$

d. $NH_4NO_2 \longrightarrow N_2 + 2H_2O$

e. $2NaHCO_3 \longrightarrow Na_2CO_3 + H_2O + CO_2$

f. $P_4O_{10} + 6H_2O \longrightarrow 4H_3PO_4$

g. $2HCl + CaCO_3 \longrightarrow CaCl_2 + H_2O + CO_2$

h. $2Al + 3H_2SO_4 \longrightarrow Al_2(SO_4)_3 + 3H_2$

i. $CO_2 + 2KOH \longrightarrow K_2CO_3 + H_2O$

j. $CH_4 + 2O_2 \longrightarrow CO_2 + 2H_2O$

k. $Be_2C + 4H_2O \longrightarrow 2Be(OH)_2 + CH_4$

l. $3Cu + 8HNO_3 \longrightarrow 3Cu(NO_3)_2 + 2NO + 4H_2O$

m. $S + 6HNO_3 \longrightarrow H_2SO_4 + 6NO_2 + 2H_2O$

n. $2NH_3 + 3CuO \longrightarrow 3Cu + N_2 + 3H_2O$

8.11   On the reactants side there are 6 A atoms and 4 B atoms. On the products side, there are 4 C atoms and 2 D atoms. Writing an equation,

$$6A + 4B \longrightarrow 4C + 2D$$

Chemical equations are typically written with the smallest set of whole number coefficients. Dividing the equation by 2 gives,

$$3A + 2B \longrightarrow 2C + D$$

The correct answer is choice **(d)**.

8.13   a. **combustion**   b. **combination**   c. **decomposition**

8.17   **Strategy:**   The process of combustion analysis involves the following steps:

(1) Convert the mass of CO_2 to moles of CO_2—this is the same as the number of moles of C in the sample because every C atom in the sample becomes the C atom in a CO_2 molecule;

(2) Convert moles of C to grams of C;

(3) Convert the mass of H_2O to moles of H_2O. Twice this number is the number of moles of H in the sample (1 mol H_2O = 2 mol H) because for every *two* H atoms in the sample, *one* water molecule is produced;

(4) Convert moles of H to grams of H;

(5) Subtract the combined mass of C and H from the sample mass; if the result is zero, then only C and H are present in the sample; if the result is greater than zero, then this is the mass of the other element in the sample (in this example, oxygen); convert the mass of this element to moles;

(6) The mole ratio leads to the empirical formula, as in percent composition problems.

Solution:

$$? \text{ mol C} = 28.16 \text{ mg CO}_2 \times \frac{1\text{g}}{1000\,\text{mg}} \times \frac{1 \text{ mol CO}_2}{44.01 \text{ g CO}_2} \times \frac{1 \text{ mol C}}{1 \text{ mol CO}_2} = 0.0006399 \text{ mol C}$$

and

$$0.0006399 \text{ mol C} \times \frac{12.01\,\text{g C}}{1\,\text{mol C}} = 0.007685 \text{ g C} = 7.685 \text{ mg C}$$

$$? \text{ mol H} = 11.53 \text{ mg H}_2\text{O} \times \frac{1\text{g}}{1000 \text{ mg}} \times \frac{1 \text{ mol H}_2\text{O}}{18.02\,\text{g H}_2\text{O}} \times \frac{2 \text{ mol H}}{1 \text{ mol H}_2\text{O}} = 0.001280 \text{ mol H}$$

and

$$0.001280 \text{ mol H} \times \frac{1.008\,\text{g H}}{1\,\text{mol H}} = 0.001290 \text{ g H} = 1.290 \text{ mg H}$$

$$\text{mg O} = 10.00 \text{ mg} - (7.685 \text{ mg} + 1.290 \text{ mg}) = 1.025 \text{ mg O}$$

$$1.025 \text{ mg O} \times \frac{1\text{g}}{1000 \text{ mg}} \times \frac{1 \text{ mol O}}{16.00 \text{ g O}} = 0.00006406 \text{ mol O}$$

$$\text{C : H : O} = 0.0006399 : 0.001280 : 0.00006406 = 9.99 : 19.98 : 1.000 \approx 10 : 20 : 1.$$

The empirical formula for menthol is **C$_{10}$H$_{20}$O.**

8.19

$$? \text{ mol C} = 14.7 \text{ g CO}_2 \times \frac{1 \text{ mol CO}_2}{44.01\,\text{g CO}_2} \times \frac{1 \text{ mol C}}{1 \text{ mol CO}_2} = 0.334 \text{ mol C}$$

$$0.334 \text{ mol C} \times \frac{12.01\,\text{g C}}{1\,\text{mol C}} = 4.01 \text{ g C}$$

$$? \text{ mol H} = 6.00 \text{ g H}_2\text{O} \times \frac{1 \text{ mol H}_2\text{O}}{18.02\,\text{g H}_2\text{O}} \times \frac{2 \text{ mol H}}{1 \text{ mol H}_2\text{O}} = 0.666 \text{ mol H}$$

$$0.666 \text{ mol H} \times \frac{1.008\,\text{g H}}{1\,\text{mol H}} = 0.671 \text{ g H}$$

$$\text{mass O} = \text{mass sample} - (\text{mass C} + \text{mass H}) = 10.00 \text{ g} - (4.01 \text{ g} + 0.671 \text{ g}) = 5.32 \text{ g O}$$

$$5.32 \text{ g O} \times \frac{1 \text{ mol O}}{16.00\,\text{g O}} = 0.333 \text{ mol O}$$

molar ratios: C : H : O = 0.334:0.666:0.333 = $\dfrac{0.334}{0.333} : \dfrac{0.666}{0.333} : \dfrac{0.333}{0.333}$ = 1.00:2.00:1.00

The empirical formula is CH$_2$O.

8.21

$$? \text{ mol C} = 3.02 \text{ g CO}_2 \times \frac{1 \text{ mol CO}_2}{44.01 \text{ g CO}_2} \times \frac{1 \text{ mol C}}{1 \text{ mol CO}_2} = 0.0686 \text{ mol C}$$

$$0.0686 \text{ mol C} \times \frac{12.01 \text{ g C}}{1 \text{ mol C}} = 0.824 \text{ g C}$$

$$? \text{ mol H} = 0.412 \text{ g H}_2\text{O} \times \frac{1 \text{ mol H}_2\text{O}}{18.02 \text{ g H}_2\text{O}} \times \frac{2 \text{ mol H}}{1 \text{ mol H}_2\text{O}} = 0.0457 \text{ mol H}$$

$$0.0457 \text{ mol H} \times \frac{1.008 \text{ g H}}{1 \text{ mol H}} = 0.0461 \text{ g H}$$

mass Cl = mass sample – (mass C + mass H) = 1.68 g – (0.824 g + 0.0461 g) = 0.810 g Cl

$$0.810 \text{ g Cl} \times \frac{1 \text{ mol Cl}}{35.45 \text{ g Cl}} = 0.0228 \text{ mol Cl}$$

molar ratios: C : H : Cl = 0.0686:0.0457:0.0228 = $\dfrac{0.0686}{0.0228} : \dfrac{0.0457}{0.0228} : \dfrac{0.0228}{0.0228}$ = 3.00:2.00:1.00

The empirical formula is C$_3$H$_2$Cl.

The empirical formula weight is 73.5, while the molar mass is 147, so $x = \dfrac{147}{73.5} = 2$.

The correct molecular formula is thus (C$_3$H$_2$Cl)$_2$ = **C$_6$H$_4$Cl$_2$.**

8.23　a. All the C and H in the combustion products come from the acetylene. In acetylene, the ratio of C to H is 1:1. Thus, the ratio of C to H in the combustion products must also be 1:1. Since each water molecule contains two atoms of H and each carbon dioxide molecule contains one atom of C, there must be twice the number of carbon dioxide molecules as water molecules. The answer is **diagram (b).**

　　b. In ethylene, the ratio of C to H is 1:2. Since this is the same ratio as the number of C atoms in each carbon dioxide molecule to the number of H atoms in each water molecule, combustion would produce equal numbers of CO$_2$ and H$_2$O molecules. The answer is **diagram (a).**

8.27

$$\text{Si}(s) + 2\text{Cl}_2(g) \longrightarrow \text{SiCl}_4(l)$$

Strategy: Looking at the balanced equation, how do we compare the amounts of Cl$_2$ and SiCl$_4$? We can compare them based on the mole ratio from the balanced equation.

Solution: Because the balanced equation is given in the problem, the mole ratio between Cl$_2$ and SiCl$_4$ is known: 2 moles Cl$_2$ ≏ 1 mole SiCl$_4$. From this relationship, we have two conversion factors.

$$\frac{2 \text{ mol Cl}_2}{1 \text{ mol SiCl}_4} \quad \text{and} \quad \frac{1 \text{ mol SiCl}_4}{2 \text{ mol Cl}_2}$$

Which conversion factor is needed to convert from moles of $SiCl_4$ to moles of Cl_2? The conversion factor on the left is the correct one. Moles of $SiCl_4$ will cancel, leaving units of "mol Cl_2" for the answer. We calculate moles of Cl_2 reacted as follows:

$$? \text{ mol } Cl_2 \text{ reacted } = 0.507 \text{ mol } SiCl_4 \times \frac{2 \text{ mol } Cl_2}{1 \text{ mol } SiCl_4} = \textbf{1.01 mol Cl}_2$$

8.29 Starting with the 5.0 moles of C_4H_{10}, we can use the mole ratio from the balanced equation to calculate the moles of CO_2 formed.

$$2C_4H_{10}(g) + 13O_2(g) \longrightarrow 8CO_2(g) + 10H_2O(l)$$

$$? \text{ mol } CO_2 = 5.0 \text{ mol } C_4H_{10} \times \frac{8 \text{ mol } CO_2}{2 \text{ mol } C_4H_{10}} = 20 \text{ mol } CO_2 = \textbf{2.0} \times \textbf{10}^{\textbf{1}} \textbf{ mol CO}_2$$

8.31 a. $\textbf{2NaHCO}_3 \longrightarrow \textbf{Na}_2\textbf{CO}_3 + \textbf{CO}_2 + \textbf{H}_2\textbf{O}$

b.

$$\text{molar mass } NaHCO_3 = 22.99 \text{ g} + 1.008 \text{ g} + 12.01 \text{ g} + 3(16.00 \text{ g}) = 84.008 \text{ g}$$
$$\text{molar mass } CO_2 = 12.01 \text{ g} + 2(16.00 \text{ g}) = 44.01 \text{ g}$$

The balanced equation shows one mole of CO_2 formed from two moles of $NaHCO_3$.

$$\text{mass } NaHCO_3 = 20.5 \text{ g } CO_2 \times \frac{1 \text{ mol } CO_2}{44.01 \text{ g } CO_2} \times \frac{2 \text{ mol } NaHCO_3}{1 \text{ mol } CO_2} \times \frac{84.008 \text{ g } NaHCO_3}{1 \text{ mol } NaHCO_3}$$

$$= \textbf{78.3 g NaHCO}_3$$

8.33

$$C_6H_{12}O_6 \longrightarrow 2C_2H_5OH + 2CO_2$$
$$\text{glucose} \qquad\qquad \text{ethanol}$$

Strategy: We compare glucose and ethanol based on the *mole ratio* in the balanced equation. Before we can determine moles of ethanol produced, we need to convert to moles of glucose. What conversion factor is needed to convert from grams of glucose to moles of glucose? Once moles of ethanol are obtained, another conversion factor is needed to convert from moles of ethanol to grams of ethanol.

Solution: The molar mass of glucose will allow us to convert from grams of glucose to moles of glucose. The molar mass of glucose = $6(12.01 \text{ g}) + 12(1.008 \text{ g}) + 6(16.00 \text{ g}) = 180.16 \text{ g}$. The balanced equation is given, so the mole ratio between glucose and ethanol is known—that is, 1 mole glucose $\simeq$ 2 moles ethanol. Finally, the molar mass of ethanol will convert moles of ethanol to grams of ethanol. This sequence of three conversions is summarized as follows:

grams of glucose → moles of glucose → moles of ethanol → grams of ethanol

$$? \text{g } C_2H_5OH = 500.4 \text{ g } C_6H_{12}O_6 \times \frac{1 \text{ mol } C_6H_{12}O_6}{180.16 \text{ g } C_6H_{12}O_6} \times \frac{2 \text{ mol } C_2H_5OH}{1 \text{ mol } C_6H_{12}O_6} \times \frac{46.068 \text{ g } C_2H_5OH}{1 \text{ mol } C_2H_5OH}$$

$$= \textbf{255.9 g C}_2\textbf{H}_5\textbf{OH}$$

The liters of ethanol can be calculated from the density and the mass of ethanol.

$$\text{volume} = \frac{\text{mass}}{\text{density}}$$

$$\text{volume of ethanol obtained} = \frac{255.9 \text{ g}}{0.789 \text{ g/mL}} = 324 \text{ mL} = \mathbf{0.324 \text{ L}}$$

8.35   The balanced equation shows that eight moles of KCN are needed to combine with four moles of Au.

$$? \text{ mol KCN} = 29.0 \text{ g Au} \times \frac{1 \text{ mol Au}}{197.0 \text{ g Au}} \times \frac{8 \text{ mol KCN}}{4 \text{ mol Au}} = 0.2944 \text{ mol KCN}$$

Converting to mass in grams gives:

$$0.2944 \text{ mol KCN} \times \frac{65.12 \text{ g KCN}}{1 \text{ mol KCN}} = \mathbf{19.2 \text{ g KCN}}$$

8.37   a. $\mathbf{NH_4NO_3(s) \longrightarrow N_2O(g) + 2H_2O(g)}$

b.   Starting with moles of NH_4NO_3, we can use the mole ratio from the balanced equation to find moles of N_2O. Once we have moles of N_2O, we can use the molar mass of N_2O to convert to grams of N_2O. Combining the two conversions into one calculation, we have:

$$\text{mol } NH_4NO_3 \rightarrow \text{ mol } N_2O \rightarrow \text{g } N_2O$$

$$? \text{ g } N_2O = 0.46 \text{ mol } NH_4NO_3 \times \frac{1 \text{ mol } N_2O}{1 \text{ mol } NH_4NO_3} \times \frac{44.02 \text{ g } N_2O}{1 \text{ mol } N_2O} = \mathbf{2.0 \times 10^1 \text{ g } N_2O}$$

8.39   The balanced equation for the decomposition is :

$$2KClO_3(s) \longrightarrow 2KCl(s) + 3O_2(g)$$

$$? \text{ g } O_2 = 46.0 \text{ g } KClO_3 \times \frac{1 \text{ mol } KClO_3}{122.55 \text{ g } KClO_3} \times \frac{3 \text{ mol } O_2}{2 \text{ mol } KClO_3} \times \frac{32.00 \text{ g } O_2}{1 \text{ mol } O_2} = \mathbf{18.0 \text{ g } O_2}$$

8.45

$$N_2 + 3H_2 \longrightarrow 2NH_3$$

The number of N_2 molecules shown in the diagram is 3. The balanced equation shows that 3 moles H_2 are stoichiometrically equivalent to 1 mole N_2. Therefore, we need 9 molecules of H_2 to react completely with 3 molecules of N_2. There are 10 molecules of H_2 present in the diagram. H_2 is in excess. N_2 is the limiting reactant.

9 molecules of H_2 will react with 3 molecules of N_2, leaving 1 molecule of H_2 in excess. The mole ratio between N_2 and NH_3 is 1:2. When 3 molecules of N_2 react, 6 molecules of NH_3 will be produced. Because each model represents one mole, we find that **6 mol of NH_3 are produced and 1 mol H_2 is left.**

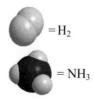

$= H_2$

$= NH_3$

8.47   According to the diagram, the reaction consumes one O_2 molecule and four NO_2 molecules, and two N_2O_5
        molecules are produced. The balanced equation is **$4NO_2 + O_2 \longrightarrow 2N_2O_5$. The limiting reagent is NO_2**
        since it was consumed fully with no leftover molecules remaining. There was some O_2 left over, so O_2 is in
        excess.

8.49   **Strategy:**   Note that this reaction gives the amounts of both reactants, so it is likely to be a limiting reactant
        problem. The reactant that produces fewer moles of product is the limiting reactant because it
        limits the amount of product that can be produced. How do we convert from the amount of
        reactant to amount of product? Perform this calculation for each reactant, then compare the
        moles of product, NO_2, formed by the given amounts of O_3 and NO to determine which reactant
        is the limiting reactant.

        Working backwards, we can determine the amount of NO that reacted to produce 0.01542 mole
        of NO_2. The amount of NO left over is the difference between the initial amount and the amount
        reacted.

Solution: We carry out two separate calculations. First, starting with 0.740 g O_3, we calculate the number
        of moles of NO_2 that could be produced if all the O_3 reacted. We complete the following
        conversions.

$$\text{grams of } O_3 \rightarrow \text{moles of } O_3 \rightarrow \text{moles of } NO_2$$

        Combining these two conversions into one calculation, we write

$$? \text{ mol } NO_2 = 0.740 \text{ g } O_3 \times \frac{1 \text{ mol } O_3}{48.00 \text{ g } O_3} \times \frac{1 \text{ mol } NO_2}{1 \text{ mol } O_3} = 0.0154 \text{ mol } NO_2$$

        Second, starting with 0.670 g of NO, we complete similar conversions.

$$\text{grams of } NO \rightarrow \text{moles of } NO \rightarrow \text{moles of } NO_2$$

        Combining these two conversions into one calculation, we write

$$? \text{ mol } NO_2 = 0.670 \text{ g } NO \times \frac{1 \text{ mol } NO}{30.01 \text{ g } NO} \times \frac{1 \text{ mol } NO_2}{1 \text{ mol } NO} = 0.0223 \text{ mol } NO_2$$

        The initial amount of O_3 limits the amount of product that can be formed; therefore, **O_3 is the
        limiting reactant**.

        The problem asks for grams of NO_2 produced. We already know the moles of NO_2 produced,
        0.0154 mole. Use the molar mass of NO_2 as a conversion factor to convert to grams (molar mass
        $NO_2 = 46.01$ g).

$$? \text{ g NO}_2 = 0.0154 \text{ mol NO}_2 \times \frac{46.01 \text{ g NO}_2}{1 \text{ mol NO}_2} = \textbf{0.709 g NO}_2 \textbf{ produced}$$

Starting with 0.01542 mole of NO_2, we can determine the moles of NO that reacted using the mole ratio from the balanced equation. We can calculate the initial moles of NO starting with 0.670 g and using molar mass of NO as a conversion factor.

$$\text{mol NO reacted} = 0.0154 \text{ mol NO}_2 \times \frac{1 \text{ mol NO}}{1 \text{ mol NO}_2} = 0.0154 \text{ mol NO}$$

$$\text{mol NO initial} = 0.670 \text{ g NO} \times \frac{1 \text{ mol NO}}{30.01 \text{ g NO}} = 0.0223 \text{ mol NO}$$

$$\text{mol NO remaining} = \text{mol NO initial} - \text{mol NO reacted}$$

$$\text{mol NO remaining} = 0.0223 \text{ mol NO} - 0.0154 \text{ mol NO} = \textbf{0.0069 mol NO remaining}$$

8.51 This is a limiting reactant problem. Let's calculate the moles of Cl_2 produced assuming complete reaction for each reactant.

$$0.86 \text{ mol MnO}_2 \times \frac{1 \text{ mol Cl}_2}{1 \text{ mol MnO}_2} = 0.86 \text{ mol Cl}_2$$

$$48.2 \text{ g HCl} \times \frac{1 \text{ mol HCl}}{36.458 \text{ g HCl}} \times \frac{1 \text{ mol Cl}_2}{4 \text{ mol HCl}} = 0.3305 \text{ mol Cl}_2$$

HCl is the limiting reactant; it limits the amount of product produced. It will be used up first. The amount of product produced is 0.3305 mole Cl_2. Let's convert this to grams.

$$? \text{ g Cl}_2 = 0.3305 \text{ mol Cl}_2 \times \frac{70.90 \text{ g Cl}_2}{1 \text{ mol Cl}_2} = \textbf{23.4 g Cl}_2$$

8.53 a. Start with a balanced chemical equation. It's given in the problem. We use NG as an abbreviation for nitroglycerin. The molar mass of NG = 227.1 g/mol.

$$4C_3H_5N_3O_9 \longrightarrow 6N_2 + 12CO_2 + 10H_2O + O_2$$

Map out the following strategy to solve this problem.

$$\text{g NG} \rightarrow \text{mol NG} \rightarrow \text{mol O}_2 \rightarrow \text{g O}_2$$

Calculate the grams of O_2 using the strategy above.

$$? \text{ g O}_2 = 2.00 \times 10^2 \text{ g NG} \times \frac{1 \text{ mol NG}}{227.1 \text{ g NG}} \times \frac{1 \text{ mol O}_2}{4 \text{ mol NG}} \times \frac{32.00 \text{ g O}_2}{1 \text{ mol O}_2} = \textbf{7.05 g O}_2$$

b. The theoretical yield was calculated in part (a), and the actual yield is given in the problem (6.55 g). The percent yield is:

$$\% \ \text{yield} \ = \ \frac{\text{actual yield}}{\text{theoretical yield}} \times 100\%$$

$$\% \ \text{yield} \ = \ \frac{6.55 \ \text{g O}_2}{7.05 \ \text{g O}_2} \times 100\% \ = \ \textbf{92.9\%}$$

8.55   The actual yield of ethylene is 481 g. Let's calculate the yield of ethylene if the reaction is 100 percent efficient. We can calculate this from the definition of percent yield. We can then calculate the mass of hexane needed.

$$\% \ \text{yield} = \frac{\text{actual yield}}{\text{theoretical yield}} \times 100\%$$

$$42.5\% \ = \ \frac{481 \ \text{g C}_2\text{H}_4}{\text{theoretical yield}} \times 100\%$$

$$\frac{42.5\%}{100\%} = \frac{481 \ \text{g C}_2\text{H}_4}{\text{theoretical yield}}$$

$$\text{theoretical yield} \times 42.5\% = 481 \ \text{g C}_2\text{H}_4 \times 100\%$$

$$\text{theoretical yield} = 481 \ \text{g C}_2\text{H}_4 \times \frac{100\%}{42.5\%}$$

$$\text{theoretical yield} = 1.132 \times 10^3 \ \text{g C}_2\text{H}_4$$

The mass of hexane needed is:

$$(1.132 \times 10^3 \ \text{g C}_2\text{H}_4) \times \frac{1 \ \text{mol C}_2\text{H}_4}{28.052 \ \text{g C}_2\text{H}_4} \times \frac{1 \ \text{mol C}_6\text{H}_{14}}{1 \ \text{mol C}_2\text{H}_4} \times \frac{86.172 \ \text{g C}_6\text{H}_{14}}{1 \ \text{mol C}_6\text{H}_{14}} \ = \ \textbf{3.48} \times \textbf{10}^3 \ \textbf{g C}_6\textbf{H}_{14}$$

8.57   This is a limiting reactant problem. Let's calculate the moles of S_2Cl_2 produced assuming complete reaction for each reactant.

$$S_8(l) + 4Cl_2(g) \longrightarrow 4S_2Cl_2(l)$$

$$4.06 \ \text{g S}_8 \times \frac{1 \ \text{mol S}_8}{256.56 \ \text{g S}_8} \times \frac{4 \ \text{mol S}_2\text{Cl}_2}{1 \ \text{mol S}_8} \ = \ 0.0633 \ \text{mol S}_2\text{Cl}_2$$

$$6.24 \ \text{g Cl}_2 \times \frac{1 \ \text{mol Cl}_2}{70.90 \ \text{g Cl}_2} \times \frac{4 \ \text{mol S}_2\text{Cl}_2}{4 \ \text{mol Cl}_2} \ = \ 0.0880 \ \text{mol S}_2\text{Cl}_2$$

S_8 is the limiting reactant; it limits the amount of product produced. The amount of product produced is 0.0633 mole S_2Cl_2. Let's convert this to grams.

$$? \ \text{g S}_2\text{Cl}_2 \ = \ 0.0633 \ \text{mol S}_2\text{Cl}_2 \times \frac{135.04 \ \text{g S}_2\text{Cl}_2}{1 \ \text{mol S}_2\text{Cl}_2} \ = \ \textbf{8.55 g S}_2\textbf{Cl}_2$$

This is the theoretical yield of S_2Cl_2. The actual yield is given in the problem (6.55 g). The percent yield is:

$$\% \ \text{yield} \ = \ \frac{\text{actual yield}}{\text{theoretical yield}} \times 100\% \ = \ \frac{6.55 \ \text{g}}{8.55 \ \text{g}} \times 100\% \ = \ \textbf{76.6\%}$$

8.59   **Strategy:**   The atom economy can be determined using Equation 8.2.

Setup:
$$\text{atom economy} = \frac{\text{mass of atoms in desired product(s)}}{\text{mass of atoms in all reactants}} \times 100\%$$

Solution: The atom economy of this process is

$$\frac{\text{mass of } CH_3CO_2H}{\text{mass of } CH_3OH \text{ and } CO} \times 100\% = \frac{60.05 \text{ g}}{60.05 \text{ g}} \times 100\% = \textbf{100\%}$$

8.63   **The electron configuration of helium is $1s^2$ and that of the other noble gases is ns^2np^6. The completely filled subshell represents great stability. Consequently, these elements are chemically unreactive.**

8.65
$$2H_2(g) + O_2(g) \longrightarrow 2H_2O(g)$$

We start with 8 molecules of H_2 and 3 molecules of O_2. The balanced equation shows 2 moles H_2 are stoichiometrically equivalent to 1 mole O_2. If 3 molecules of O_2 react, 6 molecules of H_2 will react, leaving 2 molecules of H_2 in excess. The balanced equation also shows 1 mole O_2 is stoichiometrically equivalent to 2 moles H_2O. If 3 molecules of O_2 react, 6 molecules of H_2O will be produced.

After complete reaction, there will be 2 molecules of H_2 and 6 molecules of H_2O. The correct choice is **diagram (b)**.

8.67   We assume that all the Cl in the compound ends up as HCl and all the O ends up as H_2O. Therefore, we need to find the number of moles of Cl in HCl and the number of moles of O in H_2O.

$$\text{mol Cl} = 0.233 \text{ g HCl} \times \frac{1 \text{ mol HCl}}{36.458 \text{ g HCl}} \times \frac{1 \text{ mol Cl}}{1 \text{ mol HCl}} = 0.006391 \text{ mol Cl}$$

$$\text{mol O} = 0.403 \text{ g } H_2O \times \frac{1 \text{ mol } H_2O}{18.016 \text{ g } H_2O} \times \frac{1 \text{ mol O}}{1 \text{ mol } H_2O} = 0.02237 \text{ mol O}$$

Dividing by the smallest number of moles (0.006391 mole) gives the formula, $ClO_{3.50}$. Multiplying both subscripts by two gives the empirical formula, **Cl_2O_7**.

8.69   The symbol "O" refers to moles of oxygen atoms, not oxygen molecule (O_2). Look at the molecular formulas given in parts (a) and (b). What do they tell you about the relative amounts of carbon and oxygen?

a. $0.212 \text{ mol C} \times \dfrac{1 \text{ mol O}}{1 \text{ mol C}} = \textbf{0.212 mol O}$

b. $0.212 \text{ mol C} \times \dfrac{2 \text{ mol O}}{1 \text{ mol C}} = \textbf{0.424 mol O}$

8.71   MM(nitroglycerin) = 227.10 g; MM(NO) = 30.01 g;

$$\text{mol(NO)} : \text{mol(nitroglycerin)} = 3:1 \text{ or } \frac{(3)(30.01 \text{ g})}{227.10 \text{ g}} \times 100\% = \textbf{39.64\% NO}$$

8.73   The amount of Fe that reacted is: $\dfrac{1}{8} \times 664 \text{ g} = 83.0 \text{ g}$

The amount of Fe remaining is: $664 \text{ g} - 83.0 \text{ g} = 581 \text{ g}$

Thus, 83.0 g of Fe reacts to form the compound Fe_2O_3, which has two moles of Fe atoms per 1 mole of compound. The mass of Fe_2O_3 produced is:

$$83.0 \text{ g Fe} \times \frac{1 \text{ mol Fe}}{55.85 \text{ g Fe}} \times \frac{1 \text{ mol Fe}_2O_3}{2 \text{ mol Fe}} \times \frac{159.7 \text{ g Fe}_2O_3}{1 \text{ mol Fe}_2O_3} = 119 \text{ g Fe}_2O_3$$

The final mass of the iron bar and rust is: $581 \text{ g Fe} + 119 \text{ g Fe}_2O_3 = \textbf{700 g}$

8.75   a. $\textbf{Zn}(s) + \textbf{H}_2\textbf{SO}_4(aq) \longrightarrow \textbf{ZnSO}_4(aq) + \textbf{H}_2(g)$

b. We assume that a pure sample would produce the theoretical yield of H_2. The balanced equation shows a mole ratio of 1 mole H_2 : 1 mole Zn. The theoretical yield of H_2 is:

$$3.86 \text{ g Zn} \times \frac{1 \text{ mol Zn}}{65.41 \text{ g Zn}} \times \frac{1 \text{ mol H}_2}{1 \text{ mol Zn}} \times \frac{2.016 \text{ g H}_2}{1 \text{ mol H}_2} = 0.119 \text{ g H}_2$$

$$\text{percent purity} = \frac{0.0764 \text{ g H}_2}{0.119 \text{ g H}_2} \times 100\% = \textbf{64.2\%}$$

c. **We assume that the impurities are inert and do not react with the sulfuric acid to produce hydrogen.**

8.77   a. $0.400 \text{ g aspirin} \times \dfrac{1 \text{ mol aspirin}}{180.15 \text{ g aspirin}} \times \dfrac{1 \text{ mol salicylic acid}}{1 \text{ mol aspirin}} \times \dfrac{138.12 \text{ g salicylic acid}}{1 \text{ mol salicylic acid}}$

$$= \textbf{0.307 g salicylic acid}$$

b. $0.307 \text{ g salicylic acid} \times \dfrac{1}{0.749} = \textbf{0.410 g salicylic acid}$

If you have trouble deciding whether to multiply or divide by 0.749 in the calculation, remember that if only 74.9% of salicylic acid is converted to aspirin, a larger amount of salicylic acid will be needed to yield the same amount of aspirin.

c. $9.26 \text{ g salicylic acid} \times \dfrac{1 \text{ mol salicylic acid}}{138.12 \text{ g salicylic acid}} \times \dfrac{1 \text{ mol aspirin}}{1 \text{ mol salicylic acid}} = 0.06704 \text{ mol aspirin}$

$8.54 \text{ g acetic anhydride} \times \dfrac{1 \text{ mol acetic anhydride}}{102.09 \text{ g acetic anhydride}} \times \dfrac{1 \text{ mol aspirin}}{1 \text{ mol acetic anhydride}} = 0.08365 \text{ mol aspirin}$

The limiting reactant is salicylic acid. The theoretical yield of aspirin is:

$$0.06704 \text{ mol aspirin} \times \frac{180.15 \text{ g aspirin}}{1 \text{ mol aspirin}} = \textbf{12.08 g aspirin}$$

The percent yield is:

$$\% \text{ yield} = \frac{10.9 \text{ g}}{12.08 \text{ g}} \times 100\% = \textbf{90.2\%}$$

8.79   The mass of oxygen in MO is 39.46 g – 31.70 g = 7.76 g O. Therefore, for every 31.70 g of M, there is 7.76 g of O in the compound MO. The molecular formula shows a mole ratio of 1 mole M : 1 mole O. First, calculate moles of M that react with 7.76 g O.

$$\text{mol M} = 7.76 \text{ g O} \times \frac{1 \text{ mol O}}{16.00 \text{ g O}} \times \frac{1 \text{ mol M}}{1 \text{ mol O}} = 0.485 \text{ mol M}$$

$$\text{molar mass M} = \frac{31.70 \text{ g M}}{0.485 \text{ mol M}} = 65.4 \text{ g/mol}$$

Thus, the atomic mass of M is **65.4 amu**. The metal is most likely **Zn**.

8.81   We assume that the increase in mass results from the element nitrogen. The mass of nitrogen is:

$$0.378 \text{ g} - 0.273 \text{ g} = 0.105 \text{ g N}$$

The empirical formula can now be calculated. Convert to moles of each element.

$$0.273 \text{ g Mg} \times \frac{1 \text{ mol Mg}}{24.31 \text{ g Mg}} = 0.0112 \text{ mol Mg}$$

$$0.105 \text{ g N} \times \frac{1 \text{ mol N}}{14.01 \text{ g N}} = 0.00749 \text{ mol N}$$

Dividing by the smallest number of moles gives $Mg_{1.5}N$. Recall that an empirical formula must have whole number coefficients. Multiplying by a factor of 2 gives the empirical formula $\textbf{Mg}_{\textbf{3}}\textbf{N}_{\textbf{2}}$. The name of this compound is **magnesium nitride**.

8.83   *Step 1:*   Calculate the mass of C in 55.90 g CO_2, and the mass of H in 28.61 g H_2O. This is a dimensional analysis problem. To calculate the mass of each component, you need the molar masses and the correct mole ratio.

You should come up with the following strategy:

$$\text{g } CO_2 \rightarrow \text{mol } CO_2 \rightarrow \text{mol C} \rightarrow \text{g C}$$

Step 2: $$? \text{ g C} = 55.90 \text{ g } CO_2 \times \frac{1 \text{ mol } CO_2}{44.01 \text{ g } CO_2} \times \frac{1 \text{ mol C}}{1 \text{ mol } CO_2} \times \frac{12.01 \text{ g C}}{1 \text{ mol C}} = 15.25 \text{ g C}$$

Similarly,

$$? \text{ g H} = 28.61 \text{ g } H_2O \times \frac{1 \text{ mol } H_2O}{18.02 \text{ g } H_2O} \times \frac{2 \text{ mol H}}{1 \text{ mol } H_2O} \times \frac{1.008 \text{ g H}}{1 \text{ mol H}} = 3.201 \text{ g H}$$

Since the compound contains C, H, and Pb, we can calculate the mass of Pb by difference.

$$51.36 \text{ g} = \text{mass C} + \text{mass H} + \text{mass Pb}$$

$$51.36 \text{ g} = 15.25 \text{ g} + 3.201 \text{ g} + \text{mass Pb}$$

$$\text{mass Pb} = 32.91 \text{ g Pb}$$

Step 3: Calculate the number of moles of each element present in the sample. Use molar mass as a conversion factor.

$$? \text{ mol C} = 15.25 \text{ g C} \times \frac{1 \text{ mol C}}{12.01 \text{ g C}} = 1.270 \text{ mol C}$$

Similarly,

$$? \text{ mol H} = 3.201 \text{ g H} \times \frac{1 \text{ mol H}}{1.008 \text{ g H}} = 3.176 \text{ mol H}$$

$$? \text{ mol Pb} = 32.91 \text{ g Pb} \times \frac{1 \text{ mol Pb}}{207.2 \text{ g Pb}} = 0.1588 \text{ mol Pb}$$

Thus, we arrive at the formula $Pb_{0.1588}C_{1.270}H_{3.176}$, which gives the identity and the ratios of atoms present. However, chemical formulas are written with whole numbers.

Step 4: Try to convert to whole numbers by dividing all the subscripts by the smallest subscript.

$$\text{Pb:} \frac{0.1588}{0.1588} = 1.00 \quad \text{C:} \frac{1.270}{0.1588} \approx 8 \quad \text{H:} \frac{3.176}{0.1588} \approx 20$$

This gives the empirical formula, **PbC_8H_{20}**.

8.85 a. The balanced chemical equation is:

$$\mathbf{C_3H_8}\textbf{\textit{(g)}} + \mathbf{3H_2O}\textbf{\textit{(g)}} \longrightarrow \mathbf{3CO}\textbf{\textit{(g)}} + \mathbf{7H_2}\textbf{\textit{(g)}}$$

b. We need to complete the following conversions:

$$\text{kg C}_3\text{H}_8 \rightarrow \text{g C}_3\text{H}_8 \rightarrow \text{mol C}_3\text{H}_8 \rightarrow \text{mol H}_2 \rightarrow \text{ g H}_2 \rightarrow \text{kg H}_2$$

We show the conversion steps in two parts. First, we convert from kilograms C_3H_8 to moles H_2:

$$? \text{ mol H}_2 = (2.84 \times 10^3 \text{ kg C}_3\text{H}_8) \times \frac{1000 \text{ g C}_3\text{H}_8}{1 \text{ kg C}_3\text{H}_8} \times \frac{1 \text{ mol C}_3\text{H}_8}{44.09 \text{ g C}_3\text{H}_8} \times \frac{7 \text{ mol H}_2}{1 \text{ mol C}_3\text{H}_8} = 4.51 \times 10^5 \text{ mol H}_2$$

Then we convert from moles H_2 to kilograms H_2:

$$? \text{ kg H}_2 = (4.51 \times 10^5 \text{ mol H}_2) \times \frac{2.016 \text{ g H}_2}{1 \text{ mol H}_2} \times \frac{1 \text{ kg H}_2}{1000 \text{ g H}_2} = \mathbf{909 \text{ kg H}_2}$$

8.87   a. We need to compare the mass % of K in both KCl and K_2SO_4.

$$\%K \text{ in KCl} = \frac{39.10 \text{ g}}{74.55 \text{ g}} \times 100\% = 52.45\% \text{ K}$$

$$\%K \text{ in } K_2SO_4 = \frac{2(39.10 \text{ g})}{174.27 \text{ g}} \times 100\% = 44.87\% \text{ K}$$

The price depends on the %K.

$$\frac{\text{Price of } K_2SO_4}{\text{Price of KCl}} = \frac{\%K \text{ in } K_2SO_4}{\%K \text{ in KCl}}$$

$$\text{Price of } K_2SO_4 = \text{Price of KCl} \times \frac{\%K \text{ in } K_2SO_4}{\%K \text{ in KCl}}$$

$$\text{Price of } K_2SO_4 = \frac{\$0.55}{\text{kg}} \times \frac{44.87\%}{52.45\%} = \mathbf{\$0.47 / kg}$$

b. First, calculate the number of moles of K in 1.00 kg of KCl.

$$1.00 \text{ kg KCl} \times \frac{1000 \text{ g KCl}}{1 \text{ kg KCl}} \times \frac{1 \text{ mol KCl}}{74.55 \text{ g KCl}} \times \frac{1 \text{ mol K}}{1 \text{ mol KCl}} = 13.4 \text{ mol K}$$

Next, calculate the amount of K_2O needed to supply 13.4 mol K.

$$13.4 \text{ mol K} \times \frac{1 \text{ mol } K_2O}{2 \text{ mol K}} \times \frac{94.20 \text{ g } K_2O}{1 \text{ mol } K_2O} \times \frac{1 \text{ kg}}{1000 \text{ g}} = \mathbf{0.631 \text{ kg } K_2O}$$

8.89   a. **NH_4NO_2 is the *only* reactant**, so, no matter how much of it is present initially, NH_4NO_2 will always be the reactant that determines the amount of product formed. **When a reaction involves only a single reactant, we usually do not describe that reactant as the "limiting reactant" because it is understood that this is the case.**

b. **In principle, two or more reactants can be limiting reactants if they are exhausted simultaneously. In practice, it is virtually impossible to measure the reactants so precisely that two or more run out at the same time.**

8.91   **Strategy:**   Begin by writing an unbalanced equation to represent the combination of reactants and formation of products as stated in the problem, and then balance the equation.

To determine the mass of water produced, first calculate the number of moles of O_2 used in the reaction. Use the balanced chemical equation to determine the correct stoichiometric conversion factor and calculate the mass of water produced.

Setup: There are 30.0 g of air, which is 20.0% (or 20.0/100) oxygen by mass. This means there are $0.200 \times 30.0 \text{ g} = 6.00 \text{ g O}_2$.

Write a balanced equation for the combustion of C_4H_{10} in O_2 (combustion of a hydrocarbon such as C_4H_{10} produces carbon dioxide and water):

$$C_4H_{10}(g) + O_2(g) \longrightarrow CO_2(g) + H_2O(l)$$

Balance the number of C atoms by changing the coefficient for CO_2 from 1 to 4.

$$C_4H_{10}(g) + O_2(g) \longrightarrow 4CO_2(g) + H_2O(l)$$

Balance the number of H atoms by changing the coefficient for H_2O from 1 to 5.

$$C_4H_{10}(g) + O_2(g) \longrightarrow 4CO_2(g) + 5H_2O(l)$$

Balance the number of O atoms by changing the coefficient for O_2 from 1 to $\dfrac{13}{2}$.

$$C_4H_{10}(g) + \frac{13}{2}O_2(g) \longrightarrow 4CO_2(g) + 5H_2O(l)$$

There are equal numbers of each kind of atom on both sides of the equation; however, to write the equation with the smallest *whole* number coefficients, multiply each coefficient by 2. The final balanced equation is:

$$2C_4H_{10}(g) + 13O_2(g) \longrightarrow 8CO_2(g) + 10H_2O(l)$$

The required molar masses are 32.00 g/mol for O_2 and 18.02 g/mol for H_2O. From the balanced equation we have 13 mol O_2 and 10 mol H_2O. The necessary stoichiometric conversion factor is therefore:

$$\frac{10 \text{ mol H}_2\text{O}}{13 \text{ mol O}_2}$$

Solution: To determine the mass of water produced, first calculate the number of moles of O_2 consumed:

$$6.00 \text{ g O}_2 \times \frac{1 \text{ mol O}_2}{32.00 \text{ g O}_2} = 0.1875 \text{ mol O}_2$$

Calculate the number of moles of H_2O produced from 0.185 mol O_2 using the conversion factor:

$$0.1875 \text{ mol O}_2 \times \frac{10 \text{ mol H}_2\text{O}}{13 \text{ mol O}_2} = 0.1442 \text{ mol H}_2\text{O}$$

Convert this amount to grams as follows:

$$0.1442 \text{ mol H}_2\text{O} \times \frac{18.02 \text{ g H}_2\text{O}}{1 \text{ mol H}_2\text{O}} = \textbf{2.60 g H}_2\textbf{O}$$

8.93   The figure shows 6 molecules of NO combining with 5 molecules of SO_2 to produce 2 molecules of SO_2, 3 molecules of SO_3, and 3 molecules of N_2O.

The unbalanced equation is:

$$6NO(g) + 5SO_2(g) \longrightarrow 2SO_2(g) + 3SO_3(g) + 3N_2O(g)$$

Since there are at least 2 mol $SO_2(g)$ on both sides of the equation, they cancel out like algebraic quantities. We are left with 3 mol $SO_2(g)$ on the reactant side (5 mol – 2 mol = 3 mol $SO_2(g)$) and 0 mol $SO_2(g)$ on the product side of the equation (2 mol – 2 mol = 0 mol $SO_2(g)$).

The equation becomes:

$$6NO(g) + 3SO_2(g) \longrightarrow 3SO_3(g) + 3N_2O(g)$$

The remaining coefficients are 6, 3, 3, and 3. To obtain the smallest possible *whole* number coefficient, divide each coefficient by 3 to give the final balanced equation:

$$\mathbf{2NO(g) + SO_2(g) \longrightarrow SO_3(g) + N_2O(g)}$$

8.95   **Strategy:**   Use the balanced equation to write the necessary stoichiometric conversion factor and determine which reactant is limiting. Again, using the balanced equations, write the stoichiometric conversion factors to determine the number of moles of $NH_4V_3O_8$ that can be produced.

        Setup: There are 2.5 moles each of N_2 and H_2. From the balanced equations, we have 3 mol $H_2 \triangleq 1$ mol N_2, 3 mol $H_2 \triangleq 2$ mol NH_3, 2 mol $NH_3 \triangleq 2$ mol NH_4VO_3, and 2 mol $NH_4VO_3 \triangleq 1$ mol $NH_4V_3O_8$. The necessary stoichiometric conversion factors are therefore:

$$\frac{3 \text{ mol } H_2}{1 \text{ mol } N_2} \qquad \frac{2 \text{ mol } NH_3}{3 \text{ mol } H_2} \qquad \frac{2 \text{ mol } NH_4VO_3}{2 \text{ mol } NH_3} \qquad \frac{1 \text{ mol } NH_4V_3O_8}{3 \text{ mol } NH_4VO_3}$$

        Solution: To determine which reactant is limiting, calculate the amount of H_2 needed to react completely with 2.5 mole of N_2.

$$2.5 \text{ mol } N_2 \times \frac{3 \text{ mol } H_2}{1 \text{ mol } N_2} = 7.5 \text{ mol } H_2$$

The amount of H_2 required to react with 2.5 mol N_2 is more than is present. Therefore, H_2 is the limiting reactant and N_2 is the excess reactant.

Calculate the number of moles of $NH_4V_3O_8$ produced from the number of moles of limiting reactant (H_2) consumed:

$$2.5 \text{ mol } H_2 \times \frac{2 \text{ mol } NH_3}{3 \text{ mol } H_2} \times \frac{2 \text{ mol } NH_4VO_3}{2 \text{ mol } NH_3} \times \frac{1 \text{ mol } NH_4V_3O_8}{3 \text{ mol } NH_4VO_3} = \mathbf{0.56 \text{ mol } NH_4V_3O_8}$$

8.97   **Strategy:**   The atom economy can be determined using Equation 8.2.

        Setup:

$$\text{atom economy} = \frac{\text{mass of atoms in desired product(s)}}{\text{mass of atoms in all reactants}} \times 100\%$$

Solution: The atom economy of the traditional process is:

$$\frac{\text{mass of } CH_2OCH_2}{\text{mass of } CH_2CH_2, Cl_2, \text{ and } Ca(OH)_2} \times 100\% = \frac{44.05 \text{ g}}{173.05 \text{ g}} \times 100\% = 25.46\%$$

The atom economy of the recent synthesis is

$$\frac{\text{mass of } CH_2OCH_2}{\text{mass of } CH_2CH_2 \text{ and } O_2} \times 100\% = \frac{44.05 \text{ g}}{60.05 \text{ g}} \times 100\% = 73.36\%$$

The atom economy for the recent synthesis, 73.36%, is nearly three times higher than it is for the traditional preparation, 25.46%.

8.99   Both compounds contain only Mn and O. When the first compound is heated, oxygen gas is evolved. Let's calculate the empirical formulas for the two compounds, then we can write a balanced equation.

a. **compound X:** Assume 100 g of compound.

$$63.3 \text{ g Mn} \times \frac{1 \text{ mol Mn}}{54.94 \text{ g Mn}} = 1.15 \text{ mol Mn}$$

$$36.7 \text{ g O} \times \frac{1 \text{ mol O}}{16.00 \text{ g O}} = 2.29 \text{ mol O}$$

Dividing by the smallest number of moles (1.15 moles) gives the empirical formula, **MnO$_2$**.
compound Y: Assume 100 g of compound.

$$72.0 \text{ g Mn} \times \frac{1 \text{ mol Mn}}{54.94 \text{ g Mn}} = 1.31 \text{ mol Mn}$$

$$28.0 \text{ g O} \times \frac{1 \text{ mol O}}{16.00 \text{ g O}} = 1.75 \text{ mol O}$$

Dividing by the smallest number of moles gives MnO$_{1.33}$. Recall that an empirical formula must have whole number coefficients. Multiplying by a factor of three gives the empirical formula **Mn$_3$O$_4$**.

b. The balanced equation is: **3MnO$_2$(s) $\longrightarrow$ Mn$_3$O$_4$(s) + O$_2$(g)**

Chapter 9
Chemical Reactions in Aqueous Solutions

Visualizing Chemistry

VC 9.1 b   VC 9.2 c   VC 9.3 c   VC 9.4 a

Key Skills

9.1   b   9.2   c   9.3   c   9.4   d

Problems

9.9   **Strategy:**   Strong electrolytes dissociate completely upon dissolving in water. The more ions in solution, the stronger the electrolyte.

Setup: Compound A_2E: The compound partially dissociates. All four of the molecules are partially ionized.

Compound A_2F: The compound exists in solution as molecules that are not ionized. The compound does not dissociate in aqueous solution.

Compound A_2G: The compound exists in solution predominately as molecules that are not ionized. Three of the five molecules are intact, while the remaining two are only partially ionized.

Solution: In order of increasing electrolyte strength:

$$A_2F < A_2G < A_2E$$

9.11   When NaCl dissolves in water, it dissociates into Na^+ and Cl^- ions. When the ions are hydrated, the water molecules will be oriented so that the negative end of the water dipole interacts with the positive sodium ion, and the positive end of the water dipole interacts with the negative chloride ion. The negative end of the water dipole is near the oxygen atom, and the positive end of the water dipole is near the hydrogen atoms. The diagram that best represents the hydration of NaCl when dissolved in water is **diagram (c)**.

9.13   Ionic compounds, strong acids, and strong bases (metal hydroxides) are strong electrolytes (completely broken up into ions of the compound). Weak acids and weak bases are weak electrolytes. Molecular substances other than acids or bases are nonelectrolytes.

a. **strong electrolyte** (ionic)

b. **nonelectrolyte**

c. **weak electrolyte** (weak base)

d. **strong electrolyte** (strong base)

e. **weak electrolyte** (weak acid)

9.17   **Strategy:** Write the molecular equation for the reaction that occurs when aqueous solutions of $AgNO_3$ and NaCl are combined. Determine which product will precipitate based on the solubility guidelines in Tables 9.2 and 9.3.

Setup: Predict the products by exchanging the ions and balancing the equation.

Solution: Molecular equation:

$$AgNO_3(aq) + NaCl(aq) \longrightarrow AgCl(s) + NaNO_3(aq)$$

According to Table 4.2, AgCl is insoluble in water and will precipitate from solution. $NaNO_3$ contains a metal cation and is therefore an ionic compound. It will dissociate completely in water and remain as Na^+ and NO_3^- ions in solution.

Diagram (c) best represents the mixture.

9.19   **Strategy:** Predict the products by exchanging the ions and balancing the equation. Determine which product will precipitate based on the solubility guidelines in Tables 9.2 and 9.3. Rewrite the equation showing strong electrolytes as ions. Identify and cancel spectator ions.

Setup: Molecular equation:

$$2LiOH(aq) + Cu(NO_3)_2(aq) \longrightarrow 2LiNO_3(aq) + Cu(OH)_2(s)$$

Ionic equation:

$$2Li^+(aq) + 2OH^-(aq) + Cu^{2+}(aq) + 2NO_3^-(aq) \longrightarrow 2Li^+(aq) + 2NO_3^-(aq) + Cu(OH)_2(s)$$

Net ionic equation:

$$Cu^{2+}(aq) + 2OH^-(aq) \longrightarrow Cu(OH)_2(s)$$

Solution: According to the net ionic equation, Cu^{2+} combines with OH^- in a 1:2 ratio to give $Cu(OH)_2$. **Reaction (d)** represents the combination of aqueous solutions of LiOH and $Cu(NO_3)_2$.

9.21   **Strategy:** Although it is not necessary to memorize the solubilities of compounds, you should keep in mind the following useful rules: all ionic compounds containing alkali metal cations, the ammonium ion, and the nitrate, bicarbonate, and chlorate ions are soluble. For other compounds, refer to Tables 9.2 and 9.3 of the text.

Solution: a. $Ca_3(PO_4)_2$ is **insoluble**. Most phosphate compounds are insoluble.

b. $Mn(OH)_2$ is **insoluble**. Most hydroxide compounds are insoluble.

c. $AgClO_3$ is **soluble**. All chlorate compounds are soluble.

d. K_2S is **soluble**. All compounds containing alkali metal cations are soluble.

e. $Pb_3(PO_4)_2$ is **insoluble**. Most phosphate compounds are insoluble.

9.23   **Strategy:** Recall that an *ionic equation* shows dissolved ionic compounds in terms of their free ions. What ions do the aqueous solutions contain before they are combined? A *net ionic equation* shows only the species that actually take part in the reaction.

Solution: a. In solution, $AgNO_3$ dissociates into Ag^+ and NO_3^- ions and Na_2SO_4 dissociates into Na^+ and SO_4^{2-} ions. According to Tables 9.2 and 9.3 of the text, silver ions (Ag^+) and sulfate ions (SO_4^{2-}) will form an insoluble compound, silver sulfate (Ag_2SO_4), while the other product, $NaNO_3$, is soluble and remains in solution. This is a precipitation reaction. The balanced molecular equation is:

$$2AgNO_3(aq) + Na_2SO_4(aq) \longrightarrow Ag_2SO_4(s) + 2NaNO_3(aq)$$

The ionic and net ionic equations are:

Ionic: $2Ag^+(aq) + 2NO_3^-(aq) + 2Na^+(aq) + SO_4^{2-}(aq) \longrightarrow Ag_2SO_4(s) + 2Na^+(aq) + 2NO_3^-(aq)$

Net ionic: $2Ag^+(aq) + SO_4^{2-}(aq) \longrightarrow Ag_2SO_4(s)$

b. In solution, $BaCl_2$ dissociates into Ba^{2+} and Cl^- ions and $ZnSO_4$ dissociates into Zn^{2+} and SO_4^{2-} ions. According to Tables 9.2 and 9.3 of the text, barium ions (Ba^{2+}) and sulfate ions (SO_4^{2-}) will form an insoluble compound, barium sulfate ($BaSO_4$), whereas the other product, $ZnCl_2$, is soluble and remains in solution. This is a precipitation reaction. The balanced molecular equation is:

$$BaCl_2(aq) + ZnSO_4(aq) \longrightarrow BaSO_4(s) + ZnCl_2(aq)$$

The ionic and net ionic equations are:

Ionic: $Ba^{2+}(aq) + 2Cl^-(aq) + Zn^{2+}(aq) + SO_4^{2-}(aq) \longrightarrow BaSO_4(s) + Zn^{2+}(aq) + 2Cl^-(aq)$

Net ionic: $Ba^{2+}(aq) + SO_4^{2-}(aq) \longrightarrow BaSO_4(s)$

c. In solution, $(NH_4)_2CO_3$ dissociates into NH_4^+ and CO_3^{2-} ions and $CaCl_2$ dissociates into Ca^{2+} and Cl^- ions. According to Tables 9.2 and 9.3 of the text, calcium ions (Ca^{2+}) and carbonate ions (CO_3^{2-}) will form an insoluble compound, calcium carbonate ($CaCO_3$), whereas the other product, NH_4Cl, is soluble and remains in solution. This is a precipitation reaction. The balanced molecular equation is:

$$(NH_4)_2CO_3(aq) + CaCl_2(aq) \longrightarrow CaCO_3(s) + 2NH_4Cl(aq)$$

The ionic and net ionic equations are:

Ionic: $2NH_4^+(aq) + CO_3^{2-}(aq) + Ca^{2+}(aq) + 2Cl^-(aq) \longrightarrow CaCO_3(s) + 2NH_4^+(aq) + 2Cl^-(aq)$

Net ionic: $Ca^{2+}(aq) + CO_3^{2-}(aq) \longrightarrow CaCO_3(s)$

9.25 **Strategy:** Precipitation reactions usually involve ionic compounds, but only certain combinations of electrolyte solutions result in the formation of a precipitate. Whether or not a precipitate forms when two solutions are mixed depends on the solubility of the products.

Setup: Predict the products by exchanging the ions and balancing the equation. Determine which product will precipitate based on the solubility guidelines in Tables 9.2 and 9.3.

    a. $2NaNO_3(aq) + CuSO_4(aq) \longrightarrow Na_2SO_4(aq) + Cu(NO_3)_2(aq)$

    b. $BaCl_2(aq) + K_2SO_4(aq) \longrightarrow BaSO_4(s) + 2KCl(aq)$

    c. $AgNO_3(aq) + LiC_2H_3O_2(aq) \longrightarrow AgC_2H_3O_2(aq) + LiNO_3(aq)$

Solution: a. Ionic equation:

$$2Na^+(aq) + 2NO_3^-(aq) + Cu^+(aq) + SO_4^{2-}(aq) \longrightarrow 2Na^+(aq) + SO_4^{2-}(aq) + Cu^+(aq) + 2NO_3^-(aq)$$

All the ions in solution are spectator ions; therefore, there is **no net ionic equation**. Not a precipitation reaction.

b. Ionic equation:

$$Ba^{2+}(aq) + 2Cl^-(aq) + 2K^+(aq) + SO_4^{2-}(aq) \longrightarrow BaSO_4(s) + 2K^+(aq) + 2Cl^-(aq)$$

Net ionic equation:
$$\mathbf{Ba^{2+}(aq) + SO_4^{2-}(aq) \longrightarrow BaSO_4(s)}$$

A precipitation reaction will likely result from mixing a BaCl₂ solution with a K_2SO_4 solution. $BaSO_4$ will precipitate.

c. Ionic equation:

$$Ag^+(aq) + NO_3^-(aq) + Li^+(aq) + C_2H_3O_2^-(aq) \longrightarrow Ag^+(aq) + C_2H_3O_2^-(aq) + Li^+(aq) + NO_3^-(aq)$$

All the ions in the solution are spectator ions; therefore, there is **no net ionic equation**. Not a precipitation reaction.

9.33 **Strategy:** What are the characteristics of a Brønsted acid? Does it contain at least an H atom? With the exception of ammonia, most Brønsted bases that you will encounter at this stage are anions.

Solution: a. PO_4^{3-} in water can accept a proton to become HPO_4^{2-} and is thus a **Brønsted base**.

    b. ClO_2^- in water can accept a proton to become $HClO_2$ and is thus a **Brønsted base**.

    c. NH_4^+ dissolved in water can donate a proton H^+, thus behaving as a **Brønsted acid**.

d. HCO_3^- can either donate a proton to yield H^+ and CO_3^{2-}, thus behaving as a **Brønsted acid**. Or, HCO_3^- can accept a proton to become H_2CO_3, thus behaving as a **Brønsted base**.

9.35   **Strategy:**   Recall that strong acids and strong bases are strong electrolytes. They are completely ionized in solution. An *ionic equation* will show strong electrolytes, including strong acids and strong bases as separate ions. Weak acids and weak bases are weak electrolytes. They ionize only to a small extent . Weak acids and weak bases are shown as molecules in ionic and net ionic equations. A *net ionic equation* shows only the species that actually take part in the reaction.

    Solution: a. $HC_2H_3O_2$ is a weak acid. It will be shown as a molecule in the ionic equation. KOH is a strong base. It completely dissociates to K^+ and OH^- ions. Since $HC_2H_3O_2$ is an acid, it donates an H^+ to the base, OH^-, producing water. The other product is the salt, $KC_2H_3O_2$, which is soluble and remains in solution. **The balanced molecular equation is:**

$$HC_2H_3O_2(aq) + KOH(aq) \longrightarrow KC_2H_3O_2(aq) + H_2O(l)$$

The ionic and net ionic equations are:

$$\textbf{Ionic: } HC_2H_3O_2(aq) + K^+(aq) + OH^-(aq) \longrightarrow C_2H_3O_2^-(aq) + K^+(aq) + H_2O(l)$$

$$\textbf{Net ionic: } HC_2H_3O_2(aq) + OH^-(aq) \longrightarrow C_2H_3O_2^-(aq) + H_2O(l)$$

b. H_2CO_3 is a weak acid. It will be shown as a molecule in the ionic equation. NaOH is a strong base. It completely dissociates to Na^+ and OH^- ions. Since H_2CO_3 is an acid, it donates an H^+ to the base, OH^-, producing water. The other product is the salt, Na_2CO_3, which is soluble and remains in solution. **The balanced molecular equation is:**

$$H_2CO_3(aq) + 2NaOH(aq) \longrightarrow Na_2CO_3(aq) + 2H_2O(l)$$

The ionic and net ionic equations are:

$$\textbf{Ionic: } H_2CO_3(aq) + 2Na^+(aq) + 2OH^-(aq) \longrightarrow 2Na^+(aq) + CO_3^{2-}(aq) + 2H_2O(l)$$

$$\textbf{Net ionic: } H_2CO_3(aq) + 2OH^-(aq) \longrightarrow CO_3^{2-}(aq) + 2H_2O(l)$$

c. HNO_3 is a strong acid. It completely ionizes to H^+ and NO_3^- ions. $Ba(OH)_2$ is a strong base. It completely dissociates to Ba^{2+} and OH^- ions. Since HNO_3 is an acid, it donates an H^+ to the base, OH^-, producing water. The other product is the salt, $Ba(NO_3)_2$, which is soluble and remains in solution. **The balanced molecular equation is:**

$$2HNO_3(aq) + Ba(OH)_2(aq) \longrightarrow Ba(NO_3)_2(aq) + 2H_2O(l)$$

The ionic and net ionic equations are:

$$\textbf{Ionic: } 2H^+(aq) + 2NO_3^-(aq) + Ba^{2+}(aq) + 2OH^-(aq) \longrightarrow Ba^{2+}(aq) + 2NO_3^-(aq) + 2H_2O(l)$$

$$\textbf{Net ionic: } 2H^+(aq) + 2OH^-(aq) \longrightarrow 2H_2O(l) \text{ or } H^+(aq) + OH^-(aq) \longrightarrow H_2O(l)$$

9.43 **Strategy:** In order to break a redox reaction down into an oxidation half-reaction and a reduction half-reaction, you should first assign oxidation numbers to all the atoms in the reaction. In this way, you can determine which element is oxidized (loses electrons) and which element is reduced (gains electrons).

Solution: In each part, the reducing agent is the reactant in the first half-reaction and the oxidizing agent is the reactant in the second half-reaction.

a. The product is an ionic compound whose ions are Sr^{2+} and O^{2-}. **The half-reactions are:**

$$2Sr \longrightarrow 2Sr^{2+} + 4e^- \text{ and } O_2 + 4e^- \longrightarrow 2O^{2-}$$

O_2 is the oxidizing agent. Sr is the reducing agent.

b. The product is an ionic compound whose ions are Li^+ and H^-. **The half-reactions are:**

$$2Li \longrightarrow 2Li^+ + 2e^- \text{ and } H_2 + 2e^- \longrightarrow 2H^-$$

H_2 is the oxidizing agent. Li is the reducing agent.

c. The product is an ionic compound whose ions are Cs^+ and Br^-. **The half-reactions are:**

$$2Cs \longrightarrow 2Cs^+ + 2e^- \text{ and } Br_2 + 2e^- \longrightarrow 2Br^-$$

Br_2 is the oxidizing agent. Cs is the reducing agent.

d. The product is an ionic compound whose ions are Mg^{2+} and N^{3-}. **The half-reactions are:**

$$3Mg \longrightarrow 3Mg^{2+} + 6e^- \text{ and } N_2 + 6e^- \longrightarrow 2N^{3-}$$

N_2 is the oxidizing agent. Mg is the reducing agent.

9.45 **Strategy:** In general, we follow the rules listed in Section 9.4 of the text for assigning oxidation numbers.

Solution: The oxidation number for hydrogen is +1 and for oxygen is −2. The oxidation number for sulfur in S_8 is zero (rule 1). Remember that in a molecule, the sum of the oxidation numbers of all the atoms must be zero, and in a polyatomic ion the sum of oxidation numbers of all elements in the ion must equal the net charge of the ion (rule 2).

$$H_2S (-2), S^{2-} (-2), HS^- (-2) < S_8 (0) < SO_2 (+4) < SO_3 (+6), H_2SO_4 (+6)$$

The number in parentheses denotes the oxidation number of sulfur.

9.47 See the guidelines for assigning oxidation numbers in Section 9.4 of the text.

a. $\underline{Cl}F$: F (−1), $\underline{Cl}$ (**+1**)

b. $\underline{I}F_7$: F (−1), $\underline{I}$ (**+7**)

c. $\underline{C}H_4$: H (+1), $\underline{C}$ (**−4**)

d. $\underline{C}_2H_2$: H (+1), $\underline{C}$ (**−1**)

e. $\underline{C_2}H_4$: H (+1), $\underline{C}$ (**−2**)

f. $K_2\underline{Cr}O_4$: K (+1), O (−2), $\underline{Cr}$ (**+6**)

g. $K_2\underline{Cr}_2O_7$: K (+1), O (−2), $\underline{Cr}$ (**+6**)

h. $K\underline{Mn}O_4$: K (+1), O (−2), $\underline{Mn}$ (**+7**)

i. $NaH\underline{C}O_3$: Na (+1), H (+1), O (−2), $\underline{C}$ (**+4**)

j. $\underline{Li}_2$: by Rule 1, $\underline{Li}$ (**0**)

k. $Na\underline{I}O_3$: Na (+1), O (−2), $\underline{I}$ (**+5**)

l. $K\underline{O}_2$: K (+1), $\underline{O}$ (**−1/2**)

m. $\underline{P}F_6^-$: F (−1), $\underline{P}$ (**+5**)

n. $K\underline{Au}Cl_4$: K (+1), Cl (−1), $\underline{Au}$ (**+3**)

9.49   a. $\underline{Cs}_2O$: O (−2), $\underline{Cs}$ (**+1**)

b. $Ca\underline{I}_2$: Ca (+2), $\underline{I}$ (**−1**)

c. $\underline{Al}_2O_3$: O (−2), $\underline{Al}$ (**+3**)

d. $H_3\underline{As}O_3$: H (+1), O (−2), $\underline{As}$ (**+3**)

e. $\underline{Ti}O_2$: O (−2), $\underline{Ti}$ (**+4**)

f. $\underline{Mo}O_4^{2-}$: O (−2), $\underline{Mo}$ (**+6**)

g. $\underline{Pt}Cl_4^{2-}$: Cl (−1), $\underline{Pt}$ (**+2**)

h. $\underline{Pt}Cl_6^{2-}$: Cl (−1), $\underline{Pt}$ (**+4**)

i. $\underline{Sn}F_2$: F (−1), $\underline{Sn}$ (**+2**)

j. $\underline{Cl}F_3$: F (−1), $\underline{Cl}$ (**+3**)

k. $\underline{Sb}F_6^-$: F (−1), $\underline{Sb}$ (**+5**)

9.51   **If nitric acid is a strong oxidizing agent and zinc is a strong reducing agent, then zinc metal will probably reduce nitric acid when the two react; that is, N will gain electrons and the oxidation number of N must decrease. Since the oxidation number of nitrogen in nitric acid is +5, then the nitrogen-containing product must have a smaller oxidation number for nitrogen. The only compound in the list that doesn't have a nitrogen oxidation number less than +5 is N_2O_5 (what is the oxidation number of N in N_2O_5?) This is never a product of the reduction of nitric acid.**

9.53   In order to solve this problem, you need to assign the oxidation numbers to all the elements in the compounds. In each case oxygen has an oxidation number of −2. These oxidation numbers should then be compared to the range of possible oxidation numbers that each element can have. (See Figure 9.7 in the text.) **Molecular oxygen is a powerful oxidizing agent. In SO_3, the oxidation number of the element bound to oxygen (S) is at its maximum value (+6); the sulfur cannot be oxidized further. The other elements bound to oxygen in this problem have less than their maximum oxidation number and can undergo further oxidation. Only SO_3 does not react with molecular oxygen.**

9.55   a. **decomposition** (This is a special type of decomposition reaction called a "**disproportionation**," in which a single species undergoes both oxidation and reduction.)

       b. **displacement**

       c. **decomposition**

       d. **combination**

9.63   **Strategy:** First, calculate the moles of KI needed to prepare the solution, and then convert from moles to grams. How many moles of KI does 5.00×10^2 mL of a 2.80 M solution contain?

       Solution: From the molarity (2.80 M) and the volume (5.00×10^2 mL), we can calculate the moles of KI needed to prepare the solution.

First, calculate the moles of KI needed to prepare the solution.

$$\text{mol of KI} = \frac{2.80 \text{ mol KI}}{1000 \text{ mL solution}} \times (5.00 \times 10^2 \text{ mL solution}) = 1.40 \text{ mol KI}$$

Converting to grams of KI:

$$1.40 \text{ mol KI} \times \frac{166.0 \text{ g KI}}{1 \text{ mol KI}} = \mathbf{232\,g\,KI}$$

9.65   Since the problem asks for moles of solute ($MgCl_2$), you should be thinking that you can calculate moles of solute from the molarity and volume of solution.

$$\text{mol} = M \times L$$

$$60.0 \text{ mL} = 0.0600 \text{ L}$$

$$\text{mol MgCl}_2 = \frac{0.100 \text{ mol MgCl}_2}{1 \text{ L soln}} \times 0.0600 \text{ L solution} = \mathbf{6.00 \times 10^{-3} \text{ mol MgCl}_2}$$

9.67   a. molar mass of C_2H_5OH = 46.068 g/mol

$$? \text{mol C}_2\text{H}_5\text{OH} = 29.0 \text{ g C}_2\text{H}_5\text{OH} \times \frac{1 \text{ mol C}_2\text{H}_5\text{OH}}{46.068 \text{ g C}_2\text{H}_5\text{OH}} = 0.630 \text{ mol C}_2\text{H}_5\text{OH}$$

$$\text{Molarity} = \frac{\text{mol solute}}{\text{L soln}} = \frac{0.630 \text{ mol C}_2\text{H}_5\text{OH}}{0.545 \text{ L solution}} = \mathbf{1.16\,M}$$

b. molar mass of $C_{12}H_{22}O_{11}$ = 342.3 g/mol

$$? \, mol \, C_{12}H_{22}O_{11} = 15.4 \, g \, C_{12}H_{22}O_{11} \times \frac{1 \, mol \, C_{12}H_{22}O_{11}}{342.3 \, g \, C_{12}H_{22}O_{11}} = 0.0450 \, mol \, C_{12}H_{22}O_{11}$$

$$Molarity = \frac{mol \, solute}{L \, soln} = \frac{0.0450 \, mol \, C_{12}H_{22}O_{11}}{74.0 \times 10^{-3} \, L \, solution} = \boldsymbol{0.608 \, M}$$

c. molar mass of NaCl = 58.44 g/mol

$$? \, mol \, NaCl = 9.00 \, g \, NaCl \times \frac{1 \, mol \, NaCl}{58.44 \, g \, NaCl} = 0.154 \, mol \, NaCl$$

$$Molarity = \frac{mol \, solute}{L \, soln} = \frac{0.154 \, mol \, NaCl}{86.4 \times 10^{-3} \, L \, solution} = \boldsymbol{1.78 \, M}$$

9.69   First, calculate the moles of each solute. Then, you can calculate the volume (in L) from the molarity and the number of moles of solute.

a.

$$? \, mol \, NaCl = 2.14 \, g \, NaCl \times \frac{1 \, mol \, NaCl}{58.44 \, g \, NaCl} = 0.03662 \, mol \, NaCl$$

$$L \, soln = \frac{mol \, solute}{Molarity} = \frac{0.03662 \, mol \, NaCl}{0.270 \, mol/L} = 0.136 \, L = \boldsymbol{136 \, mL \, soln}$$

b.

$$? \, mol \, C_2H_5OH = 4.30 \, g \, C_2H_5OH \times \frac{1 \, mol \, C_2H_5OH}{46.068 \, g \, C_2H_5OH} = 0.09334 \, mol \, C_2H_5OH$$

$$L \, soln = \frac{mol \, solute}{Molarity} = \frac{0.09334 \, mol \, C_2H_5OH}{1.50 \, mol/L} = 0.0622 \, L = \boldsymbol{62.2 \, mL \, soln}$$

c.

$$? \, mol \, HC_2H_3O_2 = 0.85 \, g \, HC_2H_3O_2 \times \frac{1 \, mol \, HC_2H_3O_2}{60.052 \, g \, HC_2H_3O_2} = 0.0142 \, mol \, HC_2H_3O_2$$

$$L \, soln = \frac{mol \, solute}{Molarity} = \frac{0.0142 \, mol \, HC_2H_3O_2}{0.30 \, mol/L} = 0.047 \, L = \boldsymbol{47 \, mL \, soln}$$

9.71   **Strategy:**   Because the volume of the final solution is greater than the original solution, this is a dilution process. Keep in mind that in a dilution, the concentration of the solution decreases, but the number of moles of the solute remains the same (Equation 9.2).

Solution: We prepare for the calculation by tabulating our data. The subscripts c and d stand for concentrated and dilute.

$$M_c = 2.00 \, M \qquad\qquad M_d = 0.646 \, M$$

$$L_c = ? \qquad\qquad L_d = 1.00 \, L$$

We substitute the data into Equation (9.3) of the text.

$$M_c L_c = M_d L_d$$

You can solve the equation algebraically for L_c. Then substitute in the given quantities to solve for the volume of 2.00 M HCl needed to prepare 1.00 L of a 0.646 M HCl solution.

$$L_c = \frac{M_d \times L_d}{M_c} = \frac{0.646\ M \times 1.00\ L}{2.00\ M} = \textbf{0.323 L = 323 mL}$$

To prepare the 0.646 M solution, you would **dilute 323 mL of the 2.00 M HCl solution to a final volume of 1.00 L.**

9.73   You can solve the equation algebraically for L_c. Then substitute in the given quantities to solve for the volume of 4.00 M HNO$_3$ needed to prepare 60.0 mL of a 0.200 M HNO$_3$ solution.

$$L_c = \frac{M_d \times L_d}{M_c} = \frac{0.200\ M \times 60.0\ mL}{4.00\ M} = \textbf{3.00 mL}$$

To prepare the 0.200 M solution, you would **dilute 3.00 mL of the 4.00 M HNO$_3$ solution to a final volume of 60.0 mL.**

9.75   Moles of KMnO$_4$ from the 1.66 M solution:

$$(0.0352\ L)\left(\frac{1.66\ mol\ KMnO_4}{1\ L}\right) = 0.0584\ mol\ KMnO_4$$

Moles of KMnO$_4$ from the 0.892 M solution:

$$(0.0167\ L)\left(\frac{0.892\ mol\ KMnO_4}{1\ L}\right) = 0.0149\ mol\ KMnO_4$$

Total moles of KMnO$_4$ = 0.0584 + 0.0149 = 0.0733 mol.

Assume the volumes are additive. Then, the volume of the final solution will be 0.0352 + 0.0167 = 0.0519 L. The concentration is 0.0733 mol / 0.0519 L = **1.41 M KMnO$_4$.**

9.77 $$\left(\frac{126\ mg}{dL}\right)\left(\frac{1\ g}{1000\ mg}\right)\left(\frac{10\ dL}{1\ L}\right)\left(\frac{1\ mol\ C_6H_{12}O_6}{180.16\ g\ C_6H_{12}O_6}\right) = \textbf{0.00699 } \boldsymbol{M}\ \textbf{C}_\textbf{6}\textbf{H}_\textbf{12}\textbf{O}_\textbf{6}$$

9.79   Given $[H_3O]^+$, using the equation $[H_3O]^+ = 10^{-pH}$; that is, pH $= -\log(2.7 \times 10^{-3})$

a.   pH $= -\log(2.7 \times 10^{-3}) = \textbf{2.57}$
b.   pH $= -\log(1.9 \times 10^{-9}) = \textbf{8.72}$
c.   pH $= -\log(5.6 \times 10^{-2}) = \textbf{1.25}$

9.81   a. $-\log[H_3O^+] = 2.42$

$$[H_3O^+] = 10^{-2.42}$$
$$= \textbf{3.8} \times \textbf{10}^{\textbf{-3}}\ \boldsymbol{M}$$

b. $[H_3O^+] = 10^{-11.21}$

$= 6.2 \times 10^{-12}\ M$

c. $[H_3O^+] = 10^{-6.96}$

$= 1.1 \times 10^{-7}\ M$

d. $[H_3O^+] = 10^{-15.00}$

$= 1.0 \times 10^{-15}\ M$

9.83

pH	$[H^+]$	Solution is
<7	$>1.0 \times 10^{-7}\ M$	**acidic**
>7	$<1.0 \times 10^{-7}\ M$	**basic**
=7	$=1.0 \times 10^{-7}\ M$	**neutral**

9.85   a.

$$\mathbf{BaCl_2:}\left(\frac{0.150\ \text{mol BaCl}_2}{1\ \text{L}}\right)\left(\frac{2\ \text{mol Cl}^-}{1\ \text{mol BaCl}_2}\right) = \mathbf{0.300\ \textit{M}\ Cl^-}$$

$$\mathbf{NaCl:}\left(\frac{0.566\ \text{mol NaCl}}{1\ \text{L}}\right)\left(\frac{1\ \text{mol Cl}^-}{1\ \text{mol NaCl}}\right) = \mathbf{0.566\ \textit{M}\ Cl^-}$$

$$\mathbf{AlCl_3:}\left(\frac{1.202\ \text{mol AlCl}_3}{1\ \text{L}}\right)\left(\frac{3\ \text{mol Cl}^-}{1\ \text{mol AlCl}_3}\right) = \mathbf{3.606\ \textit{M}\ Cl^-}$$

b.

$$\left(\frac{2.55\ \text{mol NO}_3^{2-}}{1\ \text{L}}\right)\left(\frac{1\ \text{mol Sr(NO}_3)_2}{2\ \text{mol NO}_3^{2-}}\right) = \mathbf{1.28\ \textit{M}\ Sr(NO_3)_2}$$

9.87   Moles of nitrate in the 0.992 *M* solution:

$$(0.0950\ \text{L})\left(\frac{0.992\ \text{mol KNO}_3}{1\ \text{L}}\right)\left(\frac{1\ \text{mol NO}_3^-}{1\ \text{mol KNO}_3}\right) = 0.0942\ \text{mol NO}_3^-$$

Moles of nitrate in the 1.570 *M* solution:

$$(0.1555\ \text{L})\left(\frac{1.570\ \text{mol Ca(NO}_3)_2}{1\ \text{L}}\right)\left(\frac{2\ \text{mol NO}_3^-}{1\ \text{mol Ca(NO}_3)_2}\right) = 0.4883\ \text{mol NO}_3^-$$

Total nitrate = 0.0942 mol + 0.4883 mol = 0.5825 mol

Assume the volumes are additive. Then, the volume of the final solution is 0.0950 L + 0.1555 L = 0.2505 L. The concentration of nitrate in the final solution is 0.5825 mol / 0.2505 L = **2.325 *M*.**

9.89   **Strategy:**   Using either a graphing calculator or an Excel spreadsheet, we graph the tabulated data to verify that the relationship between absorbance and concentration is linear and that the line goes through the origin. We then use the line and the measured absorbance to determine the concentration of the unknown solution.

The molar absorptivity is a measure of how strongly the species absorbs light of a particular wavelength. The quantitative relationship between absorbance and a solution's concentration is called the Beer-Lambert law and is expressed as $A = \varepsilon bc$.

Setup: a. Graphing the data tabulated in Problem 9.88 with Excel gives the following plot:

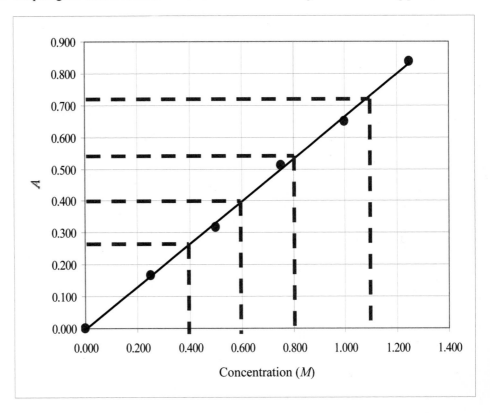

b. $A = \varepsilon bc$

where ε = molar absorptivity
         b = path length of solution (in cm)
         c = molar concentration of the solution

$b = 1.00$ cm

Absorbance is a unitless quantity.

Solution: a. Solving for the unknown absorbance for each of the following concentrations gives:

(M)	A
0.4	**0.26**
0.6	**0.40**
0.8	**0.53**
1.1	**0.73**

b. Rearranging to solve for ε, gives $\varepsilon = \dfrac{A}{bc}$.

0.250 *M*:

$$\varepsilon = \frac{A}{bc} = \frac{0.165}{(1 \text{ cm})(0.250 \text{ mol/L})} = 0.660 \text{ L/mol·cm}$$

0.500 *M*:

$$\varepsilon = \frac{A}{bc} = \frac{0.317}{(1 \text{ cm})(0.500 \text{ mol/L})} = 0.634 \text{ L/mol·cm}$$

0.750 *M*:

$$\varepsilon = \frac{A}{bc} = \frac{0.510}{(1 \text{ cm})(0.750 \text{ mol/L})} = 0.680 \text{ L/mol·cm}$$

1.000 *M*:

$$\varepsilon = \frac{A}{bc} = \frac{0.650}{(1 \text{ cm})(1.000 \text{ mol/L})} = 0.650 \text{ L/mol·cm}$$

1.250 *M*:

$$\varepsilon = \frac{A}{bc} = \frac{0.837}{(1 \text{ cm})(1.250 \text{ mol/L})} = 0.670 \text{ L/mol·cm}$$

Calculating the average molar absorptivity gives:

$$[(0.660 + 0.634 + 0.680 + 0.650 + 0.670)/5] = \mathbf{0.659 \text{ L/mol·cm}}$$

9.91   **Strategy:** First, write the formula for sodium sulfate. Then, identify the compound as ionic or molecular.

Setup: Na^+ and SO_4^{2-} combine in a 2:1 ratio to give Na_2SO_4. Na_2SO_4 contains a metal cation and is therefore an ionic compound. Ionic compounds dissociate completely in water and are strong electrolytes.

Solution: **Diagram (e)** best represents an aqueous solution of sodium sulfate. There are twice as many cations as anions (2:1 ion ratio) and the ions have completely dissociated in water as expected for a strong electrolyte.

9.99   The balanced equation is:

$$CaCl_2(aq) + 2AgNO_3(aq) \longrightarrow Ca(NO_3)_2(aq) + 2AgCl(s)$$

We need to determine the limiting reactant. Ag^+ and Cl^- combine in a 1:1 mole ratio to produce AgCl. Let's calculate the amount of Ag^+ and Cl^- in solution.

$$\text{mol Ag}^+ = \frac{0.100 \, \text{mol Ag}^+}{1000 \, \text{mL soln}} \times 15.0 \, \text{mL soln} = 1.50 \times 10^{-3} \, \text{mol Ag}^+$$

$$\text{mol Cl}^- = \frac{0.150 \, \text{mol CaCl}_2}{1000 \, \text{mL soln}} \times \frac{2 \, \text{mol Cl}^-}{1 \, \text{mol CaCl}_2} \times 30.0 \, \text{mL soln} = 9.00 \times 10^{-3} \, \text{mol Cl}^-$$

Since Ag^+ and Cl^- combine in a 1:1 mole ratio, $AgNO_3$ is the limiting reactant. Only 1.50×10^{-3} mole of AgCl can form. Converting to grams of AgCl:

$$1.50 \times 10^{-3} \, \text{mol AgCl} \times \frac{143.35 \, \text{g AgCl}}{1 \, \text{mol AgCl}} = \textbf{0.215 g AgCl}$$

9.101   The net ionic equation is: $\textbf{Ag}^+\textbf{(aq)} + \textbf{Cl}^-\textbf{(aq)} \longrightarrow \textbf{AgCl(s)}$

One mole of Cl^- is required per mole of Ag^+. First, find the number of moles of Ag^+.

$$\text{mol Ag}^+ = \frac{0.0113 \, \text{mol Ag}^+}{1000 \, \text{mL soln}} \times (2.50 \times 10^2 \, \text{mL soln}) = 2.825 \times 10^{-3} \, \text{mol Ag}^+$$

Now, calculate the mass of NaCl using the mole ratio from the balanced equation.

$$(2.825 \times 10^{-3} \, \text{mol Ag}^+) \times \frac{1 \, \text{mol Cl}^-}{1 \, \text{mol Ag}^+} \times \frac{1 \, \text{mol NaCl}}{1 \, \text{mol Cl}^-} \times \frac{58.44 \, \text{g NaCl}}{1 \, \text{mol NaCl}} = \textbf{0.165 g NaCl}$$

9.103   **Strategy:**   Using the mass given and the molar mass of KHP, determine the number of moles of KHP. Recognize that the number of moles of KOH in the volume given is equal to the number of moles of KHP. Divide moles of KOH by volume (in liters) to get molarity.

             Setup: The molar mass of KHP ($KHC_8H_4O_4$) = [39.1 g + 5(1.008 g) + 8(12.01 g) + 4(16.00 g)] = 204.2 g/mol.

             Solution:

$$\text{moles of KHP} = \frac{0.4218 \, \text{g}}{204.2 \, \text{g/mol}} = 0.002066 \, \text{mol KHP}$$

Because moles of KHP = moles of KOH, it follows that moles of KOH = 0.002066 mol.

$$\text{molarity of KOH} = \frac{0.002066 \, \text{mol}}{0.01868 \, \text{L}} = \textbf{0.1106} \, \textbf{\textit{M}}$$

9.105   a. In order to have the correct mole ratio to solve the problem, you must start with a balanced chemical equation.

$$\text{HCl}(aq) + \text{NaOH}(aq) \longrightarrow \text{NaCl}(aq) + \text{H}_2\text{O}(l)$$

From the molarity and volume of the HCl solution, you can calculate moles of HCl. Then, using the mole ratio from the balanced equation above, you can calculate moles of NaOH.

$$? \, \text{mol NaOH} = 25.00 \, \text{mL} \times \frac{2.430 \, \text{mol HCl}}{1000 \, \text{mL soln}} \times \frac{1 \, \text{mol NaOH}}{1 \, \text{mol HCl}} = 6.075 \times 10^{-2} \, \text{mol NaOH}$$

Solving for the volume of NaOH:

$$\text{liters of solution} = \frac{\text{moles of solute}}{M}$$

$$\text{volume of NaOH} = \frac{6.075 \times 10^{-2} \text{ mol NaOH}}{1.420 \text{ mol/L}} = 4.278 \times 10^{-2} \text{ L} = \mathbf{42.78\,mL}$$

b. This problem is similar to part (a). The difference is that the mole ratio between base and acid is 2:1.

$$H_2SO_4(aq) + 2NaOH(aq) \longrightarrow Na_2SO_4(aq) + \quad 2H_2O(l)$$

$$? \text{mol NaOH} = 25.00 \text{ mL} \times \frac{4.500 \text{ mol } H_2SO_4}{1000 \text{ mL soln}} \times \frac{2 \text{ mol NaOH}}{1 \text{ mol } H_2SO_4} = 0.2250 \text{ mol NaOH}$$

$$\text{volume of NaOH} = \frac{0.2250 \text{ mol NaOH}}{1.420 \text{ mol/L}} = 0.1585 \text{ L} = \mathbf{158.5\ mL}$$

c. This problem is similar to parts (a) and (b). The difference is that the mole ratio between base and acid is 3:1.

$$H_3PO_4(aq) + 3NaOH(aq) \longrightarrow Na_3PO_4(aq) + 3H_2O(l)$$

$$? \text{mol NaOH} = 25.00 \text{ mL} \times \frac{1.500 \text{ mol } H_3PO_4}{1000 \text{ mL soln}} \times \frac{3 \text{ mol NaOH}}{1 \text{ mol } H_3PO_4} = 0.1125 \text{ mol NaOH}$$

$$\text{volume of NaOH} = \frac{0.1125 \text{ mol NaOH}}{1.420 \text{ mol/L}} = 0.07923 \text{ L} = \mathbf{79.23\,mL}$$

9.107   **Strategy:** Write the molecular equation for the reaction that occurs when aqueous solutions of NaCl and $Pb(NO_3)_2$ are combined. Determine which product will precipitate based on the solubility guidelines in Tables 9.2 and 9.3.

Setup: Predict the products by exchanging the ions and balancing the equation.

Solution: Molecular equation:

$$2NaCl(aq) + Pb(NO_3)_2(aq) \longrightarrow 2NaNO_3(aq) + PbCl_2(s)$$

According to Table 9.2, $PbCl_2$ is insoluble in water and will precipitate from solution. Pb^{2+} combines with Cl^- in a 1:2 ratio. Therefore, there are twice as many anions as cations in the precipitate.

$NaNO_3$ contains a metal cation and is therefore an ionic compound. It will dissociate completely in water and remain as Na^+ and NO_3^- ions in solution.

Assuming 1 L of solution, we have 0.10 moles of NaCl and 0.10 moles of $Pb(NO_3)_2$. According to the balanced equation, there are 2 moles of NaCl for every mole of $Pb(NO_3)_2$.

To determine which reactant is limiting, calculate the amount of $Pb(NO_3)_2$ necessary to react completely with 0.10 mol NaCl.

$$0.10 \text{ mol NaCl} \times \frac{1 \text{ mol Pb}(NO_3)_2}{2 \text{ mol NaCl}} = 0.05 \text{ mol Pb}(NO_3)_2$$

0.05 mol of $Pb(NO_3)_2$ are required to react completely with 0.10 mol of NaCl. Since there are 0.10 mol of $Pb(NO_3)_2$, all of the NaCl will react before the $Pb(NO_3)_2$ is consumed. Therefore, NaCl is the limiting reactant. As a result, there will be excess $Pb(NO_3)_2$, which will be present in solution as Pb^{2+} and NO_3^- ions.

Diagram (c) best represents the mixture.

9.109   **Strategy:** Count the number of hydrogen ions consumed in each case to determine how many are required per formula unit of base. If one H^+ ion is consumed per base, the base is monobasic. If two H^+ ions are consumed per base, the base is dibasic.

Solution: **Diagram (d) corresponds to the neutralization of NaOH. Diagram (c) corresponds to the neutralization of Ba(OH)₂.**

9.111   **Choice (d), 0.20 *M* Mg(NO₃)₂,** should be the best conductor of electricity because it **has the greatest concentration of ions;** the total ion concentration in this solution is 0.60 *M*. The total ion concentrations for solutions (a) and (c) are 0.40 *M* and 0.50 *M*, respectively. We can rule out choice (b), because acetic acid is a weak electrolyte.

9.113   The balanced equation for the displacement reaction is:

$$Zn(s) + CuSO_4(aq) \longrightarrow ZnSO_4(aq) + Cu(s)$$

The moles of $CuSO_4$ that react with 7.89 g of zinc are:

$$7.89 \text{ g Zn} \times \frac{1 \text{ mol Zn}}{65.41 \text{ g Zn}} \times \frac{1 \text{ mol CuSO}_4}{1 \text{ mol Zn}} = 0.1206 \text{ mol CuSO}_4$$

The volume of the 0.156 *M* $CuSO_4$ solution needed to react with 7.89 g Zn is:

$$\text{L of soln} = \frac{\text{mole solute}}{M} = \frac{0.1206 \text{ mol CuSO}_4}{0.156 \text{ mol/L}} = 0.773 \text{ L} = \textbf{773 mL}$$

9.115   a. **Weak electrolyte.** Ethanolamine ($C_2H_5ONH_2$) is a weak base.

b. **Strong electrolyte.** Potassium fluoride (KF) is a soluble ionic compound.

c. **Strong electrolyte.** Ammonium nitrate (NH_4NO_3) is a soluble ionic compound.

d. **Nonelectrolyte.** Isopropanol (C_3H_7OH) is a molecular compound that is neither an acid nor a base.

9.117   a. Weak electrolytes exist predominantly as molecules in solution. **C₂H₅ONH₂ molecules.**

b. Strong electrolytes exist predominantly as ions in solution. **K⁺ and F⁻ ions.**

c. Strong electrolytes exist predominantly as ions in solution. **NH_4^+ and NO_3^- ions.**

d. Nonelectrolytes exist entirely as molecules in solution. **C_3H_7OH molecules.**

9.119   The neutralization reaction is: $HA(aq) + NaOH(aq) \longrightarrow NaA(aq) + H_2O(l)$

The mole ratio between the acid and NaOH is 1:1. The moles of HA that react with NaOH are:

$$20.27 \text{ mL soln} \times \frac{0.1578 \text{ mol NaOH}}{1000 \text{ mL soln}} \times \frac{1 \text{ mol HA}}{1 \text{ mol NaOH}} = 3.1986 \times 10^{-3} \text{ mol HA}$$

3.664 g of the acid reacted with the base. The molar mass of the acid is:

$$\text{Molar mass} = \frac{3.664 \text{ g HA}}{3.1986 \times 10^{-3} \text{ mol HA}} = \textbf{1146 g/mol}$$

9.121   Let's call the original solution, soln 1; the first dilution, soln 2; and the second dilution, soln 3. Start with the concentration of soln 3, 0.00383 M. From the concentration and volume of soln 3, we can find the concentration of soln 2. Then, from the concentration and volume of soln 2, we can find the concentration of soln 1, the original solution.

$$M_2V_2 = M_3V_3$$

$$M_2 = \frac{M_3V_3}{V_2} = \frac{(0.00383 \, M)(1.000 \times 10^3 \text{ mL})}{25.00 \text{ mL}} = 0.153 \, M$$

$$M_1V_1 = M_2V_2$$

$$M_1 = \frac{M_2V_2}{V_1} = \frac{(0.153 \, M)(125.0 \text{ mL})}{15.00 \text{ mL}} = \textbf{1.28 } \boldsymbol{M}$$

9.123   The balanced equation is: $Ba(OH)_2(aq) + Na_2SO_4(aq) \longrightarrow BaSO_4(s) + 2NaOH(aq)$

moles $Ba(OH)_2$:   (2.27 L)(0.0820 mol/L) = 0.186 mol $Ba(OH)_2$
moles Na_2SO_4:   (3.06 L)(0.0664 mol/L) = 0.203 mol Na_2SO_4

Since the mole ratio between $Ba(OH)_2$ and Na_2SO_4 is 1:1, $Ba(OH)_2$ is the limiting reactant. The mass of $BaSO_4$ formed is:

$$0.186 \text{ mol Ba(OH)}_2 \times \frac{1 \text{ mol BaSO}_4}{1 \text{ mol Ba(OH)}_2} \times \frac{233.37 \text{ g BaSO}_4}{1 \text{ mol BaSO}_4} = \textbf{43.4 g BaSO}_4$$

9.125   In redox reactions, the oxidation numbers of elements change. To test whether an equation represents a redox process, assign the oxidation numbers to each of the elements in the reactants and products. If oxidation numbers change, it is a redox reaction.

a. On the left the oxidation number of chlorine in Cl_2 is zero (rule 1). On the right it is −1 in Cl^- (rule 2) and +1 in OCl^-. Since chlorine is both oxidized and reduced, this is a disproportionation **redox** reaction.

b. The oxidation numbers of calcium and carbon do not change. This is not a redox reaction; it is a **precipitation** reaction.

c. The oxidation numbers of nitrogen and hydrogen do not change. This is not a redox reaction; it is an **acid-base** reaction.

d. The oxidation numbers of carbon, chlorine, chromium, and oxygen do not change. This is not a redox reaction; it doesn't fit easily into any category, but could be considered as a type of **combination** reaction.

e. The oxidation number of calcium changes from 0 to +2, and the oxidation number of fluorine changes from 0 to −1. This is a combination **redox reaction**.

The remaining parts (f) through (j) can be worked the same way.

f. **redox**

g. **precipitation**

h. **redox**

i. **redox**

j. **redox**

9.127   a. **Magnesium hydroxide is insoluble in water. It can be prepared by mixing a solution containing Mg^{2+} ions such as $MgCl_2(aq)$ or $Mg(NO_3)_2(aq)$ with a solution containing hydroxide ions such as $NaOH(aq)$. $Mg(OH)_2$ will precipitate and can be collected by filtration.** The net ionic reaction is:

$$Mg^{2+}(aq) + 2OH^-(aq) \longrightarrow Mg(OH)_2(s)$$

b. The balanced equation is:

$$2HCl + Mg(OH)_2 \longrightarrow MgCl_2 + 2H_2O$$

The moles of $Mg(OH)_2$ in 10 mL of milk of magnesia are:

$$10 \text{ mL soln} \times \frac{0.080 \text{ g Mg(OH)}_2}{1 \text{ mL soln}} \times \frac{1 \text{ mol Mg(OH)}_2}{58.326 \text{ g Mg(OH)}_2} = 0.0137 \text{ mol Mg(OH)}_2$$

$$\text{Moles of HCl reacted} = 0.0137 \text{ mol Mg(OH)}_2 \times \frac{2 \text{ mol HCl}}{1 \text{ mol Mg(OH)}_2} = 0.0274 \text{ mol HCl}$$

$$\text{Volume of HCl} = \frac{\text{mol solute}}{M} = \frac{0.0274 \text{ mol HCl}}{0.035 \text{ mol/L}} = \textbf{0.78 L}$$

9.129   First, calculate the number of moles of glucose present.

$$\frac{0.513 \text{ mol glucose}}{1000 \text{ mL soln}} \times 60.0 \text{ mL} = 0.03078 \text{ mol glucose}$$

$$\frac{2.33 \text{ mol glucose}}{1000 \text{ mL soln}} \times 120.0 \text{ mL} = 0.2796 \text{ mol glucose}$$

Add the moles of glucose, and then divide by the total volume of the combined solutions to calculate the molarity.

$$60 \text{ mL} + 120.0 \text{ mL} = 180.0 \text{ mL} = 0.180 \text{ L}$$

$$\text{Molarity of final solution} = \frac{(0.03078 + 0.2796) \text{ mol glucose}}{0.180 \text{ L}} = 1.72 \text{ mol/L} = \textbf{1.72 } \textbf{\textit{M}}$$

9.131   The three chemical tests might include:

(1)   electrolysis to ascertain if hydrogen and oxygen were produced;
(2)   the reaction with an alkali metal to see if a base and hydrogen gas were produced;
(3)   the dissolution of a metal oxide to see if a base was produced (or a nonmetal oxide to see if an acid was produced).

9.133   **Diagram (a)**, which shows Ag^+ and NO_3^- ions. The balanced molecular equation for the reaction is:

$$\textbf{AgOH}(aq) + \textbf{HNO}_3(aq) \longrightarrow \textbf{H}_2\textbf{O}(l) + \textbf{AgNO}_3(aq)$$

9.135   a. **Check with litmus paper; combine with carbonate or bicarbonate to see if CO_2 gas is produced; combine with a base and check for neutralization with an indicator.**

          b. **Titrate a known quantity of acid with a standard NaOH solution.** Since it is a monoprotic acid, the moles of NaOH reacted equals the moles of the acid. Dividing the mass of acid by the number of moles gives the molar mass of the acid.

          c. **Visually compare the conductivity of the acid with a standard NaCl solution of the same molar concentration.** A strong acid will have a similar conductivity to the NaCl solution. The conductivity of a weak acid will be considerably less than the NaCl solution.

9.137   a.
$$\textbf{Pb(NO}_3)_2(aq) + \textbf{Na}_2\textbf{SO}_4(aq) \longrightarrow \textbf{PbSO}_4(s) + 2\textbf{NaNO}_3(aq)$$

$$\textbf{Pb}^{2+}(aq) + \textbf{SO}_4^{2-}(aq) \longrightarrow \textbf{PbSO}_4(s)$$

          b. First, calculate the moles of Pb^{2+} in the polluted water.

$$0.00450 \text{ g Na}_2\text{SO}_4 \times \frac{1 \text{ mol Na}_2\text{SO}_4}{142.05 \text{ g Na}_2\text{SO}_4} \times \frac{1 \text{ mol Pb(NO}_3)_2}{1 \text{ mol Na}_2\text{SO}_4} \times \frac{1 \text{ mol Pb}^{2+}}{1 \text{ mol Pb(NO}_3)_2} = 3.17 \times 10^{-5} \text{ mol Pb}^{2+}$$

          The volume of the polluted water sample is 500 mL (0.500 L). The molar concentration of Pb^{2+} is:

$$[Pb^{2+}] = \frac{\text{mol Pb}^{2+}}{\text{L soln}} = \frac{3.17 \times 10^{-5} \text{ mol Pb}^{2+}}{0.500 \text{ L soln}} = \textbf{6.34} \times \textbf{10}^{-5} \textbf{ \textit{M}}$$

9.139   a. An acid and a base react to form water and a salt. Potassium iodide is a salt; therefore, the acid and base are chosen to produce this salt.

$$\textbf{HI}(aq) + \textbf{KOH}(aq) \longrightarrow \textbf{KI}(aq) + \textbf{H}_2\textbf{O}(l)$$

          The resulting solution could be evaporated to dryness to isolate the KI.

b. Acids react with carbonates to form carbon dioxide gas. Again, chose the acid and carbonate salt so that KI is produced.

$$2HI(aq) + K_2CO_3(aq) \longrightarrow 2KI(aq) + CO_2(g) + H_2O(l)$$

Again, **the resulting solution could be evaporated to dryness to isolate the KI**.

9.141   All the three products are water insoluble. Use this information in formulating your answer.

a. **Combine any soluble magnesium salt with a soluble hydroxide; filter the precipitate.**

Example: $MgCl_2(aq) + 2NaOH(aq) \longrightarrow$ **$Mg(OH)_2(s)$** $+ 2NaCl(aq)$

b. **Combine any soluble silver salt with any soluble iodide salt; filter the precipitate.**

Example: $AgNO_3(aq) + NaI(aq) \longrightarrow$ **$AgI(s)$** $+ NaNO_3(aq)$

c. **Combine any soluble barium salt with any soluble phosphate salt; filter the precipitate.**

Example: $3Ba(OH)_2(aq) + 2H_3PO_4(aq) \longrightarrow$ **$Ba_3(PO_4)_2(s)$** $+ 6H_2O(l)$

9.143   a. **Add Na_2SO_4,** or any soluble sulfate salt. Barium sulfate would precipitate leaving sodium ions in solution.

b. **Add KOH,** or any soluble compound containing hydroxide, carbonate, phosphate, or sulfide ion to precipitate magnesium cations. In the case of adding KOH, $Mg(OH)_2$ would precipitate leaving potassium cations in solution.

c. **Add $AgNO_3$,** or any soluble silver salt. AgBr would precipitate, leaving nitrate ions in solution.

d. **Add $Ca(NO_3)_2$,** or any soluble compound containing a cation other than ammonium or a Group 1A cation to precipitate the phosphate ions; the nitrate ions will remain in solution.

e. **Add $Mg(NO_3)_2$,** or any solution containing a cation other than ammonium or a Group 1A cation to precipitate the carbonate ions; the nitrate ions will remain in solution.

9.145   **reaction 1:** $SO_3^{2-}(aq) + H_2O_2(aq) \longrightarrow SO_4^{2-}(aq) + H_2O(l)$

reaction 2: $SO_4^{2-}(aq) + Ba^{2+}(aq) \longrightarrow BaSO_4(s)$

9.147   We assume that O has an oxidation state of −2. From this, we determine the ratio of combination of chlorine and oxygen necessary to make a neutral formula for each of the specified oxidation states of Cl.

Cl_2O (Cl = +1)   **Cl_2O_3 (Cl = +3)**   **ClO_2 (Cl = +4)**   **Cl_2O_6 (Cl = +6)**

Cl_2O_7 (Cl = +7)

9.149   Since aspirin is a monoprotic acid, it will react with NaOH in a 1:1 mole ratio.

First, calculate the moles of aspirin in the tablet.

$$12.25 \text{ mL soln} \times \frac{0.1466 \text{ mol NaOH}}{1000 \text{ mL soln}} \times \frac{1 \text{ mol aspirin}}{1 \text{ mol NaOH}} = 1.796 \times 10^{-3} \text{ mol aspirin}$$

Next, convert from moles of aspirin to grains of aspirin.

$$1.796 \times 10^{-3} \text{ mol aspirin} \times \frac{180.15 \text{ g aspirin}}{1 \text{ mol aspirin}} \times \frac{1 \text{ grain}}{0.0648 \text{ g}} = \textbf{4.99 grains of aspirin in one tablet}$$

9.151   a.
$$\textbf{CaF}_2\textbf{(}s\textbf{)} + \textbf{H}_2\textbf{SO}_4\textbf{(}aq\textbf{)} \longrightarrow \textbf{CaSO}_4\textbf{(}s\textbf{)} + \textbf{2HF(}g\textbf{)}$$

$$\textbf{2NaCl(}s\textbf{)} + \textbf{H}_2\textbf{SO}_4\textbf{(}aq\textbf{)} \longrightarrow \textbf{Na}_2\textbf{SO}_4\textbf{(}aq\textbf{)} + \textbf{2HCl(}g\textbf{)}$$

b. HBr and HI **cannot** be prepared similarly, because **the sulfuric acid would oxidize Br⁻ and I⁻ ions to Br₂ and I₂.**

c.
$$\textbf{PBr}_3\textbf{(}l\textbf{)} + \textbf{3H}_2\textbf{O(}l\textbf{)} \longrightarrow \textbf{3HBr(}g\textbf{)} + \textbf{H}_3\textbf{PO}_3\textbf{(}aq\textbf{)}$$

9.153   **Electric furnace method:**

$$\textbf{P}_4\textbf{(}s\textbf{)} + \textbf{5O}_2\textbf{(}g\textbf{)} \longrightarrow \textbf{P}_4\textbf{O}_{10}\textbf{(}s\textbf{)} \qquad \textbf{(redox)}$$

$$\textbf{P}_4\textbf{O}_{10}\textbf{(}s\textbf{)} + \textbf{6H}_2\textbf{O(}l\textbf{)} \longrightarrow \textbf{4H}_3\textbf{PO}_4\textbf{(}aq\textbf{)} \qquad \textbf{(acid-base)}$$

Wet process:
$$\textbf{Ca}_5\textbf{(PO}_4\textbf{)}_3\textbf{F(}s\textbf{)} + \textbf{5H}_2\textbf{SO}_4\textbf{(}aq\textbf{)} \longrightarrow \textbf{3H}_3\textbf{PO}_4\textbf{(}aq\textbf{)} + \textbf{HF(}aq\textbf{)} + \textbf{5CaSO}_4\textbf{(}s\textbf{)}$$

(acid-base and precipitation)

9.155   a.
$$\textbf{4KO}_2\textbf{(}s\textbf{)} + \textbf{2CO}_2\textbf{(}g\textbf{)} \longrightarrow \textbf{2K}_2\textbf{CO}_3\textbf{(}s\textbf{)} + \textbf{3O}_2\textbf{(}g\textbf{)}$$

b. The oxidation number of oxygen in the O_2^{2-} ion is **−1/2.**

c.
$$? \text{ L air} = 7.00 \text{ g KO}_2 \times \frac{1 \text{ mol KO}_2}{71.10 \text{ g KO}_2} \times \frac{2 \text{ mol CO}_2}{4 \text{ mol KO}_2} \times \frac{44.01 \text{ g CO}_2}{1 \text{ mol CO}_2} \times \frac{1 \text{ L air}}{0.063 \text{ g CO}_2} = \textbf{34 L air}$$

9.157   **No.** In a redox reaction, electrons must be transferred between reacting species. In other words, oxidation numbers must change in a redox reaction. **In both O₂ (molecular oxygen) and O₃ (ozone), the oxidation number of oxygen is zero.** This is *not* a redox reaction.

9.159   a. **acid: H₃O⁺, base: OH⁻**

$$\text{OH}^- \qquad\qquad \text{H}_3\text{O}^+ \qquad\qquad\qquad \text{H}_2\text{O} \qquad\qquad \text{H}_2\text{O}$$

b. **acid: NH_4^+ , base: NH_2^-**

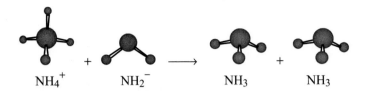

$$NH_4^+ \qquad NH_2^- \qquad NH_3 \qquad NH_3$$

9.161   **When a solid dissolves in solution, the volume of the solution usually changes.**

9.163   a. **The precipitate $CaSO_4$ formed over Ca preventing the Ca from reacting with the sulfuric acid.**

b. **Aluminum is protected by a tenacious oxide layer with the composition Al_2O_3.**

c. **These metals react more readily with water.**

$$2Na(s) + 2H_2O(l) \longrightarrow 2NaOH(aq) + H_2(g)$$

d. **The metal should be placed below Fe and above H.**

e. **Any metal above Al in the activity series will react with Al^{3+}. Metals from Mg to Li will work.**

9.165   a. The balanced equations are:

(1) $Cu(s) + 4HNO_3(aq) \longrightarrow Cu(NO_3)_2(aq) + 2NO_2(g) + 2H_2O(l)$ **(redox)**

(2) $Cu(NO_3)_2(aq) + 2NaOH(aq) \longrightarrow Cu(OH)_2(s) + 2NaNO_3(aq)$ **(precipitation)**

(3) $Cu(OH)_2(s) \longrightarrow CuO(s) + H_2O(g)$ **(decomposition)**

(4) $CuO(s) + H_2SO_4(aq) \longrightarrow CuSO_4(aq) + H_2O(l)$ **(acid-base)**

(5) $CuSO_4(aq) + Zn(s) \longrightarrow Cu(s) + ZnSO_4(aq)$ **(redox)**

(6) $Zn(s) + 2HCl(aq) \longrightarrow ZnCl_2(aq) + H_2(g)$ **(redox)**

b. We start with 65.6 g Cu, which is $65.6 \text{ g Cu} \times \dfrac{1 \text{ mol Cu}}{63.55 \text{ g Cu}} = 1.032 \text{ mol Cu}$. The mole ratio between the product and the reactant in each reaction is 1:1. Therefore, the theoretical yield in each reaction is 1.032 moles.

(1) $1.032 \text{ mol} \times \dfrac{187.57 \text{ g Cu(NO}_3)_2}{1 \text{ mol Cu(NO}_3)_2} = \mathbf{194 \text{ g Cu(NO}_3)_2}$

(2) $1.032 \text{ mol} \times \dfrac{97.566 \text{ g Cu(OH)}_2}{1 \text{ mol Cu(OH)}_2} = \mathbf{101 \text{ g Cu(OH)}_2}$

(3) $1.032 \text{ mol} \times \dfrac{79.55 \text{ g CuO}}{1 \text{ mol CuO}} = \mathbf{82.1 \text{ g CuO}}$

(4) $$1.032 \text{ mol} \times \frac{159.62 \text{ g CuSO}_4}{1 \text{ mol CuSO}_4} = \textbf{165 g CuSO}_4$$

(5) $$1.032 \text{ mol} \times \frac{63.55 \text{ g Cu}}{1 \text{ mol Cu}} = \textbf{65.6 g Cu}$$

(6) **The amount of copper recovered in Step 6 will vary with the washing technique used.**

c. **All the reaction steps are clean and almost quantitative; therefore, the recovery yield should be high**.

9.167   The reaction between $Mg(NO_3)_2$ and NaOH is:

$$Mg(NO_3)_2(aq) + 2NaOH(aq) \longrightarrow Mg(OH)_2(s) + 2NaNO_3(aq)$$

Magnesium hydroxide, $Mg(OH)_2$, precipitates from solution. Na^+ and NO_3^- are spectator ions. This is most likely a limiting reactant problem as the amounts of both reactants are given. Let's first determine which reactant is the limiting reactant before we try to determine the concentration of ions remaining in the solution.

$$1.615 \text{ g Mg(NO}_3)_2 \times \frac{1 \text{ mol Mg(NO}_3)_2}{148.33 \text{ g Mg(NO}_3)_2} = 0.010888 \text{ mol Mg(NO}_3)_2$$

$$1.073 \text{ g NaOH} \times \frac{1 \text{ mol NaOH}}{39.998 \text{ g NaOH}} = 0.026826 \text{ mol NaOH}$$

From the balanced equation, we need twice as many moles of NaOH compared to $Mg(NO_3)_2$. We have more than twice as much NaOH (2×0.010888 mol $= 0.021776$ mol) and therefore $Mg(NO_3)_2$ is the limiting reactant. NaOH is in excess and ions of Na^+, OH^-, and NO_3^- will remain in solution. Because Na^+ and NO_3^- are spectator ions, the number of moles after reaction will equal the initial number of moles. The excess moles of OH^- need to be calculated based on the amount that reacts with Mg^{2+}. The combined volume of the two solutions is: 22.02 mL $+ 28.64$ mL $= 50.66$ mL $= 0.05066$ L.

$$[\textbf{Na}^+] = 0.026826 \text{ mol NaOH} \times \frac{1 \text{ mol Na}^+}{1 \text{ mol NaOH}} \times \frac{1}{0.05066 \text{ L}} = \textbf{0.5295 }\boldsymbol{M}$$

$$[\textbf{NO}_3^-] = 0.010888 \text{ mol Mg(NO}_3)_2 \times \frac{2 \text{ mol NO}_3^-}{1 \text{ mol Mg(NO}_3)_2} \times \frac{1}{0.05066 \text{ L}} = \textbf{0.4298 }\boldsymbol{M}$$

The moles of OH^- reacted are:

$$0.010888 \text{ mol Mg}^{2+} \times \frac{2 \text{ mol OH}^-}{1 \text{ mol Mg}^{2+}} = 0.021776 \text{ mol OH}^- \text{ reacted}$$

The moles of excess OH^- are:

$$0.026826 \text{ mol} - 0.021776 \text{ mol} = 0.005050 \text{ mol OH}^-$$

$$[\textbf{OH}^-] = \frac{0.005050 \text{ mol}}{0.05066 \text{ L}} = \textbf{0.09968 }\boldsymbol{M}$$

$$[\textbf{Mg}^{2+}] \approx \textbf{0 }\boldsymbol{M}$$

[The concentration of Mg^{2+} is approximately zero as almost all of it will precipitate as $Mg(OH)_2$.]

Chapter 10
Energy Changes in Chemical Reactions

Visualizing Chemistry

VC 10.1	c	VC 10.2	a	VC 10.3	b	VC 10.4	c
VC 10.5	c	VC 10.6	a	VC 10.7	b	VC 10.8	c
VC 10.9	b	VC 10.10	c	VC 10.11	c	VC 10.12	b

Key Skills

10.1 e   10.2 a   10.3 c   10.4 b

Problems

10.5 **law of conservation of energy**

10.7 **Energy is needed to break chemical bonds, whereas energy is released when bonds are formed.**

10.11 Using Equation 10.1 and the sign conventions for q and w,

$$\Delta U = q + w$$

$$\Delta U = (-26 \text{ J}) + (74 \text{ J}) = \textbf{48 J}$$

10.13 Using Equation 10.1 and the sign conventions for q and w,

$$\Delta U = q + w$$

$$510 \text{ J} = (-67 \text{ J}) + w$$

$$w = \textbf{577 J (work done } \textit{on} \textbf{ the system)}$$

10.15 a. Since work is done on the system, volume decreases. **Diagram (iii)** fits.

b. Rearrange Equation 10.1 to solve for work.
$$\Delta U = q + w$$

$$w = \Delta U - q$$

Since $\Delta U > 0$ and $q < 0$, w must be greater than 0. This corresponds to **diagram (iii)**.

c. Since $\Delta U > 0$ and $q > 0$, w depends on the magnitude of the internal energy and heat. Thus it is not possible to determine which system is correct. **Diagrams (i), (ii), and (iii)** are possible.

10.21 a. Because the external pressure is zero, no work is done in the expansion.

$$w = -P\Delta V = -(0)(89.3 - 26.7) \text{ mL}$$

$$w = \textbf{0 J}$$

b. The external, opposing pressure is 1.5 atm, so

$$w = -P\Delta V = -(1.5 \text{ atm})(89.3 - 26.7) \text{ mL}$$

$$w = -94 \text{ mL} \cdot \text{atm} \times \frac{0.001 \text{ L}}{1 \text{ mL}} = -0.094 \text{ L} \cdot \text{atm}$$

To convert the answer to joules, we write:

$$w = -0.094 \text{ L} \cdot \text{atm} \times \frac{101.3 \text{ J}}{1 \text{ L} \cdot \text{atm}} = \mathbf{-9.5 \text{ J}}$$

c. The external, opposing pressure is 2.8 atm, so

$$w = -P\Delta V = -(2.8 \text{ atm})(89.3 - 26.7) \text{ mL}$$

$$w = \left(-1.8 \times 10^2 \text{ mL} \cdot \text{atm}\right) \times \frac{0.001 \text{ L}}{1 \text{ mL}} = -0.18 \text{ L} \cdot \text{atm}$$

To convert the answer to joules, we write:

$$w = -0.18 \text{ L} \cdot \text{atm} \times \frac{101.3 \text{ J}}{1 \text{ L} \cdot \text{atm}} = \mathbf{-18 \text{ J}}$$

10.23 **Strategy:** The thermochemical equation shows that for every 2 moles of ZnS roasted, 879 kJ of heat are given off (note the negative sign). We can write a conversion factor from this information.

$$\frac{-879 \text{ kJ}}{2 \text{ mol ZnS}}$$

How many moles of ZnS are in 1 g of ZnS? What conversion factor is needed to convert from grams to moles?

Solution: We need to first calculate the number of moles of ZnS in 1 g of the compound. Then, we can convert to kilojoules of heat produced from the exothermic reaction. The sequence of conversions is:

grams of ZnS → moles of ZnS → kilojoules of heat generated

Therefore, the enthalpy change for the roasting of 1 g of ZnS is:

$$\Delta H = \frac{-879 \text{ kJ}}{2 \text{ mol ZnS}} \times \frac{1 \text{ mol ZnS}}{97.48 \text{ g ZnS}} = -4.51 \text{ kJ/g ZnS}$$

Because the question asked is *how much heat* is evolved, it is not necessary to include the negative sign in the answer. The roasting of ZnS produces **4.51 kJ/g.**

10.25 We can calculate ΔU by rearranging Equation 10.8 of the text.

$$\Delta U = \Delta H - P\Delta V$$

Substituting into the above equation:

$$\Delta U = 483.6 \text{ kJ} - (1.00 \text{ atm})(32.7 \text{ L})\left(\frac{101.3 \text{ J}}{\text{L} \cdot \text{atm}}\right)\left(\frac{1 \text{ kJ}}{1000 \text{ J}}\right)$$

$$\Delta U = \mathbf{4.80 \times 10^2 \text{ kJ}}$$

10.27 The second reaction is the reverse of the first reaction. Consequently, the magnitude of ΔH remains the same, but its sign changes.

$$\Delta H = \mathbf{595.8 \text{ kJ/mol}}$$

10.31 See Table 10.2 of the text for the specific heat of Cu.

$$q = m_{Cu}s_{Cu}\Delta T = (6.22 \times 10^3 \text{ g})(0.385 \text{ J/g} \cdot {}^\circ\text{C})(324.3^\circ\text{C} - 20.5^\circ\text{C}) = 7.28 \times 10^5 \text{ J} = \mathbf{728 \text{ kJ}}$$

10.33 **Strategy:** We know the masses of gold and iron as well as the initial temperatures of each. We can look up the specific heats of gold and iron in Table 10.2 of the text. Assuming no heat is lost to the surroundings, we can equate the heat lost by the iron sheet to the heat gained by the gold sheet. With this information, we can solve for the final temperature of the combined metals.

Solution: Treating the calorimeter as an isolated system (no heat lost to the surroundings), we can write:

$$q_{Au} + q_{Fe} = 0$$

$$\text{or } q_{Au} = -q_{Fe}$$

The heat gained by the gold sheet is given by:

$$q_{Au} = m_{Au}s_{Au}\Delta T = (10.0 \text{ g})(0.129 \text{ J/g} \cdot {}^\circ\text{C})(T_f - 18.0)^\circ\text{C}$$

where m and s are the mass and specific heat, and $\Delta T = T_{final} - T_{initial}$.

The heat lost by the iron sheet is given by:

$$q_{Fe} = m_{Fe}s_{Fe}\Delta T = (20.0 \text{ g})(0.444 \text{ J/g} \cdot {}^\circ\text{C})(T_f - 55.6)^\circ\text{C}$$

Substituting into the equation derived above, we can solve for T_f.

$$q_{Au} = -q_{Fe}$$

$$(10.0 \text{ g})(0.129 \text{ J/g} \cdot {}^\circ\text{C})(T_f - 18.0)^\circ\text{C} = -(20.0 \text{ g})(0.444 \text{ J/g} \cdot {}^\circ\text{C})(T_f - 55.6)^\circ\text{C}$$

$$1.29 \, T_f - 23.2 = -8.88 \, T_f + 494$$

$$10.2 \, T_f = 517$$

$$T_f = \mathbf{50.7^\circ\text{C}}$$

10.35 **Strategy:** The neutralization reaction is exothermic. 56.2 kJ of heat is released when 1 mole of H^+ reacts with 1 mole of OH^-. Assuming no heat is lost to the surroundings, we can equate the heat lost by the reaction to the heat gained by the combined solution. How do we calculate the heat released during the reaction? Are we combining 1 mole of H^+ with 1 mole of OH^-? How do we calculate the heat absorbed by the combined solution?

Solution: Assuming no heat is lost to the surroundings, we can write:

$$q_{soln} + q_{rxn} = 0$$

$$\text{or } q_{soln} = -q_{rxn}$$

First, let's set up how we would calculate the heat gained by the solution,

$$q_{soln} = m_{soln}s_{soln}\Delta T$$

where m and s are the mass and specific heat of the solution and $\Delta T = T_f - T_i$.

We assume that the specific heat of the solution is the same as the specific heat of water, and we assume that the density of the solution is the same as the density of water (1.00 g/mL). Since the density is 1.00 g/mL, the mass of 400 mL of solution (200 mL + 200 mL) is 400 g.

Substituting into the equation above, the heat gained by the solution can be represented as:

$$q_{soln} = (4.00 \times 10^2 \text{ g})(4.184 \text{ J/g·°C})(T_f - 20.48°C)$$

Next, let's calculate q_{rxn}, the heat released when 200 mL of 0.862 M HCl are mixed with 200 mL of 0.431 M $Ba(OH)_2$. The equation for the neutralization is:

$$2HCl(aq) + Ba(OH)_2(aq) \longrightarrow 2H_2O(l) + BaCl_2(aq)$$

There is exactly enough $Ba(OH)_2$ to neutralize all the HCl. Note that 2 moles HCl is stoichiometrically equivalent to 1 mole $Ba(OH)_2$ and that the concentration of HCl is double the concentration of $Ba(OH)_2$. The number of moles of HCl is:

$$\left(2.00 \times 10^2 \text{ mL}\right) \times \frac{0.862 \text{ mol HCl}}{1000 \text{ mL}} = 0.172 \text{ mol HCl}$$

The amount of heat released when 1 mole of H^+ is neutralized is given in the problem (−56.2 kJ/mol). The amount of heat liberated when 0.172 mole of H^+ is neutralized is:

$$q_{rxn} = 0.172 \text{ mol} \times \frac{-56.2 \times 10^3 \text{ J}}{1 \text{ mol}} = -9.67 \times 10^3 \text{ J}$$

Finally, knowing that the heat lost by the reaction equals the heat gained by the solution, we can solve for the final temperature of the mixed solution.

$$q_{soln} = -q_{rxn}$$

$$(4.00 \times 10^2 \text{ g})(4.184 \text{ J/g·°C})(T_f - 20.48°C) = -(-9.67 \times 10^3 \text{ J})$$

$$(1.67 \times 10^3)T_f - (3.43 \times 10^4) = 9.67 \times 10^3 \text{ J}$$

$$T_f = \mathbf{26.3°C}$$

10.37 Assuming that all the heat produced by the body heats the water contained in the body and using Equation 10.11 from the text,

$$q = ms\Delta T$$

In this case, q is the heat produced by the body, m is the mass of water contained in the body, and ΔT is the rise in temperature over the course of a day.

$$1.0 \times 10^4 \text{ kJ/day} = (50 \text{ kg})(4.184 \text{ kJ/kg·°C})(\Delta T)$$

$$\Delta T = \frac{1.0 \times 10^4 \text{ kJ/day}}{(50 \text{ kg})(4.184 \text{ kJ/kg·°C})} = \mathbf{48 \text{ °C/day}}$$

Next, determine what mass of water could be vaporized by that amount of heat.

$$\text{mass of water} = \left(1.0 \times 10^4 \text{ kJ/day}\right)\left(\frac{1 \text{ g}}{2.41 \text{ kJ}}\right)\left(\frac{1 \text{ kg}}{1000 \text{ g}}\right) = \mathbf{4.1 \text{ kg water/day}}$$

A person would have to drink approximately this much water each day to maintain normal body temperature, so the human body is not an isolated system. Other external factors, such as the choice of clothing that allows evaporation, also must play a role in maintaining proper body temperature.

10.39 **Strategy:** The first law of thermodynamics dictates that the energy given off by a system must be absorbed by the surroundings. In this case, the surroundings consist of ethanol and the calorimeter, both of which absorb the heat given off by methanol, which we will call the system. Because $q_{surr} = -q_{sys}$, we know that the sum of $q_{ethanol}$ and $q_{calorimeter}$ must equal $-q_{methanol}$. According to Equation 10.11, $q = ms\Delta T$.

Setup: The specific heat of ethanol, from Table 10.2, is 2.46 J/g °C. $\Delta T_{methanol}$ is

$$28.5°C - 35.6°C = -7.1°C.$$

$\Delta T_{ethanol}$ and $\Delta T_{calorimeter}$ are both

$$28.5°C - 24.7°C = 3.8°C.$$

Solution: Since $q_{ethanol} + q_{calorimeter} = -q_{methanol}$, we can write:

$$38.65 \text{ g} \times \frac{2.46 \text{ J}}{\text{g}°C} \times 3.8°C + \frac{19.3 \text{ J}}{°C} \times 3.8°C = -\left[25.95 \text{ g} \times s \times (-7.1°C)\right]$$

$$361 \text{ J} + 73.3 \text{ J} = (184 \text{ g·°C})s$$

$$s = \mathbf{2.36 \text{ J/g·°C}}$$

10.41 choice **(d)**

10.45 Making the necessary changes to the reactions given in the problem, and making the corresponding changes to the $\Delta H°$ values:

Reaction	$\Delta H°$ (kJ/mol)
$CO_2(g) + 2H_2O(l) \longrightarrow CH_3OH(l) + \frac{3}{2}O_2(g)$	726.4
$C(graphite) + O_2(g) \longrightarrow CO_2(g)$	−393.5
$2H_2(g) + O_2(g) \longrightarrow 2H_2O(l)$	2(−285.8)
$C(graphite) + 2H_2(g) + \frac{1}{2}O_2(g) \longrightarrow CH_3OH(l)$	$\Delta H°_{rxn} = \textbf{−238.7 kJ/mol}$

We have just calculated an enthalpy at standard conditions, which we abbreviate as $\Delta H°_{rxn}$. In this case, the reaction in question was for the formation of *one* mole of CH_3OH *from its elements* in their standard state.

Therefore, the $\Delta H°_{rxn}$ that we calculated is also, by definition, the standard heat of formation $\Delta H°_f$ of CH_3OH (**−238.7 kJ/mol**).

10.47 **Strategy:** Our goal is to calculate the enthalpy change for the formation of monoclinic sulfur from rhombic sulfur. To do so, we must arrange the equations that are given in the problem in such a way that they will sum to the desired overall equation. This requires reversing the second equation and changing the sign of its $\Delta H°$ value.

Solution:

Reaction	$\Delta H°$ (kJ/mol)
$S(rhombic) + O_2(g) \longrightarrow SO_2(g)$	−296.06
$SO_2(g) \longrightarrow S(monoclinic) + O_2(g)$	296.36
$S(rhombic) \longrightarrow S(monoclinic)$	$\Delta H°_{rxn} = \textbf{0.30 kJ/mol}$

10.49 The reaction can be written as:

$$6 \text{ green} + 8 \text{ gray} \longrightarrow 4 \text{ blue}$$

Arrange the given equations so that their sum is the desired equation.

$2 \text{ green} + 8 \text{ purple} \longrightarrow 4 \text{ blue}$	$\Delta H = -100 \text{ kJ/mol}$
$8 \text{ gray} + 4 \text{ green} \longrightarrow 8 \text{ purple}$	$\Delta H = \frac{4}{3}(-60 \text{ kJ/mol}) = -80 \text{ kJ/mol}$
$6 \text{ green} + 8 \text{ gray} \longrightarrow 4 \text{ blue}$	$\Delta H = \textbf{−180 kJ/mol}$

10.55 The standard enthalpy of formation of any element in its most stable form is zero. Therefore, since $\Delta H°_f(O_2) = 0$, $\textbf{O}_2$ is the more stable form of the element oxygen at this temperature.

10.57 **Strategy:** Use Hess's law: the enthalpy of a reaction is the difference between the sum of the enthalpies of the products and the sum of the enthalpies of the reactants. The enthalpy of each species (reactant or product) is given by the product of the stoichiometric coefficient and the standard enthalpy of formation, $\Delta H°_f$, of the species.

Solution: We use the ΔH_f° values in Appendix 2 and Equation 10.18 of the text.

$$\Delta H_{rxn}^{\circ} = [\Delta H_f^{\circ}(CaO) + \Delta H_f^{\circ}(CO_2)] - \Delta H_f^{\circ}(CaCO_3)$$

$$\Delta H^{\circ} = [(1)(-635.6 \text{ kJ/mol}) + (1)(-393.5 \text{ kJ/mol})] - (1)(-1206.9 \text{ kJ/mol}) = \textbf{177.8 kJ/mol}$$

10.59   a.
$$\Delta H^{\circ} = 2\Delta H_f^{\circ}(H_2O) - 2\Delta H_f^{\circ}(H_2) - \Delta H_f^{\circ}(O_2)$$

$$\Delta H^{\circ} = (2)(-285.8 \text{ kJ/mol}) - (2)(0) - (1)(0) = \textbf{-571.6 kJ/mol}$$

b.
$$\Delta H^{\circ} = 4\Delta H_f^{\circ}(CO_2) + 2\Delta H_f^{\circ}(H_2O) - 2\Delta H_f^{\circ}(C_2H_2) - 5\Delta H_f^{\circ}(O_2)$$

$$\Delta H^{\circ} = (4)(-393.5 \text{ kJ/mol}) + (2)(-285.8 \text{ kJ/mol}) - (2)(226.6 \text{ kJ/mol}) - (5)(0) = \textbf{-2598.8 kJ/mol}$$

10.61   The given enthalpies are in units of kJ/g. We must convert them to units of kJ/mol.

a. $\dfrac{-22.6 \text{ kJ}}{1 \text{ g}} \times \dfrac{32.04 \text{ g}}{1 \text{ mol}} = \textbf{-724 kJ/mol}$

b. $\dfrac{-29.7 \text{ kJ}}{1 \text{ g}} \times \dfrac{46.07 \text{ g}}{1 \text{ mol}} = \textbf{-1.37} \times \textbf{10}^{\textbf{3}} \textbf{ kJ/mol}$

c. $\dfrac{-33.4 \text{ kJ}}{1 \text{ g}} \times \dfrac{60.09 \text{ g}}{1 \text{ mol}} = \textbf{-2.01} \times \textbf{10}^{\textbf{3}} \textbf{ kJ/mol}$

10.63
$$\Delta H_{rxn}^{\circ} = 6\Delta H_f^{\circ}(CO_2) + 6\Delta H_f^{\circ}(H_2O) - [\Delta H_f^{\circ}(C_6H_{12}) + 9\Delta H_f^{\circ}(O_2)]$$

$$\Delta H_{rxn}^{\circ} = (6)(-393.5 \text{ kJ/mol}) + (6)(-285.8 \text{ kJ/mol}) - (1)(-151.9 \text{ kJ/mol}) - (1)(0) = \textbf{-3923.9 kJ/mol}$$

10.65   The amount of heat given off is:

$$\left(1.26 \times 10^4 \text{ g NH}_3\right) \times \frac{1 \text{ mol NH}_3}{17.03 \text{ g NH}_3} \times \frac{-92.6 \text{ kJ}}{2 \text{ mol NH}_3} = \textbf{-3.43} \times \textbf{10}^{\textbf{4}} \textbf{ kJ}$$

10.67   We know that heat is required to convert liquid water to water vapor:

$$H_2O(l) \longrightarrow H_2O(g) \qquad \text{endothermic}$$

Therefore, we know that

$$\Delta H_{rxn}^{\circ} = \Delta H_f^{\circ}[H_2O(g)] - \Delta H_f^{\circ}[H_2O(l)] > 0$$

and

$\boldsymbol{\Delta H_f^{\circ}}$ **[H$_2$O(l)]** is more negative since $\Delta H_{rxn}^{\circ} > 0$.

10.69
$$2H_2O_2(l) \longrightarrow 2H_2O(l) + O_2(g)$$

$\boldsymbol{\Delta H_f^{\circ}}$ **[H$_2$O$_2$(l)]** is indeed negative, but $\boldsymbol{\Delta H_f^{\circ}}$ **[H$_2$0(l)]** is even more negative, meaning that **2H$_2$O(l) + O$_2$(g) is more stable than H$_2$O$_2$(l).**

10.71
$$\Delta H^{\circ}_{rxn} = \sum n\Delta H^{\circ}_f(\text{products}) - \sum m\Delta H^{\circ}_f(\text{reactants})$$

According to the diagram, the reactants are CO and NO and the products are CO_2 and N_2. The NO is in excess (some of NO is left after all the CO is consumed). The balanced equation for the reaction is:

$$2CO(g) + 2NO(g) \longrightarrow 2CO_2(g) + N_2(g)$$

$$\Delta H^{\circ}_{rxn} = \left[2\Delta H^{\circ}_f(CO_2) + \Delta H^{\circ}_f(N_2) \right] - \left[2\Delta H^{\circ}_f(CO) + 2\Delta H^{\circ}_f(NO) \right]$$

$$\Delta H^{\circ}_{rxn} = [(2)(-393.5 \text{ kJ/mol}) + (1)(0 \text{ kJ/mol})] - [(2)(-110.5 \text{ kJ/mol}) + (2)(90.4 \text{ kJ/mol})] = \mathbf{-746.8\,kJ/mol}$$

10.75 Strategy: Keep in mind that bond breaking is an energy absorbing (endothermic) process and bond making is an energy releasing (exothermic) process. Therefore, the overall energy change is the difference between these two opposing processes, as described in Equation 10.19 of the text.

Solution: There are two oxygen-to-oxygen bonds in ozone. We will represent these bonds as O–O. However, these bonds might not be true oxygen-to-oxygen single bonds. Using Equation 10.19 of the text, we write:

$$\Delta H^{\circ} = \sum BE(\text{reactants}) - \sum BE(\text{products})$$

$$\Delta H^{\circ} = BE(O=O) - 2BE(O-O)$$

In the problem, we are given ΔH° for the reaction, and we can look up the O=O bond enthalpy in Table 10.4 of the text. Solving for the average bond enthalpy in ozone,

$$2BE(O-O) = \Delta H^{\circ} - BE(O=O)$$

$$BE(O-O) = \frac{BE(O=O) - \Delta H^{\circ}}{2} = \frac{498.7 \text{ kJ/mol} + 107.2 \text{ kJ/mol}}{2} = \mathbf{303.0\,kJ/mol}$$

Considering the resonance structures for ozone, it is expected that the O–O bond enthalpy in ozone is between the single O–O bond enthalpy (142 kJ) and the double O=O bond enthalpy (498.7 kJ).

10.77 a.

Bonds Broken	Number Broken	Bond Enthalpy (kJ/mol)	Enthalpy Change (kJ/mol)
C–H	12	414	4968
C–C	2	347	694
O=O	7	498.7	3491

Bonds Formed	Number Formed	Bond Enthalpy (kJ/mol)	Enthalpy Change (kJ/mol)
C=O	8	799	6392
O–H	12	460	5520

$$\Delta H^{\circ} = \text{total energy input} - \text{total energy released}$$

$$= (4968 + 694 + 3491) - (6392 + 5520) = \mathbf{-2759\,kJ/mol}$$

b.
$$\Delta H^\circ_{rxn} = 4\Delta H^\circ_f(CO_2) + 6\Delta H^\circ_f(H_2O) - [2\Delta H^\circ_f(C_2H_6) + 7\Delta H^\circ_f(O_2)]$$

$$\Delta H^\circ_{rxn} = (4)(-393.5 \text{ kJ/mol}) + (6)(-241.8 \text{ kJ/mol}) - [(2)(-84.7 \text{ kJ/mol}) + (7)(0)] = \textbf{-2855.4 kJ/mol}$$

The answers for part (a) and (b) are different, because *average* bond enthalpies are used for part (a).

10.79 **Strategy:** Draw Lewis structures for the reactants and products to determine the types and numbers of bonds. The balanced equation for the reaction is given below. We use this equation and Table 10.4 to estimate the enthalpy of reaction.

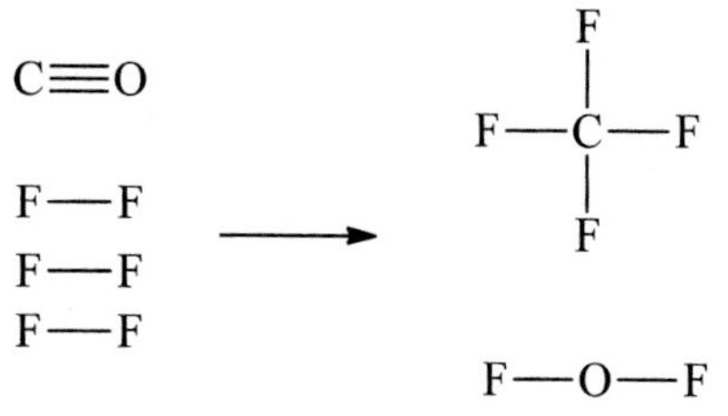

Solution:
$$\Delta H_{rxn} = BE(C \equiv O) + 3BE(F\text{–}F) - \left[4BE(C\text{–}F) + 2BE(O\text{–}F)\right]$$
$$= 1070 \text{ kJ/mol} + (3)(156.9 \text{ kJ/mol}) - \left[(4)(453 \text{ kJ/mol}) + (2)(190 \text{ kJ/mol})\right]$$
$$= \textbf{-651 kJ/mol}$$

10.83 Using data from Figures 4.8 and 4.10 as well as Appendix 2 (see the method outlined in Figure 10.14), we determine the lattice energy as follows:

Lattice energy = $\Delta H^\circ_f[Ca(g)] + 2\Delta H^\circ_f[Cl(g)] + IE_1(Ca) + IE_2(Ca) + \left|\Delta H^\circ_f[CaCl_2(s)]\right| - 2EA(Cl)$

Note that we account for the second ionization of Ca and that because there are two chloride ions in $CaCl_2$, we must multiply both the heat of formation (atomization) and the electron affinity of Cl by 2.

$Ca(s) \longrightarrow Ca(g)$	$\Delta H^\circ_f[Ca(g)] = 179.3 \text{ kJ/mol}$
$Cl_2(g) \longrightarrow 2Cl(g)$	$2 \times \Delta H^\circ_f[Cl(g)] = 243.4 \text{ kJ/mol}$
$Ca(g) \longrightarrow Ca^+(g) + e^-$	$IE_1(Ca) = 590 \text{ kJ/mol}$
$Ca^+(g) \longrightarrow Ca^{2+}(g) + e^-$	$IE_2(Ca) = 1145 \text{ kJ/mol}$
$Ca(s) + Cl_2(g) \longrightarrow CaCl_2(s)$	$\Delta H^\circ_f\left[CaCl_2^-(s)\right] = -794.96 \text{ kJ/mol}$
	$\left|\Delta H^\circ_f[CaCl_2(s)]\right| = 794.96 \text{ kJ/mol}$
$2[Cl(g) + e^- \longrightarrow Cl^-(g)]$	$2 \times EA(Cl) = 698 \text{ kJ/mol}$

$$(179.3 \text{ kJ/mol}) + (243.4 \text{ kJ/mol}) + (590 \text{ kJ/mol}) + (1145 \text{ kJ/mol})$$
$$+ (794.96 \text{ kJ/mol}) - (698 \text{ kJ/mol}) = 2255 \text{ kJ/mol}$$

The lattice energy of $CaCl_2$ is **2255 kJ/mol**.

10.85 **Strategy:** Use Equation 10.18, $\Delta H_{rxn}^{\circ} = \sum n\Delta H_f^{\circ}(\text{products}) - \sum m\Delta H_f^{\circ}(\text{reactants})$, and ΔH_f° values from Appendix 2 to calculate ΔH_f° of NF$_3$.

Setup: The ΔH_f° values for NH$_3(g)$, F$_2(g)$, and HF(g) are -46.3, 0, and -271.6 kJ/mol, respectively. The ΔH_{rxn}° value is -881.2 kJ/mol.

Solution: Using Equation 10.18,

$$\Delta H_{rxn}^{\circ} = [\Delta H_f^{\circ}(\text{NF}_3) + 3\Delta H_f^{\circ}(\text{HF})] - [\Delta H_f^{\circ}(\text{NH}_3) + 3\Delta H_f^{\circ}(\text{F}_2)]$$

$$881.2 \text{ kJ/mol} = [\Delta H_f^{\circ}(\text{NF}_3) + 3(-271.6 \text{ kJ/mol})] - [(-46.3 \text{ kJ/mol}) + 3(0 \text{ kJ/mol})]$$

$$-881.2 \text{ kJ/mol} = [\Delta H_f^{\circ}(\text{NF}_3) + (-814.8 \text{ kJ/mol})] - [(-46.3 \text{ kJ/mol}) + (0 \text{ kJ/mol})]$$

$$-881.2 \text{ kJ/mol} = [\Delta H_f^{\circ}(\text{NF}_3) - 814.8 \text{ kJ/mol})] + 46.3 \text{ kJ/mol}$$

$$-66.4 \text{ kJ/mol} = \Delta H_f^{\circ}(\text{NF}_3) + 46.3 \text{ kJ/mol}$$

$$\Delta H_f^{\circ}(\text{NF}_3) = \mathbf{-112.7 \text{ kJ / mol}}$$

10.87 The reaction corresponding to standard enthalpy of formation, ΔH_f°, of AgNO$_2(s)$ is:

$$\text{Ag}(s) + \tfrac{1}{2}\text{N}_2(g) + \text{O}_2(g) \longrightarrow \text{AgNO}_2(s)$$

Rather than measuring the enthalpy directly, we can use the enthalpy of formation of AgNO$_3(s)$ and the ΔH_{rxn}° provided.

$$\text{AgNO}_3(s) \longrightarrow \text{AgNO}_2(s) + \tfrac{1}{2}\text{O}_2(g)$$

$$\Delta H_{rxn}^{\circ} = \Delta H_f^{\circ}(\text{AgNO}_2) + \frac{1}{2}\Delta H_f^{\circ}(\text{O}_2) - \Delta H_f^{\circ}(\text{AgNO}_3)$$

$$78.67 \text{ kJ/mol} = \Delta H_f^{\circ}(\text{AgNO}_2) + 0 - (-123.02 \text{ kJ/mol})$$

$$\Delta H_f^{\circ}(\text{AgNO}_2) = \mathbf{-44.35 \text{ kJ / mol}}$$

10.89
$$w = -P\Delta V = -(1 \text{ atm})(-98 \text{ L}) = 98 \text{ L} \cdot \text{atm} \times \frac{101.3 \text{ J}}{1 \text{ L} \cdot \text{atm}} = \mathbf{9.9 \times 10^3 \text{ J}} = \mathbf{9.9 \text{ kJ}}$$

$$\Delta H = \Delta U + P\Delta V \text{ or } \Delta U = \Delta H - P\Delta V$$

Using ΔH as -185.2 kJ $= (2 \times -92.6$ kJ$)$ (because the question involves the formation of 4 moles of ammonia not 2 moles of ammonia for which the standard enthalpy is given in the question), and $-P\Delta V$ as 9.9 kJ (for which we just solved):

$$\Delta U = -185.2 \text{ kJ} + 9.9 \text{ kJ} = \mathbf{-175.3 \text{ kJ}}$$

10.91   The formation of CH_4 from its elements is:

$$C(s) + 2H_2(g) \longrightarrow CH_4(g)$$

The reaction could take place in two steps:

Step 1: $C(s) + 2H_2(g) \longrightarrow C(g) + 4H(g)$ $\Delta H^\circ_{rxn} = (716 + 872.8) \text{kJ/mol} = 1589 \text{ kJ/mol}$

Step 2: $C(g) + 4H(g) \longrightarrow CH_4(g)$ $\Delta H^\circ_{rxn} \approx -4 \times (\text{bond energy of C-H bond})$

$$= -4 \times 414 \text{ kJ/mol} = -1656 \text{ kJ/mol}$$

Therefore, $\Delta H^\circ_f(CH_4)$ would be approximately the sum of the enthalpy changes for the two steps. See Section 10.5 of the text (Hess's law).

$$\Delta H^\circ_f(CH_4) = \Delta H^\circ_{rxn}(1) + \Delta H^\circ_{rxn}(2)$$

$$\Delta H^\circ_f(CH_4) = (1589 - 1656) \text{kJ/mol} = \mathbf{-67 \text{ kJ/mol}}$$

The actual value of $\Delta H^\circ_f(CH_4) = -74.85 \text{ kJ/mol}$.

10.93   a. Using Equation 10.18 of the text,

$$\Delta H^\circ = 2\Delta H^\circ_f[HI(g)] - \left[\Delta H^\circ_f[H_2(g)] + \Delta H^\circ_f[I_2(g)] \right]$$
$$\Delta H^\circ = (2)(25.9 \text{ kJ/mol}) - [(0) + (1)(61.0 \text{ kJ/mol})] = \mathbf{-9.2 \text{ kJ/mol}}$$

b. Using Equation 10.19 of the text,

$$\Delta H = \Sigma BE(\text{reactants}) - \Sigma BE(\text{products})$$

$$\Delta H = [(436.4 + 151.0) - 2(298.3)] = \mathbf{-9.2 \text{ kJ/mol}}$$

10.95 $$H(g) + Br(g) \longrightarrow HBr(g) \quad \Delta H^\circ_{rxn}$$

Rearrange the equations as necessary so that they can be added to yield the desired equation.

$$H(g) \longrightarrow \tfrac{1}{2} H_2(g) \qquad \Delta H^\circ_{rxn} = \tfrac{1}{2}(-436.4 \text{ kJ/mol}) = -218.2 \text{ kJ/mol}$$

$$Br(g) \longrightarrow \tfrac{1}{2} Br_2(g) \qquad \Delta H^\circ_{rxn} = \tfrac{1}{2}(-192.5 \text{ kJ/mol}) = -96.25 \text{ kJ/mol}$$

$$\tfrac{1}{2} H_2(g) + \tfrac{1}{2} Br_2(g) \longrightarrow HBr(g) \quad \Delta H^\circ_{rxn} = \tfrac{1}{2}(-72.4 \text{ kJ/mol}) = -36.2 \text{ kJ/mol}$$

$$\overline{H(g) + Br(g) \longrightarrow HBr(g) \qquad \Delta H^\circ = \mathbf{-350.7 \text{ kJ/mol}}}$$

10.97 $$q_{sys} = q_{metal} + q_{water} + q_{calorimeter} = 0$$

$$m_{metal}s_{metal}(T_{final} - T_{initial}) + m_{water}s_{water}(T_{final} - T_{initial}) + C_{calorimeter}(T_{final} - T_{initial}) = 0$$

All the needed values are given in the problem. Plug in the values and solve for s_{metal}.

$$(44.0 \text{ g})(s_{metal})(28.4 - 99.0)°C + (80.0 \text{ g})(4.184 \text{ J/g·°C})(28.4 - 24.0)°C + (12.4 \text{ J/°C})(28.4 - 24.0)°C = 0$$

$$(-3.11 \times 10^3)s_{metal} \text{ (g·°C)} = -1.53 \times 10^3 \text{ J}$$

$$s_{metal} = \textbf{0.492 J/g·°C}$$

10.99   A good starting point would be to calculate the standard enthalpy for both reactions.

Calculate the standard enthalpy for the reaction: $C(s) + \frac{1}{2}O_2(g) \longrightarrow CO(g)$

This reaction corresponds to the standard enthalpy of formation of CO, so we use the value of -110.5 kJ/mol (see Appendix 2 of the text).

Calculate the standard enthalpy for the reaction: $C(s) + H_2O(g) \longrightarrow CO(g) + H_2(g)$

$$\Delta H_{rxn}^° = \left[\Delta H_f^°(CO) + \Delta H_f^°(H_2)\right] - \left[\Delta H_f^°(C) + \Delta H_f^°(H_2O)\right]$$

$$\Delta H_{rxn}^° = [(1)(-110.5 \text{ kJ/mol}) + (1)(0)] - [(1)(0) + (1)(-241.8 \text{ kJ/mol})] = 131.3 \text{ kJ/mol}$$

The first reaction, which is exothermic, can be used to promote the second reaction, which is endothermic. Thus, the two gases are produced alternately.

10.101 First, calculate the energy produced by 1 mole of octane, C_8H_{18}.

$$C_8H_{18}(l) + \frac{25}{2}O_2(g) \longrightarrow 8CO_2(g) + 9H_2O(l)$$

$$\Delta H_{rxn}^° = 8\Delta H_f^°(CO_2) + 9\Delta H_f^°[H_2O(l)] - \left[\Delta H_f^°(C_8H_{18}) + \frac{25}{2}\Delta H_f^°(O_2)\right]$$

$$\Delta H_{rxn}^° = [(8)(-393.5 \text{ kJ/mol}) + (9)(-285.8 \text{ kJ/mol})] - [(1)(-249.9 \text{ kJ/mol}) + (\tfrac{25}{2})(0)]$$

$$= -5470 \text{ kJ/mol}$$

The problem asks for the energy produced by the combustion of 1 gallon of octane. $\Delta H_{rxn}^°$ above has units of kJ/mol octane. We need to convert from kJ/mol octane to kJ/gallon octane. The heat of combustion for 1 gallon of octane is:

$$\Delta H° = \frac{-5470 \text{ kJ}}{1 \text{ mol octane}} \times \frac{1 \text{ mol octane}}{114.2 \text{ g octane}} \times \frac{2660 \text{ g}}{1 \text{ gal}} = -1.274 \times 10^5 \text{ kJ/gal}$$

The combustion of hydrogen corresponds to the standard heat of formation of water:

$$H_2(g) + \frac{1}{2}O_2(g) \longrightarrow H_2O(l)$$

Thus, ΔH_{rxn}° is the same as ΔH_f° for $H_2O(l)$, which has a value of -285.8 kJ/mol. The number of moles of hydrogen required to produce 1.274×10^5 kJ of heat is:

$$n_{H_2} = \left(1.274 \times 10^5 \text{ kJ}\right) \times \frac{1 \text{ mol } H_2}{285.8 \text{ kJ}} = \textbf{446 mol } \textbf{H}_2$$

10.103 The combustion reaction is:

$$C_2H_6(l) + \tfrac{7}{2} O_2(g) \longrightarrow 2CO_2(g) + 3H_2O(l)$$

The heat released during the combustion of 1 mole of ethane is:

$$\Delta H_{rxn}^{\circ} = [2\Delta H_f^{\circ}(CO_2) + 3\Delta H_f^{\circ}(H_2O)] - [\Delta H_f^{\circ}(C_2H_6) + \tfrac{7}{2}\Delta H_f^{\circ}(O_2)]$$

$$\Delta H_{rxn}^{\circ} = [(2)(-393.5 \text{ kJ/mol}) + (3)(-285.8 \text{ kJ/mol})] - [(1)(-84.7 \text{ kJ/mol}) + (\tfrac{7}{2})(0)]$$
$$= -1559.7 \text{ kJ/mol}$$

The heat required to raise the temperature of the water to 98°C is:

$$m_{H_2O} s_{H_2O} \Delta t = (855 \text{ g})(4.184 \text{ J/g} \cdot {}^{\circ}\text{C})(98.0 - 25.0){}^{\circ}\text{C} = 2.61 \times 10^5 \text{ J} = 261 \text{ kJ}$$

The combustion of 1 mole of ethane produces 1559.7 kJ; the number of moles required to produce 261 kJ is:

$$n_{ethane} = 261 \text{ kJ} \times \frac{1 \text{ mol ethane}}{1559.7 \text{ kJ}} = \textbf{0.167 mol ethane}$$

10.105 The heat gained by the liquid nitrogen must be equal to the heat lost by the water.

$$q_{N_2} = -q_{H_2O}$$

If we can calculate the heat lost by the water, we can calculate the heat gained by 60.0 g of the nitrogen.

$$\text{Heat lost by the water} = q_{H_2O} = m_{H_2O} s_{H_2O} \Delta T$$

$$q_{H_2O} = (2.00 \times 10^2 \text{ g})(4.184 \text{ J/g} \cdot {}^{\circ}\text{C})(41.0 - 55.3){}^{\circ}\text{C} = -1.20 \times 10^4 \text{ J}$$

The heat gained by 60.0 g nitrogen is the opposite sign of the heat lost by the water.

$$q_{N_2} = -q_{H_2O}$$
$$q_{N_2} = 1.20 \times 10^4 \text{ J}$$

The problem asks for the molar heat of vaporization of liquid nitrogen. Above, we calculated the amount of heat necessary to vaporize 60.0 g of liquid nitrogen. We need to convert from J/60.0 g N_2 to J/mol N_2.

$$\Delta H_{vap} = \frac{1.20 \times 10^4 \text{ J}}{60.0 \text{ g } N_2} \times \frac{28.02 \text{ g } N_2}{1 \text{ mol } N_2} = 5.60 \times 10^3 \text{ J/mol} = \textbf{5.60 kJ/mol}$$

10.107 Recall that the standard enthalpy of formation (ΔH_f°) is defined as the heat change that results when 1 mole of a compound is formed from its elements at a pressure of 1 atm. Only in **reaction (a)** does $\Delta H_{rxn}^\circ = \Delta H_f^\circ$. In reaction (b), C(diamond) is *not* the most stable form of elemental carbon under standard conditions; C(graphite) is the most stable form.

10.109 a. No work is done by a gas expanding in a vacuum, because the pressure exerted on the gas is zero.

b.
$$w = -P\Delta V$$

$$w = -(0.20 \text{ atm})(0.50 - 0.050)\text{L} = -0.090 \text{ L}\cdot\text{atm}$$

Converting to units of joules:
$$w = -0.090 \text{ L}\cdot\text{atm} \times \frac{101.3 \text{ J}}{\text{L}\cdot\text{atm}} = -\textbf{9.1 J}$$

10.111 We start by balancing the equation to give:

$$2\text{ZnS}(s) + 3\text{O}_2(g) \longrightarrow 2\text{ZnO}(s) + 2\text{SO}_2(g)$$

We can calculate ΔH_{rxn}° using Equation 10.18 and the standard enthalpies of formation found in Appendix 2.

$$\Delta H_{rxn}^\circ = \left[2\Delta H_f^\circ(\text{ZnO}) + 2\Delta H_f^\circ(\text{SO}_2)\right] - \left[2\Delta H_f^\circ(\text{ZnS}) + 3\Delta H_f^\circ(\text{O}_2)\right]$$

$$\Delta H_{rxn}^\circ = [(2)(-348.0 \text{ kJ/mol}) + (2)(-296.4 \text{ kJ/mol})] - [(2)(-202.9 \text{ kJ/mol}) + (0)] = -883 \text{ kJ/mol}$$

ΔH_{rxn}° has units of kJ/mol. We want the ΔH° change for 1.26×10^4 g SO_2. We need to calculate how many moles of SO_2 are in 1.26×10^4 g SO_2. Use the following strategy to solve the problem.

$$\text{g SO}_2 \rightarrow \text{mol SO}_2 \rightarrow \text{kJ } (\Delta H^\circ)$$

$$\Delta H^\circ = (1.26\times10^4 \text{ g SO}_2) \times \frac{1 \text{ mol SO}_2}{64.07 \text{ g SO}_2} \times \frac{-883 \text{ kJ}}{2 \text{ mol SO}_2} = \textbf{-8.68×10}^4 \textbf{ kJ}$$

10.113 $4\text{Fe}(s) + 3\text{O}_2(g) \longrightarrow 2\text{Fe}_2\text{O}_3(s)$. This equation represents twice the standard enthalpy of formation of Fe_2O_3. From Appendix 2, the standard enthalpy of formation of $\text{Fe}_2\text{O}_3 = -822.2$ kJ/mol. So, ΔH° for the given reaction is:

$$\Delta H_{rxn}^\circ = (2)(-822.2 \text{ kJ/mol}) = -1644 \text{ kJ/mol}$$

Looking at the balanced equation, this is the amount of heat released when four moles of Fe react. But, this is the reaction of 250 g of Fe, not four moles. We can convert from grams of Fe to moles of Fe, then use ΔH° as a conversion factor to convert to kJ.

$$250 \text{ g Fe} \times \frac{1 \text{ mol Fe}}{55.85 \text{ g Fe}} \times \frac{-1644 \text{ kJ}}{4 \text{ mol Fe}} = \textbf{-1.84×10}^3 \textbf{ kJ}$$

10.115 The heat required to raise the temperature of 1 L water by 1°C is:

$$4.184 \ \frac{J}{g \cdot °C} \times \frac{1 \ g}{1 \ mL} \times \frac{1000 \ mL}{1 \ L} \times 1°C = 4184 \ J/L$$

Next, convert the volume of the Pacific Ocean to liters.

$$(7.2 \times 10^8 \ km^3) \times \left(\frac{1000 \ m}{1 \ km}\right)^3 \times \left(\frac{100 \ cm}{1 \ m}\right)^3 \times \frac{1 \ L}{1000 \ cm^3} = 7.2 \times 10^{20} \ L$$

The amount of heat needed to raise the temperature of 7.2×10^{20} L of water is:

$$(7.2 \times 10^{20} \ L) \times \frac{4184 \ J}{1 \ L} = 3.0 \times 10^{24} \ J$$

Finally, we can calculate the number of atomic bombs needed to produce this much heat.

$$(3.0 \times 10^{24} \ J) \times \frac{1 \ atomic \ bomb}{1.0 \times 10^{15} \ J} = 3.0 \times 10^9 \ atomic \ bombs = \textbf{3.0 billion atomic bombs}$$

10.117 a. $\textbf{2LiOH}(aq) + \textbf{CO}_2(g) \longrightarrow \textbf{Li}_2\textbf{CO}_3(aq) + \textbf{H}_2\textbf{O}(l)$

b. The metabolism of glucose is the same as the combustion of glucose:

$$C_6H_{12}O_6(aq) + 6O_2(g) \longrightarrow 6CO_2(g) + 6H_2O(l)$$

Using Equation 10.18 from the text,

$$\Delta H^\circ_{rxn} = [6\Delta H^\circ_f (CO_2) + 6\Delta H^\circ_f (H_2O)] - \Delta H^\circ_f (C_6H_{12}O_6)$$

$$= [6(-393.5 \ kJ/mol) + 6(-285.8 \ kJ/mol)] - (-1274.5 \ kJ/mol) = -2801 \ kJ$$

$$? \ g \ CO_2 = 1.2 \times 10^4 \ kJ \times \frac{6 \ mol \ CO_2}{2801 \ kJ} \times \frac{44.01 \ g \ CO_2}{1 \ mol \ CO_2} = 1131 \ g = \textbf{1.1 kg CO}_2$$

$$? \ g \ LiOH = 1131 \ g \ CO_2 \times \frac{1 \ mol \ CO_2}{44.01 \ g \ CO_2} \times \frac{2 \ mol \ LiOH}{1 \ mol \ CO_2} \times \frac{23.9 \ g \ LiOH}{1 \ mol \ LiOH} = 1229 \ g = \textbf{1.2 kg LiOH}$$

10.119 Begin by calculating the standard enthalpy of reaction.

$$\Delta H^\circ_{rxn} = 2\Delta H^\circ_f (CaSO_4) - [2\Delta H^\circ_f (CaO) + 2\Delta H^\circ_f (SO_2) + \Delta H^\circ_f (O_2)]$$

$$= (2)(-1432.7 \ kJ/mol) - [(2)(-635.6 \ kJ/mol) + (2)(-296.4 \ kJ/mol) + 0]$$

$$= -1001 \ kJ/mol$$

This is the enthalpy change for every 2 moles of SO_2 that are removed. The problem asks to calculate the enthalpy change for this process if 6.6×10^5 g of SO_2 are removed.

$$(6.6 \times 10^5 \text{ g SO}_2) \times \frac{1 \text{ mol SO}_2}{64.07 \text{ g SO}_2} \times \frac{-1001 \text{ kJ}}{2 \text{ mol SO}_2} = \mathbf{-5.2 \times 10^6 \text{ kJ}}$$

10.121 The heat produced by the reaction heats the solution and the calorimeter: $q_{rxn} = -(q_{soln} + q_{cal})$

$$q_{soln} = ms\Delta T = (50.0 \text{ g})(4.184 \text{ J/g·°C})(22.17°C - 19.25°C) = 611 \text{ J}$$

$$q_{cal} = C\Delta T = (98.6 \text{ J/°C})(22.17°C - 19.25°C) = 288 \text{ J}$$

$$-q_{rxn} = (q_{soln} + q_{cal}) = (611 + 288)\text{J} = 899 \text{ J}$$

The 899 J produced was for 50.0 mL of a 0.100 M AgNO$_3$ solution.

$$50.0 \text{ mL} \times \frac{0.100 \text{ mol Ag}^+}{1000 \text{ mL soln}} = 5.00 \times 10^{-3} \text{ mol Ag}^+$$

On a molar basis, the heat produced was:

$$\frac{899 \text{ J}}{5.00 \times 10^{-3} \text{ mol Ag}^+} = 1.80 \times 10^5 \text{ J/mol Ag}^+ = 180 \text{ kJ/mol Ag}^+$$

The balanced equation involves 2 moles of Ag$^+$, so the heat produced is 2 mol × 180 kJ/mol = 360 kJ.

Since the reaction produces heat (or by noting the sign convention above), then:

$$\Delta H_{rxn} = q_{rxn} = \mathbf{-3.60 \times 10^2 \text{ kJ/mol Zn (or } -3.60 \times 10^2 \text{ kJ/2 mol Ag}^+)}$$

10.123 The standard enthalpy of formation (ΔH_f°) is the enthalpy change associated with the formation of 1 mole of a substance from its constituent elements, each in their standard states.

a. **The ΔH_{rxn}^0 is not the ΔH_f° value because more than 1 mole of a substance is formed.**

b. **The ΔH_{rxn}^0 is not the ΔH_f° value because more than 1 mole of a substance is formed.**

c. **The ΔH_{rxn}^0 is the ΔH_f° value because 1 mole of a substance is formed from its constituent elements in their standard states.**

d. **The ΔH_{rxn}^0 is not the ΔH_f° value because Cu^{2+}(aq) is not copper's standard state and SO$_4^{2-}$ is not an element.**

e. **The ΔH_{rxn}^0 is the ΔH_f° value because 1 mole of a substance is formed from its constituent elements in their standard states.**

10.125 **Water has a larger specific heat than air. Thus cold, damp air can extract more heat from the body than cold, dry air. By the same token, hot, humid air can deliver more heat to the body.**

10.127 Energy intake for mechanical work:

$$0.17 \times 500 \text{ g} \times \frac{3000 \text{ J}}{1 \text{ g}} = 2.6 \times 10^5 \text{ J}$$

$$2.6 \times 10^5 \text{ J} = mgh$$

$$1 \text{ J} = 1 \text{ kg·m}^2\text{s}^{-2}$$

$$2.6 \times 10^5 \frac{\text{kg·m}^2}{\text{s}^2} = (46 \text{ kg})\left(9.8 \text{ m/s}^2\right)h$$

$$h = \mathbf{5.8 \times 10^2 \text{ m}}$$

10.129   For Al: $(0.900 \text{ J/g·°C})(26.98 \text{ g}) = 24.3 \text{ J/°C}$

This law does not hold for Hg because it is a liquid.

10.131

$$C_6H_6(l) + \tfrac{15}{2} O_2(g) \longrightarrow 6CO_2(g) + 3H_2O(l) \qquad\qquad \Delta H° = -3267.4 \text{ kJ/mol}$$

$$C_2H_2(g) + \tfrac{5}{2} O_2(g) \longrightarrow 2CO_2(g) + H_2O(l) \qquad\qquad \Delta H° = -1299.4 \text{ kJ/mol}$$

Using the $\Delta H_f°$ values in Appendix 2,

$$C(\text{graphite}) + O_2 \longrightarrow CO_2(g) \qquad\qquad \Delta H° = -393.5 \text{ kJ/mol}$$

$$H_2(g) + \tfrac{1}{2} O_2(g) \longrightarrow H_2O(l) \qquad\qquad \Delta H° = -285.8 \text{ kJ/mol}$$

Using Hess's law, we can add the equations in the following manner to calculate the standard enthalpies of formation of C_2H_2 and C_6H_6.

C_2H_2: − (b) + 2(c) + (d)

$$2C(\text{graphite}) + H_2(g) \longrightarrow C_2H_2(g) \qquad\qquad \Delta H° = +226.6 \text{ kJ/mol}$$

Therefore, $\Delta H_f°(C_2H_2) = \mathbf{226.6 \text{ kJ/mol}}$

C_6H_6: − (a) + 6(c) + 3(d)

$$6C(\text{graphite}) + 3H_2(g) \longrightarrow C_6H_6(l) \qquad\qquad \Delta H° = +49.0 \text{ kJ/mol}$$

Therefore, $\Delta H_f°(C_6H_6) = \mathbf{49.0 \text{ kJ/mol}}$

Finally: $\qquad\qquad\qquad\qquad 3C_2H_2(g) \longrightarrow C_6H_6(l)$

$$\Delta H_{rxn} = (1)(49.0 \text{ kJ/mol}) - (3)(226.6 \text{ kJ/mol}) = \mathbf{-630.8 \text{ kJ/mol}}$$

10.133   a.   One C–H bond is being broken, and an O–H bond is being formed.

$$\Delta H = \Sigma BE(\text{reactants}) - \Sigma BE(\text{products})$$

$$\Delta H = 414 - 460 = \mathbf{-46 \text{ kJ/mol}}$$

b. $$\text{Energy of one O}-\text{H bond} = \frac{460 \times 10^3 \text{ J}}{1 \text{ mol}} \times \frac{1 \text{ mol}}{6.022 \times 10^{23} \text{ bonds}} = 7.64 \times 10^{-19} \text{ J/bond}$$

$$\lambda = \frac{hc}{\Delta H} = \frac{\left(6.63 \times 10^{-34} \text{ J} \cdot \text{s}\right)\left(3.00 \times 10^8 \text{ m/s}\right)}{7.64 \times 10^{-19} \text{ J}}$$

$$\lambda = 2.60 \times 10^{-7} \text{ m} = \textbf{260 nm}$$

10.135 a. The bond enthalpy of F_2^- is the energy required to break up F_2^- into an F atom and an F^- ion.

$$F_2^- (g) \longrightarrow F(g) + F^-(g)$$

We can arrange the equations given in the problem so that they add up to the above equation. See Section 10.5 of the text (Hess's law).

$$
\begin{aligned}
F_2^- (g) &\longrightarrow F_2(g) + e^- &\Delta H^\circ &= 290 \text{ kJ/mol} \\
F_2(g) &\longrightarrow 2F(g) &\Delta H^\circ &= 156.9 \text{ kJ/mol} \\
F(g) + e^- &\longrightarrow F^-(g) &\Delta H^\circ &= -333 \text{ kJ/mol} \\
\hline
F_2^- (g) &\longrightarrow F(g) + F^-(g)
\end{aligned}
$$

The bond enthalpy of F_2^- is the sum of the enthalpies of reaction.

$$BE(F_2^-) = [290 + 156.9 + (-333)] \text{kJ/mol} = \textbf{114 kJ/mol}$$

b. **The bond in F_2^- is weaker** (114 kJ/mol) than the bond in F_2 (156.9 kJ/mol), because the extra electron increases repulsion between the F atoms.

10.137 a.

$$2.0 \text{ g glucose} \times \frac{1 \text{ mol glucose}}{180.2 \text{ g glucose}} = 0.011 \text{ mol glucose}$$

The heat of combustion for glucose is –2801 kJ/mol (see the solution to Problem 10.117). Therefore, the energy released by the combustion of a 2.0 g tablet is:

For **glucose** ($C_6H_{12}O_6$):

$$0.011 \text{ mol glucose} \times \frac{2801 \text{ kJ}}{1 \text{ mol glucose}} = \textbf{31 kJ}$$

For **sucrose** ($C_{12}H_{22}O_{11}$):

$$2.0 \text{ g sucrose} \times \frac{1 \text{ mol sucrose}}{342.3 \text{ g sucrose}} = 0.0058 \text{ mol sucrose}$$

The heat of combustion for sucrose (using values from Appendix 2) is −5644 kJ/mol. The energy released by combustion of a 2.0 g sucrose tablet is:

$$0.0058 \text{ mol sucrose} \times \frac{5644 \text{ kJ}}{1 \text{ mol sucrose}} = \textbf{33 kJ}$$

b. For **glucose**:

$$31 \text{ kJ} \times 0.30 = 9.3 \text{ kJ} = 9.3 \times 10^3 \text{ J}$$

$$9.3 \times 10^3 \text{ J} = 65 \text{ kg} \times 9.8 \text{ m/s}^2 \times h$$

$$h = \textbf{15 m}$$

For **sucrose**:

$$33 \text{ kJ} \times 0.30 = 9.9 \text{ kJ} = 9.9 \times 10^3 \text{ J}$$

$$9.9 \times 10^3 \text{ J} = 65 \text{ kg} \times 9.8 \text{ m/s}^2 \times h$$

$$h = \textbf{16 m}$$

10.139 There are four C–H bonds in CH_4, so the average bond enthalpy of a C–H bond is:

$$\frac{1656 \text{ kJ/mol}}{4} = 414.0 \text{ kJ/mol}$$

The Lewis structure of propane is:

There are eight C–H bonds and two C–C bonds. We write:

$$8(\text{C–H}) + 2(\text{C–C}) = 4006 \text{ kJ/mol}$$

$$8(414.0 \text{ kJ/mol}) + 2(\text{C–C}) = 4006 \text{ kJ/mol}$$

$$2(\text{C–C}) = 694 \text{ kJ/mol}$$

So, the average bond enthalpy of a C–C bond is: $\dfrac{694}{2}$ kJ/mol = **347 kJ/mol**

10.141 Rearranging Equation 10.8 to solve for ΔU,

$$\Delta U = \Delta H - P\Delta V$$

$$\Delta U = -571.6 \text{ kJ} - (1.00 \text{ atm})(-73.4 \text{ L})\left(101.3 \times 10^{-3} \text{ kJ/L} \cdot \text{atm}\right) = \textbf{–564.2 kJ}$$

10.143 First, we calculate ΔH for the combustion of 1 mole of glucose using data in Appendix 2 of the text. We can then calculate the heat produced in the calorimeter. Using the heat produced along with ΔH for the combustion of 1 mole of glucose will allow us to calculate the mass of glucose in the sample. Finally, the mass % of glucose in the sample can be calculated.

$$C_6H_{12}O_6(s) + 6O_2(g) \longrightarrow 6CO_2(g) + 6H_2O(l)$$

$$\Delta H^{\circ}_{rxn} = (6)(-393.5 \text{ kJ/mol}) + (6)(-285.8 \text{ kJ/mol}) - (1)(-1274.5 \text{ kJ/mol}) = -2801.3 \text{ kJ/mol}$$

The heat produced in the calorimeter is:

$$(3.134°C)(19.65 \text{ kJ/°C}) = 61.58 \text{ kJ}$$

Let x equal the mass of glucose in the sample:

$$x \text{ g glucose} \times \frac{1 \text{ mol glucose}}{180.2 \text{ g glucose}} \times \frac{2801.3 \text{ kJ}}{1 \text{ mol glucose}} = 61.58 \text{ kJ}$$

$$x = 3.961 \text{ g}$$

$$\% \text{ glucose} = \frac{3.961 \text{ g}}{4.117 \text{ g}} \times 100\% = \mathbf{96.21\%}$$

10.145 a. Although we cannot measure ΔH°_{rxn} for this reaction, the reverse process is the combustion of glucose. **We could easily measure ΔH°_{rxn} for this combustion by burning a mole of glucose in a bomb calorimeter.**

$$C_6H_{12}O_6(s) + 6O_2(g) \longrightarrow 6CO_2(g) + 6H_2O(l)$$

b. We can calculate ΔH°_{rxn} using standard enthalpies of formation.

$$\Delta H^{\circ}_{rxn} = \Delta H^{\circ}_f[C_6H_{12}O_6(s)] + 6\Delta H^{\circ}_f[O_2(g)] - \left[6\Delta H^{\circ}_f[CO_2(g)] + 6\Delta H^{\circ}_f[H_2O(l)] \right]$$

$$\Delta H^{\circ}_{rxn} = [(1)(-1274.5 \text{ kJ/mol}) + 0] - [(6)(-393.5 \text{ kJ/mol}) + (6)(-285.8 \text{ kJ/mol})] = 2801.3 \text{ kJ/mol}$$

ΔH°_{rxn} has units of kJ/mol glucose. We want the ΔH° change for 7.0×10^{14} kg glucose. We need to calculate how many moles of glucose are in 7.0×10^{14} kg glucose. Use the following strategy to solve the problem.

kg glucose $\rightarrow$ g glucose $\rightarrow$ mol glucose $\rightarrow$ kJ (ΔH°)

$$\Delta H^{\circ} = \left(7.0 \times 10^{14} \text{ kg}\right) \times \frac{1000 \text{ g}}{1 \text{ kg}} \times \frac{1 \text{ mol } C_6H_{12}O_6}{180.2 \text{ g } C_6H_{12}O_6} \times \frac{2801.3 \text{ kJ}}{1 \text{ mol } C_6H_{12}O_6} = \mathbf{1.1 \times 10^{19} \text{ kJ}}$$

10.147 a. $$\mathbf{CaC_2(s) + 2H_2O(l) \longrightarrow Ca(OH)_2(s) + C_2H_2(g)}$$

b. The reaction for the combustion of acetylene is:

$$2C_2H_2(g) + 5O_2(g) \longrightarrow 4CO_2(g) + 2H_2O(l)$$

We can calculate the enthalpy change for this reaction from standard enthalpy of formation values given in Appendix 2 of the text.

$$\Delta H_{rxn}^{\circ} = [4\Delta H_f^{\circ}(CO_2) + 2\Delta H_f^{\circ}(H_2O)] - [2\Delta H_f^{\circ}(C_2H_5) + 5\Delta H_f^{\circ}(O_2)]$$

$$\Delta H_{rxn}^{\circ} = [(4)(-393.5 \text{ kJ/mol}) + (2)(-285.8 \text{ kJ/mol})] - [(2)(226.6 \text{ kJ/mol}) + (5)(0)]$$

$$\Delta H_{rxn}^{\circ} = -2599 \text{ kJ/mol}$$

Looking at the balanced equation, this is the amount of heat released when two moles of C_2H_2 are reacted. The problem asks for the amount of heat that can be obtained starting with 74.6 g of CaC_2. From this amount of CaC_2, we can calculate the moles of C_2H_2 produced.

$$74.6 \text{ g } CaC_2 \times \frac{1 \text{ mol } CaC_2}{64.10 \text{ g } CaC_2} \times \frac{1 \text{ mol } C_2H_2}{1 \text{ mol } CaC_2} = 1.16 \text{ mol } C_2H_2$$

Now, we can use the ΔH_{rxn}° calculated above as a conversion factor to determine the amount of heat obtained when 1.16 moles of C_2H_2 are burned.

$$1.16 \text{ mol } C_2H_2 \times \frac{2599 \text{ kJ}}{2 \text{ mol } C_2H_2} = 1.51 \times 10^3 \text{ kJ} = \mathbf{1.51 \times 10^6 \text{ J}}$$

10.149 We assume that when the car is stopped, its kinetic energy is completely converted into heat (friction of the brakes and friction between the tires and the road). Thus,

$$q = \frac{1}{2}mu^2$$

The amount of heat generated must be proportional to the braking distance, *d*:

$$\boldsymbol{d \propto q}$$

$$\boldsymbol{d \propto u^2}$$

Therefore, as *u* increases to 2*u*, *d* increases to $(2u)^2 = 4u^2$ which is proportional to 4*d*.

10.151 This is an application of Hess's law.

$2H + H^+ \longrightarrow H_3^+$		$\Delta H = -849 \text{ kJ/mol}$
$H_2 \longrightarrow 2H$		$\Delta H = 36.4 \text{ kJ/mol}$
$H^+ + H_2 \longrightarrow H_3^+$		$\Delta H = \mathbf{-413 \text{ kJ/mol}}$

The energy released in forming H_3^+ from H^+ and H_2 is almost as large as the formation of H_2 from two H atoms.

10.153 When 1.034 g of naphthalene are burned, 41.56 kJ of heat are evolved. Convert this to the amount of heat evolved on a molar basis. The molar mass of naphthalene is 128.2 g/mol.

$$q = \frac{-41.56 \text{ kJ}}{1.034 \text{ g } C_{10}H_8} \times \frac{128.2 \text{ g } C_{10}H_8}{1 \text{ mol } C_{10}H_8} = -5153 \text{ kJ/mol}$$

q has a negative sign because this is an exothermic reaction.

This reaction takes place at constant volume ($\Delta V = 0$); therefore, no work will result from the change.

$$w = -P\Delta V = \mathbf{0}$$

From Equation 10.1 of the text, it follows that the change in energy is equal to the heat change.

$$\Delta U = q + w = q = \mathbf{-5153\ kJ/mol}$$

10.155 The heat required to heat 200 g of water (assume $d = 1$ g/mL) from 20°C to 100°C is:

$$q = ms\Delta T$$

$$q = (200\ \text{g})(4.184\ \text{J/g·°C})\,(100 - 20)°C = 6.69 \times 10^4\ \text{J}$$

Since 50% of the heat from the combustion of methane is lost to the surroundings, twice the amount of heat needed must be produced during the combustion: $2(6.69 \times 10^4\ \text{J}) = 1.34 \times 10^5\ \text{J} = 1.34 \times 10^2\ \text{kJ}$.

Use standard enthalpies of formation (see Appendix 2) to calculate the heat of combustion of methane.

$$CH_4(g) + 2O_2(g) \longrightarrow CO_2(g) + 2H_2O(l) \qquad\qquad \Delta H° = -890.3\ \text{kJ/mol}$$

The number of moles of methane needed to produce 1.34×10^2 kJ of heat is:

$$\left(1.34 \times 10^2\ \text{kJ}\right) \times \frac{1\ \text{mol CH}_4}{890.3\ \text{kJ}} = 0.151\ \text{mol CH}_4$$

$$(0.151\ \text{mol CH}_4) \times \frac{\$0.27}{1\ \text{mol CH}_4} = \$0.041$$

The cost of the methane is about **4.1 cents**.

10.157 **C—C: 347 kJ/mol; N—N: 193 kJ/mol; O—O: 142 kJ/mol; Lone pairs appear to weaken the bond.**

Chapter 11
Gases

Visualizing Chemistry

VC 11.1 b   VC 11.2 a   VC 11.3 a   VC 11.4 b

Key Skills

11.1   d   11.2   c   11.3   b   11.4   e

Problems

11.9   **Strategy:**   To calculate the root-mean-square speed, we use Equation 11.1 of the text. What units should we use for R and $\mathcal{M}$ so that u_{rms} will be expressed in units of m/s?

Solution: To calculate u_{rms}, the units of R should be 8.314 J/K·mol, and because 1 J = 1 kg·m²/s², the units of molar mass must be kg/mol.

First, let's calculate the molar masses ($\mathcal{M}$) of N_2, O_2, and O_3. Remember, $\mathcal{M}$ must be in units of kg/mol.

$$\mathcal{M}_{N_2} = 2(14.01 \text{ g/mol}) = 28.02 \text{ g/mol}$$

$$28.02 \frac{\text{g}}{\text{mol}} \times \frac{1 \text{ kg}}{1000 \text{ g}} = 0.02802 \text{ kg/mol}$$

$$\mathcal{M}_{O_2} = 2(16.00 \text{ g/mol}) = 32.00 \text{ g/mol}$$

$$32.00 \frac{\text{g}}{\text{mol}} \times \frac{1 \text{ kg}}{1000 \text{ g}} = 0.03200 \text{ kg/mol}$$

$$\mathcal{M}_{O_3} = 3(16.00 \text{ g/mol}) = 48.00 \text{ g/mol}$$

$$48.00 \frac{\text{g}}{\text{mol}} \times \frac{1 \text{ kg}}{1000 \text{ g}} = 0.04800 \text{ kg/mol}$$

Now, we can substitute into Equation 11.1 of the text.

$$u_{rms} = \sqrt{\frac{3RT}{\mathcal{M}}}$$

$$u_{rms}(N_2) = \sqrt{\frac{(3)\left(8.314 \frac{\text{J}}{\text{K} \cdot \text{mol}}\right)(-23 + 273)\text{K}}{\left(0.02802 \frac{\text{kg}}{\text{mol}}\right)}}$$

$u_{rms}(N_2)$ **= 472 m/s**

Similarly, $u_{rms}(O_2)$ **= 441 m/s**, $u_{rms}(O_3)$ **= 360 m/s.**

11.11
$$\text{RMS speed} = \sqrt{\frac{\left(2.0^2 + 2.2^2 + 2.6^2 + 2.7^2 + 3.3^2 + 3.5^2\right)(m/s)^2}{6}} = \textbf{2.8 m/s}$$

$$\text{Average speed} = \frac{(2.0 + 2.2 + 2.6 + 2.7 + 3.3 + 3.5)m/s}{6} = \textbf{2.7 m/s}$$

The root-mean-square value is always greater than the average value, because squaring favors the larger values compared to just taking the average value.

11.13   The rate of effusion is the number of molecules passing through a porous barrier in a given time. The longer it takes, the slower the rate of effusion. Therefore, Equation 11.2 of the text can be written as

$$\frac{r_1}{r_2} = \frac{t_2}{t_1} = \sqrt{\frac{\mathcal{M}_2}{\mathcal{M}_1}}$$

where t_1 and t_2 are the times of effusion for gases 1 and 2, respectively.

The molar mass of N_2 is 28.02 g/mol. We write

$$\frac{15.0 \text{ min}}{12.0 \text{ min}} = \sqrt{\frac{\mathcal{M}}{28.02 \text{ g/mol}}}$$

where $\mathcal{M}$ is the molar mass of the unknown gas. Solving for $\mathcal{M}$, we obtain

$$\mathcal{M} = \left(\frac{15.0 \text{ min}}{12.0 \text{ min}}\right)^2 \times 28.02 \text{ g/mol} = \textbf{43.8 g/mol}$$

The gas is **carbon dioxide, CO_2** ($\mathcal{M}$ = 44.01 g/mol). During the fermentation of glucose, ethanol and carbon dioxide are produced.

11.15   Graham's law states that the rate of effusion of a gas is inversely proportional to the square root of its molar mass. Thus, the gas with the greater molar mass will effuse more slowly.

   a. **Since more of the yellow gas was able to escape the container in the same amount of time, the rate of effusion for the yellow gas is greater. Therefore, the molar mass of the yellow gas is lower.**

   b. **More of the red gas escaped; hence, the molar mass of the blue gas is greater.**

11.23   **Strategy:** We use the conversion factors provided in Table 11.2 in the text to convert a pressure in mmHg to atm, bar, torr, and Pa.

   Solution: Converting to atm:

$$?\text{atm} = 375 \text{ mmHg} \times \frac{133.322 \text{ Pa}}{1 \text{ mmHg}} \times \frac{1 \text{ atm}}{101{,}325 \text{ Pa}} = \textbf{0.493 atm}$$

We could also have solved this by remembering that 760 mmHg = 1 atm.

$$?\text{atm} = 375 \text{ mmHg} \times \frac{1 \text{ atm}}{760 \text{ mmHg}} = \textbf{0.493 atm}$$

Converting to bar:

$$?bar = 375 \text{ mmHg} \times \frac{133.322 \text{ Pa}}{1 \text{ mmHg}} \times \frac{1 \text{ bar}}{1 \times 10^5 \text{ Pa}} = \textbf{0.500 bar}$$

Converting to torr:

$$?torr = 375 \text{ mmHg} \times \frac{133.322 \text{ Pa}}{1 \text{ mmHg}} \times \frac{1 \text{ torr}}{133.322 \text{ Pa}} = \textbf{375 torr}$$

Note that because both 1 mmHg and 1 torr are equal to 133.322 Pa, this could be simplified by recognizing that 1 torr = 1 mmHg.

$$375 \text{ mmHg} \times \frac{1 \text{ torr}}{1 \text{ mmHg}} = \textbf{375 torr}$$

Converting to Pa:

$$?Pa = 375 \text{ mmHg} \times \frac{1 \text{ atm}}{760 \text{ mmHg}} \times \frac{101{,}325 \text{ Pa}}{1 \text{ atm}} = \textbf{5.00} \times \textbf{10}^4 \textbf{ Pa}$$

11.25   **Strategy:**   This problem is similar to Worked Example 11.2. We can use the equation derived in the sample problem to solve for the height of the column of methanol. The equation is

$$\text{pressure} = \text{height} \times \text{density} \times \text{gravitational constant}$$

The gravitational constant is 9.80665 m/s^2.

Solution: Solving the equation for height of the column gives

$$\text{height} = \frac{\text{pressure}}{\text{density} \times \text{gravitational constant}}$$

Recall that for units to cancel properly, pressure must be expressed in Pa (1 Pa = 1 kg/m·s^2) and density in kg/m^3. Converting the data given in the problem to the appropriate units,

$$?Pa = 1 \text{ atm} \times \frac{101{,}325 \text{ Pa}}{1 \text{ atm}} = 101{,}325 \text{ Pa}$$

Note that we consider atmospheric pressure to be exactly 1 atm.

$$?kg/m^3 = \frac{0.787 \text{ g}}{1 \text{ cm}^3} \times \frac{1 \text{ kg}}{1000 \text{ g}} \times \left(\frac{100 \text{ cm}}{1 \text{ m}}\right)^3 = 787 \text{ kg/m}^3$$

Remember that when a unit is raised to a power, any conversion factor used must be raised to the same power. Substituting into the equation to solve for height gives:

$$\text{height} = \frac{101{,}325 \text{ Pa}}{\left(787 \text{ kg/m}^3\right)\left(9.80665 \text{ m/s}^2\right)} = \frac{101{,}325 \text{ kg/m} \cdot \text{s}^2}{\left(787 \text{ kg/m}^3\right)\left(9.80665 \text{ m/s}^2\right)} = \textbf{13.1 m}$$

11.27 **Strategy:** This problem is very similar to Worked Example 11.2. We can use the equation derived in the sample problem to solve for the pressure exerted by the column of toluene. The equation is:

$$\text{pressure} = \text{height} \times \text{density} \times \text{gravitational constant}$$

The gravitational constant is 9.80665 m/s^2.

Solution: Recall that for units to cancel properly, density must be expressed in kg/m^3.

$$?\text{kg/m}^3 = \frac{0.867 \text{ g}}{1 \text{ cm}^3} \times \frac{1 \text{ kg}}{1000 \text{ g}} \times \left(\frac{100 \text{ cm}}{1 \text{ m}}\right)^3 = 867 \text{ kg/m}^3$$

Remember that when a unit is raised to a power, any conversion factor used must be raised to the same power. Substituting into the equation to solve for pressure gives:

Converting to atm:

$$?\text{atm} = = 87 \text{ m} \times \frac{867 \text{ kg}}{\text{m}^3} \times \frac{9.80665 \text{ m}}{\text{s}^2} = 7.4 \times 10^5 \text{ kg/m} \cdot \text{s}^2$$

$$7.4 \times 10^5 \text{ kg/m} \cdot \text{s}^2 = 7.4 \times 10^5 \text{ Pa} \times \frac{1 \text{ atm}}{101{,}325 \text{ Pa}} = \textbf{7.3 atm}$$

Converting to bar:

$$?\text{bar} = 7.3 \text{ atm} \times \frac{101{,}325 \text{ Pa}}{1 \text{ atm}} \times \frac{1 \text{ bar}}{1 \times 10^5 \text{ Pa}} = \textbf{7.4 bar}$$

11.31 **Strategy:** This is a Boyle's law problem. Both temperature and the amount of gas are constant. Therefore, we can use Equation 11.5 to solve for the final volume.

$$P_1 V_1 = P_2 V_2$$

Solution:

Initial Conditions	Final Conditions
$P_1 = 0.970$ atm	$P_2 = 0.541$ atm
$V_1 = 25.6$ mL	$V_2 = ?$

$$V_2 = \frac{P_1 V_1}{P_2} = \frac{(0.970 \text{ atm})(25.6 \text{ mL})}{0.541 \text{ atm}} = \textbf{45.9 mL}$$

11.33 **Strategy:** The amount of gas and the temperature remain constant in this problem. We can use Equation 11.5 (Boyle's law) to solve for the unknown pressure.

Solution:

Initial Conditions	Final Conditions
$P_1 = 1.00$ atm $= 760$ mmHg	$P_2 = ?$
$V_1 = 7.15$ L	$V_2 = 9.25$ L

$$P_1 V_1 = P_2 V_2$$

$$P_2 = \frac{P_1 V_1}{V_2} = \frac{(760 \text{ mmHg})(7.15 \text{ L})}{9.25 \text{ L}} = \textbf{587 mmHg}$$

11.35 **Strategy:** Pressure is held constant in this problem. Only volume and temperature change. This is a Charles's law problem. We use Equation 11.7 to solve for the unknown volume.

Solution:

Initial Conditions

$T_1 = 35°C + 273 = 308$ K

$V_1 = 28.4$ L

Final Conditions

$T_2 = 72° + 273 = 345$ K

$V_2 = ?$

$$V_2 = \frac{V_1 T_2}{T_1} = \frac{(28.4\,\text{L})(345\,\text{K})}{308\,\text{K}} = \textbf{31.8 L}$$

11.37 The balanced equation is:
$$4NH_3(g) + 5O_2(g) \longrightarrow 4NO(g) + 6H_2O(g)$$

Recall that Avogadro's law states that the volume of a gas is directly proportional to the number of moles of gas at constant temperature and pressure. The ammonia and nitric oxide coefficients in the balanced equation are the same, so **one volume** of nitric oxide must be obtained from one volume of ammonia.

11.39 a. If the final temperature of the sample is above the boiling point, it would still be in the gas phase. The choice that best represents this is **diagram (d)**.

b. If the final temperature of the sample is below its boiling point, it will condense to a liquid. The liquid will have a vapor pressure, so some of the sample will remain in the gas phase. The choice that best represents this is **diagram (b)**.

11.43 **Strategy:** This problem gives the amount, volume, and temperature of CO gas. Is the gas undergoing a change in any of its properties?

Solution: Because no changes in gas properties occur, we can use the ideal gas equation to calculate the pressure. Rearranging Equation 11.11 of the text, we write:

$$P = \frac{nRT}{V}$$

$$P = \frac{(6.9\,\text{mol})\left(0.08206\,\dfrac{\text{L}\cdot\text{atm}}{\text{K}\cdot\text{mol}}\right)(355\,\text{K})}{30.4\,\text{L}} = \textbf{6.6 atm}$$

11.45 In this problem, the moles of gas and the volume that the gas occupies are constant. Temperature and pressure change. We use Equation 11.10(b) and divide out the equal volumes V_1 and V_2:

$$\frac{P_1}{T_1} = \frac{P_2}{T_2}$$

Initial Conditions

$P_1 = 1.00$ atm

$T_1 = 273$ K

Final Conditions

$P_2 = ?$

$T_2 = 210°C + 273 = 483$ K

Solving for the final pressure gives:

$$P_2 = \frac{P_1 T_2}{T_1} = \frac{(1.00\,\text{atm})(483\,\text{K})}{273\,\text{K}} = \textbf{1.8 atm}$$

11.47 **Strategy:** In this problem, the moles of gas and the pressure on the gas are constant. Both temperature and volume change. This is a Charles's law problem, and we use Equation 11.7 to solve it. Note that the way the problem is stated, it is V_1 that is unknown.

$$\frac{V_1}{T_1} = \frac{V_2}{T_2}$$

Initial Conditions
$V_1 = ?$
$T_1 = 36.5 + 273.15 = 309.7 \text{ K}$

Final Conditions
$V_2 = 0.67 \text{ L}$
$T_2 = 22.5°\text{C} + 273 = 295.7 \text{ K}$

Solution: Solving Equation 11.7 for the original volume of the gas gives:

$$V_1 = \frac{V_2 T_1}{T_2} = \frac{(0.67 \text{ L})(309.7 \text{ K})}{295.7 \text{ K}} = \textbf{0.70 L}$$

11.49 In the problem, temperature and pressure are given. If we can determine the moles of CO_2, we can calculate the volume it occupies using the ideal gas equation.

$$? \text{ mol } CO_2 = 124.3 \text{ g } CO_2 \times \frac{1 \text{ mol } CO_2}{44.01 \text{ g } CO_2} = 2.8244 \text{ mol } CO_2$$

We now substitute into the ideal gas equation to calculate volume of CO_2.

$$V_{CO_2} = \frac{nRT}{P} = \frac{(2.8244 \text{ mol})\left(0.08206 \dfrac{\text{L} \cdot \text{atm}}{\text{K} \cdot \text{mol}}\right)(273.15 \text{ K})}{1 \text{ atm}} = \textbf{63.31 L}$$

Note that because there are four significant figures in the mass, we use more significant figures than we usually do for R and for the temperature. We also carried an extra digit in the calculated number of moles to avoid rounding error in the final result.

Alternatively, we could use the fact that 1 mole of an ideal gas occupies a volume of 22.41 L at STP. After calculating the moles of CO_2, we can use this fact as a conversion factor to convert to volume of CO_2.

$$? \text{ L } CO_2 = 2.8244 \text{ mol} \times \frac{22.41 \text{ L}}{1 \text{ mol}} = \textbf{63.29 L}$$

The slight difference in the results of our two calculations is due to rounding the volume occupied by 1 mole of an ideal gas to 22.41 L.

11.51 The molar mass of $CO_2 = 44.01$ g/mol. Since $PV = nRT$, we write:

$$P = \frac{nRT}{V}$$

$$P = \frac{\left(0.050 \text{ g} \times \dfrac{1 \text{ mol}}{44.01 \text{ g}}\right)\left(0.0821 \dfrac{\text{L} \cdot \text{atm}}{\text{K} \cdot \text{mol}}\right)(30 + 273)\text{K}}{4.6 \text{ L}} = \textbf{6.1} \times \textbf{10}^{-3} \textbf{ atm}$$

11.53   **Strategy:**   We can calculate the molar mass of a gas if we know its density, temperature, and pressure.

Solution: We need to use Equation 11.13 of the text to calculate the molar mass of the gas.

$$\mathcal{M} = \frac{dRT}{P}$$

Before substituting into the above equation, we need to calculate the density and check that the other known quantities (P and T) have the appropriate units.

$$d = \frac{7.10\,g}{5.40\,L} = 1.314\,g/L$$

$$T = 44°C + 273 = 317\,K$$

$$P = 741\,torr \times \frac{1\,atm}{760\,torr} = 0.975\,atm$$

Calculate the molar mass by substituting in the known quantities.

$$\mathcal{M} = \frac{\left(1.314\,\frac{g}{L}\right)\left(0.0821\,\frac{L \cdot atm}{K \cdot mol}\right)(317\,K)}{0.975\,atm} = \textbf{35.1\,g/mol}$$

Alternatively, we can solve for the molar mass by writing:

$$\text{molar mass of compound} = \frac{\text{mass of compound}}{\text{moles of compound}}$$

Mass of compound is given in the problem (7.10 g), so we need to solve for moles of compound to calculate the molar mass.

$$n = \frac{PV}{RT}$$

$$n = \frac{(0.975\,atm)(5.40\,L)}{\left(0.08206\,\frac{L \cdot atm}{K \cdot mol}\right)(317\,K)} = 0.202\,mol$$

Now, we can calculate the molar mass of the gas.

$$\text{molar mass of compound} = \frac{\text{mass of compound}}{\text{moles of compound}} = \frac{7.10\,g}{0.202\,mol} = \textbf{35.1\,g/mol}$$

11.55   The number of particles in 1 L of gas at STP is:

$$\text{Number of particles} = 1\,L \times \frac{1\,mol}{22.41\,L} \times \frac{6.022 \times 10^{23}\,\text{particles}}{1\,mol} = 2.69 \times 10^{22}\,\text{particles}$$

Number of N_2 molecules = $0.78 \times 2.69 \times 10^{22}$ particles = **2.1×10^{22} N_2 molecules**

Number of O_2 molecules = $0.21 \times 2.69 \times 10^{22}$ particles = **5.6×10^{22} O_2 molecules**

Number of Ar atoms = $0.01 \times 2.69 \times 10^{22}$ particles = **2.7×10^{20} Ar atoms**

11.57   The density can be calculated from the ideal gas equation.

$$d = \frac{P\mathcal{M}}{RT}$$

$$\mathcal{M} = 1.008 \text{ g/mol} + 79.90 \text{ g/mol} = 80.91 \text{ g/mol}$$

$$T = 46°C + 273 = 319 \text{ K}$$

$$P = 733 \text{ mmHg} \times \frac{1 \text{ atm}}{760 \text{ mmHg}} = 0.964 \text{ atm}$$

$$d = \frac{(0.964 \text{ atm})\left(\dfrac{80.91 \text{ g}}{1 \text{ mol}}\right)}{319 \text{ K}} \times \frac{\text{K} \cdot \text{mol}}{0.0821 \text{ L} \cdot \text{atm}} = \textbf{2.98 g / L}$$

Alternatively, we can solve for the density by writing:

$$\text{density} = \frac{\text{mass}}{\text{volume}}$$

Assuming that we have 1 mole of HBr, the mass is 80.91 g. The volume of the gas can be calculated using the ideal gas equation.

$$V = \frac{nRT}{P}$$

$$V = \frac{(1 \text{ mol})\left(0.08206 \dfrac{\text{L} \cdot \text{atm}}{\text{K} \cdot \text{mol}}\right)(319 \text{ K})}{0.964 \text{ atm}} = 27.2 \text{ L}$$

Now, we can calculate the density of HBr gas.

$$\text{density} = \frac{\text{mass}}{\text{volume}} = \frac{80.91 \text{ g}}{27.2 \text{ L}} = \textbf{2.97 g / L}$$

11.59   This is an extension of an ideal gas law calculation involving molar mass. If you determine the molar mass of the gas, you will be able to determine the molecular formula from the empirical formula.

$$\mathcal{M} = \frac{dRT}{P}$$

Calculate the density, then substitute its value into the above equation.

$$d = \frac{0.100\,\text{g}}{22.1\,\text{mL}} \times \frac{1000\,\text{mL}}{1\,\text{L}} = 4.52\,\text{g/L}$$

$$T(\text{K}) = 20°\text{C} + 273 = 293\,\text{K}$$

$$\mathcal{M} = \frac{\left(4.52\,\dfrac{\text{g}}{\text{L}}\right)\left(0.08206\,\dfrac{\text{L} \cdot \text{atm}}{\text{K} \cdot \text{mol}}\right)(293\,\text{K})}{1.02\,\text{atm}} = 107\,\text{g/mol}$$

Compare the empirical mass to the molar mass.

$$\text{empirical mass} = 32.07\,\text{g/mol} + 4(19.00\,\text{g/mol}) = 108.07\,\text{g/mol}$$

Remember, the molar mass will be a whole number multiple of the empirical mass. In this case, the $\dfrac{\text{molar mass}}{\text{empirical mass}} \approx 1$. Therefore, the molecular formula is the same as the empirical formula, **SF$_4$**.

11.61 **Strategy:** In this problem, the moles of gas are constant. Use the combined gas law (Equation 11.10b).

$$\frac{P_1 V_1}{T_1} = \frac{P_2 V_2}{T_2}$$

Because V is constant, the above equation reduces to

$$\frac{P_1}{T_1} = \frac{P_2}{T_2}$$

Solution: The final temperature is given by:

$$T_2 = \frac{T_1 P_2}{P_1}$$

$$T_2 = \frac{(298\,\text{K})(5.00\,\text{atm})}{0.800\,\text{L}} = 1.86 \times 10^3\,\text{K} = \mathbf{1590°C}$$

11.67 In this problem, we are comparing the pressure as determined by the van der Waals equation with that determined by the ideal gas equation.

van der Waals equation:

We find the pressure by first solving algebraically for P.

$$P = \frac{nRT}{V - nb} - \frac{an^2}{V^2}$$

where $n = 2.50$ mol, $V = 5.00$ L, $T = 450$ K, $a = 3.59$ atm·L^2/mol^2, and $b = 0.0427$ L/mol.

$$P = \frac{(2.50\,\text{mol})\left(0.0821\,\dfrac{\text{L} \cdot \text{atm}}{\text{K} \cdot \text{mol}}\right)(450\,\text{K})}{5.00\,\text{L} - (2.50\,\text{mol} \times 0.0427\,\text{L/mol})} - \frac{\left(3.59\,\dfrac{\text{atm} \cdot \text{L}^2}{\text{mol}^2}\right)(2.50\,\text{mol})^2}{(5.00\,\text{L})^2} = \mathbf{18.0\,atm}$$

Ideal gas equation:

$$P = \frac{nRT}{V} = \frac{(2.50\,\text{mol})\left(0.0821\dfrac{\text{L}\cdot\text{atm}}{\text{K}\cdot\text{mol}}\right)(450\,\text{K})}{(5.00\,\text{L})} = \textbf{18.5\,atm}$$

Since the pressure calculated using van der Waals equation is comparable to the pressure calculated using the ideal gas equation, we conclude that CO_2 behaves fairly ideally under these conditions.

11.71 Dalton's law states that the total pressure of the mixture is the sum of the partial pressures.

a. $$P_{\text{total}} = 0.32\ \text{atm} + 0.15\ \text{atm} + 0.42\ \text{atm} = \textbf{0.89 atm}$$

b. We know:

Initial Conditions	Final Conditions
$P_1 = 0.15$ atm $+ 0.42$ atm $= 0.57$ atm	$P_2 = 1.0$ atm
$T_1 = 15°C + 273 = 288$ K	$T_2 = 273$ K
$V_1 = 2.5$ L	$V_2 = ?$

$$\frac{P_1 V_1}{n_1 T_1} = \frac{P_2 V_2}{n_2 T_2}$$

Because $n_1 = n_2$, we can write:

$$V_2 = \frac{P_1 V_1 T_2}{P_2 T_1}$$

$$V_2 = \frac{(0.57\,\text{atm})(2.5\,\text{L})(273\,\text{K})}{(1.0\,\text{atm})(288\,\text{K})} = \textbf{1.4 L at STP}$$

11.73 $$P_{\text{total}} = P_1 + P_2 + P_3 + \ldots + P_n$$

In this case,

$$P_{\text{total}} = P_{\text{Ne}} + P_{\text{He}} + P_{\text{H}_2\text{O}}$$

$$P_{\text{Ne}} = P_{\text{total}} - P_{\text{He}} - P_{\text{H}_2\text{O}}$$

$$P_{\text{Ne}} = 745\ \text{mm Hg} - 368\ \text{mmHg} - 28.3\ \text{mmHg} = \textbf{349 mmHg}$$

11.75 $$P_i = \chi_i P_{\text{total}}$$

We need to calculate the mole fractions of each component to determine their partial pressures. To calculate mole fraction, write the balanced chemical equation to determine the correct mole ratio.

$$2NH_3(g) \longrightarrow N_2(g) + 3H_2(g)$$

The mole fractions of N_2 and H_2 are:

$$\chi_{N_2} = \frac{1\,mol}{3\,mol + 1\,mol} = 0.250$$

$$\chi_{H_2} = \frac{3\,mol}{3\,mol + 1\,mol} = 0.750$$

The partial pressures of N_2 and H_2 are:

$$\boldsymbol{P_{N_2}} = \chi_{N_2} P_T = (0.250)(866\,mmHg) = \boldsymbol{217\,mmHg}$$

$$\boldsymbol{P_{H_2}} = \chi_{H_2} P_T = (0.750)(866\,mmHg) = \boldsymbol{650\,mmHg}$$

11.77 **Strategy:** Because the gas that the patient breathes is inside the hyperbaric chamber, its total pressure is the same as the chamber pressure. Use Equation 11.18 to determine the chamber pressure.

Setup: Rearranging Equation 11.18 to solve for chamber pressure (total pressure) gives:

$$P_{total} = \frac{P_i}{\chi_i}$$

The partial pressure of O_2 is 2.8 atm. The breathing gas is 21 percent O_2, so the mole fraction of O_2 is 0.21.

Solution:

$$\boldsymbol{P_{total}} = \frac{2.8\,atm}{0.21} = \boldsymbol{13\,atm}$$

11.79 a. **box on the right**   b. **box on the right**

11.81 **Strategy:** From the moles of CH_4 that reacts, we can calculate the moles of CO_2 produced. From the balanced equation, we see that 1 mol CH_4 is stoichiometrically equivalent to 1 mol CO_2. Once moles of CO_2 are determined, we can use the ideal gas equation to calculate the volume of CO_2.

Solution: First let's calculate moles of CO_2 produced.

$$?\,mol\,CO_2 = 15.0\,mol\,CH_4 \times \frac{1\,mol\,CO_2}{1\,mol\,CH_4} = 15.0\,mol\,CO_2$$

Now, we can substitute moles, temperature, and pressure into the ideal gas equation to solve for volume of CO_2.

$$V = \frac{nRT}{P}$$

$$\boldsymbol{V_{CO_2}} = \frac{(15.0\,mol)\left(0.0821\dfrac{L \cdot atm}{K \cdot mol}\right)(23 + 273)\,K}{0.985\,atm} = \boldsymbol{3.70 \times 10^2\,L}$$

11.83   From the amount of glucose that reacts (5.97 g), we can calculate the theoretical yield of CO_2. We can then compare the theoretical yield to the actual yield given in the problem (1.44 L) to determine the percent yield.

First, let's determine the moles of CO_2 that can be produced theoretically. Then, we can use the ideal gas equation to determine the volume of CO_2.

$$? \, mol \, CO_2 = 5.97 \, g \, glucose \times \frac{1 \, mol \, glucose}{180.2 \, g \, glucose} \times \frac{2 \, mol \, CO_2}{1 \, mol \, glucose} = 0.0663 \, mol \, CO_2$$

Now, substitute moles, pressure, and temperature into the ideal gas equation to calculate the volume of CO_2.

$$V = \frac{nRT}{P}$$

$$V_{CO_2} = \frac{(0.0663 \, mol)\left(0.0821 \dfrac{L \cdot atm}{K \cdot mol}\right)(293 \, K)}{0.984 \, atm} = 1.62 \, L$$

This is the theoretical yield of CO_2. The actual yield, which is given in the problem, is 1.44 L. We can now calculate the percent yield.

$$percent \, yield = \frac{actual \, yield}{theoretical \, yield} \times 100\%$$

$$\textbf{percent yield} = \frac{1.44 \, L}{1.62 \, L} \times 100\% = \textbf{88.9\%}$$

11.85   **Strategy:**   We can calculate the moles of M consumed, and the moles of H_2 gas produced. By comparing the number of moles of M consumed to the number of moles H_2 produced, we can determine the mole ratio in the balanced equation.

Solution: First let's calculate the moles of the metal (M) consumed.

$$mol \, M = 0.225 \, g \, M \times \frac{1 \, mol \, M}{27.0 \, g \, M} = 8.33 \times 10^{-3} \, mol \, M$$

Solve the ideal gas equation algebraically for n_{H_2}. Then, calculate the moles of H_2 by substituting the known quantities into the equation.

$$P = 741 \, mmHg \times \frac{1 \, atm}{760 \, mmHg} = 0.975 \, atm$$

$$T = 17°C + 273 = 290 \, K$$

$$n_{H_2} = \frac{PV_{H_2}}{RT}$$

$$n_{H_2} = \frac{(0.975\,\text{atm})(0.303\,\text{L})}{\left(0.0821\dfrac{\text{L}\cdot\text{atm}}{\text{K}\cdot\text{mol}}\right)(290\,\text{K})} = 1.24\times10^{-2}\,\text{mol}\,H_2$$

Compare the number of moles of H_2 produced to the number of moles of M consumed.

$$\frac{1.24\times10^{-2}\,\text{mol}\,H_2}{8.33\times10^{-3}\,\text{mol}\,M} \approx 1.5$$

This means that the mole ratio of H_2 to M is 1.5:1.

We can now write the balanced equation since we know the mole ratio between H_2 and M. The unbalanced equation is:

$$M(s) + HCl(aq) \longrightarrow 1.5H_2(g) + M_xCl_y(aq)$$

We have three atoms of H on the products side of the reaction, so a 3 must be placed in front of HCl. The ratio of M to Cl on the reactants side is now 1:3. Therefore the formula of the metal chloride must be MCl_3.

The balanced equation is:

$$\mathbf{M(s) + 3HCl(aq) \longrightarrow 1.5H_2(g) + MCl_3(aq)}$$

From the formula of the metal chloride, we determine that the charge of the metal is +3. Therefore, the formulas of the metal oxide and the metal sulfate are $\mathbf{M_2O_3}$ and $\mathbf{M_2(SO_4)_3}$, respectively.

11.87 From the moles of CO_2 produced, we can calculate the amount of calcium carbonate that must have reacted. We can then determine the percent by mass of $CaCO_3$ in the 3.00 g sample.

The balanced equation is:

$$CaCO_3(s) + 2HCl(aq) \longrightarrow CO_2(g) + CaCl_2(aq) + H_2O(l)$$

The moles of CO_2 produced can be calculated using the ideal gas equation.

$$n_{CO_2} = \frac{PV_{CO_2}}{RT}$$

$$n_{CO_2} = \frac{\left(792\,\text{mmHg}\times\dfrac{1\,\text{atm}}{760\,\text{mmHg}}\right)(0.656\,\text{L})}{\left(0.0821\dfrac{\text{L}\cdot\text{atm}}{\text{K}\cdot\text{mol}}\right)(20+273\,\text{K})} = 2.84\times10^{-2}\,\text{mol}\,CO_2$$

The balanced equation shows a 1:1 mole ratio between CO_2 and $CaCO_3$. Therefore, 2.84×10^{-2} mole of $CaCO_3$ must have reacted.

$$?\,\text{g}\,CaCO_3\,\text{reacted} = (2.84\times10^{-2}\,\text{mol}\,CaCO_3)\times\frac{100.1\,\text{g}\,CaCO_3}{1\,\text{mol}\,CaCO_3} = 2.84\,\text{g}\,CaCO_3$$

The percent by mass of the $CaCO_3$ sample is:

$$\% \, CaCO_3 = \frac{2.84\,g}{3.00\,g} \times 100\% = \mathbf{94.7\%}$$

Assumption: **The impurity (or impurities) must not react with HCl to produce CO_2 gas**.

11.89   The balanced equation is:

$$C_2H_5OH(l) + 3O_2(g) \longrightarrow 2CO_2(g) + 3H_2O(l)$$

The moles of O_2 needed to react with 185 g ethanol are:

$$185\,g\,C_2H_5OH \times \frac{1\,mol\,C_2H_5OH}{46.07\,g\,C_2H_5OH} \times \frac{3\,mol\,O_2}{1\,mol\,C_2H_5OH} = 12.05\,mol\,O_2$$

12.05 moles of O_2 correspond to a volume of:

$$V_{O_2} = \frac{n_{O_2}RT}{P} = \frac{(12.05\,mol\,O_2)\left(0.0821\dfrac{L \cdot atm}{K \cdot mol}\right)(318\,K)}{\left(793\,mmHg \times \dfrac{1\,atm}{760\,mmHg}\right)} = 302\,L\,O_2$$

Since air is 21.0 percent O_2 by volume, we can write:

$$V_{air} = V_{O_2}\left(\frac{100\%\,air}{21\%\,O_2}\right) = (302\,L\,O_2)\left(\frac{100\%\,air}{21\%\,O_2}\right) = \mathbf{1.44 \times 10^3\,L\,air}$$

11.91   **Strategy:** To solve for moles of H_2 generated, we must first calculate the partial pressure of H_2 in the mixture. What gas law do we need? How do we convert from moles of H_2 to amount of Zn reacted?

Solution: Dalton's law of partial pressure states that:

$$P_{total} = P_1 + P_2 + P_3 + \ldots + P_n$$

In this case,

$$P_{total} = P_{H_2} + P_{H_2O}$$

$$P_{H_2} = P_{total} - P_{H_2O}$$

$$P_{H_2} = 0.980\,atm - (23.8\,mmHg)\left(\frac{1\,atm}{760\,mmHg}\right) = 0.949\,atm$$

Now that we know the pressure of H_2 gas, we can calculate the moles of H_2. Then, using the mole ratio from the balanced equation, we can calculate moles of Zn.

$$n_{H_2} = \frac{P_{H_2}V}{RT}$$

$$n_{H_2} = \frac{(0.949\,\text{atm})(7.80\,\text{L})}{(25+273)\,\text{K}} \times \frac{\text{K} \cdot \text{mol}}{0.0821\,\text{L} \cdot \text{atm}} = 0.303\,\text{mol}\,H_2$$

Using the mole ratio from the balanced equation and the molar mass of zinc, we can now calculate the grams of zinc consumed in the reaction.

$$\textbf{?}\,\textbf{g Zn} = 0.303\,\text{mol}\,H_2 \times \frac{1\,\text{mol Zn}}{1\,\text{mol}\,H_2} \times \frac{65.39\,\text{g Zn}}{1\,\text{mol Zn}} = \textbf{19.8 g Zn}$$

11.93   You can map out the following strategy to solve for the total volume of gas.

grams nitroglycerin → moles nitroglycerin → moles products → volume of products

$$\text{? mol products} = 2.6 \times 10^2\,\text{g nitroglycerin} \times \frac{1\,\text{mol nitroglycerin}}{227.09\,\text{g nitroglycerin}} \times \frac{29\,\text{mol product}}{4\,\text{mol nitroglycerin}}$$

$$= 8.3\,\text{mol}$$

Calculating the volume of products:

$$V_{\text{product}} = \frac{n_{\text{product}}RT}{P} = \frac{(8.3\,\text{mol})\left(0.08206\,\dfrac{\text{L} \cdot \text{atm}}{\text{K} \cdot \text{mol}}\right)(298\,\text{K})}{(1.2\,\text{atm})} = \textbf{1.7} \times \textbf{10}^2\,\textbf{L}$$

The relationship between partial pressure and P_{total} is:

$$P_i = \chi_i P_{\text{total}}$$

Calculate the mole fraction of each gaseous product, then calculate its partial pressure using the equation above.

$$\chi_{\text{component}} = \frac{\text{moles component}}{\text{total moles of all components}}$$

$$\chi_{CO_2} = \frac{12\,\text{mol}\,CO_2}{29\,\text{mol product}} = 0.41$$

Similarly, $\chi_{H_2O} = 0.34$, $\chi_{N_2} = 0.21$, and $\chi_{O_2} = 0.034$

$$P_{CO_2} = \chi_{CO_2} P_T$$

$$P_{CO_2} = (0.41)(1.2\,\text{atm}) = \textbf{0.49 atm}$$

Similarly, $\boldsymbol{P_{H_2O} = 0.41\,\textbf{atm}}$, $\boldsymbol{P_{N_2} = 0.25\,\textbf{atm}}$, and $\boldsymbol{P_{O_2} = 0.041\,\textbf{atm}}$.

11.95 a.

$$NH_4NO_2(s) \longrightarrow N_2(g) + 2H_2O(l)$$

b. Map out the following strategy to solve the problem.

$$\text{volume } N_2 \rightarrow \text{moles } N_2 \rightarrow \text{moles } NH_4NO_2 \rightarrow \text{grams } NH_4NO_2$$

First, calculate the moles of N_2 using the ideal gas equation.

$$T(K) = 22°C + 273 = 295 \text{ K}$$

$$V = 86.2 \text{ mL} \times \frac{1\text{L}}{1000\,\text{mL}} = 0.0862 \text{ L}$$

$$n_{N_2} = \frac{P_{N_2}V}{RT}$$

$$n_{N_2} = \frac{(1.20\,\text{atm})(0.0862\,\text{L})}{\left(0.0821\dfrac{\text{L}\cdot\text{atm}}{\text{K}\cdot\text{mol}}\right)(295\,\text{K})} = 4.27\times10^{-3} \text{ mol}$$

Next, calculate the mass of NH_4NO_2 needed to produce 4.27×10^{-3} mole of N_2.

$$\textbf{? g NH}_4\textbf{NO}_2 = (4.27\times10^{-3}\,\text{mol N}_2)\times\frac{1\,\text{mol NH}_4\text{NO}_2}{1\,\text{mol N}_2}\times\frac{64.05\,\text{g NH}_4\text{NO}_2}{1\,\text{mol NH}_4\text{NO}_2} = \textbf{0.273 g}$$

11.97 a. The total pressure in (i) is 2.0 atm. The gas is represented by a total of nine spheres in a volume of 2.0 L. In (ii), the volume is only 1.0 L, but the gas is represented by a total of nine spheres. The same amount of gas in half the volume will exert twice the pressure. Therefore, $P_{ii} = \textbf{4.0 atm}$. In (iii), the gas is represented by 12 spheres in a 2.0 L volume. It contains 1/3 more spheres than (i) in the same volume, so its pressure will be 1/3 greater. $P_{iii} = \textbf{2.7 atm}$.

b. When the valves are opened, the gases will distribute themselves among the flasks, and the pressure will be the same throughout. The total number of spheres is $9 + 9 + 12 = 30$ spheres. The spheres will be distributed among the three flasks in a total volume of $2.0 \text{ L} + 1.0 \text{ L} + 2.0 \text{ L} = 5.0 \text{ L}$. Therefore, there will be $30 \div 5 = 6$ spheres per liter. Six spheres per liter corresponds to a pressure of **2.7 atm**. (Flask iii originally contains six spheres per liter.) Because there are equal numbers of red spheres and blue spheres (15 each), their partial pressures will be equal. $P_A = \textbf{1.3 atm}$, $P_B = \textbf{1.3 atm}$.

11.99 a. The number of moles of $Ni(CO)_4$ formed is:

$$86.4\,\text{g Ni} \times \frac{1\,\text{mol Ni}}{58.69\,\text{g Ni}} \times \frac{1\,\text{mol Ni(CO)}_4}{1\,\text{mol Ni}} = 1.47\,\text{mol Ni(CO)}_4$$

The pressure of $Ni(CO)_4$ is:

$$P = \frac{nRT}{V} = \frac{(1.472\,\text{mol})\left(0.08206\dfrac{\text{L}\cdot\text{atm}}{\text{K}\cdot\text{mol}}\right)(316\,\text{K})}{4.00\,\text{L}} = \textbf{9.54 atm}$$

b. **If the tetracarbonylnickel gas starts to decompose significantly above 43°C, then for every mole of tetracarbonylnickel gas decomposed, four moles of carbon monoxide would be produced.**

$$Ni(CO)_4(g) \longrightarrow Ni(s) + 4CO(g)$$

11.101 Using the ideal gas equation, we can calculate the moles of gas.

$$n = \frac{PV}{RT} = \frac{(1.1 \text{ atm})\left(5.0 \times 10^2 \text{ mL} \times \dfrac{0.001 \text{ L}}{1 \text{ mL}}\right)}{\left(0.0821 \dfrac{\text{L} \cdot \text{atm}}{\text{K} \cdot \text{mol}}\right)(37 + 273)\text{K}} = 0.022 \text{ mol gas}$$

Next, use Avogadro's number to convert to molecules of gas.

$$0.022 \text{ mol gas} \times \frac{6.022 \times 10^{23} \text{ molecules}}{1 \text{ mol gas}} = \textbf{1.3} \times \textbf{10}^{\textbf{22}} \textbf{ molecules of gas}$$

The most common gases present in exhaled air are **CO_2, O_2, N_2, and H_2O.**

11.103
a. $$\textbf{2KClO}_3\textbf{(s)} \longrightarrow \textbf{2KCl(s)} + \textbf{3O}_2\textbf{(g)}$$

b. First, calculate the moles of $KClO_3$ in 20.4 g. The molar mass of $KClO_3$ is 122.55 g/mol.

$$20.4 \text{ g KClO}_3 \times \frac{1 \text{ mol KClO}_3}{122.55 \text{ g KClO}_3} = 0.1665 \text{ mol KClO}_3$$

According to the balanced equation, three moles of O_2 form for every two moles of $KClO_3$ that decompose.

Therefore,

$$0.1665 \text{ mol KClO}_3 \times \frac{3 \text{ mol O}_2}{2 \text{ mol KClO}_3} = 0.2498 \text{ mol O}_2$$

Finally, we use the ideal gas equation to determine the volume of O_2.

$$V_{O_2} = \frac{nRT}{P} = \frac{(0.2498 \text{ mol})\left(0.08206 \dfrac{\text{L} \cdot \text{atm}}{\text{K} \cdot \text{mol}}\right)(291.5 \text{ K})}{0.962 \text{ atm}} = \textbf{6.21 L}$$

11.105 To calculate the molarity of NaOH, we need moles of NaOH and volume of the NaOH solution. The volume is given in the problem; therefore, we need to calculate the moles of NaOH. The moles of NaOH can be calculated from the reaction of NaOH with HCl. The balanced equation is:

$$NaOH(aq) + HCl(aq) \longrightarrow H_2O(l) + NaCl(aq)$$

The number of moles of HCl gas is found from the ideal gas equation. $V = 0.189$ L, $T = 25°C + 273$ K = 298 K, and $P = 108 \text{ mmHg} \times \dfrac{1 \text{ atm}}{760 \text{ mmHg}} = 0.142 \text{ atm}$.

$$n_{HCl} = \frac{PV_{HCl}}{RT} = \frac{(0.142 \text{ atm})(0.189 \text{ L})}{\left(0.0821 \dfrac{\text{L} \cdot \text{atm}}{\text{K} \cdot \text{mol}}\right)(298 \text{ K})} = 1.10 \times 10^{-3} \text{ mol HCl}$$

The moles of NaOH can be calculated using the mole ratio from the balanced equation.

$$(1.10 \times 10^{-3} \text{ mol HCl}) \times \frac{1 \text{ mol NaOH}}{1 \text{ mol HCl}} = 1.10 \times 10^{-3} \text{ mol NaOH}$$

The molarity of the NaOH solution is:

$$M = \frac{\text{mol NaOH}}{\text{L of soln}} = \frac{1.10 \times 10^{-3} \text{ mol NaOH}}{0.0157 \text{ L soln}} = 0.0701 \text{ mol/L} = \textbf{0.0701 } \boldsymbol{M}$$

11.107 To calculate the partial pressures of He and Ne, the total pressure of the mixture is needed. To calculate the total pressure of the mixture, we need the total number of moles of gas in the mixture (mol He + mol Ne).

$$n_{He} = \frac{PV}{RT} = \frac{(0.63 \text{ atm})(1.2 \text{ L})}{\left(0.08206 \dfrac{\text{L} \cdot \text{atm}}{\text{K} \cdot \text{mol}}\right)(16 + 273)\text{K}} = 0.032 \text{ mol He}$$

$$n_{Ne} = \frac{PV}{RT} = \frac{(2.8 \text{ atm})(3.4 \text{ L})}{\left(0.08206 \dfrac{\text{L} \cdot \text{atm}}{\text{K} \cdot \text{mol}}\right)(16 + 273)\text{K}} = 0.40 \text{ mol Ne}$$

The total pressure is:

$$P_{total} = \frac{(n_{He} + n_{Ne})RT}{V_{total}} = \frac{(0.032 + 0.40)\text{mol}\left(0.08206 \dfrac{\text{L} \cdot \text{atm}}{\text{K} \cdot \text{mol}}\right)(16 + 273)\text{K}}{(1.2 + 3.4)\text{L}} = 2.2 \text{ atm}$$

$P_i = \mathcal{X}_i P_T$. The partial pressures of He and Ne are:

$$P_{He} = \frac{0.032 \text{ mol}}{(0.032 + 0.40)\text{mol}} \times 2.2 \text{ atm} = \textbf{0.16 atm}$$

$$P_{Ne} = \frac{0.40 \text{ mol}}{(0.032 + 0.40)\text{mol}} \times 2.2 \text{ atm} = \textbf{2.0 atm}$$

11.109 **When the water enters the flask from the dropper, some hydrogen chloride dissolves, creating a partial vacuum. Pressure from the atmosphere forces more water up the vertical tube.**

11.111 Use the ideal gas equation to calculate the moles of water produced. We carry an extra significant figure in the first step of the calculation to limit rounding errors.

$$n_{H_2O} = \frac{PV}{RT} = \frac{(24.8 \text{ atm})(2.00 \text{ L})}{\left(0.08206 \dfrac{\text{L} \cdot \text{atm}}{\text{K} \cdot \text{mol}}\right)(120 + 273)\text{K}} = 1.537 \text{ mol H}_2\text{O}$$

Next, we can determine the mass of H_2O in the 54.2 g sample. Subtracting the mass of H_2O from 54.2 g will give the mass of $MgSO_4$ in the sample.

$$1.537 \text{ mol } H_2O \times \frac{18.02 \text{ g } H_2O}{1 \text{ mol } H_2O} = 27.7 \text{ g } H_2O$$

$$\text{Mass } MgSO_4 = 54.2 \text{ g sample} - 27.7 \text{ g } H_2O = 26.5 \text{ g } MgSO_4$$

Finally, we can calculate the moles of $MgSO_4$ in the sample. Comparing moles of $MgSO_4$ to moles of H_2O will allow us to determine the correct mole ratio in the formula.

$$26.5 \text{ g } MgSO_4 \times \frac{1 \text{ mol } MgSO_4}{120.4 \text{ g } MgSO_4} = 0.220 \text{ mol } MgSO_4$$

$$\frac{\text{mol } H_2O}{\text{mol } MgSO_4} = \frac{1.54 \text{ mol}}{0.220 \text{ mol}} = 7.00$$

Therefore, the mole ratio between H_2O and $MgSO_4$ in the compound is 7:1. Thus, the value of $x = 7$, and the formula is $MgSO_4 \cdot 7H_2O$.

11.113 The circumference of the cylinder is $= 2\pi r = 2\pi\left(\dfrac{15.0 \text{ cm}}{2}\right) = 47.1 \text{ cm}$

a. The speed at which the target is moving equals:

$$\text{speed of target} = \text{circumference} \times \text{revolutions/second}$$

$$\text{speed of target} = \frac{47.1 \text{ cm}}{1 \text{ revolution}} \times \frac{130 \text{ revolutions}}{1 \text{ s}} \times \frac{0.01 \text{ m}}{1 \text{ cm}} = \mathbf{61.2 \text{ m/s}}$$

b.
$$2.80 \text{ cm} \times \frac{0.01 \text{ m}}{1 \text{ cm}} \times \frac{1 \text{ s}}{61.2 \text{ m}} = \mathbf{4.58 \times 10^{-4} \text{ s}}$$

c. The Bi atoms must travel across the cylinder to hit the target. This distance is the diameter of the cylinder, which is 15.0 cm. The Bi atoms travel this distance in 4.58×10^{-4} s.

$$\frac{15.0 \text{ cm}}{4.58 \times 10^{-4} \text{ s}} \times \frac{0.01 \text{ m}}{1 \text{ cm}} = \mathbf{328 \text{ m/s}}$$

The rms speed of Bi = 366 m/s (average of all Bi atoms). The experiment is measuring the most probable speed u_{mp} instead of the rms speed u_{rms}. The average speed is always about 81.6% of the rms speed (see Exercise 11.154a).

$$u_{rms} = \sqrt{\frac{3RT}{M}} = \sqrt{\frac{3(8.314 \text{ J/K} \cdot \text{mol})(850 + 273)\text{K}}{209.0 \times 10^{-3} \text{ kg/mol}}} = \mathbf{366 \text{ m/s}}$$

The magnitudes of the speeds are comparable, but not identical. The rms speed is derived for three-dimensional motion and is not applicable to the speed distributions in directed molecular beams. Also, the experiment is measuring a most-probable speed, not an rms speed.

11.115 The moles of O_2 can be calculated from the ideal gas equation. The mass of O_2 can then be calculated using the molar mass as a conversion factor.

$$n_{O_2} = \frac{PV}{RT} = \frac{(132 \text{ atm})(120 \text{ L})}{\left(0.08206 \frac{\text{L} \cdot \text{atm}}{\text{K} \cdot \text{mol}}\right)(22+273)\text{K}} = 654 \text{ mol } O_2$$

$$? \text{ g } O_2 = 654 \text{ mol } O_2 \times \frac{32.00 \text{ g } O_2}{1 \text{ mol } O_2} = \mathbf{2.09 \times 10^4 \text{ g } O_2}$$

The volume of O_2 gas under conditions of 1.00 atm pressure and a temperature of 22°C can be calculated using the ideal gas equation. The moles of $O_2 = 654$ moles.

$$V_{O_2} = \frac{n_{O_2}RT}{P} = \frac{(654 \text{ mol})\left(0.08206\frac{\text{L} \cdot \text{atm}}{\text{K} \cdot \text{mol}}\right)(22+273)\text{K}}{1.00 \text{ atm}} = \mathbf{1.58 \times 10^4 \text{ L } O_2}$$

11.117 **The fruit ripens more rapidly because the quantity (partial pressure) of ethylene gas inside the bag increases.**

11.119 **As the pen is used the amount of ink decreases, increasing the volume inside the pen. As the volume increases, the pressure inside the pen decreases. The hole is needed to equalize the pressure as the volume inside the pen increases.**

11.121 a. $\mathbf{NH_4NO_3(s) \longrightarrow N_2O(g) + 2H_2O(l)}$

b. $$R = \frac{PV}{nT} = \frac{\left(718 \text{ mmHg} \times \dfrac{1 \text{ atm}}{760 \text{ mmHg}}\right)(0.340 \text{ L})}{\left(0.580 \text{ g } N_2O \times \dfrac{1 \text{ mol } N_2O}{44.02 \text{ g } N_2O}\right)(24+273)\text{K}} = \mathbf{0.0821 \frac{\text{L} \cdot \text{atm}}{\text{K} \cdot \text{mol}}}$$

11.123 The value of a indicates how strongly molecules of a given type of gas attract one another. $\mathbf{C_6H_6}$ has the greatest intermolecular attractions due to its larger size compared to the other choices. Therefore, it has the largest value of a.

11.125 **The gases inside the mine were a mixture of carbon dioxide, carbon monoxide, methane, and other harmful compounds. The low atmospheric pressure caused the gases to flow out of the mine (the gases in the mine were at a higher pressure), and the man suffocated.**

11.127 a. This is a Boyle's law problem.

$$P_{\text{tire}}V_{\text{tire}} = P_{\text{air}}V_{\text{air}}$$

$$(5.0 \text{ atm})(0.98 \text{ L}) = (1.0 \text{ atm})V_{\text{air}}$$

$$V_{\text{air}} = \mathbf{4.90 \text{ L}}$$

b. $\qquad$ Pressure in the tire − atmospheric pressure = gauge pressure

Pressure in the tire − 1.0 atm = 5.0 atm

Pressure in the tire = **6.0 atm**

c. Again, this is a Boyle's law problem.

$$P_{pump}V_{pump} = P_{gauge}V_{gauge}$$

$$(1 \text{ atm})(0.33 V_{tire}) = P_{gauge}V_{gauge}$$

$$P_{gauge} = 0.33 \text{ atm}$$

This is the gauge pressure after one pump stroke. After three strokes, the gauge pressure will be $(3 \times 0.33 \text{ atm})$, or approximately **1 atm**. The assumption is that the initial gauge pressure was zero.

11.129 a. First, let's convert the concentration of hydrogen from atoms/cm^3 to mol/L. The concentration in mol/L can be substituted into the ideal gas equation to calculate the pressure of hydrogen.

$$\frac{1 \text{ H atom}}{1 \text{ cm}^3} \times \frac{1 \text{ mol H}}{6.022 \times 10^{23} \text{ H atoms}} \times \frac{1000 \text{ cm}^3}{1 \text{ L}} = \frac{1.7 \times 10^{-21} \text{ mol H}}{L}$$

The pressure of H is:

$$P_H = \left(\frac{n}{V}\right)RT = \left(\frac{1.7 \times 10^{-21} \text{ mol H}}{L}\right)\left(0.08206 \frac{L \cdot atm}{K \cdot mol}\right)(3 \text{ K}) = \mathbf{4 \times 10^{-22}} \text{ atm}$$

b. From part (a), we know that 1 L contains 1.7×10^{-21} mole of H atoms. We convert to the volume that contains 1.0 g of H atoms.

$$\frac{1 \text{ L}}{1.7 \times 10^{-21} \text{ mol H}} \times \frac{1 \text{ mol H}}{1.008 \text{ g H}} = \mathbf{6 \times 10^{20} \text{ L/g of H}}$$

11.131 From Table 11.6, the equilibrium vapor pressure at 30°C is 31.8 mmHg.

Converting 3.9×10^3 Pa to units of mmHg:

$$(3.9 \times 10^3 \text{ Pa}) \times \frac{760 \text{ mmHg}}{1.01325 \times 10^5 \text{ Pa}} = 29 \text{ mmHg}$$

$$\text{Relative Humidity} = \frac{\text{partial pressure of water vapor}}{\text{equilibrium vapor pressure}} \times 100\% = \frac{29 \text{ mmHg}}{31.8 \text{ mmHg}} \times 100\% = \mathbf{91\%}$$

11.133 The volume of one alveolus is:

$$V = \frac{4}{3}\pi r^3 = \frac{4}{3}\pi(0.0050 \text{ cm})^3 = (5.2 \times 10^{-7} \text{ cm}^3) \times \frac{1 \text{ mL}}{1 \text{ cm}^3} \times \frac{0.001 \text{ L}}{1 \text{ mL}} = 5.2 \times 10^{-10} \text{ L}$$

The number of moles of air in one alveolus can be calculated using the ideal gas equation.

$$n = \frac{PV}{RT} = \frac{(1.0 \text{ atm})(5.2 \times 10^{-10} \text{ L})}{\left(0.0821 \frac{L \cdot atm}{K \cdot mol}\right)(37 + 273)K} = 2.0 \times 10^{-11} \text{ mol of air}$$

Since the air inside the alveolus is 14 percent oxygen, the moles of oxygen in one alveolus equals:

$$(2.0 \times 10^{-11} \text{ mol of air}) \times \frac{14\% \text{ oxygen}}{100\% \text{ air}} = 2.8 \times 10^{-12} \text{ mol O}_2$$

Converting to O_2 molecules:

$$(2.8 \times 10^{-12} \text{ mol O}_2) \times \frac{6.022 \times 10^{23} \text{ O}_2 \text{ molecules}}{1 \text{ mol O}_2} = \mathbf{1.7 \times 10^{12} \text{ O}_2 \text{ molecules}}$$

11.135   The molar mass of a gas can be calculated using Equation 11.13 of the text.

$$\mathcal{M} = \frac{dRT}{P} = \frac{\left(1.33 \frac{\text{g}}{\text{L}}\right)\left(0.0821 \frac{\text{L} \cdot \text{atm}}{\text{K} \cdot \text{mol}}\right)(150 + 273)\text{K}}{\left(764 \text{ mmHg} \times \frac{1 \text{ atm}}{760 \text{ mmHg}}\right)} = 45.9 \text{ g/mol}$$

Some nitrogen oxides and their molar masses are:

NO   30 g/mol N_2O 44 g/mol NO_2 46 g/mol

The nitrogen oxide is most likely **NO_2**, although N_2O cannot be completely ruled out.

11.137   First, calculate the moles of hydrogen gas needed to fill a 4.1-L life jacket.

$$n_{H_2} = \frac{PV}{RT} = \frac{(0.97 \text{ atm})(4.1 \text{ L})}{\left(0.08206 \frac{\text{L} \cdot \text{atm}}{\text{K} \cdot \text{mol}}\right)(12 + 273)\text{K}} = 0.17 \text{ mol H}_2$$

The balanced equation shows a mole ratio between H_2 and LiH of 1:1. Therefore, 0.17 mole of LiH is needed. Converting to mass in grams:

$$? \text{ g LiH} = 0.17 \text{ mol LiH} \times \frac{7.949 \text{ g LiH}}{1 \text{ mol LiH}} = \mathbf{1.4 \text{ g LiH}}$$

11.139
$$P_{CO_2}: (0.965) \times (9.0 \times 10^6 \text{ Pa}) = \mathbf{8.7 \times 10^6 \text{ Pa}}$$

$$P_{N_2}: (0.035) \times (9.0 \times 10^6 \text{ Pa}) = \mathbf{3.2 \times 10^5 \text{ Pa}}$$

$$P_{SO_2}: (1.5 \times 10^{-4}) \times (9.0 \times 10^6 \text{ Pa}) = \mathbf{1.4 \times 10^3 \text{ Pa}}$$

11.141   a. (i)   **He atoms in both flasks have the same rms speed and average kinetic energy.**

(ii)   **The atoms in the larger volume strike the walls of their container with the same average force as those in the smaller volume, but the frequency of collisions in the larger volume will be less.**

b. (i)   **u_2 is greater than u_1 by a factor of $\sqrt{\dfrac{T_2}{T_1}}$.**

(ii)   **Both the frequency of collision and the average force of collision of the atoms at T_2 will be greater than those at T_1.**

c. (i)   **False. The rms speed of He is greater than that of Ne because He is less massive.**

(ii)   **True. The average kinetic energies of the two gases are equal.**

(iii)   **The rms speed of the whole collection of atoms is 1.47×10^3 m/s, but the speeds of the individual atoms are random and constantly changing.**

11.143   The ideal gas law can be used to calculate the moles of water vapor per liter.

$$\frac{n}{V} = \frac{P}{RT} = \frac{1.0 \text{ atm}}{\left(0.08206 \dfrac{\text{L} \cdot \text{atm}}{\text{K} \cdot \text{mol}}\right)(100+273)\text{K}} = 0.033 \text{ mol/L}$$

We eventually want to find the distance between molecules. Therefore, let's convert moles to molecules, and convert liters to a volume unit that will allow us to get to distance (m^3).

$$\left(\frac{0.033 \text{ mol}}{1 \text{ L}}\right)\left(\frac{6.022 \times 10^{23} \text{ molecules}}{1 \text{ mol}}\right)\left(\frac{1000 \text{ L}}{1 \text{ m}^3}\right) = 2.0 \times 10^{25} \text{ molecules/m}^3$$

This is the number of ideal gas molecules in a cube that is 1 m on each side. Assuming an equal distribution of molecules along the three mutually perpendicular directions defined by the cube, a linear density in one direction may be found:

$$\left(\frac{2.0 \times 10^{25} \text{ molecules}}{1 \text{ m}^3}\right)^{\frac{1}{3}} = 2.7 \times 10^8 \text{ molecules/m}^3$$

This is the number of molecules on a line 1 m in length. The distance between each **vapor** molecule is given by:

$$\frac{1 \text{ m}}{2.70 \times 10^8} = 3.7 \times 10^{-9} \text{ m} = \mathbf{3.7 \text{ nm}}$$

Assuming a water molecule to be a sphere with a diameter of 0.3 nm, the water molecules are separated by over 12 times their diameter: $\dfrac{3.7 \text{ nm}}{0.3 \text{ nm}} \approx 12$ times.

A similar calculation is done for liquid water. Starting with density, we convert to molecules per cubic meter.

$$\frac{0.96 \text{ g}}{1 \text{ cm}^3} \times \frac{1 \text{ mol H}_2\text{O}}{18.02 \text{ g H}_2\text{O}} \times \frac{6.022 \times 10^{23} \text{ molecules}}{1 \text{ mol H}_2\text{O}} \times \left(\frac{100 \text{ cm}}{1 \text{ m}}\right)^3 = 3.2 \times 10^{28} \text{ molecules/m}^3$$

This is the number of liquid water molecules in 1 m^3. From this point, the calculation is the same as that for water vapor, and the space between **liquid** molecules is found using the same assumptions.

$$\left(\frac{3.2\times10^{28}\text{ molecules}}{1\text{ m}^3}\right)^{\frac{1}{3}} = 3.2\times10^9 \text{ molecules/m}$$

$$\frac{1\text{ m}}{3.2\times10^9} = 3.1\times10^{-10}\text{ m} = \textbf{0.31 nm}$$

Assuming a water molecule to be a sphere with a diameter of 0.3 nm, to one significant figure, **water molecules are packed very closely together in the liquid, but much farther apart in the steam.**

11.145 We need to find the total moles of gas present in the flask after the reaction. Then, we can use the ideal gas equation to calculate the total pressure inside the flask. Since we are given the amounts of both reactants, we need to find which reactant is used up first. This is a limiting reactant problem. Let's calculate the moles of each reactant present.

$$5.72\text{ g C} \times \frac{1\text{ mol C}}{12.01\text{ g C}} = 0.4763\text{ mol C}$$

$$68.4\text{ g O}_2 \times \frac{1\text{ mol O}_2}{32.00\text{ g O}_2} = 2.138\text{ mol O}_2$$

The mole ratio between C and O_2 in the balanced equation is 1:1. Therefore, C is the limiting reactant. The amount of C remaining after complete reaction is 0 moles. Since the mole ratio between C and O_2 is 1:1, the amount of O_2 that reacts is 0.4763 mole. The amount of O_2 remaining after reaction is:

moles O_2 remaining = moles O_2 initial − moles O_2 consumed = 2.138 mol − 0.4763 mol = 1.662 mol O_2

The amount of CO_2 produced in the reaction is:

$$0.4763\text{ mol C} \times \frac{1\text{ mol CO}_2}{1\text{ mol C}} = 0.4763\text{ mol CO}_2$$

The total moles of gas present after reaction are:

total mol of gas = mol CO_2 + mol O_2 = 0.4763 mol + 1.662 mol = 2.138 mol

Using the ideal gas equation, we can now calculate the total pressure inside the flask.

$$P = \frac{nRT}{V} = \frac{(2.138\text{ mol})\left(0.0821\frac{\text{L}\cdot\text{atm}}{\text{K}\cdot\text{mol}}\right)(182+273)\text{K}}{8.00\text{ L}} = \textbf{9.98 atm}$$

11.147 When the drum is dented, the volume decreases, and therefore, we expect the pressure to increase. However, the pressure due to acetone vapor (400 mmHg) will not change as long as the temperature stays at 18°C (vapor pressure is constant at a given temperature). As the pressure increases, more acetone vapor will condense to liquid. Assuming that air does not dissolve in acetone, the pressure inside the drum will increase due to an increase in the pressure due to air. Initially, the total pressure is 750 mmHg, and the pressure due to air is:

$$P_{total} = P_{air} + P_{acetone}$$

$$P_{air} = P_{total} - P_{acetone} = 750\text{ mmHg} - 400\text{ mmHg} = 350\text{ mmHg}$$

The initial volume of vapor in the drum is:

$$V_1 = 25.0 \text{ gal} - 15.4 \text{ gal} = 9.6 \text{ gal}$$

When the drum is dented, the volume the vapor occupies decreases to:

$$V_2 = 20.4 \text{ gal} - 15.4 \text{ gal} = 5.0 \text{ gal}$$

The same number of air molecules now occupies a smaller volume. The pressure increases according to Boyle's Law.

$$P_1 V_1 = P_2 V_2$$

$$(350 \text{ mmHg})(9.6 \text{ gal}) = P_2(5.0 \text{ gal})$$

$$P_2 = 672 \text{ mmHg}$$

This is the pressure due to air. The pressure due to acetone vapor is still 400 mmHg. The total pressure inside the drum after the accident is:

$$P_{\text{total}} = P_{\text{air}} + P_{\text{acetone}}$$

$$P_{\text{total}} = 672 \text{ mmHg} + 400 \text{ mmHg} = 1072 \text{ mmHg}$$

$$1072 \text{ mmHg} \times \frac{1 \text{ atm}}{760 \text{ mmHg}} = \mathbf{1.41 \text{ atm}}$$

11.149 The reactions are:

$$CH_4 + 2O_2 \longrightarrow CO_2 + 2H_2O$$

$$2C_2H_6 + 7O_2 \longrightarrow 4CO_2 + 6H_2O$$

For a given volume and temperature, $n \propto P$. This means that the greater the pressure of reactant, the more moles of reactant, and hence the more product (CO_2) that will be produced. The pressure of CO_2 produced comes from the combustion of both methane and ethane. We set up an equation using the mole ratios from the balanced equation to convert to pressure of CO_2.

$$\left(P_{CH_4} \times \frac{1 \text{ mol } CO_2}{1 \text{ mol } CH_4} \right) + \left(P_{C_2H_6} \times \frac{4 \text{ mol } CO_2}{2 \text{ mol } C_2H_6} \right) = 356 \text{ mmHg } CO_2$$

$$(1) \quad P_{CH_4} + 2P_{C_2H_6} = 356 \text{ mmHg}$$

Also,

$$(2) \quad P_{CH_4} + P_{C_2H_6} = 294 \text{ mmHg}$$

Subtracting Equation (2) from Equation (1) gives:

$$P_{C_2H_6} = 356 - 294 = 62 \text{ mmHg}$$

$$P_{CH_4} = 294 - 62 = 232 \text{ mmHg}$$

Lastly, because $n \propto P$, we can solve for the mole fraction of each component using partial pressures.

$$\chi_{CH_4} = \frac{232}{294} = 0.789 \qquad \chi_{C_2H_6} = \frac{62}{294} = 0.211$$

methane   **ethane**

11.151   a. **Strategy:** The volume around each atom into which the center of another atom cannot penetrate is called the excluded volume. To determine the excluded volume, use the formula for volume of a sphere to calculate the volume of a sphere of radius $2r$.

      Setup: Assuming the atoms are spherical, the equation for the volume of a sphere is $\frac{4}{3}\pi r^3$.

      Solution: Excluded volume defined by two atoms $= \frac{4}{3}\pi(2r)^3 = 8\left(\frac{4}{3}\pi r^3\right)$

   b. **Strategy:** Multiply the excluded volume per molecule by Avogadro's number to determine the excluded volume per mole.

      Setup: $N_A = 6.022 \times 10^{23}$

      Solution: Excluded volume per mole $= \dfrac{8\left(\frac{4}{3}\pi r^3\right)}{2 \text{ molecules}} \times \dfrac{N_A \text{ molecules}}{1 \text{ mole}} = \dfrac{4N_A\left(\frac{4}{3}\pi r^3\right)}{1 \text{ mole}}$

The excluded volume per mole is four times the volume actually occupied by a mole of molecules.

11.153   From the root-mean-square speed, we can calculate the molar mass of the gaseous oxide.

$$u_{rms} = \sqrt{\frac{3RT}{\mathcal{M}}}$$

$$\mathcal{M} = \frac{3RT}{(u_{rms})^2} = \frac{3(8.314 \text{ J/K} \cdot \text{mol})(293 \text{ K})}{(493 \text{ m/s})^2} = 0.0301 \text{ kg/mol} = 30.1 \text{ g/mol}$$

The compound must be a monoxide because 2 moles of oxygen atoms would have a mass of 32.00 g. The molar mass of the other element is:

$$30.1 \text{ g/mol} - 16.00 \text{ g/mol} = 14.1 \text{ g/mol}$$

The compound is nitrogen monoxide, **NO**.

11.155   Pressure and volume are constant. The number of moles and temperature are changing. We start with the ideal gas equation, writing it for initial and final conditions, and setting the two expressions equal to each other.

$$\frac{P_1 V_1}{n_1 T_1} = \frac{P_2 V_2}{n_2 T_2}$$

Because pressure and volume are constant, this equation reduces to:

$$\frac{1}{n_1 T_1} = \frac{1}{n_2 T_2}$$

or

$$n_1 T_1 = n_2 T_2$$

Because $T_1 = 2T_2$, substituting into the above equation gives:

$$2n_1 T_2 = n_2 T_2$$

or

$$2n_1 = n_2$$

This equation indicates that the number of moles of gas after reaction is twice the number of moles of gas before reaction. Only **reaction (b)** fits this description.

11.157 **Strategy:** The density of an ideal gas is $d = \dfrac{P\mathcal{M}}{RT}$ (Equation 11.12). Plug in the given information and calculate.

Solution: $d = \dfrac{(1.10\ \text{atm})(4.003\ \text{g/mol})}{(0.08206\ \text{L} \cdot \text{atm/K} \cdot \text{mol})(273.15 + 25.0)(\text{K})} = 0.180\ \text{g/L} = \mathbf{1.80 \times 10^{-4}}\ \textbf{g/mL}$

11.159 **The kinetic energy of a particle depends on its mass and its speed (KE = ½mv^2). At a given temperature, the more massive SF$_6$ molecules move more slowly than do the Ne molecules.**

11.161 The compressibility factor Z of a gas is defined as $Z = P_{\text{real}}V/nRT$. Note that for an ideal gas, $Z = 1$ at all temperatures and pressures. To understand the Z vs. P plot of a real gas, it is useful to model the real gas using the van der Waals equation. For a van der Waals gas,

$$Z_{\text{vdW}} = \dfrac{\left(\dfrac{nRT}{V-nb} - \dfrac{an^2}{V^2} \right)V}{nRT} = \dfrac{V}{V-nb} - \left(\dfrac{an}{RT} \right)\left(\dfrac{1}{V} \right)$$

For fixed n and T, low pressures mean V is large, and high pressures mean V is small. At low pressures, the term nb is negligible compared to V, and the compressibility is approximately $Z_{\text{vdW}} \approx 1 - \left(\dfrac{an}{RT} \right)\left(\dfrac{1}{V} \right)$ (note that the compressibility is less than 1). This effect is amplified when a is large (large intermolecular forces). As pressure increases, V gets smaller and approaches nb, which means the denominator of the term $\dfrac{V}{V-nb}$ approaches zero, causing Z to get very large. This effect is amplified when b is large (larger molecules). To summarize, **intermolecular attractions tend to lower Z at low pressures and excluded volume tends to increase Z at high pressures.**

Chapter 12
Liquids and Solids

Key Skills

12.1   a, d, e   12.2   a, d   12.3   b   12.4   d

Problems

12.13 Using Equation 12.4 of the text:

$$\ln\frac{P_1}{P_2} = \frac{\Delta H_{vap}}{R}\left(\frac{1}{T_2} - \frac{1}{T_1}\right)$$

$$\ln\left(\frac{1}{2}\right) = \left(\frac{\Delta H_{vap}}{8.314 \text{ J/K}\cdot\text{mol}}\right)\left(\frac{1}{373 \text{ K}} - \frac{1}{348 \text{ K}}\right) = \Delta H_{vap}\left(\frac{-1.93\times10^{-4}}{8.314 \text{ J/mol}}\right)$$

$$\Delta H_{vap} = 2.99 \times 10^4 \text{ J/mol} = \textbf{29.9 kJ/mol}$$

12.15 **Ethylene glycol has two –OH groups, allowing it to exert strong intermolecular forces through hydrogen bonding. Its viscosity should fall between ethanol (one –OH group) and glycerol (three –OH groups).**

12.17 Using the Clausius-Clapeyron equation (Equation 12.1), we see that a plot of ln P vs. $1/T$ is a straight line whose slope is proportional to ΔH_{vap}. The relative positions of the points ($1/T$, ln P) for each liquid is shown below. (Note that higher temperature corresponds to smaller values of $1/T$.)

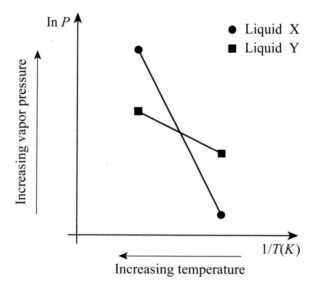

According to the graph, liquid X has a more negative slope. So, we can conclude that **liquid X has a larger ΔH_{vap} than does liquid Y**.

12.27 A corner sphere is shared equally among eight unit cells, so only one-eighth of each corner sphere "belongs" to any one unit cell. A face-centered sphere is divided equally between the two unit cells sharing the face. A body-centered sphere belongs entirely to its own unit cell.

In a *simple cubic cell*, there are eight corner spheres. One-eighth of each belongs to the individual cell giving a total of **one sphere** per cell. In a *body-centered cubic cell*, there are eight corner spheres and one body-center sphere giving a total of **two spheres** per unit cell (one-eighth from each of the eight corners and one from the body center). In a *face-centered cubic cell*, there are eight corner spheres and six face-centered spheres (six faces). The total number would be **four spheres**: one from the corners and three from the faces.

12.29 **Strategy:** First, we need to calculate the volume (in cm^3) occupied by 1 mole of Ba atoms. Next, we calculate the volume that a Ba atom occupies. Once we have these values, we can multiply them together to end up with the number of Ba atoms per mole of Ba.

$$\frac{\text{number of Ba atoms}}{cm^3} \times \frac{cm^3}{1 \text{ mol Ba}} = \frac{\text{number of Ba atoms}}{1 \text{ mol Ba}}$$

Solution: The volume that contains one mole of Ba atoms can be calculated from the density using the following strategy:

$$\frac{\text{volume}}{\text{mass of Ba}} \rightarrow \frac{\text{volume}}{\text{mol Ba}}$$

$$\frac{1 \text{ cm}^3}{3.50 \text{ g Ba}} \times \frac{137.3 \text{ g Ba}}{1 \text{ mol Ba}} = \frac{39.23 \text{ cm}^3}{1 \text{ mol Ba}}$$

We carry an extra significant figure in this calculation to limit rounding errors. Next, the volume that contains two Ba atoms is the volume of the body-centered cubic unit cell. Some of this volume is empty space because packing is only 68.0 percent efficient. But, this will not affect our calculation.

$$V = a^3$$

Let's also convert to cm^3.

$$V = (502 \text{ pm})^3 \times \left(\frac{1 \times 10^{-12} \text{ m}}{1 \text{ pm}}\right)^3 \times \left(\frac{1 \text{ cm}}{0.01 \text{ m}}\right)^3 = \frac{1.265 \times 10^{-22} \text{ cm}^3}{2 \text{ Ba atoms}}$$

We can now calculate the number of Ba atoms in 1 mole using the strategy presented above.

$$\frac{\text{number of Ba atoms}}{cm^3} \times \frac{cm^3}{1 \text{ mol Ba}} = \frac{\text{number of Ba atoms}}{1 \text{ mol Ba}}$$

$$\frac{2 \text{ Ba atoms}}{1.265 \times 10^{-22} \text{ cm}^3} \times \frac{39.23 \text{ cm}^3}{1 \text{ mol Ba}} = \mathbf{6.20 \times 10^{23} \text{ atoms / mol}}$$

This is close to Avogadro's number, 6.022×10^{23} particles/mol.

12.31 The mass of the unit cell is the mass in grams of two Eu atoms.

$$m = \frac{2 \text{ Eu atoms}}{1 \text{ unit cell}} \times \frac{1 \text{ mol Eu}}{6.022 \times 10^{23} \text{ Eu atoms}} \times \frac{152.0 \text{ g Eu}}{1 \text{ mol Eu}} = 5.048 \times 10^{-22} \text{ g Eu/unit cell}$$

$$V = \frac{5.048 \times 10^{-22} \text{ g}}{1 \text{ unit cell}} \times \frac{1 \text{ cm}^3}{5.26 \text{ g}} = 9.60 \times 10^{-23} \text{ cm}^3/\text{unit cell}$$

The edge length (a) is:

$$a = V^{1/3} = (9.60 \times 10^{-23} \text{ cm}^3)^{1/3} = 4.58 \times 10^{-8} \text{ cm} = \textbf{458 pm}$$

12.33 **Strategy:** Recall that a corner atom is shared with eight unit cells, so only one-eighth of corner atom is within a given unit cell. Also recall that a face atom is shared with two unit cells, so one-half of a face atom is within a given unit cell. See Figure 12.21 of the text.

Solution: In a face-centered cubic unit cell, there are atoms at each of the eight corners, and there is one atom in each of the six faces. Only one-half of each face-centered atom and one-eighth of each corner atom belong to the unit cell.

X atoms/unit cell = (8 corner atoms)(1/8 atom per corner) = 1 X atom/unit cell

Y atoms/unit cell = (6 face-centered atoms)(1/2 atom per face) = 3 Y atoms/unit cell

The unit cell is the smallest repeating unit in the crystal; therefore, the empirical formula is **XY$_3$**.

12.35 Rearranging the equation on page 499, we have:

$$\lambda = \frac{2d \sin \theta}{n} = \frac{2(282 \text{ pm})(\sin 23.0°)}{1} = 220 \text{ pm} = \textbf{0.220 nm}$$

12.37 The cell is face-centered cubic, determined by the positions of O^{2-} ions, so there are four O^{2-} ions. There are also four Zn^{2+} ions. Therefore, the formula of zinc oxide is **ZnO**.

12.41 See Table 12.4 of the text. The properties listed are those of a **molecular solid**.

12.43 In a molecular crystal, the lattice points are occupied by molecules. Of the solids listed, the ones that are composed of *molecules* are **Se$_8$, HBr, CO$_2$, P$_4$O$_6$**, and **SiH$_4$**. In covalent crystals, atoms are held together in an extensive three-dimensional network entirely by covalent bonds. Of the solids listed, the ones that are composed of atoms held together by *covalent bonds* are **Si** and **C**.

12.45 **Diamond: Each carbon atom is covalently bonded to four other carbon atoms. Because these bonds are strong and uniform, diamond is a very hard substance. Graphite: The carbon atoms in each layer are linked by strong bonds, but the layers are bound by weak dispersion forces. As a result, graphite may be cleaved easily between layers and is not hard.**

In graphite, all atoms are sp^2 hybridized; each atom is covalently bonded to three other atoms. The remaining unhybridized $2p$ orbital is used in pi bonding forming a delocalized molecular orbital. The electrons are free to move around in this extensively delocalized molecular orbital, making graphite a good conductor of electricity in directions along the planes of carbon atoms.

12.65   *Step 1:*   Warming ice to the melting point.

$$q_1 = ms\Delta T = (866 \text{ g H}_2\text{O})(2.03 \text{ J/g·°C})[0 - (-15)°\text{C}] = 26 \text{ kJ}$$

   Step 2:   Converting ice at the melting point to liquid water at 0°C. (See Table 12.7 of the text for the heat of fusion of water.)

$$q_2 = 866 \text{ g H}_2\text{O} \times \frac{1 \text{ mol}}{18.02 \text{ g H}_2\text{O}} \times \frac{6.01 \text{ kJ}}{1 \text{ mol}} = 289 \text{ kJ}$$

   Step 3:   Heating water from 0°C to 100°C.

$$q_3 = ms\Delta T = (866 \text{ g H}_2\text{O})(4.184 \text{ J/g·°C})[(100 - 0)°\text{C}] = 362 \text{ kJ}$$

   Step 4:   Converting water at 100°C to steam at 100°C. (See Table 12.5 of the text for the heat of vaporization of water.)

$$q_4 = 866 \text{ g H}_2\text{O} \times \frac{1 \text{ mol}}{18.02 \text{ g H}_2\text{O}} \times \frac{40.79 \text{ kJ}}{1 \text{ mol}} = 1.96 \times 10^3 \text{ kJ}$$

   Step 5: Heating steam from 100°C to 146°C.

$$q_5 = ms\Delta T = (866 \text{ g H}_2\text{O})(1.99 \text{ J/g·°C})\left[(146 - 100)°\text{C}\right] = 79.3 \text{ kJ}$$

$$q_{total} = q_1 + q_2 + q_3 + q_4 + q_5 = \textbf{2.72} \times \textbf{10}^3 \textbf{ kJ}$$

12.67   **Two phase changes occur in this process. First, the liquid is turned to solid (freezing), then the solid ice is turned to gas (sublimation).**

12.69   **When steam condenses to liquid water at 100°C, it releases a large amount of heat equal to the enthalpy of vaporization. Thus steam at 100°C exposes one to more heat than an equal amount of water at 100°C.**

12.71   **The solid ice turns to vapor (sublimation). The temperature is too low for melting to occur.**

12.75   **Initially, the ice melts because of the increase in pressure. As the wire sinks into the ice, the water abov the wire refreezes. Eventually the wire moves completely through the ice block without cutting it in half.**

12.77   Region labels: The region containing point A is the solid region. The region containing point B is the liquid region. The region containing point C is the gas region.

   a. **Ice would melt. (If heating continues, the liquid water would eventually boil and become a vapor.)**

   b. **Liquid water would vaporize.**

   c. **Water vapor would solidify without becoming a liquid.**

12.79   The properties of hardness, high melting point, poor conductivity, and so on could place boron in either the ionic or covalent categories. However, boron atoms will not alternately form positive and negative ions to achieve an ionic crystal. The structure is **covalent** because the units are single boron atoms.

12.81 a. A low surface tension means the attraction between molecules making up the surface is weak. Water has a high surface tension; water bugs could not "walk" on the surface of a liquid with a low surface tension.

b. A low critical temperature means a gas is very difficult to liquefy by cooling. This is the result of weak intermolecular attractions. Helium has the lowest known critical temperature (5.3 K).

c. A low boiling point means weak intermolecular attractions. It takes little energy to separate the particles. All ionic compounds have extremely high boiling points.

d. A low vapor pressure means it is difficult to remove molecules from the liquid phase because of high intermolecular attractions. Substances with low vapor pressures have high boiling points.

Thus, only choice **(d) indicates strong intermolecular forces in a liquid.** The other choices indicate weak intermolecular forces in a liquid.

12.83 Determine the number of base pairs and the total length of the molecule.

$$\frac{5.0 \times 10^9 \text{ g/mol}}{650 \text{ g/mol} \cdot \text{base pair}} = \textbf{7.69} \times \textbf{10}^{\textbf{6}} \textbf{ base pairs}$$

$$7.69 \times 10^6 \text{ base pairs} \times \frac{3.4 \text{ Å}}{\text{base pair}} = 2.6 \times 10^7 \text{ Å}$$

$$2.6 \times 10^7 \text{ Å} \times \frac{1 \text{ m}}{1 \times 10^{10} \text{ Å}} = 0.0026 \text{ m or } \textbf{0.26 cm}$$

12.85 The vapor pressure of mercury (as well as all other substances) is **760 mmHg** at its normal boiling point.

12.87 **It has reached the critical point; the point of critical temperature (T_c) and critical pressure (P_c).**

12.89 **Crystalline SiO_2; Its regular structure results in a more efficient packing.**

12.91 a. **False.** Permanent dipoles are usually much stronger than temporary dipoles.

b. **False.** The hydrogen atom must be bonded to N, O, or F.

c. **True**

Statements **(a) and (b)** are false.

12.93 From Figure 12.30, the sublimation temperature is −78°C or 195 K at a pressure of 1 atm.

$$\ln \frac{P_1}{P_2} = \frac{\Delta H_{sub}}{R} \left(\frac{1}{T_2} - \frac{1}{T_1} \right)$$

$$\ln \frac{1}{P_2} = \frac{25.9 \times 10^3 \text{ J/mol}}{8.314 \text{ J/K} \cdot \text{mol}} \left(\frac{1}{150 \text{ K}} - \frac{1}{195 \text{ K}} \right)$$

$$\ln \frac{1}{P_2} = 4.79$$

Taking the anti-ln of both sides gives:

$$P_2 = 8.3 \times 10^{-3} \text{ atm}$$

12.95 The standard enthalpy change for the formation of gaseous iodine from solid iodine is simply the difference between the standard enthalpies of formation of the products and the reactants in the equation:

$$I_2(s) \longrightarrow I_2(g)$$

$$\Delta H_{sub} = \Delta H_f^\circ[I_2(g)] - \Delta H_f^\circ[I_2(s)] = 62.4 \text{ kJ/mol} - 0 \text{ kJ/mol} = \mathbf{62.4 \text{ kJ / mol}}$$

12.97 CH_4 is a tetrahedral, nonpolar molecule that can only exert weak dispersion-type attractive forces. SO_2 is bent and possesses a dipole moment, which gives rise to stronger dipoledipole attractions. **SO_2 will behave less ideally because it is polar and has greater intermolecular forces**.

12.99 **Smaller ions can approach water molecules more closely, resulting in larger ion-dipole interactions. The greater the ion-dipole interaction, the larger is the heat of hydration**.

12.101
a. Using data from Appendix 2, for the process: $Br_2(l) \longrightarrow Br_2(g)$

$$\Delta H^\circ = \Delta H_f^\circ[Br_2(g)] - \Delta H_f^\circ[Br_2(l)] = (1)(30.7 \text{ kJ/mol}) - 0 = \mathbf{30.7 \text{ kJ / mol}}$$

b. Using data from Table 10.4 for the process: $Br_2(g) \longrightarrow 2Br(g)$

$$\Delta H^\circ = \mathbf{192.5 \text{ kJ/mol}}$$

As expected, the bond enthalpy represented in part (b) is much greater than the energy of vaporization represented in part (a). **It requires more energy to break the bond than to vaporize the molecule**.

12.103 a. **decreases**   b. **no change**   c. **no change**

12.105
$$CaCO_3(s) \longrightarrow CaO(s) + CO_2(g)$$

Initial state: one solid phase, final state: two solid phase components and one gas phase component. $CaCO_3$ and CaO constitute two separate solid phases because they are separated by well-defined boundaries.

12.107 **The time required to cook food depends on the boiling point of the water in which it is cooked. The boiling point of water increases when the pressure inside the cooker increases**.

12.109 **Extra heat is produced when steam condenses at 100°C. Steaming avoids the extraction of ingredients caused by boiling in water**.

12.111 a. **~2.3 K**

b. **~10 atm**

c. **~5 K**

d. **No**. There is no solid-vapor phase boundary.

12.113 a. **Pumping allows Ar atoms to escape, thus removing heat from the liquid phase. Eventually the liquid freezes.**

b. **The slope of the solid-liquid line of cyclohexane is positive. Therefore, its melting point increases with pressure.**

c. **These droplets are supercooled liquids.**

d. **When the dry ice is added to water, it sublimes. The cold CO_2 gas generated causes nearby water vapor to condense, hence the appearance of fog.**

12.115 a. **two triple points: diamond/graphite/liquid and graphite/liquid/vapor**

b. **diamond**

c. **Apply high pressure at high temperature.**

12.117 **Ethanol mixes well with water. The mixture has a lower surface tension and readily flows out of the ear channel.**

12.119 The two main reasons for spraying the trees with water are:

(1) **When water freezes it releases heat, helping keep the fruit warm enough not to freeze.**

$$H_2O(l) \longrightarrow H_2O(s) \quad -\Delta H_{fus} = -6.01 \text{ kJ/mol}$$

(2) **A layer of ice is a thermal insulator.**

12.121 **The fuel source for the Bunsen burner is most likely methane gas. When methane burns in air, carbon dioxide and water are produced.**

$$CH_4(g) + 2O_2(g) \longrightarrow CO_2(g) + 2H_2O(g)$$

The water vapor produced during the combustion condenses to liquid water when it comes in contact with the outside of the cold beaker.

12.123 First, we need to calculate the volume (in cm^3) occupied by 1 mole of Fe atoms. Next, we calculate the volume that a Fe atom occupies. Once we have these two pieces of information, we can multiply them together to end up with the number of Fe atoms per mole of Fe.

$$\frac{\text{number of Fe atoms}}{cm^3} \times \frac{cm^3}{1 \text{ mol Fe}} = \frac{\text{number of Fe atoms}}{1 \text{ mol Fe}}$$

The volume that contains 1 mole of Fe atoms can be calculated from the density using the following strategy:

$$\frac{\text{volume}}{\text{mass of Fe}} \rightarrow \frac{\text{volume}}{\text{mol Fe}}$$

$$\frac{1 \text{ cm}^3}{7.874 \text{ g Fe}} \times \frac{55.85 \text{ g Fe}}{1 \text{ mol Fe}} = \frac{7.093 \text{ cm}^3}{1 \text{ mol Fe}}$$

572

Next, the volume that contains two iron atoms is the volume of the body-centered cubic unit cell. Some of this volume is empty space because packing is only 68.0 percent efficient. But, this will not affect our calculation.

$$V = a^3$$

Let's also convert to cm^3.

$$V = (286.7 \text{ pm})^3 \times \left(\frac{1 \times 10^{-12} \text{ m}}{1 \text{ pm}}\right)^3 \times \left(\frac{1 \text{ cm}}{0.01 \text{ m}}\right)^3 = \frac{2.357 \times 10^{-23} \text{ cm}^3}{2 \text{ Fe atoms}}$$

We can now calculate the number of iron atoms in 1 mole using the strategy presented above.

$$\frac{\text{number of Fe atoms}}{\text{cm}^3} \times \frac{\text{cm}^3}{1 \text{ mol Fe}} = \frac{\text{number of Fe atoms}}{1 \text{ mol Fe}}$$

$$\frac{2 \text{ Fe atoms}}{2.357 \times 10^{-23} \text{ cm}^3} \times \frac{7.093 \text{ cm}^3}{1 \text{ mol Ba}} = \mathbf{6.019 \times 10^{23} \text{ Fe atoms / mol}}$$

The small difference between the above number and 6.022×10^{23} is the result of rounding off and using rounded values for density and other constants.

12.125 If we know the values of ΔH_{vap} and P of a liquid at one temperature, we can use the Clausius-Clapeyron equation, Equation 12.4 of the text, to calculate the vapor pressure at a different temperature. At 65.0°C, we can calculate ΔH_{vap} of methanol. Because this is the boiling point, the vapor pressure will be 1 atm (760 mmHg).

First, we calculate ΔH_{vap}. From Appendix 2 of the text, $\Delta H_f^\circ [CH_3OH(l)] = -238.7$ kJ/mol

$$CH_3OH(l) \longrightarrow CH_3OH(g)$$

$$\Delta H_{vap} = \Delta H_f^\circ[CH_3OH(g)] - \Delta H_f^\circ[CH_3OH(l)]$$

$$\Delta H_{vap} = -201.2 \text{ kJ/mol} - (-238.7 \text{ kJ/mol}) = 37.50 \text{ kJ/mol}$$

Next, we substitute into Equation 12.4 of the text to solve for the vapor pressure of methanol at 25°C.

$$\ln \frac{P_1}{P_2} = \frac{\Delta H_{vap}}{R}\left(\frac{1}{T_2} - \frac{1}{T_1}\right)$$

$$\ln \frac{P_1}{760} = \frac{37.50 \times 10^3 \text{ J/mol}}{8.314 \text{ J/K} \cdot \text{mol}}\left(\frac{1}{338 \text{ K}} - \frac{1}{298 \text{ K}}\right)$$

$$\ln \frac{P_1}{760} = -1.79$$

Taking the anti-ln of both sides gives: $\qquad P_1 = \mathbf{127 \text{ mmHg}}$

12.127 If half the water remains in the liquid phase, there is 1.0 g of water vapor. We can derive a relationship between vapor pressure and temperature using the ideal gas equation.

$$P = \frac{nRT}{V} = \frac{\left(1.0 \text{ g H}_2\text{O} \times \dfrac{1 \text{ mol H}_2\text{O}}{18.02 \text{ g H}_2\text{O}}\right)\left(0.0821 \dfrac{\text{L} \cdot \text{atm}}{\text{K} \cdot \text{mol}}\right)T}{9.6 \text{ L}} = (4.7 \times 10^{-4})T \text{ atm}$$

Converting to units of mmHg:

$$\left(4.7 \times 10^{-4}\right)T \text{ atm} \times \frac{760 \text{ mmHg}}{1 \text{ atm}} = 0.36T \text{ mmHg}$$

To determine the temperature at which only half the water remains, we set up the following table and refer to Table 11.6 of the text. The calculated value of vapor pressure that most closely matches the vapor pressure in Table 11.6 would indicate the approximate value of the temperature.

$T(K)$	$P_{\text{H}_2\text{O}}$ mmHg (from Table 11.6)	$(0.36\,T)$ mmHg
313	55.3	112.7
318	71.9	114.5
323	92.5	116.3
328	118.0	118.1 (closest match)
333	149.4	119.9
338	187.5	121.7

Therefore, the temperature is about 328 K = **55°C** at which half the water has vaporized.

12.129 Use Equation 11.10 to compute the density of the ideal gas:

$$d = \frac{P\mathcal{M}}{RT} = \frac{(101{,}325 \text{ Pa})\left(20.008 \times 10^{-3} \text{ kg/mol}\right)}{(8.314 \text{ J/K} \cdot \text{mol})(273.15 + 19.5)(\text{K})} = 0.833 \text{ kg/m}^3 = \textbf{0.833 g / L}$$

This density is nearly four times smaller than the experimental value of 3.10 g/L. **The hydrogen-bonding interactions in HF are relatively strong, and since the ideal gas equation ignores intermolecular forces, it underestimates significantly the density of HF gas near its boiling point.**

12.131 **Strategy:** Use the equation on page 499 to solve for θ:

$$2d \sin \theta = n\lambda$$

Setup: $d = 312$ pm, $\lambda = 0.154$ nm $= 154$ pm, and $n = 1$.

Solution:

$$\sin \theta = \frac{n\lambda}{2d} = \frac{1(154 \text{ pm})}{2(312 \text{ pm})} = 0.247$$

$$\theta = \sin^{-1}(0.247)$$

$$\theta = \textbf{14.3°}$$

12.133 Strategy: Cold water is warmed to body temperature in a single step: a temperature change. The melting of ice and the subsequent warming of the resulting liquid water takes place in two steps: a phase change and a temperature change. In each case, the heat transferred during a temperature change depends on the mass of the water, the specific heat of water, and the change in temperature. For the phase change, the heat transferred depends on the amount of water (in moles) and the molar heat of fusion (ΔH_{fus}). The total energy expended is the sum of the energy changes for the individual steps.

Setup: The specific heat of water is 4.184 J/g°C. The molar heat of fusion (ΔH_{fus}) of water is 6.01 kJ/mol. The molar mass of water is 18.02 g/mol.

The density of water is 1.00 g/cm^3 or 1.00 g/mL. Converting 2.84 L of water to grams gives:

$$2.84 \text{ L} \times \frac{1 \times 10^3 \text{ mL}}{1 \text{ L}} \times \frac{1.00 \text{ g}}{\text{mL}} = 2840 \text{ g}$$

Converting to moles of water gives:

$$2840 \text{ g} \times \frac{\text{mol}}{18.02 \text{ g}} = 157.60 \text{ mol}$$

Solution: Energy to warm water:

$$\Delta T = 37°C - 10°C = 27°C$$

$$q = ms\Delta T = 2840 \text{ g} \times \frac{4.184 \text{ J}}{\text{g} \cdot °C} \times 27°C = 3.2 \times 10^5 \text{ J} = \mathbf{3.2 \times 10^2 \text{ kJ}}$$

Energy to warm ice:

$$q_1 = n\Delta H_{fus} = 157.60 \text{ mol} \times \frac{6.01 \text{ kJ}}{\text{mol}} = 947.2 \text{ kJ}$$

$$\Delta T = 37°C - 0°C = 37°C$$

$$q_2 = ms\Delta T = 2840 \text{ g} \times \frac{4.184 \text{ J}}{\text{g} \cdot °C} \times 37°C = 4.40 \times 10^5 \text{ J} = 440 \text{ kJ}$$

The energy expended to warm the ice from 0°C to 37°C is the sum of q_1 and q_2:

$$947.2 \text{ kJ} + 440 \text{ kJ} = 1387 \text{ kJ or } 1.39 \times 10^3 \text{ kJ}$$

Therefore, it takes 1.39×10^3 kJ $- 3.2 \times 10^2$ kJ or $\mathbf{1.1 \times 10^3}$ **kJ** more energy if the water were consumed as ice.

Chapter 13
Physical Properties of Solutions

Key Skills

13.1   a, c   13.2   d   13.3   e   13.4   a

Problems

13.9   In predicting solubility, remember the saying: "**Like dissolves like**." A nonpolar solute will dissolve in a nonpolar solvent; ionic compounds will generally dissolve in polar solvents due to favorable ion-dipole interactions; solutes that can form hydrogen bonds with a solvent will also have high solubility in the solvent. **Naphthalene and benzene are nonpolar, whereas CsF is ionic.**

13.11   The order of increasing solubility is: $O_2 < Br_2 < LiCl < CH_3OH$. Methanol is miscible with water because of strong hydrogen bonding. LiCl is an ionic solid and is very soluble because of the high polarity of the water molecules. Both oxygen and bromine are nonpolar and exert only weak dispersion forces. Bromine is a larger molecule and is therefore more polarizable and susceptible to dipole-induced dipole attractions.

13.17   a. The molality is the number of moles of sucrose (molar mass 342.3 g/mol) divided by the mass of the solvent (water) in kg.

$$\text{mol sucrose} = 14.3 \text{ g sucrose} \times \frac{1 \text{ mol}}{342.3 \text{ g sucrose}} = 0.0418 \text{ mol}$$

$$\text{molality} = \frac{0.0418 \text{ mol sucrose}}{0.685 \text{ kg H}_2\text{O}} = \textbf{0.0610 } \boldsymbol{m}$$

   b.

$$\text{molality} = \frac{7.15 \text{ mol ethylene glycol}}{3.505 \text{ kg H}_2\text{O}} = \textbf{2.04 } \boldsymbol{m}$$

13.19   In each case we consider 1 L of solution.

$$\text{mass of solution} = \text{volume} \times \text{density}$$

   a.

$$\text{mass of sugar} = 1.22 \text{ mol sugar} \times \frac{342.3 \text{ g sugar}}{1 \text{ mol sugar}} = 418 \text{ g sugar}$$

$$418 \text{ g sugar} \times \frac{1 \text{ kg}}{1000 \text{ g}} = 0.418 \text{ kg sugar}$$

$$\text{mass of soln} = 1000 \text{ mL} \times \frac{1.12 \text{ g}}{1 \text{ mL}} = 1120 \text{ g}$$

$$1120 \text{ g} \times \frac{1 \text{ kg}}{1000 \text{ g}} = 1.120 \text{ kg}$$

$$\text{molality} = \frac{1.22 \text{ mol sugar}}{(1.120 - 0.418) \text{ kg H}_2\text{O}} = \textbf{1.73 } \boldsymbol{m}$$

b. $$\text{mass of NaOH} = 0.87 \text{ mol NaOH} \times \frac{40.00 \text{ g NaOH}}{1 \text{ mol NaOH}} = 35 \text{ g NaOH}$$

$$\text{mass solvent (H}_2\text{O)} = 1040 \text{ g} - 35 \text{ g} = 1005 \text{ g} = 1.005 \text{ kg}$$

$$\text{molality} = \frac{0.87 \text{ mol NaOH}}{1.005 \text{ kg H}_2\text{O}} = \textbf{0.87 } \boldsymbol{m}$$

c. $$\text{mass of NaHCO}_3 = 5.24 \text{ mol NaHCO}_3 \times \frac{84.01 \text{ g NaHCO}_3}{1 \text{ mol NaHCO}_3} = 440 \text{ g NaHCO}_3$$

$$\text{mass solvent (H}_2\text{O)} = 1190 \text{ g} - 440 \text{ g} = 750 \text{ g} = 0.750 \text{ kg}$$

$$\text{molality} = \frac{5.24 \text{ mol NaHCO}_3}{0.750 \text{ kg H}_2\text{O}} = \textbf{6.99 } \boldsymbol{m}$$

13.21   We find the volume of ethanol in 1.00 L of 75 proof gin. Note that 75 proof means $\left(\dfrac{75}{2}\right)\%$.

$$\text{Volume} = 1.00 \text{ L} \times \left(\frac{75}{2}\right)\% = 0.38 \text{ L} = 3.8 \times 10^2 \text{ mL}$$

$$\text{Ethanol mass} = (3.8 \times 10^2 \text{ mL}) \times \frac{0.798 \text{ g}}{1 \text{ mL}} = \textbf{3.0} \times \textbf{10}^2 \, \boldsymbol{g}$$

13.23   **Strategy:**   In this problem, the masses of both solute and solvent are given, and the density is given.

   Solution: To determine molarity, we need to know the volume of the solution. We add the masses of solute and solvent to get the total mass and use the density to determine the volume.

$$35.0 \text{ g NH}_3 + 75.0 \text{ g H}_2\text{O} = 110.0 \text{ g soln}$$

$$\text{Volume of soln} = 110.0 \text{ g soln} \times \frac{1 \text{ mL soln}}{0.982 \text{ g soln}} = 112 \text{ mL soln}$$

We then convert the mass of NH_3 to moles and divide by the volume in liters to get molarity.

$$35.0 \text{ g NH}_3 \times \frac{1 \text{ mol NH}_3}{17.03 \text{ g NH}_3} = 2.055 \text{ mol NH}_3$$

$$\text{molarity} = \frac{2.055 \text{ mol NH}_3}{0.112 \text{ L soln}} = \textbf{18.3 } \boldsymbol{M}$$

To calculate molality, we use Equation 13.1.

$$\text{molality} = \frac{2.055 \text{ mol NH}_3}{0.0750 \text{ kg H}_2\text{O}} = \textbf{27.4}$$

13.25   Assuming that N_2 and O_2 are the only dissolved gases, the mole fractions in the total dissolved gas can be determined as follows:

We multiply each partial pressure by the corresponding Henry's law constant (solubility).

For O_2, we have

$$0.20 \text{ atm} \times 1.3 \times 10^{-3} \text{ mol/L·atm} = 2.6 \times 10^{-4} \text{ mol/L}$$

For N_2, we have

$$0.80 \text{ atm} \times 6.8 \times 10^{-4} \text{ mol/L·atm} = 5.44 \times 10^{-4} \text{ mol/L}$$

In a liter of water, then, there will be 2.6×10^{-4} mol O_2 and 5.44×10^{-4} mol N_2. The mole fractions are

$$\chi_{O_2} = \frac{2.6 \times 10^{-4} \text{ mol } O_2}{\left(2.6 \times 10^{-4} + 5.44 \times 10^{-4}\right) \text{ mol}} = \mathbf{0.32}$$

$$\chi_{N_2} = \frac{5.44 \times 10^{-4} \text{ mol } N_2}{\left(2.6 \times 10^{-4} + 5.44 \times 10^{-4}\right) \text{ mol}} = \mathbf{0.68}$$

Because of the greater solubility of oxygen, it has a larger mole fraction in solution than it does in the air.

13.33   At 75°C, 155 g of KNO_3 dissolves in 100 g of water to form 255 g of solution. When cooled to 25°C, only 38.0 g of KNO_3 remain dissolved. This means that $(155 - 38.0)$ g = 117 g of KNO_3 will crystallize. The amount of KNO_3 formed when 100 g of saturated solution at 75°C is cooled to 25°C can be found by a simple unit conversion.

$$100 \text{ g saturated soln} \times \frac{117 \text{ g } KNO_3 \text{ crystallized}}{255 \text{ g saturated soln}} = \mathbf{45.9 \text{ g } KNO_3}$$

13.35   **Strategy:**   The given solubility allows us to calculate Henry's law constant (k), which can then be used to determine the concentration of CO_2 at 0.0003 atm.

   Solution: First, calculate the Henry's law constant, k, using the concentration of CO_2 in water at 1 atm.

$$k = \frac{c}{P} = \frac{0.034 \text{ mol/L}}{1 \text{ atm}} = 0.034 \text{ mol/L·atm}$$

For atmospheric conditions we write:

$$c = kP = (0.034 \text{ mol/L·atm})(0.00030 \text{ atm}) = \mathbf{1.0 \times 10^{-5} \text{ mol/L}}$$

13.37   **Strategy:**   Consider the number of polar groups, and determine whether each molecule is predominantly polar or predominantly nonpolar. Predominantly polar molecules tend to be water soluble. Predominantly nonpolar molecules tend to be fat soluble.

   Setup: Polar groups include those containing O atoms or N atoms (atoms with lone pairs). Vitamin C contains six polar groups (six O atoms) making it predominately polar. Vitamin E contains only two polar groups and is therefore predominately nonpolar.

   Solution: **Vitamin C: water soluble; Vitamin E: fat soluble**

13.39 **Strategy:** The mass of CO_2 liberated from the container is given by the difference of the two mass measurements. Assume that, at the time of the second mass measurement, all the CO_2 has been released and that the amount of CO_2 in the vapor phase of the unopened soda is negligible. Then, we can use the following conversions to calculate the pressure of CO_2 in the unopened soda.

Mass of CO_2 lost $\rightarrow$ moles of CO_2 in solution $\rightarrow$ pressure of CO_2 above solution

Solution: Mass of CO_2 lost = 853.5 g $-$ 851.3 g = 2.2 g CO_2

$$\text{moles of } CO_2 = \left(2.2 \text{ g } CO_2\right)\left(\frac{1 \text{ mol } CO_2}{44.01 \text{ g } CO_2}\right) = 0.050 \text{ mol } CO_2$$

Using Henry's law for the final conversion:

$$c = kP$$

$$\frac{0.050 \text{ mol } CO_2}{0.4524 \text{ L}} = (3.4 \times 10^{-2} \text{ mol/L} \cdot \text{atm}) P$$

$$P = \textbf{3.3 atm}$$

This pressure is only an estimate since we ignored the amount of CO_2 that was present in the unopened container in the gas phase.

13.41 The dissolution of the **red solute** is **exothermic**, because increasing the temperature causes it to precipitate.

The dissolution of the **green solute** is **endothermic**, because increasing the temperature causes more solute to dissolve.

The numerical value of ΔH_{soln} is greater for the red solute, since changing the temperature produces a greater difference in solubility. The formation of chemical bonds of the precipitate in the red solute releases energy, whereas the breaking of chemical bonds in the dissolution in the green solute absorbs energy.

13.57 **Strategy:** From the vapor pressure of water at 20°C and the change in vapor pressure for the solution (2.0 mmHg), we can solve for the mole fraction of sucrose using Equation 13.5 of the text. From the mole fraction of sucrose, we can solve for moles of sucrose. Lastly, we convert from moles to grams of sucrose.

Solution: Using Equation 13.5 of the text, we can calculate the mole fraction of sucrose that causes a 2.0 mmHg drop in vapor pressure.

$$\Delta P = \chi_2 P_1^\circ$$

$$\Delta P = \chi_{sucrose} P_{water}^\circ$$

$$\chi_{sucrose} = \frac{\Delta P}{P_{water}^\circ} = \frac{2.0 \text{ mmHg}}{17.5 \text{ mmHg}} = 0.114$$

From the definition of mole fraction, we can calculate moles of sucrose.

$$\chi_{sucrose} = \frac{n_{sucrose}}{n_{water} + n_{sucrose}}$$

$$\text{moles of water} = 552 \text{ g} \times \frac{1 \text{ mol}}{18.02 \text{ g}} = 30.63 \text{ mol } H_2O$$

$$0.114 = \frac{n_{sucrose}}{30.63 + n_{sucrose}}$$

$$n_{sucrose} = 3.94 \text{ mol sucrose}$$

Using the molar mass of sucrose as a conversion factor, we can calculate the mass of sucrose.

$$\text{mass of sucrose} = 3.94 \text{ mol sucrose} \times \frac{342.3 \text{ g sucrose}}{1 \text{ mol}} = \mathbf{1.3 \times 10^3 \text{ g sucrose}}$$

13.59 For any solution the sum of the mole fractions of the components is always 1.00, so the mole fraction of 1-propanol is 0.700. The partial pressures are:

$$P_{ethanol} = \chi_{ethanol} \times P^{\circ}_{ethanol} = (0.300)(100 \text{ mmHg}) = \mathbf{30.0 \text{ mmHg}}$$

$$P_{1\text{-propanol}} = \chi_{1\text{-propanol}} \times P^{\circ}_{1\text{-propanol}} = (0.700)(37.6 \text{ mmHg}) = \mathbf{26.3 \text{ mmHg}}$$

13.61 This problem is very similar to Problem 13.57.

$$\Delta P = \chi_{urea} P^{\circ}_{water}$$

$$2.50 \text{ mmHg} = \chi_{urea}(31.8 \text{ mmHg})$$

$$\chi_{urea} = 0.0786$$

The number of moles of water is:

$$n_{water} = 658 \text{ g } H_2O \times \frac{1 \text{ mol } H_2O}{18.02 \text{ g } H_2O} = 36.51 \text{ mol } H_2O$$

$$\chi_{urea} = \frac{n_{urea}}{n_{water} + n_{urea}}$$

$$0.0786 = \frac{n_{urea}}{36.51 + n_{urea}}$$

$$n_{urea} = 3.11 \text{ mol}$$

$$\text{mass of urea} = 3.11 \text{ mol urea} \times \frac{60.06 \text{ g urea}}{1 \text{ mol urea}} = \mathbf{187 \text{ g urea}}$$

13.63 $\quad m = \dfrac{\Delta T_f}{K_f} = \dfrac{1.1°C}{1.86°C/m} = \mathbf{0.59\ \textit{m}}$

13.65   We first find the number of moles of gas using the ideal gas equation.

$$n = \frac{PV}{RT} = \frac{\left(748 \text{ mmHg} \times \dfrac{1 \text{ atm}}{760 \text{ mmHg}}\right)(4.00 \text{ L})}{(27 + 273) \text{ K}} \times \frac{\text{K} \cdot \text{mol}}{0.08206 \text{ L} \cdot \text{atm}} = 0.160 \text{ mol}$$

$$\text{molality} = \frac{0.160 \text{ mol}}{0.0750 \text{ kg benzene}} = 2.13 \, m$$

$$\Delta T_f = K_f m = (5.12°\text{C}/m)(2.13 \, m) = 10.9°\text{C}$$

$$\text{freezing point} = 5.5°\text{C} - 10.9°\text{C} = \mathbf{-5.4°C}$$

13.67   a. NaCl is a strong electrolyte. The concentration of particles (ions) is double the concentration of NaCl. Note that because the density of water is 1 g/mL, 135 mL of water has a mass of 135 g.

The number of moles of NaCl is:

$$21.2 \text{ g NaCl} \times \frac{1 \text{ mol}}{58.44 \text{ g}} = 0.363 \text{ mol NaCl}$$

Next, we can find the changes in freezing and boiling points ($i = 2$).

$$m = \frac{0.363 \text{ mol}}{0.135 \text{ kg}} = 2.70 \, m$$

$$\Delta T_f = iK_f m = 2(1.86°\text{C}/m)(2.70 \, m) = 10.0°\text{C}$$

$$\Delta T_b = iK_b m = 2(0.52°\text{C}/m)(2.70 \, m) = 2.8°\text{C}$$

The *freezing point* is **−10.0°C**; the *boiling point* is **102.8°C**.

b. Urea is a nonelectrolyte. The particle concentration is just equal to the urea concentration.

The molality of the urea solution is:

$$\text{moles of urea} = 15.4 \text{ g urea} \times \frac{1 \text{ mol urea}}{60.06 \text{ g urea}} = 0.256 \text{ mol urea}$$

$$m = \frac{0.256 \text{ mol urea}}{0.0667 \text{ kg H}_2\text{O}} = 3.84 \, m$$

$$\Delta T_f = iK_f m = 1(1.86°\text{C}/m)(3.84 \, m) = 7.14°\text{C}$$

$$\Delta T_b = iK_b m = 1(0.52°\text{C}/m)(3.84 \, m) = 2.0°\text{C}$$

The *freezing point* is **−7.14°C**; the *boiling point* is **102.0°C**.

13.69   **Both NaCl and CaCl₂ are strong electrolytes. Urea and sucrose are nonelectrolytes. NaCl or CaCl₂ will yield more particles per mole of the solid dissolved, resulting in greater freezing point depression. Also, sucrose and urea would make a mess when the ice melts.**

13.71   The temperature and molarity of the two solutions are the same. If we divide Equation 13.11 of the text for one solution by the same equation for the other, we can find the ratio of the van't Hoff factors in terms of the osmotic pressures ($i = 1$ for urea).

$$\frac{\pi_{CaCl_2}}{\pi_{urea}} = \frac{iMRT}{MRT} = i = \frac{0.605 \text{ atm}}{0.245 \text{ atm}} = \textbf{2.47}$$

13.73   For this problem, we first need to calculate the pressure necessary to support a column of water 105 m high. We use the equation $P = hdg$, where h, d, and g are the column height in m, the density of water in kg/m^3, and the gravitational constant, respectively.

$$P = 105 \text{ m} \times 1.00 \times 10^3 \frac{\text{kg}}{\text{m}^3} \times 9.81 \frac{\text{m}}{\text{s}^2} = 1.03 \times 10^6 \text{ kg/m·s}^2 = 1.03 \times 10^6 \text{ Pa}$$

$$1.03 \times 10^6 \text{ Pa} \times \frac{1 \text{ atm}}{101,325 \text{ Pa}} = 10.16 \text{ atm}$$

Because atmospheric pressure contributes 1 atm, the osmotic pressure required is only **9.16 atm**.

13.75   CaCl$_2$ is an ionic compound and is therefore an electrolyte in water. Assuming that CaCl$_2$ is a strong electrolyte and completely dissociates (no ion pairs, van't Hoff factor $i = 3$), the total ion concentration will be 3×0.35 m = 1.05 m, which is larger than the urea (nonelectrolyte) concentration of 0.90 m.

a. The **CaCl$_2$** solution will show a larger boiling point elevation because there are more particles of solute.

b. The freezing point of the **urea** solution will be higher because the CaCl$_2$ solution will have a larger freezing point depression.

c. The **CaCl$_2$** solution will have a lower vapor pressure. **CaCl$_2$ is an ionic compound and is therefore an electrolyte in water. Assuming that CaCl$_2$ is a strong electrolyte and completely dissociates (no ion pairs, van't Hoff factor $i = 3$), the total ion concentration will be $3 \times 0.35 = 1.05$ m, which is larger than the urea (nonelectrolyte) concentration of 0.90 m.**

13.77   Assume that all the salts are completely dissociated. Calculate the molality of the ions in the solutions.

0.10 m Na$_3$PO$_4$:	0.10 $m \times 4$ ions/unit = 0.40 m
0.35 m NaCl:	0.35 $m \times 2$ ions/unit = 0.70 m
0.20 m MgCl$_2$:	0.20 $m \times 3$ ions/unit = 0.60 m
0.15 m C$_6$H$_{12}$O$_6$:	nonelectrolyte, 0.15 m
0.15 m CH$_3$COOH:	weak electrolyte, slightly greater than 0.15 m

The solution with the lowest molality will have the highest freezing point (smallest freezing point depression): **0.15 m C$_6$H$_{12}$O$_6$ > 0.15 m CH$_3$COOH > 0.10 m Na$_3$PO$_4$ > 0.20 m MgCl$_2$ > 0.35 m NaCl.**

13.79   a. **NaCl**

b. **MgCl$_2$**

c. **KBr**

13.83   First, from the freezing point depression we can calculate the molality of the solution. See Table 13.2 of the text for the normal freezing point and K_f value for benzene.

$$\Delta T_f = (5.5 - 4.3)°C = 1.2°C$$

$$m = \frac{\Delta T_f}{K_f} = \frac{1.2°C}{5.12°C/m} = 0.23 \ m$$

Multiplying the molality by the mass of solvent (in kg) gives moles of unknown solute. Then, dividing the mass of solute (in g) by the moles of solute gives the molar mass of the unknown solute.

$$? \ \text{mol of unknown solute} = \frac{0.23 \ \text{mol solute}}{1 \ \text{kg benzene}} \times 0.0250 \ \text{kg benzene} = 0.0058 \ \text{mol solute}$$

$$\text{molar mass of unknown} = \frac{2.50 \ \text{g}}{0.0058 \ \text{mol}} = \mathbf{4.3 \times 10^2 \ g/mol}$$

The empirical molar mass of C_6H_5P is 108.1 g/mol. Therefore, the molecular formula is $(C_6H_5P)_4$ or **$C_{24}H_{20}P_4$**.

13.85   **Strategy:** We are asked to calculate the molar mass of the polymer. Grams of the polymer are given in the problem, so we need to solve for moles of polymer.

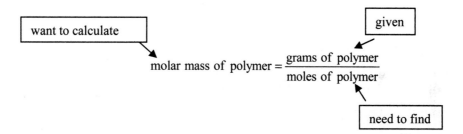

From the osmotic pressure of the solution, we can calculate the molarity of the solution. Then, from the molarity, we can determine the number of moles in 0.8330 g of the polymer.

Solution: First, we calculate the molarity using Equation 13.8 of the text.

$$\pi = MRT$$

$$M = \frac{\pi}{RT} = \frac{\left(5.20 \ \text{mmHg} \times \dfrac{1 \ \text{atm}}{760 \ \text{mmHg}}\right)}{298 \ \text{K}} \times \frac{\text{K} \cdot \text{mol}}{0.0821 \ \text{L} \cdot \text{atm}} = 2.80 \times 10^{-4} \ M$$

Multiplying the molarity by the volume of solution (in L) gives moles of solute (polymer).

$$? \ \text{mol of polymer} = (2.80 \times 10^{-4} \ \text{mol/L})(0.170 \ \text{L}) = 4.76 \times 10^{-5} \ \text{mol polymer}$$

Lastly, dividing the mass of polymer (in g) by the moles of polymer gives the molar mass of the polymer.

$$\text{molar mass of polymer} = \frac{0.8330 \text{ g polymer}}{4.76 \times 10^{-5} \text{ mol polymer}} = \mathbf{1.75 \times 10^4 \text{ g/mol}}$$

13.87 We use the osmotic pressure data to determine the molarity.

$$M = \frac{\pi}{RT} = \frac{4.61 \text{ atm}}{(20 + 273)\text{K}} \times \frac{\text{K} \cdot \text{mol}}{0.08206 \text{ L} \cdot \text{atm}} = 0.192 \text{ mol/L}$$

Next we use the density and the solution mass to find the volume of the solution.

$$\text{mass of soln} = 6.85 \text{ g} + 100.0 \text{ g} = 106.9 \text{ g soln}$$

$$\text{volume of soln} = 106.9 \text{ g soln} \times \frac{1 \text{ mL}}{1.024 \text{ g}} = 104.4 \text{ mL} = 0.1044 \text{ L}$$

Multiplying the molarity by the volume (in L) gives moles of solute (carbohydrate).

$$\text{mol of solute} = M \times \text{L} = (0.192 \text{ mol/L})(0.1044 \text{ L}) = 0.02004 \text{ mol solute}$$

Finally, dividing mass of carbohydrate by moles of carbohydrate gives the molar mass of the carbohydrate.

$$\text{molar mass} = \frac{6.85 \text{ g carbohydrate}}{0.02004 \text{ mol carbohydrate}} = \mathbf{342 \text{ g/mol}}$$

13.89 Use the osmotic pressure to calculate the molar concentration of dissolved particles in the HB solution.

$$M = \frac{\pi}{RT} = \frac{2.83 \text{ atm}}{\left(0.08206 \dfrac{\text{L} \cdot \text{atm}}{\text{K} \cdot \text{mol}}\right)(298 \text{ K})} = 0.1157 \ M$$

When an HB molecule ionizes, it produces one H_2B^+ ion and one OH^- ion.

$$HB(aq) + H_2O(l) \rightleftharpoons H_2B^+(aq) + OH^-(aq)$$

The concentration of HB is originally 0.100 M. It decreases by x and the concentrations of H_2B^+ and OH^- each increase by x. Thus, the concentration of dissolved particles is $0.100 - x + 2x = 0.100 + x$. We solve for x and determine what percentage x is of the original concentration, 0.100 M.

$$\% \text{ ionization} = \frac{(0.1157 - 0.100)}{0.100} \times 100\% = \mathbf{15.7\%}$$

13.95 Convert each concentration to moles of glucose per mL and to total moles and total grams in 5.0 L.

Levels before:

$$\frac{0.140 \text{ g glucose}}{100 \text{ mL blood}} \times \frac{1 \text{ mol glucose}}{180.2 \text{ g glucose}} \times 1 \text{ mL} = \mathbf{7.77 \times 10^{-6} \text{ mol glucose/mL blood}}$$

$$\frac{7.77 \times 10^{-6} \text{ mol glucose}}{1 \text{ mL blood}} \times \frac{1000 \text{ mL}}{1 \text{ L}} \times 5.0 \text{ L} = \mathbf{3.88 \times 10^{-3} \text{ mol glucose}}$$

$$\frac{0.140 \text{ g glucose}}{100 \text{ mL blood}} \times \frac{1000 \text{ mL}}{1 \text{ L}} \times 5.0 \text{ L} = \textbf{7.0 g glucose}$$

Levels after:

$$\frac{0.240 \text{ g glucose}}{100 \text{ mL blood}} \times \frac{1 \text{ mol glucose}}{180.2 \text{ g glucose}} \times 1 \text{ mL} = \textbf{1.332} \times \textbf{10}^{-5} \textbf{ mol glucose/mL blood}$$

$$\frac{1.332 \times 10^{-5} \text{ mol glucose}}{1 \text{ mL blood}} \times \frac{1000 \text{ mL}}{1 \text{ L}} \times 5.0 \text{ L} = \textbf{6.66} \times \textbf{10}^{-2} \textbf{ mol glucose}$$

$$\frac{0.240 \text{ g glucose}}{100 \text{ mL blood}} \times \frac{1000 \text{ mL}}{1 \text{ L}} \times 5.0 \text{ L} = \textbf{12 g glucose}$$

13.97 **Water migrates through the semipermeable cell walls of the cucumber into the concentrated salt solution.**

13.99 $$\Delta T_f = iK_f m$$

You can find the K_f for water in Table 13.2 of the text. $i = \dfrac{\Delta T_f}{K_f m} = \dfrac{2.6\,°\text{C}}{(1.86\,°\text{C}/m)(0.40\,m)} = \textbf{3.5}$

13.101 One manometer has pure water over the mercury, one manometer has a 1.0 *m* solution of NaCl, and the other manometer has a 1.0 *m* solution of urea. The pure water will have the highest vapor pressure and will thus force the mercury column down the most; column X. Both the salt and the urea will lower the overall pressure of the water. However, the salt dissociates into sodium and chloride ions (van't Hoff factor *i* = 2), whereas urea is a molecular compound with a van't Hoff factor of 1. Therefore the urea solution will lower the pressure only half as much as the salt solution. Y is the NaCl solution, and Z is the urea solution.

X: water; Y: NaCl; Z: urea

13.103 **The pill is in a hypotonic solution. Consequently, by osmosis, water moves across the semipermeable membrane into the pill. The increase in pressure pushes the elastic membrane to the right, causing the drug to exit through the small holes at a constant rate.**

13.105 **Reverse osmosis involves no phase changes and is usually cheaper than distillation or freezing.**

To reverse the osmotic migration of water across a semipermeable membrane, an external pressure exceeding the osmotic pressure must be applied. To find the osmotic pressure of 0.70 *M* NaCl solution, we must use the van't Hoff factor because NaCl is a strong electrolyte and the total ion concentration becomes 2(0.70 *M*) = 1.4 *M*.

The osmotic pressure of sea water is:

$$\pi = iMRT = 2(0.70 \text{ mol/L})(0.08206 \text{ L·atm/K·mol})(298 \text{ K}) = \textbf{34 atm}$$

To cause reverse osmosis, a pressure in excess of 34 atm must be applied.

13.107 The total concentration of dissolved species is expressed in molarity. Using Equation 13.8 of the text, we find the molarity of the solution.

$$M = \frac{\pi}{RT} = \frac{7.5 \text{ atm}}{\left(0.08206 \dfrac{\text{L} \cdot \text{atm}}{\text{K} \cdot \text{mol}}\right)(310 \text{ K})} = \textbf{0.295 } \boldsymbol{M}$$

This is the combined concentrations of all the ions. For dilute solutions, $M \approx m$. You can find the K_f for water in Table 13.2 of the text. Therefore, the freezing point of blood can be calculated using Equation 13.7.

$$\Delta T_f = K_f\, m = (1.86°\text{C}/m)(0.295\ m) = 0.55°\text{C}$$

Thus, the freezing point of blood should be about **–0.55°C**.

13.109 a. Using Equation 13.8 of the text, we find the molarity of the solution.

$$M = \frac{\pi}{RT} = \frac{0.257 \text{ atm}}{(0.08206\ \text{L} \cdot \text{atm/K} \cdot \text{mol})(298 \text{ K})} = 0.0105 \text{ mol/L}$$

This is the combined concentrations of all the ions. The amount dissolved in 10.0 mL (0.01000 L) is

$$? \text{ moles} = \frac{0.0105 \text{ mol}}{1 \text{ L}} \times 0.0100 \text{ L} = 1.05 \times 10^{-4} \text{ mol}$$

Since the mass of this amount of protein is 0.225 g, the apparent molar mass is

$$\frac{0.225 \text{ g}}{1.05 \times 10^{-4} \text{ mol}} = \textbf{2.14} \boldsymbol{\times 10^3} \textbf{ g / mol}$$

b. We need to use a van't Hoff factor to take into account the fact that the protein is a strong electrolyte. There are 20 Na^+ ions and 1 negatively charged protein P^{20-} for a total of 21 ions per protein molecule, so the van't Hoff factor will be $i = 21$.

$$M = \frac{\pi}{iRT} = \frac{0.257 \text{ atm}}{(21)(0.0821\ \text{L} \cdot \text{atm/K} \cdot \text{mol})(298 \text{ K})} = 5.00 \times 10^{-4} \text{ mol/L}$$

This is the actual concentration of the protein. The amount in 10.0 mL (0.0100 L) is

$$\frac{5.00 \times 10^{-4} \text{ mol}}{1 \text{ L}} \times 0.0100 \text{ L} = 5.00 \times 10^{-6} \text{ mol}$$

Therefore the actual molar mass is:

$$\frac{0.225 \text{ g}}{5.00 \times 10^{-6} \text{ mol}} = \textbf{4.50} \boldsymbol{\times 10^4} \textbf{ g / mol}$$

13.111

$$2H_2O_2 \longrightarrow 2H_2O + O_2$$

$$10 \text{ mL} \times \frac{3.0 \text{ g } H_2O_2}{100 \text{ mL}} \times \frac{1 \text{ mol } H_2O_2}{34.02 \text{ g } H_2O_2} \times \frac{1 \text{ mol } O_2}{2 \text{ mol } H_2O_2} = 4.4 \times 10^{-3} \text{ mol } O_2$$

a. Using the ideal gas law:

$$V = \frac{nRT}{P} = \frac{\left(4.4 \times 10^{-3} \text{ mol O}_2\right)(0.08206 \text{ L} \cdot \text{atm/K} \cdot \text{mol})(273 \text{ K})}{1.0 \text{ atm}} = \textbf{99 mL} = \textbf{0.099 L}$$

b. The ratio of the volumes:

$$\frac{99 \text{ mL}}{10 \text{ mL}} = \textbf{9.9}$$

13.113 a. **At reduced pressure, the solution is supersaturated with CO_2.**

b. **As the escaping CO_2 expands, it cools, condensing water vapor in the air to form fog.**

13.115 a. **Runoff of the salt solution into the soil increases the salinity of the soil. If the soil becomes hypertonic relative to the tree cells, osmosis would reverse, and the tree would lose water to the soil and eventually die of dehydration.**

b. **Assuming the collecting duct acts as a semipermeable membrane, water would flow from the urine into the hypertonic fluid, thus returning water to the body.**

13.117 At equilibrium, the concentrations in the two beakers are equal. Let x L be the change in volume.

$$\frac{n_A}{(0.050 - x)(\text{L})} = \frac{n_B}{(0.050 + x)(\text{L})}$$

$$\frac{(0.050 \text{ L})(1.0 \text{ } M)}{(0.050 - x)(\text{L})} = \frac{(0.050 \text{ L})(2.0 \text{ } M)}{(0.050 + x)(\text{L})}$$

$$(0.050 + x)(1.0) = (0.050 - x)(2.0) \text{ (all units cancel)}$$

$$3x = 0.050$$

$$x = 0.0167 \text{ L} = 16.7 \text{ mL} \quad (x \text{ has a unit of liters})$$

The final volumes are:

$$(50 - 16.7) \text{ mL} = \textbf{33 mL}$$

$$(50 + 16.7) \text{ mL} = \textbf{67 mL}$$

13.119 First, we calculate the number of moles of HCl in 100 g of solution.

$$n_{HCl} = 100 \text{ g soln} \times \frac{37.7 \text{ g HCl}}{100 \text{ g soln}} \times \frac{1 \text{ mol HCl}}{36.46 \text{ g HCl}} = 1.03 \text{ mol HCl}$$

Next, we calculate the volume of 100 g of solution.

$$V = 100 \text{ g} \times \frac{1 \text{ mL}}{1.19 \text{ g}} \times \frac{1 \text{ L}}{1000 \text{ mL}} = 0.0840 \text{ L}$$

Finally, the molarity (molar concentration) of the solution is:

$$\frac{1.03 \text{ mol}}{0.0840 \text{ L}} = \textbf{12.3 } \textbf{\textit{M}}$$

13.121 Let the mass of NaCl be x g. Then, the mass of sucrose is $(10.2 - x)$ g.

We know that the equation representing the osmotic pressure is:

$$\pi = MRT$$

π, R, and T are given. Using this equation and the definition of molarity, we can calculate the percentage of NaCl in the mixture.

$$\text{molarity} = \frac{\text{mol solute}}{\text{L soln}}$$

Remember that NaCl dissociates into two ions in solution; therefore, we multiply the moles of NaCl by 2.

$$\text{mol solute} = 2\left(x \text{ g NaCl} \times \frac{1 \text{ mol NaCl}}{58.44 \text{ g NaCl}} \right) + \left((10.2 - x) \text{ g sucrose} \times \frac{1 \text{ mol sucrose}}{342.3 \text{ g sucrose}} \right)$$

$$\text{mol solute} = 0.03422x + 0.02980 - 0.002921x$$

$$\text{mol solute} = 0.03130x + 0.02980$$

$$\text{Molarity of solution} = \frac{\text{mol solute}}{\text{L soln}} = \frac{(0.03130x + 0.02980) \text{ mol}}{0.250 \text{ L}}$$

Substitute molarity into the equation for osmotic pressure to solve for x.

$$\pi = MRT$$

$$7.32 \text{ atm} = \left(\frac{(0.03130x + 0.02980) \text{ mol}}{0.250 \text{ L}} \right) \left(0.08206 \frac{\text{L} \cdot \text{atm}}{\text{K} \cdot \text{mol}} \right) (296 \text{ K})$$

$$0.0753 = 0.03130x + 0.02980$$

$$x = 1.45 \text{ g} = \text{mass of NaCl}$$

$$\text{Mass \% NaCl} = \frac{1.45 \text{ g}}{10.2 \text{ g}} \times 100\% = \mathbf{14.2\%}$$

13.123 a. **Solubility decreases with increasing lattice energy.**

b. **Ionic compounds are more soluble in a polar solvent.**

c. **Solubility increases with enthalpy of hydration of the cation and anion.**

13.125 Let's assume we have 100 g of solution. The 100 g of solution will contain 70.0 g of HNO_3 and 30.0 g of H_2O.

$$\text{mol solute } (HNO_3) = 70.0 \text{ g } HNO_3 \times \frac{1 \text{ mol } HNO_3}{63.02 \text{ g } HNO_3} = 1.11 \text{ mol } HNO_3$$

$$\text{kg solvent } (H_2O) = 30.0 \text{ g } H_2O \times \frac{1 \text{ kg}}{1000 \text{ g}} = 0.0300 \text{ kg } H_2O$$

$$\text{molality} = \frac{1.11 \text{ mol } HNO_3}{0.0300 \text{ kg } H_2O} = \mathbf{37.0 \ \textit{m}}$$

To calculate the density, let's again assume we have 100 g of solution. Since,

$$d = \frac{\text{mass}}{\text{volume}}$$

We know the mass (100 g) and therefore need to calculate the volume of the solution. We know from the molarity that 15.9 mol of HNO_3 are dissolved in a solution volume of 1000 mL. In 100 g of solution, there are 1.11 moles HNO_3 (calculated above). What volume will 1.11 moles of HNO_3 occupy?

$$1.11 \text{ mol } HNO_3 \times \frac{1000 \text{ mL soln}}{15.9 \text{ mol } HNO_3} = 69.8 \text{ mL soln}$$

Dividing the mass by the volume gives the density.

$$d = \frac{100 \text{ g}}{69.8 \text{ mL}} = \mathbf{1.43 \text{ g} / \text{mL}}$$

13.127 **NH_3 can form hydrogen bonds with water; NCl_3 cannot. (Like dissolves like.)**

13.129 We can calculate the molality of the solution from the freezing point depression. See Table 13.2 of the text for the K_f value for water.

$$\Delta T_f = K_f m$$

$$0.203 = 1.86 \ m$$

$$m = \frac{0.203 \degree C}{1.86 \degree C/m} = 0.109 \ m$$

The molality of the original solution was 0.106 m. Some of the solution has ionized to H^+ and CH_3COO^-.

$$CH_3COOH \rightleftharpoons CH_3COO^- + H^+$$

Initial	0.106 m	0	0
Change	$-x$	$+x$	$+x$
Equilibrium	0.106 $m - x$	x	x

At equilibrium, the total concentration of species in solution is 0.109 m.

$$(0.106 - x) + 2x = 0.109 \ m$$

$$x = 0.003 \ m$$

The percentage of acid that has undergone ionization is:

$$\frac{0.003 \ m}{0.106 \ m} \times 100\% = \mathbf{3\%}$$

13.131 percent by mass **NaBr** $= \dfrac{\text{mass of NaBr}}{\text{mass of NaBr} + \text{mass of solvent}} \times 100\%$

$\qquad = \dfrac{5.75 \text{ g}}{67.9 \text{ g}} \times 100\% = \textbf{8.47\%}$

percent by mass **KCl** $= \dfrac{\text{mass of KCl}}{\text{mass of KCl} + \text{mass of water}} \times 100\%$

$\qquad = \dfrac{24.6 \text{ g}}{24.6 \text{ g} + 114 \text{ g}} \times 100\% = \textbf{17.7\%}$

percent by mass **toluene** $= \dfrac{\text{mass of toluene}}{\text{mass of toluene} + \text{mass of benzene}} \times 100\%$

$\qquad = \dfrac{4.8 \text{ g}}{4.8 \text{ g} + 39 \text{ g}} \times 100\% = \textbf{11\%}$

13.133 $192 \ \mu\text{g} = 192 \times 10^{-6} \text{ g}$ or $1.92 \times 10^{-4} \text{ g}$

$$\text{mass of lead/L} = \dfrac{1.92 \times 10^{-4} \text{ g}}{2.6 \text{ L}} = 7.4 \times 10^{-5} \text{ g/L}$$

Safety limit: 0.050 ppm implies a mass of 0.050 g Pb per 1×10^6 g of water. 1 liter of water has a mass of 1000 g.

$$\text{mass of lead} = \dfrac{0.050 \text{ g Pb}}{1 \times 10^6 \text{ g H}_2\text{O}} \times 1000 \text{ g H}_2\text{O} = 5.0 \times 10^{-5} \text{ g/L}$$

Yes, the concentration of lead calculated above (7.4×10^{-5} g/L) exceeds the safety limit of 5.0×10^{-5} g/L.

13.135 **As the water freezes, dissolved minerals in the water precipitate from the solution. The minerals refract light and create an opaque appearance**.

13.137 At equilibrium, the vapor pressure of benzene over each beaker must be the same. Assuming ideal solutions, this means that the mole fraction of benzene in each beaker must be identical at equilibrium. Consequently, the mole fraction of solute is also the same in each beaker, even though the solutes are different in the two solutions. Assuming the solute to be nonvolatile, equilibrium is reached by the transfer of benzene, through the vapor phase, from beaker A to beaker B.

The mole fraction of naphthalene in beaker A at equilibrium can be determined from the data given. The number of moles of naphthalene is given, and the moles of benzene can be calculated using its molar mass and knowing that 100 g – 7.0 g = 93.0 g of benzene remain in the beaker.

$$\chi_{\text{C}_{10}\text{H}_8} = \dfrac{0.15 \text{ mol}}{0.15 \text{ mol} + \left(93.0 \text{ g benzene} \times \dfrac{1 \text{ mol benzene}}{78.11 \text{ g benzene}} \right)} = 0.112$$

Now, let the number of moles of unknown compound be *n*. Assuming all the benzene lost from beaker A is transferred to beaker B, there are 100 g + 7.0 g = 107 g of benzene in the beaker. Also, recall that the mole fraction of solute in beaker B is equal to that in beaker A at equilibrium (0.112). The mole fraction of the unknown compound is:

$$\chi_{\text{unknown}} = \frac{n}{n + \left(107 \text{ g benzene} \times \dfrac{1 \text{ mol benzene}}{78.11 \text{ g benzene}}\right)}$$

$$0.112 = \frac{n}{n + 1.370}$$

$$n = 0.1728 \text{ mol}$$

There are 31 grams of the unknown compound dissolved in benzene. The molar mass of the unknown is:

$$\frac{31 \text{ g}}{0.173 \text{ mol}} = \mathbf{1.8 \times 10^2 \text{ g / mol}}$$

Temperature is assumed constant and ideal behavior is also assumed. Both solutes are assumed to be nonvolatile.

13.139 To solve for the molality of the solution, we need the moles of solute (urea) and the kilograms of solvent (water). If we assume that we have 1 mole of water, we know the mass of water. Using the change in vapor pressure, we can solve for the mole fraction of urea and then the moles of urea.

Using Equation 13.5 of the text, we solve for the mole fraction of urea.

$$\Delta P = 23.76 \text{ mmHg} - 22.98 \text{ mmHg} = 0.78 \text{ mmHg}$$

$$\Delta P = \chi_2 P_1^\circ = \chi_{\text{urea}} P_{\text{water}}^\circ$$

$$\chi_{\text{urea}} = \frac{\Delta P}{P_{\text{water}}^\circ} = \frac{0.78 \text{ mmHg}}{23.76 \text{ mmHg}} = 0.033$$

Assuming that we have 1 mole of water, we can now solve for moles of urea.

$$\chi_{\text{urea}} = \frac{\text{mol urea}}{\text{mol urea} + \text{mol water}}$$

$$0.033 = \frac{n_{\text{urea}}}{n_{\text{urea}} + 1}$$

$$0.033 n_{\text{urea}} + 0.033 = n_{\text{urea}}$$

$$0.033 = 0.967 n_{\text{urea}}$$

$$n_{\text{urea}} = 0.034 \text{ mol}$$

1 mole of water has a mass of 18.02 g or 0.01802 kg. We now know the moles of solute (urea) and the kilograms of solvent (water), so we can solve for the molality of urea in the solution.

$$m = \frac{\text{mol solute}}{\text{kg solvent}} = \frac{0.034 \text{ mol}}{0.01802 \text{ kg}} = \mathbf{1.9 \ m}$$

13.141 The total vapor pressure depends on the vapor pressures of A and B in the mixture, which in turn depends on the vapor pressures of pure A and B. With the total vapor pressure of the two mixtures known, a pair of simultaneous equations can be written in terms of the vapor pressures of pure A and B. We carry two extra significant figures throughout this calculation to avoid rounding errors.

For the solution containing 1.2 moles of A and 2.3 moles of B,

$$\chi_A = \frac{1.2 \text{ mol}}{1.2 \text{ mol} + 2.3 \text{ mol}} = 0.3429$$

$$\chi_B = 1 - 0.3429 = 0.6571$$

$$P_{total} = P_A + P_B = \chi_A P_A^\circ + \chi_B P_B^\circ$$

Substituting in P_{total} and the mole fractions calculated gives:

$$331 \text{ mmHg} = 0.3429 P_A^\circ + 0.6571 P_B^\circ$$

Solving for P_A°,

$$P_A^\circ = \frac{331 \text{ mmHg} - 0.6571 P_B^\circ}{0.3429} = 965.3 \text{ mmHg} - 1.916 P_B^\circ \qquad (1)$$

Now, consider the solution with the additional mole of B.

$$\chi_A = \frac{1.2 \text{ mol}}{1.2 \text{ mol} + 3.3 \text{ mol}} = 0.2667$$

$$\chi_B = 1 - 0.2667 = 0.7333$$

$$P_{total} = P_A + P_B = \chi_A P_A^\circ + \chi_B P_B^\circ$$

Substituting in P_{total} and the mole fractions calculated gives:

$$347 \text{ mmHg} = 0.2667 P_A^\circ + 0.7333 P_B^\circ \qquad (2)$$

Substituting Equation (1) into Equation (2) gives:

$$347 \text{ mmHg} = 0.2667(965.3 \text{ mmHg} - 1.916 P_B^\circ) + 0.7333 P_B^\circ$$

$$0.2223 P_B^\circ = 89.55 \text{ mmHg}$$

$$\boldsymbol{P_B^\circ = 402.8 \text{ mmHg} = 4.0 \times 10^2 \text{ mmHg}}$$

Substitute the value of P_B° into Equation (1) to solve for P_A°.

$$\boldsymbol{P_A^\circ = 965.3 \text{ mmHg} - 1.916(402.8 \text{ mmHg}) = 193.5 \text{ mmHg} = 1.9 \times 10^2 \text{ mmHg}}$$

13.143 To calculate the freezing point of the solution, we need the solution molality and the freezing-point depression constant for water (see Table 13.2 of the text). We can first calculate the molarity of the solution using Equation 13.8 of the text: $\pi = MRT$. The solution molality can then be determined from the molarity.

$$M = \frac{\pi}{RT} = \frac{10.50 \text{ atm}}{(0.08206 \text{ L} \cdot \text{atm/K} \cdot \text{mol})(298 \text{ K})} = 0.429 \text{ } M$$

Let's assume that we have 1 L (1000 mL) of solution. The mass of 1000 mL of solution is:

$$\frac{1.16 \text{ g}}{1 \text{ mL}} \times 1000 \text{ mL} = 1160 \text{ g soln}$$

The mass of the solvent (H_2O) is:

$$\text{mass } H_2O = \text{mass soln} - \text{mass solute}$$

$$\text{mass } H_2O = 1160 \text{ g} - \left(0.429 \text{ mol glucose} \times \frac{180.2 \text{ g glucose}}{1 \text{ mol glucose}} \right) = 1083 \text{ g} = 1.083 \text{ kg}$$

The molality of the solution is:

$$\text{molality} = \frac{\text{mol solute}}{\text{kg solvent}} = \frac{0.429 \text{ mol}}{1.083 \text{ kg}} = 0.396 \text{ } m$$

The freezing point depression is:

$$\Delta T_f = K_f m = (1.86°C/m)(0.396 \text{ } m) = 0.737°C$$

The solution will freeze at $0°C - 0.737°C = \textbf{–0.737°C}$

13.145 **Strategy:** From the figure, note that the normal boiling point of the solution is elevated by about 1°C. Use the boiling point elevation equation (Equation 13.6) with $K_b = 2.53°C/m$ (Table 13.2).

Solution:
$$\Delta T_b = K_b m$$

$$1°C = \left(2.53°C/m \right) m$$

$$m = \frac{1°C}{2.53°C/m} \approx 0.4 \text{ } m$$

The solution concentration is **about 0.4 molal**.

13.147 **Strategy:** Use Equation 13.2 and the definition of ppm to calculate the percent F^- by mass.

Use the molar mass of NaF to determine the mass of NaF in a gram of solution. Subtract the mass of NaF from the mass of the solution to determine the mass of the solvent (water). Use Equation 13.1 to determine the molality.

Setup: The molar mass of NaF is 41.99 g/mol.

$$1.0 \text{ ppm } F^- = \frac{1.0 \text{ g } F^-}{1,000,000 \text{ g solution}}$$

Consider the ionization of NaF:

$$NaF(aq) \longrightarrow Na^+(aq) + F^-(aq)$$

According to this equation, if 1.0 mol NaF molecules ionize, we get 1.0 mol Na^+ ion and 1.0 mol F^- ions.

Solution: To calculate percent mass by F^-,

$$1.0 \text{ ppm } F^- = \frac{1.0 \text{ g } F^-}{1{,}000{,}000 \text{ g solution}} \times 100\% = \mathbf{0.00010\% \ F^-} = \mathbf{1.0 \times 10^{-4}\% \ F^-}$$

To calculate molality, first determine the number of moles and mass of NaF per gram of solution.

$$\left(\frac{1.0 \text{ g } F^-}{1{,}000{,}000 \text{ g solution}}\right)\left(\frac{1 \text{ mol } F^-}{19.00 \text{ g } F^-}\right)\left(\frac{1 \text{ mol NaF}}{1 \text{ mol } F^-}\right) = \frac{5.3 \times 10^{-8} \text{ mol NaF}}{\text{g solution}}$$

$$\left(\frac{5.3 \times 10^{-8} \text{ mol NaF}}{\text{g solution}}\right)\left(\frac{41.99 \text{ g NaF}}{1 \text{ mol NaF}}\right) = \frac{2.2 \times 10^{-6} \text{ g NaF}}{\text{g solution}}$$

Finally, calculate the molality.

$$m = \left(\frac{\text{mol NaF}}{\text{kg H}_2\text{O}}\right) = \left(\frac{5.3 \times 10^{-8} \text{ mol NaF}}{1.0 \text{ g solution} - 2.2 \times 10^{-6} \text{ g NaF}}\right)\left(\frac{1000 \text{ g H}_2\text{O}}{1 \text{ kg H}_2\text{O}}\right) = \mathbf{5.3 \times 10^{-5} \ m \ NaF}$$

13.149 a. Strategy: The mole fraction of a component is the number of moles of the component divided by the total number of moles in the mixture:

$$\chi_i = \frac{n_i}{n_{total}}$$

Setup: The solution is prepared by mixing equal masses of A and B. Assume a mass of 1.00 g for each component.

The molar mass of A is 100 g/mol and the molar mass of B is 110 g/mol.

$$1.00 \text{ g A} \times \frac{1 \text{ mol A}}{100 \text{ g A}} = 0.01 \text{ mol A}$$

$$1.00 \text{ g B} \times \frac{1 \text{ mol B}}{110 \text{ g B}} = 0.0091 \text{ mol B}$$

Solution:
$$\chi_A = \frac{\text{mol A}}{\text{mol A} + \text{mol B}} = \frac{0.01}{(0.01 + 0.0091)} = \mathbf{0.52}$$

$$\chi_B = \frac{\text{mol B}}{\text{mol B} + \text{mol A}} = \frac{0.0091}{(0.0091 + 0.01)} = \mathbf{0.48}$$

b. **Strategy:** According to Raoult's law, the vapor pressure of the solution is the sum of the individual partial pressures exerted by the solution components. Use Equation 13.4 to calculate the partial pressures of A and B.

$$P_1 = \chi_1 P_1^{\circ}$$

Setup: At 55°C, $P_A^{\circ} = 98$ mmHg and $P_B^{\circ} = 42$ mmHg.

The mole fractions of A and B were calculated in part (a): $\chi_A = 0.52$ and $\chi_B = 0.48$.

Solution:
$$P_A = \chi_A P_A^{\circ} = 0.52 \times 98 \text{ mmHg} = \mathbf{51\,mmHg}$$

$$P_B = \chi_B P_B^{\circ} = 0.48 \times 42 \text{ mmHg} = \mathbf{20\,mmHg}$$

c. **Strategy:** The total pressure is given by Dalton's law of partial pressures, $P_{total} = P_A + P_B$. The mole fraction is then equal to the partial pressure of a component divided by the total pressure according to Equation 11.16:

$$\chi_i = \frac{P_i}{P_{total}}$$

Setup: Use the partial pressures of A and B (calculated in part (b)) to determine the total pressure, P_T.

$$P_{total} = 51 \text{ mmHg} + 20 \text{ mmHg} = 71 \text{ mmHg}$$

Solution:
$$\chi_A = \frac{P_A}{P_{total}} = \frac{51 \text{ mmHg}}{71 \text{ mmHg}} = \mathbf{0.72}$$

$$\chi_B = \frac{P_B}{P_{total}} = \frac{20 \text{ mmHg}}{71 \text{ mmHg}} = \mathbf{0.28}$$

d. **Strategy:** Use Equation 13.4 to calculate the partial pressures of A and B.

$$P_1 = \chi_1 P_1^{\circ}$$

Setup: The mole fraction of each component above the condensed liquid was calculated in part (c): $\chi_A = 0.72$ and $\chi_B = 0.28$.

Solution:
$$P_A = \chi_A P_A^{\circ} = 0.72 \times 98 \text{ mmHg} = \mathbf{71\,mmHg}$$

$$P_B = \chi_B P_B^{\circ} = 0.28 \times 42 \text{ mmHg} = \mathbf{12\,mmHg}$$

Chapter 14
Entropy and Free Energy

Visualizing Chemistry

VC 14.1   b   VC 14.2   a   VC 14.3   c   VC 14.4   b

Key Skills

14.1   e   14.2   c   14.3   b   14.4   b

Problems

14.7   **Strategy:** According to Equation 14.2, the number of ways of arranging N molecules in X cells is $W = X^N$. Use this equation to calculate the number of arrangements and Equation 14.1 to calculate the entropy.

       Setup: For the setup in the figure, $X = 4$ when the barrier is in place and $X = 8$ when it is absent.

       Solution: a. **with barrier:** $N = 2$; $W = 4^2 =$ **16**; **without barrier**: $N = 2$; $W = 8^2 =$ **64**

          b. From part (a), we know that **16** of the 64 arrangements have both particles in the left side of the container. Similarly, there are **16** ways for the particles to be found on the right side. The number of arrangements with one particle per side is $64 - 16 - 16 =$ **32**. **Both particles on one side:** $S = k \ln W = (1.38 \times 10^{-23}$ J/K$) \ln(16) = $**3.83 × 10^{-23} J/K**; **particles on opposite sides:** $S = (1.38 \times 10^{-23}$ J/K$) \ln(32) = $**4.78 × 10^{-23} J/K**; **The most probable state is the one with the larger entropy (particles are on opposite sides).**

14.13   **Strategy:** Equation 14.4 gives the entropy change for the isothermal expansion of an ideal gas. Substitute the given values into the equation and compute ΔS.

       Solution: a. $\Delta S_{sys} = (0.0050 \text{ mol})(8.314 \text{ J/K} \cdot \text{mol}) \ln\left(\dfrac{52.5 \text{ mL}}{112 \text{ mL}}\right) = $**−0.031 J / K**

          b. $\Delta S_{sys} = (0.015 \text{ mol})(8.314 \text{ J/K} \cdot \text{mol}) \ln\left(\dfrac{22.5 \text{ mL}}{225 \text{ mL}}\right) = $**−0.29 J / K**

          c. $\Delta S_{sys} = (22.1 \text{ mol})(8.314 \text{ J/K} \cdot \text{mol}) \ln\left(\dfrac{275 \text{ L}}{122 \text{ L}}\right) = $**1.49×10^2 J / K**

14.15   **Strategy:** To calculate the standard entropy change of a reaction, we look up the standard entropies of reactants and products in Appendix 2 of the text and apply Equation 14.6. As in the calculation of enthalpy of reaction, the stoichiometric coefficients have no units, so ΔS°_{rxn} is expressed in units of J/K·mol.

Solution: The standard entropy change for a reaction can be calculated using the following equation.

$$\Delta S^\circ_{rxn} = \Sigma n S^\circ(\text{products}) - \Sigma m S^\circ(\text{reactants})$$

a.
$$\Delta S^\circ_{rxn} = S^\circ(\text{Cu}) + S^\circ[\text{H}_2\text{O}(g)] - [S^\circ(\text{H}_2) + S^\circ(\text{CuO})]$$

$$= (1)(33.3 \text{ J/K·mol}) + (1)(188.7 \text{ J/K·mol}) - [(1)(131.0 \text{ J/K·mol}) + (1)(43.5 \text{ J/K·mol})]$$

$$= \mathbf{47.5 \text{ J/K · mol}}$$

b.
$$\Delta S^\circ_{rxn} = S^\circ(\text{Al}_2\text{O}_3) + 3S^\circ(\text{Zn}) - [2S^\circ(\text{Al}) + 3S^\circ(\text{ZnO})]$$

$$= (1)(50.99 \text{ J/K·mol}) + (3)(41.6 \text{ J/K·mol}) - [(2)(28.3 \text{ J/K·mol}) + (3)(43.9 \text{ J/K·mol})]$$

$$= \mathbf{-12.5 \text{ J/K · mol}}$$

c.
$$\Delta S^\circ_{rxn} = S^\circ(\text{CO}_2) + 2S^\circ[\text{H}_2\text{O}(l)] - [S^\circ(\text{CH}_4) + 2S^\circ(\text{O}_2)]$$

$$= (1)(213.6 \text{ J/K·mol}) + (2)(69.9 \text{ J/K·mol}) - [(1)(186.2 \text{ J/K·mol}) + (2)(205.0 \text{ J/K·mol})]$$

$$= \mathbf{-242.8 \text{ J/K · mol}}$$

14.17 In order of increasing entropy per mole at 25°C:

$$\mathbf{(c) < (d) < (e) < (a) < (b)}$$

a. **Ne(g): a monatomic gas of higher molar mass than H$_2$. (For gas phase molecules, increasing molar mass tends to increase the molar entropy. While Ne(g) is structurally less complex than H$_2$(g), its much larger molar mass more than offsets the complexity difference.)**

b. **SO$_2$(g): a polyatomic gas of higher complexity and higher molar mass than Ne(g)** (see the explanation for (a) above).

c. **Na(s): highly ordered, crystalline material.**

d. **NaCl(s): highly ordered crystalline material, but with more particles per mole than Na(s).**

e. **H$_2$(g): a diatomic gas, hence of higher entropy than a solid.**

14.21 **Strategy:** Assume all reactants and products are in their standard states. The entropy change in the surroundings is related to the enthalpy change of the system by Equation 14.7. For each reaction in Exercise 14.15, use Appendix 2 to calculate ΔH_{sys} ($=\Delta H_{rxn}$) (Section 10.6) and then use Equation 14.7 to calculate ΔS_{surr}. Finally, use Equation 14.8 to compute ΔS_{univ}. If ΔS_{univ} is positive, then according to the second law of thermodynamics, the reaction is spontaneous. The values for ΔS_{sys} are found in the solution for Problem 14.15.

Solution: a.

$$\Delta H_{sys} = \Delta H_f^\circ (Cu) + \Delta H_f^\circ [H_2O(g)] - \left[\Delta H_f^\circ (H_2) + \Delta H_f^\circ (CuO) \right]$$

$$\Delta H_{sys} = 0 + (1)(-241.8 \text{ kJ/mol}) - \left[0 + (1)(-155.2 \text{ kJ/mol}) \right]$$

$$\Delta H_{sys} = -86.6 \text{ kJ/mol}$$

$$\Delta S_{surr} = -\frac{\Delta H_{sys}}{T} = -\frac{(-86.6 \text{ kJ/mol})}{298 \text{ K}} = 0.291 \text{ kJ/K·mol}$$

$$\Delta S_{surr} = \mathbf{291 \ J/K \cdot mol}$$

$$\Delta S_{univ} = \Delta S_{sys} + \Delta S_{surr} = 47.5 \text{ J/K·mol} + 291 \text{ J/K·mol}$$

$$\Delta S_{univ} = 339 \text{ J/K·mol}$$

$$\Delta S_{univ} > 0$$

spontaneous

b.

$$\Delta H_{sys} = \Delta H_f^\circ (Al_2O_3) + 3\Delta H_f^\circ (Zn) - \left[2\Delta H_f^\circ (Al) + 3\Delta H_f^\circ (ZnO) \right]$$

$$\Delta H_{sys} = (1)(-1669.8 \text{ kJ/mol}) + 0 - \left[0 + (3)(-348.0 \text{ kJ/mol}) \right]$$

$$\Delta H_{sys} = -626 \text{ kJ/mol}$$

$$\Delta S_{surr} = -\frac{\Delta H_{sys}}{T} = -\frac{(-626 \text{ kJ/mol})}{298 \text{ K}} = 2.10 \text{ kJ/K·mol}$$

$$\Delta S_{surr} = \mathbf{2.10 \times 10^3 \ J/K \cdot mol}$$

$$\Delta S_{univ} = \Delta S_{sys} + \Delta S_{surr} = -12.5 \text{ J/K·mol} + 2.10 \times 10^3 \text{ J/K·mol}$$

$$\Delta S_{univ} = 2.09 \times 10^3 \text{ J/K·mol}$$

$$\Delta S_{univ} > 0$$

spontaneous

c.

$$\Delta H_{sys} = \Delta H_f^\circ (CO_2) + 2\Delta H_f^\circ [H_2O(l)] - \left[\Delta H_f^\circ (CH_4) + 2\Delta H_f^\circ (O_2) \right]$$

$$\Delta H_{sys} = (1)(-393.5 \text{ kJ/mol}) + (2)(-285.8 \text{ kJ/mol}) - \left[(1)(-74.85 \text{ kJ/mol}) + 0 \right]$$

$$\Delta H_{sys} = -890.3 \text{ kJ/mol}$$

$$\Delta S_{surr} = -\frac{\Delta H_{sys}}{T} = -\frac{(-890.3 \text{ kJ/mol})}{298 \text{ K}} = 2.99 \text{ kJ/K·mol}$$

$$\Delta S_{surr} = 2.99 \times 10^3 \text{ J/K} \cdot \text{mol}$$

$$\Delta S_{univ} = \Delta S_{sys} + \Delta S_{surr} = -242.8 \text{ J/K} \cdot \text{mol} + 2.99 \times 10^3 \text{ J/K} \cdot \text{mol}$$

$$\Delta S_{univ} = 2.75 \times 10^3 \text{ J/K} \cdot \text{mol}$$

$$\Delta S_{univ} > 0$$

spontaneous

14.23 **Strategy:** Assume all reactants and products are in their standard states. According to Equation 14.7, the entropy change in the surroundings for an isothermal process is:

$$\Delta S_{surr} = -\frac{\Delta H_{sys}}{T}$$

Also, Equation 14.8 states that the entropy change of the universe is:

$$\Delta S_{univ} = \Delta S_{sys} + \Delta S_{surr}$$

Use Appendix 2 to calculate ΔS_{sys} $(= \Delta S_{rxn}^{\circ})$ and ΔH_{sys} $(= \Delta H_{rxn}^{\circ})$. Substitute these results into the above equations and determine the sign of ΔS_{univ}. If ΔS_{univ} is positive, then according to the second law of thermodynamics, the reaction is spontaneous.

Solution: a.

$$\Delta S_{rxn}^{\circ} = S^{\circ}\left(PCl_5\right) - \left[S^{\circ}\left(PCl_3(g)\right) + S^{\circ}\left(Cl_2\right)\right]$$

$$\Delta S_{rxn}^{\circ} = 364.5 \text{ J/K} \cdot \text{mol} - \left[217.1 \text{ J/K} \cdot \text{mol} + 223.0 \text{ J/K} \cdot \text{mol}\right]$$

$$\Delta S_{rxn}^{\circ} = -75.6 \text{ J / K} \cdot \text{mol}$$

$$\Delta H_{rxn}^{\circ} = \Delta H_{sys} = \Delta H_f^{\circ}\left(PCl_5\right) - \left[\Delta H_f^{\circ}\left(PCl_3(g)\right) + \Delta H_f^{\circ}\left(Cl_2\right)\right]$$
$$\Delta H_{sys} = -374.9 \text{ kJ/mol} - \left[-319.7 \text{ kJ/mol} + 0\right]$$

$$\Delta H_{sys} = -55.2 \text{ kJ/mol}$$

$$\Delta S_{surr} = -\frac{\Delta H_{sys}}{T} = -\frac{(-55.2 \text{ kJ/mol})}{298 \text{ K}} = 0.185 \text{ kJ/K} \cdot \text{mol}$$

$$\Delta S_{surr} = 185 \text{ J / K} \cdot \text{mol}$$

$$\Delta S_{univ} = \Delta S_{sys} + \Delta S_{surr} = \Delta S_{rxn}^{\circ} + \Delta S_{surr} = -75.6 \text{ J/K} \cdot \text{mol} + 185 \text{ J/K} \cdot \text{mol}$$

$$\Delta S_{univ} = 109 \text{ J/K} \cdot \text{mol}$$

$$\Delta S_{univ} > 0$$

spontaneous

b. $$\Delta S^{\circ}_{\text{rxn}} = 2S^{\circ}\left(\text{Hg}\right) + S^{\circ}\left(\text{O}_2\right) - 2S^{\circ}\left(\text{HgO}\right)$$

$$\Delta S^{\circ}_{\text{rxn}} = (2)\left(77.4 \text{ J/K·mol}\right) + 205.0 \text{ J/K·mol} - (2)\left(72.0 \text{ J/K·mol}\right)$$

$$\Delta S^{\circ}_{\text{rxn}} = \textbf{215.8 J / K·mol}$$

$$\Delta H^{\circ}_{\text{rxn}} = \Delta H_{\text{sys}} = 2\Delta H^{\circ}_{\text{f}}\left(\text{Hg}\right) + \Delta H^{\circ}_{\text{f}}\left(\text{O}_2\right) - 2\Delta H^{\circ}_{\text{f}}\left(\text{HgO}\right)$$

$$\Delta H_{\text{sys}} = 0 + 0 - \left(-90.7 \text{ kJ/mol}\right)$$

$$\Delta H_{\text{sys}} = 90.7 \text{ kJ/mol}$$

$$\Delta S_{\text{surr}} = -\frac{\Delta H_{\text{sys}}}{T} = -\frac{\left(90.7 \text{ kJ/mol}\right)}{298 \text{ K}} = -0.304 \text{ kJ/K·mol}$$

$$\Delta S_{\text{surr}} = \textbf{--304 J / K·mol}$$

$$\Delta S_{\text{univ}} = \Delta S_{\text{sys}} + \Delta S_{\text{surr}} = \Delta S^{\circ}_{\text{rxn}} + \Delta S_{\text{surr}} = 215.8 \text{ J/K·mol} + \left(-304 \text{ J/K·mol}\right)$$

$$\Delta S_{\text{univ}} = -88 \text{ J/K·mol}$$

$$\Delta S_{\text{univ}} < 0$$

not spontaneous

c. $$\Delta S^{\circ}_{\text{rxn}} = 2S^{\circ}[\text{H}(g)] - S^{\circ}[\text{H}_2(g)] = (2)\left(114.6 \text{ J/K·mol}\right) - (1)\left(131.0 \text{ J/K·mol}\right)$$

$$\Delta S^{\circ}_{\text{rxn}} = \textbf{98.2 J / K·mol}$$

To calculate $\Delta H^{\circ}_{\text{rxn}}$ for $\text{H}_2 \rightarrow 2\text{H}$, use the bond enthalpy from Table 10.4.

$$\Delta H^{\circ}_{\text{rxn}} = \Delta H_{\text{sys}} = 436.4 \text{ kJ / mol}$$

Calculate ΔS_{surr} and ΔS_{univ}:

$$\Delta S_{\text{surr}} = -\frac{\Delta H_{\text{sys}}}{T} = -\frac{\left(436.4 \text{ kJ/mol}\right)}{298 \text{ K}} = -1.46 \text{ kJ/K·mol}$$

$$\Delta S_{\text{surr}} = \textbf{--1.46} \times \textbf{10}^3 \textbf{ J / K·mol}$$

$$\Delta S_{\text{univ}} = \Delta S_{\text{sys}} + \Delta S_{\text{surr}} = \Delta S^{\circ}_{\text{rxn}} + \Delta S_{\text{surr}} = 98.2 \text{ J/K·mol} + \left(-1.46 \times 10^3 \text{ J/K·mol}\right)$$

$$\Delta S_{\text{univ}} = -1.36 \times 10^3 \text{ J/K·mol}$$

$$\Delta S_{\text{univ}} < 0$$

not spontaneous

d.
$$\Delta S_{rxn}^{\circ} = S^{\circ}\left(UF_6\right) - \left[S^{\circ}\left(U\right) + 3S^{\circ}\left(F_2\right)\right]$$

$$\Delta S_{rxn}^{\circ} = 378 \text{ J/K·mol} - \left[50.21 \text{ J/K·mol} + (3)(203.34 \text{ J/K·mol})\right]$$

$$\Delta S_{rxn}^{\circ} = -282 \text{ J / K·mol}$$

$$\Delta H_{rxn}^{\circ} = \Delta H_{sys} = \Delta H_f^{\circ}\left(UF_6\right) - \left[\Delta H_f^{\circ}\left(U\right) + 3\Delta H_f^{\circ}\left(F_2\right)\right]$$

$$\Delta H_{sys} = -2147 \text{ kJ/mol} - (0 + 0)$$

$$\Delta H_{sys} = -2147 \text{ kJ/mol}$$

$$\Delta S_{surr} = -\frac{\Delta H_{sys}}{T} = -\frac{(-2147 \text{ kJ/mol})}{298 \text{ K}} = 7.20 \text{ kJ/K·mol}$$

$$\Delta S_{surr} = 7.20 \times 10^3 \text{ J / K·mol}$$

$$\Delta S_{univ} = \Delta S_{sys} + \Delta S_{surr} = \Delta S_{rxn}^{\circ} + \Delta S_{surr} = -282 \text{ J/K·mol} + 7.20 \times 10^3 \text{ J/K·mol}$$

$$\Delta S_{univ} = 6.92 \times 10^3 \text{ J/K·mol}$$

$$\Delta S_{univ} > 0$$

spontaneous

14.29 **Strategy:** Use Equation 14.10 to solve for ΔG.

Setup: From problem 14.24, $\Delta S = 1.52$ kJ/K · mol and $\Delta H = 510$ kJ/mol.

$T = (20 + 273.15) = 293.2$ K

Solution: $\Delta G = \Delta H - T\Delta S = 510$ kJ/mol $- (293.2$ K$)(1.52$ kJ/K·mol$) = \textbf{64 kJ/mol}$

14.31 **Strategy:** To calculate the standard free energy change of a reaction, we look up the standard free energies of formation of reactants and products in Appendix 2 of the text and apply Equation 14.12. Note that all the stoichiometric coefficients have no units so ΔG_{rxn}° is expressed in units of kJ/mol. The standard free energy of formation of any element in its stable allotropic form at 1 atm and 25°C is zero.

Solution: The standard free energy change for a reaction can be calculated using the following equation.

$$\Delta G_{rxn}^{\circ} = \Sigma n\Delta G_f^{\circ}(\text{products}) - \Sigma m\Delta G_f^{\circ}(\text{reactants})$$

a.
$$\Delta G^\circ = 2\Delta G_f^\circ(MgO) - [2\Delta G_f^\circ(Mg) + \Delta G_f^\circ(O_2)]$$

$$\Delta G^\circ = (2)(-569.6 \text{ kJ/mol}) - [(2)(0) + (1)(0)] = \textbf{-1139 kJ / mol}$$

b.
$$\Delta G^\circ = 2\Delta G_f^\circ(SO_3) - [2\Delta G_f^\circ(SO_2) + \Delta G_f^\circ(O_2)]$$

$$\Delta G^\circ = (2)(-370.4 \text{ kJ/mol}) - [(2)(-300.4 \text{ kJ/mol}) + (1)(0)] = \textbf{-140.0 kJ / mol}$$

c.
$$\Delta G^\circ = 4\Delta G_f^\circ\left(CO_2(g)\right) + 6\Delta G_f^\circ\left(H_2O(l)\right) - \left[2\Delta G_f^\circ\left(C_2H_6(g)\right) + 7\Delta G_f^\circ\left(O_2(g)\right)\right]$$

$$\Delta G^\circ = (4)(-394.4 \text{ kJ/mol}) + (6)(-237.2 \text{ kJ/mol}) - [(2)(-32.89 \text{ kJ/mol}) + (7)(0)]$$
$$= \textbf{-2935 kJ / mol}$$

14.33   a. Calculate ΔG from ΔH and ΔS.

$$\Delta G = \Delta H - T\Delta S = -126{,}000 \text{ J/mol} - (298 \text{ K})(84 \text{ J/K} \cdot \text{mol}) = -151{,}000 \text{ J/mol}$$

The free energy change is negative so the reaction is spontaneous at 298 K. Since ΔH is negative and ΔS is positive, the reaction is spontaneous at **all temperatures**.

b. Calculate ΔG.

$$\Delta G = \Delta H - T\Delta S = -11{,}700 \text{ J/mol} - (298 \text{ K})(-105 \text{ J/K·mol}) = +19{,}600 \text{ J}$$

The free energy change is positive at 298 K, which means the reaction is not spontaneous at that temperature. The positive sign of ΔG results from the large negative value of ΔS. At lower temperatures, the $-T\Delta S$ term will be smaller, thus allowing the free energy change to be negative.

ΔG will equal zero when $\Delta H = T\Delta S$.

Rearranging,

$$T = \frac{\Delta H}{\Delta S} = \frac{-11700 \text{ J/mol}}{-105 \text{ J/K} \cdot \text{mol}} = 111 \text{ K}$$

At temperatures **below 111 K**, ΔG will be negative and the reaction will be spontaneous.

14.35   **Strategy:** Equation 14.7 from the text relates the entropy change of the surroundings to the enthalpy change of the system and the temperature at which a phase change occurs.

$$\Delta S_{surr} = \frac{-\Delta H_{sys}}{T}$$

We know that $\Delta S_{sys} = -\Delta S_{surr}$, so we can rewrite Equation 14.7 as

$$\Delta S_{sys} = \frac{\Delta H_{sys}}{T}$$

Solution: $$\Delta S_{fus} = \frac{\Delta H_{fus}}{T_{melting}} = \frac{23.4 \text{ kJ/mol}}{234.3 \text{ K}} \times \frac{1000 \text{ J}}{1 \text{ kJ}} = \mathbf{99.9 \text{ J/K} \cdot \text{mol}}$$

$$\Delta S_{vap} = \frac{\Delta H_{vap}}{T_{boiling}} = \frac{59.0 \text{ kJ/mol}}{630 \text{ K}} \times \frac{1000 \text{ J}}{1 \text{ kJ}} = \mathbf{93.7 \text{ J/K} \cdot \text{mol}}$$

14.37 Using Equation 14.12 from the text,

$$\Delta G_{rxn}^{\circ} = \Delta G_f^{\circ}(\text{NO}_3^-) - [1/2 \ \Delta G_f^{\circ}(\text{O}_2) + \Delta G_f^{\circ}(\text{NO}_2^-)]$$

$$= (-110.5 \text{ kJ/mol}) - [(0 \text{ kJ/mol}) + (-34.6 \text{ kJ/mol}) = -75.9 \text{ kJ/mol}$$

75.9 kJ of Gibbs free energy are released.

14.41 **Strategy:** Melting is an endothermic process, $\Delta H_{fus} > 0$. Also, a liquid generally has a higher entropy than the solid at the same temperature, so $\Delta S_{fus} > 0$. To determine the sign of ΔG_{fus}, use the fact that melting is spontaneous ($\Delta G_{fus} < 0$) for temperatures above the freezing point and is not spontaneous ($\Delta G_{fus} > 0$) for temperatures below the freezing point.

Solution: a. The temperature is above the freezing point, so melting is spontaneous and $\Delta G_{fus} < 0$.

b. The temperature is at the freezing point, so the solid and the liquid are in equilibrium and $\Delta G_{fus} = 0$.

c. The temperature is below the freezing point, so melting is not spontaneous and $\Delta G_{fus} > 0$.

14.43 Setting ΔG equal to zero and solving Equation 14.10 for T gives

$$T = \frac{\Delta H}{\Delta S} = \frac{1.25 \times 10^5 \text{ J/mol}}{397 \text{ J/K} \cdot \text{mol}} = 315 \text{ K} = \mathbf{42°C}$$

14.45 If the process is *spontaneous* as well as *endothermic*, the signs of ΔG and ΔH must be negative and positive, respectively. Since $\Delta G = \Delta H - T\Delta S$, the sign of ΔS **must be positive ($\Delta S > 0$)** for ΔG to be negative.

14.47 a. Using the relationship:

$$\frac{\Delta H_{vap}}{T_{bp}} = \Delta S_{vap} \approx 90 \text{ J/K} \cdot \text{mol}$$

benzene $\Delta S_{vap} = \mathbf{87.8 \text{ J/K} \cdot \text{mol}}$

hexane $\Delta S_{vap} = \mathbf{90.1 \text{ J/K} \cdot \text{mol}}$

mercury $\Delta S_{vap} = \mathbf{93.7 \text{ J/K} \cdot \text{mol}}$

toluene $\Delta S_{vap} = \mathbf{91.8 \text{ J/K} \cdot \text{mol}}$

Trouton's rule is a statement about ΔS_{vap}. In most substances, the molecules are in constant and random motion in both the liquid and gas phases, so $\Delta S_{vap} \approx 90$ J/K $\cdot$ mol.

b. Using the data in Table 12.5 of the text, we find:

ethanol	$\Delta S_{vap} = 112$ **J/K $\cdot$ mol**
water	$\Delta S_{vap} = 109$ **J/K $\cdot$ mol**

In ethanol and water, there are fewer possible arrangements of the molecules due to the network of H-bonds, so ΔS_{vap} is greater.

14.49   a. ΔS is **positive**   b. ΔS is **negative**   c. ΔS is **positive**   d. ΔS is **positive**

14.51   At the temperature of the normal boiling point, the free energy difference between the liquid and gaseous forms of mercury (or any other substances) is zero, that is, the two phases are in equilibrium. We can therefore use Equation 14.10 of the text to find this temperature. For the equilibrium,

$$\text{Hg}(l) \rightleftharpoons \text{Hg}(g)$$

$$\Delta G = \Delta H - T\Delta S = 0$$

$$\Delta H = \Delta H_f^\circ[\text{Hg}(g)] - \Delta H_f^\circ[\text{Hg}(l)] = 60,780 \text{ J/mol} - 0 = 60,780 \text{ J/mol}$$

$$\Delta S = S^\circ[\text{Hg}(g)] - S^\circ[\text{Hg}(l)] = 174.7 \text{ J/K} \cdot \text{mol} - 77.4 \text{ J/K} \cdot \text{mol} = 97.3 \text{ J/K} \cdot \text{mol}$$

$$T_{bp} = \frac{\Delta H}{\Delta S} = \frac{60,780 \text{ J/mol}}{97.3 \text{ J/K} \cdot \text{mol}} = \textbf{625 K}$$

The calculation assumes that **ΔH° and ΔS° do not depend on temperature.**

14.53   **No. A negative ΔG° tells us that a reaction has the potential to happen but gives no indication of the rate.**

14.55   **Strategy:** We assume that the denaturation is a single-step equilibrium process. Therefore, because ΔG is zero at equilibrium, we can use Equation 14.10, substituting 0 for ΔG and solving for T, to determine the temperature at which the denaturation becomes spontaneous.

Setup: $\Delta H = 512$ kJ/mol and $\Delta S = 1.60$ kJ/K $\cdot$ mol.

Solution: **ΔH and ΔS are both positive, indicating this is an endothermic process with high entropy.**

$$T = \frac{\Delta H}{\Delta S} = \frac{512 \text{ kJ/mol}}{1.60 \text{ kJ/K} \cdot \text{mol}} = \textbf{320 K} = \textbf{47°C}$$

14.57   First convert to moles of ice.

$$74.6 \text{ g H}_2\text{O}(s) \times \frac{1 \text{ mol H}_2\text{O}(s)}{18.02 \text{ g H}_2\text{O}(s)} = 4.14 \text{ mol H}_2\text{O}(s)$$

For a phase transition:

$$\Delta S_{sys} = \frac{\Delta H_{sys}}{T}$$

$$\Delta S_{sys} = \frac{(4.14 \text{ mol})(6010 \text{ J/mol})}{273 \text{ K}} = \textbf{91.1 J/K}$$

$$\Delta S_{surr} = \frac{-\Delta H_{sys}}{T}$$

$$\Delta S_{surr} = \frac{-(4.14 \text{ mol})(6010 \text{ J/mol})}{273 \text{ K}} = \textbf{-91.1 J/K}$$

$$\Delta S_{univ} = \Delta S_{sys} + \Delta S_{surr} = \textbf{0}$$

The system is at equilibrium.

14.59 For a phase transition, $\Delta G = 0$. We write:

$$\Delta G = \Delta H - T\Delta S$$

$$0 = \Delta H - T\Delta S$$

$$\Delta S_{sub} = \frac{\Delta H_{sub}}{T}$$

Substituting ΔH and the temperature, $(-78°C + 273)K = 195$ K, gives

$$\Delta S_{sub} = \frac{\Delta H_{sub}}{T} = \frac{25.2 \times 10^3 \text{ J / mol}}{195 \text{ K}} = 129 \text{ J/K} \cdot \text{mol}$$

This value of ΔS_{sub} is for the sublimation of 1 mole of CO_2. We convert to the ΔS value for the sublimation of 84.8 g of CO_2.

$$84.8 \text{ g } CO_2 \times \frac{1 \text{ mol } CO_2}{44.01 \text{ g } CO_2} \times \frac{129 \text{ J}}{\text{K} \cdot \text{mol}} = \textbf{249 J/K}$$

14.61 We can calculate ΔS_{sys} from standard entropy values in Appendix 2 of the text. We can calculate ΔS_{surr} from the ΔH_{sys} value given in the problem. Finally, we can calculate ΔS_{univ} from the ΔS_{sys} and ΔS_{surr} values.

$$\Delta S_{sys} = (2)(69.9 \text{ J/K} \cdot \text{mol}) - [(2)(131.0 \text{ J/K} \cdot \text{mol}) + (1)(205.0 \text{ J/K} \cdot \text{mol})] = \textbf{-327.2 J/K} \cdot \textbf{mol}$$

$$\Delta S_{surr} = \frac{-\Delta H_{sys}}{T} = \frac{-\left(-571.6 \times 10^3 \text{ J/mol}\right)}{298 \text{ K}} = \textbf{1918 J / K} \cdot \textbf{mol}$$

$$\Delta S_{univ} = \Delta S_{sys} + \Delta S_{surr} = (-327.2 + 1918) \text{ J/K} \cdot \text{mol} = \textbf{1591 J/K} \cdot \textbf{mol}$$

14.63 **q and w are not state functions**. Recall that state functions represent properties that are determined by the state of the system, regardless of how that condition is achieved. Heat and work are not state functions because they are not properties of the system. They manifest themselves only during a process (during a change). Thus their values depend on the path of the process and vary accordingly.

14.65 Since the adsorption is spontaneous, **ΔG must be negative**. When hydrogen bonds to the surface of the catalyst, there is a decrease in the number of possible arrangements of atoms in the system, **ΔS must be negative**. Since there is a decrease in entropy, the adsorption must be exothermic for the process to be spontaneous, **ΔH must be negative**.

14.67 Only **U and H** are associated with the first law alone.

Chapter 15
Chemical Equilibrium

Visualizing Chemistry

VC 15.1 a   VC 15.2 c   VC 15.3 c   VC 15.4 c   VC 15.5 c   VC 15.6 b

VC 15.7 a and c   VC 15.8 a, b, and c   VC 15.9 c   VC 15.10 a   VC 15.11 b

VC 15.12 c

Key Skills

15.1   e   15.2   a   15.3   b   15.4   d

Problems

15.9   a. $\dfrac{[N_2][H_2O]^2}{[NO]^2[H_2]^2}$

     b. $K_c = \dfrac{(0.082)(4.64)^2}{(0.31)^2(0.16)^2} = 7.2 \times 10^2$

15.11   The problem states that the system is at equilibrium, so we simply substitute the equilibrium concentrations into the equilibrium expression to calculate K_c.

Step 1:   Calculate the concentrations of the components in units of mol/L. The molarities can be calculated by simply dividing the number of moles by the volume of the flask.

$$[H_2] = \frac{2.50 \text{ mol}}{12.0 \text{ L}} = 0.208 \ M$$

$$[S_2] = \frac{1.35 \times 10^{-5} \text{ mol}}{12.0 \text{ L}} = 1.13 \times 10^{-6} \ M$$

$$[H_2S] = \frac{8.70 \text{ mol}}{12.0 \text{ L}} = 0.725 \ M$$

Step 2: Once the molarities are known, K_c can be found by substituting the molarities into the equilibrium expression.

$$K_c = \frac{[H_2S]^2}{[H_2]^2[S_2]} = \frac{(0.725)^2}{(0.208)^2(1.13 \times 10^{-6})} = 1.08 \times 10^7$$

15.13   The equilibrium expression for this reaction is:

$$K_c = \frac{[B]^2}{[A]^2}$$

According to the first diagram, there are six yellow spheres (A) and four red spheres (B). The equilibrium expression becomes

$$K_c = \frac{[B]^2}{[A]^2} = \frac{(4)^2}{(6)^2} = \frac{16}{36}$$

In the second diagram, there are nine yellow spheres (A). For this reaction also to be at equilibrium,

$$\frac{16}{36} = \frac{x^2}{(9)^2} \text{ or } \frac{x^2}{81}$$

$$x^2 = 36$$

$$x = 6$$

There must be **six** red spheres.

15.23   When the equation for a reversible reaction is written in the opposite direction, the equilibrium constant becomes the reciprocal of the original equilibrium constant.

$$K_c' = \frac{1}{K_c} = \frac{1}{4.17 \times 10^{-34}} = \mathbf{2.40 \times 10^{33}}$$

15.25   **Strategy:**   The relationship between K_c and K_P is given by Equation 15.4 of the text. What is the change in the number of moles of gases from reactant to product? Recall that:

$$\Delta n = \text{moles of gaseous products} - \text{moles of gaseous reactants}$$

What unit of temperature should we use?

Solution: The relationship between K_c and K_P is given by Equation 15.4 of the text.

$$K_P = K_c \,[(0.08206 \text{ L·atm/K·mol}) \times T]^{\Delta n}$$

Rearrange the equation relating K_P and K_c, solving for K_c.

$$K_c = \frac{K_P}{(0.08206\,T)^{\Delta n}}$$

Because $T = 623$ K and $\Delta n = 3 - 2 = 1$, we have:

$$K_c = \frac{1.8 \times 10^{-5}}{(0.08206)(623 \text{ K})} = \mathbf{3.5 \times 10^{-7}}$$

15.27   The equilibrium expressions are:

a.

$$K_c = \frac{[NH_3]^2}{[N_2][H_2]^3}$$

Substituting the given equilibrium concentration gives:

$$K_c = \frac{(0.25)^2}{(0.11)(1.91)^3} = 0.082 \text{ or } 8.2 \times 10^{-2}$$

b.

$$K_c = \frac{[NH_3]}{[N_2]^{\frac{1}{2}}[H_2]^{\frac{3}{2}}}$$

Substituting the given equilibrium concentration gives:

$$K_c = \frac{(0.25)}{(0.11)^{\frac{1}{2}}(1.91)^{\frac{3}{2}}} = 0.29$$

15.29   Because pure solids do not enter into an equilibrium expression, we can calculate K_P directly from the pressure that is due solely to $CO_2(g)$.

$$K_P = P_{CO_2} = 0.105$$

Now, we can convert K_P to K_c using the following equation.

$$K_P = K_c \, [(0.08206 \text{ L·atm/K·mol}) \times T]^{\Delta n}$$

$$K_c = \frac{K_P}{(0.08206 \, T)^{\Delta n}}$$

$$K_c = \frac{0.105}{(0.08206 \times 623)^{(1-0)}} = 2.05 \times 10^{-3}$$

15.31   **Strategy:**   Because they are constant quantities, the concentrations of solids and liquids do not appear in the equilibrium expressions for heterogeneous systems. The total pressure at equilibrium that is given is due to both NH_3 and CO_2. Note that for every 1 atm of CO_2 produced, 2 atm of NH_3 will be produced due to the stoichiometry of the balanced equation. Using this ratio, we can calculate the partial pressures of NH_3 and CO_2 at equilibrium.

Solution: The equilibrium expression for the reaction is

$$K_P = \left(P_{NH_3}\right)^2 P_{CO_2}$$

The total pressure in the flask (0.363 atm) is the sum of the partial pressures of NH_3 and CO_2.

$$P_{total} = P_{NH_3} + P_{CO_2} = 0.363 \text{ atm}$$

Let the partial pressure of $CO_2 = x$. From the stoichiometry of the balanced equation, you should find that $P_{NH_3} = 2P_{CO_2}$. Therefore, the partial pressure of $NH_3 = 2x$.

Substituting into the equation for total pressure gives:

$$P_{total} = P_{NH_3} + P_{CO_2} = 2x + x = 3x$$

$$3x = 0.363 \text{ atm}$$

$$x = P_{CO_2} = 0.121 \text{ atm}$$

$$P_{NH_3} = 2x = 0.242 \text{ atm}$$

Substitute the equilibrium pressures into the equilibrium expression to solve for K_P.

$$K_P = \left(P_{NH_3}\right)^2 P_{CO_2} = (0.242)^2(0.121) = \mathbf{7.09 \times 10^{-3}}$$

15.33   Let x be the initial pressure of NOBr. Using the balanced equation, we can write expressions for the partial pressures at equilibrium.

$$P_{NOBr} = (1 - 0.34)x = 0.66x$$

$$P_{NO} = 0.34x$$

$$P_{Br_2} = 0.17x$$

The sum of these is the total pressure.

$$0.66x + 0.34x + 0.17x = 1.17x = 0.25 \text{ atm}$$

$$x = 0.214 \text{ atm}$$

The equilibrium pressures are then:

$$P_{NOBr} = 0.66(0.214) = 0.141 \text{ atm}$$

$$P_{NO} = 0.34(0.214) = 0.073 \text{ atm}$$

$$P_{Br_2} = 0.17(0.214) = 0.036 \text{ atm}$$

We find K_P by substitution.

$$K_P = \frac{\left(P_{NO}\right)^2 P_{Br_2}}{\left(P_{NOBr}\right)^2} = \frac{(0.073)^2(0.036)}{(0.141)^2} = \mathbf{9.6 \times 10^{-3}}$$

The relationship between K_P and K_c is given by:

$$K_P = K_c(RT)^{\Delta n}$$

We find K_c (for this system $\Delta n = +1$)

$$K_c = \frac{K_P}{(RT)^{\Delta n}} = \frac{K_P}{RT} = \frac{9.6 \times 10^{-3}}{(0.08206 \times 298)} = 3.9 \times 10^{-4}$$

15.35

$$K_c = K_c' K_c''$$

$$K_c = (6.5 \times 10^{-2})(6.1 \times 10^{-5})$$

$$\boldsymbol{K_c = 4.0 \times 10^{-6}}$$

15.37 To obtain $2SO_2$ as a reactant in the final equation, we must reverse the first equation and multiply by 2. For the equilibrium, $2SO_2(g) \rightleftharpoons 2S(s) + 2O_2(g)$

$$K_c''' = \left(\frac{1}{K_c'}\right)^2 = \left(\frac{1}{4.2 \times 10^{52}}\right)^2 = 5.7 \times 10^{-106}$$

Now we can add the above equation to the second equation to obtain the final equation. Since we add the two equations, the equilibrium constant is the product of the equilibrium constants for the two reactions.

$$2SO_2(g) \rightleftharpoons 2S(s) + 2O_2(g) \qquad K_c''' = 5.7 \times 10^{-106}$$

$$2S(s) + 3O_2(g) \rightleftharpoons 2SO_3(g) \qquad K_c'' = 9.8 \times 10^{128}$$

$$\overline{2SO_2(g) + O_2(g) \rightleftharpoons 2SO_3(g) \qquad \boldsymbol{K_c = K_c''' \times K_c'' = 5.6 \times 10^{23}}}$$

15.39

$$K_c = \frac{[B]}{[A]}$$

(1) With $K_c = 10$, products are favored at equilibrium. Because the coefficients for both A and B are one, we expect the concentration of B to be 10 times that of A at equilibrium. **Diagram (a)** is the best choice with 10 B molecules and 1 A molecule.

(2) With $K_c = 0.10$, reactants are favored at equilibrium. Because the coefficients for both A and B are one, we expect the concentration of A to be 10 times that of B at equilibrium. **Diagram (d)** is the best choice with 10 A molecules and 1 B molecule.

You can calculate K_c in each case without knowing the volume of the container because the mole ratio between A and B is the same. **Volume will cancel from the K_c expression.** Only moles of each component are needed to calculate K_c.

15.43 **Strategy:** We are given the concentrations of all species in a biochemical reaction and must determine whether the reaction will proceed in the forward direction or the reverse direction in order to establish equilibrium. This requires calculating the value of Q and comparing it to the value of K, which is given in the problem.

Solution: The reaction quotient is

$$Q = \frac{[\alpha-\text{ketoglutarate}][\text{L-alanine}]}{[\text{L-glutamate}][\text{pyruvate}]} = \frac{(1.6 \times 10^{-2})(6.25 \times 10^{-3})}{(3.0 \times 10^{-5})(3.3 \times 10^{-4})} = 1.0 \times 10^4$$

Because K is 1.11, Q is greater than K. Thus, the reaction will proceed to the left (the reverse reaction will occur) in order to establish equilibrium. **The forward reaction will not occur.**

15.45
$$K_{sp} = [Fe^{2+}][OH^-]^2 = 1.6 \times 10^{-14}$$

$$\Delta G° = -RT \ln K_{sp} = -(8.314 \text{ J/K·mol})(298 \text{ K})\ln(1.6 \times 10^{-14}) = 7.9 \times 10^4 \text{ J/mol} = \textbf{79 kJ/mol}$$

15.47 a. We first find the standard free energy change of the reaction.

$$\Delta G°_{rxn} = \Delta G°_f\left(PCl_3(g)\right) + \Delta G°_f\left(Cl_2(g)\right) - \Delta G°_f\left(PCl_5(g)\right)$$
$$= (-269.6 \text{ kJ/mol}) + (0) - (-305.0 \text{ kJ/mol}) = \textbf{35.4 kJ/mol}$$

We can calculate K_P by rearranging Equation 15.6 of the text.

$$K_P = e^{\frac{-\Delta G°}{RT}} = e^{\frac{-35.4 \times 10^3 \text{ J/mol}}{(8.314 \text{ J/K·mol})(298 \text{ K})}} = e^{-14.29} = \textbf{6.2} \times \textbf{10}^{-7}$$

b. We are finding the free energy difference between the reactants and the products at their non-equilibrium values. The result tells us the direction of and the potential for further chemical change. We use the given non-equilibrium pressures to compute Q_P.

$$Q_P = \frac{P_{PCl_3} P_{Cl_2}}{P_{PCl_5}} = \frac{(0.27)(0.40)}{0.0029} = 37$$

The value of ΔG (notice that this is not the standard free energy difference) can be found using Equation 15.5 of the text and the result from part (a).

$$\Delta G = \Delta G° + RT \ln Q_P = (35.4 \times 10^3 \text{ J/mol}) + (8.314 \text{ J/K·mol})(298 \text{ K})\ln(37) = \textbf{44.6 kJ/mol}$$

15.49 The expression of K_P is: $K_P = P_{CO_2}$

Thus you can predict the equilibrium pressure directly from the value of the equilibrium constant. The only task at hand is computing the values of K_P using Equations 14.10 and 15.6 of the text.

a. At 25°C, $\Delta G° = \Delta H° - T\Delta S° = (177.8 \times 10^3 \text{ J/mol}) - (298 \text{ K})(160.5 \text{ J/K·mol})$

$$= 130.0 \times 10^3 \text{ J/mol}$$

$$P_{CO_2} = K_P = e^{\frac{-\Delta G°}{RT}} = e^{\frac{-130.0 \times 10^3 \text{ J/mol}}{(8.314 \text{ J/K·mol})(298 \text{ K})}} = e^{-52.47} = \textbf{1.6} \times \textbf{10}^{-23} \textbf{ atm}$$

b. At 800°C, $\Delta G° = \Delta H° - T\Delta S° = (177.8 \times 10^3 \text{ J/mol}) - (1073 \text{ K})(160.5 \text{ J/K·mol})$

$$= 5.584 \times 10^3 \text{ J/mol}$$

$$P_{CO_2} = K_P = e^{\frac{-\Delta G°}{RT}} = e^{\frac{-5.58 \times 10^3 \text{ J/mol}}{(8.314 \text{ J/K·mol})(1073 \text{ K})}} = e^{-0.625} = \textbf{0.535 atm}$$

15.51   The equilibrium constant expression is: $K_P = P_{H_2O}$

We are actually finding the equilibrium vapor pressure of water (compare Problem 15.49). We use Equation 15.6 of the text.

$$P_{H_2O} = K_P = e^{\frac{-\Delta G^\circ}{RT}} = e^{\frac{-8.6 \times 10^3 \text{ J/mol}}{(8.314 \text{ J/K·mol})(298 \text{ K})}} = e^{-3.47} = \mathbf{3.1 \times 10^{-2} \text{ atm or } 23.6 \text{ mmHg}}$$

15.53   **Strategy:**   According to Equation 15.6 of the text, the equilibrium constant for the reaction is related to the standard free energy change; that is, $\Delta G^\circ = -RT\ln K$. Since we are given ΔG° in the problem, we can solve for the equilibrium constant. What temperature unit should be used?

Solution: Solving Equation 15.6 for K gives

$$K_p = e^{\frac{-\Delta G^\circ}{RT}} = e^{\frac{-2.60 \times 10^3 \text{ J/mol}}{(8.314 \text{ J/K·mol})(298 \text{ K})}} = e^{-1.05} = \mathbf{0.35}$$

15.55   **Strategy:**   Use the law of mass action to write the equilibrium expression. The equilibrium expression must yield $K_c = 2$ for the reaction to be at equilibrium.

Setup: The equilibrium expression has the form of concentrations of products over concentrations of reactants, each raised to the appropriate power.

$$K_c = \frac{[AB]^2}{[A_2][B_2]}$$

Plug in the concentrations of all three species to evaluate K_c.

Solution: Diagram (a):

$$K_c = \frac{[2]^2}{[2][1]} = 2$$

Diagram (b):

$$K_c = \frac{[2]^2}{[1][2]} = 2$$

Diagram (c):

$$K_c = \frac{[4]^2}{[2][4]} = 2$$

Diagram (d):

$$K_c = \frac{[4]^2}{[2][2]} = 4$$

Since $K_c = 2$, **diagrams (a), (b), and (c)** represent an equilibrium mixture of A_2, B_2, and AB.

15.57 **Strategy:** The equilibrium constant K_c is given, and we start with a mixture of CO and H_2O. From the stoichiometry of the reaction, we can determine the concentration of H_2 at equilibrium. Knowing the partial pressure of H_2 and the volume of the container, we can calculate the number of moles of H_2.

Solution: The balanced equation shows that 1 mole of water will combine with 1 mole of carbon monoxide to form 1 mole of hydrogen and 1 mole of carbon dioxide. (This is the *reverse* reaction.) Let x be the depletion in the concentration of either CO or H_2O at equilibrium. The equilibrium concentration of hydrogen must then also be equal to x. The changes are summarized as shown in the table.

	H_2	$+$ CO_2	$\rightleftharpoons$ H_2O	$+$ CO
Initial (M):	0	0	0.0300	0.0300
Change (M):	$+x$	$+x$	$-x$	$-x$
Equilibrium (M):	x	x	$(0.0300 - x)$	$(0.0300 - x)$

The equilibrium constant is:

$$K_c = \frac{[H_2O][CO]}{[H_2][CO_2]} = 0.534$$

Substituting,

$$\frac{(0.0300 - x)^2}{x^2} = 0.534$$

Taking the square root of both sides, we obtain:

$$\frac{(0.0300 - x)}{x} = \sqrt{0.534}$$

$$x = 0.0173\ M$$

The number of moles of H_2 formed is:

$$0.0173\ \text{mol/L} \times 10.0\ \text{L} = \textbf{0.173 mol } \mathbf{H_2}$$

15.59 Notice that the balanced equation requires that for every 2 moles of HBr consumed, 1 mole of H_2 and 1 mole of Br_2 must be formed. Let $2x$ be the depletion in the concentration of HBr at equilibrium. The equilibrium concentrations of H_2 and Br_2 must therefore each be x. The changes are shown in the table.

	H_2 $+$	Br_2	$\rightleftharpoons$ 2HBr
Initial (M):	0	0	0.267
Change (M):	$+x$	$+x$	$-2x$
Equilibrium (M):	x	x	$(0.267 - 2x)$

The equilibrium constant relationship is given by:

$$K_c = \frac{[HBr]^2}{[H_2][Br_2]}$$

Substitution of the equilibrium concentration expressions gives

$$K_c = \frac{(0.267 - 2x)^2}{x^2} = 2.18 \times 10^6$$

Taking the square root of both sides we obtain:

$$\frac{0.267 - 2x}{x} = 1.48 \times 10^3$$

$$x = 1.80 \times 10^{-4}$$

The equilibrium concentrations are:

$$[H_2] = [Br_2] = 1.80 \times 10^{-4} \, M$$

$$[HBr] = 0.267 - 2(1.80 \times 10^{-4}) = 0.267 \, M$$

15.61 Since equilibrium pressures are desired, we calculate K_P.

$$K_P = K_c \, [(0.08206 \, \text{L·atm/K·mol}) \times T]^{\Delta n} = (4.63 \times 10^{-3})(0.08206 \times 800)^1 = 0.304$$

	$COCl_2(g)$	$\rightleftharpoons$	$CO(g)$	+	$Cl_2(g)$
Initial (atm):	0.760		0.000		0.000
Change (atm):	$-x$		$+x$		$+x$
Equilibrium (atm):	$(0.760 - x)$		x		x

$$\frac{x^2}{0.760 - x} = 0.304$$

$$x^2 + 0.304x - 0.231 = 0$$

$$x = 0.352 \, \text{atm}$$

At equilibrium:

$$P_{COCl_2} = (0.760 - 0.352)\text{atm} = \mathbf{0.408 \ atm}$$

$$P_{CO} = P_{Cl_2} = \mathbf{0.352 \ atm}$$

15.63 The equilibrium expression for the system is:

$$K_P = \frac{(P_{CO})^2}{P_{CO_2}}$$

The total pressure can be expressed as:

$$P_{total} = P_{CO_2} + P_{CO}$$

If we let the partial pressure of CO be x, then the partial pressure of CO_2 is:

$$P_{CO_2} = P_{total} - x = (4.50 - x)\text{atm}$$

Substitution gives the equation:

$$K_P = \frac{(P_{CO})^2}{P_{CO_2}} = \frac{x^2}{(4.50 - x)} = 1.52$$

This can be rearranged to the quadratic:

$$x^2 + 1.52x - 6.84 = 0$$

The solutions are $x = 1.96$ and $x = -3.48$; only the positive result has physical significance. The equilibrium pressures are

$$\boldsymbol{P_{CO} = x = 1.96\ \text{atm}}$$

$$\boldsymbol{P_{CO_2}} = (4.50 - 1.96) = \boldsymbol{2.54\ \text{atm}}$$

15.69   **Strategy:** According to Le Châtelier's principle, increasing the volume causes a shift toward the side with the *larger* number of moles of gas. Decreasing the volume of an equilibrium mixture causes a shift toward the side of the equation with the *smaller* number of moles of gas. For equilibrium with equal numbers of gaseous moles on both sides, a change in volume does not cause the equilibrium to shift in either direction.

   Setup: Determine the numbers of moles of gas in both the reactants and the products for each reaction.

   Solution: There are equal number of moles of gas in the reactants and products for **reaction (b)**. A change in volume will not affect the position of its equilibrium.

15.71   **Strategy:** Use Le Châtelier's principle to determine which of the following scenarios will cause the reaction equilibrium to shift to the right.

   Solution: a. Decreasing the volume of an equilibrium mixture causes a shift toward the side of the equation with the *smaller* number of moles of gas. The equilibrium will **shift to the left**.

   b. Increasing the volume of an equilibrium mixture causes a shift toward the side with the *larger* number of moles of gas. The equilibrium will **shift to the right**.

   c. Altering the amount of a solid **does not change the position of the equilibrium** because doing so does not change the value of Q.

   d. The addition of more reactant will **shift the equilibrium to the right**.

   e. The removal of a product will **shift the equilibrium to the right**.

   (b), (d), and (e) will cause the equilibrium to shift to the right.

15.73   a. Removal of $CO_2(g)$ from the system: **The equilibrium would shift to the right**.

   b. Addition of more solid Na_2CO_3: **The equilibrium would be unaffected**. $[Na_2CO_3]$ does not appear in the equilibrium expression.

c. Removal of some of the solid NaHCO₃: **The equilibrium would be unaffected**. Same reason as (b).

15.75   **Strategy:**   A change in pressure can affect only the volume of a gas but not that of a solid or liquid because solids and liquids are much less compressible. The stress applied is an increase in pressure. According to Le Châtelier's principle, the system will adjust to partially offset this stress. In other words, the system will adjust to decrease the pressure. This can be achieved by shifting to the side of the equation that has fewer moles of gas. Recall that pressure is directly proportional to moles of gas: $PV = nRT$ so $P \propto n$.

   Solution: a. Changes in pressure ordinarily do not affect the concentrations of reacting species in condensed phases because liquids and solids are virtually incompressible. Pressure change should have **no effect** on this system.

   b.   **No effect**. Same situation as (a).

   c. Only the product is in the gas phase. An increase in pressure should favor the reaction that decreases the total number of moles of gas. The equilibrium should **shift to the left**, that is, the amount of B should decrease and that of A should increase.

   d. In this equation there are equal moles of gaseous reactants and products. A shift in either direction will have no effect on the total number of moles of gas present. There will be **no effect** on the position of the equilibrium when the pressure is increased.

   e. A shift in the direction of the reverse reaction (**shift to the left**) will have the result of decreasing the total number of moles of gas present.

15.77   **Strategy:**   (a) What does the sign of $\Delta H°$ indicate about the heat change (endothermic or exothermic) for the forward reaction? (b) The stress is the addition of Cl₂ gas. How will the system adjust to partially offset the stress? (c) The stress is the removal of PCl₃ gas. How will the system adjust to partially offset the stress? (d) The stress is an increase in pressure. The system will adjust to decrease the pressure. Remember, pressure is directly proportional to moles of gas. (e) What is the function of a catalyst? How does it affect a reacting system that is not at equilibrium?

   Solution: a. The stress applied is the heat added to the system. Note that the reaction is endothermic ($\Delta H° > 0$). Endothermic reactions absorb heat from the surroundings; therefore, we can think of heat as a reactant.

$$\text{heat} + PCl_5(g) \rightleftharpoons PCl_3(g) + Cl_2(g)$$

   The system will adjust to remove some of the added heat by undergoing a decomposition reaction (**shift to the right**).
   b. The stress is the addition of Cl₂ gas. The system will shift in the direction to remove some of the added Cl₂. The system **shifts to the left** until equilibrium is re-established.

   c. The stress is the removal of PCl₃ gas. The system will shift to replace some of the PCl₃ th was removed. The system **shifts to the right** until equilibrium is re-established.

   d. The stress applied is an increase in pressure. The system will adjust to remove the stress by decreasing the pressure. Recall that pressure is directly proportional to the number of moles of gas. In the balanced equation we see 1 mole of gas on the reactants side and 2 moles of gas on the products side. The pressure can be decreased by shifting to the side with the fewer moles of gas. The system **shifts to the left** to re-establish equilibrium.

e. The function of a catalyst is to increase the rate of a reaction. If a catalyst is added to the reacting system not at equilibrium, the system will reach equilibrium faster than if left undisturbed. If a system is already at equilibrium, as in this case, the addition of a catalyst will not affect either the concentrations of reactant and product or the equilibrium constant. **A catalyst has no effect on equilibrium position**.

15.79 For this system,

$$K_P = P_{CO_2}$$

This means that to remain at equilibrium, the pressure of CO_2 must stay at a fixed value as long as the temperature remains the same.

a. If the volume is increased, the pressure of CO_2 will drop (according to Boyle's law, pressure and volume are inversely proportional). Some $CaCO_3$ will break down to form more CO_2 and CaO (so the system **shifts to right**).

b. Assuming that the amount of added solid CaO is not so large that the volume of the system is altered significantly, there should be **no effect**. If a huge amount of CaO was added, this would have the effect of reducing the volume of the container.

c. Assuming that the amount of $CaCO_3$ removed doesn't alter the volume of the container significantly, there should be **no effect**. Removing a huge amount of $CaCO_3$ will have the effect of increasing the container volume. The result in that case will be the same as in part (a).

d. The pressure of CO_2 will be greater and will exceed the value of K_P. Some CO_2 will combine with CaO to form more $CaCO_3$. (The system **shifts to the left**.)

e. CO_2 combines with aqueous NaOH according to the equation

$$CO_2(g) + NaOH(aq) \longrightarrow NaHCO_3(aq)$$

This will have the effect of reducing the CO_2 pressure and causing more $CaCO_3$ to break down to CO_2 and CaO (so the system **shifts to the right**).

f. CO_2 does not react with HCl, but $CaCO_3$ does.

$$CaCO_3(s) + 2HCl(aq) \longrightarrow CaCl_2(aq) + CO_2(g) + H_2O(l)$$

The CO_2 produced by the action of the acid will combine with CaO as discussed in (d) above (so the system **shifts to the left**).

g. This is a decomposition reaction. Decomposition reactions are endothermic. Increasing the temperature will favor this reaction and produce more CO_2 and CaO (the system **shifts to the right**).

15.81 a. $2O_3(g) \rightleftharpoons 3O_2(g)$ $\Delta H^\circ_{rxn} = -284.4$ kJ/mol

b. Equilibrium would shift to the left. **The number of O_3 molecules would increase and the number of O_2 molecules would decrease.**

15.83 **Strategy:** According to Le Châtelier's principle, adding a product to an equilibrium mixture shifts the equilibrium toward the reactant side (to the left) of the equation.

Solution: The addition of acid (product) will shift the equilibrium to the left. As a result, **the concentration of oxyhemoglobin will decrease as acidosis occurs**.

15.85 The equilibrium expression for the top diagram is:

$$K_c = \frac{[C]^2}{[A][B]}$$

According to the diagram, there are three spheres of A, four spheres of B, and four spheres of C. The equilibrium expression becomes:

$$K_c = \frac{[C]^2}{[A][B]} = \frac{4^2}{[3][4]} = \frac{16}{12}$$

For a system to be at equilibrium, the reaction quotient Q_c must be equal to the equilibrium constant K_c. To determine which of the remaining systems are also at equilibrium, we must calculate the value of Q_c and compare it to the K_c calculated above.

For (a), the reaction quotient is:

$$Q_c = \frac{[C]^2}{[A][B]} = \frac{(3)^2}{(5)(3)} = \frac{9}{15}$$

For (b), the reaction quotient is:

$$Q_c = \frac{[C]^2}{[A][B]} = \frac{(4)^2}{(4)(3)} = \frac{16}{12}$$

For (c), the reaction quotient is:

$$Q_c = \frac{[C]^2}{[A][B]} = \frac{(3)^2}{(4)(4)} = \frac{9}{16}$$

For (d), the reaction quotient is:

$$Q_c = \frac{[C]^2}{[A][B]} = \frac{(4)^2}{(6)(2)} = \frac{16}{12}$$

Since Q_c equals K_c for **diagrams (b) and (d)**, they must also be at equilibrium.

15.87 a. The equation that relates K_P and K_c is:

$$K_P = K_c\,[(0.08206\ \text{L·atm/K·mol}) \times T]^{\Delta n}$$

For this reaction, $\Delta n = 3 - 2 = 1$

$$K_c = \frac{K_P}{(0.08206\,T)} = \frac{2 \times 10^{-42}}{(0.08206 \times 298)} = 8 \times 10^{-44}$$

    b. **A mixture of H_2 and O_2 can be kept at room temperature because of a very large activation energy.** The reaction of hydrogen with oxygen is infinitely slow without a catalyst or an initiator. The action of a single spark on a mixture of these gases results in the explosive formation of water.

15.89   a.
$$2CO + 2NO \longrightarrow 2CO_2 + N_2$$

    b. **The oxidizing agent is NO; the reducing agent is CO.**

    c.
$$\Delta G^\circ = 2\Delta G_f^\circ(CO_2) + \Delta G_f^\circ(N_2) - 2\Delta G_f^\circ(CO) - 2\Delta G_f^\circ(NO)$$

$$\Delta G^\circ = (2)(-394.4 \text{ kJ/mol}) + (0) - (2)(-137.3 \text{ kJ/mol}) - (2)(86.7 \text{ kJ/mol}) = -687.6 \text{ kJ/mol}$$

$$\Delta G^\circ = -RT \ln K_P$$

$$\ln K_P = \frac{6.876 \times 10^5 \text{ J/mol}}{(8.314 \text{ J/K} \cdot \text{mol})(298 \text{ K})} = 277.5$$

$$K_P = 3.29 \times 10^{120}$$

    d.
$$Q_P = \frac{(P_{N_2})(P_{CO_2})^2}{(P_{CO})^2 (P_{NO})^2} = \frac{(0.80)(3.0 \times 10^{-4})^2}{(5.0 \times 10^{-5})^2 (5.0 \times 10^{-7})^2} = 1.2 \times 10^{14}$$

Since $Q_P \ll K_P$, the reaction will proceed **to the right**.

    e. Using data from Appendix 2 we calculate the enthalpy change for the reaction.

$$\Delta H^\circ = 2\Delta H_f^\circ(CO_2) + \Delta H_f^\circ(N_2) - 2\Delta H_f^\circ(CO) - 2\Delta H_f^\circ(NO)$$

$$\Delta H^\circ = (2)(-393.5 \text{ kJ/mol}) + (0) - (2)(-110.5 \text{ kJ/mol}) - (2)(90.4 \text{ kJ/mol}) = -746.8 \text{ kJ/mol}$$

Since ΔH° is negative, raising the temperature will decrease K_P (Le Châtelier's principle), thereby increasing the amount of reactants and decreasing the amount of products. **No, the formation of N_2 and CO_2 is not favored by raising the temperature.**

15.91   a. Calculate the value of K_P by substituting the equilibrium partial pressures into the equilibrium expression.

$$K_P = \frac{P_B}{(P_A)^2} = \frac{(0.60)}{(0.60)^2} = 1.7$$

    b. The total pressure is the sum of the partial pressures for the two gaseous components, A and B. We can write:

$$P_A + P_B = 1.5 \text{ atm}$$

and

$$P_B = 1.5 - P_A$$

Substituting into the expression for K_P gives:

$$K_P = \frac{(1.5 - P_A)}{(P_A)^2} = 1.7$$

$$1.7(P_A)^2 + P_A - 1.5 = 0$$

Solving the quadratic equation, we obtain:

$$P_A = 0.69 \text{ atm}$$

and by difference,

$$P_B = 0.81 \text{ atm}$$

Check that substituting these equilibrium concentrations into the equilibrium expression gives the equilibrium constant calculated in part (a).

$$K_P = \frac{P_B}{(P_A)^2} = \frac{0.81}{(0.69)^2} = 1.7$$

15.93   Total number of moles of gas is:

$$0.020 + 0.040 + 0.96 = 1.02 \text{ mol of gas}$$

You can calculate the partial pressure of each gaseous component from the mole fraction and the total pressure.

$$P_{NO} = \chi_{NO} P_{total} = \frac{0.040}{1.02} \times 0.20 = 0.0078 \text{ atm}$$

$$P_{O_2} = \chi_{O_2} P_{total} = \frac{0.020}{1.02} \times 0.20 = 0.0039 \text{ atm}$$

$$P_{NO_2} = \chi_{NO_2} P_{total} = \frac{0.96}{1.02} \times 0.20 = 0.19 \text{ atm}$$

Calculate K_P by substituting the partial pressures into the equilibrium expression.

$$K_P = \frac{(P_{NO_2})^2}{(P_{NO})^2 P_{O_2}} = \frac{(0.19)^2}{(0.0078)^2 (0.0039)} = \mathbf{1.5 \times 10^5}$$

15.95   a. If $\Delta G°$ for the reaction is 173.4 kJ/mol, then

$$\Delta G_f° = \frac{173.4 \text{ kJ/mol}}{2} = \mathbf{86.7 \text{ kJ / mol}}$$

b.
$$\Delta G^\circ = -RT \ln K_P$$

$$173.4 \times 10^3 \text{ J/mol} = -(8.314 \text{ J/K·mol})(298 \text{ K}) \ln K_P$$

$$K_P = 4.0 \times 10^{-31}$$

c. ΔH° for the reaction is $2 \times \Delta H_f^\circ \text{ (NO)} = (2)(90.4 \text{ kJ/mol}) = 180.8 \text{ kJ/mol}$

Using the equation in Problem 15.122:

$$\ln \frac{K_2}{4.0 \times 10^{-31}} = \frac{180.8 \times 10^3 \text{ J/mol}}{8.314 \text{ J/K·mol}} \left(\frac{1373 \text{ K} - 298 \text{ K}}{(1373 \text{ K})(298 \text{ K})} \right)$$

$$K_2 = 2.62 \times 10^{-6}$$

d. **Lightning supplies the energy necessary to drive this reaction, converting the two most abundant gases in the atmosphere into NO(g). The NO gas dissolves in the rain, which carries it into the soil where it is converted into nitrate and nitrite by bacterial action. This "fixed" nitrogen is a necessary nutrient for plants.**

15.97   Set up a table that contains the initial concentrations, the change in concentrations, and the equilibrium concentration. Assume that the vessel has a volume of 1 L.

	H_2	$+ Cl_2$	$\rightleftharpoons$	$2HCl$
Initial (*M*):	0.47	0		3.59
Change (*M*):	$+x$	$+x$		$-2x$
Equilibrium (*M*):	$(0.47 + x)$	x		$(3.59 - 2x)$

Substitute the equilibrium concentrations into the equilibrium expression, then solve for x. Since $\Delta n = 0$, $K_c = K_P$.

$$K_c = \frac{[\text{HCl}]^2}{[\text{H}_2][\text{Cl}_2]} = \frac{(3.59 - 2x)^2}{(0.47 + x)x} = 193$$

Solving the quadratic equation,

$$x = 0.103$$

Having solved for x, calculate the equilibrium concentrations of all species.

$$[\text{H}_2] = 0.57 \ M \qquad [\text{Cl}_2] = 0.103 \ M \qquad [\text{HCl}] = 3.384 \ M$$

Since we assumed that the vessel had a volume of 1 L, the above molarities also correspond to the number of moles of each component.

From the mole fraction of each component and the total pressure, we can calculate the partial pressure of each component.

$$\text{Total number of moles} = 0.57 + 0.103 + 3.384 = 4.06 \text{ mol}$$

$$P_{H_2} = \frac{0.57}{4.06} \times 2.00 = \mathbf{0.28\ atm}$$

$$P_{Cl_2} = \frac{0.103}{4.06} \times 2.00 = \mathbf{0.0507\ atm}$$

$$P_{HCl} = \frac{3.384}{4.06} \times 2.00 = \mathbf{1.67\ atm}$$

15.99

$$2O_3 \rightleftharpoons 3O_2$$

You can look up the standard free energy of formation values in Appendix 2 of the text.

$$\Delta G° = 3\Delta G_f°(O_2) - 2\Delta G_f°(O_3) = 0 - (2)(163.4\ kJ/mol)$$

$$\Delta G° = -326.8\ kJ/mol$$

$$-326.8 \times 10^3\ J/mol = -(8.314\ J/K \cdot mol)(243\ K)\ln K_P$$

$$K_P = \mathbf{1.78 \times 10^{70}}$$

Because of the large magnitude of *K*, you would expect this reaction to be spontaneous in the forward direction. However, this reaction has a large activation energy, so the rate of reaction is extremely slow.

15.101   We carry an additional significant figure throughout this calculation to minimize rounding errors. The initial molarity of SO_2Cl_2 is:

$$[SO_2Cl_2] = \frac{6.75\ g\ SO_2Cl_2 \times \dfrac{1\ mol\ SO_2Cl_2}{135.0\ g\ SO_2Cl_2}}{2.00\ L} = 0.02500\ M$$

The concentration of SO_2 at equilibrium is:

$$[SO_2] = \frac{0.0345\ mol}{2.00\ L} = 0.01725\ M$$

Since there is a 1:1 mole ratio between SO_2 and SO_2Cl_2, the concentration of SO_2 at equilibrium (0.01725 *M*) equals the concentration of SO_2Cl_2 reacted. The concentrations of SO_2Cl_2 and Cl_2 at equilibrium are:

	$SO_2Cl_2(g) \rightleftharpoons$	$SO_2(g)$ +	$Cl_2(g)$
Initial (*M*):	0.02500	0	0
Change (*M*):	−0.01725	+0.01725	+0.01725
Equilibrium (*M*):	0.00775	0.01725	0.01725

Substitute the equilibrium concentrations into the equilibrium expression to calculate K_c.

$$K_c = \frac{[SO_2][Cl_2]}{[SO_2Cl_2]} = \frac{(0.01725)(0.01725)}{(0.00775)} = 3.84 \times 10^{-2}\ or\ \mathbf{0.0384}$$

15.103   This is a difficult problem. Express the equilibrium number of moles in terms of the initial moles and the change in number of moles (x). Next, calculate the mole fraction of each component. Using the mole fraction, you should come up with a relationship between partial pressure and total pressure for each component. Substitute the partial pressures into the equilibrium expression to solve for the total pressure, P_T.

The reaction is:

	N_2	$+$	$3H_2$	$\rightleftharpoons$	$2NH_3$
Initial (mol):	1		3		0
Change (mol):	$-x$		$-3x$		$2x$
Equilibrium (mol):	$(1-x)$		$(3-3x)$		$2x$

$$\text{Mole fraction of NH}_3 = \frac{\text{mol of NH}_3}{\text{total number of moles}}$$

$$\chi_{NH_3} = \frac{2x}{(1-x)+(3-3x)+2x} = \frac{2x}{4-2x}$$

$$0.21 = \frac{2x}{4-2x}$$

$$x = 0.35 \text{ mol}$$

Substituting x into the following mole fraction equations, the mole fractions of N_2 and H_2 can be calculated.

$$\chi_{N_2} = \frac{1-x}{4-2x} = \frac{1-0.35}{4-2(0.35)} = 0.20$$

$$\chi_{H_2} = \frac{3-3x}{4-2x} = \frac{3-3(0.35)}{4-2(0.35)} = 0.59$$

The partial pressures of each component are equal to the mole fraction multiplied by the total pressure.

$$P_{NH_3} = 0.21 P_{total} \qquad P_{N_2} = 0.20 P_{total} \qquad P_{H_2} = 0.59 P_{total}$$

Substitute the partial pressures above (in terms of P_T) into the equilibrium expression, and solve for P_T.

$$K_P = \frac{\left(P_{NH_3}\right)^2}{\left(P_{H_2}\right)^3 P_{N_2}}$$

$$4.31 \times 10^{-4} = \frac{(0.21)^2 \left(P_{total}\right)^2}{\left(0.59 P_{total}\right)^3 \left(0.20 P_{total}\right)}$$

$$4.31 \times 10^{-4} = \frac{1.07}{\left(P_{total}\right)^2}$$

$$P_{total} = 5.0 \times 10^1 \text{ atm}$$

15.105 Of the original 1.05 moles of Br_2, 1.20% has dissociated. The amount of Br_2 dissociated in molar concentration is:

$$[Br_2] = 0.0120 \times \frac{1.05 \text{ mol}}{0.980 \text{ L}} = 0.0129 \text{ } M$$

Setting up a table:

	$Br_2(g)$	$\rightleftharpoons$	$2Br(g)$
Initial (*M*):	$\dfrac{1.05 \text{ mol}}{0.980 \text{ L}} = 1.07 \text{ } M$		0
Change (*M*):	-0.0129		$+2(0.0129)$
Equilibrium (*M*):	1.06		0.0258

$$K_c = \frac{[Br]^2}{[Br_2]} = \frac{(0.0258)^2}{1.06} = 6.28 \times 10^{-4}$$

15.107 For a 100% yield, 2.00 moles of SO_3 would be formed. An 80% yield means 2.00 moles $\times$ (0.80) = 1.60 moles SO_3 is formed.

The amount of SO_2 remaining at equilibrium = (2.00 − 1.60)mol = 0.40 mol

The amount of O_2 reacted = $\dfrac{1}{2} \times$ (amount of SO_2 reacted) = $\left(\dfrac{1}{2} \times 1.60\right)$ mol = 0.80 mol

The amount of O_2 remaining at equilibrium = (2.00 − 0.80)mol = 1.20 mol

Total moles at equilibrium = moles SO_2 + moles O_2 + moles SO_3 = (0.40 + 1.20 + 1.60)mol = 3.20 moles

$$P_{SO_2} = \frac{0.40}{3.20} P_{total} = 0.125 \, P_{total}$$

$$P_{O_2} = \frac{1.20}{3.20} P_{total} = 0.375 \, P_{total}$$

$$P_{SO_3} = \frac{1.60}{3.20} P_{total} = 0.500 \, P_{total}$$

$$K_P = \frac{\left(P_{SO_3}\right)^2}{\left(P_{SO_2}\right)^2 P_{O_2}}$$

$$0.13 = \frac{\left(0.500 \, P_{total}\right)^2}{\left(0.125 \, P_{total}\right)^2 \left(0.375 \, P_{total}\right)}$$

$$P_{total} = 3.3 \times 10^2 \text{ atm}$$

15.109   a.   It is the reverse of a **disproportionation redox reaction**.

b. $$\Delta G^\circ = (2)(-228.6 \text{ kJ/mol}) - (2)(-33.0 \text{ kJ/mol}) - (1)(-300.4 \text{ kJ/mol})$$

$$\Delta G^\circ = -90.8 \text{ kJ/mol}$$

$$-90.8 \times 10^3 \text{ J/mol} = -(8.314 \text{ J/K·mol})(298 \text{ K}) \ln K$$

$$K = 8.25 \times 10^{15}$$

Because of the large value of K, **this method is feasible for removing SO$_2$.**

c. $$\Delta H^\circ = (2)(-241.8 \text{ kJ/mol}) + (3)(0) - (2)(-20.15 \text{ kJ/mol}) - (1)(-296.4 \text{ kJ/mol})$$

$$\Delta H^\circ = -146.9 \text{ kJ/mol}$$

$$\Delta S^\circ = (2)(188.7 \text{ J/K·mol}) + (3)(31.88 \text{ J/K·mol}) - (2)(205.64 \text{ J/K·mol}) - (1)(248.5 \text{ J/K·mol})$$

$$\Delta S^\circ = -186.7 \text{ J/K·mol}$$

$$\Delta G^\circ = \Delta H^\circ - T\Delta S^\circ$$

Because of the negative entropy change, ΔS°, the free energy change, ΔG°, will become positive at higher temperatures. Therefore, the reaction will be **less effective** at high temperatures.

15.111   a. $$\Delta G^\circ = 2\Delta G_f^\circ(\text{HBr}) - \Delta G_f^\circ(\text{H}_2) - \Delta G_f^\circ(\text{Br}_2) = (2)(-53.2 \text{ kJ/mol}) - (1)(0) - (1)(0)$$

$$\Delta G^\circ = -106.4 \text{ kJ/mol}$$

$$\ln K_P = \frac{-\Delta G^\circ}{RT} = \frac{106.4 \times 10^3 \text{ J/mol}}{(8.314 \text{ J/K · mol})(298 \text{ K})} = 42.9$$

$$K_P = 3.87 \times 10^{18}$$

b. $$\Delta G^\circ = \Delta G_f^\circ(\text{HBr}) - \tfrac{1}{2}\Delta G_f^\circ(\text{H}_2) - \tfrac{1}{2}\Delta G_f^\circ(\text{Br}_2) = (1)(-53.2 \text{ kJ/mol}) - \left(\tfrac{1}{2}\right)(0) - \left(\tfrac{1}{2}\right)(0)$$

$$\Delta G^\circ = -53.2 \text{ kJ/mol}$$

$$\ln K_P = \frac{-\Delta G^\circ}{RT} = \frac{53.2 \times 10^3 \text{ J/mol}}{(8.314 \text{ J/K · mol})(298 \text{ K})} = 21.5$$

$$K_P = 2.17 \times 10^9$$

The K_P in (a) is the square of the K_P in (b). Both ΔG° and K_P depend on the number of moles of reactants and products specified in the balanced equation.

15.113   According to the ideal gas law, pressure is directly proportional to the concentration of a gas in mol/L if the reaction is at constant volume and temperature. Therefore, pressure may be used as a concentration unit.

The reaction is:

$$N_2 \quad + \quad 3H_2 \;\rightleftharpoons\; 2NH_3$$

Initial (atm):	0.862	0.373	0
Change (atm):	$-x$	$-3x$	$+2x$
Equilibrium (atm):	$(0.862-x)$	$(0.373-3x)$	$2x$

$$K_P = \frac{\left(P_{NH_3}\right)^2}{\left(P_{H_2}\right)^3 P_{N_2}}$$

$$4.31 \times 10^{-4} = \frac{(2x)^2}{(0.373-3x)^3(0.862-x)}$$

At this point, we need to make two assumptions that $3x$ is very small compared to 0.373 and that x is very small compared to 0.862. Hence,

$$0.373 - 3x \approx 0.373$$

and

$$0.862 - x \approx 0.862$$

$$4.31 \times 10^{-4} \approx \frac{(2x)^2}{(0.373)^3(0.862)}$$

Solving for x.

$$x = 2.20 \times 10^{-3} \text{ atm}$$

The equilibrium pressures are:

$$P_{N_2} = \left[0.862 - \left(2.20 \times 10^{-3}\right)\right]\text{atm} = \mathbf{0.860\ atm}$$

$$P_{H_2} = \left[0.373 - (3)\left(2.20 \times 10^{-3}\right)\right]\text{atm} = \mathbf{0.366\ atm}$$

$$P_{NH_3} = (2)\left(2.20 \times 10^{-3}\text{ atm}\right) = \mathbf{4.40 \times 10^{-3}\ atm}$$

15.115   a. The equation is:

$$\text{fructose} \;\rightleftharpoons\; \text{glucose}$$

Initial (M):	0.244	0
Change (M):	-0.131	$+0.131$
Equilibrium (M):	0.113	0.131

Calculating the equilibrium constant,

$$K_c = \frac{[\text{glucose}]}{[\text{fructose}]} = \frac{0.131}{0.113} = \mathbf{1.16}$$

b.

$$\text{Percent converted} = \frac{\text{amount of fructose converted}}{\text{original amount of fructose}} \times 100\%$$

$$= \frac{0.131}{0.244} \times 100\% = \mathbf{53.7\%}$$

15.117 **Strategy:** Use Equation 15.6 to solve for the equilibrium constant, K.

Setup: $R = 8.314 \times 10^{-3}$ kJ/K·mol and $T = (20 + 273)\text{K} = 293$ K

From problem 14.29, $\Delta G = 64$ kJ/mol.

Solution: Solving Equation 15.6 for K gives

$$K = e^{\frac{-\Delta G^\circ}{RT}} = e^{\frac{-64\,\text{kJ/mol}}{(8.314\,\text{J/K·mol})(293\,\text{K})}} = e^{-0.026} = \mathbf{0.97}$$

15.119 a. There is only one gas phase component, O_2. The equilibrium constant is simply

$$K_P = P_{O_2} = \mathbf{0.49\ atm}$$

b. From the ideal gas equation, we can calculate the moles of O_2 produced by the decomposition of CuO.

$$n_{O_2} = \frac{PV}{RT} = \frac{(0.49\ \text{atm})(2.0\ \text{L})}{(0.08206\ \text{L·atm/K·mol})(1297\ \text{K})} = 9.2 \times 10^{-3}\ \text{mol}\ O_2$$

From the balanced equation,

$$\left(9.2 \times 10^{-3}\ \text{mol}\ O_2\right) \times \frac{4\ \text{mol CuO}}{1\ \text{mol}\ O_2} = 3.7 \times 10^{-2}\ \text{mol CuO decomposed}$$

$$\text{Fraction of CuO decomposed} = \frac{\text{amount of CuO lost}}{\text{original amount of CuO}}$$

$$= \frac{3.7 \times 10^{-2}\ \text{mol}}{0.16\ \text{mol}} = \mathbf{0.23}\ (\text{or 23\%})$$

c. If a 1.0 mol sample were used, the pressure of oxygen would still be the same (0.49 atm) and it would be due to the same quantity of O_2. Remember, a pure solid does not affect the equilibrium position. The moles of CuO lost would still be 3.7×10^{-2} mol. Thus the fraction decomposed would be:

$$\frac{0.037}{1.0} = \mathbf{0.037}\ (\text{or 3.7\%})$$

d. If the number of moles of CuO were less than 0.037 mol, the equilibrium could not be established because the pressure of O_2 would be less than 0.49 atm. Therefore, **the smallest number of moles of CuO needed to establish equilibrium must be slightly greater than 0.037 mol.**

15.121   We must first find the initial concentrations of all the species in the system.

$$[H_2]_0 = \frac{0.714 \text{ mol}}{2.40 \text{ L}} = 0.298 \ M$$

$$[I_2]_0 = \frac{0.984 \text{ mol}}{2.40 \text{ L}} = 0.410 \ M$$

$$[HI]_0 = \frac{0.886 \text{ mol}}{2.40 \text{ L}} = 0.369 \ M$$

Calculate the reaction quotient by substituting the initial concentrations into the appropriate equation.

$$Q_c = \frac{[HI]_0^2}{[H_2]_0[I_2]_0} = \frac{(0.369)^2}{(0.298)(0.410)} = 1.11$$

We find that Q_c is less than K_c. The equilibrium will shift to the right, decreasing the concentrations of H_2 and I_2 and increasing the concentration of HI.

We set up the usual table. Let x be the decrease in concentration of H_2 and I_2.

	H_2	+	I_2	$\rightleftharpoons$	$2HI$
Initial (M):	0.298		0.410		0.369
Change (M):	$-x$		$-x$		$+2x$
Equilibrium (M):	$(0.298 - x)$		$(0.410 - x)$		$(0.369 + 2x)$

The equilibrium expression is:

$$K_c = \frac{[HI]^2}{[H_2][I_2]} = \frac{(0.369 + 2x)^2}{(0.298 - x)(0.410 - x)} = 54.3$$

This becomes the quadratic equation

$$50.3x^2 - 39.9x + 6.49 = 0$$

The smaller root is $x = 0.228 \ M$. (The larger root is physically impossible.)

Having solved for x, calculate the equilibrium concentrations.

$$[H_2] = (0.298 - 0.228) \ M = \textbf{0.070} \ \boldsymbol{M}$$

$$[I_2] = (0.410 - 0.228) \ M = \textbf{0.182} \ \boldsymbol{M}$$

$$[HI] = [0.369 + 2(0.228)] \ M = \textbf{0.825} \ \boldsymbol{M}$$

15.123   The gas cannot be (a) because the color became lighter while heating. Heating (a) to 150°C would produce some HBr, which is colorless and would lighten rather than darken the gas.

The gas cannot be (b) because Br_2 does not dissociate into Br atoms at 150°C, so the color should not change.

The gas must be (c). N_2O_4(colorless) $\longrightarrow$ $2NO_2$(brown) is consistent with the observations. The reaction is endothermic, so heating darkens the color. Above 150°C, the NO_2 breaks up into colorless NO and O_2:

$$2NO_2(g) \longrightarrow 2NO(g) + O_2(g)$$

An increase in pressure shifts the equilibrium back to the left, restoring the color by producing NO_2.

15.125   Given the following:

$$K_c = \frac{[NH_3]^2}{[N_2][H_2]^3} = 1.2$$

a. Temperature must have units of Kelvin.

$$K_P = K_c \left[(0.08206 \text{ L·atm/K·mol}) \times T\right]^{\Delta n}$$

$$\boldsymbol{K_P = (1.2)(0.08206 \times 648)^{(2-4)} = 4.2 \times 10^{-4}}$$

b. Recalling that,

$$K_{\text{forward}} = \frac{1}{K_{\text{reverse}}}$$

Therefore,

$$\boldsymbol{K_c' = \frac{1}{1.2} = 0.83}$$

c. Since the equation

$$\tfrac{1}{2} N_2(g) + \tfrac{3}{2} H_2(g) \rightleftharpoons NH_3(g)$$

is equivalent to

$$\tfrac{1}{2} [N_2(g) + 3H_2(g) \rightleftharpoons 2NH_3(g)]$$

then, K_c' for the reaction:

$$\tfrac{1}{2} N_2(g) + \tfrac{3}{2} H_2(g) \rightleftharpoons NH_3(g)$$

equals $(K_c)^{\frac{1}{2}}$ for the reaction:

$$N_2(g) + 3H_2(g) \rightleftharpoons 2NH_3(g)$$

Thus,

$$K_c' = (K_c)^{\frac{1}{2}} = \sqrt{1.2} = 1.1$$

d. For K_P in part (b):

$$K_P = (0.83)(0.08206 \times 648)^{+2} = 2.3 \times 10^3$$

and for K_P in part (c):

$$K_P = (1.1)(0.08206 \times 648)^{-1} = 2.1 \times 10^{-2}$$

15.127   a. **color deepens**   b. **increases**   c. **decreases**   d. **increases**   e. **unchanged**

15.129   **Potassium is more volatile than sodium. Therefore, its removal shifts the equilibrium from left to right.**

15.131   In this problem, you are asked to calculate K_c.

Step 1: Calculate the initial concentration of NOCl. We carry an extra significant figure throughout this calculation to minimize rounding errors.

$$[NOCl]_0 = \frac{2.50 \text{ mol}}{1.50 \text{ L}} = 1.667 \ M$$

Step 2: Let's represent the change in concentration of NOCl as $-2x$. Setting up a table:

	$2NOCl(g)$	$\rightleftharpoons$	$2NO(g)$	$+$	$Cl_2(g)$
Initial (*M*):	1.667		0		0
Change (*M*):	$-2x$		$+2x$		$+x$
Equilibrium (*M*):	$1.667 - 2x$		$2x$		x

If 28.0 percent of the NOCl has dissociated at equilibrium, the amount consumed is:

$$(0.280)(1.667 \ M) = 0.4668 \ M$$

In the table above, we have represented the amount of NOCl that reacts as $2x$. Therefore,

$$2x = 0.4668 \ M$$

$$x = 0.2334 \ M$$

The equilibrium concentrations of NOCl, NO, and Cl_2 are:

$$[NOCl] = (1.667 - 2x)M = (1.667 - 0.4668)M = 1.200 \ M$$

$$[NO] = 2x = 0.4668 \ M$$

$$[Cl_2] = x = 0.2334 \ M$$

Step 3: The equilibrium constant K_c can be calculated by substituting the above concentrations into the equilibrium expression.

$$K_c = \frac{[NO]^2[Cl_2]}{[NOCl]^2} = \frac{(0.4668)^2(0.2334)}{(1.200)^2} = 0.035 \text{ or } \mathbf{3.5 \times 10^{-2}}$$

15.133   a. **shifts to right**

       b. **shifts to right**

       c. **no change**

       d. **no change**

       e. **no change**

       f. **shifts to left**

15.135   The equilibrium is: $\qquad\qquad\qquad\qquad N_2O_4(g) \rightleftharpoons 2NO_2(g)$

$$K_P = \frac{(P_{NO_2})^2}{P_{N_2O_4}} = \frac{0.15^2}{0.20} = 0.113$$

Volume is doubled so pressure is halved. Let's calculate Q_P and compare it to K_P.

$$Q_P = \frac{\left(\dfrac{0.15}{2}\right)^2}{\left(\dfrac{0.20}{2}\right)} = 0.0563 < K_P$$

Equilibrium will shift to the right. Some N_2O_4 will react, and some NO_2 will be formed. Let x = amount of N_2O_4 reacted.

$$N_2O_4(g) \rightleftharpoons 2NO_2(g)$$

	N_2O_4	$2NO_2$
Initial (atm):	0.10	0.075
Change (atm):	$-x$	$+2x$
Equilibrium (atm):	$0.10 - x$	$0.075 + 2x$

Substitute into the K_P expression to solve for x.

$$K_P = 0.113 = \frac{(0.075 + 2x)^2}{0.10 - x}$$

$$4x^2 + 0.413x - 5.68 \times 10^{-3} = 0$$

$$x = 0.0123$$

At equilibrium:

$$P_{N_2O_4} = 0.10 - 0.0123 = \textbf{0.09 atm}$$

$$P_{NO_2} = 0.075 + 2(0.0123) = 0.0996 \approx \textbf{0.10 atm}$$

15.137 a. Molar mass of PCl_5 = 208.2 g/mol

$$P = \frac{nRT}{V} = \frac{\left(2.50 \text{ g} \times \dfrac{1 \text{ mol}}{208.2 \text{ g}}\right)\left(0.08206 \dfrac{\text{L} \cdot \text{atm}}{\text{K} \cdot \text{mol}}\right)(523 \text{ K})}{0.500 \text{ L}} = \textbf{1.03 atm}$$

b.

	PCl_5	$\rightleftharpoons$	PCl_3	+	Cl_2
Initial (atm)	1.03		0		0
Change (atm)	$-x$		$+x$		$+x$
Equilibrium (atm)	$1.03 - x$		x		x

$$K_P = 1.05 = \frac{x^2}{1.03 - x}$$

$$x^2 + 1.05x - 1.08 = 0$$

$$x = 0.639$$

At equilibrium:

$$P_{PCl_5} = 1.03 - 0.639 = \textbf{0.39 atm}$$

c. $$P_{total} = (1.03 - x) + x + x = 1.03 + 0.639 = \textbf{1.67 atm}$$

d. $$\frac{0.639 \text{ atm}}{1.03 \text{ atm}} = \textbf{0.620} \, (62.0\%)$$

15.139 a. $\textbf{\textit{K}}_P = P_{Hg} = 0.0020$ mmHg $= 2.6 \times 10^{-6}$ atm $= \textbf{2.6} \times \textbf{10}^{-6}$ (equilibrium constants are expressed without units)

$$\textbf{\textit{K}}_c = \frac{K_P}{(0.08206\,T)^{\Delta n}} = \frac{2.6 \times 10^{-6}}{(0.08206 \times 299)^1} = \textbf{1.1} \times \textbf{10}^{-7}$$

b. Volume of lab = (6.1 m)(5.3 m)(3.1 m) = 100 m^3

$$[Hg] = K_c$$

$$\textbf{Total mass of Hg vapor} = \frac{1.1 \times 10^{-7} \text{ mol}}{1 \text{ L}} \times \frac{200.6 \text{ g}}{1 \text{ mol}} \times \frac{1 \text{ L}}{1000 \text{ cm}^3} \times \left(\frac{1 \text{ cm}}{0.01 \text{ m}}\right)^3 \times 100 \text{ m}^3 = \textbf{2.2 g}$$

The concentration of mercury vapor in the room is:

$$\frac{2.2 \text{ g}}{100 \text{ m}^3} = 0.022 \text{ g/m}^3 = \mathbf{22 \text{ mg/m}^3}$$

Yes. This concentration exceeds the safety limit of 0.05 mg/m^3.

15.141   a.

$$\Delta G° = \Delta G_f°(\text{H}_2) + \Delta G_f°(\text{Fe}^{2+}) - \Delta G_f°(\text{Fe}) - 2\Delta G_f°(\text{H}^+)]$$

$$\Delta G° = (1)(0) + (1)(-84.9 \text{ kJ/mol}) - (1)(0) - (2)(0)$$

$$\Delta G° = -84.9 \text{ kJ/mol}$$

$$\Delta G° = -RT \ln K$$

$$-84.9 \times 10^3 \text{ J/mol} = -(8.314 \text{ J/K·mol})(298 \text{ K}) \ln K$$

$$\boldsymbol{K = 7.62 \times 10^{14}}$$

   b.

$$\Delta G° = \Delta G_f°(\text{H}_2) + \Delta G_f°(\text{Cu}^{2+}) - \Delta G_f°(\text{Cu}) - 2\Delta G_f°(\text{H}^+)$$

$$\Delta G° = 64.98 \text{ kJ/mol}$$

$$\Delta G° = -RT \ln K$$

$$64.98 \times 10^3 \text{ J/mol} = -(8.314 \text{ J/K·mol})(298 \text{ K}) \ln K$$

$$\boldsymbol{K = 4.07 \times 10^{-12}}$$

The activity series is correct. The very large value of *K* for reaction (a) indicates that *products* are highly favored, whereas the very small value of *K* for reaction (b) indicates that *reactants* are highly favored.

15.143   **There is a temporary dynamic equilibrium between the melting ice cubes and the freezing of water between the ice cubes.**

15.145   First, let's calculate the initial concentration of ammonia.

$$[\text{NH}_3] = \frac{14.6 \text{ g} \times \dfrac{1 \text{ mol NH}_3}{17.03 \text{ g NH}_3}}{4.00 \text{ L}} = 0.214 \ M$$

Let's set up a table to represent the equilibrium concentrations. We represent the amount of NH$_3$ that reacts as $2x$.

	2NH$_3$(*g*)	$\rightleftharpoons$	N$_2$(*g*) +	3H$_2$(*g*)
Initial (*M*):	0.214		0	0
Change (*M*):	$-2x$		$+x$	$+3x$
Equilibrium (*M*):	$0.214 - 2x$		x	$3x$

Substitute into the equilibrium expression to solve for *x*.

$$K_c = \frac{[N_2][H_2]^3}{[NH_3]^2}$$

$$0.83 = \frac{(x)(3x)^3}{(0.214 - 2x)^2} = \frac{27x^4}{(0.214 - 2x)^2}$$

Taking the square root of both sides of the equation gives:

$$0.91 = \frac{5.20x^2}{0.214 - 2x}$$

Rearranging,

$$5.20x^2 + 1.82x - 0.195 = 0$$

Solving the quadratic equation gives the solutions:

$$x = 0.086 \ M \text{ and } x = -0.44 \ M$$

The positive root is the correct answer. The equilibrium concentrations are:

$$[\textbf{NH}_3] = 0.214 - 2(0.086) = \textbf{0.042 } \textbf{\textit{M}}$$

$$[\textbf{N}_2] = \textbf{0.086 } \textbf{\textit{M}}$$

$$[\textbf{H}_2] = 3(0.086) = \textbf{0.26 } \textbf{\textit{M}}$$

15.147   a. From the balanced equation

	N$_2$O$_4$	$\rightleftharpoons$	2NO$_2$
Initial (mol):	1		0
Change (mol):	$-x$		$+2x$
Equilibrium (mol):	$(1-x)$		$2x$

The total moles in the system = (moles N$_2$O$_4$ + moles NO$_2$) = $[(1-x) + 2x] = 1 + x$. If the total pressure in the system is P, then:

$$P_{N_2O_4} = \frac{1-x}{1+x}P \text{ and } P_{NO_2} = \frac{2x}{1+x}P$$

$$K_P = \frac{\left(P_{NO_2}\right)^2}{P_{N_2O_4}} = \frac{\left(\dfrac{2x}{1+x}\right)^2 P^2}{\left(\dfrac{1-x}{1+x}\right)P}$$

$$K_P = \frac{\left(\dfrac{4x^2}{1+x}\right)P}{1-x} = \frac{4x^2}{1-x^2}P$$

b. Rearranging the K_P expression:

$$4x^2P = K_P - x^2K_P$$

$$x^2(4P + K_P) = K_P$$

$$x^2 = \frac{K_P}{4P + K_P}$$

$$x = \sqrt{\frac{K_P}{4P + K_P}}$$

K_P is a constant (at constant temperature). If P increases, the fraction $\dfrac{4x^2}{1-x^2}$ (and therefore x) must decrease. Equilibrium shifts to the left to produce less NO_2 and more N_2O_4, as predicted.

15.149   Initially, the pressure of SO_2Cl_2 is 9.00 atm. The pressure is held constant, so after the reaction reaches equilibrium, $P_{SO_2Cl_2} + P_{SO_2} + P_{Cl_2} = 9.00$ atm . The amount (pressure) of SO_2Cl_2 reacted must equal the pressure of SO_2 and Cl_2 produced for the pressure to remain constant. If we let $P_{SO_2} + P_{Cl_2} = x$, then the pressure of SO_2Cl_2 reacted must be $2x$. We set up a table showing the initial pressures, the change in pressures, and the equilibrium pressures.

$$SO_2Cl_2(g) \rightleftharpoons SO_2(g) + Cl_2(g)$$

	SO_2Cl_2	SO_2	Cl_2
Initial (atm):	9.00	0	0
Change (atm):	$-2x$	$+x$	$+x$
Equilibrium (atm):	$9.00 - 2x$	x	x

Again, note that the change in pressure for SO_2Cl_2 $(-2x)$ does not match the stoichiometry of the reaction, because we are expressing changes in pressure. The total pressure is kept at 9.00 atm throughout.

$$K_P = \frac{P_{SO_2} P_{Cl_2}}{P_{SO_2Cl_2}}$$

$$2.05 = \frac{(x)(x)}{9.00 - 2x}$$

$$x^2 + 4.10x - 18.45 = 0$$

Solving the quadratic equation, $x = 2.71$ atm. At equilibrium,

$$P_{SO_2Cl_2} = 9.00 - 2(2.71) = \textbf{3.58 atm}$$

$$P_{SO_2} = P_{Cl_2} = x = \textbf{2.71 atm}$$

15.151   We start with a table.

$$
\begin{array}{ccccc}
 & A_2 & + & B_2 & \rightleftharpoons & 2AB \\
\text{Initial (mol):} & 1 & & 3 & & 0 \\
\text{Change (mol):} & -\dfrac{x}{2} & & -\dfrac{x}{2} & & +x \\
\hline
\text{Equilibrium (mol):} & 1-\dfrac{x}{2} & & 3-\dfrac{x}{2} & & x
\end{array}
$$

After the addition of two moles of A,

$$
\begin{array}{ccccc}
 & A_2 & + & B_2 & \rightleftharpoons & 2AB \\
\text{Initial (mol):} & 3-\dfrac{x}{2} & & 3-\dfrac{x}{2} & & x \\
\text{Change (mol):} & -\dfrac{x}{2} & & -\dfrac{x}{2} & & +x \\
\hline
\text{Equilibrium (mol):} & 3-x & & 3-x & & 2x
\end{array}
$$

We write two different equilibrium constants expressions for the two tables.

$$
K = \frac{[AB]^2}{[A_2][B_2]}
$$

$$
K = \frac{x^2}{\left(1-\dfrac{x}{2}\right)\left(3-\dfrac{x}{2}\right)} \quad \text{and} \quad K = \frac{(2x)^2}{(3-x)(3-x)}
$$

We equate the equilibrium expressions and solve for x.

$$
\frac{x^2}{\left(1-\dfrac{x}{2}\right)\left(3-\dfrac{x}{2}\right)} = \frac{(2x)^2}{(3-x)(3-x)}
$$

$$
\frac{1}{\dfrac{1}{4}(x^2-8x+12)} = \frac{4}{x^2-6x+9}
$$

$$
-6x+9 = -8x+12
$$

$$
x = 1.5
$$

We substitute x back into one of the equilibrium expressions to solve for K.

$$
\boldsymbol{K} = \frac{(2x)^2}{(3-x)(3-x)} = \frac{(3)^2}{(1.5)(1.5)} = \mathbf{4.0}
$$

Substitute x into the other equilibrium expression to see if you obtain the same value for K. Note that we used moles rather than molarity for the concentrations because the volume, V cancels in the equilibrium expressions.

15.153 **Strategy:** The concentrations of solids and pure liquids do not appear in K_C. So, even though the given equilibrium is heterogeneous, Equation 15.4 still applies.

 Solution: There are 3 moles of gas molecules on the product side and 6 moles on the reactant side. So, $\Delta n = 3 - 6 = -3.$

15.155 **Strategy:** According to Equation 15.4, K_P and K_C are the same when $\Delta n = 0$:

$$K_p = K_c \left[(0.08206 \text{ L} \cdot \text{atm} / \text{K} \cdot \text{mol}) \times T \right]^{\Delta n} = K_c \left(0.08206T \right)^0 = K_c \cdot 1 = K_c$$

So, look for reactions that show no change in the number of gas molecules as the reaction occurs. Also, note that Equation 15.4 does not apply to *heterogeneous* equilibria if liquid phases are *solutions*.

 Solution: a. $K_P \neq K_C$ ($\Delta n \neq 0$)

 b. $K_P \neq K_C$ ($\Delta n \neq 0$)

 c. **Equation 15.4 is not applicable** (some species are aqueous).

 d. $K_P = K_C$ ($\Delta n = 0$)

 e. $K_P \neq K_C$ ($\Delta n \neq 0$)

 f. **Equation 15.4 is not applicable** (some species are aqueous).

 g. $K_P = K_C$ ($\Delta n = 0$)

 h. $K_P \neq K_C$ ($\Delta n \neq 0$)

15.157 a. **We start by writing the van't Hoff equation at two different temperatures.**

$$\ln K_1 = \frac{-\Delta H^\circ}{RT_1} + C$$

$$\ln K_2 = \frac{-\Delta H^\circ}{RT_2} + C$$

$$\ln K_1 - \ln K_2 = \frac{-\Delta H^\circ}{RT_1} - \frac{-\Delta H^\circ}{RT_2}$$

$$\ln \frac{K_1}{K_2} = \frac{\Delta H^\circ}{R} \left(\frac{1}{T_2} - \frac{1}{T_1} \right)$$

Assume an endothermic reaction, $\Delta H° > 0$ and $T_2 > T_1$. Then, $\dfrac{\Delta H°}{R}\left(\dfrac{1}{T_2} - \dfrac{1}{T_1}\right) < 0$, meaning that

$\ln\dfrac{K_1}{K_2} < 0$ or $K_1 < K_2$. A larger K_2 indicates that there are more products at equilibrium as the temperature is raised. This agrees with Le Châtelier's principle that an increase in temperature favors the forward endothermic reaction. The opposite of the above discussion holds for an exothermic reaction.

b. Treating

$$H_2O(l) \rightleftharpoons H_2O(g) \qquad \Delta H_{vap} = ?$$

as a heterogeneous equilibrium, $K_P = P_{H_2O}$.

We substitute into the equation derived in part (a) to solve for ΔH_{vap}.

$$\ln\frac{K_1}{K_2} = \frac{\Delta H°}{R}\left(\frac{1}{T_2} - \frac{1}{T_1}\right)$$

$$\ln\frac{31.82 \text{ mmHg}}{92.51 \text{ mmHg}} = \frac{\Delta H°}{8.314 \text{ J/K}\cdot\text{mol}}\left(\frac{1}{323 \text{ K}} - \frac{1}{303 \text{ K}}\right)$$

$$-1.067 = \Delta H°\left(-2.46 \times 10^{-5}\right)$$

$$\Delta H° = 4.34 \times 10^4 \text{ J/mol} = \textbf{434 kJ/mol}$$

Chapter 16
Acids, Bases, and Salts

Visualizing Chemistry

VC 16.1 b   VC 16.2 a   VC 16.3 a   VC 16.4 b

Key Skills

16.1   b   16.2   d   16.3   d   16.4   b

Problems

16.3   Tables 16.1 and 16.2 of the text list important Brønsted acids and bases and their respective conjugates.

 a. **both**

 b. **base**

 c. **acid**

 d. **base**

 e. **acid**

 f. **base**

 g. **base**

 h. **base**

 i. **acid**

 j. **acid**

16.5   Recall that the conjugate base of a Brønsted acid is the species that remains when *one* proton has been removed from the acid.

 a. nitrite ion: NO_2^-

 b. hydrogen sulfate ion (also called bisulfate ion): HSO_4^-

 c. hydrogen sulfide ion (also called bisulfide ion): HS^-

 d. cyanide ion: CN^-

 e. formate ion: $HCOO^-$

16.7   The conjugate base of any acid is simply the acid minus one proton.

   a. CH_2ClCOO^-

   b. IO_4^-

   c. $H_2PO_4^-$

   d. HPO_4^{2-}

   e. PO_4^{3-}

   f. HSO_4^-

   g. SO_4^{2-}

   h. IO_3^-

   i. SO_3^{2-}

   j. NH_3

   k. HS^-

   l. S^{2-}

   m. OCl^-

16.11   All the listed pairs are oxoacids that contain different central atoms whose elements are in the same group of the periodic table and have the same oxidation number. In this situation the acid with the most electronegative central atom will be the strongest.

   a. $H_2SO_4 > H_2SeO_4$ b. $H_3PO_4 > H_3AsO_4$

16.13   **The conjugate bases are $C_6H_5O^-$ from phenol and CH_3O^- from methanol. The $C_6H_5O^-$ is stabilized by resonance:**

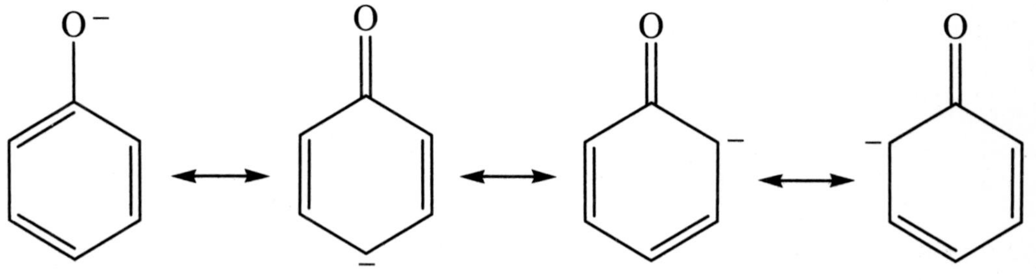

The CH_3O^- ion has no such resonance stabilization. A more stable conjugate base means an increase in the strength of the acid.

16.19 **Strategy:** The equilibrium concentrations of H^+ and OH^- must satisfy Equation 16.1.

Setup: $K_W = [H^+][OH^-] = 1.0 \times 10^{-14}$ (25°C). Substitute the given concentrations and solve for $[H^+]$:

$$[H^+] = \frac{1.0 \times 10^{-14}}{[OH^-]}$$

Solution: a.

$$[H^+] = \frac{1.0 \times 10^{-14}}{2.50 \times 10^{-2}} = \mathbf{4.0 \times 10^{-13}\, M}$$

b.

$$[H^+] = \frac{1.0 \times 10^{-14}}{1.67 \times 10^{-5}} = \mathbf{6.0 \times 10^{-10}\, M}$$

c.

$$[H^+] = \frac{1.0 \times 10^{-14}}{8.62 \times 10^{-3}} = \mathbf{1.2 \times 10^{-12}\, M}$$

d.

$$[H^+] = \frac{1.0 \times 10^{-14}}{1.75 \times 10^{-12}} = \mathbf{5.7 \times 10^{-3}\, M}$$

16.21 **Strategy:** Use Equation 16.1 to calculate the $[H_3O^+]$ concentrations.

$$K_w = [H_3O^+][OH^-]$$

Setup: $K_w = 5.13 \times 10^{-13}$ at 100°C. Rearranging Equation 16.1 to solve for $[H_3O^+]$ gives:

$$[H_3O^+] = \frac{5.13 \times 10^{-13}}{[OH^-]}$$

Solution: a. $[H_3O^+] = \dfrac{5.13 \times 10^{-13}}{2.50 \times 10^{-2}} = \mathbf{2.05 \times 10^{-11}\, M}$

b. $[H_3O^+] = \dfrac{5.13 \times 10^{-13}}{1.67 \times 10^{-5}} = \mathbf{3.07 \times 10^{-8}\, M}$

c. $[H_3O^+] = \dfrac{5.13 \times 10^{-13}}{8.62 \times 10^{-3}} = \mathbf{5.95 \times 10^{-11}\, M}$

d. $[H_3O^+] = \dfrac{5.13 \times 10^{-13}}{1.75 \times 10^{-12}} = \mathbf{2.93 \times 10^{-1}\, M}$

16.25

$$[H^+] = 1.4 \times 10^{-3}\, M$$

$$[OH^-] = \frac{K_w}{[H^+]} = \frac{1.0 \times 10^{-14}}{1.4 \times 10^{-3}} = \mathbf{7.1 \times 10^{-12}\, M}$$

16.27　a. HCl is a strong acid, so the concentration of hydrogen ion is also 0.0010 M. We use the definition of pH.

$$pH = -\log[H^+] = -\log(0.0010) = \mathbf{3.00}$$

b. KOH is an ionic compound and completely dissociates into ions. We first find the concentration of hydrogen ion.

$$[H^+] = \frac{K_w}{[OH^-]} = \frac{1.0 \times 10^{-14}}{0.76} = 1.3 \times 10^{-14} \ M$$

The pH is then found from its defining equation

$$pH = -\log[H^+] = -\log[1.3 \times 10^{-14}] = \mathbf{13.89}$$

16.29　The pH can be found by using Equation 16.4 of the text.

$$pH = 14.00 - pOH = 14.00 - 9.40 = 4.60$$

The hydrogen ion concentration can be found as follows.

$$4.60 = -\log[H^+]$$

Taking the antilog of both sides:

$$[H^+] = \mathbf{2.5 \times 10^{-5} \ M}$$

16.31　We can calculate the OH⁻ concentration from the pOH.

$$pOH = 14.00 - pH = 14.00 - 10.00 = 4.00$$

$$[OH^-] = 10^{-pOH} = 1.0 \times 10^{-4} \ M$$

Since NaOH is a strong base, it ionizes completely. The OH⁻ concentration equals the initial concentration of NaOH.

$$[NaOH] = 1.0 \times 10^{-4} \ mol/L$$

So, we need to prepare 546 mL of $1.0 \times 10^{-4} \ M$ NaOH.

This is a dimensional analysis problem. We need to perform the following unit conversions.

$$mol/L \rightarrow mol \ NaOH \rightarrow grams \ NaOH$$

$$546 \ mL = 0.546 \ L$$

$$? \, g \ NaOH = 546 \ mL \times \frac{1.00 \times 10^{-4} \ mol \ NaOH}{1000 \ mL \ soln} \times \frac{40.00 \ g \ NaOH}{1 \ mol \ NaOH} = \mathbf{2.2 \times 10^{-3} \ g \ NaOH}$$

16.35　a. **−0.009**　　　　　b. **1.46**　　　　　c. **5.82**

16.37　**Strategy:**　HNO₃ is a strong acid. Therefore, the concentration of HNO₃ is equal to the concentration of hydronium ion. We use Equation 9.6 to determine $[H_3O^+]$:

Solution: a. $[HNO_3] = [H_3O^+] = 10^{-4.21} = \mathbf{6.2 \times 10^{-5}}$ $\textit{\textbf{M}}$

b. $[HNO_3] = [H_3O^+] = 10^{-3.55} = \mathbf{2.8 \times 10^{-4}}$ $\textit{\textbf{M}}$

c. $[HNO_3] = [H_3O^+] = 10^{-0.98} = \mathbf{0.10}$ $\textit{\textbf{M}}$

16.39 **Strategy:** We use Equation 16.2 to determine pOH and Equation 16.4 to determine pH. The hydroxide ion concentration of a monobasic strong base, such as LiOH and NaOH, is equal to the base concentration. In the case of a dibasic base such as $Ba(OH)_2$, the hydroxide concentration is twice the base concentration.

Solution: a.
$$[OH^-] = [LiOH] = 1.24\ M$$

$$pOH = -\log[OH^-] = -\log(1.24) = \mathbf{-0.093}$$

$$pH = 14.00 - pOH = 14.00 - (-0.093) = \mathbf{14.09}$$

b.
$$[OH^-] = 2[Ba(OH)_2] = 2(0.22\ M) = 0.44\ M$$

$$pOH = -\log[OH^-] = -\log(0.44) = \mathbf{0.36}$$

$$pH = 14.00 - pOH = 14.00 - 0.36 = \mathbf{13.64}$$

c.
$$[OH^-] = [NaOH] = 0.085\ M$$

$$pOH = -\log[OH^-] = -\log(0.085) = \mathbf{1.07}$$

$$pH = 14.00 - pOH = 14.00 - 1.07 = \mathbf{12.93}$$

16.41 **Strategy:** We use Equation 16.4 to determine pOH and Equation 16.3 to determine hydroxide ion concentration. The concentration of a monobasic strong base such as LiOH is equal to the hydroxide concentration. In the case of a dibasic base such as $Ba(OH)_2$, the base concentration is *half* the hydroxide concentration.

Solution: a.
$$pOH = 14.00 - pH = 14.00 - 11.04 = 2.96$$

$$[KOH] = [OH^-] = 10^{-pOH} = 10^{-2.96} = \mathbf{1.1 \times 10^{-3}}\ \textit{\textbf{M}}$$

b.
$$[OH^-] = 10^{-pOH} = 10^{-2.96} = 1.1 \times 10^{-3}\ M$$

$$[Ba(OH)_2] = [OH^-] \times \frac{1\ \text{mol Ba(OH)}_2}{2\ \text{mol OH}^-} = \frac{1.1 \times 10^{-3}\ M}{2} = \mathbf{5.5 \times 10^{-4}}\ \textit{\textbf{M}}$$

16.43 **Strategy:** HNO_3 is a strong acid. Given pH, use Equation 9.6 to solve for $[H_3O^+]$.

$$[H_3O^+] = 10^{-pH}$$

Use reaction stoichiometry to determine HNO_3 concentration.

Solution:
$$[H_3O^+] = 10^{-4.65} = 2.24 \times 10^{-5}\ M$$

The hydronium ion concentration is equal that of HNO_3 according to the balanced equation:

$$HNO_3(aq) + H_2O(aq) \longrightarrow H_3O^+(aq) + NO_3^-(aq)$$

Therefore, $[H_3O^+] = [HNO_3] = \textbf{2.24} \times \textbf{10}^{-5} \textbf{\textit{M}}$

16.51 **Strategy:** Recall that a weak acid only partially ionizes in water. We are given the initial quantity of a weak acid (CH_3COOH) and asked to calculate the pH. For this we will need to calculate $[H^+]$, so we follow the procedure outlined in Section 16.3 of the text.

Solution: We set up a table for the dissociation.

	$C_6H_5COOH(aq)$	$\rightleftharpoons$	$H^+(aq)$	$+ C_6H_5COO^-(aq)$
Initial (M):	0.10		0.00	0.00
Change (M):	$-x$		$+x$	$+x$
Equilibrium (M):	$(0.10 - x)$		x	x

$$K_a = \frac{[H^+][C_6H_5COO^-]}{[C_6H_5COOH]}$$

$$6.5 \times 10^{-5} = \frac{x^2}{(0.10 - x)}$$

Assuming that x is small compared to 0.10, we neglect it in the denominator:

$$6.5 \times 10^{-5} = \frac{x^2}{0.10}$$

$$x = 2.55 \times 10^{-3} \, M = [H^+]$$

$$pH = -\log(2.55 \times 10^{-3}) = \textbf{2.59}$$

This problem could also have been solved using the quadratic equation. The difference in pH obtained using the approximation $x \approx 0$ is very small. (Using the quadratic equation would have given pH = 2.60, a difference of less than 0.4%.)

16.53 We set up a table for the dissociation.

	$HCN(aq)$	$\rightleftharpoons$	$H^+(aq)$	$+ CN^-(aq)$
Initial (M):	0.095		0.00	0.00
Change (M):	$-x$		$+x$	$+x$
Equilibrium (M):	$(0.095 - x)$		x	x

$$K_a = \frac{[H^+][CN^-]}{[HCN]}$$

$$4.9 \times 10^{-10} = \frac{x^2}{(0.095 - x)}$$

Assuming that x is small compared to 0.095, we neglect it in the denominator:

$$4.9 \times 10^{-10} = \frac{x^2}{0.095}$$

$$x = 6.82 \times 10^{-6} \ M = [H^+]$$

$$pH = -\log(6.82 \times 10^{-6}) = \mathbf{5.17}$$

16.55 **Strategy:** Formic acid is a weak acid, and some of it will ionize in solution according to the equilibrium. You can find the K_a value in Table 16.5 of the text.

$$HF(aq) + H_2O(l) \rightleftharpoons F^-(aq) + H_3O^+(aq)$$

$$K_a = 1.7 \times 10^{-4} = \frac{[H^+][F^-]}{[HF]}$$

The percent ionization of a weak acid is given by Equation 16.5:

$$\% \ \text{ionization} = \frac{[H^+]}{[HA]_0} \times 100\%$$

where $[HA]_0$ is the initial concentration of the acid (i.e., the concentration assuming there is *no* ionization). Using the K_a expression above, set up a concentration table and solve for $[H^+]$. Then, use this value to compute the percent ionization.

Solution: a.

	HF(aq) +	H$_2$O(l)	$\rightleftharpoons$	F$^-$(aq)	+ H$_3$O$^+$(aq)
Initial (*M*):	0.016			0	0
Change (*M*):	$-x$			$+x$	$+x$
Equilibrium (*M*):	$0.016 - x$			x	x

$$1.7 \times 10^{-4} = \frac{(x)(x)}{(0.016 - x)} = \frac{x^2}{(0.016 - x)} \qquad (1)$$

As a first approximation, assume that x is small compared to 0.016 so that $0.016 - x \approx 0.016$.

$$1.7 \times 10^{-4} \approx \frac{x^2}{0.016}$$

$$x \approx \sqrt{(1.7 \times 10^{-4})(0.016)} = 0.0016$$

Note that this approximate value of x is 10% of 0.016, so it is probably not appropriate to have assumed that $0.016 - x \approx 0.016$. Instead, attempt to solve for x without making any approximations. Start by clearing the denominator of Equation (1) and then expanding the result to get:

$$x^2 + 1.7 \times 10^{-4} x - 2.72 \times 10^{-6} = 0$$

This equation is a *quadratic equation* of the form $ax^2 + bx + c = 0$, the solutions of which are given by (see Appendix 1):

$$x = \frac{-b \pm \sqrt{b^2 - 4ac}}{2a}$$

For this case, $a = 1$, $b = 1.7 \times 10^{-4}$, and $c = -2.72 \times 10^{-6}$. Substituting these values into the quadratic formula gives:

$$x = \frac{-1.7\times10^{-4} \pm \sqrt{\left(1.7\times10^{-4}\right)^2 - 4(1)\left(-2.72\times10^{-6}\right)}}{(2)(1)}$$

$$= \frac{-1.7\times10^{-4} \pm 3.30\times10^{-3}}{2}$$

$$x = 1.6\times10^{-3}\ M \quad \text{or} \quad x = -1.7\times10^{-3}\ M$$

The negative solution is not physically reasonable since there is no such thing as a negative concentration. The solution is $x = 1.6\times10^{-3}\ M$.

Finally,

$$\text{\% ionization} = \frac{[\text{H}^+]}{[\text{HF}]_0}\times100\% = \frac{x}{0.016}\times100\% = \frac{0.0016}{0.016}\times100\% = \mathbf{10\%}$$

b. Proceed in the same manner as part (a). Note though that the initial concentration is even less than it was in part (a), making it even more necessary to avoid the approximate solution.

	HF(aq) +	H$_2$O(l)	$\rightleftharpoons$	F$^-$(aq)	+ H$_3$O$^+$(aq)
Initial (M):	5.7×10^{-4}			0	0
Change (M):	$-x$			$+x$	$+x$
Equilibrium (M):	$5.7\times10^{-4} - x$			x	x

$$1.7\times10^{-4} = \frac{x^2}{\left(5.7\times10^{-4} - x\right)}$$

$$x^2 + 1.7\times10^{-4}x - 9.69\times10^{-8} = 0$$

$$x = \frac{-1.7\times10^{-4} \pm \sqrt{\left(1.7\times10^{-4}\right)^2 - 4(1)\left(-9.69\times10^{-8}\right)}}{(2)(1)}$$

$$= \frac{-1.7\times10^{-4} \pm 6.45\times10^{-4}}{2}$$

$$x = 4.8\times10^{-4}\ M \quad \text{or} \quad x = -8.2\times10^{-4}\ M$$

The negative solution is rejected, giving $x = 6.3\times10^3\ M$. Thus,

$$\text{\% ionization} = \frac{[\text{H}^+]}{[\text{HF}]_0}\times100\% = \frac{x}{5.7\times10^{-4}}\times100\% = \frac{4.8\times10^{-4}}{5.7\times10^{-4}}\times100\% = \mathbf{84\%}$$

c. Proceed in the same manner as part (a). Note though that the initial concentration is very large, and we may safely seek an approximate solution.

	HF(aq) +	H$_2$O(l)	$\rightleftharpoons$	F$^-$(aq)	+ H$_3$O$^+$(aq)
Initial (M):	1.75			0	0
Change (M):	$-x$			$+x$	$+x$
Equilibrium (M):	$1.75-x$			x	x

$$1.7 \times 10^{-4} = \frac{x^2}{(1.75-x)} \approx \frac{x^2}{1.75}$$

$$x \approx \sqrt{\left(1.7 \times 10^{-4}\right)(1.75)} = 0.017 \ M$$

$$\% \text{ ionization} = \frac{[\text{H}^+]}{[\text{HF}]_0} \times 100\% = \frac{x}{1.75} \times 100\% = \frac{0.017}{1.75} \times 100\% = \mathbf{0.97\%}$$

16.57 **Strategy:** We are asked to compute K_a for the equilibrium

$$\text{HA}(aq) + \text{H}_2\text{O}(l) \rightleftharpoons \text{A}^-(aq) + \text{H}_3\text{O}^+(aq)$$

$$K_a = \frac{[\text{H}^+][\text{A}^-]}{[\text{HA}]} \quad (1)$$

We are given the percent ionization of the acid, which is related to [H$^+$] and [HA]$_0$ by Equation 16.5:

$$\% \text{ ionization} = \frac{[\text{H}^+]}{[\text{HA}]_0} \times 100\% \quad \text{or} \quad [\text{H}^+] = \left(\frac{\% \text{ ionization}}{100\%}\right)[\text{HA}]_0 \quad (2)$$

where [HA]$_0$ is the initial concentration of the acid (that is, the concentration assuming there is *no* ionization). Since the percent ionization (0.92%) is very small, it is safe to assume that [HA] $\approx$ [HA]$_0$. Using Equation (2) above for [H$^+$], along with the fact that [H$^+$] = [A–], substitute into Equation (1) and evaluate K_a.

Solution:

$$K_a = \frac{[\text{H}^+][\text{A}^-]}{[\text{HA}]}$$

$$\approx \frac{\left[\left(\dfrac{\% \text{ ionization}}{100\%}\right)[\text{HA}]_0\right]^2}{[\text{HA}]_0}$$

$$\approx \left(\frac{\% \text{ ionization}}{100\%}\right)^2 [\text{HA}]_0$$

$$\approx (0.92/100)^2 (0.015)$$

$$K_a \approx 1.3 \times 10^{-6}$$

16.59 **Strategy:** We use the pH to determine $[H^+]_{eq}$ and set up a table for the dissociation of the weak acid, filling in what we know. The initial concentration of weak acid is 0.19 M. The initial concentrations of H^+ and the anion are both zero; and because they both are products of the ionization of the weak acid, their equilibrium concentrations are equal: $[H^+]_{eq} = [A^-]_{eq}$.

Solution: $[H^+] = 10^{-4.52} = 3.02 \times 10^{-5}\ M$

	HA(aq)	$\rightleftharpoons$	H^+(aq)	+ A$^-$(aq)
Initial (M):	0.19		0	0
Change (M):	-3.02×10^{-5}		$+3.02 \times 10^{-5}$	$+3.02 \times 10^{-5}$
Equilibrium (M):	$(0.19 - 3.02 \times 10^{-5})$		3.02×10^{-5}	3.02×10^{-5}

The amount of HA ionized is very small compared to the initial concentration:

$$0.19 - 3.02 \times 10^{-5} \approx 0.19$$

Therefore, $K_a = \dfrac{[H^+]_{eq}[A^-]_{eq}}{[HA]_{eq}} = \dfrac{(3.02 \times 10^{-5})^2}{0.19} = 4.8 \times 10^{-9}$

16.61 A pH of 3.26 corresponds to a $[H^+]$ of $5.5 \times 10^{-4}\ M$. Let the original concentration of formic acid be x. If the concentration of $[H^+]$ is $5.5 \times 10^{-4}\ M$, it means that $5.5 \times 10^{-4}\ M$ of HCOOH ionized because of the 1:1 mole ratio between HCOOH and H^+.

	HCOOH(aq)	$\rightleftharpoons$	H^+(aq)	+ HCOO$^-$(aq)
Initial (M):	x		0	0
Change (M):	-5.5×10^{-4}		$+5.5 \times 10^{-4}$	$+5.5 \times 10^{-4}$
Equilibrium (M):	$x - (5.5 \times 10^{-4})$		5.5×10^{-4}	5.5×10^{-4}

Substitute K_a and the equilibrium concentrations into the ionization constant expression to solve for x.

$$K_a = \frac{[H^+][HCOO^-]}{[HCOOH]}$$

$$1.7 \times 10^{-4} = \frac{(5.5 \times 10^{-4})^2}{x - (5.5 \times 10^{-4})}$$

$$x = [HCOOH] = 2.3 \times 10^{-3}\ M$$

16.63   At 37°C, $K_w = 2.5 \times 10^{-14}$ so $[H^+][OH^-] = 2.5 \times 10^{-14}$. Because in neutral water the concentrations of hydronium and hydroxide are equal, we can solve for pH as follows:

$$[H^+] = [OH^-] = x$$

$$x^2 = 2.5 \times 10^{-14}$$

$$x = \sqrt{2.5 \times 10^{-14}} = 1.6 \times 10^{-7} \; M$$

$$pH = -\log(1.6 \times 10^{-7}) = \mathbf{6.80}$$

16.65   a. **strong base**   b. **weak base**   c. **weak base**   d. **weak base**   e. **strong base**

16.69   **Strategy:**   Weak bases only partially ionize in water.

$$B(aq) + H_2O(l) \rightleftharpoons BH^+(aq) + OH^-(aq)$$

Note that the concentration of the weak base given refers to the initial concentration before ionization has started. The pH of the solution, on the other hand, refers to the situation at equilibrium. To calculate K_b, we need to know the concentrations of all three species [B], [BH$^+$], and [OH$^-$] at equilibrium. We ignore the ionization of water as a source of OH$^-$ ions.

Solution: We proceed as follows.

Step 1:   The major species in solution are B, OH$^-$, and the conjugate acid BH$^+$.

Step 2:   First, we need to calculate the hydroxide ion concentration from the pH value. Calculate the pOH from the pH. Then, calculate the OH$^-$ concentration from the pOH.

$$pOH = 14.00 - pH = 14.00 - 10.66 = 3.34$$

$$pOH = -\log[OH^-]$$

$$-pOH = \log[OH^-]$$

Taking the antilog of both sides of the equation,

$$10^{-pOH} = [OH^-]$$

$$[OH^-] = 10^{-3.34} = 4.57 \times 10^{-4} \; M$$

Step 3:   If the concentration of OH$^-$ is 4.57×10^{-4} M at equilibrium, it must mean that 4.57×10^{-4} M of the base is ionized. We summarize the changes.

	$B(aq) +$	$H_2O(l) \rightleftharpoons$	$BH^+(aq)$	$+ OH^-(aq)$
Initial (M):	0.30		0	0
Change (M):	-4.57×10^{-4}		$+4.57 \times 10^{-4}$	$+4.57 \times 10^{-4}$
Equilibrium (M):	$0.30 - (4.57 \times 10^{-4})$		4.57×10^{-4}	4.57×10^{-4}

Step 4: Substitute the equilibrium concentrations into the ionization constant expression to solve for K_b.

$$K_b = \frac{[BH^+][OH^-]}{[B]}$$

$$K_b = \frac{\left(4.57 \times 10^{-4}\right)^2}{\left(0.30 - 4.57 \times 10^{-4}\right)} = \mathbf{7.0 \times 10^{-7}}$$

16.71 We set up an equilibrium table to solve for $[OH^-]$ and use pOH to get pH. We use B and BH^+ to represent the weak base and its protonated form, respectively.

	B(aq) +	H₂O(l) ⇌	BH⁺(aq)	+ OH⁻(aq)
Initial (M):	0.61		0.00	0.00
Change (M):	−x		+x	+x
Equilibrium (M):	0.61 − x		x	x

$$K_b = \frac{[BH^+][OH^-]}{[B]}$$

$$[OH^-] = \frac{1.5 \times 10^{-4}[B]}{[BH^+]}$$

$$x = \frac{1.5 \times 10^{-4}(0.61 - x)}{x}$$

Assuming x is small relative to 0.61, then

$$x = \frac{\left(1.5 \times 10^{-4}\right)(0.61)}{x}$$

$$x = 0.0096\ M = [OH^-]$$

$$pOH = -\log(0.0096) = 2.02$$

$$pH = 14.00 - 2.02 = \mathbf{11.98}$$

16.73 We set up an equilibrium table to solve for $[OH^-]$ and use pOH to get pH. We use B and BH^+ to represent the weak base and its protonated form, respectively.

	B(aq) +	H₂O(l) ⇌	BH⁺(aq)	+ OH⁻(aq)
Initial (M):	0.045		0.00	0.00
Change (M):	−x		+x	+x
Equilibrium (M):	0.045 − x		x	x

$$K_b = \frac{\left[OH^-\right]\left[B^+\right]}{[BOH]}$$

$$[OH^-] = \frac{4.2 \times 10^{-10}[B^+]}{[BH^+]}$$

$$x = \frac{4.2 \times 10^{-10}(0.045 - x)}{x}$$

Assuming x is small relative to 0.045, then

$$x = \frac{4.2 \times 10^{-10}(0.045)}{x}$$

$$x = 4.3 \times 10^{-6} M = [OH^-]$$

$$pOH = -\log(4.3 \times 10^{-6}) = 5.36$$

$$pH = 14.00 - 5.36 = \mathbf{8.64}$$

16.77 **Strategy:** We calculate the K_b value for a conjugate base using Equation 16.6.

Solution: You can find the K_a values in Table 16.5 and Table 16.7 of the text.

$$\mathbf{K_b(CN^-)} = \frac{K_w}{K_a(HCN)} = \frac{1.0 \times 10^{-14}}{4.9 \times 10^{-10}} = \mathbf{2.0 \times 10^{-5}}$$

$$\mathbf{K_b(F^-)} = \frac{K_w}{K_a(HF)} = \frac{1.0 \times 10^{-14}}{7.1 \times 10^{-4}} = \mathbf{1.4 \times 10^{-11}}$$

$$\mathbf{K_b(CH_3COO^-)} = \frac{K_w}{K_a(CH_3COOH)} = \frac{1.0 \times 10^{-14}}{1.8 \times 10^{-5}} = \mathbf{5.6 \times 10^{-10}}$$

$$\mathbf{K_b(HCO_3^-)} = \frac{K_w}{K_a(H_2CO_3)} = \frac{1.0 \times 10^{-14}}{4.2 \times 10^{-7}} = \mathbf{2.4 \times 10^{-8}}$$

16.79 a. HA has the largest K_a value. Therefore, **A⁻** has the smallest K_b value.

b. HB has the smallest K_a value. Therefore, **B⁻** is the strongest base.

16.83 The pH of a 0.040 M HCl solution (strong acid) is: pH = $-\log(0.040)$ = **1.40**. Follow the procedure for calculating the pH of a diprotic acid to calculate the pH of the sulfuric acid solution.

Strategy: Determining the pH of a diprotic acid in aqueous solution is more involved than for a monoprotic acid. The first stage of ionization for H_2SO_4 goes to completion. We follow the procedure for determining the pH of a strong acid for this stage. The conjugate base produced in the first ionization (HSO_4^-) is a weak acid. We follow the procedure for determining the pH of a weak acid for this stage.

Solution: We proceed according to the following steps.

Step 1: H_2SO_4 is a strong acid. The first ionization stage goes to completion. The ionization of H_2SO_4 is

$$H_2SO_4(aq) \longrightarrow H^+(aq) + HSO_4^-(aq)$$

The concentrations of all the species (H_2SO_4, H^+, and HSO_4^-) before and after ionization can be represented as follows.

	$H_2SO_4(aq)$	$\rightarrow$	$H^+(aq)$	$+ HSO_4^-(aq)$
Initial (*M*):	0.040		0	0
Change (*M*):	−0.040		+0.040	+0.040
Equilibrium (*M*):	0		0.040	0.040

Step 2: Now, consider the second stage of ionization. HSO_4^- is a weak acid. Set up a table showing the concentrations for the second ionization stage. Let x be the change in concentration. Note that the initial concentration of H^+ is 0.040 *M* from the first ionization.

	$HSO_4^-(aq)$	$\rightleftarrows$	$H^+(aq)$	$+ SO_4^{2-}(aq)$
Initial (*M*):	0.040		0.040	0
Change (*M*):	−x		+x	+x
Equilibrium (*M*):	$0.040 - x$		$0.040 + x$	x

Write the ionization constant expression for K_a. Then, solve for x. You can find the K_a value in Table 16.7 of the text.

$$K_a = \frac{[H^+][SO_4^{2-}]}{[HSO_4^-]}$$

$$1.3 \times 10^{-2} = \frac{(0.040 + x)(x)}{(0.040 - x)}$$

Since K_a is quite large, we cannot make the assumptions that

$$0.040 - x \approx 0.040 \text{ and } 0.040 + x \approx 0.040$$

Therefore, we must solve a quadratic equation.

$$x^2 + 0.053x - (5.2 \times 10^{-4}) = 0$$

$$x = \frac{-0.053 \pm \sqrt{(0.053)^2 - 4(1)(-5.2 \times 10^{-4})}}{2(1)}$$

$$x = \frac{-0.053 \pm 0.070}{2}$$

$$x = 8.5 \times 10^{-3}\ M \text{ or } x = -0.062\ M$$

The second solution is physically impossible because you cannot have a negative concentration. The first solution is the correct answer.

Step 3: Having solved for x, we can calculate the H^+ concentration at equilibrium. We can then calculate the pH from the H^+ concentration.

$$[H^+] = 0.040\ M + x = [0.040 + (8.5 \times 10^{-3})]M = 0.049\ M$$

$$pH = -\log(0.049) = \textbf{1.31}$$

16.85   For the first stage of ionization:

$$H_2CO_3(aq) \;\rightleftharpoons\; H^+(aq) \;+\; HCO_3^-(aq)$$

Initial (M):	0.025	0.00	0.00
Change (M):	$-x$	$+x$	$+x$
Equilibrium (M):	$(0.025 - x)$	$+x$	$+x$

$$K_{a_1} = \frac{[H^+][HCO_3^-]}{[H_2CO_3]}$$

$$4.2 \times 10^{-7} = \frac{x^2}{(0.025 - x)} \approx \frac{x^2}{0.025}$$

$$x = 1.0 \times 10^{-4}\ M$$

For the second ionization:

$$HCO_3^-(aq) \;\rightleftharpoons\; H^+(aq) \;+\; CO_3^{2-}(aq)$$

Initial (M):	1.0×10^{-4}	1.0×10^{-4}	0.00
Change (M):	$-y$	$+y$	$+y$
Equilibrium (M):	$(1.0 \times 10^{-4}) - y$	$(1.0 \times 10^{-4}) + y$	y

$$K_{a_2} = \frac{[H^+][CO_3^{2-}]}{[HCO_3^-]}$$

$$4.8 \times 10^{-11} = \frac{\left[(1.0 \times 10^{-4}) + y\right](y)}{1.0 \times 10^{-4} - y} \approx \frac{(1.0 \times 10^{-4})(y)}{(1.0 \times 10^{-4})}$$

$$y = 4.8 \times 10^{-11}\ M$$

Since HCO_3^- is a very weak acid, there is little ionization at this stage. Therefore we have:

$$\textbf{[H}^+\textbf{] = [HCO}_3^-\textbf{] = 1.0} \times \textbf{10}^{-4}\ \textbf{\textit{M} and [CO}_3^{2-}\textbf{] =} \textit{y} = \textbf{4.8} \times \textbf{10}^{-11}\ \textbf{\textit{M}}$$

16.87　Oxalic acid is diprotic. Because its first and second ionization constants differ only by a factor of about 1000, we consider both the first and second ionizations to determine $[H^+]$. We set up an equilibrium table:

$$H_2C_2O_4(aq) \;\rightleftharpoons\; H^+(aq) \;+\; HC_2O_4^-(aq)$$

	Initial (*M*):	0.25	0.00	0.00
	Change (*M*):	$-x$	$+x$	$+x$
	Equilibrium (*M*):	$(0.25 - x)$	$+x$	$+x$

$$6.5 \times 10^{-2} = \frac{x^2}{0.25 - x}$$

Using the quadratic equation to solve for x, we find

$$x = 0.099$$

For the second ionization:

$$HC_2O_4^-(aq) \;\rightleftharpoons\; H^+(aq) \;+\; C_2O_4^{2-}(aq)$$

	Initial (*M*):	0.099	0.099	0.00
	Change (*M*):	$-y$	$+y$	$+y$
	Equilibrium (*M*):	$(0.099 - y)$	$0.099 + y$	y

$$6.1 \times 10^{-5} = \frac{(0.099 + y)y}{0.099 - y}$$

We neglect y with respect to 0.099 and find

$$y = 6.1 \times 10^{-5}$$

The concentration of hydronium ion after the second ionization is therefore

$$0.099 + 6.1 \times 10^{-5} \approx 0.099 \ M$$

and

$$pH = -\log(0.099) = \mathbf{1.00}$$

Note that the second ionization did not contribute enough to the hydronium concentration to change the value of pH.

16.89　1.　The two steps in the ionization of a weak diprotic acid are:

$$H_2A(aq) + H_2O(l) \;\rightleftharpoons\; H_3O^+(aq) + HA^-(aq)$$

$$HA^-(aq) + H_2O(l) \;\rightleftharpoons\; H_3O^+(aq) + A^{2-}(aq)$$

The diagram that represents a weak diprotic acid is **diagram (c)**. In this diagram, we only see the first step of the ionization, because HA^- is a much weaker acid than H_2A.

2. Both **diagrams (b) and (d)** are chemically implausible situations. Because HA^- is a much weaker acid than H_2A, you would not see a higher concentration of A^{2-} compared to HA^-.

16.95   The salt ammonium chloride completely ionizes upon dissolution, producing $0.42\ M\ [NH_4^+]$ and $0.42\ M$ $[Cl^-]$ ions. NH_4^+ will undergo hydrolysis because it is a weak acid (NH_4^+ is the conjugate acid of the weak base, NH_3).

Step 1:   Express the equilibrium concentrations of all species in terms of initial concentrations and a single unknown x, which represents the change in concentration. Let $(-x)$ be the depletion in concentration (mol/L) of NH_4^+. From the stoichiometry of the reaction, it follows that the increase in concentration for both H_3O^+ and NH_3 must be x. Complete the table that lists the initial concentrations, the change in concentrations, and the equilibrium concentrations.

$$NH_4^+(aq)\ +\ H_2O(l)\ \rightleftharpoons\ NH_3(aq)\ +\ H_3O^+(aq)$$

Step 2:

Initial (M):	0.42	0.00	0.00
Change (M):	$-x$	$+x$	$+x$
Equilibrium (M):	$(0.42 - x)$	$+x$	$+x$

You can calculate the K_a value for NH_4^+ from the K_b value of NH_3. The relationship is:

$$K_a \times K_b = K_w$$

or

$$K_a = \frac{K_w}{K_b} = \frac{1.0 \times 10^{-14}}{1.8 \times 10^{-5}} = 5.6 \times 10^{-10}$$

Step 3:   Write the ionization constant expression in terms of the equilibrium concentrations. Knowing the value of the equilibrium constant (K_a), solve for x.

$$K_a = \frac{[NH_3][H_3O^+]}{[NH_4^+]}$$

$$5.6 \times 10^{-10} = \frac{x^2}{0.42 - x} \approx \frac{x^2}{0.42}$$

$$x = [H^+] = 1.5 \times 10^{-5}\ M$$

$$pH = -\log(1.5 \times 10^{-5}) = \mathbf{4.82}$$

Since NH_4Cl is the salt of a weak base (aqueous ammonia) and a strong acid (HCl), we expect the solution to be slightly acidic, which is confirmed by the calculation.

16.97   **Strategy:**   We are asked to determine the pH of a salt solution. We must first determine the identities of the ions in solution and decide which, if any, of the ions will hydrolyze. In this case, the ions in solution are $C_2H_5NH_3^+$ and I^-. The iodide ion, I^-, is the anion of the strong acid, HI. Therefore, it will not hydrolyze. The $C_2H_5NH_3^+$ ion, however, is the conjugate acid of the weak base $C_2H_5NH_2$ (ethylamine). It hydrolyzes to produce H^+ and the weak base $C_2H_5NH_2$:

$$C_2H_5NH_3^+(aq) + H_2O(l) \rightleftharpoons C_2H_5NH_2(aq) + H_3O^+(aq)$$

In order to determine the pH, we must first determine the concentration of hydronium ion produced by the hydrolysis. For this, we need the ionization constant, K_a, of the $C_2H_5NH_3^+$ ion.

Solution: Using the K_b value for ethylamine and Equation 16.6, we calculate K_a for $C_2H_5NH_3^+$:

$$K_a = \frac{1.0 \times 10^{-14}}{5.6 \times 10^{-4}} = 1.8 \times 10^{-11}$$

We then construct an equilibrium table and fill in what we know to find $[H_3O^+]$:

	$C_2H_5NH_3^+(aq)$	$H_2O(l)$	$\rightleftharpoons$	$C_2H_5NH_2(aq)$	+	$H_3O^+(aq)$
Initial (*M*):	0.91			0.00		0.00
Change (*M*):	$-x$			$+x$		$+x$
Equilibrium (*M*):	$(0.91 - x)$			x		x

$$1.8 \times 10^{-11} = \frac{[C_2H_5NH_2][H_3O^+]}{[C_2H_5NH_3^+]} = \frac{x^2}{(0.91 - x)}$$

Assuming that x is very small compared to 0.91, we can write

$$1.8 \times 10^{-11} = \frac{x^2}{0.91}$$

and

$$x = [H_3O^+] = 4.05 \times 10^{-6} \ M$$

$$pH = -\log(4.05 \times 10^{-6}) = \mathbf{5.39}$$

16.99 **Strategy:** In deciding whether a salt will undergo hydrolysis, ask yourself the following questions: Is the cation a highly charged metal ion or an ammonium ion? Is the anion the conjugate base of a weak acid? If yes to either question, then hydrolysis will occur. In cases where both the cation and the anion react with water, the pH of the solution will depend on the relative magnitudes of K_a for the cation and K_b for the anion.

Solution: We first break up the salt into its cation and anion components and then examine the possible reaction of each ion with water.

a. The Na^+ cation does not hydrolyze. The Br^- anion is the conjugate base of the strong acid HBr. Therefore, Br^- will not hydrolyze either, and the solution is **neutral**.

b. The K^+ cation does not hydrolyze. The SO_3^{2-} anion is the conjugate base of the weak acid HSO_3^- and will hydrolyze to give HSO_3^- and OH^-. The solution will be **basic**.

c. Both the NH_4^+ and NO_2^- ions will hydrolyze. NH_4^+ is the conjugate acid of the weak base NH_3, and NO_2^- is the conjugate base of the weak acid HNO_2. Using ionization constants from Tables 16.5 and 16.6, and Equation 16.6, we find that the K_a of NH_4^+ ($1.0 \times 10^{-14}/1.8 \times 10^{-5} = 5.6 \times 10^{-10}$) is greater than the K_b of NO_2^- ($1.0 \times 10^{-14}/4.5 \times 10^{-4} = 2.2 \times 10^{-11}$). Therefore, the solution will be **acidic**.

d. Cr^{3+} is a small metal cation with a high charge, which hydrolyzes to produce H^+ ions. The NO_3^- anion does not hydrolyze. It is the conjugate base of the strong acid, HNO_3. The solution will be **acidic**.

16.101 There is an inverse relationship between acid strength and conjugate base strength. As acid strength decreases, the proton accepting power of the conjugate base increases. In general the weaker the acid, the stronger the conjugate base. All three of the potassium salts ionize completely to form the conjugate base of the respective acid. The greater the pH, the stronger the conjugate base and, therefore, the weaker the acid.

The order of increasing acid strength is **HZ < HY < HX.**

16.103

$$HCO_3^- \rightleftharpoons H^+ + CO_3^{2-} \qquad\qquad K_a = 4.8 \times 10^{-11}$$

$$HCO_3^- + H_2O \rightleftharpoons H_2CO_3 + OH^- \qquad K_b = \frac{K_w}{K_a} = \frac{1.0 \times 10^{-14}}{4.2 \times 10^{-7}} = 2.4 \times 10^{-8}$$

HCO_3^- has a greater tendency to hydrolyze than to ionize ($K_b > K_a$). The solution will be basic (**pH > 7**).

16.107 The most basic oxides occur with metal ions having the lowest positive charges (or lowest oxidation numbers).

a. **Al_2O_3 < BaO < K_2O**   b. **CrO_3 < Cr_2O_3 < CrO**

16.109 $Al(OH)_3$ is an amphoteric hydroxide. The reaction is:

$$Al(OH)_3(s) + OH^-(aq) \longrightarrow Al(OH)_4^-(aq)$$

This is a Lewis acid-base reaction.

16.111 a. The basic metallic oxides react with water to form metal hydroxides:

$$Li_2O(s) + H_2O(l) \longrightarrow 2LiOH(aq)$$

b. As in part (a) the basic metal oxide reacts to form a metal hydroxide:

$$CaO(s) + H_2O(l) \longrightarrow Ca(OH)_2(aq)$$

c. The reaction between an acidic oxides and water is:

$$SO_3(g) + H_2O(l) \longrightarrow H_2SO_4(aq)$$

16.115 **$AlCl_3$ is a Lewis acid with an incomplete octet of electrons and Cl^- is the Lewis base donating a pair of electrons.**

16.117 By definition Brønsted acids are proton donors; therefore, such compounds must contain at least one hydrogen atom. In Problem 16.114, Lewis acids that do not contain hydrogen, and therefore are not Brønsted acids, are **CO_2, SO_2, and BCl_3.**

16.119 a. **$AlBr_3$ is the Lewis acid; Br^- is the Lewis base.**

b. **Cr is the Lewis acid; CO is the Lewis base.**

c. **Cu^{2+} is the Lewis acid; CN^- is the Lewis base.**

16.121 A **strong acid**, such as HCl, will be completely ionized, **diagram (b)**.

A **weak acid** will only ionize to a lesser extent compared to a strong acid, **diagram (c)**.

A **very weak acid** will remain almost exclusively as the acid molecule in solution; **diagram (d)** is the best choice.

16.123 In theory, **the products will be $CH_3COO^-(aq)$ and $HCl(aq)$, but this reaction will *not* occur to any measurable extent** because Cl^- is the conjugate base of the strong acid, HCl. It is a negligibly weak base and has no affinity for protons.

16.125 HNO_2 is a weak acid with $K_a = 4.5 \times 10^{-4}$. (See Table 16.5.)

$$HNO_2(aq) \rightleftarrows H^+(aq) + NO_2^-(aq)$$

Initial (*M*):	0.88	0.00	0.00
Change (*M*):	$-x$	$+x$	$+x$
Equilibrium (*M*):	$(0.88 - x)$	x	x

$$K_b = \frac{[H^+][NO_2^-]}{[HNO_2]}$$

$$4.5\times10^{-4} = \frac{x^2}{0.88 - x}$$

We neglect x in the denominator and find

$$x = \sqrt{\left(4.5\times10^{-4}\right)(0.88)} = 0.020$$

$$pH = -\log(0.020) = \mathbf{1.70}$$

$$\text{\% ionization} = \frac{0.020}{0.88} \times 100\% = 2.3\%$$

16.127 **Choice (c)** because 0.70 M KOH has a higher pH than 0.60 M NaOH. Adding an equal volume of 0.60 M NaOH lowers the [OH$^-$] to 0.65 M, hence lowering the pH.

16.129 a. **For the forward reaction NH_4^+ and NH_3 are the conjugate acid and base pair** (Brønsted acid and base pair)**, respectively. For the reverse reaction NH_3 and NH_2^- are the conjugate acid and base pair** (Brønsted acid and base pair)**, respectively.**

b. **H$^+$ corresponds to NH_4^+ ; OH$^-$ corresponds to NH_2^- . For the neutral solution, $[NH_4^+] = [NH_2^-]$.**

16.131

$$K_a = \frac{[H^+][A^-]}{[HA]}$$

$$[HA] \approx 0.1\ M$$

$$[A^-] \approx 0.1\ M$$

Therefore,

$$K_a = [H^+] = \frac{K_w}{[OH^-]}$$

$$[OH^-] = \frac{K_w}{K_a}$$

16.133

$$HCOOH \rightleftharpoons HCOO^- + H^+ \qquad K_a = 1.7 \times 10^{-4}$$

$$H^+ + OH^- \rightleftharpoons H_2O \qquad K_w' = \frac{1}{K_w} = \frac{1}{1.0 \times 10^{-14}} = 1.0 \times 10^{14}$$

$$HCOOH + OH^- \rightleftharpoons HCOO^- + H_2O \qquad K = K_a K_w' = (1.7 \times 10^{-4})(1.0 \times 10^{14}) = \mathbf{1.7 \times 10^{10}}$$

16.135 a.
$$\begin{array}{ccccccc} H^- & + & H_2O & \longrightarrow & OH^- & + & H_2 \\ \text{base}_1 & & \text{acid}_2 & & \text{base}_2 & & \text{acid}_1 \end{array}$$

b. **H$^-$ is the reducing agent, and H$_2$O is the oxidizing agent.**

16.137

$$K_b = 8.91 \times 10^{-6}$$

$$K_a = \frac{K_w}{K_b} = \frac{1.0 \times 10^{-14}}{8.91 \times 10^{-6}} = 1.1 \times 10^{-9}$$

$$pH = 7.40$$

$$[H^+] = 10^{-7.40} = 3.98 \times 10^{-8}$$

$$K_a = \frac{[\text{H}^+][\text{conjugate base}]}{[\text{acid}]}$$

Therefore,

$$\frac{[\text{conjugate base}]}{[\text{acid}]} = \frac{K_a}{[\text{H}^+]} = \frac{1.1 \times 10^{-9}}{3.98 \times 10^{-8}} = 0.028 \text{ or } \mathbf{2.8 \times 10^{-2}}$$

16.139 Because the P–H bond is weaker, there is a greater tendency for PH_4^+ to ionize. Therefore, **PH$_3$ is a weaker base than NH$_3$.**

16.141 a. **HNO$_2$**

b. **HF**

c. **BF$_3$**

d. **NH$_3$**

e. **H$_2$SO$_3$**

f. **HCO$_3^-$ and CO$_3^{2-}$**

The reactions for (f) are: $\text{HCO}_3^- (aq) + \text{H}^+(aq) \longrightarrow \text{CO}_2(g) + \text{H}_2\text{O}(l)$

$\text{CO}_3^{2-} (aq) + 2\text{H}^+(aq) \longrightarrow \text{CO}_2(g) + \text{H}_2\text{O}(l)$

16.143 a. The Lewis structure of H_3O^+ is:

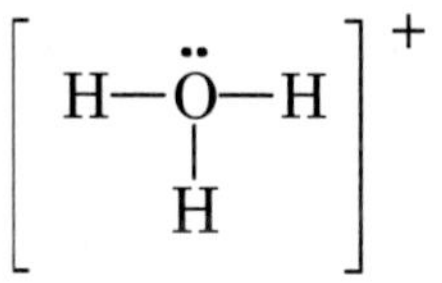

$$\left[\begin{array}{c} \text{H}-\overset{\displaystyle ..}{\text{O}}-\text{H} \\ | \\ \text{H} \end{array} \right]^+$$

Note that this structure is very similar to the Lewis structure of NH$_3$. The geometry is **trigonal pyramidal**.

b. **H$_4$O^{2+} does *not* exist because the positively charged H$_3$O$^+$ has no affinity to accept the positive H$^+$ ion. If H$_4$O^{2+} existed, it would have a tetrahedral geometry.**

16.145 The equations are: **Cl$_2$(g) + H$_2$O(l) $\rightleftharpoons$ HCl(aq) + HClO(aq)**

HCl(aq) + AgNO$_3$(aq) $\rightleftharpoons$ AgCl(s) + HNO$_3$(aq)

In the presence of OH$^-$ ions, the first equation is shifted to the right:

H$^+$ (from HCl) + OH$^-$ $\longrightarrow$ H$_2$O

Therefore, the concentration of HClO increases. (The "bleaching action" is due to ClO$^-$ ions.)

16.147 We examine the hydrolysis of the cation and anion separately.

$$NH_4CN(aq) \longrightarrow NH_4^+(aq) + CN^-(aq)$$

Cation: $NH_3(aq) + H_2O(l) \rightleftharpoons NH_4^+(aq) + H_3O^+(aq)$

	NH_3		NH_4^+	H_3O^+
Initial (M):	2.00		0.00	0.00
Change (M):	$-x$		$+x$	$+x$
Equilibrium (M):	$2.00 - x$		x	x

Use the K_b value for NH_3 (1.8×10^{-5} from Table 16.6) and Equation 16.6 to determine K_a for NH_4^+.

$$K_a = \frac{K_w}{K_b} = \frac{1.0 \times 10^{-14}}{1.8 \times 10^{-5}} = 5.6 \times 10^{-10}$$

$$K_a = \frac{[NH_3][H_3O^+]}{[NH_4^+]}$$

$$5.6 \times 10^{-10} = \frac{x^2}{2.00 - x} \approx \frac{x^2}{2.00}$$

$$x = 3.35 \times 10^{-5} \, M = [H_3O^+]$$

Use the K_a value for HCN (4.9×10^{-10} from Table 16.5) and Equation 16.6 to determine K_b for CN^-.

$$K_b = \frac{K_w}{K_a} = \frac{1.0 \times 10^{-14}}{4.9 \times 10^{-10}} = 2.0 \times 10^{-5}$$

$$K_b = \frac{[HCN][OH^-]}{[CN^-]}$$

$$2.0 \times 10^{-5} = \frac{y^2}{2.00 - y} \approx \frac{y^2}{2.00}$$

$$y = 6.32 \times 10^{-3} \, M = [OH^-]$$

CN^- is stronger as a base than NH_4^+ is as an acid. Some OH^- produced from the hydrolysis of CN^- will be neutralized by H_3O^+ produced from the hydrolysis of NH_4^+.

	$H_3O^+(aq) +$	$OH^-(aq)$	$\rightarrow$	$2H_2O(l)$
Initial (M):	3.35×10^{-5}	6.32×10^{-3}		
Change (M):	-3.35×10^{-5}	-3.35×10^{-5}		
Equilibrium (M):	0	6.29×10^{-3}		

$$[OH^-] = 6.29 \times 10^{-3} \ M$$

$$pOH = -\log[6.29 \times 10^{-3}] = 2.20$$

$$\mathbf{pH} = 14.00 - 2.20 = \mathbf{11.80}$$

16.149 a. We carry an additional significant figure throughout this calculation to minimize rounding errors.

$$\text{Number of moles NaOH} = M \times \text{vol (L)} = 0.0568 \ M \times 0.0138 \ L = 7.838 \times 10^{-4} \ \text{mol}$$

If the acid were all dimer, then:

$$\text{mol of dimer} = \frac{\text{mol NaOH}}{2} = \frac{7.838 \times 10^{-4} \ \text{mol}}{2} = 3.919 \times 10^{-4} \ \text{mol}$$

If the acetic acid were all dimer, the pressure that would be exerted would be:

$$P = \frac{nRT}{V} = \frac{(3.919 \times 10^{-4} \ \text{mol})(0.08206 \ \text{L} \cdot \text{atm/K} \cdot \text{mol})(324 \ \text{K})}{0.360 \ \text{L}} = 0.02896 \ \text{atm}$$

However, the actual pressure is 0.0342 atm. If α mol of dimer dissociates to monomers, then 2α monomer forms.

$$(CH_3COOH)_2 \ \rightleftharpoons \ 2CH_3COOH$$
$$1 - \alpha \qquad\qquad 2\alpha$$

The total moles of acetic acid is:

$$\text{moles dimer} + \text{monomer} = (1 - \alpha) + 2\alpha = 1 + \alpha$$

Using partial pressures:

$$P_{\text{observed}} = P(1 + \alpha)$$

$$0.0342 \ \text{atm} = (0.02896 \ \text{atm})(1 + \alpha)$$

$$\boldsymbol{\alpha = 0.181 \ \text{or} \ 18.1\%}$$

b. The equilibrium constant is:

$$K_P = \frac{\left(P_{CH_3COOH}\right)^2}{P_{(CH_3COOH)_2}} = \frac{\left(\dfrac{2\alpha}{1+\alpha}\right)^2 \left(P_{\text{observed}}\right)^2}{\left(\dfrac{1-\alpha}{1+\alpha}\right)P_{\text{observed}}} = \frac{4\alpha^2 P_{\text{observed}}}{1 - \alpha^2} = \mathbf{4.63 \times 10^{-3}}$$

16.151

$$[CO_2] = kP = (2.28 \times 10^{-3} \ \text{mol/L} \cdot \text{atm})(3.20 \ \text{atm}) = 7.30 \times 10^{-3} \ M$$

$$CO_2(aq) + H_2O(l) \ \rightleftharpoons \ H^+(aq) + HCO_3^-(aq)$$
$$(7.30 \times 10^{-3} - x) \ M \qquad\qquad x \ M \qquad x \ M$$

$$K_a = \frac{[H^+][HCO_3^-]}{[CO_2]}$$

$$4.2 \times 10^{-7} = \frac{x^2}{\left(7.30 \times 10^{-3}\right) - x} \approx \frac{x^2}{7.30 \times 10^{-3}}$$

$$x = 5.5 \times 10^{-5}\ M = [H^+]$$

pH = 4.26

16.153 When the pH is 10.00, the pOH is 4.00 and the concentration of hydroxide ion is $1.0 \times 10^{-4}\ M$. The concentration of HCN must be the same. If the concentration of NaCN is x, the table looks like:

	$CN^-(aq) +$	$H_2O(l)$	$\rightleftarrows$	$HCN(aq)$	$+\ OH^-(aq)$
Initial (*M*):	x			0.00	0.00
Change (*M*):	-1.0×10^{-4}			$+1.0 \times 10^{-4}$	$+1.0 \times 10^{-4}$
Equilibrium (*M*):	$(x - 1.0 \times 10^{-4})$			(1.0×10^{-4})	(1.0×10^{-4})

$$K_b = \frac{[HCN][OH^-]}{[CN^-]}$$

$$2.04 \times 10^{-5} = \frac{\left(1.0 \times 10^{-4}\right)^2}{x - 1.0 \times 10^{-4}}$$

$$x = 5.9 \times 10^{-4}\ M = [CN^-]_0$$

$$\text{Amount of NaCN} = 250\ \text{mL} \times \frac{5.9 \times 10^{-4}\ \text{mol NaCN}}{1000\ \text{mL}} \times \frac{49.01\ \text{g NaCN}}{1\ \text{mol NaCN}} = \mathbf{7.2 \times 10^{-3}\ g\ NaCN}$$

16.155 The equilibrium is established:

	$CH_3COOH(aq)$	$\rightleftarrows$	$CH_3COO^-(aq)$	$+\ H^+(aq)$
Initial (*M*):	0.150		0.00	0.100
Change (*M*):	$-x$		$+x$	$+x$
Equilibrium (*M*):	$(0.150 - x)$		x	$(0.100 + x)$

According to Table 16.5, K_a for acetic acid is 1.8×10^{-5}.

$$K_a = \frac{[CH_3COO^-][H^+]}{[CH_3COOH]}$$

$$1.8 \times 10^{-5} = \frac{x(0.100 + x)}{0.150 - x} \approx \frac{0.100x}{0.150}$$

$$x = 2.7 \times 10^{-5}\ M$$

2.7×10^{-5} *M* is the [H$^+$] contributed by CH$_3$COOH. HCl is a strong acid that completely ionizes. It contributes a [H$^+$] of 0.100 *M* to the solution.

$$[\text{H}^+]_{\text{total}} = [0.100 + (2.7 \times 10^{-5})] \ M \approx 0.100 \ M$$

$$\text{pH} = \textbf{1.000}$$

To the correct number of significant digits, the contribution to the pH from CH$_3$COOH is negligible.

16.157 a. **The pH of the solution of HA would be lower**.

b. **The electrical conductance of the HA solution would be greater**.

c. **The rate of hydrogen evolution from the HA solution would be greater**. Presumably, the rate of the reaction between the metal and hydrogen ion would depend on the hydrogen ion concentration (i.e., this would be part of the rate law). The hydrogen ion concentration will be greater in the HA solution.

16.159

	HCOOH (*aq*)	$\rightleftarrows$	H$^+$(*aq*)	+ HCOO$^-$(*aq*)
Initial (*M*):	0.400		0.00	0.00
Change (*M*):	$-x$		$+x$	$+x$
Equilibrium (*M*):	$0.400 - x$		x	x

Total concentration of particles in solution: $(0.400 - x) + x + x = 0.400 + x$

Assuming the molarity of the solution is equal to the molality, we can write:

$$\Delta T_{\text{f}} = K_{\text{f}}m$$

$$0.758 = (1.86)(0.400 + x)$$

$$x = 0.0075 = [\text{H}^+] = [\text{HCOO}^-]$$

$$K_{\text{a}} = \frac{[\text{H}^+][\text{HCOO}^-]}{[\text{HCOOH}]} = \frac{(0.0075)(0.0075)}{0.400 - 0.0075} = \textbf{1.4} \times \textbf{10}^{-4}$$

16.161

$$\text{pH} = 10.64$$

$$\text{pOH} = 3.36$$

$$[\text{OH}^-] = 4.4 \times 10^{-4} \ M$$

$$\text{CH}_3\text{NH}_2(aq) + \text{H}_2\text{O}(l) \rightleftarrows \text{CH}_3\text{NH}_3^+ (aq) + \text{OH}^-(aq)$$

$$(x - 4.4 \times 10^{-4}) \ M \quad 4.4 \times 10^{-4} \ M \quad 4.4 \times 10^{-4} \ M$$

$$K_{\text{b}} = \frac{[\text{CH}_3\text{NH}_3^+][\text{OH}^-]}{[\text{CH}_3\text{NH}_2]}$$

$$4.4 \times 10^{-4} = \frac{(4.4 \times 10^{-4})(4.4 \times 10^{-4})}{x - (4.4 \times 10^{-4})}$$

$$4.4 \times 10^{-4}x - 1.9 \times 10^{-7} = 1.9 \times 10^{-7}$$

$$x = 8.6 \times 10^{-4} \, M$$

The molar mass of CH_3NH_2 is 31.06 g/mol.

The mass of CH_3NH_2 in 100.0 mL is:

$$100.0 \text{ mL} \times \frac{8.6 \times 10^{-4} \text{ mol } CH_3NH_2}{1000 \text{ mL}} \times \frac{31.06 \text{ g } CH_3NH_2}{1 \text{ mol } CH_3NH_2} = \mathbf{2.7 \times 10^{-3} \text{ g } CH_3NH_2}$$

16.163 a.

$$NH_2^- \text{ (base)} + H_2O \text{ (acid)} \longrightarrow NH_3 + OH^-$$

$$N^{3-} \text{ (base)} + 3H_2O \text{ (acid)} \longrightarrow NH_3 + 3OH^-$$

b. N^{3-} is the stronger base since each ion produces three OH^- ions.

16.165 **When smelling salt is inhaled, some of the powder dissolves in the basic solution. The ammonium ions react with the base as follows:**

$$NH_4^+ \text{ (aq)} + OH^- \text{(aq)} \longrightarrow NH_3(aq) + H_2O(l)$$

It is the pungent odor of ammonia that prevents a person from fainting.

16.167 **Reaction (c)** does not represent a Lewis acid-base reaction. In this reaction, the F–F single bond is broken and single bonds are formed between P and each F atom. For a Lewis acid-base reaction, the Lewis acid is an electron-pair acceptor and the Lewis base is an electron-pair donor.

16.169 From the given pH's, we can calculate the $[H^+]$ in each solution.

Solution (1): $[H^+] = 10^{-pH} = 10^{-4.12} = 7.6 \times 10^{-5} \, M$
Solution (2): $[H^+] = 10^{-5.76} = 1.7 \times 10^{-6} \, M$
Solution (3): $[H^+] = 10^{-5.34} = 4.6 \times 10^{-6} \, M$

We are adding solutions (1) and (2) to make solution (3). The volume of solution (2) is 0.528 L. We are going to add a given volume of solution (1) to solution (2). Let this volume be x. The moles of H^+ in solutions (1) and (2) will equal the moles of H^+ in solution (3).

$$\text{mol } H^+ \text{ soln (1)} + \text{mol } H^+ \text{ soln (2)} = \text{mol } H^+ \text{ soln (3)}$$

Recall that mol = $M \times$ vol (L). We have:

$$(7.6 \times 10^{-5} \text{ mol/L})(x \text{ L}) + (1.7 \times 10^{-6} \text{ mol/L})(0.528 \text{ L}) = (4.6 \times 10^{-6} \text{ mol/L})(0.528 + x)\text{L}$$

$$(7.6 \times 10^{-5})x + (9.0 \times 10^{-7}) = (2.4 \times 10^{-6}) + (4.6 \times 10^{-6})x$$

$$(7.1 \times 10^{-5})x = 1.5 \times 10^{-6}$$

$$x = 0.021 \text{ L} = \mathbf{21 \text{ mL}}$$

16.171 The balanced equations for the two reactions are:

$$MCO_3(s) + 2HCl(aq) \longrightarrow MCl_2(aq) + CO_2(g) + H_2O(l)$$

$$HCl(aq) + NaOH(aq) \longrightarrow NaCl(aq) + H_2O(l)$$

First, let's find the number of moles of excess acid from the reaction with NaOH.

$$0.03280 \text{ L} \times \frac{0.588 \text{ mol NaOH}}{1 \text{ L soln}} \times \frac{1 \text{ mol HCl}}{1 \text{ mol NaOH}} = 0.0193 \text{ mol HCl}$$

The original number of moles of acid was:

$$0.500 \text{ L} \times \frac{1.00 \text{ mol HCl}}{1 \text{ L soln}} = 0.500 \text{ mol HCl}$$

The amount of hydrochloric acid that reacted with the metal carbonate is:

$$(0.500 \text{ mol HCl}) - (0.0193 \text{ mol HCl}) = 0.4807 \text{ mol HCl}$$

The mole ratio from the balanced equation is 1 mole MCO_3 : 2 mole HCl. The moles of MCO_3 that reacted are:

$$0.4807 \text{ mol HCl} \times \frac{1 \text{ mol MCO}_3}{2 \text{ mol HCl}} = 0.2404 \text{ mol MCO}_3$$

We can now determine the molar mass of MCO_3, which will allow us to identify the metal.

$$\text{molar mass MCO}_3 \times \frac{20.27 \text{ g MCO}_3}{0.2404 \text{ mol MCO}_3} = 84.3 \text{ g/mol}$$

We subtract the mass of CO_3^{2-} to identify the metal.

$$\text{molar mass of M} = 84.3 \text{ g/mol} - 60.01 \text{ g/mol} = 24.3 \text{ g/mol}$$

The metal is **magnesium**.

16.173 **Both NaF and SnF$_2$ provide F⁻ ions in solution.**

$$NaF \longrightarrow Na^+ + F^-$$

$$SnF_2 \longrightarrow Sn^{2+} + 2F^-$$

Because HF is a much stronger acid than H_2O, it follows that F⁻ is a much weaker base than OH⁻. **The F⁻ ions replace OH⁻ ions during the remineralization process.**

$$\textbf{5Ca}^{2+} + \textbf{3PO}_4^{3-} + \textbf{F}^- \longrightarrow \textbf{Ca}_5\textbf{(PO}_4\textbf{)}_3\textbf{F (fluorapatite)}$$

because OH⁻ has a much greater tendency to combine with H⁺

$$OH^- + H^+ \longrightarrow H_2O$$

than F^- does.

$$F^- + H^+ \rightleftharpoons HF$$

Because F^- is a weaker base than OH^-, fluorapatite is more resistant to attacks by acids compared to hydroxyapatite.

16.175 **Strategy:** All alkali metal oxides and all alkaline earth metal oxides except BeO are basic. Beryllium oxide and several metallic oxides in Groups 3A and 4A are amphoteric. Nonmetal oxides in which the oxidation number of the main group elements is high are acidic.

   Setup: Figure 16.6 shows the oxides of the main group elements in their highest oxidation states.

   Solution: **Li_2O, lithium oxide, basic**

   BeO, beryllium oxide, amphoteric

   B_2O_3, diboron trioxide, acidic

   CO_2, carbon dioxide, acidic

   N_2O_5, dinitrogen pentoxide, acidic

16.177 **Strategy:** The first ionization is complete. The conjugate base resulting from the first ionization is the acid for the second ionization, and its starting concentration is the concentration of the sulfuric acid.

   Setup: The initial concentration of H_2SO_4 is 4.00×10^{-5} M. Since the first ionization is complete, the final concentration of H_2SO_4 is 0 and $[H^+] = HSO_4^- = 4.00 \times 10^{-5}$ M.

   The ionization constant for the second ionization is $K_{a_1} = 1.3 \times 10^{-2}$.

   Construct an equilibrium table for the second ionization using x as the unknown.

	HSO (aq)	$\rightleftharpoons$	$H^+(aq)$	$+$ $SO_4^{2-}(aq)$
Initial (M):	4.00×10^{-5}		4.00×10^{-5}	0.00
Change (M):	$-x$		$+x$	$+x$
Equilibrium (M):	$4.00 \times 10^{-5} - x$		$4.00 \times 10^{-5} + x$	x

   Solution: These equilibrium concentrations are then substituted into the equilibrium expression to give:

$$K_{a_1} = \frac{\left(4.00 \times 10^{-5} + x\right)(x)}{\left(4.00 \times 10^{-5} - x\right)} \quad (1)$$

   Because K_a is relatively large, solve for x without making any approximations. Start by clearing the denominator of equation (1) and then expanding the result to get:

$$x^2 + 1.30 \times 10^{-2} x - \left(5.2 \times 10^{-7}\right) = 0$$

This equation is a *quadratic equation* of the form $ax^2 + bx + c = 0$, the solutions of which are given by (see Appendix 1):

$$x = \frac{-b \pm \sqrt{b^2 - 4ac}}{2a}.$$

For this case, $a = 1$, $b = 1.30 \times 10^{-2}$, and $c = -5.2 \times 10^{-7}$. Substituting these values into the quadratic formula gives:

$$x = \frac{-1.3 \times 10^{-2} \pm \sqrt{\left(1.3 \times 10^{-2}\right)^2 - 4(1)\left(-5.2 \times 10^{-7}\right)}}{(2)(1)}$$

$$= \frac{-1.3 \times 10^{-2} \pm 1.308 \times 10^{-2}}{2}$$

$$x = 4.0 \times 10^{-5} \ M \quad \text{or} \quad x = -1.3 \times 10^{-2} \ M$$

The negative solution is not physically reasonable since there is no such thing as a negative concentration. The solution is $x = 4.0 \times 10^{-5}$ M.

According to the equilibrium table, $x = [\,SO_4^{2-}\,] = [H^+]$, however there is also $[H^+]$ produced from the first ionization. The final H^+ concentration is equal to the sum of both concentrations.

At equilibrium the final concentrations of all species are:

$$[\,SO_4^{2-}\,] = 4.0 \times 10^{-5} \ M$$

$$[H^+] = 4.0 \times 10^{-5} \ M + 4.0 \times 10^{-5} \ M = 8.0 \times 10^{-5} \ M$$

$$[\,HSO_4^-\,] = 4.0 \times 10^{-5} \ M - 4.0 \times 10^{-5} \ M = 0 \ M$$

$$[H_2SO_4] = 0 \ M$$

Chapter 17
Acid-Base Equilibria and Solubility Equilibria

Visualizing Chemistry

VC 17.1	b	VC 17.2	c	VC 17.3	a	VC 17.4	a
VC 17.5	c	VC 17.6	a	VC 17.7	c	VC 17.8	a

Key Skills

17.1	e	17.2	d	17.3	e	17.4	a

Problems

17.5   a. This is a weak acid problem. Setting up the standard equilibrium table:

$$CH_3COOH(aq) \rightleftharpoons H^+(aq) + CH_3COO^-(aq)$$

	$CH_3COOH(aq)$	$H^+(aq)$	$CH_3COO^-(aq)$
Initial (M):	0.40	0.00	0.00
Change (M):	$-x$	$+x$	$+x$
Equilibrium (M):	$(0.40 - x)$	x	x

According to Table 16.5, K_a for CH_3COOH is 1.8×10^{-5}.

$$K_a = \frac{[H^+][CH_3COO^-]}{[CH_3COOH]}$$

$$1.8 \times 10^{-5} = \frac{x^2}{(0.40 - x)} \approx \frac{x^2}{0.40}$$

$$x = [H^+] = 2.7 \times 10^{-3}\ M$$

$$\mathbf{pH} = -\log[H^+] = -\log(2.7 \times 10^{-3}) = \mathbf{2.57}$$

  b. In addition to the acetate ion formed from the ionization of acetic acid, we also have acetate ion formed from the dissolution of sodium acetate.

$$CH_3COONa(aq) \longrightarrow CH_3COO^-(aq) + Na^+(aq)$$

Dissolving 0.20 M sodium acetate initially produces 0.20 M CH_3COO^- and 0.20 M Na^+. The sodium ions are not involved in any further equilibrium, but the acetate ions must be added to the equilibrium in part (a).

$$CH_3COOH(aq) \rightleftharpoons H^+(aq) + CH_3COO^-(aq)$$

	$CH_3COOH(aq)$	$H^+(aq)$	$CH_3COO^-(aq)$
Initial (M):	0.40	0.00	0.20
Change (M):	$-x$	$+x$	$+x$
Equilibrium (M):	$(0.40 - x)$	x	$(0.20 + x)$

$$K_a = \frac{[H^+][CH_3COO^-]}{[CH_3COOH]}$$

$$1.8 \times 10^{-5} = \frac{(x)(0.20 + x)}{(0.40 - x)} \approx \frac{x(0.20)}{0.40}$$

$$x = [H^+] = 3.6 \times 10^{-5}\ M$$

$$\textbf{pH} = -\log[H^+] = -\log(3.6 \times 10^{-5}) = \textbf{4.44}$$

An alternate way to solve part (b) of this problem is to use the Henderson-Hasselbalch equation.

$$pH = pK_a + \log\frac{[\text{conjugate base}]}{[\text{acid}]}$$

$$pH = -\log(1.8 \times 10^{-5}) + \log\frac{0.20\ M}{0.40\ M} = 4.74 - 0.30 = \textbf{4.44}$$

17.9 Strategy: The pH of a buffer system can be calculated using the Henderson-Hasselbalch equation (Equation 17.3). The K_a of a conjugate acid such as NH_4^+ is calculated using the K_b of its weak base (in this case, NH_3) from Table 16.6 and Equation 16.6.

Solution:

$$NH_4^+(aq) \rightleftharpoons NH_3(aq) + H^+(aq)$$

$$K_a = \frac{K_w}{1.8 \times 10^{-5}} = \frac{1.0 \times 10^{-14}}{1.8 \times 10^{-5}} = 5.56 \times 10^{-10}$$

$$pK_a = -\log K_a = 9.26$$

$$\textbf{pH} = pK_a + \log\frac{[NH_3]}{[NH_4^+]} = 9.26 + \log\frac{0.15\ M}{0.35\ M} = \textbf{8.89}$$

17.11

$$H_2CO_3(aq) \rightleftharpoons HCO_3^-(aq) + H^+(aq)$$

$$K_{a_1} = 4.2 \times 10^{-7}\ \text{(See Table 16.7)}$$

$$pK_{a_1} = -\log K_{a_1} = -\log(4.2 \times 10^{-7}) = 6.38$$

$$pH = pK_a + \log\frac{[HCO_3^-]}{[H_2CO_3]}$$

$$8.00 = 6.38 + \log\frac{[HCO_3^-]}{[H_2CO_3]}$$

$$\log \frac{[HCO_3^-]}{[H_2CO_3]} = 1.62$$

$$\frac{[HCO_3^-]}{[H_2CO_3]} = 42$$

$$\frac{[H_2CO_3]}{[HCO_3^-]} = \mathbf{0.024}$$

17.13   Using the Henderson-Hasselbalch equation:

$$pH = pK_a + \log \frac{[CH_3COO^-]}{[CH_3COOH]}$$

$$4.50 = 4.74 + \log \frac{[CH_3COO^-]}{[CH_3COOH]}$$

Thus,

$$\frac{[CH_3COO^-]}{[CH_3COOH]} = \mathbf{0.58}$$

17.15   For the first part we use K_a for ammonium ion. The Henderson-Hasselbalch equation is:

$$pH = -\log\left(5.6 \times 10^{-10}\right) + \log \frac{(0.20 \ M)}{(0.20 \ M)} = \mathbf{9.25}$$

For the second part, the acid-base reaction is:

$$NH_3(g) + H^+(aq) \longrightarrow NH_4^+(aq)$$

We find the number of moles of HCl added:

$$10.0 \ \text{mL} \times \frac{0.10 \ \text{mol HCl}}{1000 \ \text{mL soln}} = 0.0010 \ \text{mol HCl}$$

The number of moles of NH_3 and NH_4^+ originally present are:

$$65.0 \ \text{mL} \times \frac{0.20 \ \text{mol}}{1000 \ \text{mL soln}} = 0.013 \ \text{mol}$$

Using the acid-base reaction, we find the number of moles of NH_3 and NH_4^+ after the addition of HCl.

	$NH_3(aq)+$	$H^+(aq)$ $\rightleftarrows$	$NH_4^+(aq)$
Initial (mol):	0.013	0.0010	0.013
Change (mol):	−0.0010	−0.0010	+0.0010
Final (mol):	0.012	0	0.014

We find the new pH:

$$pH = 9.25 + \log\frac{(0.012)}{(0.014)} = \textbf{9.18}$$

17.17 **Strategy:** What constitutes a buffer system? Which of the solutions described in the problem contains a weak acid and its salt (containing the weak conjugate base)? Which contains a weak base and its salt (containing the weak conjugate acid)? Why is the conjugate base of a strong acid not able to neutralize an added acid?

Solution: The criteria for a buffer system are that we must have a weak acid and its salt (containing the weak conjugate base) or a weak base and its salt (containing the weak conjugate acid).

a. HCl (hydrochloric acid) is a strong acid. A buffer is a solution containing both a weak acid and a weak base. Therefore, this is *not* a buffer system.

b. H_2SO_4 (sulfuric acid) is a strong acid. A buffer is a solution containing both a weak acid and a weak base. Therefore, this is *not* a buffer system.

c. This solution contains both a weak acid, $H_2PO_4^-$, and its conjugate base, HPO_4^{2-}. Therefore, this is a buffer system.

d. HNO_2 (nitrous acid) is a weak acid, and its conjugate base, NO_2^- (nitrite ion, the anion of the salt KNO_2) is a weak base. Therefore, this is a buffer system.

Only **solutions (c) and (d)** are buffer systems.

17.19 In order for the buffer solution to function effectively, the pK_a of the acid component must be close to the desired pH. We write:

$$K_{a_1} = 1.1 \times 10^{-3} \qquad pK_{a_1} = -\log\left(1.1 \times 10^{-3}\right) = 2.96$$

$$K_{a_2} = 2.5 \times 10^{-6} \qquad pK_{a_2} = -\log\left(2.5 \times 10^{-6}\right) = 5.60$$

Therefore, the proper buffer system is **Na₂A/NaHA**.

17.21 1. In order to function as a buffer, a solution must contain species that will consume both acid and base. Solution (a) contains HA^-, which can consume either acid or base:

$$HA^- + H^+ \longrightarrow H_2A$$

$$HA^- + OH^- \longrightarrow A^{2-} + H_2O$$

and A^{2-}, which can consume acid:

$$A^{2-} + H+ \longrightarrow HA^-$$

Solution (b) contains HA^-, which can consume either acid or base, and H_2A, which can consume base:

$$H_2A + OH^- \longrightarrow HA^- + H_2O$$

Solution (c) contains only H_2A and consequently can consume base but not acid. Solution (c) cannot function as a buffer.
Solution (d) contains HA^- and A^{2-}.

Solutions (a), (b), and (c) can function as buffers.

2. **Solution (a)** should be the most effective buffer because it has the highest concentrations of acid and base-consuming species.

17.27   Since the acid is monoprotic, the number of moles of KOH is equal to the number of moles of acid.

$$\text{Moles acid} = 16.4 \text{ mL} \times \frac{0.08133 \text{ mol}}{1000 \text{ mL}} = 0.00133 \text{ mol}$$

$$\textbf{Molar mass} = \frac{0.2688 \text{ g}}{0.00133 \text{ mol}} = \textbf{202 g / mol}$$

17.29   The neutralization reaction is:

$$H_2SO_4(aq) + 2NaOH(aq) \longrightarrow Na_2SO_4(aq) + 2H_2O(l)$$

Since 1 mole of sulfuric acid combines with 2 moles of sodium hydroxide, we write:

$$\text{mol NaOH} = 12.5 \text{ mL } H_2SO_4 \times \frac{0.500 \text{ mol } H_2SO_4}{1000 \text{ mL soln}} \times \frac{2 \text{ mol NaOH}}{1 \text{ mol } H_2SO_4} = 0.0125 \text{ mol NaOH}$$

$$\text{concentration of NaOH} = \frac{0.0125 \text{ mol NaOH}}{50.0 \times 10^{-3} \text{ L soln}} = \textbf{0.250 } \textbf{\textit{M}}$$

17.31   a. Since the acid is monoprotic, the moles of acid equals the moles of base added.

$$HA(aq) + NaOH(aq) \longrightarrow NaA(aq) + H_2O(l)$$

$$\text{Moles acid} = 18.4 \text{ mL} \times \frac{0.0633 \text{ mol}}{1000 \text{ mL soln}} = 0.00116 \text{ mol}$$

We know the mass of the unknown acid in grams and the number of moles of the unknown acid.

$$\text{Molar mass} = \frac{0.1276 \text{ g}}{0.00116 \text{ mol}} = \textbf{1.10} \times \textbf{10}^2 \textbf{ g / mol}$$

b. The number of moles of NaOH in 10.0 mL of solution is

$$10.0 \text{ mL} \times \frac{0.0633 \text{ mol}}{1000 \text{ mL soln}} = 6.33 \times 10^{-4} \text{ mol}$$

The neutralization reaction is:

$$HA(aq) \quad + NaOH(aq) \quad \rightarrow \quad NaA(aq) \quad + H_2O(l)$$

	HA(aq)	+ NaOH(aq)	NaA(aq)	+ H₂O(l)
Initial (mol):	0.00116	6.33×10^{-4}	0	
Change (mol):	-6.33×10^{-4}	-6.33×10^{-4}	$+6.33 \times 10^{-4}$	
Final (mol):	5.2×10^{-4}	0	6.33×10^{-4}	

Now, the weak acid equilibrium will be re-established. The total volume of the solution is 35.0 mL.

$$[HA] = \frac{5.2 \times 10^{-4} \text{ mol}}{0.0350 \text{ L}} = 0.015 \ M$$

$$[A^-] = \frac{6.33 \times 10^{-4} \text{ mol}}{0.0350 \text{ L}} = 0.0181 \ M$$

We can calculate the $[H^+]$ from the pH.

$$[H^+] = 10^{-pH} = 10^{-5.87} = 1.35 \times 10^{-6} \ M$$

	HA(aq)	$\rightleftharpoons$	H^+(aq) +	A^-(aq)
Initial (*M*):	0.015		0	0.0181
Change (*M*):	-1.35×10^{-6}		$+1.35 \times 10^{-6}$	$+1.35 \times 10^{-6}$
Equilibrium (*M*):	0.015		1.35×10^{-6}	0.0181

Substitute the equilibrium concentrations into the equilibrium constant expression to solve for K_a.

$$K_a = \frac{[H^+][A^-]}{[HA]} = \frac{\left(1.35 \times 10^{-6}\right)(0.0181)}{0.015} = \mathbf{1.6 \times 10^{-6}}$$

17.33

$$HCl(aq) + CH_3NH_2(aq) \rightleftharpoons CH_3NH_3^+ (aq) + Cl^-(aq)$$

Since the concentrations of acid and base are equal, equal volumes of each solution will need to be added to reach the equivalence point. Therefore, the solution volume is doubled at the equivalence point, and the concentration of the conjugate acid from the salt, $CH_3NH_3^+$, is:

$$\frac{0.20 \ M}{2} = 0.10 \ M$$

The conjugate acid undergoes hydrolysis.

$$CH_3NH_3^+(aq) \quad + H_2O(l) \rightleftharpoons H_3O^+(aq) \ + \ CH_3NH_2(aq)$$

	$CH_3NH_3^+$(aq)	+ H₂O(l)	H_3O^+(aq) +	CH_3NH_2(aq)
Initial (*M*):	0.10		0	0
Change (*M*):	$-x$		$+x$	$+x$
Equilibrium (*M*):	$0.10 - x$		x	x

$$K_a = \frac{[H_3O^+][CH_3NH_2]}{[CH_3NH_3^+]}$$

$$2.3 \times 10^{-11} = \frac{x^2}{0.10 - x}$$

Assuming that $0.10 - x \approx 0.10$,

$$x = [H_3O^+] = 1.5 \times 10^{-6} \, M$$

$$\mathbf{pH = 5.82}$$

17.35   The reaction between CH_3COOH and KOH is:

$$CH_3COOH(aq) + KOH(aq) \longrightarrow CH_3COOK(aq) + H_2O(l)$$

We see that 1 mol CH_3COOH is stoichiometrically equivalent to 1 mol KOH. Therefore, at every stage of titration, we can calculate the number of moles of acid reacting with base, and the pH of the solution is determined by the excess acid or base left over. At the equivalence point, however, the neutralization is complete, and the pH of the solution will depend on the extent of the hydrolysis of the salt formed, which is CH_3COOK.

a. No KOH has been added. This is a weak acid calculation.

	$CH_3COOH(aq)$	$+ H_2O(l)$	$\rightleftharpoons$	$H_3O^+(aq)$	$+ CH_3COO^-(aq)$
Initial (*M*):	0.100			0	0
Change (*M*):	$-x$			$+x$	$+x$
Equilibrium (*M*):	$0.100 - x$			x	x

According to Table 16.5, K_a for acetic acid is 1.8×10^{-5}.

$$1.8 \times 10^{-5} = \frac{(x)(x)}{0.100 - x} \approx \frac{x^2}{0.100}$$

$$x = 1.3 \times 10^{-3} \, M = [H_3O^+]$$

$$\mathbf{pH} = -\log[H_3O^+] = -\log(1.3 \times 10^{-3}) = \mathbf{2.87}$$

b. The number of moles of CH_3COOH originally present in 25.0 mL of solution is:

$$25.0 \text{ mL} \times \frac{0.100 \text{ mol } CH_3COOH}{1000 \text{ mL } CH_3COOH \text{ soln}} = 2.50 \times 10^{-3} \text{ mol}$$

The number of moles of KOH in 5.0 mL is:

$$5.0 \text{ mL} \times \frac{0.200 \text{ mol KOH}}{1000 \text{ mL KOH soln}} = 1.0 \times 10^{-3} \text{ mol}$$

We work with moles at this point because when two solutions are mixed, the solution volume increases. As the solution volume increases, molarity will change, but the number of moles will remain the same. The changes in the number of moles are summarized.

	$CH_3COOH(aq)$	$+ KOH(aq)$	$\rightarrow$ $CH_3COOK(aq)$	$+ H_2O(l)$
Initial (mol):	2.50×10^{-3}	1.0×10^{-3}	0	
Change (mol):	-1.0×10^{-3}	-1.0×10^{-3}	$+1.0 \times 10^{-3}$	
Final (mol):	1.5×10^{-3}	0	1.0×10^{-3}	

At this stage, we have a buffer system made up of CH_3COOH and CH_3COO^- (from the salt, CH_3COOK). We use the Henderson-Hasselbalch equation to calculate the pH.

$$pH = pK_a + \log \frac{[\text{conjugate base}]}{[\text{acid}]}$$

$$pH = -\log(1.8 \times 10^{-5}) + \log\left(\frac{1.0 \times 10^{-3}}{1.5 \times 10^{-3}}\right)$$

pH = 4.56

c. This part is solved similarly to part (b).

The number of moles of KOH in 10.0 mL is:

$$10.0 \text{ mL} \times \frac{0.200 \text{ mol KOH}}{1000 \text{ mL KOH soln}} = 2.00 \times 10^{-3} \text{ mol}$$

The changes in the number of moles are summarized.

	$CH_3COOH(aq)$	$+ KOH(aq)$	$\rightarrow$ $CH_3COOK(aq)$	$+ H_2O(l)$
Initial (mol):	2.50×10^{-3}	2.00×10^{-3}	0	
Change (mol):	-2.00×10^{-3}	-2.00×10^{-3}	$+2.00 \times 10^{-3}$	
Final (mol):	0.500×10^{-3}	0	2.00×10^{-3}	

At this stage, we have a buffer system made up of CH_3COOH and CH_3COO^- (from the salt, CH_3COOK). We use the Henderson-Hasselbalch equation to calculate the pH.

$$pH = pK_a + \log\frac{[\text{conjugate base}]}{[\text{acid}]}$$

$$pH = -\log\left(1.8 \times 10^{-5}\right) + \log\left(\frac{2.00 \times 10^{-3}}{0.500 \times 10^{-3}}\right)$$

$$pH = 5.34$$

d. We have reached the equivalence point of the titration. 2.50×10^{-3} mole of CH_3COOH reacts with 2.50×10^{-3} mole of KOH to produce 2.50×10^{-3} mole of CH_3COOK. The only major species present in the solution at the equivalence point is the salt, CH_3COOK, which contains the conjugate base, CH_3COO^-. Let's calculate the molarity of CH_3COO^-. The volume of the solution is: (25.0 mL + 12.5 mL = 37.5 mL = 0.0375 L).

$$M (CH_3COO^-) = \frac{2.50 \times 10^{-3} \text{ mol}}{0.0375 \text{ L}} = 0.0667 \ M$$

We set up the hydrolysis of CH_3COO^-, which is a weak base.

	$CH_3COO^-(aq)$	$+ H_2O(l)$	$\rightleftarrows$	$CH_3COOH(aq)$	$+ OH^-(aq)$
Initial (*M*):	0.0667			0	0
Change (*M*):	$-x$			$+x$	$+x$
Equilibrium (*M*):	$0.0667 - x$			x	x

Use the K_a value for CH_3COOH (1.8×10^{-5} from Table 16.5) and Equation 16.6 to determine K_b for CH_3COO^-.

$$K_b = \frac{K_w}{K_a} = \frac{1.0 \times 10^{-14}}{1.8 \times 10^{-5}} = 5.56 \times 10^{-10}$$

$$K_b = \frac{[CH_3COOH][OH^-]}{[CH_3COO^-]}$$

$$5.56 \times 10^{-10} = \frac{(x)(x)}{0.0667 - x} \approx \frac{x^2}{0.0667}$$

$$x = 6.09 \times 10^{-6} \ M = [OH^-]$$

$$pOH = -\log[OH^-] = -\log(6.09 \times 10^{-6}) = 5.22$$

$$pH = 14.00 - 5.22 = 8.78$$

e. We have passed the equivalence point of the titration. The excess strong base, KOH, will determine the pH at this point. The moles of KOH in 15.0 mL are:

$$15.0 \text{ mL} \times \frac{0.200 \text{ mol KOH}}{1000 \text{ mL KOH soln}} = 3.00 \times 10^{-3} \text{ mol}$$

The changes in number of moles are summarized.

$$CH_3COOH(aq) \quad + KOH(aq) \quad \rightarrow \quad CH_3COOK(aq) \quad + H_2O(l)$$

	CH₃COOH(aq)	KOH(aq)	CH₃COOK(aq)
Initial (mol):	2.50×10^{-3}	3.00×10^{-3}	0
Change (mol):	-2.50×10^{-3}	-2.50×10^{-3}	$+2.50 \times 10^{-3}$
Final (mol):	0	0.500×10^{-3}	2.50×10^{-3}

Let's calculate the molarity of the KOH in solution. The volume of the solution is now 40.0 mL = 0.0400

$$M\ (KOH) = \frac{0.500 \times 10^{-3}\ mol}{0.0400\ L} = 0.0125\ M$$

KOH is a strong base. The pOH is:

$$pOH = -\log(0.0125) = 1.90$$

$$\mathbf{pH = 12.10}$$

17.37 a. HCOOH is a weak acid and NaOH is a strong base. Suitable indicators are **cresol red and phenolphthalein**.

b. HCl is a strong acid and KOH is a strong base. **Most of the indicators in Table 17.3, except thymol blue and, to a lesser extent, bromophenol blue and methylorange**, are suitable for a strong acid–strong base titration.

c. HNO_3 is a strong acid and CH_3NH_2 is a weak base. Suitable indicators are **bromophenol blue, methyl orange, methyl red, and chlorophenol blue**.

17.39 The weak acid equilibrium is

$$HIn(aq) \rightleftharpoons H^+(aq) + In^-(aq)$$

We can write a K_a expression for this equilibrium.

$$K_a = \frac{\left[H^+\right][In^-]}{[HIn]}$$

Rearranging,

$$\frac{[HIn]}{[In^-]} = \frac{\left[H^+\right]}{K_a}$$

From the pH, we can calculate the H^+ concentration.

$$[H^+] = 10^{-pH} = 10^{-4.00} = 1.0 \times 10^{-4}\ M$$

$$\frac{[HIn]}{[In^-]} = \frac{\left[H^+\right]}{K_a} = \frac{1.0 \times 10^{-4}}{1.0 \times 10^{-6}} = 1.0 \times 10^2$$

Since the concentration of HIn is 100 times greater than the concentration of In^-, the color of the solution will be that of HIn, the nonionized formed. The color of the solution will be **red**.

679

17.41   1. **diagram (c)**

2. **diagram (b)**

3. **diagram (d)**

4. **Diagram (a); the pH at the equivalence point is below 7 (acidic)** because the species in the solution at the equivalence point is the conjugate acid of a weak base.

17.49   a. The solubility equilibrium is given by the equation:

$$AgI(s) \rightleftharpoons Ag^+(aq) + I^-(aq)$$

The expression for K_{sp} is given by:

$$K_{sp} = [Ag^+][I^-]$$

The value of K_{sp} can be found in Table 17.4 of the text. If the equilibrium concentration of silver ion is the value given, the concentration of iodide ion must be:

$$[I^-] = \frac{K_{sp}}{[Ag^+]} = \frac{8.3 \times 10^{-17}}{9.1 \times 10^{-9}} = \mathbf{9.1 \times 10^{-9}}\ \boldsymbol{M}$$

b. The value of K_{sp} for aluminum hydroxide can be found in Table 17.4 of the text. The equilibrium expressions are:

$$Al(OH)_3(s) \rightleftharpoons Al^{3+}(aq) + 3OH^-(aq)$$

$$K_{sp} = [Al^{3+}][OH^-]^3$$

Using the given value of the hydroxide ion concentration, the equilibrium concentration of aluminum ion is:

$$[Al^{3+}] = \frac{K_{sp}}{[OH^-]^3} = \frac{1.8 \times 10^{-33}}{\left(2.9 \times 10^{-9}\right)^3} = \mathbf{7.4 \times 10^{-8}}\ \boldsymbol{M}$$

17.51   For $MnCO_3$ dissolving, we write:

$$MnCO_3(s) \rightleftharpoons Mn^{2+}(aq) + CO_3^{2-}(aq)$$

For every mole of $MnCO_3$ that dissolves, 1 mole of Mn^{2+} will be produced and 1 mole of CO_3^{2-} will be produced. If the molar solubility of $MnCO_3$ is s mol/L, then the concentrations of Mn^{2+} and CO_3^{2-} are:

$$[Mn^{2+}] = [CO_3^{2-}] = s = 4.2 \times 10^{-6}\ M$$

$$\boldsymbol{K_{sp}} = [Mn^{2+}][CO_3^{2-}] = s^2 = (4.2 \times 10^{-6})^2 = \mathbf{1.8 \times 10^{-11}}$$

17.53   The charges of the M and X ions are +3 and −2, respectively. We first calculate the number of moles of M_2X_3 that dissolve in 1.0 L of water. We carry an additional significant figure throughout this calculation to minimize rounding errors.

$$\text{Moles } M_2X_3 = \left(3.6 \times 10^{-17}\text{ g}\right) \times \frac{1\text{ mol}}{288\text{ g}} = 1.25 \times 10^{-19}\text{ mol}$$

The molar solubility, s, of the compound is therefore 1.3×10^{-19} M. At equilibrium the concentration of M^{3+} must be $2s$ and that of X^{2-} must be $3s$.

$$K_{sp} = [M^{3+}]^2[X^{2-}]^3 = [2s]^2[3s]^3 = 108s^5$$

Since these are equilibrium concentrations, the value of K_{sp} can be found by simple substitution.

$$\boldsymbol{K_{sp}} = 108s^5 = 108(1.25 \times 10^{-19})^5 = \boldsymbol{3.3 \times 10^{-93}}$$

17.55   Let s be the molar solubility of $Zn(OH)_2$. The equilibrium concentrations of the ions are then

$$[Zn^{2+}] = s \text{ and } [OH^-] = 2s$$

$$K_{sp} = [Zn^{2+}][OH^-]^2 = (s)(2s)^2 = 4s^3 = 1.8 \times 10^{-14}$$

$$s = \sqrt[3]{\left(\frac{1.8 \times 10^{-14}}{4}\right)} = 1.65 \times 10^{-5}$$

$$[OH^-] = 2s = 3.30 \times 10^{-5} \text{ } M \text{ and pOH} = 4.48$$

$$\textbf{pH} = 14.00 - 4.48 = \textbf{9.52}$$

17.57   According to the solubility rules, the only precipitate that might form is $BaCO_3$.

$$Ba^{2+}(aq) + CO_3^{2-}(aq) \longrightarrow BaCO_3(s)$$

The number of moles of Ba^{2+} present in the original 20.0 mL of $Ba(NO_3)_2$ solution is:

$$20.0\text{ mL} \times \frac{0.10\text{ mol }Ba^{2+}}{1000\text{ mL soln}} = 2.0 \times 10^{-3}\text{ mol }Ba^{2+}$$

The total volume after combining the two solutions is 70.0 mL. The concentration of Ba^{2+} in 70 mL is:

$$[Ba^{2+}] = \frac{2.0 \times 10^{-3}\text{ mol }Ba^{2+}}{70.0 \times 10^{-3}\text{ L}} = 2.9 \times 10^{-2}\text{ }M$$

The number of moles of CO_3^{2-} present in the original 50.0 mL Na_2CO_3 solution is:

$$50.0\text{ mL} \times \frac{0.10\text{ mol }CO_3^{2-}}{1000\text{ mL soln}} = 5.0 \times 10^{-3}\text{ mol }CO_3^{2-}$$

The concentration of CO_3^{2-} in the 70.0 mL of combined solution is:

$$[CO_3^{2-}] = \frac{5.0 \times 10^{-3} \text{ mol } CO_3^{2-}}{70.0 \times 10^{-3} \text{ L}} = 7.1 \times 10^{-2} \text{ } M$$

Now we must compare Q and K_{sp}. From Table 17.4 of the text, the K_{sp} for $BaCO_3$ is 8.1×10^{-9}. As for Q,

$$Q = [Ba^{2+}]_0[CO_3^{2-}]_0 = (2.9 \times 10^{-2})(7.1 \times 10^{-2}) = 2.1 \times 10^{-3}$$

Since $(2.1 \times 10^{-3}) > (8.1 \times 10^{-9})$, then $Q > K_{sp}$. Therefore, **yes**, $BaCO_3$ will precipitate.

17.59 Strategy: The addition of soluble salts to the solution changes the concentrations of M^{2+} and A^-. The salt which causes the largest change in concentration (and therefore a change in the reaction quotient, Q) will cause the precipitation of the greatest quantity of MA_2.

Setup: The balanced equation is:

$$MA_2 \rightleftharpoons M^{2+} + 2A^-$$

Therefore,

$$K_{sp} = [M^{2+}][A^-]^2$$

Solution: Since the anion, A^-, in the K_{sp} equation is squared, increasing its value will have the greater impact on the Q value than an equivalent change in the cation, M^{2+}.

AlA_3 contains three anions, whereas NaA and BaA_2, contain only one or two. $M_3(PO_4)_2$ contains three cations, whereas $M(NO_3)_2$ contains only one cation. Since the $[A^-]$ is squared in the K_{sp} equation, changing its concentration will have a greater impact on the precipitation than the threefold change in $[M^{2+}]$. Therefore, **AlA_3** will cause the greatest quantity of MA_2 to precipitate.

17.65 Strategy: In parts (b) and (c), this is a common-ion problem. In part (b), the common ion is Br^-, which is supplied by both $PbBr_2$ and KBr. Remember that the presence of a common ion will affect only the solubility of $PbBr_2$ but not the K_{sp} value because it is an equilibrium constant. In part (c), the common ion is Pb^{2+}, which is supplied by both $PbBr_2$ and $Pb(NO_3)_2$.

Solution: a. Set up a table to find the equilibrium concentrations in pure water.

	$PbBr_2(s)$	$\rightleftharpoons$	$Pb^{2+}(aq)$	$+ 2Br^-(aq)$
Initial (M):			0	0
Change (M):	$-s$		$+s$	$+2s$
Equilibrium (M):			s	$2s$

$$K_{sp} = [Pb^{2+}][Br^-]^2$$

$$8.9 \times 10^{-6} = (s)(2s)^2$$

$$s = \text{molar solubility} = \textbf{0.013 } \textbf{\textit{M}} \text{ or } \textbf{1.3} \times \textbf{10}^{-2} \textbf{\textit{M}}$$

b. Set up a table to find the equilibrium concentrations in 0.20 M KBr. KBr is a soluble salt that ionizes completely giving an initial concentration of $Br^- = 0.20$ M.

$$PbBr_2(s) \quad \rightleftharpoons \quad Pb^{2+}(aq) \quad + 2Br^-(aq)$$

Initial (*M*):		0	0.20
Change (*M*):	$-s$	$+s$	$+2s$
Equilibrium (*M*):		s	$0.20 + 2s$

$$K_{sp} = [Pb^{2+}][Br^-]^2$$

$$8.9 \times 10^{-6} = (s)(0.20 + 2s)^2$$

$$8.9 \times 10^{-6} \approx (s)(0.20)^2$$

$$s = \text{molar solubility} = \mathbf{2.2 \times 10^{-4} \ M}$$

Thus, the molar solubility of $PbBr_2$ is reduced from 0.013 *M* to 2.2×10^{-4} *M* as a result of the common ion (Br^-) effect.

c. Set up a table to find the equilibrium concentrations in 0.20 *M* $Pb(NO_3)_2$. $Pb(NO_3)_2$ is a soluble salt that dissociates completely giving an initial concentration of $[Pb^{2+}] = 0.20$ *M*.

$$PbBr_2(s) \quad \rightleftharpoons \quad Pb^{2+}(aq) \quad + 2Br^-(aq)$$

Initial (*M*):	0.20	0	
Change (*M*):	$-s$	$+s$	$+2s$
Equilibrium (*M*):		$0.20 + s$	$2s$

$$K_{sp} = [Pb^{2+}][Br^-]^2$$

$$8.9 \times 10^{-6} = (0.20 + s)(2s)^2$$

$$8.9 \times 10^{-6} \approx (0.20)(2s)^2$$

$$s = \text{molar solubility} = \mathbf{3.3 \times 10^{-3} \ M}$$

Thus, the molar solubility of $PbBr_2$ is reduced from 0.013 *M* to 3.3×10^{-3} *M* as a result of the common ion (Pb^{2+}) effect.

17.67   a. The equilibrium equation is:

$$BaSO_4(s) \quad \rightleftharpoons \quad Ba^{2+}(aq) \quad + \quad SO_4^{2-}(aq)$$

Initial (*M*):		0	0
Change (*M*):	$-s$	$+s$	$+s$
Equilibrium (*M*):		s	s

From Table 17.4, the value of K_{sp} for barium sulfate is 1.1×10^{-10}.

$$K_{sp} = [Ba^{2+}][SO_4^{2-}]$$

$$1.1 \times 10^{-10} = s^2$$

$$s = \mathbf{1.0 \times 10^{-5} \ M}$$

The molar solubility of $BaSO_4$ in pure water is 1.0×10^{-5} mol/L.

b. The initial concentration of SO_4^{2-} is 1.0 M.

$$BaSO_4(s) \rightleftarrows Ba^{2+}(aq) + SO_4^{2-}(aq)$$

		Ba^{2+}	SO_4^{2-}
Initial (M):		0	1.0
Change (M):	$-s$	$+s$	$+s$
Equilibrium (M):		s	$1.0 + s$

$$K_{sp} = [Ba^{2+}][SO_4^{2-}]$$

$$1.1 \times 10^{-10} = (s)(1.0 + s) \approx (s)(1.0)$$

$$\mathbf{s = 1.1 \times 10^{-10} \, M}$$

Because of the common ion effect, the molar solubility of $BaSO_4$ decreases to 1.1×10^{-10} mol/L in 1.0 M $SO_4^{2-}(aq)$ compared to 1.0×10^{-5} mol/L in pure water.

17.69   From Table 17.4, the value of K_{sp} for iron(II) hydroxide is 1.6×10^{-14}.

a. At pH = 8.00, pOH = 14.00 − 8.00 = 6.00, and $[OH^-] = 10^{-6.00} = 1.0 \times 10^{-6} \, M$

$$[Fe^{2+}] = \frac{K_{sp}}{[OH^-]^2} = \frac{1.6 \times 10^{-14}}{\left(1.0 \times 10^{-6}\right)^2} = 0.016 \, M$$

The *molar solubility* of iron(II) hydroxide at pH = 8.00 is **0.016 M or 1.6×10^{-2} M**

b. At pH = 10.00, pOH = 14.00 − 10.00 = 4.00, and $[OH^-] = 10^{-4.00} = 1.0 \times 10^{-4} \, M$

$$[Fe^{2+}] = \frac{K_{sp}}{[OH^-]^2} = \frac{1.6 \times 10^{-14}}{\left(1.0 \times 10^{-4}\right)^2} = 1.6 \times 10^{-6} \, M$$

The *molar solubility* of iron(II) hydroxide at pH = 10.00 is **1.6×10^{-6} M**.

17.71   We first determine the effect of the added ammonia. Let's calculate the concentration of NH_3. This is a dilution problem.

$$M_i V_i = M_f V_f$$

$$(0.60 \, M)(2.00 \text{ mL}) = M_f(1002 \text{ mL})$$

$$M_f = 0.0012 \, M \, NH_3$$

Ammonia is a weak base ($K_b = 1.8 \times 10^{-5}$).

$$NH_3(aq) + H_2O(l) \rightleftarrows NH_4^+(aq) + OH^-(aq)$$

	NH_3		NH_4^+	OH^-
Initial (M):	0.0012		0	0
Change (M):	$-x$		$+x$	$+x$
Equilibrium (M):	$0.0012 - x$		x	x

$$K_b = \frac{[NH_4^+][OH^-]}{[NH_3]}$$

$$1.8 \times 10^{-5} = \frac{x^2}{(0.0012 - x)}$$

Solving the resulting quadratic equation gives $x = 0.00014$ or $[OH^-] = 0.00014\ M$.

This is a solution of iron(II) sulfate, which contains Fe^{2+} ions. These Fe^{2+} ions could combine with OH^- to precipitate $Fe(OH)_2$. Therefore, we must use K_{sp} for iron(II) hydroxide. We compute the value of Q for this solution.

$$Fe(OH)_2(s) \rightleftharpoons Fe^{2+}(aq) + 2OH^-(aq)$$

$$Q = [Fe^{2+}]_0[OH^-]_0^2 = (1.0 \times 10^{-3})(0.00014)^2 = 2.0 \times 10^{-11}$$

Note that when adding 2.00 mL of NH_3 to 1.0 L of $FeSO_4$, the concentration of $FeSO_4$ will decrease slightly. However, rounding off to two significant figures, the concentration of $1.0 \times 10^{-3}\ M$ does not change. Q is larger than K_{sp} $[Fe(OH)_2] = 1.6 \times 10^{-14}$. The concentrations of the ions in the solution are greater than the equilibrium concentrations; the solution is saturated. The system will shift left to re-establish equilibrium; therefore, **a precipitate of $Fe(OH)_2$ will form**.

17.73 **Strategy:** The addition of $Cd(NO_3)_2$ to the NaCN solution results in complex ion formation. In solution, Cd^{2+} ions will complex with CN^- ions. The concentration of Cd^{2+} will be determined by the following equilibrium:

$$Cd^{2+}(aq) + 4CN^-(aq) \rightleftharpoons Cd(CN)\,Cd(CN)_4^{2-}$$

From Table 17.5 of the text, we see that the formation constant (K_f) for this reaction is very large $(K_f = 7.1 \times 10^{16})$. Because K_f is so large, the reaction lies mostly to the right. At equilibrium, the concentration of Cd^{2+} will be very small. As a good approximation, we can assume that essentially all the dissolved Cd^{2+} ions end up as $Cd(CN)_4^{2-}$ ions. What is the initial concentration of Cd^{2+} ions? A very small amount of Cd^{2+} will be present at equilibrium. Set up the K_f expression for the above equilibrium to solve for $[Cd^{2+}]$.

Solution: Calculate the initial concentration of Cd^{2+} ions.

$$[Cd^{2+}]_0 = \frac{0.50\ g \times \dfrac{1\ mol\ Cd(NO_3)_2}{236.42\ g\ Cd(NO_3)_2} \times \dfrac{1\ mol\ Cd^{2+}}{1\ mol\ Cd(NO_3)_2}}{0.50\ L} = 4.2 \times 10^{-3}\ M$$

If we assume that the above equilibrium goes to completion, we can write

	$Cd^{2+}(aq) +$	$4CN^-(aq)$	$\rightarrow$	$Cd(CN)_4^{2-}(aq)$
Initial (M):	4.2×10^{-3}	0.50		0
Change (M):	-4.2×10^{-3}	$-4(4.2 \times 10^{-3})$		$+4.2 \times 10^{-3}$
Equilibrium (M):	0	0.48		4.2×10^{-3}

To find the concentration of free Cd^{2+} at equilibrium, use the formation constant expression.

$$K_f = \frac{[Cd(CN)_4^{2-}]}{[Cd^{2+}][CN^-]^4}$$

Rearranging,

$$[Cd^{2+}] = \frac{[Cd(CN)_4^{2-}]}{K_f[CN^-]^4}$$

Substitute the equilibrium concentrations calculated above into the formation constant expression to calculate the equilibrium concentration of Cd^{2+}.

$$\mathbf{[Cd^{2+}]} = \frac{[Cd(CN)_4^{2-}]}{K_f[CN^-]^4} = \frac{4.2 \times 10^{-3}}{\left(7.1 \times 10^{16}\right)(0.48)^4} = \mathbf{1.1 \times 10^{-18}\ M}$$

$$\mathbf{[Cd(CN)_4^{2-}]} = (4.2 \times 10^{-3}\ M) - (1.1 \times 10^{-18}) = \mathbf{4.2 \times 10^{-3}\ M}$$

$$\mathbf{[CN^-]} = 0.48\ M + 4(1.1 \times 10^{-18}\ M) = \mathbf{0.48\ M}$$

17.75 Silver iodide is only slightly soluble. It dissociates to form a small amount of Ag^+ and I^- ions. The Ag^+ ions then complex with NH_3 in the solution to form the complex ion $Ag(NH_3)_2^+$. The balanced equations are:

$$AgI(s) \rightleftharpoons Ag^+(aq) + I^-(aq) \qquad K_{sp} = [Ag^+][I^-] = 8.3 \times 10^{-17} \text{ (See Table 17.4)}$$

$$Ag^+(aq) + 2NH_3(aq) \rightleftharpoons Ag(NH_3)_2^+(aq) \quad K_f = \frac{[Ag(NH_3)_2^+]}{[Ag^+][NH_3]^2} = 1.5 \times 10^7 \text{ (See Table 17.5)}$$

Overall: $AgI(s) + 2NH_3(aq) \rightleftharpoons Ag(NH_3)_2^+(aq) + I^-(aq) \quad K = K_{sp} \times K_f = 1.2 \times 10^{-9}$

If s is the molar solubility of AgI then,

	AgI(s) +	2NH₃(aq)	⇌	Ag(NH₃)₂⁺(aq)	+	I⁻(aq)
Initial (*M*):		1.0		0.0		0.0
Change (*M*):	−s	−2s		+s		+s
Equilibrium (*M*):		(1.0 − 2s)		s		s

Because K_f is large, we can assume that all the silver ions exist as $Ag(NH_3)_2^+$. Thus,

$$[Ag(NH_3)_2^+] = [I^-] = s$$

We can write the equilibrium constant expression for the above reaction and then solve for s.

$$K = 1.2 \times 10^{-9} = \frac{(s)(s)}{(1.0 - 2s)^2} \approx \frac{(s)(s)}{(1.0)^2}$$

$$s = 3.5 \times 10^{-5}\,M$$

At equilibrium, 3.5×10^{-5} moles of AgI dissolves in 1 L of 1.0 M NH$_3$ solution.

17.77 **Strategy:** Substitute the molar solubility into the equilibrium expression to determine K_{sp}.

Setup: $s = 7 \times 10^{-8}\,M$

The equation and the equilibrium expression for the dissociation of Ca$_5$(PO$_4$)$_3$F are:

$$Ca_5(PO_4)_3F(s) \rightleftharpoons 5Ca^{2+}(aq) + 3PO_4^{3-}(aq) + F^-(aq)$$

and $\quad K_{sp} = [Ca^{2+}]^5[\,PO_4^{3-}\,]^3[F^-]$

Solution: a. Substituting the molar solubility into the equilibrium expression gives:

$$K_{sp} = [Ca^{2+}]^5[PO_4^{3-}]^3[F^-] = [5(7 \times 10^{-8})]^5[3(7 \times 10^{-8})]^3(7 \times 10^{-8})$$

$$K_{sp} = (5.3 \times 10^{-33})(9.3 \times 10^{-21})(7 \times 10^{-8})$$

$$K_{sp} = 3 \times 10^{-60}$$

b. In an aqueous solution with $[F^-] = 0.10\,M$

$$3 \times 10^{-60} = (5s)^5(3s)^3(0.10)$$

$$3 \times 10^{-59} = 84{,}375s^8$$

$$3.6 \times 10^{-64} = s^8$$

$$s = 1.2 \times 10^{-8}\,M$$

17.79 a. I$^-$ is the conjugate base of the strong acid HI.

b. SO$_4^{2-}$ (aq) is a weak base.

c. OH$^-$(aq) is a strong base.

d. C$_2$O$_4^{2-}$ (aq) is a weak base.

e. PO$_4^{3-}$ (aq) is a weak base.

The solubilities of the above (**b, c, d, and e**) will increase in acidic solution. Only (a), which contains an extremely weak base (I$^-$ is the conjugate base of the strong acid HI) is unaffected by the acid solution.

17.83 For Fe(OH)$_3$, $K_{sp} = 1.1 \times 10^{-36}$. When $[Fe^{3+}] = 0.010\,M$, the [OH$^-$] value is:

$$K_{sp} = [Fe^{3+}][OH^-]^3$$

or

$$[OH^-] = \left(\frac{K_{sp}}{[Fe^{3+}]}\right)^{\frac{1}{3}}$$

$$[OH^-] = \left(\frac{1.1 \times 10^{-36}}{0.010}\right)^{\frac{1}{3}} = 4.8 \times 10^{-12} \ M$$

This [OH⁻] corresponds to a pH of 2.68. In other words, Fe(OH)₃ will begin to precipitate from this solution at a pH of 2.68.

For Zn(OH)₂, $K_{sp} = 1.8 \times 10^{-14}$. When $[Zn^{2+}] = 0.010 \ M$, the [OH⁻] value is:

$$[OH^-] = \left(\frac{K_{sp}}{[Zn^{2+}]}\right)^{\frac{1}{2}}$$

$$[OH^-] = \left(\frac{1.8 \times 10^{-14}}{0.010}\right)^{\frac{1}{2}} = 1.3 \times 10^{-6} \ M$$

This corresponds to a pH of 8.11. In other words Zn(OH)₂ will begin to precipitate from the solution at a pH of 8.11. These results show that Fe(OH)₃ will precipitate when the pH just exceeds 2.68 and that Zn(OH)₂ will precipitate when the pH just exceeds 8.11. Therefore, to selectively remove iron as Fe(OH)₃, the **pH must be *greater than* 2.68 but *less than* 8.11**.

17.85 Since some PbCl₂ precipitates, the solution is saturated. From Table 17.4, the value of K_{sp} for lead(II) chloride is 2.4×10^{-4}. The equilibrium is:

$$PbCl_2(aq) \rightleftharpoons Pb^{2+}(aq) + 2Cl^-(aq)$$

We can write the solubility product expression for the equilibrium.

$$K_{sp} = [Pb^{2+}][Cl^-]^2$$

K_{sp} and [Cl⁻] are known. Solving for the Pb²⁺ concentration,

$$\mathbf{[Pb^{2+}]} = \frac{K_{sp}}{[Cl^-]^2} = \frac{2.4 \times 10^{-4}}{(0.15)^2} = \mathbf{0.011 \ M}$$

17.87 **Chloride ion will precipitate Ag⁺ but not Cu²⁺. Hence, dissolve some solid in H₂O and add HCl. If a precipitate forms, the salt is AgNO₃. A flame test will also work. Cu²⁺ gives a green flame test.**

17.89 First, calculate the pH of the 2.00 *M* weak acid (HNO₂) solution before any NaOH is added. According to Table 16.5, K_a for nitrous acid, HNO₂(*aq*), is $4.5 \times 10^{-4} \ M$.

	$HNO_2(aq)$	$\rightleftharpoons$	$H^+(aq)$	+	$NO_2^-(aq)$
Initial (*M*):	2.00		0		0
Change (*M*):	−x		+x		+x
Equilibrium (*M*):	2.00 − x		x		x

$$K_a = \frac{[H^+][NO_2^-]}{[HNO_2]}$$

$$4.5 \times 10^{-4} = \frac{x^2}{2.00 - x} \approx \frac{x^2}{2.00}$$

$$x = [H^+] = 0.030 \ M$$

$$pH = -\log(0.030) = 1.52 \quad \cdot$$

Since the pH after the addition is 1.5 pH units greater, the new pH = $1.52 + 1.50 = 3.02$.

From this new pH, we can calculate the $[H^+]$ in solution.

$$[H^+] = 10^{-pH} = 10^{-3.02} = 9.55 \times 10^{-4} \ M$$

When NaOH is added, we dilute our original 2.00 M HNO_2 solution to:

$$M_i V_i = M_f V_f$$

$$(2.00 \ M)(400 \ mL) = M_f(600 \ mL)$$

$$M_f = 1.33 \ M$$

Since we have not reached the equivalence point, we have a buffer solution. The reaction between HNO_2 and NaOH is:

$$HNO_2(aq) + NaOH(aq) \longrightarrow NaNO_2(aq) + H_2O(l)$$

or

$$HNO_2(aq) + OH^-(aq) \longrightarrow NO_2^-(aq) + H_2O(l)$$

Since the mole ratio between HNO_2 and NaOH is 1:1, the decrease in $[HNO_2]$ is the same as the decrease in [NaOH].

We can calculate the decrease in $[HNO_2]$ by setting up the weak acid equilibrium. From the pH of the solution, we know that the $[H^+]$ at equilibrium is $9.55 \times 10^{-4} \ M$.

	$HNO_2(aq)$	$\rightleftarrows$	$H^+(aq)$	$+$	$NO_2^-(aq)$
Initial (M):	1.33		0		0
Change (M):	$-x$				$+x$
Equilibrium (M):	$1.33 - x$		9.55×10^{-4}		x

We can calculate x from the equilibrium constant expression.

$$K_a = \frac{[H^+][NO_2^-]}{[HNO_2]}$$

$$4.5 \times 10^{-4} = \frac{\left(9.55 \times 10^{-4}\right)(x)}{1.33 - x}$$

$$x = 0.43 \ M$$

Thus, x is the decrease in $[HNO_2]$, which equals the concentration of added OH^-. However, this is the concentration of NaOH after it has been diluted to 600 mL. We need to correct for the dilution from 200 mL to 600 mL to calculate the concentration of the original NaOH solution.

$$M_i V_i = M_f V_f$$

$$M_i(200 \ \text{mL}) = (0.43 \ M)(600 \ \text{mL})$$

$$\textbf{[NaOH]} = M_i = \textbf{1.3} \ \textbf{\textit{M}}$$

17.91   The resulting solution is not a buffer system. There is excess NaOH, and the neutralization is well past the equivalence point.

$$\text{Moles NaOH} = 0.500 \ \text{L} \times \frac{0.167 \ \text{mol}}{1 \ \text{L}} = 0.0835 \ \text{mol}$$

$$\text{Moles HCOOH} = 0.500 \ \text{L} \times \frac{0.100 \ \text{mol}}{1 \ \text{L}} = 0.0500 \ \text{mol}$$

	HCOOH(*aq*)	+ NaOH(*aq*)	→	HCOONa(*aq*)
Initial (mol):	0.0500	0.0835		0
Change (mol):	−0.0500	−0.0500		+0.0500
Final (mol):	0	0.0335		0.0500

The volume of the resulting solution is 1.00 L (500 mL + 500 mL = 1000 mL).

$$\textbf{[Na}^+\textbf{]} = \frac{(0.0335 + 0.0500)\,\text{mol}}{1.00 \ \text{L}} = \textbf{0.0835} \ \textbf{\textit{M}}$$

$$\textbf{[HCOO}^-\textbf{]} = \frac{0.0500 \ \text{mol}}{1.00 \ \text{L}} = \textbf{0.0500} \ \textbf{\textit{M}}$$

$$\textbf{[OH}^-\textbf{]} = \frac{0.0335 \ \text{mol}}{1.00 \ \text{L}} = \textbf{0.0335} \ \textbf{\textit{M}}$$

$$\textbf{[H}^+\textbf{]} = \frac{K_w}{[OH^-]} = \frac{1.0 \times 10^{-14}}{0.0335} = \textbf{3.0} \times \textbf{10}^{-13} \ \textbf{\textit{M}}$$

	HCOO$^-$(*aq*)	+ H$_2$O(*l*)	⇌	HCOOH(*aq*)	+ OH$^-$(*aq*)
Initial (*M*):	0.0500			0	0.0335
Change (*M*):	−*x*			+*x*	+*x*
Equilibrium (*M*):	0.0500 − *x*			*x*	0.0335 + *x*

$$K_b = \frac{[\text{HCOOH}][\text{OH}^-]}{[\text{HCOO}^-]}$$

$$5.9 \times 10^{-11} = \frac{(x)(0.0335 + x)}{(0.0500 - x)} \approx \frac{(x)(0.0335)}{(0.0500)}$$

$$x = [\text{HCOOH}] = 8.8 \times 10^{-11} \ M$$

17.93 Most likely the increase in solubility is due to complex ion formation:

$$\text{Cd(OH)}_2(s) + 2\text{OH}^-(aq) \rightleftharpoons \text{Cd(OH)}_4^{2-} \ (aq)$$

This is a **Lewis acid-base reaction** in which the metal cation Cd^{2+} combines with the Lewis base NaOH results in the formation of complex ion $Cd(OH)_4^{2-}$.

17.95 A solubility equilibrium is an equilibrium between a solid (reactant) and its components (products: ions, neutral molecules, etc.) in solution. Only **reaction (d)** represents a solubility equilibrium.

17.97 Since equal volumes of the two solutions were used, the initial molar concentrations will be halved.

$$[\text{Ag}^+] = \frac{0.12 \ M}{2} = 0.060 \ M$$

$$[\text{Cl}^-] = \frac{2(0.14 \ M)}{2} = 0.14 \ M$$

Let's assume that the Ag^+ ions and Cl^- ions react completely to form AgCl(s). Then, we will re-establish the equilibrium between AgCl, Ag^+, and Cl^-.

Now, setting up the equilibrium,

	$\text{Ag}^+(aq)$	$+ \ \text{Cl}^-(aq)$	$\rightarrow$	$\text{AgCl}(s)$
Initial (*M*):	0.060	0.14		0
Change (*M*):	−0.060	−0.060		+0.060
Final (*M*):	0	0.080		0.060

	$\text{AgCl}(s)$	$\rightleftharpoons$	$\text{Ag}^+(aq)$	$+ \ \text{Cl}^-(aq)$
Initial (*M*):	0.060		0	0.080
Change (*M*):	−s		+s	+s
Equilibrium (*M*):	0.060 − s		s	0.080 + s

Set up the K_{sp} expression to solve for s. From Table 17.4, the value of K_{sp} for silver chloride is 1.6×10^{-10}.

$$K_{sp} = [\text{Ag}^+][\text{Cl}^-]$$

$$1.6 \times 10^{-10} = (s)(0.080 + s)$$

$$s = 2.0 \times 10^{-9} \ M$$

$$[Ag^+] = s = 2.0 \times 10^{-9} \ M$$

$$[Cl^-] = 0.080 \ M + s = 0.080 \ M$$

$$[Zn^{2+}] = \frac{0.14 \ M}{2} = 0.070 \ M$$

$$[NO_3^-] = \frac{0.12 \ M}{2} = 0.060 \ M$$

17.99 First we find the molar solubility and then convert moles to grams. The solubility equilibrium for silver carbonate is:

$$Ag_2CO_3(s) \ \rightleftarrows \ 2Ag^+(aq) \ + \ CO_3^{2-}(aq)$$

Initial (*M*):		0	0
Change (*M*):	−s	+2s	+s
Equilibrium (*M*):		2s	s

From Table 17.4, the value of K_{sp} for silver carbonate is 8.1×10^{-12}.

$$K_{sp} = [Ag^+]^2[CO_3^{2-}] = (2s)^2(s) = 4s^3 = 8.1 \times 10^{-12}$$

$$s = \sqrt[3]{\left(\frac{8.1 \times 10^{-12}}{4}\right)} = 1.27 \times 10^{-4} \ M$$

Converting from mol/L to g/L:

$$\textbf{concentration in g/L} = \frac{1.27 \times 10^{-4} \ mol}{1 \ L \ soln} \times \frac{275.8 \ g}{1 \ mol} = \textbf{0.035 g/L}$$

17.101 The equilibrium reaction is:

$$Pb(IO_3)_2 \ (aq) \ \rightleftarrows \ Pb^{2+}(aq) \ + \ 2IO_3^-(aq)$$

Initial (*M*):		0	0.10
Change (*M*):	-2.4×10^{-11}	$+2.4 \times 10^{-11}$	$+2(2.4 \times 10^{-11})$
Equilibrium (*M*):		2.4×10^{-11}	≈ 0.10

Substitute the equilibrium concentrations into the solubility product expression to calculate K_{sp}.

$$K_{sp} = [Pb^{2+}][IO_3^-]^2$$

$$\textbf{K}_{sp} = (2.4 \times 10^{-11})(0.10)^2 = \textbf{2.4} \times \textbf{10}^{-13}$$

17.103

$$BaSO_4(s) \rightleftarrows Ba^{2+}(aq) + SO_4^{2-}(aq)$$

$$K_{sp} = [Ba^{2+}][SO_4^{2-}] = 1.1 \times 10^{-10} \ \text{(See Table 17.4)}$$

$$[Ba^{2+}] = 1.0 \times 10^{-5} \ M$$

In 5.0 L, the number of moles of Ba^{2+} is

$$(5.0 \text{ L})\left(\frac{1.0\times10^{-5} \text{ mol Ba}^{2+}}{1 \text{ L}}\right) = 5.0\times10^{-5} \text{ mol Ba}^{2+}.$$

In practice, even less $BaSO_4$ will dissolve because the $BaSO_4$ is not in contact with the entire volume of blood. **$Ba(NO_3)_2$ is too soluble to be used for this purpose**.

17.105 a. The solubility product expressions for both substances have exactly the same mathematical form and are therefore directly comparable. The substance having the smaller K_{sp} (**AgBr**) will precipitate first.

   b. When CuBr just begins to precipitate the solubility product expression will just equal K_{sp} (saturated solution). The concentration of Cu^+ at this point is 0.010 M (given in the problem), so the concentration of bromide ion must be:

$$K_{sp} = [Cu^+][Br^-] = (0.010)[Br^-] = 4.2 \times 10^{-8} \text{ (See Table 17.4)}$$

$$[Br^-] = \frac{4.2 \times 10^{-8}}{0.010} = 4.2 \times 10^{-6} \ M$$

   Using this value of $[Br^-]$, we find the silver ion concentration:

$$[Ag^+] = \frac{K_{sp}}{[Br^-]} = \frac{7.7 \times 10^{-13}}{4.2 \times 10^{-6}} = \mathbf{1.8 \times 10^{-7} \ M}$$

   c. The percent of silver ion remaining in solution is:

$$\% \ Ag^+(aq) = \frac{1.8 \times 10^{-7} \ M}{0.010 \ M} \times 100\% = \mathbf{0.0018\% \text{ or } 1.8 \times 10^{-3}\%}$$

17.107 The initial number of moles of Ag^+ is

$$\text{mol Ag}^+ = 50.0 \text{ mL} \times \frac{0.010 \text{ mol Ag}^+}{1000 \text{ mL soln}} = 5.0 \times 10^{-4} \text{ mol Ag}^+$$

We can use the counts of radioactivity as being proportional to concentration. Thus, we can use the ratio to determine the quantity of Ag^+ still in solution. However, since our original 50 mL of solution has been diluted to 500 mL, the counts per mL will be reduced by 10. Our diluted solution would then produce 7402.5 counts per minute if no removal of Ag^+ had occurred.

The number of moles of Ag^+ that correspond to 44.4 counts are:

$$44.4 \text{ counts} \times \frac{5.0 \times 10^{-4} \text{ mol Ag}^+}{7402.5 \text{ counts}} = 3.0 \times 10^{-6} \text{ mol Ag}^+$$

$$\text{Original mol of IO}_3^- = 100 \text{ mL} \times \frac{0.030 \text{ mol IO}_3^-}{1000 \text{ mL soln}} = 3.0 \times 10^{-3} \text{ mol}$$

The quantity of IO_3^- remaining after reaction with Ag^+:

(original moles − moles reacted with Ag^+) = $(3.0 \times 10^{-3} \text{ mol}) - [(5.0 \times 10^{-4} \text{ mol}) - (3.0 \times 10^{-6} \text{ mol})]$

$$= 2.5 \times 10^{-3} \text{ mol } IO_3^-$$

The total final volume is 500 mL or 0.50 L.

$$[Ag^+] = \frac{3.0 \times 10^{-6} \text{ mol } Ag^+}{0.50 \text{ L}} = 6.0 \times 10^{-6} \text{ M}$$

$$[IO_3^-] = \frac{2.5 \times 10^{-3} \text{ mol } IO_3^-}{0.50 \text{ L}} = 5.0 \times 10^{-3} \text{ M}$$

$$AgIO_3(s) \rightleftharpoons Ag^+(aq) + IO_3^- \ (aq)$$

$$K_{sp} = [Ag^+][IO_3^-] = (6.0 \times 10^{-6})(5.0 \times 10^{-3}) = \mathbf{3.0 \times 10^{-8}}$$

17.109 a.
$$H^+ + OH^- \longrightarrow H_2O \qquad K = \mathbf{1.0 \times 10^{14}}$$

b.
$$H^+ + NH_3 \longrightarrow NH_4^+ \quad K = \frac{1}{K_a} = \frac{1}{5.6 \times 10^{-10}} = \mathbf{1.8 \times 10^9}$$

c.
$$CH_3COOH + OH^- \longrightarrow CH_3COO^- + H_2O$$

Broken into two equations:

$$CH_3COOH \longrightarrow CH_3COO^- + H^+ \qquad\qquad K_a$$

$$H^+ + OH^- \longrightarrow H_2O \qquad\qquad 1/K_w$$

$$K = \frac{K_a}{K_w} = \frac{1.8 \times 10^{-5}}{1.0 \times 10^{-14}} = \mathbf{1.8 \times 10^9}$$

d.
$$CH_3COOH + NH_3 \longrightarrow CH_3COONH_4$$

Broken into two equations:

$$CH_3COOH \longrightarrow CH_3COO^- + H^+ \qquad\qquad K_a$$

$$NH_3 + H^+ \longrightarrow NH_4^+ \qquad\qquad \frac{1}{K_a'}$$

$$K = \frac{K_a}{K_a'} = \frac{1.8 \times 10^{-5}}{5.6 \times 10^{-10}} = \mathbf{3.2 \times 10^4}$$

17.111 a. **Mix 500 mL of 0.40 *M* CH₃COOH with 500 mL of 0.40 *M* CH₃COONa.** Since the final volume is 1.00 L, the concentrations of the two solutions that were mixed must be one-half of their initial concentrations.

b. **Mix 500 mL of 0.80 *M* CH₃COOH with 500 mL of 0.40 *M* NaOH.** (Note: half of the acid reacts with all of the base to make a solution identical to that in part (a) above.)

$$CH_3COOH + NaOH \longrightarrow CH_3COONa + H_2O$$

c. **Mix 500 mL of 0.80 *M* CH₃COONa with 500 mL of 0.40 *M* HCl.** (Note: half of the salt reacts with all of the acid to make a solution identical to that in part (a) above.)

$$CH_3COO^- + H^+ \longrightarrow CH_3COOH$$

17.113 a. When a strong acid H_3O^+ is added to a buffer solution containing the weak acid HA and its conjugate base A^-, the strong acid H_3O^+ protonates the conjugate base A^-. This corresponds to **figure (b)**.

b. When a strong base OH^- is added to a buffer solution containing the weak acid HA and its conjugate base A^-, the strong base OH^- de-protonates the conjugate acid HA. This corresponds to **figure (a)**.

17.115 a. **Add Na₂CO₃.** Na_2CO_3 is soluble but $BaCO_3$ is not.

b. **Add HCl.** KCl is soluble but $PbCl_2$ is not.

c. **Add H₂S.** ZnS is soluble but HgS is not.

17.117 The amphoteric oxides cannot be used to prepare buffer solutions because **they are insoluble in water**.

17.119 **The ionized polyphenols have a dark color. In the presence of citric acid from lemon juice, the anions are converted to the lighter-colored acids.**

17.121 Assuming the density of water to be 1.00 g/mL, 0.05 g Pb^{2+} per 10^6 g water is equivalent to 5×10^{-5} g Pb^{2+}/L

$$\frac{0.05 \text{ g Pb}^{2+}}{1 \times 10^6 \text{ g H}_2O} \times \frac{1 \text{ g H}_2O}{1 \text{ mL H}_2O} \times \frac{1000 \text{ mL H}_2O}{1 \text{ L H}_2O} = 5 \times 10^{-5} \text{ g Pb}^{2+} / \text{L}$$

$$PbSO_4(s) \rightleftarrows Pb^{2+}(aq) + SO_4^{2-}(aq)$$

Initial (*M*):		0	0.10
Change (*M*):	−s	+s	+s
Equilibrium (*M*):		s	S

$$K_{sp} = [Pb^{2+}][SO_4^{2-}]$$

$$1.6 \times 10^{-8} = s^2$$

$$s = 1.3 \times 10^{-4} \text{ } M$$

The solubility of $PbSO_4$ in g/L is:

$$\frac{1.3 \times 10^{-4} \text{ mol}}{1 \text{ L}} \times \frac{303.3 \text{ g}}{1 \text{ mol}} = 4.0 \times 10^{-2} \text{ g / L}$$

Yes. The $[Pb^{2+}]$ exceeds the safety limit of 5×10^{-5} g Pb^{2+}/L.

17.123 **(c)** has the highest $[H^+]$.

$$F^- + SbF_5 \longrightarrow SbF_6^-$$

Removal of F^- promotes further ionization of HF.

17.125 Because oxalate is the anion of a weak acid, increasing the hydrogen ion concentration (decreasing the pH) would consume oxalate ion to produce hydrogen oxalate and oxalic acid:

$$C_2O_4^{2-}(aq) + H^+(aq) \longrightarrow HC_2O_4^-(aq)$$

$$HC_2O_4^-(aq) + H^+(aq) \longrightarrow H_2C_2O_4(aq)$$

Decreasing the concentration of oxalate ion would, by Le Châtelier's principle, increase the solubility of calcium oxalate. **Decreasing the pH would increase the solubility of calcium oxalate and should help minimize the formation of calcium oxalate kidney stones.**

Note: Although this makes sense from the standpoint of principles presented in Chapter 17, the actual mechanism of kidney stone formation is more complex than that can be described using only these equilibria.

17.127 $\text{pH} = \text{p}K_a + \log \dfrac{[\text{conjugate base}]}{[\text{acid}]}$

At pH = 1.0:

−COOH

$$1.0 = 2.3 + \log \frac{[-COO^-]}{[-COOH]}$$

$$\frac{[-COOH]}{[-COO^-]} = 20$$

−NH$_3$$^+$

$$1.0 = 9.6 + \log \frac{[-NH_2]}{[-NH_3^+]}$$

$$\frac{[-NH_3^+]}{[-NH_2]} = 4.0 \times 10^8$$

Therefore, the predominant species is: **$^+$NH$_3$ − CH$_2$ − COOH**

At pH = 7.0:

$$-COOH \qquad\qquad 7.0 = 2.3 + \log\frac{[-COO^-]}{[-COOH]}$$

$$\frac{[-COO^-]}{[-COOH]} = 5.0 \times 10^4$$

$$-NH_3^+ \qquad\qquad 7.0 = 9.6 + \log\frac{[-NH_2]}{[-NH_3^+]}$$

$$\frac{[-NH_3^+]}{[-NH_2]} = 4.0 \times 10^2$$

Predominant species: $^+NH_3 - CH_2 - COO^-$

At pH = 12.0:

$$-COOH \qquad\qquad 12.0 = 2.3 + \log\frac{[-COO^-]}{[-COOH]}$$

$$\frac{[-COO^-]}{[-COOH]} = 5.0 \times 10^9$$

$$-NH_3^+ \qquad\qquad 12.0 = 9.6 + \log\frac{[-NH_2]}{[-NH_3^+]}$$

$$\frac{[-NH_2]}{[-NH_3^+]} = 2.5 \times 10^2$$

Predominant species: $NH_2 - CH_2 - COO^-$

17.129 a. **Before dilution:**

$$pH = pK_a + \log\frac{[CH_3COO^-]}{[CH_3COOH]}$$

$$pH = 4.74 + \log\frac{[0.500]}{[0.500]} = \textbf{4.74}$$

After a 10-fold dilution:

$$pH = 4.74 + \log\frac{[0.0500]}{[0.0500]} = \textbf{4.74}$$

There is no change in the pH of a buffer upon dilution.

b. **Before dilution**:

$$CH_3COOH(aq) \quad + H_2O(l) \quad \rightleftharpoons \quad H_3O^+(aq) \quad + CH_3COO^-(aq)$$

Initial (*M*):	0.500		0	0
Change (*M*):	−*x*		+*x*	+*x*
Equilibrium (*M*):	0.500 − *x*		*x*	*x*

$$K_a = \frac{[H_3O^+][CH_3COO^-]}{[CH_3COOH]}$$

$$1.8 \times 10^{-5} = \frac{x^2}{0.500 - x} \approx \frac{x^2}{0.500}$$

$$x = 3.0 \times 10^{-3} \ M = [H_3O^+]$$

$$pH = -\log(3.0 \times 10^{-3}) = \textbf{2.52}$$

After dilution:

$$1.8 \times 10^{-5} = \frac{x^2}{0.0500 - x} \approx \frac{x^2}{0.0500}$$

$$x = 9.5 \times 10^{-4} \ M = [H_3O^+]$$

$$pH = -\log(9.5 \times 10^{-4}) = \textbf{3.02}$$

17.131 **Strategy:** The possible precipitates are $AgBr(s)$, $Ag_2CO_3(s)$, and $Ag_2SO_4(s)$. Using the K_{sp} values from Table 17.4 and the given anion concentrations, calculate the concentration of $Ag^+(aq)$ that will just begin to precipitate each anion.

Solution: $AgBr(s)$: $K_{sp} = 7.7 \times 10^{-13}$

$$AgBr(s) \quad \rightarrow \quad Ag^+(aq) \quad + Br^-(aq)$$

Initial (*M*):			0	0.1
Change (*M*):	−*x*		+*x*	+*x*
Equilibrium (*M*):			*x*	0.1 + *x*

$$K_{sp} = (x)(0.1 + x) \approx 0.1x$$

$$7.7 \times 10^{-13} = 0.1x$$

$$x = 7.7 \times 10^{-12}$$

$$[Ag^+] = 7.7 \times 10^{-12} \ M$$

$Ag_2CO_3(s)$: $K_{sp} = 8.1 \times 10^{-12}$

$$Ag_2CO_3(s) \rightarrow 2Ag^+(aq) + CO_3^{2-}(aq)$$

		$2Ag^+(aq)$	$CO_3^{2-}(aq)$
Initial (*M*):		0.00	0.1
Change (*M*):	$-x$	$+2x$	$+x$
Equilibrium (*M*):		$2x$	$0.1 + x$

$$K_{sp} = (2x)^2(0.1+x) \approx 0.4x^2$$

$$8.1 \times 10^{-12} = 0.4x^2$$

$$x = 4.5 \times 10^{-6}$$

$$[Ag^+] = 2x = 9.0 \times 10^{-6} \ M$$

$Ag_2SO_4(s)$: $K_{sp} = 1.5 \times 10^{-5}$

$$Ag_2SO_4(s) \rightarrow 2Ag^+(aq) + SO_4^{2-}(aq)$$

		$2Ag^+(aq)$	$SO_4^{2-}(aq)$
Initial (*M*):		0.00	0.1
Change (*M*):	$-x$	$+2x$	$+x$
Equilibrium (*M*):		$2x$	$0.1 + x$

$$K_{sp} = (2x)^2(0.1+x) \approx 0.4x^2$$

$$1.5 \times 10^{-5} = 0.4x^2$$

$$x = 6.1 \times 10^{-3}$$

$$[Ag^+] = 2x = 1.2 \times 10^{-2} \ M$$

AgBr(s) will precipitate first (when $[Ag^+] = 7.7 \times 10^{-12}$ M), **Ag$_2$CO$_3$(s) will precipitate last** (when $[Ag^+] = 9.0 \times 10^{-6}$ M), **then Ag$_2$SO$_4$(s)** (when $[Ag^+] = 1.2 \times 10^{-2}$ M).

17.133 According to the solubility guidelines (Tables 9.2 and 9.3), the cation $Cd^{2+}(aq)$ will not precipitate with $NO_3^-(aq)$, $SO_4^{2-}(aq)$, or $ClO_3^-(aq)$. Likewise, the anion $S^{2-}(aq)$ will not precipitate with $Li^+(aq)$, $Na^+(aq)$, or $K^+(aq)$. However, $Cd^{2+}(aq)$ forms a complex with CN^-(aq) to produce $Cd(CN)_4^{2-}(aq)$. This reduces $[Cd^{2+}]$ in the solution and, by Le Châtelier's principle, promotes further dissolving of the CdS(s). Hence, **compound (c)** will increase the solubility.

17.135 **Strategy:** Let s be the molar solubility of Ag_2CO_3. We wish to calculate

$$K_{sp} = [Ag^+]^2[CO_3^{2-}] = (2s)^2(s) = 4s^3$$

Each mole of $CO_3^{2-}(aq)$ produces one mole of $CO_2(g)$:

$$CO_3^{2-}(aq) + 2HCl(aq) \longrightarrow H_2O(l) + 2Cl^-(aq) + CO_2(g)$$

so the number of moles of $CO_3^{2-}(aq)$ in the saturated solution is equal to the number of moles of CO_2 (g) collected. Use the ideal gas law and the given information to calculate $n(CO_2) = n(CO_3^{2-})$, then use this result to calculate s and K_{sp}.

Solution: The amount of CO_2 is:

$$n(CO_2) = \frac{PV}{RT} = \frac{(114 \text{ mmHg})\left(\dfrac{1 \text{ atm}}{760 \text{ mmHg}}\right)(0.019 \text{ L})}{(0.08206 \text{ L} \cdot \text{atm/K} \cdot \text{mol})(273+5)(K)} = 1.3 \times 10^{-4} \text{ mol}$$

Since $n(CO_3^{2-}) = n(CO_2) = 1.3 \times 10^{-4}$ mol and the volume of the saturated solution is 1.0 L, then

$$s = \frac{1.3 \times 10^{-4} \text{ mol } CO_3^{2-}}{1.0 \text{ L}} = 1.3 \times 10^{-4} \text{ } M$$

Thus,

$$K_{sp} = 4\left(1.3 \times 10^{-4}\right)^3 = 8.8 \times 10^{-12}$$

17.137 The reaction is:

$$NH_3 + HCl \longrightarrow NH_4Cl$$

First, we calculate moles of HCl and NH_3.

$$n_{HCl} = \frac{PV}{RT} = \frac{\left(372 \text{ mmHg} \times \dfrac{1 \text{ atm}}{760 \text{ mmHg}}\right)(0.96 \text{ L})}{\left(0.08206 \dfrac{\text{L} \cdot \text{atm}}{\text{K} \cdot \text{mol}}\right)(295 \text{ K})} = 0.0194 \text{ mol}$$

$$n_{NH_3} = \frac{0.57 \text{ mol } NH_3}{1 \text{ L soln}} \times 0.034 \text{ L} = 0.0194 \text{ mol}$$

The mole ratio between NH_3 and HCl is 1:1, so we have complete neutralization.

	$NH_3(aq)$	+ HCl (aq)	$\rightarrow$	NH_4Cl (s)
Initial (mol):	0.0194	0.0194		0
Change (mol):	−0.0194	−0.0194		+0.0194
Final (mol):	0	0		0.0194

NH_4^+ is a weak acid. We set up the reaction representing the hydrolysis of NH_4^+.

	$NH_4^+(aq)$	+ $H_2O(l)$	$\rightleftharpoons$	$H_3O^+(aq)$	+ $NH_3(aq)$
Initial (M):	0.0194 mol/0.034 L			0	0
Change (M):	−x			+x	+x
Equilibrium (M):	$0.57 - x$			x	x

$$K_a = \frac{[H_3O^+][NH_3]}{[NH_4^+]}$$

$$5.6 \times 10^{-10} = \frac{x^2}{0.57 - x} \approx \frac{x^2}{0.57}$$

$$x = 1.79 \times 10^{-5}\ M = [H_3O^+]$$

$$\textbf{pH} = -\log(1.79 \times 10^{-5}) = \textbf{4.75}$$

17.139 The balanced equation is:

$$HA(aq) + NaOH(aq) \longrightarrow NaA(aq) + H_2O(l)$$

This equation can be simplified to:

$$HA(aq) + OH^-(aq) \longrightarrow A^-(aq) + H_2O(l)$$

a. Before the addition of any base, only HA is present. This corresponds to **diagram (b)**.

b. At the equivalence point, all the HA has been neutralized and we are left with A^-. This corresponds to **diagram (e)**.

c. After the equivalence point, all the HA has been consumed, so there is nothing left to consume additional OH^-. The diagram will have only OH^- and A^-. This corresponds to **diagram (d)**.

d. The pH is equal to the pK_a at the half-equivalence point. At this point, $[HA] = [A^-]$. This corresponds to **diagram (a)**, where there is an equal number of HA and A^- ions.

17.141 We can use the Henderson-Hasselbalch equation to solve for the pH when the indicator is 90% acid / 10% conjugate base and when the indicator is 10% acid / 90% conjugate base.

$$pH = pK_a + \log\frac{[\text{conjugate base}]}{[\text{acid}]}$$

Solving for the pH with 90% of the indicator in the HIn form:

$$pH = 3.46 + \log\frac{[10]}{[90]} = 3.46 - 0.95 = 2.51$$

Next, solving for the pH with 90% of the indicator in the In^- form:

$$pH = 3.46 + \log\frac{[90]}{[10]} = 3.46 + 0.95 = 4.41$$

Thus, the pH range varies from **2.51 to 4.41** as the [HIn] varies from 90% to 10%.

Chapter 18
Electrochemistry

Visualizing Chemistry

VC 18.1 a   VC 18.2 b   VC 18.3 b   VC 18.4 a

Key Skills

18.1   b   18.2   a   18.3   d   18.4   d

Problems

18.1   **Strategy:**   We follow the stepwise procedure for balancing redox reactions presented in Section 18.1 of the text.

       Solution: a. In *Step 1,* we separate the half-reactions.

$$\text{oxidation:} \qquad Fe^{2+} \longrightarrow Fe^{3+}$$
$$\text{reduction:} \qquad H_2O_2 \longrightarrow H_2O$$

Step 2 is unnecessary because the half-reactions are already balanced with respect to iron.

Step 3: We balance each half-reaction for O by adding H_2O. The oxidation half-reaction is already balanced in this regard (it contains no O atoms). The reduction half-reaction requires the addition of one H_2O on the product side.

$$H_2O_2 \longrightarrow H_2O + H_2O \text{ or simply } H_2O_2 \longrightarrow 2H_2O$$

Step 4: We balance each half-reaction for H by adding H^+. Again, the oxidation half-reaction is already balanced in this regard. The reduction half-reaction requires the addition of two H^+ ions on the reactant side.

$$H_2O_2 + 2H^+ \longrightarrow 2H_2O$$

Step 5: We balance both half-reactions for charge by adding electrons. In the oxidation half-reaction, there is a total charge of +2 on the left and a total charge of +3 on the right. Adding one electron to the product side makes the total charge on each side +2.

$$Fe^{2+} \longrightarrow Fe^{3+} + e^-$$

In the reduction half-reaction, there is a total charge of +2 on the left and a total charge of 0 on the right. Adding two electrons to the reactant side makes the total charge on each side 0.

$$H_2O_2 + 2H^+ + 2e^- \longrightarrow 2H_2O$$

Step 6: Because the number of electrons is not the same in both half-reactions, we multiply the oxidation half-reaction by 2.

$$2\times(Fe^{2+} \longrightarrow Fe^{3+} + e^-) = 2Fe^{2+} \longrightarrow 2Fe^{3+} + 2e^-$$

Step 7: Finally, we add the resulting half-reactions, cancelling electrons and any other identical species to get the overall balanced equation.

$$2Fe^{2+} \rightarrow 2Fe^{3+} + 2e^-$$

$$H_2O_2 + 2H^+ + 2e^- \rightarrow 2H_2O$$

$$\overline{2H^+ + H_2O_2 + 2Fe^{2+} \rightarrow 2Fe^{3+} + 2H_2O}$$

b. In *Step 1*, we separate the half-reactions.

oxidation: $\quad Cu \longrightarrow Cu^{2+}$

reduction: $\quad HNO_3 \longrightarrow NO$

Step 2 is unnecessary because the half-reactions are already balanced with respect to copper and nitrogen.

Step 3: We balance each half-reaction for O by adding H_2O. The oxidation half-reaction is already balanced in this regard (it contains no O atoms). The reduction half-reaction requires the addition of two H_2O molecules on the product side.

$$HNO_3 \longrightarrow NO + 2H_2O$$

Step 4: We balance each half-reaction for H by adding H^+. Again, the oxidation half-reaction is already balanced in this regard. The reduction half-reaction requires the addition of three H^+ ions on the reactant side.

$$3H^+ + HNO_3 \longrightarrow NO + 2H_2O$$

Step 5: We balance both half-reactions for charge by adding electrons. In the oxidation half-reaction, there is a total charge of 0 on the left and a total charge of +2 on the right. Adding two electrons to the product side makes the total charge on each side 0.

$$Cu \longrightarrow Cu^{2+} + 2e^-$$

In the reduction half-reaction, there is a total charge of +3 on the left and a total charge of 0 on the right. Adding three electrons to the reactant side makes the total charge on each side 0.

$$3H^+ + HNO_3 + 3e^- \longrightarrow NO + 2H_2O$$

Step 6: Because the number of electrons is not the same in both half-reactions, we multiply the oxidation half-reaction by 3, and the reduction half-reaction by 2.

$$3\times(Cu \rightarrow Cu^{2+} + 2e^-) = 3Cu \longrightarrow 3Cu^{2+} + 6e^-$$

$$\overline{2\times(3H^+ + HNO_3 + 3e^- \longrightarrow NO + 2H_2O) = 6H^+ + 2HNO_3 + 6e^- \rightarrow 2NO + 4H_2O}$$

Step 7: Finally, we add the resulting half-reactions, cancelling electrons and any other identical species to get the overall balanced equation.

$$3Cu \rightarrow 3Cu^{2+} + \boxed{6e^-}$$
$$6H^+ + 2HNO_3 + \boxed{6e^-} \rightarrow 2NO + 4H_2O$$
$$\overline{\mathbf{6H^+ + 2HNO_3 + 3Cu \rightarrow 3Cu^{2+} + 2NO + 4H_2O}}$$

c. In *Step 1*, we separate the half-reactions.

$$\text{oxidation:} \qquad CN^- \longrightarrow CNO^-$$
$$\text{reduction:} \qquad MnO_4^- \longrightarrow MnO_2$$

Step 2 is unnecessary because the half-reactions are already balanced with respect to carbon, nitrogen, and manganese.

Step 3: We balance each half-reaction for O by adding H_2O. The oxidation half-reaction requires the addition of one H_2O molecule to the reactant side. The reduction half-reaction requires the addition of two H_2O molecules on the product side.

$$H_2O + CN^- \longrightarrow CNO^-$$

$$MnO_4^- \longrightarrow MnO_2 + 2H_2O$$

Step 4: We balance each half-reaction for H by adding H^+. The oxidation half-reaction requires the addition of two H^+ ions to the product side. The reduction half-reaction requires the addition of four H^+ ions on the reactant side.

$$H_2O + CN^- \longrightarrow CNO^- + 2H^+$$
$$4H^+ + MnO_4^- \longrightarrow MnO_2 + 2H_2O$$

Step 5: We balance both half-reactions for charge by adding electrons. In the oxidation half-reaction, there is a total charge of -1 on the left and a total charge of $+1$ on the right. Adding two electrons to the product side makes the total charge on each side -1.

$$H_2O + CN^- \longrightarrow CNO^- + 2H^+ + 2e^-$$

In the reduction half-reaction, there is a total charge of $+3$ on the left and a total charge of 0 on the right. Adding three electrons to the reactant side makes the total charge on each side 0.

$$3e^- + 4H^+ + MnO_4^- \longrightarrow MnO_2 + 2H_2O$$

Step 6: Because the number of electrons is not the same in both half-reactions, we multiply the oxidation half-reaction by 3, and the reduction half-reaction by 2.

$$3\times(H_2O + CN^- \rightarrow CNO^- + 2H^+ + 2e^-) = 3H_2O + 3CN^- \rightarrow 3CNO^- + 6H^+ + 6e^-$$

$$2\times(3e^- + 4H^+ + MnO_4^- \rightarrow MnO_2 + 2H_2O) = 6e^- + 8H^+ + 2MnO_4^- \rightarrow 2MnO_2 + 4H_2O$$

Step 7: Finally, we add the resulting half-reactions, cancelling electrons and any other identical species to get the overall balanced equation.

$$3H_2O + 3CN^- \rightarrow 3CNO^- + 6H^+ + \boxed{6e^-}$$

$$\underline{\boxed{6e^-} + 6H^+ + 2MnO_4^- \rightarrow 2MnO_2 + H_2O}$$

$$3CN^- + 2MnO_4^- + 2H^+ \rightarrow 3CNO^- + 2MnO_2 + 4H_2O$$

Balancing a redox reaction in basic solution requires two additional steps.

Step 8: For each H^+ ion in the final equation, we add one OH^- ion to each side of the equation, combining the H^+ and OH^- ions to produce H_2O.

$$3CN^- + 2MnO_4^- + 2H^+ \longrightarrow 3CNO^- + 2MnO_2 + H_2O$$
$$\underline{ + 2OH^-   + 2OH^-}$$

$$2H_2O + 2MnO_4^- + 3CN^- \longrightarrow 2MnO_2 + 3CNO^- + H_2O + 2OH^-$$

Step 9: Lastly, we cancel H_2O molecules that result from Step 8.

$$3CN^- + 2MnO_4^- + H_2O \longrightarrow 3CNO^- + 2MnO_2 + 2OH^-$$

d. Parts (d) and (e) are solved using the methods outlined in (a) through (c).

$$6OH^- + 3Br_2 \longrightarrow BrO_3^- + 3H_2O + 5Br^-$$

e.

$$2S_2O_3^{2-} + I_2 \longrightarrow S_4O_6^{2-} + 2I^-$$

18.11 **Strategy:** At first, it may not be clear how to assign the electrodes in the galvanic cell. From Table 18.1 of the text, we write the standard reduction potentials of Al and Ag and compare their values to determine which is the anode half-reaction and which is the cathode half-reaction.

Solution: The standard reduction potentials are:

$$Ag^+(1.0\ M) + e^- \longrightarrow Ag(s) \qquad E° = 0.80\ V$$
$$Al^{3+}(1.0\ M) + 3e^- \longrightarrow Al(s) \qquad E° = -1.66\ V$$

The silver half-reaction, with the more positive $E°$ value, will occur at the cathode (as a reduction). The aluminum half-reaction will occur at the anode (as an oxidation). To balance the numbers of electrons in the two half-reactions, we multiply the reduction by 3 before summing the two half-reactions. Note that multiplying a half-reaction by 3 does not change its $E°$ value because reduction potential is an intensive property.

$$Al(s) \longrightarrow Al^{3+}(1.0\ M) + 3e^-$$

$$\underline{3Ag^+(1.0\ M) + 3e^- \longrightarrow 3Ag(s)}$$

Overall: $$Al(s) + 3Ag^+(1.0\ M) \longrightarrow Al^{3+}(1.0\ M) + 3Ag(s)$$

We find the emf of the cell using Equation 18.1.

$$E^\circ_{cell} = E^\circ_{cathode} - E^\circ_{anode} = E^\circ_{Ag^+/Ag} - E^\circ_{Al^{3+}/Al}$$

$$E^\circ_{cell} = 0.80 \text{ V} - (-1.66 \text{ V}) = \textbf{2.46 V}$$

18.13   The half-reaction for oxidation is:

$$2H_2O(l) \xrightarrow{\text{oxidation (anode)}} O_2(g) + 4H^+(aq) + 4e^- \qquad E^\circ_{anode} = +1.23 \text{ V}$$

The species that can oxidize water to molecular oxygen must have an E°_{red} more positive than +1.23 V. From Table 18.1 of the text, we see that only $\textbf{Cl}_2\textbf{(g)}$ and $\textbf{MnO}_4^-\textbf{(aq)}$ in acid solution can oxidize water to oxygen.

18.15   **Strategy:**   In each case, we can calculate the standard cell emf from the potentials for the two half-reactions.

$$E^\circ_{cell} = E^\circ_{cathode} - E^\circ_{anode}$$

Solution: Using E° values from Table 18.1:

a. $E^\circ = -0.40 \text{ V} - (-2.87 \text{ V}) = 2.47 \text{ V}$. The reaction is **spontaneous**.

b. $E^\circ = -0.14 \text{ V} - 1.07 \text{ V} = -1.21 \text{ V}$. The reaction is **not spontaneous**.

c. $E^\circ = -0.25 \text{ V} - 0.80 \text{ V} = -1.05 \text{ V}$. The reaction is **not spontaneous**.

d. $E^\circ = 0.77 \text{ V} - 0.15 \text{ V} = 0.62 \text{ V}$. The reaction is **spontaneous**.

18.17   **Strategy:**   The greater the tendency for the substance to be oxidized, the stronger its tendency to act as a reducing agent. The species that has a stronger tendency to be oxidized will have a smaller reduction potential.

Solution: In each pair, look for the one with the smaller reduction potential. This indicates a greater tendency for the substance to be oxidized.

a.   From Table 18.1, we have the following reduction potentials:

Na: −2.71 V
Li: −3.05 V

Li has the smaller (more negative) reduction potential and is therefore more easily oxidized. The more easily oxidized substance is the better reducing agent. **Li** is the better reducing agent.

Following the same logic for parts (b) through (d) gives

b.   **H₂**

c.   **Fe²⁺**

d.   **Br⁻**

18.21 **Strategy:** The relationship between the equilibrium constant, K, and the standard emf is given by Equation 18.5 of the text: $E_{cell}^{\circ} = \dfrac{0.0592 \text{ V}}{n} \log K$. Thus, knowing E_{cell}° and n (the moles of electrons transferred), we can then determine equilibrium constant. We find the standard reduction potentials in Table 18.1 of the text.

$$E_{cell}^{\circ} = E_{cathode}^{\circ} - E_{anode}^{\circ} = -0.76 \text{ V} - (-2.37 \text{ V}) = 1.61 \text{ V}$$

Solution: We must rearrange Equation 18.5 to solve for K:

$$E_{cell}^{\circ} = \dfrac{0.0592 \text{ V}}{n} \log K$$

$$\log K = \dfrac{n E_{cell}^{\circ}}{0.0592 \text{ V}}$$

$$K = 10^{n E_{cell}^{\circ}/0.0592 \text{ V}}$$

We see in the reaction that Mg becomes Mg^{2+} and Zn^{2+} becomes Zn. Therefore, 2 moles of electrons are transferred during the redox reaction; i.e., $n = 2$.

$$K = 10^{(2)(1.61 \text{ V})/0.0592 \text{ V}}$$

$$\boldsymbol{K = 10^{54.4} = 3 \times 10^{54}}$$

18.23 In each case we use standard reduction potentials from Table 18.1 together with Equation 18.5 of the text.

a. We break the equation into two half-reactions:

$$Br_2(l) + 2e^- \xrightarrow{\text{ reduction (cathode) }} 2Br^-(aq) \qquad\qquad E_{cathode}^{\circ} = 1.07 \text{ V}$$
$$2I^-(aq) \xrightarrow{\text{ oxidation (anode) }} I_2(s) + 2e^- \qquad\qquad E_{anode}^{\circ} = 0.53 \text{ V}$$

The standard emf is

$$E_{cell}^{\circ} = E_{cathode}^{\circ} - E_{anode}^{\circ} = 1.07 \text{ V} - 0.53 \text{ V} = 0.54 \text{ V}$$

Next, we can calculate K using Equation 18.5 of the text.

$$K = 10^{n E_{cell}^{\circ}/0.0592 \text{ V}}$$

$$K = 10^{(2)(0.54 \text{ V})/0.0592 \text{ V}}$$

$$\boldsymbol{K = 10^{18.2} = 2 \times 10^{18}}$$

b. We break the equation into two half-reactions:

$$2Ce^{4+}(aq) + 2e^- \xrightarrow{\text{ reduction (cathode) }} 2Ce^{3+}(aq) \qquad\qquad E_{cathode}^{\circ} = 1.61 \text{ V}$$
$$2Cl^-(aq) \xrightarrow{\text{ oxidation (anode) }} Cl_2(g) + 2e^- \qquad\qquad E_{anode}^{\circ} = 1.36 \text{ V}$$

The standard emf is

$$E^\circ_{cell} = E^\circ_{cathode} - E^\circ_{anode} = 1.61\ V - 1.36\ V = 0.25\ V$$

$$K = 10^{(2)(0.25\ V)/0.0592\ V}$$

$$\boldsymbol{K = 10^{8.4} = 3 \times 10^8}$$

c. We break the equation into two half-reactions:

$$MnO_4^-(aq) + 8H^+(aq) + 5e^- \xrightarrow{\text{reduction (cathode)}} Mn^{2+}(aq) + 4H_2O(l) \qquad E^\circ_{cathode} = 1.51\ V$$

$$5Fe^{2+}(aq) \xrightarrow{\text{oxidation (anode)}} 5Fe^{3+}(aq) + 5e^- \qquad E^\circ_{anode} = 0.77\ V$$

The standard emf is

$$E^\circ_{cell} = E^\circ_{cathode} - E^\circ_{anode} = 1.51\ V - 0.77\ V = 0.74\ V$$

$$K = 10^{(5)(0.74\ V)/0.0592\ V}$$

$$\boldsymbol{K = 10^{62.5} = 3 \times 10^{62}}$$

18.25 **Strategy:** The spontaneous reaction that occurs must include one reduction and one oxidation. We examine the reduction potentials of the species present to determine which half-reaction will occur as the reduction and which will occur as the oxidation. The relationship between the standard free-energy change and the standard emf of the cell is given by Equation 18.3 of the text: $\Delta G^\circ = -nFE^\circ_{cell}$. The relationship between the equilibrium constant, K, and the standard emf is given by Equation 18.5 of the text: $E^\circ_{cell} = \dfrac{0.0592\ V}{n} \log K$. Thus, once we determine E°_{cell}, we can calculate ΔG° and K. We find the standard reduction potentials in Table 18.1 of the text.

Solution: The half-reactions (both written as reductions) are:

$$Fe^{3+}(aq) + e^- \longrightarrow Fe^{2+}(aq) \qquad E^\circ_{anode} = 0.77\ V$$

$$Ce^{4+}(aq) + e^- \longrightarrow Ce^{3+}(aq) \qquad E^\circ_{cathode} = 1.61\ V$$

Because it has the larger reduction potential, Ce^{4+} is the more easily reduced and will oxidize Fe^{2+} to Fe^{3+}. This makes the Fe^{2+}/Fe^{3+} half-reaction the anode. The spontaneous reaction is:

$$\boldsymbol{Ce^{4+}(aq) + Fe^{2+}(aq) \longrightarrow Ce^{3+}(aq) + Fe^{3+}(aq)}$$

The standard cell emf is found using Equation 18.1 of the text.

$$E^\circ_{cell} = E^\circ_{cathode} - E^\circ_{anode} = 1.61\ V - 0.77\ V = 0.84\ V$$

The values of $\Delta G°$ and K_c are found using Equations 18.3 and 18.5 of the text.

$$\mathbf{\Delta G°} = -nFE°_{cell} = -(1)(96,500 \text{ J/V} \cdot \text{mol})(0.84 \text{ V}) = \mathbf{-81 \text{ kJ / mol}}$$

$$\log K = \frac{nE°_{cell}}{0.0592 \text{ V}}$$

$$K = 10^{nE°_{cell}/0.0592 \text{ V}}$$

$$\mathbf{K} = 10^{(1)(0.84 \text{ V})/0.0592 \text{ V}} = \mathbf{2 \times 10^{14}}$$

18.27 Strategy: According to Section 18.4, a simplified unbalanced equation for oxidation of tin from an amalgam filling is:

$$Sn(s) + O_2(g) \longrightarrow Sn^{2+}(aq) + H_2O(l)$$

Apply Steps 1 through 7 for balancing redox equations to balance for mass and charge. Then use Equation 18.1 to calculate the standard cell potential for the reaction.

Solution: *Step 1:* Separate the unbalanced reaction into half-reactions.

$$\text{oxidation:} \qquad Sn \longrightarrow Sn^{2+}$$
$$\text{reduction:} \qquad O_2 \longrightarrow H_2O$$

Step 2: This step is unnecessary because the half-reactions are balanced for all elements, excluding O and H.

Step 3: Balance both half-reactions for O by adding H_2O.

$$Sn \longrightarrow Sn^{2+}$$
$$O_2 \longrightarrow \mathbf{2}H_2O$$

Step 4: Balance both half-reactions for H by adding H^+.

$$Sn \longrightarrow Sn^{2+}$$
$$4H^+ + O_2 \longrightarrow 2H_2O$$

Step 5: Balance the total charge of both half-reactions by adding electrons.

$$Sn \longrightarrow Sn^{2+} + 2e^-$$
$$4e^- + 4H^+ + O_2 \longrightarrow 2H_2O$$

Step 6: Multiply the half-reactions to make the numbers of electrons the same in both.

$$2(Sn \longrightarrow Sn^{2+} + 2e^-)$$
$$4e^- + 4H^+ + O_2 \longrightarrow 2H_2O$$

Step 7: Add the half-reactions back together, cancelling electrons.

$$2Sn \rightarrow 2Sn^{2+} + 4e^-$$
$$4e^- + 4H^+ + O_2 \rightarrow 2H_2O$$
$$\mathbf{2Sn + 4H^+ + O_2 \rightarrow 2Sn^{2+} + 2H_2O}$$

The standard cell potential is:

$$E^\circ_{cell} = E^\circ_{cathode} - E^\circ_{anode} = 1.23 - (-0.14) = \mathbf{1.37\ V}$$

18.31 **Strategy:** The standard emf (E°) can be calculated using the standard reduction potentials in Table 18.1 of the text. Because the reactions are not run under standard-state conditions (concentrations are not 1 M), we need the Nernst equation (Equation 18.7) of the text to calculate the emf (E) of a hypothetical galvanic cell. Remember that solids do not appear in the reaction quotient (Q) term in the Nernst equation. We can calculate ΔG from E using Equation 18.2 of the text: $\Delta G = -nFE_{cell}$.

Solution: The half-cell reactions are:

$$Cu^{2+}(aq) + 2e^- \longrightarrow Cu(s)$$
$$Zn^{2+}(aq) + 2e^- \longrightarrow Zn(s)$$

The standard emf is:

$$E^\circ = E^\circ_{cathode} - E^\circ_{anode} = 0.34 - (-0.76) = 1.10\ V$$

$$E = E^\circ - \frac{0.0592\ V}{n} \log Q$$

$$E = 1.10\ V - \frac{0.0592\ V}{2} \log \frac{[Zn^{2+}]}{[Cu^{2+}]}$$

$$\mathbf{E = 1.10\ V - \frac{0.0592\ V}{2} \log \frac{0.25}{0.15} = 1.09\ V}$$

18.33 Use E° values from Table 18.1 to determine E° for the reaction.

The overall reaction is: $\quad Zn(s) + 2H^+(aq) \longrightarrow Zn^{2+}(aq) + H_2(g)$

$$\mathbf{E^\circ_{cell}} = E^\circ_{cathode} - E^\circ_{anode} = 0.00\ V - (-0.76\ V) = \mathbf{0.76\ V}$$

$$E = E^\circ - \frac{0.0592\ V}{n} \log \frac{[Zn^{2+}]P_{H_2}}{[H^+]^2}$$

$$\mathbf{E = 0.76\ V - \frac{0.0592\ V}{2} \log \frac{(0.45)(2.0)}{(1.8)^2} = 0.78\ V}$$

18.35   As written, the reaction is not spontaneous under standard-state conditions; the cell emf is negative.

$$E^\circ_{cell} = E^\circ_{cathode} - E^\circ_{anode} = -0.76 \text{ V} - 0.34 \text{ V} = -1.10 \text{ V}$$

The reaction will become spontaneous when the concentrations of zinc(II) and copper(II) ions are such as to make the emf positive. The turning point is when the emf is zero. We solve the Nernst equation (Equation 18.6) for the $[Cu^{2+}]/[Zn^{2+}]$ ratio at this point.

$$E_{cell} = E^\circ - \frac{RT}{nF} \ln Q$$

At 25°C:

$$0 = -1.1 \text{ V} - \frac{(8.314 \text{ J/K} \cdot \text{mol})(298 \text{ K})}{(2 \text{ mol } e^-)(96,500 \text{ J/V} \cdot \text{mol } e^-)} \ln \frac{[Cu^{2+}]}{[Zn^{2+}]}$$

$$\ln \frac{[Cu^{2+}]}{[Zn^{2+}]} = -85.7$$

$$\frac{[Cu^{2+}]}{[Zn^{2+}]} = e^{-85.7} = \textbf{6} \times \textbf{10}^{-\textbf{38}}$$

In other words for the reaction to be spontaneous, the $[Cu^{2+}]/[Zn^{2+}]$ ratio must be less than 6×10^{-38}.

18.41   We can calculate the standard free-energy change, ΔG°, from the standard free energies of formation, ΔG°_f using Equation 14.12 and Appendix 2 of the text. Then, we can calculate the standard cell emf, E°_{cell}, from ΔG°.

The overall reaction is:

$$C_3H_8(g) + 5O_2(g) \longrightarrow 3CO_2(g) + 4H_2O(l)$$

$$\Delta G^\circ_{rxn} = 3\Delta G^\circ_f \left(CO_2(g)\right) + 4\Delta G^\circ_f \left(H_2O(l)\right) - \left[\Delta G^\circ_f \left(C_3H_8(g)\right) + 5\Delta G^\circ_f \left(O_2(g)\right)\right]$$

$$\Delta G^\circ_{rxn} = (3)(-394.4 \text{ kJ/mol}) + (4)(-237.2 \text{ kJ/mol}) - [(1)(-23.5 \text{ kJ/mol}) + (5)(0)] = -2108.5 \text{ kJ/mol}$$

We can now calculate the standard emf using the following equation:

$$\Delta G^\circ = -nFE^\circ_{cell}$$

or

$$E^\circ_{cell} = \frac{-\Delta G^\circ}{nF}$$

Check the half-reactions of the text to determine that 20 moles of electrons are transferred during this redox reaction.

$$E^\circ_{cell} = \frac{-(-2108.5 \times 10^3 \text{ J/mol})}{(20)(96,500 \text{ J/V} \cdot \text{mol})} = \textbf{1.09 V}$$

18.45 **Strategy:** A faraday is a mole of electrons. Knowing how many moles of electrons are needed to reduce a mole of magnesium ions, we can determine how many moles of magnesium will be produced. The half-reaction shows that 2 moles of e^- are required per mole of Mg.

$$Mg^{2+}(aq) + 2e^- \longrightarrow Mg(s)$$

Solution: $\quad$ **Mass Mg** $= 1.00\ F \times \dfrac{1\ \text{mol Mg}}{2\ \text{mol}\ e^-} \times \dfrac{24.31\ \text{g Mg}}{1\ \text{mol Mg}} = \textbf{12.2 g Mg}$

18.47 The half-reactions are:
$$Na^+ + e^- \longrightarrow Na$$
$$Al^{3+} + 3e^- \longrightarrow Al$$

As long as we are comparing equal masses, we can use any mass that is convenient. In this case, we will use 1 g.

$$1\ \text{g Na} \times \dfrac{1\ \text{mol}}{22.99\ \text{g Na}} \times 1\ e^- = 0.043\ \text{mol}\ e^-$$

$$1\ \text{g Al} \times \dfrac{1\ \text{mol}}{26.98\ \text{g Al}} \times 3\ e^- = 0.11\ \text{mol}\ e^-$$

It is cheaper to prepare 1 ton of sodium by electrolysis.

18.49 Find the amount of oxygen using the ideal gas equation.

$$n = \dfrac{PV}{RT} = \dfrac{\left(755\ \text{mmHg} \times \dfrac{1\ \text{atm}}{760\ \text{mmHg}}\right)(0.076\ \text{L})}{(0.08206\ \text{L} \cdot \text{atm/K} \cdot \text{mol})(298\ \text{K})} = 3.1 \times 10^{-3}\ \text{mol O}_2$$

Since the half-reaction shows that 1 mole of oxygen requires 4 faradays of electric charge, we write:

$$(3.1 \times 10^{-3}\ \text{mol O}_2) \times \dfrac{4\ F}{1\ \text{mol O}_2} = \textbf{0.012}\ \textbf{\textit{F}}$$

18.51 The half-reactions are:
$$Cu^{2+}(aq) + 2e^- \longrightarrow Cu(s)$$

$$2Br^-(aq) \longrightarrow Br_2(l) + 2e^-$$

The mass of copper produced is:

$$4.50\ \text{A} \times 1\ \text{h} \times \dfrac{3600\ \text{s}}{1\ \text{h}} \times \dfrac{1\ \text{C}}{1\ \text{A} \cdot \text{s}} \times \dfrac{1\ \text{mol}\ e^-}{96{,}500\ \text{C}} \times \dfrac{1\ \text{mol Cu}}{2\ \text{mol}\ e^-} \times \dfrac{63.55\ \text{g Cu}}{1\ \text{mol Cu}} = \textbf{5.33 g Cu}$$

The mass of bromine produced is:

$$4.50\ \text{A} \times 1\ \text{h} \times \dfrac{3600\ \text{s}}{1\ \text{h}} \times \dfrac{1\ \text{C}}{1\ \text{A} \cdot \text{s}} \times \dfrac{1\ \text{mol}\ e^-}{96{,}500\ \text{C}} \times \dfrac{1\ \text{mol Br}_2}{2\ \text{mol}\ e^-} \times \dfrac{159.8\ \text{g Br}_2}{1\ \text{mol Br}_2} = \textbf{13.4 g Br}_2$$

18.53   The half-reaction is:

$$Co^{2+} + 2e^- \longrightarrow Co$$

The half-reaction tells us that 2 moles of electrons are needed to reduce 1 mol of Co^{2+} to Co metal. We can set up the following strategy to calculate the quantity of electricity (in C) needed to deposit 2.35 g of Co.

$$2.35 \text{ g Co} \times \frac{1 \text{ mol Co}}{58.93 \text{ g Co}} \times \frac{2 \text{ mol } e^-}{1 \text{ mol Co}} \times \frac{96,500 \text{ C}}{1 \text{ mol } e^-} = \mathbf{7.70 \times 10^3 \text{ C}}$$

18.55   The half-reaction for the oxidation of chloride ion is:

$$2Cl^-(aq) \longrightarrow Cl_2(g) + 2e^-$$

First, let's calculate the moles of e^- flowing through the cell in 1 hr.

$$1500 \text{ A} \times \frac{1 \text{ C}}{1 \text{ A} \cdot \text{s}} \times \frac{3600 \text{ s}}{1 \text{ h}} \times \frac{1 \text{ mol } e^-}{96,500 \text{ C}} = 55.96 \text{ mol } e^-$$

Next, let's calculate the hourly production rate of chlorine gas (in kg). Note that the anode efficiency is 93.0%.

$$55.96 \text{ mol } e^- \times \frac{1 \text{ mol Cl}_2}{2 \text{ mol } e^-} \times \frac{0.07090 \text{ kg Cl}_2}{1 \text{ mol Cl}_2} \times \frac{93.0\%}{100\%} = \mathbf{1.84 \text{ kg Cl}_2 \text{ / h}}$$

18.57   The quantity of charge passing through the solution is:

$$0.750 \text{ A} \times \frac{1 \text{ C}}{1 \text{ A} \cdot \text{s}} \times \frac{60 \text{ s}}{1 \text{ min}} \times \frac{1 \text{ mol } e-}{96,500 \text{ C}} \times 25.0 \text{ min} = 1.166 \times 10^{-2} \text{ mol } e^-$$

Since the charge of the copper ion is +2, the number of moles of copper formed must be:

$$(1.166 \times 10^{-2} \text{ mol } e^-) \times \frac{1 \text{ mol Cu}}{2 \text{ mol } e^-} = 5.83 \times 10^{-3} \text{ mol Cu}$$

The unit of molar mass is grams per mole. The molar mass of copper is:

$$\frac{0.369 \text{ g}}{5.83 \times 10^{-3} \text{ mol}} = \mathbf{63.3 \text{ g/mol}}$$

18.59   The number of faradays supplied is:

$$1.44 \text{ g Ag} \times \frac{1 \text{ mol Ag}}{107.9 \text{ g Ag}} \times \frac{1 \text{ mol } e^-}{1 \text{ mol Ag}} = 0.01335 \text{ mol } e^-$$

Since we need three faradays to reduce 1 mole of X^{3+}, the molar mass of X must be:

$$\frac{0.120 \text{ g X}}{0.01335 \text{ mol } e^-} \times \frac{3 \text{ mol } e^-}{1 \text{ mol X}} = \mathbf{27.0 \text{ g/mol}}$$

18.65   a. (i) The **half-reactions** are:

$$H_2(g) \longrightarrow 2H^+(aq) + 2e^-$$

$$Ni^{2+}(aq) + 2e^- \longrightarrow Ni(s)$$

(ii) The complete **balanced equation** is: $H_2(g) + Ni^{2+}(aq) \longrightarrow 2H^+(aq) + Ni(s)$

(iii) Ni(s) is below and to the right of $H^+(aq)$ in Table 18.1 of the text (see the half-reactions at −0.25 and 0.00 V). Therefore, the spontaneous reaction is the reverse of the above reaction; therefore, **the reaction will proceed to the left.**

b. (i) The **half-reactions** are:

$$5e^- + 8H^+(aq) + MnO_4^-(aq) \longrightarrow Mn^{2+}(aq) + 4H_2O(l)$$

$$2Cl^-(aq) \longrightarrow Cl_2(g) + 2e^-$$

(ii) The complete **balanced equation** is:

$$16H^+(aq) + 2MnO_4^-(aq) + 10Cl^-(aq) \longrightarrow 2Mn^{2+}(aq) + 8H_2O(l) + 5Cl_2(g)$$

(iii) In Table 18.1 of the text, $Cl^-(aq)$ is below and to the right of $MnO_4^-(aq)$; therefore, the spontaneous reaction is as written. **The reaction will proceed to the right.**

c. (i) The **half-reactions** are: $Cr(s) \longrightarrow Cr^{3+}(aq) + 3e^-$

$$Zn^{2+}(aq) + 2e^- \longrightarrow Zn(s)$$

(ii) The complete **balanced equation** is:

$$2Cr(s) + 3Zn^{2+}(aq) \longrightarrow 2Cr^{3+}(aq) + 3Zn(s)$$

(iii) In Table 18.1 of the text, Zn(s) is below and to the right of $Cr^{3+}(aq)$; therefore, the spontaneous reaction is the reverse of the reaction as written. **The reaction will proceed to the left.**

18.67   The balanced equation is:

$$5SO_2(g) + 2MnO_4^-(aq) + 2H_2O(l) \longrightarrow 5SO_4^{2-}(aq) + 2Mn^{2+}(aq) + 4H^+(aq)$$

The mass of SO_2 in the water sample is given by

$$7.37 \text{ mL} \times \frac{0.00800 \text{ mol}}{1000 \text{ mL soln}} \times \frac{5 \text{ mol SO}_2}{2 \text{ mol KMnO}_4} \times \frac{64.07 \text{ g SO}_2}{1 \text{ mol SO}_2} = \textbf{0.00944 g SO}_2$$

18.69   a. The balanced equation is:

$$2MnO_4^- + 6H^+ + 5H_2O_2 \longrightarrow 2Mn^{2+} + 8H_2O + 5O_2$$

b. The number of moles of potassium permanganate in 36.44 mL of the solution is:

$$36.44 \text{ mL} \times \frac{0.01652 \text{ mol}}{1000 \text{ mL soln}} = 6.020 \times 10^{-4} \text{ mol of KMnO}_4$$

From the balanced equation it can be seen that in this particular reaction 2 moles of permanganate is stoichiometrically equivalent to 5 moles of hydrogen peroxide. The number of moles of H_2O_2 oxidized is therefore

$$(6.020 \times 10^{-4} \text{ mol MnO}_4^-) \times \frac{5 \text{ mol H}_2O_2}{2 \text{ mol MnO}_4^-} = 1.505 \times 10^{-3} \text{ mol H}_2O_2$$

The molarity of H_2O_2 is:

$$[\mathbf{H_2O_2}] = \frac{1.505 \times 10^{-3} \text{ mol}}{25.00 \times 10^{-3} \text{ L}} = 0.06020 \text{ mol/L} = \mathbf{0.06020 \ M}$$

18.71   The balanced equation is:

$$2MnO_4^- + 5C_2O_4^{2-} + 16H^+ \longrightarrow 2Mn^{2+} + 10CO_2 + 8H_2O$$

Therefore, 2 mol MnO_4^- reacts with 5 mol $C_2O_4^{2-}$

$$\text{Moles of MnO reacted } = 24.2 \text{ mL} \times \frac{9.56 \times 10^{-4} \text{ mol MnO}_4^-}{1000 \text{ mL soln}} = 2.314 \times 10^{-5} \text{ mol MnO}_4^-$$

Recognize that the mole ratio of Ca^{2+} to $C_2O_4^{2-}$ is 1:1 in CaC_2O_4. The mass of Ca^{2+} in 10.0 mL is:

$$\left(2.314 \times 10^{-5} \text{ mol MnO}_4^-\right) \times \frac{5 \text{ mol Ca}^{2+}}{2 \text{ mol MnO}_4^-} \times \frac{40.08 \text{ g Ca}^{2+}}{1 \text{ mol Ca}^{2+}} = 2.319 \times 10^{-3} \text{ g Ca}^{2+}$$

Finally, converting to mg/mL, we have:

$$\frac{2.319 \times 10^{-3} \text{ g Ca}^{2+}}{10.0 \text{ mL}} \times \frac{1000 \text{ mg}}{1 \text{ g}} = \mathbf{0.232 \text{ mg Ca}^{2+}/mL \text{ blood}}$$

18.73   The solubility equilibrium of AgBr is:

$$AgBr(s) \rightleftharpoons Ag^+(aq) + Br^-(aq)$$

By reversing the first given half-reaction and adding it to the first, we obtain:

$$Ag(s) \longrightarrow Ag^+(aq) + e^- \qquad E_{anode}^\circ = 0.80 \text{ V}$$
$$\underline{AgBr(s) + e^- \longrightarrow Ag(s) + Br^-(aq) \qquad E_{cathode}^\circ = 0.07 \text{ V}}$$
$$AgBr(s) \rightleftharpoons Ag^+(aq) + Br^-(aq)$$

$$E_{cell}^\circ = E_{cathode}^\circ - E_{anode}^\circ = 0.07 \text{ V} - 0.80 \text{ V} = -0.73 \text{ V}$$

At equilibrium, we have:

$$E = E° - \frac{0.0592 \text{ V}}{n} \log[\text{Ag}^+][\text{Br}^-]$$

$$0 = -0.73 \text{ V} - \frac{0.0592 \text{ V}}{1} \log K_{sp}$$

$$\log K_{sp} = -12.3$$

$$\boldsymbol{K_{sp} = 10^{-12.3} = 5 \times 10^{-13}}$$

(Note that this value differs from that given in Table 17.4 of the text, since the data quoted here were obtained from a student's lab report.)

18.75 a. If this were a standard cell, the concentrations would all be 1.00 M, and the voltage would just be the standard emf calculated from Table 18.1 of the text. Since cell emf's depend on the concentrations of the reactants and products, we must use the Nernst equation (Equation 18.7 of the text) to find the emf of a nonstandard cell.

The standard emf is

$$E° = E°_{cathode} - E°_{anode} = 0.80 \text{ V} - (-2.37 \text{ V}) = 3.17 \text{ V}$$

$$E = E° - \frac{0.0592 \text{ V}}{n} \log Q$$

$$E = 3.17 \text{ V} - \frac{0.0592 \text{ V}}{2} \log \frac{[\text{Mg}^{2+}]}{[\text{Ag}^+]^2}$$

$$\boldsymbol{E} = 3.17 \text{ V} - \frac{0.0592 \text{ V}}{2} \log \frac{0.100}{(0.100)^2} = \boldsymbol{3.14 \text{ V}}$$

b. First we calculate the concentration of silver ion remaining in solution after the deposition of 1.20 g of silver metal.

Ag originally in solution: $\dfrac{0.100 \text{ mol Ag}^+}{1 \text{ L}} \times 0.346 \text{ L} = 3.46 \times 10^{-2} \text{ mol Ag}^+$

Ag deposited: $1.20 \text{ g Ag} \times \dfrac{1 \text{ mol}}{107.9 \text{ g}} = 1.11 \times 10^{-2} \text{ mol Ag}$

Ag remaining in solution: $(3.46 \times 10^{-2} \text{ mol Ag}) - (1.11 \times 10^{-2} \text{ mol Ag}) = 2.35 \times 10^{-2} \text{ mol Ag}$

$$[\text{Ag}^+] = \frac{2.35 \times 10^{-2} \text{ mol}}{0.346 \text{ L}} = 6.79 \times 10^{-2} \text{ M}$$

The overall reaction is: $\text{Mg}(s) + 2\text{Ag}^+(aq) \longrightarrow \text{Mg}^{2+}(aq) + 2\text{Ag}(s)$

We use the balanced equation to find the amount of magnesium metal suffering oxidation and dissolving.

$$(1.11 \times 10^{-2} \text{ mol Ag}) \times \frac{1 \text{ mol Mg}}{2 \text{ mol Ag}} = 5.55 \times 10^{-3} \text{ mol Mg}$$

The amount of magnesium originally in solution was:

$$0.288 \text{ L} \times \frac{0.100 \text{ mol}}{1 \text{ L}} = 2.88 \times 10^{-2} \text{ mol}$$

The new magnesium ion concentration is:

$$\frac{[(5.55 \times 10^{-3}) + (2.88 \times 10^{-2})] \text{mol}}{0.288 \text{ L}} = 0.119 \ M$$

The new cell emf is:

$$E = E^\circ - \frac{0.0592 \text{ V}}{n} \log Q$$

$$\boldsymbol{E} = 3.17 \text{ V} - \frac{0.0592 \text{ V}}{2} \log \frac{0.119}{\left(6.79 \times 10^{-2}\right)^2} = \textbf{3.13 V}$$

18.77 The cell voltage is given by:

$$E_{cell} = E^\circ - \frac{0.0592 \text{ V}}{2} \log \frac{[\text{Cu}^{2+}]_{\text{dilute}}}{[\text{Cu}^{2+}]_{\text{concentrated}}}$$

$$\boldsymbol{E_{cell}} = 0 \text{ V} - \frac{0.0592 \text{ V}}{2} \log \frac{0.080}{1.2} = \textbf{0.035 V}$$

18.79 Since this is a concentration cell, the standard emf is zero. Using Equation 18.6, we can write equations to calculate the cell voltage for the two cells.

$$(1) \qquad E_{cell} = -\frac{RT}{nF} \ln Q = -\frac{RT}{2F} \ln \frac{[\text{Hg}_2^{2+}]\text{soln A}}{[\text{Hg}_2^{2+}]\text{soln B}}$$

$$(2) \qquad E_{cell} = -\frac{RT}{nF} \ln Q = -\frac{RT}{1F} \ln \frac{[\text{Hg}^+]\text{soln A}}{[\text{Hg}^+]\text{soln B}}$$

In the first case, two electrons are transferred per mercury ion ($n = 2$), while in the second only one is transferred ($n = 1$). Note that the concentration ratio will be 1:10 in both cases. The voltages calculated at 18°C are:

$$(1) \qquad E_{cell} = \frac{-(8.314 \text{ J/K} \cdot \text{mol})(291 \text{ K})}{2\left(96,500 \text{ J} \cdot \text{V}^{-1}\text{mol}^{-1}\right)} \ln 10^{-1} = 0.0289 \text{ V}$$

(2) $$E_{cell} = \frac{-(8.314 \text{ J/K} \cdot \text{mol})(291 \text{ K})}{1\left(96,500 \text{ J} \cdot \text{V}^{-1}\text{mol}^{-1}\right)} \ln 10^{-1} = 0.0577 \text{ V}$$

Since the calculated cell potential for cell (1) agrees with the measured cell emf, we conclude that the **mercury(I) is Hg_2^{2+}.**

18.81   We begin by treating this like an ordinary stoichiometry problem (see Chapter 9).

Step 1:   Calculate the number of moles of Mg and Ag^+.

The number of moles of magnesium is:

$$1.56 \text{ g Mg} \times \frac{1 \text{ mol Mg}}{24.31 \text{ g Mg}} = 0.0642 \text{ mol Mg}$$

The number of moles of silver ion in the solution is:

$$\frac{0.100 \text{ mol Ag}^+}{1 \text{ L}} \times 0.1000 \text{ L} = 0.0100 \text{ mol Ag}^+$$

Step 2:   Calculate the mass of Mg remaining by determining how much Mg reacts with Ag^+.

The balanced equation for the reaction is:

$$2Ag^+(aq) + Mg(s) \longrightarrow 2Ag(s) + Mg^{2+}(aq)$$

Since you need twice as much Ag^+ compared to Mg for complete reaction, Ag^+ is the limiting reagent. The amount of Mg consumed is:

$$0.0100 \text{ mol Ag}^+ \times \frac{1 \text{ mol Mg}}{2 \text{ mol Ag}^+} = 0.00500 \text{ mol Mg}$$

The amount of magnesium remaining is:

$$(0.0642 - 0.00500) \text{ mol Mg} \times \frac{24.31 \text{ g Mg}}{1 \text{ mol Mg}} = \textbf{1.44 g Mg}$$

Step 3:   Assuming complete reaction, calculate the concentration of Mg^{2+} ions produced.

Since the mole ratio between Mg and Mg^{2+} is 1:1, the mol of Mg^{2+} formed will equal the mol of Mg reacted. The concentration of Mg^{2+} is:

$$[Mg^{2+}] = \frac{0.00500 \text{ mol}}{0.100 \text{ L}} = 0.0500 \ M$$

Step 4:   First, calculate the standard emf of the cell from the standard reduction potentials in Table 18.1 of the text. Then, we can calculate the equilibrium constant for the reaction from the standard cell emf using Equation 18.5 of the text.

$$E^{\circ}_{cell} = E^{\circ}_{cathode} - E^{\circ}_{anode} = 0.80 \text{ V} - (-2.37 \text{ V}) = 3.17 \text{ V}$$

We can then compute the equilibrium constant.

$$K = e^{\frac{nE^\circ_{cell}}{0.0257}}$$

$$K = e^{\frac{(2)(3.17)}{0.0257}} = 1 \times 10^{107}$$

Step 5: To find equilibrium concentrations of Mg^{2+} and Ag^+, we have to solve an equilibrium problem.

Let x be the small amount of Mg^{2+} that reacts to achieve equilibrium. The concentration of Ag^+ will be $2x$ at equilibrium. Assume that essentially all Ag^+ has been reduced so that the initial concentration of Ag^+ is zero.

	$2Ag^+ (aq)$	$+ Mg(s)$	$\rightleftarrows$	$2Ag(s)$	$+ Mg^{2+} (aq)$
Initial (M):	0.0000				0.0500
Change (M):	+2x				$-x$
Equilibrium (M):	2x				$(0.0500 - x)$

$$K = \frac{[Mg^{2+}]}{[Ag^+]^2}$$

$$1 \times 10^{107} = \frac{(0.0500 - x)}{(2x)^2}$$

We can assume $0.0500 - x \approx 0.0500$.

$$1 \times 10^{107} \approx \frac{0.0500}{(2x)^2}$$

$$(2x)^2 = \frac{0.0500}{1 \times 10^{107}} = 0.0500 \times 10^{-107}$$

$$(2x)^2 = 5.00 \times 10^{-109} = 50.0 \times 10^{-110}$$

$$2x = 7 \times 10^{-55} \, M$$

$$\mathbf{[Ag^+] = 2x = 7 \times 10^{-55} \, M}$$

$$\mathbf{[Mg^{2+}] = 0.0500 - x = 0.0500 \, M}$$

18.83 a. Since this is an acidic solution, the gas must be hydrogen gas ($\mathbf{H_2}$) from the reduction of hydrogen ion. The two electrode reactions and the overall cell reaction are:

anode: $Cu(s) \longrightarrow Cu^{2+}(aq) + 2e^-$

cathode: $2H^+(aq) + 2e^- \longrightarrow H_2(g)$

$$Cu(s) + 2H^+(aq) \longrightarrow Cu^{2+}(aq) + H_2(g)$$

Since 0.584 g of copper was consumed, the amount of hydrogen gas produced is:

$$0.584 \text{ g Cu} \times \frac{1 \text{ mol Cu}}{63.55 \text{ g Cu}} \times \frac{1 \text{ mol H}_2}{1 \text{ mol Cu}} = 9.20 \times 10^{-3} \text{ mol H}_2$$

At STP, 1 mole of an ideal gas occupies a volume of 22.41 L. Thus, the volume of hydrogen gas at STP is:

$$V_{H_2} = (9.20 \times 10^{-3} \text{ mol H}_2) \times \frac{22.41 \text{ L}}{1 \text{ mol}} = \textbf{0.206 L}$$

b. From the current and the time, we can calculate the amount of charge:

$$1.18 \text{ A} \times \frac{1 \text{ C}}{1 \text{ A} \cdot \text{s}} \times (1.52 \times 10^3 \text{ s}) = 1.79 \times 10^3 \text{ C}$$

Since we know the charge of an electron, we can compute the number of electrons.

$$(1.79 \times 10^3 \text{ C}) \times \frac{1 \, e^-}{1.6022 \times 10^{-19} \text{ C}} = 1.12 \times 10^{22} \, e^-$$

Using the amount of copper consumed in the reaction and the fact that 2 mol of e^- are produced for every 1 mole of copper consumed, we can calculate Avogadro's number.

$$\frac{1.12 \times 10^{22} \, e^-}{9.20 \times 10^{-3} \text{ mol Cu}} \times \frac{1 \text{ mol Cu}}{2 \text{ mol } e^-} = \textbf{6.09} \times \textbf{10}^{\textbf{23}} \, \boldsymbol{e^-}\textbf{/mol } \boldsymbol{e^-}$$

In practice, Avogadro's number can be determined by electrochemical experiments like this. The charge of the electron can be found independently by Millikan's experiment.

18.85   a. We can calculate $\Delta G°$ from standard free energies of formation using Appendix 2 of the text.

$$\Delta G° = 2\Delta G_f°(\text{N}_2) + 6\Delta G_f°(\text{H}_2\text{O}) - [4\Delta G_f°(\text{NH}_3) + 3\Delta G_f°(\text{O}_2)]$$

$$\Delta G° = 0 + (6)(-237.2 \text{ kJ/mol}) - [(4)(-16.6 \text{ kJ/mol}) + 0]$$

$$\boldsymbol{\Delta G° = -1356.8 \text{ kJ/mol}}$$

b. The half-reactions are:

$$4\text{NH}_3(g) \longrightarrow 2\text{N}_2(g) + 12\text{H}^+(aq) + 12e^-$$
$$3\text{O}_2(g) + 12\text{H}^+(aq) + 12e^- \longrightarrow 6\text{H}_2\text{O}(l)$$

The overall reaction is a 12-electron process. We can calculate the standard cell emf from the standard free-energy change, $\Delta G°$.

$$\Delta G° = -nFE°_{cell}$$

$$E^\circ_{cell} = \frac{-\Delta G^\circ}{nF} = \frac{-\left(\frac{-1356.8 \text{ kJ}}{1 \text{ mol}} \times \frac{1000 \text{ J}}{1 \text{ kJ}}\right)}{(12)(96{,}500 \text{ J/V} \cdot \text{mol})} = \textbf{1.17 V}$$

18.87   The reduction of Ag^+ to Ag metal is:

$$Ag^+(aq) + e^- \longrightarrow Ag$$

We can calculate both the moles of Ag deposited and the moles of Au deposited.

$$? \text{ mol Ag} = 2.64 \text{ g Ag} \times \frac{1 \text{ mol Ag}}{107.9 \text{ g Ag}} = 2.45 \times 10^{-2} \text{ mol Ag}$$

$$? \text{ mol Au} = 1.61 \text{ g Au} \times \frac{1 \text{ mol Au}}{197.0 \text{ g Au}} = 8.17 \times 10^{-3} \text{ mol Au}$$

We do not know the oxidation state of Au ions, so we will represent the ions as Au^{n+}. If we divide the mol of Ag by the mol of Au, we can determine the ratio of Ag^+ reduced compared to Au^{n+} reduced.

$$\frac{2.45 \times 10^{-2} \text{ mol Ag}}{8.17 \times 10^{-3} \text{ mol Au}} = 3$$

That is, the same number of electrons that reduced the Ag^+ ions to Ag reduced only one-third the number of moles of the Au^{n+} ions to Au. Thus, each Au^{n+} required three electrons per ion for every one electron for Ag^+. The oxidation state for the gold ion is **+3**; the ion is Au^{3+}.

$$Au^{3+}(aq) + 3e^- \longrightarrow Au$$

18.89   We reverse the first half-reaction and add it to the second to come up with the overall balanced equation.

$$Hg_2^{2+} \longrightarrow 2Hg^{2+} + 2e^- \qquad\qquad E^\circ_{anode} = +0.92 \text{ V}$$

$$\underline{Hg_2^{2+} + 2e^- \longrightarrow 2Hg \qquad\qquad E^\circ_{cathode} = +0.85 \text{ V}}$$

$$2Hg_2^{2+} \longrightarrow 2Hg^{2+} + 2Hg$$

$$E^\circ_{cell} = 0.85 \text{ V} - 0.92 \text{ V} = -0.07 \text{ V}$$

Since the standard cell potential is an intensive property,

$$Hg_2^{2+}(aq) \longrightarrow Hg^{2+}(aq) + Hg(l) \quad E^\circ_{cell} = -0.07 \text{ V}$$

We calculate ΔG° from E°.

$$\boldsymbol{\Delta G^\circ = -nFE^\circ} = -(1)(96{,}500 \text{ J/V} \cdot \text{mol})(-0.07 \text{ V}) = \textbf{6.8 kJ/mol}$$

The corresponding equilibrium constant is:

$$K = \frac{[Hg^{2+}]}{[Hg_2^{2+}]}$$

We calculate K from $\Delta G°$.

$$\Delta G° = -RT \ln K$$

$$\ln K = \frac{-6.8 \times 10^3 \text{ J/mol}}{(8.314 \text{ J/K} \cdot \text{mol})(298 \text{ K})}$$

$K = 0.064$

18.91   The reactions for the electrolysis of NaCl(aq) are:

Anode: $2Cl^-(aq) \longrightarrow Cl_2(g) + 2e^-$

Cathode: $2H_2O(l) + 2e^- \longrightarrow H_2(g) + 2OH^-(aq)$

Overall: $2H_2O(l) + 2Cl^-(aq) \longrightarrow H_2(g) + Cl_2(g) + 2OH^-(aq)$

From the pH of the solution, we can calculate the OH^- concentration. From the [OH^-], we can calculate the moles of OH^- produced. Then, from the moles of OH^- we can calculate the average current used.

$$\text{pH} = 12.24$$

$$\text{pOH} = 14.00 - 12.24 = 1.76$$

$$[OH^-] = 1.7 \times 10^{-2} \; M$$

The moles of OH^- produced are:

$$\frac{1.7 \times 10^{-2} \text{ mol}}{1 \text{ L}} \times 0.300 \text{ L} = 5.1 \times 10^{-3} \text{ mol } OH^-$$

From the balanced equation, it takes 1 mole of e^- to produce 1 mole of OH^- ions.

$$(5.1 \times 10^{-3} \text{ mol } OH^-) \times \frac{1 \text{ mol } e^-}{1 \text{ mol } OH^-} \times \frac{96,500 \text{ C}}{1 \text{ mol } e^-} = 490 \text{ C}$$

Recall that $1 \text{ C} = 1 \text{ A·s}$

$$490 \text{ C} \times \frac{1 \text{ A·s}}{1 \text{ C}} \times \frac{1 \text{ min}}{60 \text{ s}} \times \frac{1}{6.00 \text{ min}} = \textbf{1.4 A}$$

18.93   The reaction is:

$$Pt^{n+} + ne^- \longrightarrow Pt$$

Thus, we can calculate the charge of the platinum ions by realizing that n mol of e^- is required per mol of Pt formed.

The moles of Pt formed are:

$$9.09 \text{ g Pt} \times \frac{1 \text{ mol Pt}}{195.1 \text{ g Pt}} = 0.0466 \text{ mol Pt}$$

Next, calculate the charge passed in C.

$$C = 2.00 \text{ h} \times \frac{3600 \text{ s}}{1 \text{ h}} \times \frac{2.50 \text{ C}}{1 \text{ s}} = 1.80 \times 10^4 \text{ C}$$

Convert to moles of electrons.

$$? \text{ mol } e^- = \left(1.80 \times 10^4 \text{ C}\right) \times \frac{1 \text{ mol } e^-}{96,500 \text{ C}} = 0.187 \text{ mol } e^-$$

We now know the number of moles of electrons (0.187 mol e^-) needed to produce 0.0466 mol of Pt metal. We can calculate the number of moles of electrons needed to produce 1 mole of Pt metal.

$$\frac{0.187 \text{ mol } e^-}{0.0466 \text{ mol Pt}} = 4.01 \text{ mol } e^- / \text{mol Pt}$$

Since we need 4 moles of electrons to reduce 1 mole of Pt ions, the charge on the Pt ions must be **+4**.

18.95   The half-reaction for the oxidation of water to oxygen is:

$$2H_2O(l) \xrightarrow{\text{oxidation (anode)}} O_2(g) + 4H^+(aq) + 4e^-$$

Knowing that 1 mole of any gas at STP occupies a volume of 22.41 L, we find the number of moles of oxygen.

$$4.26 \text{ L } O_2 \times \frac{1 \text{ mol}}{22.41 \text{ L}} = 0.190 \text{ mol } O_2$$

Since four electrons are required to form one oxygen molecule, the number of electrons must be:

$$0.190 \text{ mol } O_2 \times \frac{4 \text{ mol } e^-}{1 \text{ mol } O_2} \times \frac{6.022 \times 10^{23} e^-}{1 \text{ mol}} = 4.58 \times 10^{23} e^-$$

The amount of charge passing through the solution is:

$$6.00 \text{ A} \times \frac{1 \text{ C}}{1 \text{ A} \cdot \text{s}} \times \frac{3600 \text{ s}}{1 \text{ h}} \times 3.40 \text{ h} = 7.34 \times 10^4 \text{ C}$$

We find the electron charge by dividing the amount of charge by the number of electrons.

$$\frac{7.34 \times 10^4 \text{ C}}{4.58 \times 10^{23} e^-} = \mathbf{1.60 \times 10^{-19} \text{ C}/e^-}$$

In actual fact, this sort of calculation can be used to find Avogadro's number, not the electron charge. The latter can be measured independently, and one can use this charge together with electrolytic data like the above to calculate the number of objects in 1 mole. See also Problem 18.81.

18.97   **Cells of higher voltage require very reactive oxidizing and reducing agents, which are difficult to handle. (From Table 18.1 of the text, we see that 5.92 V is the theoretical limit of a cell made up of Li^+/Li and F_2/F^- electrodes under standard-state conditions.) Batteries made up of several cells in series are easier to use.**

18.99 The half-reactions are:

$$Zn(s) + 4OH^-(aq) \longrightarrow Zn(OH)_4^{2-}(aq) + 2e^- \qquad E^\circ_{anode} = -1.36 \text{ V}$$

$$Zn^{2+}(aq) + 2e^- \longrightarrow Zn(s) \qquad E^\circ_{cathode} = -0.76 \text{ V}$$

$$\overline{Zn^{2+}(aq) + 4OH^-(aq) \longrightarrow Zn(OH)_4^{2-}(aq)}$$

$$E^\circ_{cell} = -0.76 \text{ V} - (-1.36 \text{ V}) = 0.60 \text{ V}$$

$$E^\circ_{cell} = -\frac{0.0592 \text{ V}}{n} \log K_f$$

$$K_f = 10^{nE^\circ/0.0592 \text{ V}} = 10^{(2)(0.60 \text{ V})/0.0592 \text{ V}} = 2 \times 10^{20}$$

18.101 a. Since electrons flow from X to SHE, E°_{red} **for X is negative. Thus** E°_{red} **for Y is positive.**

b.
$$Y^{2+} + 2e^- \longrightarrow Y \qquad E^\circ_{cathode} = 0.34 \text{ V}$$

$$X \longrightarrow X^{2+} + 2e^- \qquad E^\circ_{anode} = -0.25 \text{ V}$$

$$\overline{X + Y^{2+} \longrightarrow X^{2+} + Y} \quad E^\circ_{cell} = 0.34 \text{ V} - (-0.25 \text{ V}) = \textbf{0.59 V}$$

18.103 a. **Gold does not tarnish in air because the reduction potential for oxygen is not sufficiently positive to result in the oxidation of gold.**

$$O_2 + 4H^+ + 4e^- \longrightarrow 2H_2O \qquad E^\circ_{cathode} = 1.23 \text{ V}$$

That is, $E^\circ_{cell} = E^\circ_{cathode} - E^\circ_{anode} < 0,$ for either oxidation by O_2 to Au^+ or Au^{3+}.

$$E^\circ_{cell} = 1.23 \text{ V} - 1.50 \text{ V} < 0$$

or

$$E^\circ_{cell} = 1.23 \text{ V} - 1.69 \text{ V} < 0$$

b.
$$3(Au^+ + e^- \longrightarrow Au) \qquad E^\circ_{cathode} = 1.69 \text{ V}$$

$$Au \longrightarrow Au^{3+} + 3e^- \qquad E^\circ_{anode} = 1.50 \text{ V}$$

$$\overline{3Au^+ \longrightarrow 2Au + Au^{3+}} \qquad E^\circ_{cell} = 1.69 \text{ V} - 1.50 \text{ V} = 0.19 \text{ V}$$

Calculating ΔG,

$$\Delta G^\circ = -nFE^\circ = -(3)(96,500 \text{ J/V·mol})(0.19 \text{ V}) = -55.0 \text{ kJ/mol}$$

For spontaneous electrochemical equations, ΔG° must be negative. **Yes**, the disproportionation occurs spontaneously.

c. Since the most stable oxidation state for gold is Au^{3+}, the predicted reaction is:

$$\textbf{2Au} + \textbf{3F}_2 \longrightarrow \textbf{2AuF}_3$$

18.105 The balanced equation is:

$$5Fe^{2+} + MnO_4^- + 8H^+ \longrightarrow Mn^{2+} + 5Fe^{3+} + 4H_2O$$

Calculate the amount of iron(II) in the original solution using the mole ratio from the balanced equation.

$$23.0 \text{ mL} \times \frac{0.0200 \text{ mol KMnO}_4}{1000 \text{ mL soln}} \times \frac{5 \text{ mol Fe}^{2+}}{1 \text{ mol KMnO}_4} = 0.00230 \text{ mol Fe}^{2+}$$

The concentration of iron(II) must be:

$$[Fe^{2+}] = \frac{0.00230 \text{ mol}}{0.0250 \text{ L}} = \textbf{0.0920 } \textbf{\textit{M}}$$

The total iron concentration can be found by simple proportion because the same sample volume (25.0 mL) and the same $KMnO_4$ solution were used.

$$[Fe]_{total} = \frac{40.0 \text{ mL KMnO}_4}{23.0 \text{ mL KMnO}_4} \times 0.0920 \text{ } M = 0.160 \text{ } M$$

$$[Fe^{3+}] = [Fe]_{total} - [Fe^{2+}] = \textbf{0.0680 } \textbf{\textit{M}}$$

18.107 From Table 18.1 of the text.

$$H_2O_2(aq) + 2H^+(aq) + 2e^- \longrightarrow 2H_2O(l) \qquad E^\circ_{cathode} = 1.77 \text{ V}$$

$$H_2O_2(aq) \longrightarrow O_2(g) + 2H^+(aq) + 2e^- \qquad E^\circ_{anode} = 0.68 \text{ V}$$

$$2H_2O_2(aq) \longrightarrow 2H_2O(l) + O_2(g)$$

$$E^\circ_{cell} = E^\circ_{cathode} - E^\circ_{anode} = 1.77 \text{ V} - (0.68 \text{ V}) = 1.09 \text{ V}$$

Because E° is positive, the decomposition is spontaneous.

18.109 a. **unchanged**   b. **unchanged**   c. **squared**   d. **doubled**   e. **doubled**

18.111

$$F_2(g) + 2H^+(aq) + 2e^- \longrightarrow 2HF(g)$$

$$E = E^\circ - \frac{RT}{2F} \ln \frac{P_{HF}^2}{P_{F_2}[H^+]^2}$$

With increasing $[H^+]$, E will be larger. **As $[H^+]$ increases, F_2 does become a stronger oxidizing agent.**

18.113

$$Pb \longrightarrow Pb^{2+} + 2e^- \qquad\qquad E^\circ_{anode} = -0.13 \text{ V}$$

$$2H^+ + 2e^- \longrightarrow H_2 \qquad\qquad E^\circ_{cathode} = 0.00 \text{ V}$$

$$Pb + 2H^+ \longrightarrow Pb^{2+} + H_2 \qquad\qquad E^\circ_{cell} = 0.00 \text{ V} - (-0.13 \text{ V}) = 0.13 \text{ V}$$

$$pH = 1.60$$

$$[H^+] = 10^{-1.60} = 0.025 \text{ } M$$

$$E = E^\circ - \frac{RT}{nF}\ln\frac{[Pb^{2+}]P_{H_2}}{[H^+]^2}$$

$$0 = 0.13 - \frac{0.0592\text{ V}}{2}\log\frac{(0.035)P_{H_2}}{(0.025)^2}$$

$$4.39 = \log\frac{(0.035)P_{H_2}}{(0.025)^2}$$

$$P_{H_2} = 4.4 \times 10^2\text{ atm}$$

18.115 a. The half-reactions are:

Anode:	$\textbf{Zn} \longrightarrow \textbf{Zn}^{2+} + \textbf{2}e^-$
Cathode:	$\frac{1}{2}\textbf{O}_2 + \textbf{2}e^- \longrightarrow \textbf{O}^{2-}$

Overall:	$\text{Zn} + \frac{1}{2}\text{O}_2 \longrightarrow \text{ZnO}$

To calculate the standard emf, we first need to calculate ΔG° for the reaction. From Appendix 2 of the text we write:

$$\Delta G^\circ = \Delta G_f^\circ(\text{ZnO}) - \left[\Delta G_f^\circ(\text{Zn}) + \frac{1}{2}\Delta G_f^\circ(\text{O}_2)\right]$$

$$\Delta G^\circ = -318.2\text{ kJ/mol} - [0 + 0]$$

$$\Delta G^\circ = -318.2\text{ kJ/mol}$$

$$\Delta G^\circ = -nFE^\circ$$

$$-318.2 \times 10^3\text{ J/mol} = -(2)(96{,}500\text{ J/V·mol})E^\circ$$

$$E^\circ_{cell} = \textbf{1.65 V}$$

b. We use Equation 18.7 from the text:

$$E = E^\circ - \frac{0.0592\text{ V}}{n}\log Q$$

$$E = 1.65\text{ V} - \frac{0.0592\text{ V}}{2}\log\frac{1}{P_{O_2}}$$

$$E = 1.65\text{ V} - \frac{0.0592\text{ V}}{2}\log\frac{1}{0.21}$$

$$E = 1.65 \text{ V} - 0.020 \text{ V}$$

$$\boldsymbol{E = 1.63 \text{ V}}$$

c. Since the free-energy change represents the maximum work that can be extracted from the overall reaction, the maximum amount of energy that can be obtained from this reaction is the free-energy change. To calculate the energy density, we multiply the free-energy change by the number of moles of Zn present in 1 kg of Zn.

$$\textbf{energy density} = \frac{318.2 \text{ kJ}}{1 \text{ mol Zn}} \times \frac{1 \text{ mol Zn}}{65.41 \text{ g Zn}} \times \frac{1000 \text{ g Zn}}{1 \text{ kg Zn}} = \boldsymbol{4.86 \times 10^3 \text{ kJ / kg Zn}}$$

d. One ampere is 1 C/s. The charge drawn every second is given by nF.

$$\text{charge} = nF$$

$$2.1 \times 10^5 \text{ C} = n(96,500 \text{ C/mol } e^-)$$

$$n = 2.2 \text{ mol } e^-$$

From the overall balanced reaction, we see that 4 moles of electrons will reduce 1 mole of O_2; therefore, the number of moles of O_2 reduced by 2.2 moles of electrons is:

$$\text{mol } O_2 = 2.2 \text{ mol } e^- \times \frac{1 \text{ mol } O_2}{4 \text{ mol } e^-} = 0.55 \text{ mol } O_2$$

The volume of oxygen at 1.0 atm partial pressure can be obtained by using the ideal gas equation.

$$V_{O_2} = \frac{nRT}{P} = \frac{(0.55 \text{ mol})(0.0821 \text{ L} \cdot \text{atm/K} \cdot \text{mol})(298 \text{ K})}{(1.0 \text{ atm})} = 13.5 \text{ L}$$

Since air is 21 percent oxygen by volume, the volume of air required every second is:

$$V_{\text{air}} = 13.5 \text{ L} \times \frac{100\% \text{ air}}{21\% \text{ } O_2} = \boldsymbol{64 \text{ L air}}$$

18.117 We can calculate $\Delta G^\circ_{\text{rxn}}$ using the following equation (Equation 14.12).

$$\Delta G^\circ_{\text{rxn}} = \Sigma n \Delta G^\circ_{\text{f}}(\text{products}) - \Sigma m \Delta G^\circ_{\text{f}}(\text{reactants})$$

$$\Delta G^\circ_{\text{rxn}} = 0 + 0 - [(1)(-293.8 \text{ kJ/mol}) + 0] = 293.8 \text{ kJ/mol}$$

Next, we can calculate E° using the equation

$$\Delta G^\circ = -nFE^\circ$$

We use a more accurate value for Faraday's constant.

$$293.8 \times 10^3 \text{ J/mol} = -(1)(96485.338 \text{ J/V} \cdot \text{mol})E^\circ$$

$$\boldsymbol{E^\circ = -3.05 \text{ V}}$$

18.119 First, we need to calculate $E°_{cell}$, then we can calculate K from the cell potential.

$$H_2(g) \longrightarrow 2H^+(aq) + 2e^- \qquad\qquad E°_{anode} = 0.00 \text{ V}$$

$$2H_2O(l) + 2e^- \longrightarrow H_2(g) + 2OH^- \qquad E°_{cathode} = -0.83 \text{ V}$$

$$\overline{2H_2O(l) \longrightarrow 2H^+(aq) + 2OH^-(aq) \qquad E°_{cell} = -0.83 \text{ V} - 0.00 \text{ V} = -0.83 \text{ V}}$$

We want to calculate K for the reaction: $H_2O(l) \longrightarrow H^+(aq) + OH^-(aq)$. The cell potential for this reaction will be the same as the above reaction, but the moles of electrons transferred, n, will equal 1.

$$E°_{cell} = \frac{0.0592 \text{ V}}{n}\log K_w$$

$$\log K_w = \frac{nE°_{cell}}{0.0592 \text{ V}}$$

$$K_w = 10^{nE°_{cell}/0.0592 \text{ V}}$$

$$K_w = 10^{(1)(-0.83 \text{ V})/0.0592 \text{ V}} = 1 \times 10^{-14}$$

18.121 a.
$$\mathbf{1 \text{ Ah} = 1 \text{ A} \times 3600 \text{ s} = 3600 \text{ C}}$$

b. Anode:
$$Pb + SO_4^{2-} \longrightarrow PbSO_4 + 2e^-$$

Two moles of electrons are produced by 1 mole of Pb. Recall that the charge of 1 mol e^- is 96,500 C. We can set up the following conversions to calculate the capacity of the battery.

$$\text{mol Pb} \rightarrow \text{mol } e^- \rightarrow \text{coulombs} \rightarrow \text{ampere-hours}$$

$$406 \text{ g} \times \frac{1 \text{ mol Pb}}{207.2 \text{ g Pb}} \times \frac{2 \text{ mol } e^-}{1 \text{ mol Pb}} \times \frac{96,500 \text{ C}}{1 \text{ mol } e^-} = \left(3.78 \times 10^5 \text{ C}\right) \times \frac{1 \text{ h}}{3600 \text{ s}} = \mathbf{105 \text{ Ah}}$$

This ampere-hour cannot be fully realized because the concentration of H_2SO_4 keeps decreasing.

c.
$$E°_{cell} = 1.70 \text{ V} - (-0.31 \text{ V}) = \mathbf{2.01 \text{ V}} \qquad\qquad \text{(from Table 18.1 of the text)}$$

$$\Delta G° = -nFE°$$

$$\mathbf{\Delta G° = -(2)(96,500 \text{ J/V·mol})(2.01 \text{ V}) = -3.88 \times 10^5 \text{ J/mol} = -388 \text{ kJ/mol}}$$

18.123 The surface area of an open cylinder is $2\pi rh$. The surface area of the culvert is:

$$2\pi(0.900 \text{ m})(40.0 \text{ m}) \times 2 \text{ (for both sides of the iron sheet)} = 452 \text{ m}^2$$

Converting to units of cm^2,

$$452 \text{ m}^2 \times \left(\frac{100 \text{ cm}}{1 \text{ m}}\right)^2 = 4.52 \times 10^6 \text{ cm}^2$$

The volume of the Zn layer is:

$$0.200 \text{ mm} \times \frac{1 \text{ cm}}{10 \text{ mm}} \times \left(4.52 \times 10^6 \text{ cm}^2\right) = 9.04 \times 10^4 \text{ cm}^3$$

The mass of Zn needed is:

$$\left(9.04 \times 10^4 \text{ cm}^3\right) \times \frac{7.14 \text{ g}}{1 \text{ cm}^3} = 6.45 \times 10^5 \text{ g Zn}$$

$$Zn^{2+} + 2e^- \longrightarrow Zn$$

$$Q = \left(6.45 \times 10^5 \text{ g Zn}\right) \times \frac{1 \text{ mol Zn}}{65.41 \text{ g Zn}} \times \frac{2 \text{ mol } e^-}{1 \text{ mol Zn}} \times \frac{96,500 \text{ C}}{1 \text{ mol } e^-} = 1.90 \times 10^9 \text{ C}$$

$$1 \text{ J} = 1 \text{ C} \times 1 \text{ V}$$

$$\text{Total energy} = \frac{\left(1.90 \times 10^9 \text{ C}\right)(3.26 \text{ V})}{0.95} = 6.5 \times 10^9 \text{ J}$$

$$\text{Cost} = \left(6.5 \times 10^9 \text{ J}\right) \times \frac{1 \text{ kW}}{1000 \frac{\text{J}}{\text{s}}} \times \frac{1 \text{ h}}{3600 \text{ s}} \times \frac{\$0.12}{1 \text{ kWh}} = \mathbf{\$220}$$

18.125 It might appear that because the sum of the first two half-reactions gives Equation (3), E_3° is given by $E_1^\circ + E_2^\circ = 0.33$ V. This is not the case, however, because emf is not an extensive property. We cannot set $E_3^\circ = E_1^\circ + E_2^\circ$. On the other hand, the Gibbs energy is an extensive property, so we can add the separate Gibbs energy changes to obtain the overall Gibbs energy change.

$$\Delta G_3^\circ = \Delta G_1^\circ + \Delta G_2^\circ$$

Substituting the relationship $\Delta G^\circ = -nFE^\circ$, we obtain:

$$n_3 F E_3^\circ = n_1 F E_1^\circ + n_2 F E_2^\circ$$

$$E_3^\circ = \frac{n_1 E_1^\circ + n_2 E_2^\circ}{n_3}$$

$n_1 = 2$, $n_2 = 1$, and $n_3 = 3$.

$$\boldsymbol{E_3^\circ} = \frac{(2)(-0.44 \text{ V}) + (1)(0.77 \text{ V})}{3} = \mathbf{-0.037 \text{ V}}$$

18.127 First, calculate the standard emf of the cell from the standard reduction potentials in Table 18.1 of the text. Then, calculate the equilibrium constant from the standard emf using Equation 18.5 of the text. Consider the half-reactions as :

$$\text{Oxidation:} \qquad Zn \longrightarrow Zn^{2+} + 2e^- \qquad E^\circ_{\text{anode}} = -0.76 \text{ V}$$

$$\text{Reduction:} \qquad Cu^{2+} + 2e^- \longrightarrow Cu \qquad E^\circ_{\text{cathode}} = 0.34 \text{ V}$$

$$E^\circ_{\text{cell}} = E^\circ_{\text{cathode}} - E^\circ_{\text{anode}} = 0.34 \text{ V} - (-0.76 \text{ V}) = 1.10 \text{ V}$$

$$\log K = \frac{nE^\circ_{\text{cell}}}{0.0592 \text{ V}}$$

$$K = 10^{nE^\circ_{\text{cell}}/0.0592 \text{ V}} = 10^{(2)(1.10 \text{ V})/0.0592 \text{ V}}$$

$$\boldsymbol{K = 10^{37.2} = 2 \times 10^{37}}$$

The very large equilibrium constant means that the oxidation of Zn by Cu^{2+} is virtually complete.

18.129 **Strategy:** According to Section 14.6 of the text, 31 kJ of free energy are required to convert 1 mole ADP to ATP:

$$ADP + H_3PO_4 \longrightarrow ATP + H_2O \quad \Delta G^\circ = 31 \text{ kJ/mol}$$

To determine the number of moles of ATP that can be produced, calculate the amount of free energy from the oxidation of 1 mole of nitrite and divide this energy by 31 kJ/mol.

Solution: Assume that the nitrite, nitrate, ADP, and ATP are present in their standard states. The overall reaction for converting the nitrite to nitrate is

$$2NO_2^-(aq) + O_2(g) \longrightarrow 2NO_3^-(aq).$$

Use Equation 18.3 to relate ΔE° to Δ

$$\Delta G^\circ = -nF\Delta E^\circ = -(4 \text{ mol } e^-)(96,500 \text{ J/V} \cdot \text{mol } e^-)(1.23 \text{ V} - 0.42 \text{ V})$$

$$\Delta G^\circ = -310 \text{ kJ}$$

Since the oxidation of 2 moles of nitrite provide 310 kJ of free energy, then

$$\text{yield} = \left(\frac{310 \text{ kJ}}{2 \text{ mol NO}_2^-(aq)} \right) \left(\frac{1 \text{ mol ATP}}{31 \text{ kJ}} \right) = \textbf{5.0 mol ATP / mol NO}_2^-.$$

18.131 The reduction potential of a half-cell is temperature-dependent. **A small non-zero emf will appear if the temperatures of the two half-cells are different.**

Chapter 19
Chemical Kinetics

Visualizing Chemistry

VC 19.1   c   VC 19.2   b   VC 19.3   a   VC 19.4   c

Key Skills

19.1   d   19.2   e   19.3   b   19.4   a

Problems

19.9   **Strategy:**   The rate is defined as the change in concentration of a reactant or product with time. Each change-in-concentration term is divided by the corresponding stoichiometric coefficient. Terms involving *reactants* are preceded by a minus sign because reactant concentrations decrease as a reaction progresses—and *reaction* rates are always expressed as positive quantities.

$$\text{Rate} = -\frac{1}{2}\frac{\Delta[\text{NO}]}{\Delta t} \qquad \frac{\Delta[\text{NO}]}{\Delta t} = -0.066 \ M/s$$

Solution: a.   If the concentration of NO is changing at the rate of –0.066 *M*/s, the rate at which NO_2 is being formed is

$$-\frac{1}{2}\frac{\Delta[\text{NO}]}{\Delta t} = \frac{1}{2}\frac{\Delta[\text{NO}_2]}{\Delta t}$$

$$\frac{\Delta[\text{NO}_2]}{\Delta t} = 0.066 \ \text{M}/\text{s}$$

The rate at which NO_2 is forming is **0.066 *M*/s**.

b.

$$-\frac{1}{2}\frac{\Delta[\text{NO}]}{\Delta t} = -\frac{\Delta[\text{O}_2]}{\Delta t}$$

$$\frac{\Delta[\text{O}_2]}{\Delta t} = \frac{-0.066 \ M/s}{2} = -0.033 \ \text{M}/\text{s}$$

The negative sign indicates that the concentration of molecular oxygen is decreasing as the reaction progresses. The rate at which molecular oxygen is reacting is **0.033 *M*/s**.

19.11   In general for a reaction $a\text{A} + b\text{B} \longrightarrow c\text{C} + d\text{D}$

$$\text{rate} = -\frac{1}{a}\frac{\Delta[\text{A}]}{\Delta t} = -\frac{1}{b}\frac{\Delta[\text{B}]}{\Delta t} = \frac{1}{c}\frac{\Delta[\text{C}]}{\Delta t} = \frac{1}{d}\frac{\Delta[\text{D}]}{\Delta t}$$

a.

$$\textbf{rate} = -\frac{\Delta[\text{H}_2]}{\Delta t} = -\frac{\Delta[\text{I}_2]}{\Delta t} = \frac{1}{2}\frac{\Delta[\text{HI}]}{\Delta t}$$

b.

$$\text{rate} = -\frac{1}{5}\frac{\Delta[Br^-]}{\Delta t} = -\frac{\Delta[BrO_3^-]}{\Delta t} = -\frac{1}{6}\frac{\Delta[H^+]}{\Delta t} = \frac{1}{3}\frac{\Delta[Br_2]}{\Delta t}$$

19.19 $\text{rate} = k[NH_4^+][NO_2^-] = (3.0 \times 10^{-4} / M \cdot s)(0.36 \, M)(0.075 \, M) = \mathbf{8.1 \times 10^{-6} \, M/s}$

19.21   **Strategy:**   We are given a set of concentrations and rate data and asked to determine the order of the reaction and the value of the rate constant. To determine the order of the reaction, we need to find the rate law for the reaction. We assume that the rate law takes the form

$$\text{rate} = k[A]^x[B]^y$$

How do we use the data to determine x and y? Once the orders of the reactants are known, we can calculate k using the set of rate and concentrations from any one of the experiments.

Solution: By comparing the first and second sets of data, we see that changing [B] does not affect the rate of the reaction. Therefore, the reaction is zero order in B. By comparing the first and third sets of data, we see that doubling [A] doubles the rate of the reaction.

$$\frac{\text{rate}_3}{\text{rate}_1} = \frac{6.40 \times 10^{-1} \, M/s}{3.20 \times 10^{-1} \, M/s} = 2 = \frac{k\,(3.00)^x\,(1.50)^y}{k\,(1.50)^x\,(1.50)^y}$$

Therefore,

$$\frac{(3.00)^x}{(1.50)^x} = 2^x = 2$$

and $x = 1$. This shows that **the reaction is first-order in A and first-order overall**.

$$\text{rate} = k[A]$$

From the set of data for the first experiment:

$$3.20 \times 10^{-1} \, M/s = k(1.50 \, M)$$

$$\mathbf{k = 0.213 \, s^{-1}}$$

19.23   a. **2**   b. **0**   c. **2.5**   d. **3**

19.25   The graph below is a plot of ln *P* versus time. Since the plot is linear, the reaction is **first-order**.

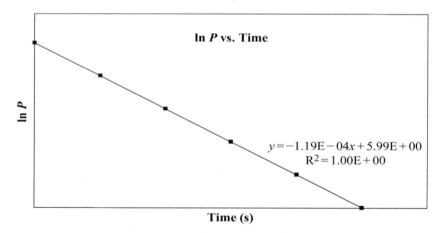

$y = -1.19E - 04x + 5.99E + 00$
$R^2 = 1.00E + 00$

slope $= -k$

$k = 1.19 \times 10^{-4} \text{ s}^{-1}$

19.31 We know that half of the substance decomposes in a time equal to the half-life, $t_{1/2}$. This leaves half of the compound. Half of what is left decomposes in a time equal to another half-life, so that only one-quarter of the original compound remains. We see that 75% of the original compound has decomposed after two half-lives. Thus two half-lives equal 1 hour, or the half-life of the decay is 30 min.

$$100\% \text{ starting compound} \xrightarrow{\ t_{1/2}\ } 50\% \text{ starting compound} \xrightarrow{\ t_{1/2}\ } 25\% \text{ starting compound}$$

Using first-order kinetics, we can solve for k using Equation 19.3 of the text, with $[A]_0 = 100$ and $[A] = 25$,

$$\ln \frac{[A]_t}{[A]_0} = -kt$$

$$\ln \frac{25}{100} = -k(60 \text{ min})$$

$$k = -\frac{\ln(0.25)}{60 \text{ min}} = 0.023 \text{ min}^{-1}$$

Then, substituting k into Equation 19.5 of the text, you arrive at the same answer for $t_{1/2}$.

$$t_{1/2} = \frac{0.693}{k} = \frac{0.693}{0.023 \text{ min}^{-1}} = \mathbf{30 \text{ min}}$$

19.33 a. Since the reaction is known to be second-order, the relationship between reactant concentration and time is given by Equation 19.6 of the text. The problem supplies the rate constant and the initial (time = 0) concentration of NOBr. The concentration after 22 s can be found easily.

$$\frac{1}{[\text{NOBr}]_t} = kt + \frac{1}{[\text{NOBr}]_0}$$

$$\frac{1}{[\text{NOBr}]_t} = (0.80/M \cdot \text{s})(22 \text{ s}) + \frac{1}{0.086 \ M}$$

$$\frac{1}{[\text{NOBr}]_t} = 29 \ M^{-1}$$

[NOBr] = 0.034 *M*

b. The half-life for a second-order reaction is dependent on the initial concentration. The half-lives can be calculated using Equation 19.7 of the text.

$$t_{1/2} = \frac{1}{k[A]_0}$$

$$t_{1/2} = \frac{1}{(0.80 / M \cdot \text{s})(0.072 \ M)}$$

$$t_{1/2} = \mathbf{17\,s}$$

For an initial concentration of 0.054 *M*, you should find $t_{1/2} = \mathbf{23\,s}$. Note that the half-life of a second-order reaction is inversely proportional to the initial reactant concentration.

19.35 a. The relative rates of the reaction in the three containers i, ii, and iii are **4:3:6**.

b. **The relative rates would be unaffected; each absolute rate would decrease by 50%.**

c. Because half-life of a first-order reaction does not depend on reactant concentration, the relative half-lives are **1:1:1**.

19.37 **Strategy:** For a zeroth-order reaction, we get a straight line if we plot reactant concentration [A] versus time. For a first-order reaction, we get a straight line if we plot the natural log of reactant concentration (ln [A]) versus time. For a second-order reaction, we obtain a straight line when we plot the reciprocal of reactant concentration (1/[A]) against time.

Solution: a. **first-order** reaction

b. **second-order** reaction

c. **zeroth-order** reaction

19.43 Using Equation 19.11,

$$\ln\frac{1}{2} = \frac{E_a}{R}\left(\frac{1}{T_2} - \frac{1}{T_1}\right)$$

Remember to convert temperatures to the Kelvin scale.

$$-0.693 = \frac{E_a}{8.314\times10^{-3}\ \text{kJ/K}\cdot\text{mol}}\left(\frac{1}{275.4\ \text{K}} - \frac{1}{272.1\ \text{K}}\right)$$

$$E_a = \frac{(-0.693)\left(8.314\times10^{-3}\ \text{kJ/K}\cdot\text{mol}\right)}{-4.4\times10^{-5}\text{K}^{-1}} = \mathbf{1.3 \times 10^2\ kJ/mol}$$

For maximum freshness, fish should be frozen immediately after capture and kept frozen until cooked.

19.45 **Strategy:** Equation 19.8 relates the rate constant to the frequency factor, the activation energy, and the temperature. Remember to make sure the units of *R* and E_a are consistent.

Solution: The appropriate value of *R* is 8.314 J/K mol, not 0.0821 L·atm/K·mol. You must also use the activation energy value of 63,000 J/mol. Once the temperature has been converted to Kelvin, the rate constant is:

$$k = Ae^{-E_a/RT} = \left(8.7\times10^{12}\ \text{s}^{-1}\right)e^{-\left[\frac{63{,}000\ \text{J/mol}}{(8.314\ \text{J/K·mol})(348\ \text{K})}\right]} = \left(8.7\times10^{12}\ \text{s}^{-1}\right)\left(3.5\times10^{-10}\right)$$

$$k = \mathbf{3.0 \times 10^3\ s^{-1}}$$

19.47   Let k_1 be the rate constant at 295 K and $2k_1$ the rate constant at 305 K. Using Equation 19.11, we write:

$$\ln\frac{k_1}{2k_1} = \frac{E_a}{R}\left(\frac{1}{T_2}-\frac{1}{T_1}\right)$$

$$-0.693 = \frac{E_a}{8.314 \text{ J/K}\cdot\text{mol}}\left(\frac{1}{305 \text{ K}}-\frac{1}{295 \text{ K}}\right)$$

$$\boldsymbol{E_a = 5.18\times10^4 \text{ J/mol} = \textbf{51.8 kJ/mol}}$$

19.49   Graphing Equation 19.10 of the text requires plotting ln k versus $1/T$. The graph is shown below.

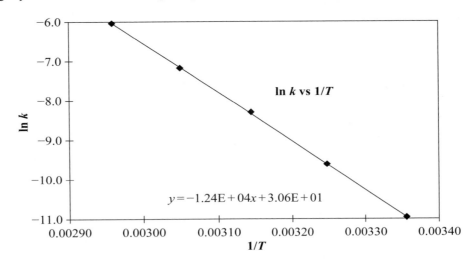

The slope of the line is -1.24×10^4 K, which is $-E_a/R$. The activation energy is:

$$-E_a = \text{slope}\times R = \left(-1.24\times10^4 \text{ K}\right)\times\left(8.314 \text{ J}/\text{K}\cdot\text{mol}\right)$$

$$\boldsymbol{E_a = 1.03\times10^5 \text{ J/mol} = \textbf{103 kJ/mol}}$$

19.51   **Strategy:**   Increasing the surface area of a solid reactant increases the reaction rate. The more finely divided a solid reactant is, the more surface area is exposed, and the more collisions that can take place with the aqueous reactant molecules.

         Increasing temperature also increases the reaction rate. As the temperature increases, so does the average kinetic energy of a sample of molecules. As a result, more molecules in the sample have sufficient kinetic energy to exceed the activation energy.

        Setup: The smaller the particle size, the greater the surface area. The crushed tablets will have a higher surface area and react faster than the tablets broken into pieces or the whole tablets.

         Tablets placed in very warm water will react faster than tablets placed in lukewarm or cold water.

    Solution: **In order of increasing time required for the effervescence to stop:**

$$\textbf{e} < \textbf{d} < \textbf{b} < \textbf{a} < \textbf{c}$$

19.61 a. The order of the reaction is simply the sum of the exponents in the rate law (Section 19.2 of the text). The reaction is **second-order**.

   b. The rate law reveals the identity of the substances participating in the slow or rate-determining step of a reaction mechanism. This rate law implies that the slow step involves the reaction of a molecule of NO with a molecule of Cl_2. If this is the case, then **the first step is the slower (rate-determining) step**.

19.63 The experimentally determined rate law is first-order in H_2 and second order in NO. In Mechanism I, the slow step is bimolecular and the rate law would be:

$$rate = k[H_2][NO]$$

Mechanism I can be discarded.

The rate-determining step in Mechanism II involves the simultaneous collision of two NO molecules with one H_2 molecule. The rate law would be:

$$rate = k[H_2][NO]^2$$

Mechanism II is a possibility.

In Mechanism III, we assume the forward and reverse reactions in the first fast step are in dynamic equilibrium, so their rates are equal:

$$k_f[NO]^2 = k_r[N_2O_2]$$

The slow step is bimolecular and involves collision of a hydrogen molecule with a molecule of N_2O_2. The rate would be:

$$rate = k_2[H_2][N_2O_2]$$

If we solve the dynamic equilibrium equation of the first step for $[N_2O_2]$ and substitute into the above equation, we have the rate law:

$$rate = \frac{k_2 k_f}{k_r}[H_2][NO]^2 = k[H_2][NO]^2$$

Mechanism III is also a possibility.

19.71 An enzyme is typically a large protein molecule that contains one or more active sites where interactions with substrates take place. **At higher temperatures the enzyme becomes denatured; that is, it loses its activity due to a change in the overall structure.** The rate of reaction will increase with increasing temperature until the enzyme denatures. Once this happens, the reaction rate drops off abruptly.

19.73 a. **termolecular**

   b. **unimolecular**

   c. **bimolecular**

19.75 Strictly, **the temperature must be specified** whenever the rate or rate constant of a reaction is quoted.

19.77 Using the diagrams, we see that the half-life of the reaction is 20 s. We rearrange Equation 19.5 to solve for rate constant:

$$k = \frac{0.693}{20\ s} = 0.035\ s^{-1}$$

19.79 **Most transition metals have several stable oxidation states. This allows the metal atoms to act as either a source or a receptor of electrons in a broad range of reactions.**

19.81 **Since the methanol contains no oxygen-18, the oxygen atom must come from the phosphate group and not the water. The mechanism must involve a bond-breaking process like:**

$$
\begin{array}{c}
\qquad\quad\ \ \overset{\displaystyle O}{\overset{\|}{}} \\
CH_3\!-\!O\!-\!\!\!\xi\!\!\!-\!\!\!P\!-\!O\!-\!H \\
\qquad\quad\ \overset{|}{O}\!-\!H
\end{array}
$$

19.83 **Temperature, energy of activation, concentration of reactants, and a catalyst.**

19.85 a. To determine the rate law, we must determine the exponents in the equation

$$\text{rate} = k[CH_3COCH_3]^x[Br_2]^y[H^+]^z$$

To determine the order of the reaction with respect to CH_3COCH_3, find two experiments in which the $[Br_2]$ and $[H^+]$ are held constant. Compare the data from experiments (1) and (5). When the concentration of CH_3COCH_3 is increased by a factor of 1.33, the reaction rate increases by a factor of 1.33. Thus, the reaction is <u>first-order</u> in CH_3COCH_3.

To determine the order with respect to Br_2, compare experiments (1) and (2). When the Br_2 concentration is doubled, the reaction rate does not change. Thus, the reaction is <u>zero order</u> in Br_2.

To determine the order with respect to H^+, compare experiments (1) and (3). When the H^+ concentration is doubled, the reaction rate doubles. Thus, the reaction is <u>first-order</u> in H^+.

The rate law is:

$$\textbf{rate} = \textbf{\textit{k}}[\textbf{CH}_3\textbf{COCH}_3][\textbf{H}^+]$$

b. Rearrange the rate law from part (a), solving for k.

$$k = \frac{\text{rate}}{[CH_3COCH_3][H^+]}$$

Substitute the data from any one of the experiments to calculate k. Using the data from Experiment (1),

$$\textbf{\textit{k}} = \frac{5.7 \times 10^{-5}\ M/s}{(0.30\ M)(0.050\ M)} = \textbf{3.8} \times \textbf{10}^{-3} / \textbf{\textit{M}} \cdot \textbf{s}$$

(The unit /M·s can also be expressed as $M^{-1}s^{-1}$.)

c. Let k_2 be the rate constant for the slow step:

$$\text{rate} = k_2[CH_3\!-\!\overset{\overset{\displaystyle +OH}{\|}}{C}\!-\!CH_3][H_2O] \qquad (1)$$

Let k_1 and k_{-1} be the rate constants for the forward and reverse steps in the fast equilibrium.

$$k_1[CH_3COCH_3][H_3O^+] = k_{-1}[CH_3\overset{\overset{+OH}{\|}}{C}\!-\!CH_3][H_2O] \qquad (2)$$

Therefore, Equation (1) becomes

$$\text{rate} = \frac{k_1 k_2}{k_{-1}}[CH_3COCH_3][H_3O^+]$$

which is the same as (a), where $k = k_1 k_2/k_{-1}$.

19.87

Fe^{3+} oxidizes I^-:	$2Fe^{3+} + 2I^- \longrightarrow 2Fe^{2+} + I_2$
Fe^{2+} reduces $S_2O_8^{2-}$:	$2Fe^{2+} + S_2O_8^{2-} \longrightarrow 2Fe^{3+} + 2SO_4^{2-}$
Overall reaction:	$2I^- + S_2O_8^{2-} \longrightarrow I_2 + 2SO_4^{2-}$

(Fe^{3+} undergoes a redox cycle: $\qquad Fe^{3+} \longrightarrow Fe^{2+} \longrightarrow Fe^{3+}$ **)**

The uncatalyzed reaction is slow because both I^- and $S_2O_8^{2-}$ are negatively charged, which makes their mutual approach unfavorable.

19.89   For a rate law, *zero order* means that the exponent is zero. In other words, the reaction rate is just equal to a constant; it does not change as time passes.

a. **(i):** The rate law would be:

$$\text{rate} = k[A]^0 = k$$

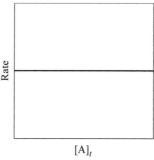

(ii): The integrated zero-order rate law is: $[A]_t = -kt + [A]_0$. Therefore, a plot of $[A]_t$ versus time should be a straight line with a slope equal to $-k$.

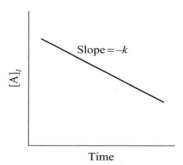

b.
$$[A]_t = [A]_0 - kt$$

At $t_{1/2}$, $[A]_t = \dfrac{[A]_0}{2}$. Substituting into the above equation:

$$\frac{[A]_0}{2} = [A]_0 - kt_{1/2}$$

$$t_{1/2} = \frac{[A]_0}{2k}$$

$$k = \frac{[A]_0}{2t_{1/2}}$$

c. When $[A]_t = 0$,

$$[A]_0 = kt$$

$$t = \frac{[A]_0}{k}$$

Substituting for k,

$$t = \frac{[A]_0}{[A]_0 / 2t_{1/2}}$$

$$t = 2t_{1/2}$$

This indicates that the integrated rate law is no longer valid after *two* half-lives.

19.91 **There are three gases present and we can measure only the total pressure of the gases. To measure the partial pressure of azomethane at a particular time, we must withdraw a sample of the mixture, analyze and determine the mole fractions. Then,**

$$\boldsymbol{P_{\text{azomethane}} = P_{\text{T}}X_{\text{azomethane}}}$$

This is a rather tedious process if many measurements are required. A mass spectrometer will help (see Section 2.5 of the text).

19.93

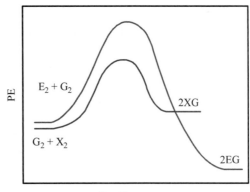

Reaction progress

19.95   a. **A catalyst works by changing the reaction mechanism, thus lowering the activation energy**.

      b. **A catalyst changes the reaction mechanism.**

      c. **A catalyst does not change the enthalpy of reaction.**

      d. **A catalyst increases the forward rate of reaction.**

      e. **A catalyst increases the reverse rate of reaction.**

19.97   **At very high $[H_2]$,**

$$k_2[H_2] \gg 1$$

$$\text{rate} = \frac{k_1[NO]^2[H_2]}{k_2[H_2]} = \frac{k_1}{k_2}[NO]^2$$

At very low $[H_2]$,

$$k_2[H_2] \ll 1$$

$$\text{rate} = \frac{k_1[NO]^2[H_2]}{1} = k_1[NO]^2[H_2]$$

The result from Problem 19.80 agrees with the rate law determined for low $[H_2]$.

19.99   First we plot the data for the reaction: $2N_2O_5 \longrightarrow 4NO_2 + O_2$

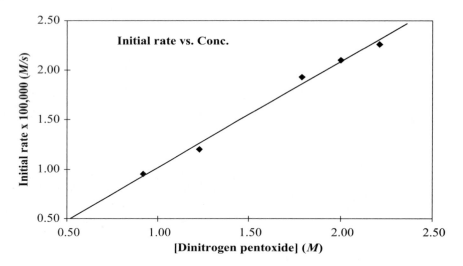

The data are linear, which means that the initial rate is directly proportional to the concentration of N_2O_5.

Thus, the rate law is:

$$\text{rate} = k[N_2O_5]$$

The rate constant k can be determined from the slope of the graph $\left(\dfrac{\Delta(\text{Initial rate})}{\Delta[N_2O_5]} \right)$ or by using any set of data.

$$k = 1.0 \times 10^{-5} \text{ s}^{-1}$$

Note that the rate law is *not* Rate = $k[N_2O_5]^2$, as we might expect from the balanced equation. In general, the order of a reaction must be determined by experiment; it cannot be deduced from the coefficients in the balanced equation.

19.101 **The red bromine vapor absorbs photons of blue light and dissociates to form bromine atoms.**

$$Br_2 \longrightarrow 2Br\cdot$$

The bromine atoms collide with methane molecules and abstract hydrogen atoms.

$$Br\cdot + CH_4 \longrightarrow HBr + \cdot CH_3$$

The methyl radical then reacts with Br$_2$, giving the observed product and regenerating a bromine atom to start the process over again:

$$\cdot CH_3 + Br_2 \longrightarrow CH_3Br + Br\cdot$$

$$Br\cdot + CH_4 \longrightarrow HBr + \cdot CH_3 \text{ and so on.}$$

19.103 **Lowering the body temperature would slow down all chemical reactions, which would be especially important for those that might damage the brain.**

19.105 a. We can write the rate law for an elementary step directly from the stoichiometry of the balanced reaction. In this rate-determining elementary step, three molecules must collide simultaneously (one X and two Y's). This makes the reaction termolecular, and consequently the rate law must be third-order: first-order in X and second-order in Y.

The rate law is:

$$\text{rate} = k[X][Y]^2$$

b. The value of the rate constant can be found by solving algebraically for k.

$$k = \frac{\text{rate}}{[X][Y]^2} = \frac{3.8 \times 10^{-3} \, M/s}{(0.26 \, M)(0.88 \, M)^2} = 1.9 \times 10^{-2} \text{ M}^{-2}\text{s}^{-1} \text{ or } \mathbf{0.019 \, M^{-2}s^{-1}}$$

19.107

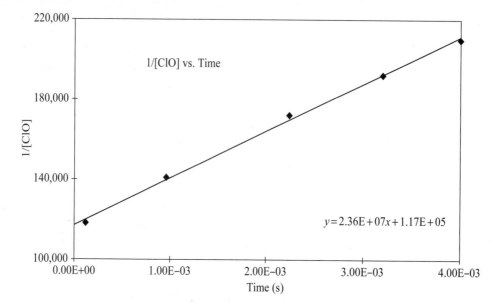

Reaction is **second-order** because a plot of 1/[ClO] versus time is a straight line. The slope of the line equals the rate constant, k.

k = slope = 2.4×10^7 /M·s or **$2.4 \times 10^7\ M^{-1}s^{-1}$**

19.109 **During the first five minutes or so the engine is relatively cold, so the exhaust gases will not fully react with the components of the catalytic converter. Remember, for almost all reactions, the rate of reaction increases with temperature.**

19.111 a. The dependence of the rate constant of a reaction on temperature can be expressed by the Arrhenius

equation (Equation 19.8), $k = Ae^{-E_a/RT}$. If the rate, k, changes significantly with a small change in temperature, T, then **the activation energy must be high.**

b. If a bimolecular reaction occurs every time an A molecule and a B molecule collide, **the molecules must have the appropriate orientation, and the energy that the colliding particles possess must be greater than the activation energy.**

19.113 First, solve for the rate constant, k, from the half-life of the decay.

$$t_{1/2} = 2.44 \times 10^5\ \text{yr} = \frac{0.693}{k}$$

$$k = \frac{0.693}{2.44 \times 10^5\ \text{yr}} = 2.84 \times 10^{-6}\ \text{yr}^{-1}$$

Now, we can calculate the time for the plutonium to decay from 5.0×10^2 g to 1.0×10^2 g using the equation for a first-order reaction relating concentration and time.

$$\ln \frac{[A]_t}{[A]_0} = -kt$$

$$\ln\frac{1.0\times10^2}{5.0\times10^2} = -\left(2.84\times10^{-6}\ yr^{-1}\right)t$$

$$-1.61 = -\left(2.84\times10^{-6}\ yr^{-1}\right)t$$

$$t = 5.7\times10^5\ yr$$

19.115 a. **Catalyst: Mn^{2+}; intermediates: Mn^{3+}, Mn^{4+}**

   First step is rate-determining.

   b. **Without the catalyst, the reaction would be a termolecular one involving three cations! (one Tl^+ and two Ce^{4+}). The reaction would be slow.**

   c. **The catalysis is homogeneous because the product has the same phase (aqueous) as the reactants.**

19.117 Initially, the number of moles of gas in terms of the volume is:

$$n = \frac{PV}{RT} = \frac{(0.350\ atm)V}{\left(0.08206\dfrac{L\cdot atm}{K\cdot mol}\right)(450 + 273)K} = 5.90\times10^{-3}\ V$$

We can calculate the concentration of dimethyl ether from the following equation.

$$\ln\frac{[(CH_3)_2O]_t}{[(CH_3)_2O]_0} = -kt$$

$$\frac{[(CH_3)_2O]_t}{[(CH_3)_2O]_0} = e^{-kt}$$

Since, the volume is held constant, it will cancel out of the equation. The concentration of dimethyl ether after 8.0 min (480 s) is:

$$[(CH_3)_2O]_t = \left(\frac{5.90\times10^{-3}\ V}{V}\right) e^{-\left(3.2\times10^{-4}\frac{1}{s}\right)(480\ s)}$$

$$[(CH_3)_2O]_t = 5.06\times10^{-3}\ M$$

After 8.0 min, the concentration of $(CH_3)_2O$ has decreased by $(5.90\times10^{-3} - 5.06\times10^{-3})\ M$ or $8.4\times10^{-4}\ M$. Since three moles of product form for each mole of dimethyl ether that reacts, the concentrations of the products are $(3)(8.4\times10^{-4}\ M) = 2.5\times10^{-3}\ M$.

The pressure of the system after 8.0 min is:

$$P = \frac{nRT}{V} = \left(\frac{n}{V}\right)RT = MRT$$

$$P = \left[\left(5.06\times10^{-3}\right) + \left(2.5\times10^{-3}\right)\right]M\times(0.08206\ L\cdot atm\ /\ K\cdot mol)(723\ K)$$

$$P = 0.45\ atm$$

19.119 a.
$$\frac{\Delta[B]}{\Delta t} = k_1[A] - k_2[B]$$

b.
If $\frac{\Delta[B]}{\Delta t} = 0$, then, from part (a) of this problem:

$$k_1[A] = k_2[B]$$

$$[B] = \frac{k_1}{k_2}[A]$$

19.121 a. The first-order rate constant can be determined from the half-life.

$$t_{1/2} = \frac{0.693}{k}$$

$$k = \frac{0.693}{t_{1/2}} = \frac{0.693}{28.1 \text{ yr}} = \mathbf{0.0247 \text{ yr}^{-1}}$$

b. See Problem 19.94. Mathematically, the amount left after ten half-lives is:

$$\left(\frac{1}{2}\right)^{10} = \mathbf{9.8 \times 10^{-4}}$$

c. If 99.0% has disappeared, then 1.0% remains. The ratio of $[A]_t/[A]_0$ is 1.0%/100% or 0.010/1.00. Substitute into the first-order integrated rate law, Equation 19.3 of the text, to determine the time.

$$\ln\frac{[A]_t}{[A]_0} = -kt$$

$$\ln\frac{0.010}{1.0} = -\left(0.0247 \text{ yr}^{-1}\right)t$$

$$-4.61 = -\left(0.0247 \text{ yr}^{-1}\right)t$$

$$\mathbf{t = 187 \text{ yr}}$$

19.123 a. There are **three** elementary steps: $A \longrightarrow B$, $B \longrightarrow C$, and $C \longrightarrow D$.

b. There are **two** intermediates: B and C.

c. **The third step, $C \longrightarrow D$, is rate determining** because it has the largest activation energy.

d. The overall reaction is **exothermic**.

19.125 Let $k_{cat} = k_{uncat}$

Then,

$$Ae^{\frac{-E_a(cat)}{RT_1}} = Ae^{\frac{-E_a(uncat)}{RT_2}}$$

Since the frequency factor is the same, we can write:

$$e^{\frac{-E_a(cat)}{RT_1}} = e^{\frac{-E_a(uncat)}{RT_2}}$$

Taking the natural log (*ln*) of both sides of the equation gives:

$$\frac{-E_a(cat)}{RT_1} = \frac{-E_a(uncat)}{RT_2}$$

or

$$\frac{E_a(cat)}{T_1} = \frac{E_a(uncat)}{T_2}$$

Substituting the given values into the equation:

$$\frac{7.0 \text{ kJ/mol}}{293 \text{ K}} = \frac{42 \text{ kJ/mol}}{T_2}$$

$$T_2 = 1.8 \times 10^3 \text{ K}$$

This temperature is too much high to be practical.

19.127 a. The rate law for the reaction is:

$$\text{rate} = k[\text{Hb}][\text{O}_2]$$

We are given the rate constant and the concentration of Hb and O_2, so we can substitute in these quantities to solve for rate of formation of HbO_2

$$\text{rate} = \left(2.1 \times 10^6 \, / \, M \cdot s\right)\left(8.0 \times 10^{-6} \, M\right)\left(1.5 \times 10^{-6} \, M\right)$$
$$\text{rate} = 2.5 \times 10^{-5} \, M/s$$

b. If HbO_2 is being formed at the rate of 2.5×10^{-5} *M*/s, then O_2 is being consumed at the same rate, **2.5×10^{-5} *M*/s**. Note the 1:1 mole ratio between O_2 and HbO_2.

c. The rate of formation of HbO_2 increases, but the concentration of Hb remains the same. Assuming that temperature is constant, we can use the same rate constant as in part (a). We substitute rate, [Hb], and the rate constant into the rate law to solve for O_2 concentration.

$$\text{rate} = k[\text{Hb}][\text{O}_2]$$

$$1.4 \times 10^{-4} \, M \, / \, s = \left(2.1 \times 10^6 \, / \, M \cdot s\right)\left(8.0 \times 10^{-6} \, M\right)[\text{O}_2]$$
$$[\text{O}_2] = 8.3 \times 10^{-6} \, M$$

19.129

$$t_{1/2} \propto \frac{1}{[A]_0^{n-1}}$$

$$t_{1/2} = C \frac{1}{[A]_0^{n-1}}, \text{ where } C \text{ is a proportionality constant.}$$

Substituting in for zero-, first-, and second-order reactions gives:

$$n = 0, \qquad t_{1/2} = C \frac{1}{[A]_0^{-1}} = C[A]_0$$

$$n = 1, \qquad t_{1/2} = C \frac{1}{[A]_0^{0}} = C$$

$$n = 2, \qquad t_{1/2} = C \frac{1}{[A]_0}$$

19.131 a. The units of the rate constant show the reaction to be second-order, meaning the rate law is most likely:

$$\text{rate} = k[H_2][I_2]$$

We can use the ideal gas equation to solve for the concentrations of H_2 and I_2. We can then solve for the initial rate in terms of H_2 and I_2 and then convert to the initial rate of formation of HI. We carry an extra significant figure throughout this calculation to minimize rounding errors.

$$n = \frac{PV}{RT}$$

$$\frac{n}{V} = M = \frac{P}{RT}$$

Since the total pressure is 1658 mmHg and there are equimolar amounts of H_2 and I_2 in the vessel, the partial pressure of each gas is 829 mmHg.

$$[H_2] = [I_2] = \frac{\left(829 \text{ mmHg} \times \dfrac{1 \text{ atm}}{760 \text{ mmHg}}\right)}{\left(0.0821 \dfrac{L \cdot atm}{K \cdot mol}\right)(400 + 273)\,K} = 0.01974 \ M$$

Let's convert the units of the rate constant to /M·min, and then we can substitute into the rate law to solve for rate.

$$k = 2.42 \times 10^{-2} \frac{1}{M \cdot s} \times \frac{60 \text{ s}}{1 \text{ min}} = 1.452 \frac{1}{M \cdot min}$$

$$\text{rate} = k[H_2][I_2]$$

$$\text{rate} = \left(1.452 \frac{1}{M \cdot min}\right)(0.01974 \ M)(0.01974 \ M) = 5.658 \times 10^{-4} \ M/min$$

We know that,

$$\text{rate} = \frac{1}{2}\frac{\Delta[\text{HI}]}{\Delta t}$$

or

$$\frac{\Delta[\text{HI}]}{\Delta t} = 2 \times \text{rate} = (2)(5.658 \times 10^{-4}\ M/\text{min}) = \mathbf{1.13 \times 10^{-3}\ M / min}$$

b. We can use the second-order integrated rate law to calculate the concentration of H_2 after 10.0 min. We can then substitute this concentration back into the rate law to solve for rate.

$$\frac{1}{[\text{H}_2]_t} = kt + \frac{1}{[\text{H}_2]_0}$$

$$\frac{1}{[\text{H}_2]_t} = \left(1.452\frac{1}{M \cdot \text{min}}\right)(10.0\ \text{min}) + \frac{1}{0.01974\ M}$$

$$[\text{H}_2]_t = 0.01534\ M$$

We can now substitute this concentration back into the rate law to solve for rate. The concentration of I_2 after 10.0 min will also equal 0.01534 *M*.

$$\text{rate} = k[\text{H}_2][\text{I}_2]$$

$$\text{rate} = \left(1.452\frac{1}{M \cdot \text{min}}\right)(0.01534\ M)(0.01534\ M) = 3.417 \times 10^{-4}\ M/\text{min}$$

We know that,

$$\text{rate} = \frac{1}{2}\frac{\Delta[\text{HI}]}{\Delta t}$$

or

$$\frac{\Delta[\text{HI}]}{\Delta t} = 2 \times \text{rate} = (2)\left(3.417 \times 10^{-4}\ M/\text{min}\right) = \mathbf{6.83 \times 10^{-4}\ M / min}$$

The concentration of HI after 10.0 min is:

$$[\text{HI}]_t = ([\text{H}_2]_0 - [\text{H}_2]_t) \times 2$$

$$\mathbf{[HI]}_t = (0.01974\ M - 0.01534\ M) \times 2 = \mathbf{8.80 \times 10^{-3}\ M}$$

19.133 The half-life is related to the initial concentration of A by

$$t_{1/2} \propto \frac{1}{[\text{A}]_0^{n-1}}$$

According to the given data, the half-life is doubled when $[A]_0$ is halved. This is possible only if the half-life is inversely proportional to $[A]_0$. Substituting $n = 2$ into the above equation gives:

$$t_{1/2} \propto \frac{1}{[A]_0}$$

Looking at this equation, it is clear that if $[A]_0$ is halved, the half-life would double. The reaction is **second-order**.

We use Equation 19.7 of the text to calculate the rate constant.

$$t_{1/2} = \frac{1}{k[A]_0}$$

$$k = \frac{1}{[A]_0 t_{1/2}} = \frac{1}{(1.20\ M)(2.0\ \text{min})} = \mathbf{0.42\ /\ M \cdot \text{min}}$$

19.135   From Equation 19.11 of the text,

$$\ln \frac{k_1}{k_2} = \frac{E_a}{R}\left(\frac{1}{T_2} - \frac{1}{T_1}\right)$$

$$\ln\left(\frac{k_1}{k_2}\right) = \frac{2.4 \times 10^5\ \text{J/mol}}{8.314\ \text{J/K} \cdot \text{mol}}\left(\frac{1}{606\ \text{K}} - \frac{1}{600\ \text{K}}\right)$$

$$\ln\left(\frac{k_1}{k_2}\right) = -0.48$$

$$\frac{k_2}{k_1} = e^{0.48} = 1.6$$

The rate constant at 606 K is 1.6 times greater than that at 600 K. This is a **60% increase** in the rate constant for a 1% increase in temperature! **The result shows the profound effect of an exponential dependence**.

19.137   **Strategy:**   Increasing the surface area of a solid reactant increases the reaction rate. The more finely divided a solid reactant is, the more surface area is exposed, and the more collisions that can take place with the aqueous reactant molecules.

Setup: The form of solid magnesium with the largest surface area (smallest particle size) will react fastest with the acid.

Solution: Form (b)

Chapter 20
Nuclear Chemistry

Visualizing Chemistry

VC 20.1   c   VC 20.2   b   VC 20.3   a   VC 20.4   a

Problems

20.5   **Strategy:**   In balancing nuclear equations, note that the sum of atomic numbers and mass numbers must match on both sides of the equation.

Solution: a.   On the left side of this equation the sum of the atomic numbers is 13 (12 + 1) and the sum of the mass numbers is 27 (26 + 1). These sums must be the same on the right side. Remember that the atomic and mass numbers of an alpha particle are 2 and 4, respectively. The atomic number of X is therefore 11 (13 − 2), and the mass number is 23 (27 − 4). X is sodium-23 ($^{23}_{11}\text{Na}$).

b.   On the left side of this equation the sum of the atomic numbers is 28 (27 + 1) and the sum of the mass numbers is 61 (59 + 2). These sums must be the same on the right side. The atomic number of X is therefore 1 (28 − 27), and the mass number is also 1 (61 − 60). X is a proton ($^{1}_{1}\text{p}$ or $^{1}_{1}\text{H}$).

c.   On the left side of this equation the sum of the atomic numbers is 92 (92 + 0) and the sum of the mass numbers is 236 (235 + 1). These sums must be the same on the right side. The atomic number of X is therefore 0 $\left[\dfrac{92-(36+56)}{3}\right]$, and the mass number is 1 $\left[\dfrac{236-(94+139)}{3}\right]$. X is a neutron ($^{1}_{0}\text{n}$).

d.   On the left side of this equation the sum of the atomic numbers is 26 (24 + 2) and the sum of the mass numbers is 57 (53 + 4). These sums must be the same on the right side. The atomic number of X is therefore 26 (26 − 0), and the mass number is 56 (57 − 1). X is iron-56 ($^{56}_{26}\text{Fe}$).

e.   On the left side of this equation the sum of the atomic numbers is 8 and the sum of the mass numbers is 20. The sums must be the same on the right side. The atomic number of X is therefore −1 (8 − 9), and the mass number is 0 (20 − 20). X is a β particle ($^{0}_{-1}\beta$).

20.13   **Strategy:**   We can use Equation 20.1, $\Delta E = (\Delta m)c^2$, to solve the problem. Recall the following conversion factor:

$$1\text{ J} = \frac{1\text{ kg}\cdot\text{m}^2}{\text{s}^2}$$

Solution: The mass change is:

$$\Delta m = \frac{\Delta E}{c^2} = \frac{-436,400 \text{ J/mol}}{(3.00 \times 10^8 \text{ m/s})^2} = -4.85 \times 10^{-12} \text{ kg / mol } \mathbf{H_2}$$

20.15 Strategy: To calculate the nuclear binding energy, we first determine the difference between the mass of the atom and the mass of all the protons, neutrons, and electrons, which gives us the mass defect. Next, we apply Einstein's mass-energy relationship [$\Delta E = (\Delta m)c^2$] (Equation 20.1), and use the procedure shown in Worked Example 20.2.

Solution: a. There are 3 protons and 4 neutrons in the lithium nucleus. The mass of 3 protons is

$$3 \times 1.00728 \text{ amu} = 3.02184 \text{ amu}$$

the mass of 3 electrons is

$$3 \times 5.4858 \times 10^{-4} \text{ amu} = 0.0016457 \text{ amu}$$

and the mass of 4 neutrons is

$$4 \times 1.008665 \text{ amu} = 4.034660 \text{ amu}$$

Therefore, the predicted mass for ^7_3Li is

$$3.02184 \text{ amu} + 0.0016457 \text{ amu} + 4.034660 \text{ amu} = 7.05815 \text{ amu}$$

The mass defect—that is, the difference between the predicted mass and the measured mass, is

$$\Delta m = 7.01600 \text{ amu} - 7.05815 \text{ amu} = -0.04215 \text{ amu}$$

$$-0.04215 \text{ amu} \times \frac{1 \text{ kg}}{6.0221418 \times 10^{26} \text{ amu}} = -6.999 \times 10^{-29} \text{ kg}$$

The mass that is converted in energy—that is, the energy released, is

$$\Delta E = (\Delta m)c^2$$

$$\Delta E = (-6.999 \times 10^{-29} \text{ kg})(2.99792458 \times 10^8 \text{ m/s})^2$$

$$\Delta E = -6.290 \times 10^{-12} \text{ kg·m}^2/\text{s}^2 = -6.290 \times 10^{-12} \text{ J}$$

Note that we report the binding energy without referring to its sign. Therefore, the nuclear binding energy is **6.290×10^{-12} J**. When comparing the stability of any two nuclei we must account for the fact that they have different numbers of nucleons. For this reason, it is more meaningful to use the *nuclear binding energy per nucleon*, defined as

$$\text{nuclear binding energy per nucleon} = \frac{\text{nuclear binding energy}}{\text{number of nucleons}}$$

The binding energy per nucleon for ^7Li is:

$$\frac{6.290 \times 10^{-12} \text{ J}}{7 \text{ nucleons}} = \textbf{8.986} \times \textbf{10}^{-13} \textbf{ J / nucleon}$$

b. Using the same procedure as in (a), using 1.00728 amu for ^1_1p, 5.4858×10^{-4} amu for e^-, and 1.008665 amu for ^1_0n, we can show that for chlorine-35:

Nuclear binding energy = $\textbf{4.779} \times \textbf{10}^{-11} \textbf{ J}$

Nuclear binding energy per nucleon = $\textbf{1.365} \times \textbf{10}^{-12} \textbf{ J/nucleon}$

20.17 **Strategy:** The mass of one ^{48}Cr atom is the sum of the masses of its constituents (24 protons, 24 electrons, and 24 neutrons) plus the mass defect. We can use the given nuclear binding energy and Equation 20.1 to compute the mass defect. Remember that $1 \text{ J} = 1 \text{ kg·m}^2/\text{s}^2$.

Solution: Calculate the mass defect:

$$\Delta E = \Delta mc^2$$

$$\left[-1.37340 \times 10^{-12} \ (\text{kg·m}^2/\text{s}^2)/\text{nucleon} \right](48 \text{ nucleons}) = (2.99792458 \times 10^8 \text{ m/s})^2 \Delta m$$

$$\Delta m = -7.33495 \times 10^{-28} \text{ kg}$$

Calculate the mass of the ^{48}Cr atom:

$$\text{mass}^{48}Cr \text{ atom} = (24 \text{ protons})(1.00728 \text{ amu/proton})$$
$$+ (24 \text{ electrons})(5.4858 \times 10^{-4} \text{ amu / electron})$$
$$+ (24 \text{ neutrons})(1.008665 \text{ amu/neutron})$$
$$- 0.44172 \text{ amu}$$

$$= 48.3958 \text{ amu} - 0.44172 \text{ amu}$$

$$= \textbf{47.9541 amu}$$

20.21 **Strategy:** Alpha emission decreases the atomic number by 2 and the mass number by 4. Beta emission increases the atomic number by 1 and has no effect on the mass number.

Solution: a.

$$^{232}_{90}\text{Th} \xrightarrow{\alpha} {}^{228}_{88}\text{Ra} \xrightarrow{\beta} {}^{228}_{89}\text{Ac} \xrightarrow{\beta} {}^{228}_{90}\text{Th}$$

b.

$$^{235}_{92}\text{U} \xrightarrow{\alpha} {}^{231}_{90}\text{Th} \xrightarrow{\beta} {}^{231}_{91}\text{Pa} \xrightarrow{\alpha} {}^{227}_{89}\text{Ac}$$

c.

$$^{237}_{93}\text{Np} \xrightarrow{\alpha} {}^{233}_{91}\text{Pa} \xrightarrow{\beta} {}^{233}_{92}\text{U} \xrightarrow{\alpha} {}^{229}_{90}\text{Th}$$

20.23 The number of atoms decreases by half for each half-life. For ten half-lives we have:

$$(5.00 \times 10^{22} \text{ atoms}) \times \left(\frac{1}{2}\right)^{10} = \textbf{4.88} \times \textbf{10}^{19} \textbf{ atoms}$$

20.25 **Strategy:** We use the equation from Section 20.3 (a variation of Equation 19.3) to solve for t, the age of the artifact. Recall that the ratio in this equation can be expressed as the number of radioactive nuclei or as the activity (disintegrations per unit time) of the sample.

$$\ln\frac{N_t}{N_0} = -kt$$

The problem gives the half-life. We must first solve for k using the equation.

$$t_{1/2} = \frac{0.693}{k}$$

$$k = \frac{0.693}{t_{1/2}} = \frac{0.693}{5715 \text{ yr}} = 1.213 \times 10^{-4} \text{ yr}^{-1}$$

Solution: Rearranging the equation to solve for t gives:

$$t = -\frac{\left(\ln\dfrac{N_t}{N_0}\right)}{k} = -\frac{\left(\ln\dfrac{18.9 \text{ dis/min}}{27.5 \text{ dis/min}}\right)}{1.213 \times 10^{-4} \text{ yr}^{-1}} = \mathbf{3.09 \times 10^3 \text{ yr}}$$

20.27   Let's consider the decay of A first.

$$k = \frac{0.693}{t_{1/2}} = \frac{0.693}{4.50 \text{ s}} = 0.154 \text{ s}^{-1}$$

Let's convert k to units of day^{-1}.

$$0.154\frac{1}{\text{s}} \times \frac{3600 \text{ s}}{1 \text{ h}} \times \frac{24 \text{ h}}{1 \text{ d}} = 1.33 \times 10^4 \text{ d}^{-1}$$

Next, use the first-order rate equation to calculate the amount of A left after 30 days.

$$\ln\frac{N_t}{N_0} = -kt$$

Let x be the amount of A left after 30 days.

$$\ln\frac{x}{1.00} = -(1.33 \times 10^4 \text{ d}^{-1})(30 \text{ d}) = -3.99 \times 10^5$$

$$\frac{x}{1.00} = e^{(-3.99 \times 10^5)}$$

$$x \approx 0$$

Thus, **no A remains.**

For B: As calculated above, all of A is converted to B in less than 30 days. In fact, essentially all of A is gone in less than 1 day! This means that at the beginning of the 30-day period, there is 1.00 mole of B present. The half-life of B is 15 days, so that after two half-lives (30 days), there should be **0.25 mole of B** left.

For C: As in the case of A, the half-life of C is also very short. Therefore, at the end of the 30-day period, **no C is left**.

For D: D is not radioactive. 0.75 mole of B reacted in 30 days; therefore, due to a 1:1 mole ratio between B and D, there should be **0.75 mole of D** present after 30 days.

20.29 Strategy: We use the integrated rate law to solve for the activity at time t, given that time $t = 8.4 \times 10^3$ yr.

$$\ln \frac{N_t}{N_0} = -kt$$

Solution: Taking the inverse ln of both sides of the equation, we get:

$$\frac{N_t}{N_0} = e^{-kt} = e^{-\left(1.213\times10^{-4} \text{ yr}^{-1}\right)\left(8.4\times10^3 \text{ yr}\right)} = e^{-1.02} = 0.361$$

$$N_t = N_0(0.361) = (15.3 \text{ dpm})(0.361) = \textbf{5.5 dpm}$$

20.31 Strategy: We begin by using the integrated rate law and the rate constant, k (determined in Problem 20.30), to determine the ratio of ^{238}U at $t = 0$ to ^{238}U at $t = 1.7 \times 10^8$ yr.

Solution:

$$\ln \frac{^{238}\text{U}_t}{^{238}\text{U}_0} = -\left(1.54\times10^{-10} \text{ yr}^{-1}\right)\left(1.7\times10^8 \text{ yr}\right) = -0.02618$$

It simplifies the problem to put the larger mass in the numerator. Doing this simply changes the sign of the result:

$$\ln \frac{^{238}\text{U}_0}{^{238}\text{U}_t} = 0.02618$$

Taking the inverse natural log of both sides gives:

$$\frac{^{238}\text{U}_0}{^{238}\text{U}_t} = e^{0.02618} = 1.027$$

The mass ratio of ^{238}U at $t = 0$ to ^{238}U now (at $t = 1.7 \times 10^8$ yr) is 1.027. Since an infinite number of mass combinations that would give this ratio, we must assume a value for one of the masses. It makes the mass easier if we assume that the mass of ^{238}U now is 1.000 g. This gives a mass of 1.027 g ^{238}U at $t = 0$. The difference between these masses is the mass of ^{238}U that has decayed to ^{206}Pb. However, because a nucleus loses mass as it decays, the mass of ^{206}Pb will be less than the mass of ^{238}U that decayed. We determine the mass of ^{206}Pb as follows:

$$0.027 \text{ g } ^{238}\text{U} \times \frac{206 \text{ g } ^{206}\text{Pb}}{238 \text{ g } ^{238}\text{U}} = 0.023 \text{ g } ^{206}\text{Pb}$$

Finally, if there are currently a mass of 1.000 g ^{238}U and a mass of 0.023 g ^{206}Pb, the ratio of ^{238}U to ^{206}Pb is:

$$\frac{1.000 \text{ g } ^{238}\text{U}}{0.023 \text{ g } ^{206}\text{Pb}} = \textbf{43:1}$$

20.35   a. ^{14}N$(\alpha,p)^{17}$O
b. ^{9}Be$(\alpha,n)^{12}$C
c. ^{238}U$(d,2n)^{238}$Np

20.37   a. ^{40}Ca$(d,p)^{41}$Ca
b. ^{32}S$(n,p)^{32}$P
c. ^{239}Pu$(\alpha,n)^{242}$Cm

Remember that it is unnecessary to include the subscripted atomic number. The elemental symbol and the atomic mass are sufficient to identify an isotope unambiguously.

20.39   Upon bombardment with neutrons, mercury-198 is first converted to mercury-199, which then emits a proton. The reaction is:

$$^{198}_{80}\text{Hg} + {}^{1}_{0}\text{n} \longrightarrow {}^{199}_{80}\text{Hg} \longrightarrow {}^{198}_{79}\text{Au} + {}^{1}_{1}\text{p}$$

20.51   **Strategy:**   Use Equation 19.5 to solve for k.

$$t_{1/2} = \frac{0.693}{k}$$

Then use Equation 19.3 to solve for time.

$$\ln\frac{[\text{A}]_t}{[\text{A}]_0} = -kt$$

Setup: At 5.00 percent of the original value, we assume $[\text{A}]_0 = 100$ and $[\text{A}]_t = 5$.

Solution: Solving Equation 19.5 for k gives:

$$k = \frac{0.693}{t_{1/2}} = \frac{0.693}{59.4 \text{ d}} = 0.01167 \text{ d}^{-1}$$

Solving Equation 19.3 for t gives:

$$\ln\frac{5.00}{100} = -(0.01167)t$$

$$t = \textbf{257 days}$$

20.53   **The fact that the radioisotope appears only in the I$_2$ shows that the IO$_3^-$ is formed only from the IO$_4^-$.**

20.55   **Add iron-59 to the person's diet, and allow a few days for the iron-59 isotope to be incorporated into the person's body. Isolate red blood cells from a blood sample and monitor radioactivity from the hemoglobin molecules present in the red blood cells.**

20.61   We start with the integrated first-order rate law (Equation 19.3):

$$\ln\frac{N_t}{N_0} = -kt$$

We can calculate the rate constant, k, from the half-life using Equation 19.5 of the text, and then substitute into Equation 19.3 to solve for the time.

$$t_{1/2} = \frac{0.693}{k}$$

$$k = \frac{0.693}{t_{1/2}} = \frac{0.693}{28.1 \text{ yr}} = 0.02466 \text{ yr}^{-1}$$

Substituting:

$$\ln\left(\frac{0.200}{1.00}\right) = -\left(0.02466 \text{ yr}^{-1}\right) t$$

$$t = \textbf{65.3 yr}$$

20.63 a. The balanced equation is:

$$_1^3\text{H} \longrightarrow {}_2^3\text{He} + {}_{-1}^{0}\beta$$

b. The number of tritium (T) atoms in 1.00 kg of water is:

$$(1.00 \times 10^3 \text{ g H}_2\text{O}) \times \frac{1 \text{ mol H}_2\text{O}}{18.02 \text{ g H}_2\text{O}} \times \frac{6.022 \times 10^{23} \text{ molecules H}_2\text{O}}{1 \text{ mol H}_2\text{O}} \times \frac{2 \text{ H atoms}}{1 \text{ H}_2\text{O}} \times \frac{1 \text{ T atom}}{1.0 \times 10^{17} \text{ H atoms}}$$

$$= 6.68 \times 10^8 \text{ T atoms}$$

The number of disintegrations per minute (dpm) will be:

$$\text{rate} = k \text{ (number of T atoms)} = kN = \frac{0.693}{t_{1/2}} N$$

$$\text{rate} = \left(\frac{0.693}{12.5 \text{ yr}} \times \frac{1 \text{ yr}}{365 \text{ d}} \times \frac{1 \text{ d}}{24 \text{ h}} \times \frac{1 \text{ h}}{60 \text{ min}}\right)\left(6.68 \times 10^8 \text{ T atoms}\right)$$

$$\textbf{rate} = \textbf{70.5 T atoms/min} = \textbf{70.5 dpm}$$

20.65 a. $_{92}^{235}\text{U} + {}_0^1\text{n} \longrightarrow {}_{56}^{140}\text{Ba} + 3\,{}_0^1\text{n} + {}_{36}^{93}\textbf{Kr}$

b. $_{92}^{235}\text{U} + {}_0^1\text{n} \longrightarrow {}_{55}^{144}\text{Cs} + {}_{37}^{90}\text{Rb} + 2\,{}_0^1\textbf{n}$

c. $_{92}^{235}\text{U} + {}_0^1\text{n} \longrightarrow {}_{35}^{87}\text{Br} + {}_{57}^{146}\textbf{La} + 3\,{}_0^1\text{n}$

d. $_{92}^{235}\text{U} + {}_0^1\text{n} \longrightarrow {}_{62}^{160}\text{Sm} + {}_{30}^{72}\text{Zn} + 4\,{}_0^1\textbf{n}$

20.67   The balanced nuclear equations are:

a.
$$\,^3_1\text{H} \longrightarrow \,^3_2\text{He} + \,^0_{-1}\beta$$

b.
$$\,^{242}_{94}\text{Pu} \longrightarrow \,^4_2\alpha + \,^{238}_{92}\text{U}$$

c.
$$\,^{131}_{53}\text{I} \longrightarrow \,^{131}_{54}\text{Xe} + \,^0_{-1}\beta$$

d.
$$\,^{251}_{98}\text{Cf} \longrightarrow \,^{247}_{96}\text{Cm} + \,^4_2\alpha$$

20.69   **Because both Ca and Sr belong to Group 2A, radioactive strontium that has been ingested into the human body becomes concentrated in bones (replacing Ca) and can damage blood cell production.**

Since radioactive decay kinetics are first-order, the decay rate is given by $R = kN$, where k is the decay constant and N is the number or radioactive nuclei.

$$R = kN$$

$$R = \left(\frac{0.693}{t_{1/2}}\right)N$$

$$R = \left(\frac{0.693}{29.1 \text{ yr}} \times \frac{1 \text{ yr}}{365 \text{ d}} \times \frac{1 \text{ d}}{24 \text{ h}} \times \frac{1 \text{ h}}{3600 \text{ s}}\right)\left(0.0156 \text{ g }^{90}\text{Sr} \times \frac{1 \text{ mol}}{90 \text{ g }^{90}\text{Sr}} \times \frac{6.022 \times 10^{23} \text{ atoms}}{1 \text{ mol}}\right)$$

$$R = \left(7.55 \times 10^{-10} \text{ s}^{-1}\right)\left(1.04 \times 10^{20} \text{ atoms}\right)$$

$$R = 7.85 \times 10^{10} \text{ atoms/s} = 7.85 \times 10^{10} \text{ disintegrations/s}$$

A curie (1 Ci) is defined as 3.70×10^{10} disintegrations per second. So, the activity in curies is:

$$\text{Radioactivity in millicuries (mCi)} = \left(7.85 \times 10^{10} \text{ disintegrations / s}\right)\left(\frac{1 \text{ Ci}}{3.70 \times 10^{10} \text{ disintegrations / s}}\right)$$

$$= 2.12 \text{ Ci} = \mathbf{2.12 \times 10^3 \text{ mCi}}$$

20.71   **Normally the human body concentrates iodine in the thyroid gland. The purpose of the large doses of KI is to displace radioactive iodine from the thyroid and allow its excretion from the body.**

20.73   a. $\,^{209}_{83}\text{Bi} + \,^4_2\alpha \longrightarrow \,^{211}_{85}\text{At} + 2\,^1_0\text{n}$ b. $\,^{209}_{83}\text{Bi}(\alpha, 2n)\,^{211}_{85}\text{At}$

20.75   *Step 1:*   The half-life of carbon-14 is 5715 years. From the half-life, we can calculate the rate constant, k.

$$k = \frac{0.693}{t_{1/2}} = \frac{0.693}{5715 \text{ yr}} = 1.21 \times 10^{-4} \text{ yr}^{-1}$$

Step 2: The age of the object can now be calculated using the following equation.

$$\ln \frac{N_t}{N_0} = -kt$$

N = the number of radioactive nuclei. In the problem, we are given disintegrations per second per gram. The number of disintegrations is directly proportional to the number of radioactive nuclei. We can write,

$$\ln \frac{\text{decay rate of old sample}}{\text{decay rate of fresh sample}} = -kt$$

$$\ln \frac{0.186 \text{ dps/g C}}{0.260 \text{ dps/g C}} = -(1.21 \times 10^{-4} \text{ yr}^{-1})t$$

$$t = \mathbf{2.77 \times 10^3 \text{ yr}}$$

20.77

$$\ln \frac{N_t}{N_0} = -kt$$

$$\ln \frac{\text{mass of old sample}}{\text{mass of fresh sample}} = -\left(1.21 \times 10^{-4} \text{ yr}^{-1}\right)\left(60,000 \text{ yr}\right)$$

$$\ln \frac{1.0 \text{ g}}{x \text{ g}} = -7.26$$

$$\frac{1.0}{x} = e^{-7.26}$$

$$x = 1422$$

$$\text{Percent of C-14 left} = \frac{1.0}{1422} \times 100\% = \mathbf{0.070\%}$$

20.79 a. In the ^{90}Sr decay, the mass defect is:

$$\Delta m = (\text{mass } ^{90}\text{Y} + \text{mass } e^-) - \text{mass } ^{90}\text{Sr}$$

$$= [(89.907152 \text{ amu} + 5.4857 \times 10^{-4} \text{ amu}) - 89.907738 \text{ amu}] = -3.7430 \times 10^{-5} \text{ amu}$$

$$= (-3.7430 \times 10^{-5} \text{ amu}) \times \frac{1 \text{ g}}{6.022 \times 10^{23} \text{ amu}} = -6.216 \times 10^{-29} \text{ g} = -6.216 \times 10^{-32} \text{ kg}$$

The energy change is given by:

$$\Delta E = (\Delta m)c^2$$

$$= (-6.216 \times 10^{-32} \text{ kg})(3.00 \times 10^8 \text{ m/s})^2$$

$$= -5.59 \times 10^{-15} \text{ kg m}^2/\text{s}^2 = -5.59 \times 10^{-15} \text{ J}$$

Similarly, for the ^{90}Y decay, we have:

$$\Delta m = (\text{mass } ^{90}\text{Zr} + \text{mass } e^-) - \text{mass } ^{90}\text{Y}$$

$$= [(89.904703 \text{ amu} + 5.4857 \times 10^{-4} \text{ amu}) - 89.907152 \text{ amu}] = -1.9004 \times 10^{-3} \text{ amu}$$

$$= \left(-1.9004 \times 10^{-3} \text{ amu}\right) \times \frac{1 \text{ g}}{6.022 \times 10^{23} \text{ amu}} = -3.156 \times 10^{-27} \text{ g} = -3.156 \times 10^{-30} \text{ kg}$$

and the energy change is:

$$\Delta E = (-3.156 \times 10^{-30} \text{ kg})(3.00 \times 10^8 \text{ m/s})^2 = -2.84 \times 10^{-13} \text{ J}$$

The energy released in the above two decays is **5.59×10^{-15} J** and **2.84×10^{-13} J**. The total amount of energy released is:

$$(5.59 \times 10^{-15} \text{ J}) + (2.84 \times 10^{-13} \text{ J}) = 2.90 \times 10^{-13} \text{ J}$$

b. This calculation requires that we know the rate constant for the decay. From the half-life, we can calculate k.

$$k = \frac{0.693}{t_{1/2}} = \frac{0.693}{28.1 \text{ yr}} = 0.0247 \text{ yr}^{-1}$$

To calculate the number of moles of ^{90}Sr decaying in a year, we apply the following equation:

$$\ln \frac{N_t}{N_0} = -kt$$

$$\ln \frac{x}{1} = -\left(0.0247 \text{ yr}^{-1}\right)(1 \text{ yr})$$

where x is the number of moles of ^{90}Sr nuclei left over. Solving, we obtain:

$$x = 0.9756 \text{ mol } ^{90}\text{Sr}$$

Thus the number of moles of nuclei which decay in a year is:

$$(1 - 0.9756) \text{ mol} = \textbf{0.0244 mol}$$

This is a reasonable number since it takes 28.1 years for 0.5 mole of ^{90}Sr to decay.

c. Since the half-life of ^{90}Y is much shorter than that of ^{90}Sr, we can safely assume that *all* the ^{90}Y formed from ^{90}Sr will be converted to ^{90}Zr. The energy changes calculated in part (a) refer to the decay of individual nuclei. In 0.0244 mole, the number of nuclei that have decayed is:

$$0.0244 \text{ mol} \times \frac{6.022 \times 10^{23} \text{ nuclei}}{1 \text{ mol}} = 1.47 \times 10^{22} \text{ nuclei}$$

Realize that there are two decay processes occurring, so we need to add the energy released for each process calculated in part (a). Thus, the heat released from 1 mole of ^{90}Sr waste in a year is given by:

$$\text{heat released} = \left(1.47 \times 10^{22} \text{ nuclei}\right) \times \frac{2.90 \times 10^{-13} \text{ J}}{1 \text{ nucleus}} = 4.26 \times 10^9 \text{ J} = \mathbf{4.26 \times 10^6 \text{ kJ}}$$

This amount is roughly equivalent to the heat generated by burning 50 tons of coal! Although the heat is released slowly during the course of a year, effective ways must be devised to prevent heat damage to the storage containers and subsequent leakage of radioactive material to the surroundings.

20.81 One curie represents 3.70×10^{10} disintegrations/s. The rate of decay of the isotope is given by the rate law: rate = kN, where N is the number of atoms in the sample and k is the first-order rate constant. We find the value of k in units of s^{-1}:

$$k = \frac{0.693}{t_{1/2}} = \frac{0.693}{1.6 \times 10^3 \text{ yr}} = 4.33 \times 10^{-4} \text{ yr}^{-1}$$

$$\frac{4.33 \times 10^{-4} \text{ yr}^{-1}}{1 \text{ yr}} \times \frac{1 \text{ yr}}{365 \text{ d}} \times \frac{1 \text{ d}}{24 \text{ h}} \times \frac{1 \text{ h}}{3600 \text{ s}} = 1.37 \times 10^{-11} \text{ s}^{-1}$$

Now, we can calculate N, the number of Ra atoms in the sample.

$$\text{rate} = kN$$

$$3.70 \times 10^{10} \text{ disintegrations/s} = (1.37 \times 10^{-11} \text{ s}^{-1})N$$

$$N = 2.70 \times 10^{21} \text{ Ra atoms}$$

By definition, 1 curie corresponds to exactly 3.7×10^{10} nuclear disintegrations per second, which is the decay rate equivalent to that of *1 g of radium*. Thus, the mass of 2.70×10^{21} Ra atoms is 1 g.

$$\frac{2.7 \times 10^{21} \text{ Ra atoms}}{1.0 \text{ g Ra}} \times \frac{226.03 \text{ g Ra}}{1 \text{ mol Ra}} = \mathbf{6.1 \times 10^{23} \text{ atoms/mol}} = N_A$$

20.83 a.

$$^{238}_{92}\text{U} \longrightarrow {}^{234}_{90}\text{Th} + {}^{4}_{2}\alpha$$

$$\Delta m = 234.03596 \text{ amu} + 4.002603 \text{ amu} - 238.05078 \text{ amu} = -0.01222 \text{ amu}$$

$$-0.01222 \text{ amu} \times \frac{1 \text{ kg}}{6.022 \times 10^{26} \text{ amu}} = 2.029 \times 10^{-29} \text{ kg}$$

$$\Delta E = 2.029 \times 10^{-29} \text{ kg} (3.00 \times 10^8 \text{ m/s})^2 = \mathbf{1.83 \times 10^{-12} \text{ J}}$$

b. **The α particle will move away faster because it is smaller**.

20.85 **U-238, $t_{1/2} = 4.5 \times 10^9$ yr and Th-232, $t_{1/2} = 1.4 \times 10^{10}$ yr.**

They are still present because of their long half-lives.

20.87

$$E = \frac{hc}{\lambda}$$

$$\lambda = \frac{hc}{E} = \frac{\left(6.63 \times 10^{-34}\ \text{J} \cdot \text{s}\right)\left(3.00 \times 10^{8}\ \text{m/s}\right)}{2.4 \times 10^{-13}\ \text{J}} = 8.3 \times 10^{-13}\ \text{m} = \mathbf{8.3 \times 10^{-4}\ nm}$$

This wavelength is clearly in the γ-ray region of the electromagnetic spectrum.

20.89 **Only ^{3}H has a suitable half-life. The other half-lives are either too long or too short to determine the time span of 6 years accurately.**

20.91 **A small-scale chain reaction (fission of ^{235}U) took place. Copper played the crucial role of reflecting neutrons from the splitting uranium-235 atoms back into the uranium sphere to trigger the chain reaction. Note that a sphere has the most appropriate geometry for such a chain reaction. In fact, during the implosion process prior to an atomic explosion, fragments of uranium-235 are pressed roughly into a sphere for the chain reaction to occur (see Section 20.5 of the text).**

20.93 In this problem, we are asked to calculate the molar mass of a radioactive isotope. Grams of sample are given in the problem, so if we can find moles of sample we can calculate the molar mass. The rate constant can be calculated from the half-life. Then, from the rate of decay and the rate constant, the number of radioactive nuclei can be calculated. The number of radioactive nuclei can be converted to moles.

First, we convert the half-life to units of minutes because the rate is given in dpm (disintegrations per minute). Then, we calculate the rate constant from the half-life.

$$(1.3 \times 10^9\ \text{yr}) \times \frac{365\ \text{d}}{1\ \text{yr}} \times \frac{24\ \text{h}}{1\ \text{d}} \times \frac{60\ \text{min}}{1\ \text{h}} = 6.8 \times 10^{14}\ \text{min}$$

$$k = \frac{0.693}{t_{1/2}} = \frac{0.693}{6.8 \times 10^{14}\ \text{min}} = 1.0 \times 10^{-15}\ \text{min}^{-1}$$

Next, we calculate the number of radioactive nuclei from the rate and the rate constant.

$$\text{rate} = kN$$

$$2.9 \times 10^4\ \text{dpm} = (1.0 \times 10^{-15}\ \text{min}^{-1})N$$

$$N = 2.9 \times 10^{19}\ \text{nuclei}$$

Convert to moles of nuclei, and then determine the molar mass.

$$\left(2.9 \times 10^{19}\ \text{nuclei}\right) \times \frac{1\ \text{mol}}{6.022 \times 10^{23}\ \text{nuclei}} = 4.8 \times 10^{-5}\ \text{mol}$$

$$\text{molar mass} = \frac{\text{g of substance}}{\text{mol of substance}} = \frac{0.0100\ \text{g}}{4.8 \times 10^{-5}\ \text{mol}} = \mathbf{2.1 \times 10^2\ g/mol}$$

20.95   a. The volume of a sphere is

$$V = \frac{4}{3}\pi r^3$$

Volume is proportional to the number of nucleons. Therefore,

$$V \propto A \text{ (mass number)}$$

$$r^3 \propto A$$

$$r \propto A^{1/3}$$

$r = r_0 A^{1/3}$, where r_0 is a proportionality constant.

b. We can calculate the volume of the ^{238}U nucleus by substituting the equation derived in part (a) into the equation for the volume of a sphere.

$$V = \frac{4}{3}\pi r^3 = \frac{4}{3}\pi r_0^3 A$$

$$V = \frac{4}{3}\pi \left(1.2 \times 10^{-15} \text{ m}\right)^3 (238) = \mathbf{1.7 \times 10^{-42} \text{ m}^3}$$

20.97 $k = 0.693/45 = 0.0154$

$$\ln(N_{20} / 15) = -0.0154(20) = -0.308$$

$$N_{20} / 15 = 0.735$$

The solution is $N_{20} \approx 11$. **Diagram (d)** most closely represents the sample of X after 20 days.

20.99   **Strategy:**   We are asked to find the volume of helium formed at STP from the radioactive decay of radium-226. The volume is easily calculated using the ideal gas law:

$$V = n_{He}RT/P$$

The overall decay process produces 5 mol He for each mole of ^{226}Ra that decays:

$$\text{mol He} = 5(\text{mol } ^{226}\text{Ra decayed}) = 5(\text{initial mol } ^{226}\text{Ra} - \text{final mol } ^{226}\text{Ra})$$

The initial amount of ^{226}Ra can be found easily from the initial mass (1.00 g) and the molar mass (226 g/mol) of radium-226. But, calculating the final amount of ^{226}Ra does not appear to be straightforward since the decay sequence from radium-226 to lead-206 involves several intermediate radioisotopes, each with its own half-life. Fortunately, though, the half-life of the first step in the decay sequence (radium-226 $\longrightarrow$ radon-222, half-life 1600 years) is significantly longer than any other half-life in the sequence, so we can approximate the overall decay process using first-order kinetics and the rate constant of the slow (first) step:

$$^{226}\text{Ra} \longrightarrow {}^{206}\text{Pb} + 5\,{}^{4}\text{He} \qquad\qquad t_{1/2} \approx 1600 \text{ yr}$$

$$\ln \frac{\text{final mol } ^{226}\text{Ra}}{\text{initial mol } ^{226}\text{Ra}} \approx -kt$$

Plugging the known values into the above equation, we can solve it for the final moles of ^{226}Ra, which in turn gives us the moles of He produced and the volume of He produced.

Solution: Calculate the initial moles of ^{226}Ra:

$$\text{initial mol } ^{226}\text{Ra} = \left(1.00 \text{ g } ^{226}\text{Ra}\right)\left(\frac{1 \text{ mol}}{226 \text{ g } ^{226}\text{Ra}}\right) = 4.42\times10^{-3} \text{ mol } ^{226}\text{Ra}$$

Next, find the final moles of ^{226}Ra:

$$\ln\frac{\text{final mol } ^{226}\text{Ra}}{0.00442 \text{ mol } ^{226}\text{Ra}} \approx -\left(\frac{0.693}{1600 \text{ yr}}\right)(125 \text{ yr})$$

$$\text{final mol } ^{226}\text{Ra} \approx \left(0.00442 \text{ mol } ^{226}\text{Ra}\right)e^{-(0.693)(125)/1600}$$

$$\text{final mol } ^{226}\text{Ra} \approx 0.00419 \text{ mol } ^{226}\text{Ra}$$

The number of moles of He produced is:

$$\text{mol He} = 5(0.00422 \text{ mol} - 0.00419 \text{ mol}) = 1.5\times10^{-4} \text{ mol He}$$

Finally, the volume of He at STP is:

$$V = \frac{\left(1.5\times10^{-4} \text{ mol}\right)(0.08206 \text{ L}\cdot\text{atm/K}\cdot\text{mol})(273 \text{ K})}{1 \text{ atm}}$$

$$V = 3.4\times10^{-3}\text{L} = \mathbf{3.4\,mL}$$

20.101 a. **Strategy:** The species written first is a reactant. The species written last is a product. Within the parenthesis, the bombarding particle (a reactant) is written first, followed by the emitted particle (a product).

Determine the atomic number for the unknown species, X, by summing the atomic numbers on both sides of the reaction.

$$\Sigma \text{ reactant atomic numbers} = \Sigma \text{ product atomic numbers}$$

Use the atomic number to determine the identity of the unknown species.

Setup: The bombarding and emitted particles are represented by $^{1}_{0}$n and γ, respectively.

The atomic number of Ir is 77.

Gamma emission has no effect on either the atomic number or the mass number. Neutron activation has no effect on the atomic number and increases the mass number by 1. Therefore, the atomic number of X = (77 + 0) = 77.

Solution:
$$X = {}^{191}_{77}Ir$$

The balanced nuclear equation is:

$$^{191}_{77}Ir + {}^{1}_{0}n \longrightarrow {}^{192}_{77}Ir + \gamma$$

b. **Strategy:** To calculate the nuclear binding energy, we first determine the difference between the mass of the nucleus and the mass of all the protons and neutrons, which yields the mass defect. Next, we must apply Einstein's mass-energy relationship $[\Delta E = (\Delta m)c^2]$. The nuclear binding energy per nucleon is given by the following equation:

$$\text{nuclear binding energy per nucleon} = \frac{\text{nuclear binding energy}}{\text{number of nucleons}}$$

Solution: There are 53 protons, 53 electrons, and 72 neutrons in the ^{125}I atom. The mass of 53 protons is

$$53 \times 1.00728 \text{ amu} = 53.38584 \text{ amu}$$

the mass of 53 electrons is

$$53 \times 5.4858 \times 10^{-4} \text{ amu} = 0.029075 \text{ amu}$$

and the mass of 72 neutrons is

$$72 \times 1.008665 \text{ amu} = 72.62388 \text{ amu}$$

Therefore, the predicted mass for $^{125}_{53}I$ is

$$53.38584 \text{ amu} + 0.029075 \text{ amu} + 72.62388 = 126.0388 \text{ amu}$$

and the mass defect is

$$\Delta m = 124.904624 - 126.0388 \text{ amu} = -1.1342 \text{ amu}$$

The energy released is

$$\Delta E = (\Delta m)c^2 = \left(-1.1342 \text{ amu} \times \frac{1 \text{ kg}}{6.0221418 \times 10^{26} \text{ amu}} \right) \left(2.99792458 \times 10^8 \text{ m/s} \right)^2$$

$$= -1.6927 \times 10^{-10} \text{ J}$$

Note that we report the binding energy without referring to its sign. Therefore, the nuclear binding energy is **1.6927×10^{-10} J**.

The binding energy per nucleon for ^{125}I is:

$$\frac{1.6927 \times 10^{-10} \text{J}}{125 \text{ nucleons}} = \mathbf{1.3541 \times 10^{-12} \text{ J / nucleon}}$$

Chapter 21
Metallurgy and the Chemistry of Metals

Problems

21.3 a. $CaCO_3$

   b. $CaCO_3 \cdot MgCO_3$

   c. CaF_2

   d. $NaCl$

   e. Al_2O_3

   f. Fe_3O_4

   g. $Be_3Al_2Si_6O_{18}$

   h. PbS

   i. $MgSO_4 \cdot 7H_2O$

   j. $CaSO_4$

21.13 For the given reaction we can calculate the standard free energy change from the standard free energies of formation. Then, we can calculate the equilibrium constant, K_P, from the standard free energy change.

$$\Delta G° = \Delta G_f°[Ni(CO)_4] - [4\Delta G_f°(CO) + \Delta G_f°(Ni)]$$
$$\Delta G° = (1)(-587.4 \text{ kJ/mol}) - [(4)(-137.3 \text{ kJ/mol}) + (1)(0)] = -38.2 \text{ kJ/mol} = -3.82 \times 10^4 \text{ J/mol}$$

Substitute $\Delta G°$, R, and T (in K) into the following equation to solve for K_P.

$$\Delta G° = -RT \ln K_P$$

$$\ln K_P = \frac{-\Delta G°}{RT} = \frac{-\left(-3.82 \times 10^4 \text{ J/mol}\right)}{(8.314 \text{ J/K} \cdot \text{mol})(353 \text{ K})}$$

$$K_P = 4.5 \times 10^5$$

21.15 a. We first find the mass of ore containing 2.0×10^8 kg of copper.

$$\left(2.0 \times 10^8 \text{ kg Cu}\right) \times \frac{100\% \text{ ore}}{0.80\% \text{ Cu}} = 2.5 \times 10^{10} \text{ kg ore}$$

We can then compute the volume from the density of the ore.

$$\left(2.5 \times 10^{10} \text{ kg}\right) \times \frac{1000 \text{ g}}{1 \text{ kg}} \times \frac{1 \text{ cm}^3}{2.8 \text{ g}} = 8.9 \times 10^{12} \text{ cm}^3$$

b. From the formula of chalcopyrite it is clear that two moles of sulfur dioxide will be formed per mole of copper. The mass of sulfur dioxide formed will be:

$$\left(2.0 \times 10^8 \text{ kg Cu}\right) \times \frac{1 \text{ mol Cu}}{0.06355 \text{ kg Cu}} \times \frac{2 \text{ mol SO}_2}{1 \text{ mol Cu}} \times \frac{0.06407 \text{ kg SO}_2}{1 \text{ mol SO}_2} = \mathbf{4.0 \times 10^8 \text{ kg SO}_2}$$

21.17 **Ag, Pt, and Au will not be oxidized but the other metals will.** (See Table 18.1 of the text.)

21.19 Very electropositive metals (i.e., very strong reducing agents) can only be isolated from their compounds by electrolysis. No chemical reducing agent is strong enough. In the given list, preparation of **Al, Na, and Ca** would require electrolysis.

21.33 a.
$$2\text{Na}(s) + 2\text{H}_2\text{O}(l) \longrightarrow 2\text{NaOH}(aq) + \text{H}_2(g)$$

b.
$$2\text{NaOH}(aq) + \text{CO}_2(g) \longrightarrow \text{Na}_2\text{CO}_3(aq) + \text{H}_2\text{O}(l)$$

c.
$$\text{Na}_2\text{CO}_3(s) + 2\text{HCl}(aq) \longrightarrow 2\text{NaCl}(aq) + \text{CO}_2(g) + \text{H}_2\text{O}(l)$$

d.
$$\text{NaHCO}_3(aq) + \text{HCl}(aq) \longrightarrow \text{NaCl}(aq) + \text{CO}_2(g) + \text{H}_2\text{O}(l)$$

e.
$$2\text{NaHCO}_3(s) \longrightarrow \text{Na}_2\text{CO}_3(s) + \text{CO}_2(g) + \text{H}_2\text{O}(g)$$

f.
$$\text{Na}_2\text{CO}_3(s) \longrightarrow \textbf{no reaction}$$

Unlike $\text{CaCO}_3(s)$, $\text{Na}_2\text{CO}_3(s)$ is not decomposed by moderate heating.

21.35 All of these reactions are discussed in Section 21.5 of the text.

a.
$$2\text{K}(s) + 2\text{H}_2\text{O}(l) \longrightarrow 2\text{KOH}(aq) + \text{H}_2(g)$$

b.
$$\text{NaH}(s) + \text{H}_2\text{O}(l) \longrightarrow \text{NaOH}(aq) + \text{H}_2(g)$$

c.
$$2\text{Na}(s) + \text{O}_2(g) \longrightarrow \text{Na}_2\text{O}_2(s)$$

d.
$$\text{K}(s) + \text{O}_2(g) \longrightarrow \text{KO}_2(s)$$

21.39 First magnesium is treated with concentrated nitric acid (redox reaction) to obtain magnesium nitrate.

$$3\text{Mg}(s) + 8\text{HNO}_3(aq) \longrightarrow 3\text{Mg(NO}_3)_2(aq) + 4\text{H}_2\text{O}(l) + 2\text{NO}(g)$$

The magnesium nitrate is recovered from solution by evaporation, dried, and heated in air to obtain magnesium oxide:

$$2\text{Mg(NO}_3)_2(s) \longrightarrow 2\text{MgO}(s) + 4\text{NO}_2(g) + \text{O}_2(g)$$

21.41 **The electron configuration of magnesium is $[\text{Ne}]3s^2$. The 3s electrons are outside the neon core (shielded), so they have relatively low ionization energies. Removing the third electron means separating an electron from the neon (closed shell) core, which requires a great deal more energy.**

21.43 **Even though helium and the Group 2A metals have ns^2 outer electron configurations, helium has a closed shell noble gas configuration and the Group 2A metals do not. The electrons in He are much closer to and more strongly attracted by the nucleus. Hence, the electrons in He are not easily removed. Helium is inert.**

21.45 a. quicklime: **CaO(*s*)**   b. slaked lime: **Ca(OH)$_2$(*s*)**

21.49 a. The relationship between cell voltage and free energy difference is:

$$\Delta G = -nFE$$

In the given reaction $n = 6$. We write:

$$E = \frac{-\Delta G}{nF} = \frac{-594 \times 10^3 \text{ J/mol}}{(6)(96,500 \text{ J/V} \cdot \text{mol})} = -1.03 \text{ V}$$

The balanced equation shows 2 moles of aluminum. Is this the voltage required to produce 1 mole of aluminum? If we divide everything in the equation by 2, we obtain:

$$\tfrac{1}{2}\text{Al}_2\text{O}_3(s) + \tfrac{3}{2}\text{C}(s) \longrightarrow \text{Al}(l) + \tfrac{3}{2}\text{CO}(g)$$

For the new equation $n = 3$ and ΔG is $\left(\tfrac{1}{2}\right)(594 \text{ kJ/mol}) = 297 \text{ kJ/mol}$. We write:

$$E = \frac{-\Delta G}{nF} = \frac{-297 \times 10^3 \text{ J/mol}}{(3)(96,500 \text{ J/V} \cdot \text{mol})} = -1.03 \text{ V}$$

The minimum voltage that must be applied is **1.03 V** (a negative sign in the above answer means that 1.03 V is required to produce the Al). The voltage required to produce 1 mole or 1000 moles of aluminum is the same; the amount of *current* will be different in each case.

b. First we convert 1.00 kg (1000 g) of Al to moles.

$$\left(1.00 \times 10^3 \text{ g Al}\right) \times \frac{1 \text{ mol Al}}{26.98 \text{ g Al}} = 37.1 \text{ mol Al}$$

The reaction in part (a) shows us that 3 moles of electrons are required to produce 1 mole of aluminum. The voltage is three times the minimum calculated above (namely, −3.09 V or −3.09 J/C). We can find the electrical energy by using the same equation with the other voltage.

$$\mathbf{\Delta G} = -nFE = -(37.1 \text{ mol Al})\left(\frac{3 \text{ mol } e^-}{1 \text{ mol Al}} \times \frac{96,500 \text{ C}}{1 \text{ mol } e^-}\right)\left(\frac{-3.09 \text{ J}}{1 \text{ C}}\right)$$

$$= 3.32 \times 10^7 \text{ J / mol} = \mathbf{3.32 \times 10^4 \text{ kJ / mol}}$$

This equation can be used because electrical work can be calculated by multiplying the voltage by the amount of charge transported through the circuit (joules = volts × coulombs). The nF term in Equation 18.2 of the text used above represents the amount of charge.

21.51 The two complex ions can be classified as AB$_4$ and AB$_6$ structures (no unshared electron pairs in Al and 4 or 6 attached atoms, respectively). Their VSEPR geometries are **tetrahedral** and **octahedral**.

The accepted explanation for the nonexistence of $AlCl_6^{3-}$ is that the chloride ion is too big to form an octahedral cluster around a very small Al^{3+} ion.

21.53 $4Al(NO_3)_3(s) \longrightarrow 2Al_2O_3(s) + 12NO_2(g) + 3O_2(g)$

21.55   The "bridge" bonds in Al_2Cl_6 break at high temperature: $Al_2Cl_6(g) \rightleftharpoons 2AlCl_3(g)$.

This increases the number of molecules in the gas phase and causes the pressure to be higher than expected for pure Al_2Cl_6.

21.57   In Al_2Cl_6, each aluminum atom is surrounded by four bonding pairs of electrons (AB_4-type molecule), and therefore each aluminum atom is sp^3 hybridized. VSEPR analysis shows $AlCl_3$ to be an AB_3-type molecule (no lone pairs on the central atom). The geometry should be trigonal planar, and the aluminum atom should therefore be sp^2 hybridized.

21.59   The formulas of the metal oxide and sulfide are MO and MS. The balanced equation must therefore be:

$$2MS(s) + 3O_2(g) \longrightarrow 2MO(s) + 2SO_2(g)$$

The number of moles of MO and MS are equal. We let x be the molar mass of metal. The number of moles of metal oxide is:

$$0.972 \text{ g} \times \frac{1 \text{ mol}}{(x + 16.00) \text{ g}}$$

The number of moles of metal sulfide is:

$$1.164 \text{ g} \times \frac{1 \text{ mol}}{(x + 32.07) \text{ g}}$$

The moles of metal oxide equal the moles of metal sulfide.

$$\frac{0.972}{(x + 16.00)} = \frac{1.164}{(x + 32.07)}$$

We solve for x.

$$0.972(x + 32.07) = 1.164(x + 16.00)$$

$$x = 65.4 \text{ g/mol}$$

21.61   **Copper(II) ion is more easily reduced than either water or hydrogen ion, so there should be no reduction of water or hydrogen ion at the cathode. (See Section 18.3 of the text.) Copper metal is more easily oxidized than water, so water should not be affected by the copper purification process under standard conditions and it should not be oxidized at the anode.**

21.63   Using Equation 14.12 from the text:

    a.

$$\Delta G_{rxn}^\circ = 4\Delta G_f^\circ(\text{Fe}) + 3\Delta G_f^\circ(O_2) - 2\Delta G_f^\circ(\text{Fe}_2O_3)$$

$$\Delta G_{rxn}^\circ = (4)(0) + (3)(0) - (2)(-741.0 \text{ kJ/mol}) = \textbf{1482 kJ / mol}$$

b.
$$\Delta G^{\circ}_{rxn} = 4\Delta G^{\circ}_f(Al) + 3\Delta G^{\circ}_f(O_2) - 2\Delta G^{\circ}_f(Al_2O_3)$$

$$\Delta G^{\circ}_{rxn} = (4)(0) + (3)(0) - (2)(-1576.4 \text{ kJ/mol}) = \textbf{3152.8 kJ / mol}$$

21.65 **Mg(*s*) reacts with N$_2$(*g*) at high temperatures to produce Mg$_3$N$_2$(*s*). Ti(*s*) also reacts with N$_2$(*g*) at high temperatures to produce TiN(*s*).**

21.67 a. **In water, the aluminum(III) ion causes an increase in the concentration of hydrogen ion (lower pH). This results from the effect of the small diameter and high charge (3+) of the aluminum ion on surrounding water molecules. The aluminum ion draws electrons in the O–H bonds to itself, thus allowing easy formation of H$^+$ ions.**

   b. **Al(OH)$_3$ is an amphoteric hydroxide. It will dissolve in strong base with the formation of a complex ion:**

$$Al(OH)_3(s) + OH^-(aq) \longrightarrow Al(OH)_4^-(aq)$$

   The concentration of OH$^-$ in aqueous ammonia is too low for this reaction to occur.

21.69 Calcium oxide is a base. The reaction is a neutralization.

$$CaO(s) + 2HCl(aq) \longrightarrow CaCl_2(aq) + H_2O(l)$$

21.71 **Metals have closely spaced energy levels and (referring to Figure 21.10 of the text) a very small energy gap between filled and empty levels. Consequently, many electronic transitions can take place with absorption and subsequent emission of light continually occurring. Some of these transitions fall in the visible region of the spectrum and give rise to the flickering appearance.**

21.73 **NaF: cavity prevention (F$^-$)**

   Li$_2$CO$_3$: antidepressant (Li$^+$)

   Mg(OH)$_2$: laxative (Milk of Magnesia$^{\circledR}$)

   CaCO$_3$: calcium supplement, antacid

   BaSO$_4$: radiocontrast agent

21.75 **Both Li and Mg form oxides (Li$_2$O and MgO). Other Group 1A metals (Na, K, etc.) also form peroxides and superoxides. In Group 1A, only Li forms nitride (Li$_3$N), like Mg (Mg$_3$N$_2$).**

   Li resembles Mg in that its carbonate, fluoride, and phosphate have low solubilities.

21.77 We know that Ag, Cu, Au, and Pt are found as free elements in nature, which leaves **Zn** by the process of elimination. Table 18.1 of the text indicates that the standard oxidation potential of Zn is +0.76 V. The positive value indicates that Zn is easily oxidized to Zn^{2+} and will not exist as a free element in nature.

21.79 There are 10.00 g of Na in 13.83 g of the mixture (the mass of sodium in the reactants and products is equal). This amount of Na is equal to the mass of Na in Na$_2$O plus the mass of Na in Na$_2$O$_2$.

   10.00 g Na = mass of Na in Na$_2$O + mass of Na in Na$_2$O$_2$

To calculate the mass of Na in each compound, the grams of compound need to be converted to grams of Na using the mass percentage of Na in the compound. If x equals the mass of Na_2O, then the mass of Na_2O_2 is $13.83 - x$. We set up the following expression and solve for x. We carry an additional significant figure throughout the calculation to minimize rounding errors.

$$10.00 \text{ g Na} = \text{mass of Na in } Na_2O + \text{mass of Na in } Na_2O_2$$

$$10.00 \text{ g Na} = \left[x \text{ g } Na_2O \times \frac{(2)(22.99 \text{ g Na})}{61.98 \text{ g } Na_2O} \right] + \left[(13.83 - x) \text{ g } Na_2O_2 \times \frac{(2)(22.99 \text{ g Na})}{77.98 \text{ g } Na_2O_2} \right]$$

$$10.00 = 0.74185x + 8.1547 - 0.58964x$$

$$0.15221x = 1.8453$$

$x = 12.123$ g, which equals the mass of Na_2O.

The mass of Na_2O_2 is $13.83 - x$, which equals 1.707 g.

The mass percent of each compound in the mixture is:

$$\textbf{\% Na}_2\textbf{O} = \frac{12.123 \text{ g}}{13.83 \text{ g}} \times 100 = \textbf{87.66\%}$$

$$\textbf{\%Na}_2\textbf{O}_2 = 100\% - 87.66\% = \textbf{12.34\%}$$

21.81 First, we calculate the density of O_2 in KO_2 using the mass percentage of O_2 in the compound.

$$\frac{32.00 \text{ g } O_2}{71.10 \text{ g } KO_2} \times \frac{2.15 \text{ g } KO_2}{1 \text{ cm}^3} = 0.968 \text{ g } O_2/\text{cm}^3$$

$$\frac{0.968 \text{ g } O_2}{1 \text{ cm}^3} \times \frac{1000 \text{ cm}^3}{1 \text{ L}} = 968 \text{ g } O_2/\text{L}$$

Now, we can use Equation 11.12 of the text to calculate the pressure of oxygen gas that would have the same density as that provided by KO_2.

$$d = \frac{P\mathcal{M}}{RT}$$

$$P = \frac{dRT}{\mathcal{M}} = \frac{\left(\dfrac{968 \text{ g}}{1 \text{ L}} \right) \left(0.0821 \dfrac{\text{L} \cdot \text{atm}}{\text{K} \cdot \text{mol}} \right) (293 \text{ K})}{\left(\dfrac{32.00 \text{ g } O_2}{1 \text{ mol}} \right)} = \textbf{727 atm}$$

Obviously, using O_2 instead of KO_2 is not practical.

Chapter 22
Coordination Chemistry

Problems

22.11   a. The oxidation number of Cr is **+3**.

   b. The coordination number of Cr is **6**.

   c. **Oxalate ion** $(C_2O_4^{2-})$ is a bidentate ligand.

22.13   **Strategy:**   The oxidation number of the metal atom is equal to its charge. First we look for known charges in the species. Recall that alkali metals are +1 and alkaline earth metals are +2. Also determine if the ligand is a charged or neutral species. From the known charges, we can deduce the net charge of the metal and hence its oxidation number.

   Solution: a. Since **Na** is always +1 and the oxygen is –2, **Mo** must have an oxidation number of **+6**.

   b. **Mg** is **+2** and oxygen –2; therefore, **W** is **+6**.

   c. CO ligands are neutral species, so the iron atom bears no net charge. The oxidation number of **Fe** is **0**.

22.15   **Strategy:**   We follow the procedure for naming coordination compounds outlined in Section 22.1 of the text and refer to Tables 22.4 and 22.5 of the text for names of ligands and anions containing metal atoms.

   Solution: a. Ethylenediamine is a neutral ligand, and each chloride has a –1 charge. Therefore, cobalt has an oxidation number of +3.
   The correct name for the ion is ***cis*-dichlorobis(ethylenediamine)cobalt(III)**. The prefix *bis* mean two; we use this instead of *di* because *di* already appears in the name ethylenediamine.

   b. There are four chlorides each with a –1 charge; therefore, Pt has a +4 charge.
   The correct name for the compound is **pentaamminechloroplatinum(IV) chloride**.

   c. There are three chlorides each with a –1 charge; therefore, Co has a +3 charge.
   The correct name for the compound is **pentaamminechlorocobalt(III) chloride**.

22.17   The atoms in a ligand that bind directly to the metal atom are known as the donor atoms. Ligands play the role of Lewis bases and the transition metals act as Lewis acids, accepting (and sharing) pairs of electrons from the Lewis bases.

   In DMSA, all four oxygen atoms, as well as both sulfur atoms, contain at least one unshared pair of valence electrons. Therefore, this ligand has **six** possible donor atoms.

22.23   a. In general for any MA$_2$B$_4$ octahedral molecule, only **two** geometric isomers are possible. The only real distinction is whether the two A-ligands are *cis* or *trans*.

   b. A model or a careful drawing is very helpful to understand the MA$_3$B$_3$ octahedral structure. There are only **two** possible geometric isomers. The first has all A's (and all B's) *cis*; this is called the facial isomer. The second has two A's (and two B's) at opposite ends of the molecule (*trans*). Try to make or draw other possibilities.

22.25 a. There are *cis* and *trans* geometric isomers (see Problem 22.23). No optical isomers.

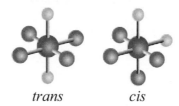

trans   *cis*

b. There are two optical isomers. The three bidentate ethylenediamine ligands are represented by the curved lines. (See Figure 22.3 of the text.)

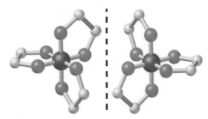

22.31 When a substance appears to be yellow, it is absorbing light from the blue-violet, high energy end of the visible spectrum. Often this absorption is just the tail of a strong absorption in the ultraviolet. Substances that appear green or blue to the eye are absorbing light from the lower energy red or orange part of the spectrum.

Cyanide ion (CN^-) is a very strong field ligand. It causes a larger crystal field splitting than water (H_2O), resulting in the absorption of higher energy (shorter wavelength) radiation when a *d* electron is excited to a higher energy *d* orbital.

22.33 a. Wavelengths of 470 nm fall between blue and blue-green, corresponding to an observed color in the **orange** part of the spectrum.

b. We convert wavelength to photon energy using the Planck relationship.

$$E = \frac{hc}{\lambda} = \frac{\left(6.63 \times 10^{-34}\ J \cdot s\right)\left(3.00 \times 10^{8}\ m/s\right)}{470 \times 10^{-9}\ m} = 4.23 \times 10^{-19}\ J$$

Here $E = \Delta$, so we have crystal field splitting (Δ) in kJ/mol is:

$$\frac{4.23 \times 10^{-19}\ J}{1\ photon} \times \frac{6.022 \times 10^{23}\ photons}{1\ mol} \times \frac{1\ kJ}{1000\ J} = \textbf{255 kJ / mol}$$

22.35 *Step 1:* The equation for freezing-point depression is:

$$\Delta T_f = K_f m$$

Solve this equation algebraically for molality (*m*), then substitute ΔT_f and K_f into the equation to calculate the molality.

$$m = \frac{\Delta T_f}{K_f} = \frac{0.56°C}{1.86°C/m} = 0.30\ m \text{ (see Table 13.2)}$$

Step 2: Multiplying the molality by the mass of solvent (in kg) gives moles of unknown solute. Then, dividing the mass of solute (in g) by the moles of solute gives the molar mass of the unknown solute.

$$? \text{ mol of unknown solute} = \frac{0.30 \text{ mol solute}}{1 \text{ kg water}} \times 0.0250 \text{ kg water} = 0.0075 \text{ mol solute}$$

$$\text{molar mass of unknown} = \frac{0.875 \text{ g}}{0.0075 \text{ mol}} = 117 \text{ g/mol}$$

The molar mass of $Co(NH_3)_4Cl_3$ is 233.4 g/mol, which is twice the computed molar mass. This implies dissociation into two ions in solution; hence, there are **two moles** of ions produced per mole of $Co(NH_3)_4Cl_3$. The formula must be:

$$\textbf{[Co(NH}_3\textbf{)}_4\textbf{Cl}_2\textbf{]Cl}$$

which contains the complex ion $[Co(NH_3)_4Cl_2]^+$ and a chloride ion, Cl^-. **Refer to Problem 22.25 (a) for a diagram of the structure** of the complex ion.

22.37 The crystal field splitting (Δ) **would be greater for the higher oxidation state.**

22.41 **Use a radioactive label such as $^{14}CN^-$ (in NaCN). Add NaCN to a solution of $K_3Fe(CN)_6$. Isolate some of the $K_3Fe(CN)_6$ and check its radioactivity. If the complex shows radioactivity, then it must mean that the CN^- ion has participated in the exchange reaction.**

22.43 **$Cu(CN)_2$ is the white precipitate.**

$$Cu^{2+}(aq) + 2CN^-(aq) \longrightarrow Cu(CN)_2(s)$$

It is soluble in KCN(aq) due to formation of $[Cu(CN)_4]^{2-}$,

$$Cu(CN)_2(s) + 2CN^-(aq) \longrightarrow [Cu(CN)_4]^{2-}(aq)$$

so the concentration of Cu^{2+} is too small for Cu^{2+} ions to precipitate with sulfide.

22.45 The formation constant expression is:

$$K_f = \frac{\left[[Fe(H_2O)_5NCS]^{2+}\right]}{\left[[Fe(H_2O)_6]^{3+}\right]\left[SCN^-\right]}$$

Notice that the original volumes of the Fe(III) and SCN^- solutions were both 1.0 mL and that the final volume is 10.0 mL. This represents a 10-fold dilution, and the concentrations of Fe(III) and SCN^- become 0.020 M and 1.0×10^{-4} M, respectively.

We make a table.

	$[Fe(H_2O)_6]^{3+}$ +	SCN^-	$\rightleftarrows$	$[Fe(H_2O)_5NCS]^{2+}$	+ H_2O
Initial (M):	0.020	1.0×10^{-4}		0	0.00
Change (M):	-7.3×10^{-5}	-7.3×10^{-5}		$+7.3 \times 10^{-5}$	$+x$
Equilibrium (M):	0.020	2.7×10^{-5}		7.3×10^{-5}	x

$$K_f = \frac{7.3 \times 10^{-5}}{(0.020)\left(2.7 \times 10^{-5}\right)} = 1.4 \times 10^2$$

22.47 The purple color is caused by the buildup of deoxyhemoglobin. When either oxyhemoglobin or deoxyhemoglobin takes up CO, the carbonylhemoglobin takes on a red color, the same as oxyhemoglobin.

22.49 Mn^{3+} is $3d^4$ and Cr^{3+} is $3d^5$. Therefore, Mn^{3+} has a greater tendency to accept an electron and is a stronger oxidizing agent. The $3d^5$ electron configuration of Cr^{3+} is a stable configuration.

22.51 Ti^{3+}; Fe^{3+}

22.53 A 100.00 g sample of hemoglobin contains 0.34 g of iron. In moles this is:

$$0.34 \text{ g Fe} \times \frac{1 \text{ mol}}{55.85 \text{ g}} = 6.1 \times 10^{-3} \text{ mol Fe}$$

The amount of hemoglobin that contains 1 mole of iron must be:

$$\frac{100.00 \text{ g hemoglobin}}{6.1 \times 10^{-3} \text{ mol Fe}} = 1.6 \times 10^4 \text{ g hemoglobin / mol Fe}$$

We compare this to the actual molar mass of hemoglobin:

$$\frac{6.5 \times 10^4 \text{ g hemoglobin}}{1 \text{ mol hemoglobin}} \times \frac{1 \text{ mol Fe}}{1.6 \times 10^4 \text{ g hemoglobin}} = 4.1 \text{ mol Fe/1 mol hemoglobin}$$

The discrepancy between our minimum value and the actual value can be explained by realizing that there are four iron atoms per mole of hemoglobin.

22.55 a. $[Cr(H_2O)_6]Cl_3$, number of ions: 4

b. $[Cr(H_2O)_5Cl]Cl_2 \cdot H_2O$, number of ions: 3

c. $[Cr(H_2O)_4Cl_2]Cl \cdot 2H_2O$, number of ions: 2

Compare the compounds with equal molar amounts of $NaCl$, $MgCl_2$, and $FeCl_3$ in an electrical conductance experiment. The solution that has conductance similar to the $NaCl$ solution contains (c); the solution with the conductance similar to $MgCl_2$ contains (b); and the solution with conductance similar to $FeCl_3$ contains (a).

22.57

$$Zn(s) \longrightarrow Zn^{2+}(aq) + 2e^- \qquad E^\circ_{anode} = -0.76 \text{ V}$$
$$2[Cu^{2+}(aq) + e^- \longrightarrow Cu^+(aq)] \qquad E^\circ_{cathode} = 0.15 \text{ V}$$
$$\overline{Zn(s) + 2Cu^{2+}(aq) \longrightarrow Zn^{2+}(aq) + 2Cu^+(aq)}$$

$$E^\circ_{cell} = E^\circ_{cathode} - E^\circ_{anode} = 0.15 \text{ V} - (-0.76 \text{ V}) = 0.91 \text{ V}$$

We carry additional significant figures throughout part of this calculation to minimize rounding errors.

$$\Delta G° = -nFE° = -(2)(96{,}500 \text{ J/V·mol})(0.91 \text{ V}) = -1.756 \times 10^5 \text{ J/mol} = \mathbf{-1.8 \times 10^2 \text{ kJ/mol}}$$

$$\Delta G° = -RT\ln K$$

$$\ln K = \frac{-\Delta G°}{RT} = \frac{-\left(-1.756 \times 10^5 \text{ J/mol}\right)}{(8.314 \text{ J/K·mol})(298 \text{ K})}$$

$$\ln K = 70.9$$

$$\boldsymbol{K = e^{70.9} = 6 \times 10^{30}}$$

22.59 **Iron is much more abundant than cobalt**.

22.61 **Oxyhemoglobin absorbs higher energy light than deoxyhemoglobin. Oxyhemoglobin is diamagnetic (low spin), while deoxyhemoglobin is paramagnetic (high spin). These differences occur because oxygen (O_2) is a strong-field ligand.** The crystal field splitting diagrams are:

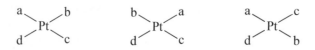

deoxyhemoglobin   oxyhemoglobin

22.63 **Complexes are expected to be colored when the highest occupied orbitals have between one and nine**
d **electrons. Such complexes can therefore have** $d \longrightarrow d$ **transitions (that are usually in the visible part of the electromagnetic radiation spectrum). Complexes with** d^0 **or** d^{10} **configurations are colorless. Zn^{2+}, Cu^+, and Pb^{2+} are** d^{10} **ions. V^{5+}, Ca^{2+}, and Sc^{3+} are** d^0 **ions.**

22.65 **Dipole moment measurement. Only the *cis* isomer has a dipole moment**.

22.67 **EDTA sequesters metal ions (like Ca^{2+} and Mg^{2+}) which are essential for the growth and function of bacteria.**

22.69 The square planar complex shown in the problem has **three** geometric isomers. They are:

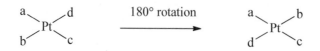

Note that in the first structure a is *trans* to c, in the second a is *trans* to d, and in the third a is *trans* to b. Make sure you realize that if we switch the positions of b and d in structure 1, we do not obtain another geometric isomer. A 180° rotation about the a–Pt–c axis gives structure 1.

22.71 The reaction is:

$$Ag^+(aq) + 2CN^-(aq) \rightleftharpoons [Ag(CN)_2]^-(aq)$$

$$K_f = 1.0 \times 10^{21} = \frac{\left[[Ag(CN)_2]^- \right]}{[Ag^+][CN^-]^2}$$

First, we calculate the initial concentrations of Ag^+ and CN^-. Then, because K_f is very large, we assume that the reaction goes to completion. This assumption will allow us to solve for the concentration of Ag^+ at equilibrium. The initial concentrations of Ag^+ and CN^- are:

$$[CN^-] = \frac{\dfrac{5.0 \text{ mol}}{1 \text{ L}} \times 9.0 \text{ L}}{99.0 \text{ L}} = 0.455 \ M$$

$$[Ag^+] = \frac{\dfrac{0.20 \text{ mol}}{1 \text{ L}} \times 90.0 \text{ L}}{99.0 \text{ L}} = 0.182 \ M$$

We set up a table to determine the concentrations after complete reaction.

	$Ag^+(aq) +$	$2CN^-(aq)$	$\rightleftharpoons$	$[Ag(CN)_2]^-(aq)$
Initial (*M*):	0.182	0.455		0
Change (*M*):	−0.182	−(2)(0.182)		+0.182
Equilibrium (*M*):	0	0.0910		0.182

$$K_f = \frac{\left[[Ag(CN)_2]^- \right]}{[Ag^+][CN^-]^2}$$

$$1.0 \times 10^{21} = \frac{0.182 \ M}{[Ag^+](0.0910 \ M)^2}$$

$$\boxed{[Ag^+] = 2.2 \times 10^{-20} \ M}$$

22.73 a. The equilibrium constant can be calculated from $\Delta G°$. We can calculate $\Delta G°$ from the cell potential.

From Table 18.1 of the text,

$$Cu^{2+} + 2e^- \longrightarrow Cu \quad E° = 0.34 \text{ V and } \Delta G° = -(2)(96,500 \text{ J/V·mol})(0.34 \text{ V}) = -6.562 \times 10^4 \text{ J/mol}$$

$$Cu^{2+} + e^- \longrightarrow Cu^+ \quad E° = 0.15 \text{ V and } \Delta G° = -(1)(96,500 \text{ J/V·mol})(0.15 \text{ V}) = -1.448 \times 10^4 \text{ J/mol}$$

These two equations need to be arranged to give the disproportionation reaction in the problem. We keep the first equation as written and reverse the second equation and multiply by 2.

$$Cu^{2+} + 2e^- \longrightarrow Cu \quad \Delta G° = -6.562 \times 10^4 \text{ J/mol}$$

$$2Cu^+ \longrightarrow 2Cu^{2+} + 2e^- \quad \Delta G° = +(2)(1.448 \times 10^4 \text{ J/mol})$$

$$2Cu^+ \longrightarrow Cu^{2+} + Cu \quad \Delta G° = -6.562 \times 10^4 \text{ J/mol} + 2.896 \times 10^4 \text{ J/mol} = -3.666 \times 10^4 \text{ J/mol}$$

We use Equation 15.6 of the text to calculate the equilibrium constant.

$$\Delta G^\circ = -RT \ln K$$

$$K = e^{-\Delta G^\circ / RT}$$

$$K = e^{-(-3.666 \times 10^4 \, \text{J/mol})/(8.314 \, \text{J/K·mol})(298 \, \text{K})}$$

$$\boldsymbol{K = 2.7 \times 10^6}$$

b. **Free Cu$^+$ ions are unstable in solution [as shown in part (a)]. Therefore, the only stable compounds containing Cu$^+$ ions are insoluble.**

22.75 When a substance appears blue, it is absorbing light from the orange part of the spectrum. Substances that appear yellow are absorbing light from the higher energy blue-violet part of the spectrum. Thus, **Y** causes a larger crystal field splitting than X, so it is a stronger field ligand.

22.77 To find the concentration at equilibrium, we make the following table.

	Pb^{2+} +	$EDTA^{4-}$	$\rightleftarrows$	$[Pb(EDTA)]^{2-}$
Initial (M):	1.0×10^{-3}	2.0×10^{-3}		0
Change (M):	$-x$	$-x$		$+x$
Equilibrium (M):	$1.0 \times 10^{-3} - x$	$2.0 \times 10^{-3} - x$		x

$$K_f = \frac{\left[Pb[EDTA]^{2-} \right]}{[Pb^{2+}][EDTA^{4-}]} = 1.0 \times 10^{18}$$

$$1.0 \times 10^{18} = \frac{x}{(1.0 \times 10^{-3} - x)(2.0 \times 10^{-3} - x)}$$

Because K_f is very large, we can either assume that the reaction goes to completion or solve the equation. *If the reaction goes to completion, Pb^{2+} is the limiting reagent.* The equilibrium concentration of $[Pb(EDTA)]^{2-}$ would be 1.0×10^{-3} M. Solving the equation would be done as follows.

$$1.0 \times 10^{18} = \frac{x}{\left(2.0 \times 10^{-6}\right) - \left(3.0 \times 10^{-3}\right)x + x^2}$$

$$0 = \left(1.0 \times 10^{18}\right)x^2 - \left(3.0 \times 10^{15}\right)x + \left(2.0 \times 10^{12}\right)$$

Use the quadratic equation to solve for x.

$$x = \frac{-b \pm \sqrt{b^2 - 4ac}}{2a}$$

$$= \frac{\left(3.0 \times 10^{15}\right) \pm \left(1.0 \times 10^{15}\right)}{2.0 \times 10^{18}}$$

$$= 2.0 \times 10^{-3} \quad \text{or} \quad 1.0 \times 10^{-3}$$

Since the equilibrium concentration of Pb^{2+} is $1.0 \times 10^{-3} - x$, the first answer is not possible. Thus, the equilibrium concentration of $[Pb(EDTA)]^{2-}$ is **0.0 M**.

22.79   The oxidation number of the metal atom is equal to its charge. Low-spin means that the crystal field splitting is large, so the lower orbitals will fill completely before the upper orbitals are populated.

$[Mn(CN)_6]^{5-}$

Each cyanide ligand is −1, so **Mn is +1**. Its electron configuration is $[Ar]4s^1 3d^5$. Thus, it has **one unpaired d electron**.

$$\overline{d_{z^2}} \qquad \overline{d_{x^2-y^2}}$$

$$\frac{\uparrow\downarrow}{d_{xy}} \qquad \frac{\uparrow\downarrow}{d_{xz}} \qquad \frac{\uparrow}{d_{yz}}$$

$[Mn(CN)_6]^{4-}$

Each cyanide ligand is −1, so **Mn is +2**. Its electron configuration is $[Ar]3d^5$. Thus, it has **one unpaired d electron**.

$$\overline{d_{z^2}} \qquad \overline{d_{x^2-y^2}}$$

$$\frac{\uparrow\downarrow}{d_{xy}} \qquad \frac{\uparrow\downarrow}{d_{xz}} \qquad \frac{\uparrow}{d_{yz}}$$

$[Mn(CN)_6]^{3-}$

Each cyanide ligand is −1, so **Mn is +3**. Its electron configuration is $[Ar]3d^4$. Thus, it has **two unpaired d electrons**.

$$\overline{d_{z^2}} \qquad \overline{d_{x^2-y^2}}$$

$$\frac{\uparrow\downarrow}{d_{xy}} \qquad \frac{\uparrow}{d_{xz}} \qquad \frac{\uparrow}{d_{yz}}$$

Chapter 23
Organic Chemistry

Problems

23.7   a. **amine**

    b. **aldehyde**

    c. **ketone**

    d. **carboxylic acid**

    e. **alcohol**

23.9   a.

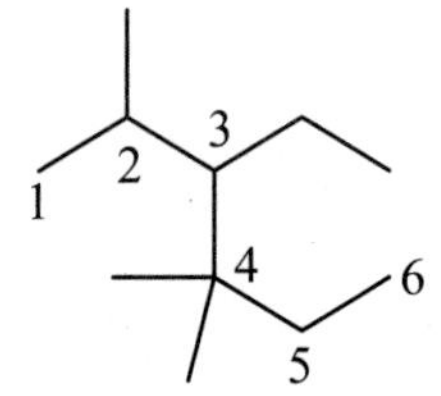

3-ethyl-2,4,4-trimethylhexane

The longest continuous chain has six carbons, so the alkane is named as a derivative of *hexane*. Its substituents, in alphabetical order, are an ethyl group and three methyl groups. It is an ethyl trimethyl derivative of hexane. When the chain is numbered beginning at the end nearest to the first branch, the substituted carbons are C-2, 3, 4, and 4.

    b.

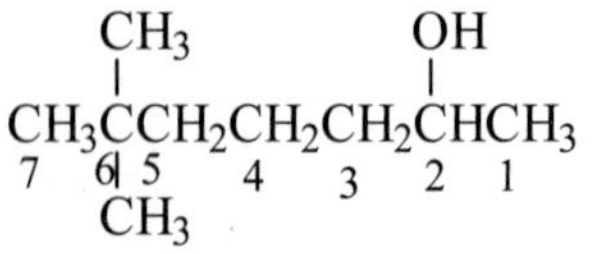

$$\underset{7\ \ \ 6\ \ 5\quad 4\quad 3\quad 2\quad 1}{CH_3CCH_2CH_2CH_2CHCH_3}$$

with CH_3 above and OH above, CH_3 below

6,6-dimethyl-2-heptanol

This compound is named as an alcohol. The longest continuous chain that bears the –OH group has seven carbons. Numbering starts from the end nearest to the –OH group. The compound is a 6,6-dimethyl derivative of 2-heptanol.

    c.

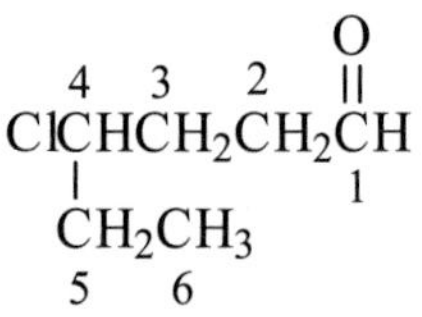

$$\underset{4\quad 3\quad 2\quad 1}{ClCHCH_2CH_2CH}\ \ \overset{O}{\underset{}{\parallel}}$$

with CH_2CH_3 below $\underset{5\quad 6}{}$

4-chlorohexanal

The compound is an aldehyde. When naming aldehydes, we drop the –*e* ending of the alkane name and replace it by –*al*. The chain is numbered beginning at the carbonyl group and is six carbons in length. This *hexanal* derivative has a chlorine substituent at C-4.

23.11   Notice that the longest continuous chain has eight carbons, not seven, and that the chain is numbered right-to-left so that the first-appearing substituent is at C-3, not C-4. The alkane is **3,5-dimethyloctane**.

23.13   a. 2,2,4-trimethylpentane ("isooctane")   **$(CH_3)_3CCH_2CH(CH_3)_2$**

    b. 3-methyl-1-butanol ("isoamyl alcohol")   **$HO(CH_2)_2CH(CH_3)_2$**

    c. hexanamide ("caproamide")   **$CH_3(CH_2)_4C(O)NH_2$**

    d. 2,2,2-trichloroethanal ("chloral")   **Cl_3CCHO**

23.15

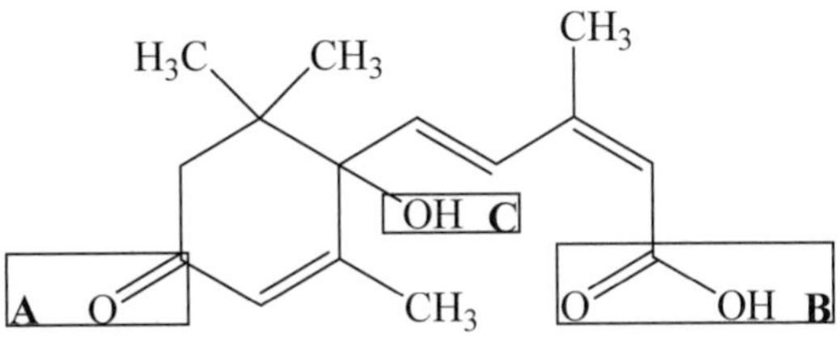

A = Carbonyl (Ketone)
B = Carboxy (Carboxylic acid)
C = Hydroxy (Alcohol)

23.17 Primary amino / Carboxy

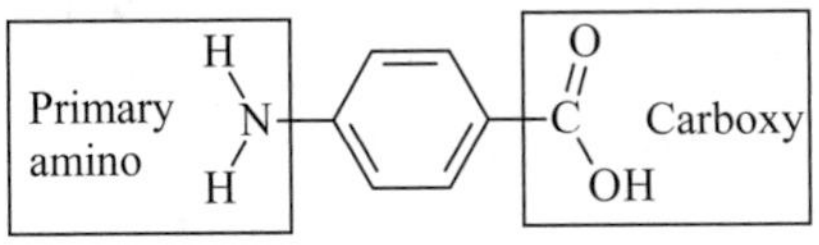

23.19 a. CH₃CH₂CHCH₂CH₂CH₃
              |
              CH₃

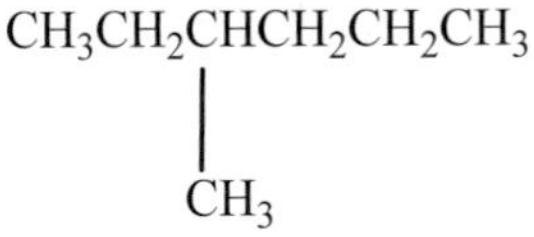

b.   CH₃
          |
   CH₃CHCHCH₂CH₃
          |
          CH₃

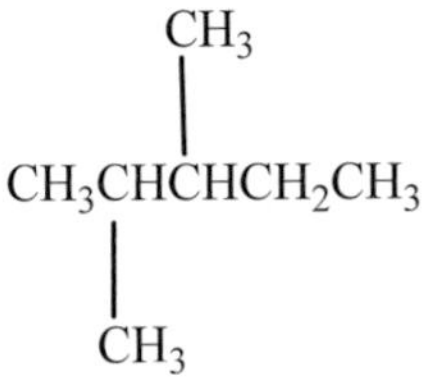

c.   Br
       |
   CH₃CHCH₂CHCH₃
            |
            C₆H₅

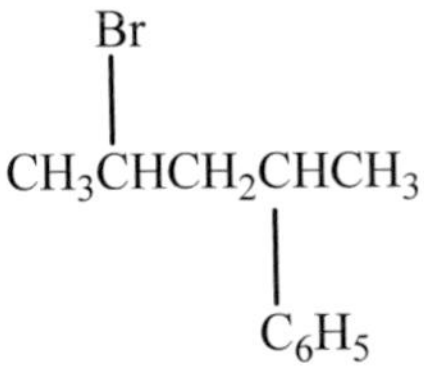

d.   CH₃ CH₃
      | |
   CH₃CH₂CHCHCHCH₂CH₂CH₃
          |
          CH₃

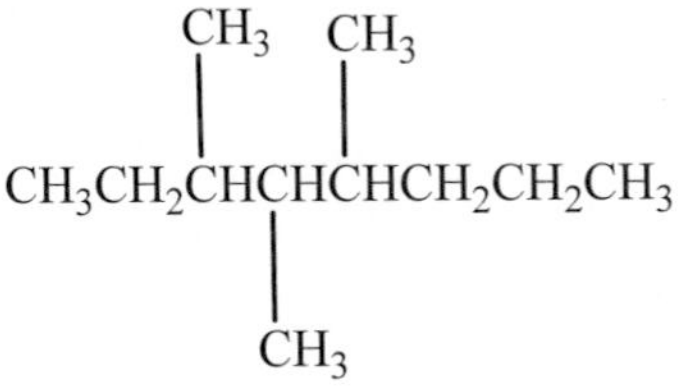

23.21 a. **Kekulé:**

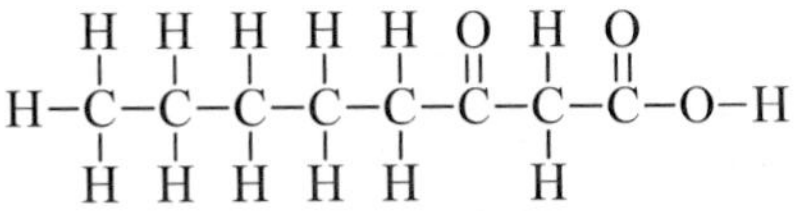

skeletal (line):

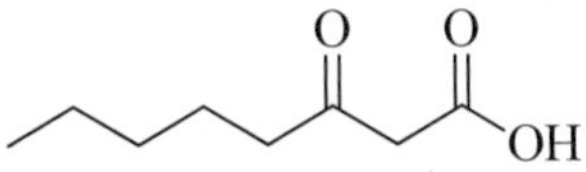

b. **condensed:** (C₂H₅)₂CHCH₂CO₂C(CH₃)₃

skeletal (line):

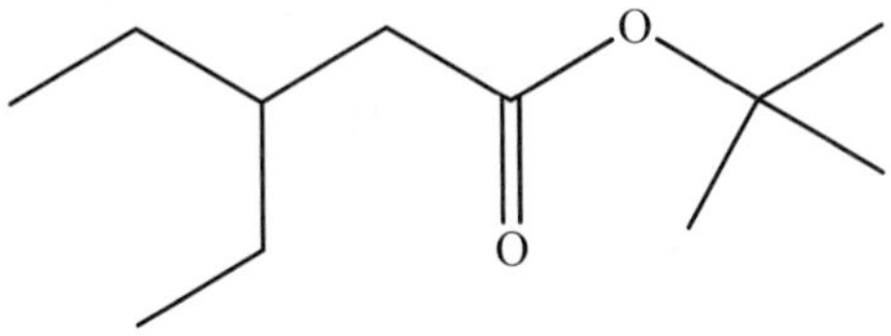

c. **condensed:** **(CH₃)₂CHCH₂NHCH(CH₃)₂**

Kekulé:

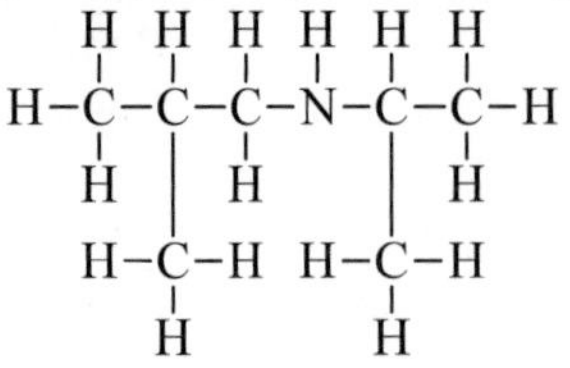

```
        H  H  H    H  H  H
        |  |  |    |  |  |
    H − C− C− C− N− C− C− H
        |  |  |       |
        H  |  H       H
           |          |
        H− C− H    H− C− H
           |          |
           H          H
```

23.23   a.   C₃H₇NO: DMF is

condensed structural: (CH₃)₂NCHO

Kekulé:

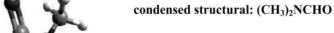

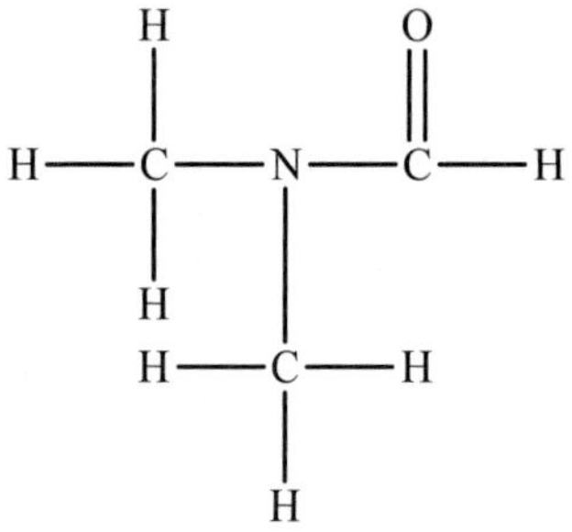

line:

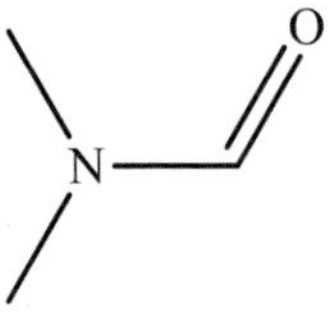

b.   C₆H₈O₇: citric acid is

condensed structural:

(CH₂COOH)₂C(OH)COOH

Kekulé:

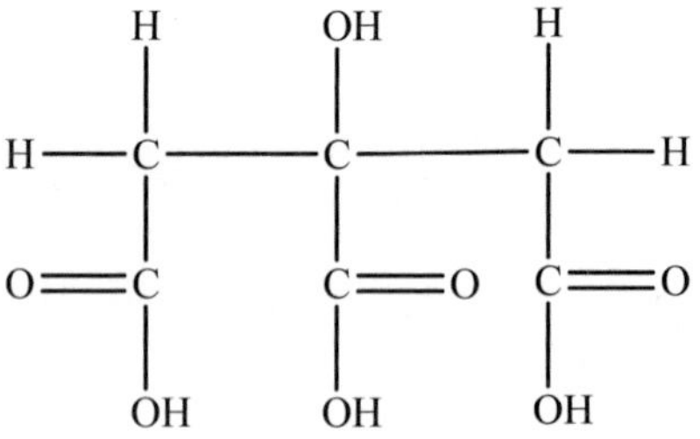

line:

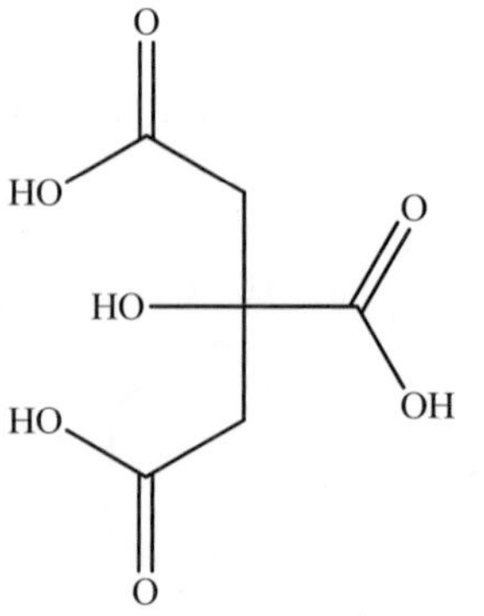

c.

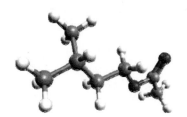

C$_7$H$_{14}$O$_2$: isoamyl acetate is

condensed structural:
$$(CH_3)_2CH(CH_2)_2OC(O)CH_3$$

Kekulé:

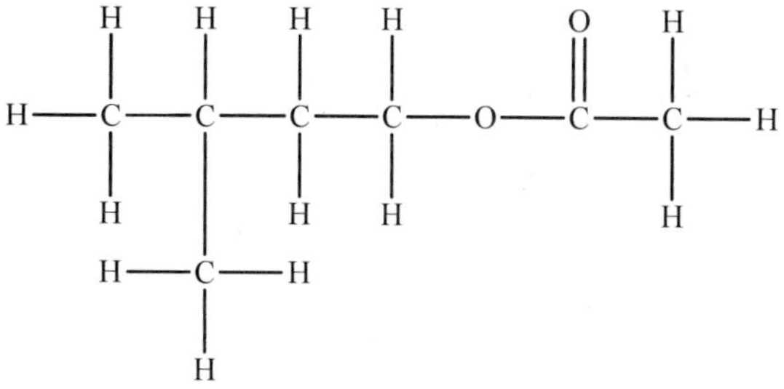

line:

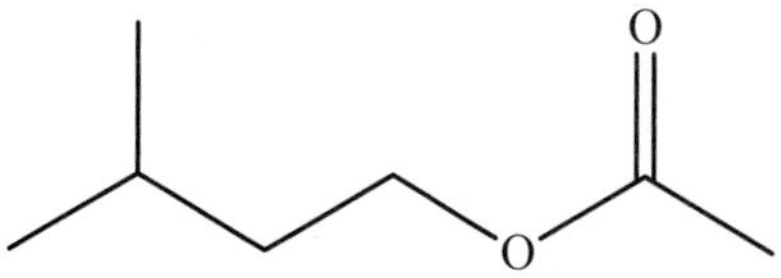

23.25

a. **line:**

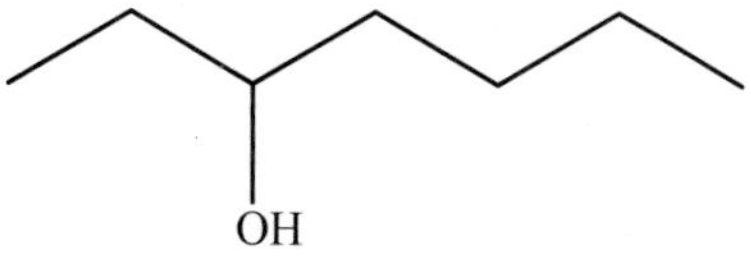

b.

$$(CH_3)_3CCH_2CHCH_2CH$$

structural:

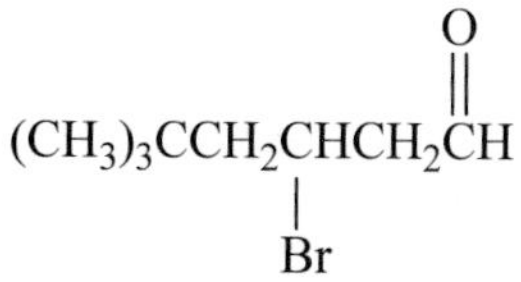

c.

line:

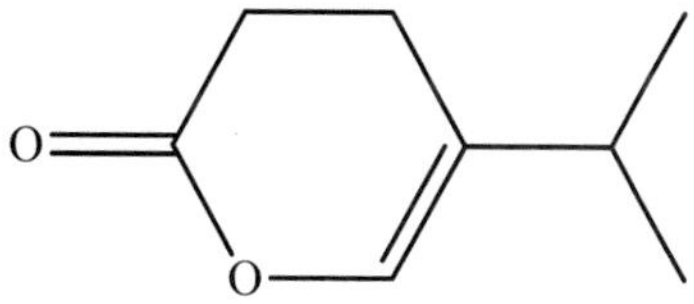

23.27

a. CH$_3$—C≡N: ⟷ CH$_3$—C=N:

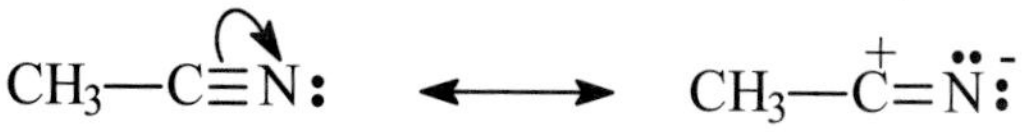

b.

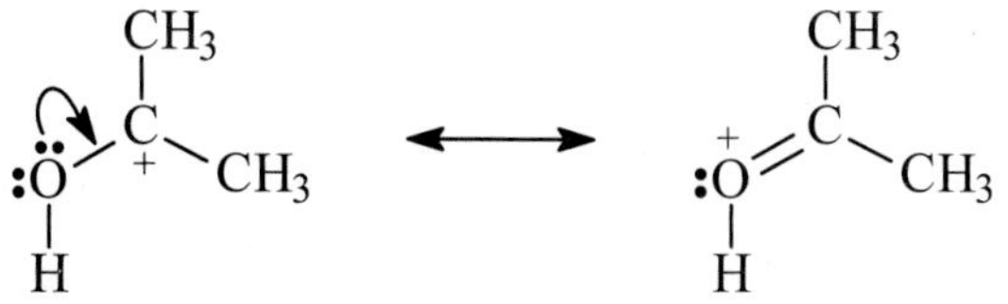

c.

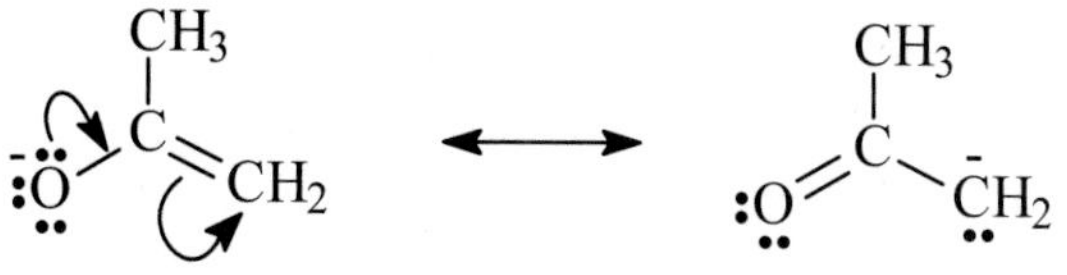

23.29 a.

b.

c.

23.37 There are nine possible structural isomers:

23.39 There are five possible isomers:

23.41 There are four possible structural isomers, that contain one benzene ring:

23.43 a.

$$CH_3-CH_2-\underset{\underset{NH_2}{|}}{\overset{\overset{CH_3}{|}}{CH}}-CH-\overset{\overset{O}{||}}{C}-NH_2 \quad CH_3-CH_2-\underset{\underset{NH_2}{\overset{*}{|}}}{\overset{\overset{CH_3}{|}}{CH}}-\overset{*}{CH}-\overset{\overset{O}{||}}{C}-NH_2$$

b.

Although the cyclopropane has two asymmetric carbon atoms, the molecule is not chiral because it has a plane of symmetry.

23.45

$$CH_3CH_2\overset{\overset{O}{||}}{C}-H$$

23.47 a.

b.

c.

d.

23.55

23.57 a. The aldehyde C has a partial positive charge since O is more electronegative than C; this positive charge attracts the negative end of the acetylide (nucleophilic attack), and as the C—C coordinate covalent bond forms, one of the C—O bond pairs flows onto the O atom and becomes a lone pair. The now negative O atom then attracts the positive (H) end of a water molecule, and as the O—H coordinate covalent bond forms, the H—O bond pair flows onto the O atom of the water molecule to form the hydroxyl ion leaving group.

b. The C atom attached to Br has a partial positive charge because Br is more electronegative than C. As the flexible C_4 chain wiggles and bends, the negative S atom can approach and be attracted to the positive C atom (nucleophilic attack), and as the S—C coordinate covalent bond forms, the C—Br bond pair flows onto the Br atom to produce the bromide leaving group.

23.59 Cl in an aromatic ring is an electron-withdrawing group, causing partial positive charges to appear on the H atom(s) next to it. The amide anion is attracted to this charge (nucleophilic attack), and as the N—H coordinate covalent bond forms, the C—H bond pair is forced onto the ring as a pi bond, which in turn forces the Cl—C bond pair onto the Cl to form the chloride leaving group. This is an **elimination** reaction.

23.61 **No. This is not an oxidation-reduction reaction. There are no changes in oxidation states for any of the atoms.**

23.63 a. **The sulfuric acid releases protons, which then protonate the terminal CH₂ group. This creates a cation; nucleophilic attack by water produces the alcohol with release of another proton. Sulfuric acid is thus a catalyst.**

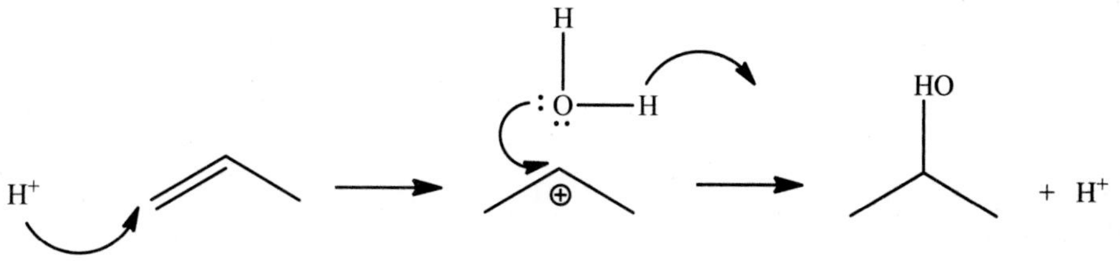

b. ***n*-propanol:**

c. **No.** Isopropanol is not chiral since the central C atom is bonded to two identical (CH₃) groups.

23.65 **Strategy:** Count the C atoms represented and the heteroatoms shown. Determine how many H atoms are present using the octet rule.

Setup: The structures are given on page 975.

Each line represents a bond. (Double lines represent double bonds.) Count one C atom at the end of each line unless another atom is shown there. Count the number of H atoms necessary to complete the octet of each C atom. Count the number of heteroatoms.

Solution: **thalidomide: $C_{13}H_{10}N_2O_4$**

lenalidomide: $C_{13}H_{13}N_3O_3$

CC-4047: $C_{13}H_{11}N_3O_4$

23.73

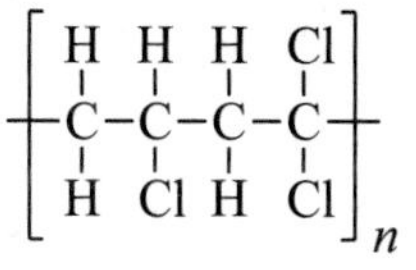

$$\left[\begin{matrix} \text{H} & \text{H} & \text{H} & \text{Cl} \\ | & | & | & | \\ \text{C} - \text{C} - \text{C} - \text{C} \\ | & | & | & | \\ \text{H} & \text{Cl} & \text{H} & \text{Cl} \end{matrix} \right]_n$$

23.75 a. $CH_2=CHCH=CH_2$

b.

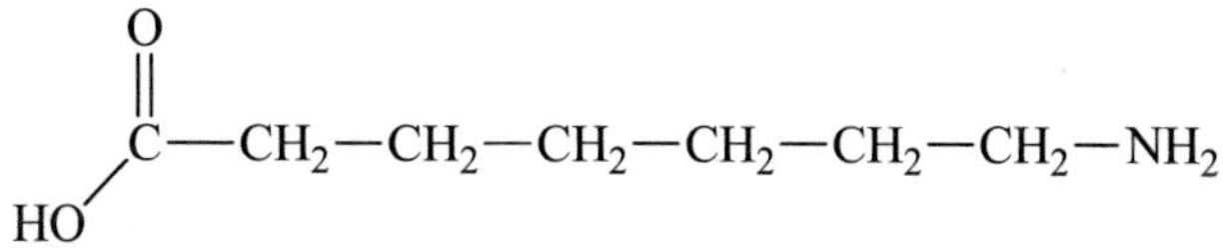

23.77 There are two: glycine-lysine and lysine-glycine.

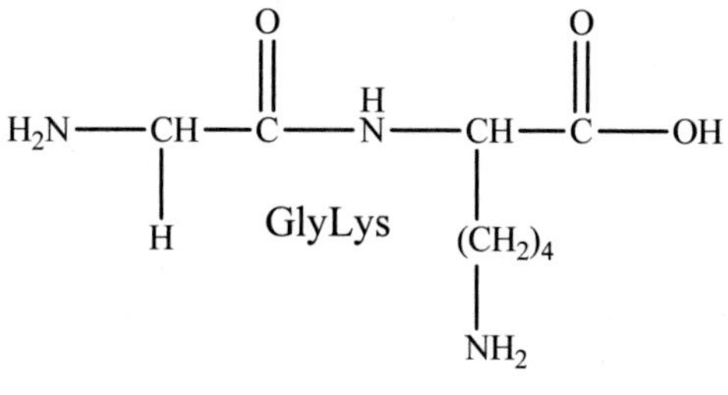

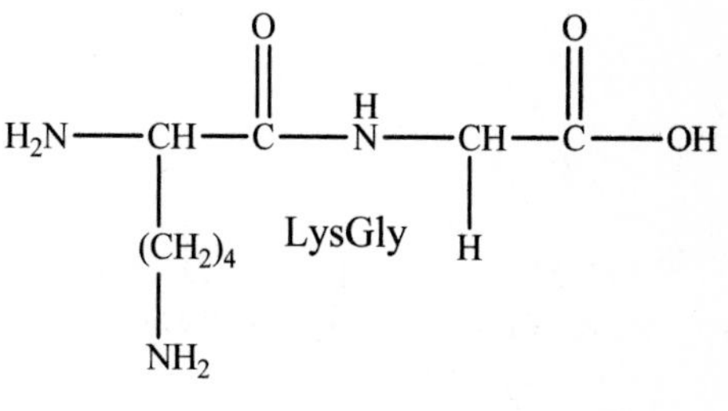

23.79 The structures and names of the constitutionally isomeric C_4H_9 alkyl groups are:

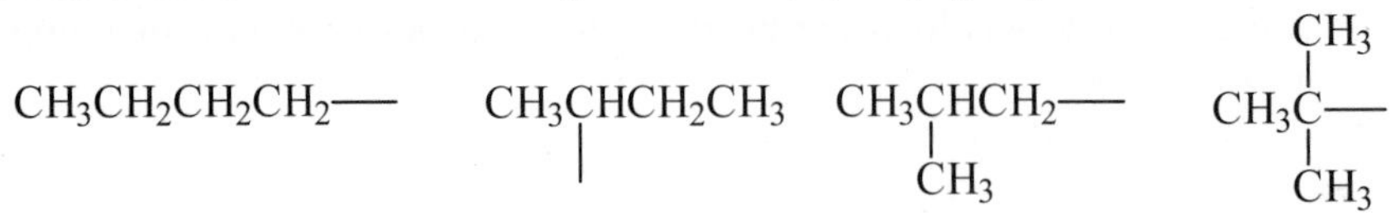

$$CH_3CH_2CH_2CH_2— \qquad CH_3CHCH_2CH_3 \qquad CH_3CHCH_2— \qquad CH_3C—$$

| Butyl | *sec*-Butyl | Isobutyl | *tert*-Butyl |

23.81 The two common α-amino acids with $R = C_4H_9$ are **leucine (Leu, isobutyl) and isoleucine (Ile, 2-*n*-butyl):**

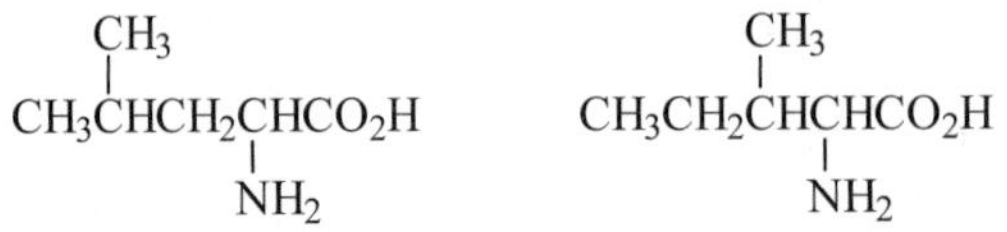

$$CH_3CHCH_2CHCO_2H \qquad CH_3CH_2CHCHCO_2H$$

Leucine   Isoleucine

23.83 a. 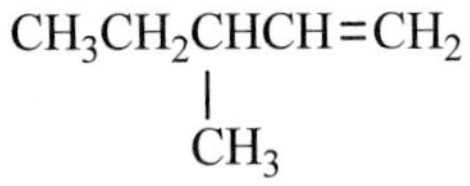 $CH_3CH_2CHCH=CH_2$
$\qquad\qquad\quad CH_3$

b.

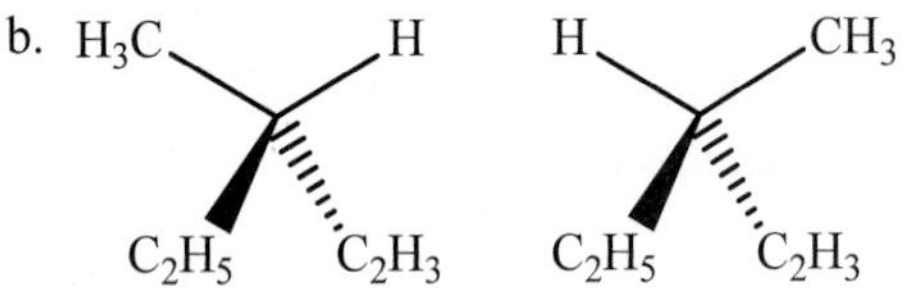

23.85 a. *cis/trans* **stereoisomers**

b. **constitutional isomers**

c. **resonance structures**

d. **different representations of the same structure**

23.87

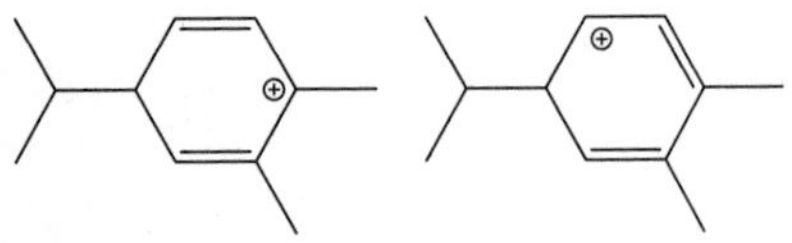

23.89 **Statement (b):** Two enantiomers of 2-chlorobutane were formed in equal amounts.

23.91 There is one noncyclic isomer of C_4H_6 with two double bonds (1,3-butadiene) and two isomers with one triple bond (1-butyne and 2-butyne).

$$sp^3 \; sp^3 \qquad\qquad sp^3 \qquad sp^3$$
$$CH_3CH_2C{\equiv}CH \qquad CH_3C{\equiv}CCH_3 \qquad H_2C=CHCH=CH_2$$
$$sp \; sp \qquad\qquad sp \; sp \qquad\qquad \text{all } sp^2$$

23.93 **Since N is less electronegative than O, electron donation in the amide would be more pronounced.**

23.95 a. **The more negative $\Delta H°$ implies stronger alkane bonds; branching decreases the total bond enthalpy (and overall stability) of the alkane.**

b. **The least highly branched isomer (*n*-octane)** would have the greatest total bond energy and should thus produce the most heat. This prediction corresponds to that determined experimentally.

23.97 a. **15.81 mg C; 1.32 mg H; 3.49 mg O** (see Chapter 5).

b. empirical formula: C_6H_6O

c. Two can be found in the *Handbook of Chemistry and Physics*:

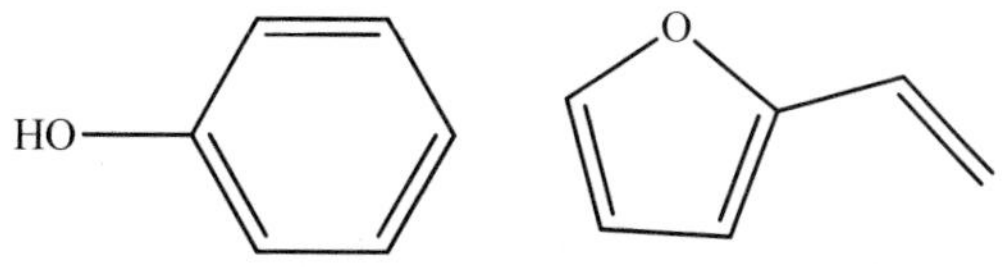

23.99

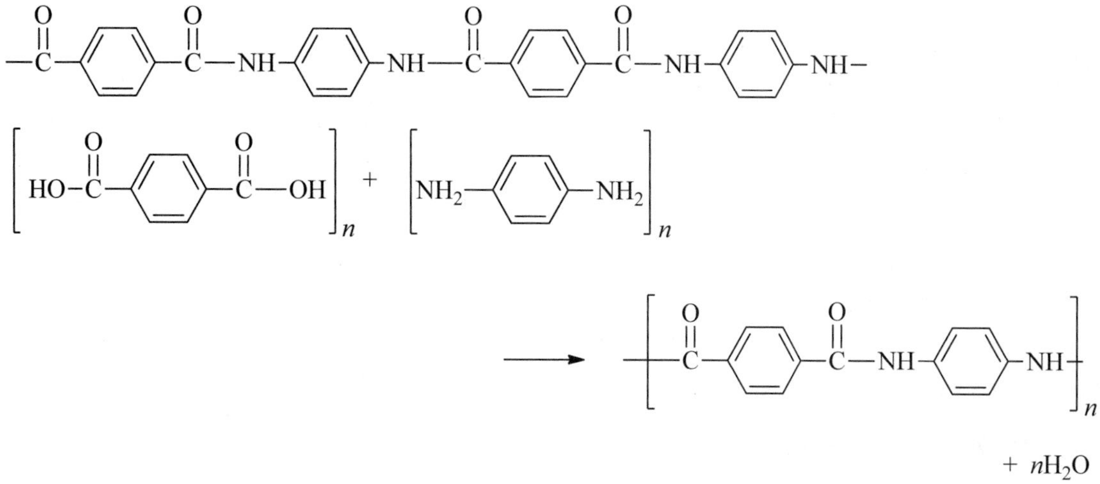

23.101

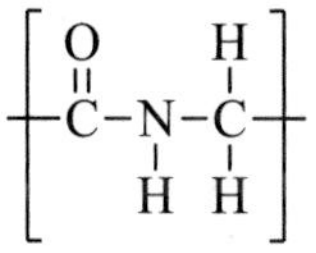

23.103 **There are $2^3 = 8$ combinations** of these two letters (amino acids):

Lys–Lys–Lys, Lys–Lys–Ala, Lys–Ala–Lys, Ala–Lys–Lys, Lys–Ala–Ala, Ala–Lys–Ala, Ala–Ala–Lys, Ala–Ala–Ala.

If it is specified that each tripeptide must contain both amino acids, then there are six such combinations.

23.105 **Strategy:** Starting with the structures of thalidomide and CC-4047, determine whether and where electrons can be repositioned to produce one or more additional structures. Indicate the movement of electrons with curved arrows, and draw all possible resonance structures. Calculate the formal charge on each atom to determine the placement of charges.

Setup: The first step of the reaction involves the substitution of an amino group for one of the H atoms on the aromatic portion of the thalidomide molecule. The next step involves a rearrangement of carbocation to form a more stable structure. A hydrogenation reaction replaces the O atoms on the nitrogen with H atoms yielding the final structure of CC-4047.

The positive charge on the carbocation can be delocalized to yield the resonance structures.

Solution: The mechanism for this reaction is:

The resonance structures for the carbocation intermediate are:

Chapter 24
Modern Materials

Problems

24.3 **The monomer must have a triple bond.**

24.5 There are two possible isomers, depending on whether the two H atoms **in the double bond are *cis*- or *trans*-.**

24.9 **(1)** **Produce the alkoxide:**

$$Sc(s) + 3C_2H_5OH(l) \longrightarrow Sc(OC_2H_5)_3(alc) + 3H^+(alc) \quad (\text{"alc" indicates a solution in alcohol})$$

(2) **Hydrolyze to produce hydroxide pellets:**

$$Sc(OC_2H_5)_3(alc) + 3H_2O(l) \longrightarrow Sc(OH)_3(s) + 3C_2H_5OH(alc)$$

(3) **Sinter pellets to produce ceramic:**

$$2Sc(OH)_3(s) \longrightarrow Sc_2O_3(s) + 3H_2O(g)$$

24.11 **No. Bakelite is not a simple polymer; its permanence is due to cross-linking of the polymer chains. Bakelite is best described as a thermosetting composite polymer.**

24.15 **No. These molecules are too flexible, and liquid crystals require long, relatively rigid molecules.**

24.19 **As shown, it is an alternating condensation random copolymer of the polyester class.**

24.21 **Metal amalgams expand with age; composite fillings tend to shrink.**

24.25 Each carbon atom is attached to three other carbon atoms in a (near) plane. Thus, sp^2 hybrids are used.

24.27 The sheets of carbon atoms in graphite are held together by **dispersion forces**. The carbon nanotubes in strands and fibers are also held together by **dispersion forces**.

24.31 a. $4 + 5$: ***n*-type** (one more e^-)   b. $4 + 3$: ***p*-type** (one less e^-)

24.35 BSCCO-2212 is $Bi_2Sr_2CaCu_2O_8$, composed of the ions Bi^{3+}, Sr^{2+}, Ca^{2+}, Cu^{2+}, and O^{2-}.

BSCCO-2201 would therefore be $Bi_2Sr_2Ca_0Cu_1O_6$ or **$Bi_2Sr_2CuO_6$**.

24.37 Two are +2 ([Ar]3d^9), one is +3 ([Ar]3d^8). The +3 oxidation state is unusual for copper.

24.39

$$H_2N-\left(CH_2\right)_8 NH_2 \qquad\qquad HO-\overset{\overset{O}{\|}}{C}\left(CH_2\right)_4\overset{\overset{O}{\|}}{C}-OH$$

24.41 In a plastic (organic) polymer, there are covalent bonds, disulfide (covalent) bonds, hydrogen bonds, and dispersion forces. In ceramics, there are mostly ionic and network covalent bonds.

24.43 Fluoroapatite is less soluble than hydroxyapatite, particularly in acidic solutions. Dental fillings must also be insoluble.

24.45 The molecule is long, flat, and rigid, so it should form a liquid crystal.

24.47 Strategy: Additional polymers, whether they are synthesized from alkynes or alkenes, form through a free-radical reaction in which one pair of pi electrons in the carbon-carbon multiple bond of a monomer molecule is used to form carbon-carbon single bonds to other monomer molecules. Use the structural formulas of poly(1-butyne) and poly(1-butene) to show that the triple and double bonds can be "opened up" to form single bonds between consecutive monomer units.

Setup: By drawing three adjacent poly(1-butyne) and poly(1-butene) molecules, we can rearrange the bonds to show three repeating units:

poly(1-butyne):

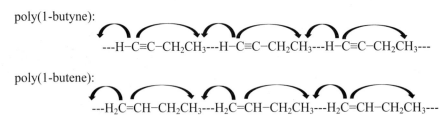

poly(1-butene):

The presence of electrons in delocalized orbitals is the key to electrical conductivity; therefore, the monomer must have a triple bond.

Solution: The structure of poly(1-butyne) is

$$-CH=C(CH_2CH_3)-CH=C(CH_2CH_3)-CH=C(CH_2CH_3)-$$

The structure of poly(1-butene) is

$$-CH_2-CH(CH_2CH_3)-CH_2-CH(CH_2CH_3)-CH_2-CH(CH_2CH_3)-$$

The two structures are similar in that they both have ethyl groups on alternating carbons. They are different in that poly(1-butyne) is a conjugated system and poly(1-butene) is not. The first structure is more likely to be an electrically conducting polymer.

APPENDIX

Answers to
Practice Questions
& Practice Quizzes

Chapter 1
Chemistry: The Central Science

PRACTICE QUESTIONS

1. a. mixture   b. pure substance   c. mixture
2. a. melting of the chocolate   b. evaporation of water
3. a. heterogenous   b. heterogeneous   c. homogeneous
4. a. phycial   b. physical   c. chemical
5. a, d, and e are physical changes; b, c, and f are chemical changes
6. a, b, c are intensive; d. is extensive
7. 37.0°C; 310.2 K
8. 3 K; more
9. 3300 cm^3
9. 11 g/cm^3
10. a. 3   b. 2   c. 3   d. 4   e. 4
11. Body fat is less dense than muscle and bone tissue. The person's body with high percentage of body fat is less dense and will float more easily.
12. a. 0.609   b. 1.0×103 c. 0.000222   d. 238.0
    e. 1.3
13. a. 26,000   b. 0.4   c. 3100   d. 13.92   e. 4.80×10^3
14. Set #1 is more precise, but set #2 is more accurate.
15. Eggs are not measured. Eggs are counted; their exact number is known for sure.
16. 6.022×10^{23} molecules = 18 g
17. 6.022×10^{23} atoms Au = 196.97 g Au
18. Estimating, there are 250 hairs/eyebrow, 2 eyebrows/person
19. 1.6×10^{15} pm
20. 14,000 g Pb

PRACTICE QUIZ

1. a. 7×10^{-6} g   b. 8.0×10^{-9} m   c. 1.4×10^5 L   d. 1.0×10^3 s
2. 13.5 g/cm^3
3. 5×10^6 cm^3
4. 4.10 h
5. 16.7 mL
6. 5.7 L
7. a. 0.125 m   b. 80 nm   c. 4.45×10^{-3} km   d. 3.25×10^{-5} km
    e. 5.73×10^{-3} mm

Chapter 2
Atom and the Periodic Table

PRACTICE QUESTIONS

1. 109 neutrons
2. $^{170}_{68}$Er
3. 68 electrons
4. 88 and 90 neutrons
5. 87.6166 amu; strontium
6. Se
7. Hg
8. 1.96×10^{24} eggs

9. 2.25 mol of shrimp
10. 1.5×10^{23} cups
11. 88.9 g/mol
12. 4.476×10^{21} atoms Pt
13. Suzie's diamond is larger.

PRACTICE QUIZ

1.

Name	Symbol	Number of protons	Number of electrons	Number of neutrons	Mass Number
sodium	^{23}Na	11	11	12	23
argon	^{40}Ar	18	18	22	40
arsenic	^{75}As	33	33	42	75
lead	^{202}Pb	82	82	120	202

2. 26 p, 30 n, 26 e
3. 64
4. a. 8 p, 9 n, 8 e   b. 47 p, 60 n, 47 e   c. 86 p, 136 n, 86 e
5. $^{46}_{20}X$ and $^{43}_{20}X$ are isotopes.
6. 0.77 g
7. 2×10^{-11} g C
8. 3.155×10^{-23} g

Chapter 3
Quantum Theory and the Electronic Structure of Atoms

PRACTICE QUESTIONS

1. 600 nm
2. 4.0×10^{14} Hz
3. 4.1×10^{-19} J
4. 2.97×10^{-19} J/photon, 179 kJ/mol
5. a. 6   b. n = 4 to n = 1   c. n = 4 to n = 3
6. 95.0 nm, ultraviolet
7. 5×10^{14} /s
8. 1.32×10^{-5} nm
9. a, b
10. a. 1   b. 5   c. 3   d. 9
11. a. 2   b. 10   c. 6   d. 18
12. Sb: $1s^2 2s^2 2p^6 3s^2 3p^6 4s^2 3d^{10} 4p^6 5s^2 4d^{10} 5p^3$;
    V: $1s^2 2s^2 2p^6 3s^2 3p^6 4s^2 3d^3$;
    Pb: $1s^2 2s^2 2p^6 3s^2 3p^6 4s^2 3d^{10} 4p^6 5s^2 4d^{10} 5p^6\ 6s^2 4f^{14} 5d^{10} 6p^2$
13. a. $[Ar]4s^2 3d^2$ b. $[Xe]6s^2 4f^{14} 5d^6$ c. $[Ar]4s^2 3d^{10} 4p^5$
14. P
15. There are three elements with atoms having three unpaired electrons in the fourth period: V, Co, and As.

PRACTICE QUIZ

1. 9.70×10^{-7} m
2. 1200 nm
3. $n = 5$

4. 4.09×10^{-19} J
5. b
6. a, c
7. a. 3d, $m_l = -2, -1, 0, 1, 2$, 5 orbitals

    b. 4f, $m_l = -3, -2, -1, 0, 1, 2, 3$, 7 orbitals

    c. 5p, $m_l = -1, 0, 1$, 3 orbitals
8. a. $n = 2$ $l = 0$, 1 orbital

    b. $n = 3$ $l = 1$, 3 orbitals

    c. $n = 4$ $l = 3$, 7 orbitals
9. Pb
10. $1s^2 2s^2 2p^6 3s^2 3p^6 4s^2 33d^{10} 4p^6 5s^1$, [Kr]$5s^1$
11. Cr
12. 2

Chapter 4
Periodic Trends of the Elements

PRACTICE QUESTIONS

1. This would be any element in main group 3: Al, Ga, In, or Tl.
2. $1s^2 2s^2 2p^6 3s^2 3p^6 4s^2 3d^{10} 4p^1$
3. a. Ar   b. O   c. C
4. O
5. Y > Se > Li > Ar
6. Although Na has valence in $n = 3$ while Tl is in $n = 6$, Na has nuclear charge of only +11 while Tl +81.
7. Fr has lowest first ionization energy; Ra has lowest second ionization energy.
8. IE_5
9. K
10. 6.5°C
11. a. K^+: $1s^2 2s^2 2p^6 3s^2 3p^6$

    b. I^-: $1s^2 2s^2 2p^6 3s^2 3p^6 4s^2 3d^{10} 4p^6 5s^2 4d^{10} 5p^6$

    c. Sr^{2+}: $1s^2 2s^2 2p^6 3s^2 3p^6 4s^2 3d^{10} 4p^6$

    d. Se^{2-}: $1s^2 2s^2 2p^6 3s^2 3p^6 4s^2 3d^{10} 4p^6$
12. $K^+ < Sr^{2+} < Se^{2-} < I^-$
13. a. Rh^{3+}: $1s^2 2s^2 2p^6 3s^2 3p^6 4s^2 3d^{10} 4p^6 4d^6$

    b. Ni^{2+}: $1s^2 2s^2 2p^6 3s^2 3p^6 3d^8$

    c. Co^{3+}: $1s^2 2s^2 2p^6 3s^2 3p^6 3d^6$
14. $Al^{3+} < Mg^{2+} < Na^+ < Ne < F^-$

PRACTICE QUIZ

1. a. noble gases   b. alkali metals
2. K
3. Cl
4. Electron affinity is the energy realeased when an atom in the gas phase accepts an electron. Alkaline earth metals have negative electron affinities.
5. Br
6. a. Na^+: $1s^2 2s^2 2p^6$

    b. N^{3-}: $1s^2 2s^2 2p^6$

    c. Ba^{2+}: $1s^2 2s^2 2p^6 3s^2 3p^6 4s^2 3d^{10} 4p^6 5s^2 4d^{10} 5p^6$

    d. Br^-: $1s^2 2s^2 2p^6 3s^2 3p^6 4s^2 3d^{10} 4p^6$

    e. Li^+: $1s^2$

    f. Al^{3+}: $1s^2 2s^2 2p^6$
7. a. Sn^{2+}: $1s^2 2s^2 2p^6 3s^2 3p^6 4s^2 3d^{10} 4p^6 5s^2 4d^{10}$

b. Sn^{4+}: $1s^22s^22p^63s^23p^64s^23d^{10}4p^64d^{10}$
c. Cu^{2+}: $1s^22s^22p^63s^23p^63d^9$
d. Cu^+: $1s^22s^22p^63s^23p^63d^{10}$
e. Ti^{2+}: $1s^22s^22p^63s^23p^63d^2$
f. Ti^{4+}: $1s^22s^22p^63s^23p^6$

8. a. Sc^{3+}: $1s^22s^22p^63s^23p^6$
   b. V^{5+}: $1s^22s^22p^63s^23p^6$
   c. Pb^{2+}: $1s^22s^22p^63s^23p^64s^23d^{10}4p^65s^24d^{10}5p^66s^24f^{14}5d^{10}$
   d. Pb^{4+}: $1s^22s^22p^63s^23p^64s^23d^{10}4p^65s^24d^{10}5p^64f^{14}5d^{10}$

9. Isoelectronic pairs: Na^+ and Ne; Ar and Cl^-

10. Se^{2-}

11. a. Co^{2+} b. S^{2-}

Chapter 5
Ionic and Covalent Compounds

PRACTICE QUESTIONS

1. a. ·Mg·   b. ·Al·   c. :Br·   d. :Xe:

2. a. one   b. seven

3. a. K^+ b. $\left[:\ddot{S}: \right]^{2-}$ c. $\left[:\ddot{N}: \right]^{3-}$

   d. $\left[:\ddot{I}: \right]^-$ e. Sr^{2+}

4. $MgO(s) \rightarrow Mg^{2+}(g) + O^{2-}(g)$ $\Delta H = $ lattice energy

5. a. CaO   b. KCl

6. a, c, e

7. a. $BaCl_2$ b. Mg_3N_2 c. Fe_2O_3 d. FeF_2

8. a. potassium nitride   b. barium fluoride   c. aluminum oxide
   d. silver chloride   e. cobalt(II) chloride

9. a. yes   b. no   c. no   d. yes   e. yes

10. CH

11. a. NO_2 b. CH_2 c. $AlCl_3$ d. Fe_2O_3 e. SF_5

12. a. phosphorus trichloride   b. dichlorine heptoxide   c. CS_2 d. B_2O_3
    e. tetraphosphorus decaoxide   f. N_2O g. nitrogen dioxide

13. a. HBr(*aq*)   b. HF(*aq*)

14. a. NH_4Cl b. Na_3PO_4 c. K_2SO_4 d. $CaCO_3$
    e. $KHCO_3$ f. $Mg(NO_2)_2$ g. $NaNO_3$ h. $Sr(OH)_2$
    i. $Cu(CN)_2$

15. a. silver carbonate   b. magnesium hydroxide   c. sodium cyanide
    d. ammonium iodide   e. iron(II) nitrate

16. a. nitric acid   b. nitrous acid   c. hydrocyanic acid   d. perchloric acid

17. a. $Ni(NO_3)\cdot6H_2O$ b. $CuSO_4\cdot5H_2O$

18. a. 114.22 amu   b. 30.03 amu   c. 169.3 amu

19. a. 142.05 g   b. 162.20 g   c. 171.3 g   d. 80.07

20. a. 27.29% C, 72.71% O
    b. 2.43% H, 52.8% As, 45.1% O

    c. 10.06% C, 0.844% H, 89.09% Cl

    d. 33.32% Na, 20.30% N, 46.38% O

    e. 2.06% H, 32.70% S, 65.24% O

21. 37.0 g

22. a. 63.55 g   b. 79.55 g   c. 159.62 g   d. 249.7 g

23 a. 0.0463 mol Ag   b. 0.217 mol Na

24. a. 2.8×10^{22} atoms   b. 1.2×10^{24} molecules   c. 9.38×10^{23} molecules

25. a. 4.0 g   b. 5.00 g   c. 4.48×10^{-23} g

26. a. 0.103 mol $NaNO_3$ b. 0.308 mol O

27. a. 95 g of $LiNO_3$ b. 150 g of MgI_2

28. 3×10^{24} Ag atoms

29. a. 327 amu   b. male, 1.6×10^{23} molecules; female, 1.3×10^{23} molecules

30. a. NO   b. N_2O c. K_3PO_4 d. $K_2Cr_2O_7$

31. Fe_2S_3

32. C_6H_{12}

PRACTICE QUIZ

1. $RbI > CaBr_2 > MgO$

2.   a. $C_3H_4O_3$ b. CH   c. HgCl   d. HO
     e. CHO_2 f. $MgCl_2$

3.   a. $Ca(ClO)_2$ b. $HgSO_4$ c. $BaSO_3$ d. ZnO
     e. N_2O f. Na_2CO_3 g. CuS   h. PbO_2

4.   a. sodium hydrogen phosphate   b. hydrogen monoiodide
     c. tetraphosphorus hexoxide   d. lithium nitrate
     e. hydroiodic acid   f. strontium nitrite
     g. sodium hydrogen carbonate   h. potassium sulfite
     i. sodium phosphate   j. aluminum hydroxide

5.   a. sulfurous acid   b. hypochlorous acid   c. chloric acid
     d. phosphorous acid   e. hydrosulfuric acid

6.   The symbol O_3 represents a molecule of ozone, a molecular form of oxygen containing 3 oxygen atoms. The symbol 3O represents 3 separate oxygen atoms.

7.   The empirical formula of a compound gives the simplest whole-number ratio of the atoms of the elements making up the compound. Ionic compounds consist of positive and negative ions in the ratio needed to give an electrically neutral substance. It is this ratio of positive to negative ions that is essential to the formation of a neutral ionic compound. Any crystal of the compound having the proper ratio of positive and negative ions will always be electrically neutral, no matter what its size.

8.   5.2 g

9. 2×10^{-11} g

10. 3.16×10^{-23} g

11. a. 2×10^{24} atoms   b. 6×10^{23} atoms   c. 4.3×10^{23} atoms

12. 51.7 g

13. a. 4.41 mol $CaSO_4$ b. 17.6 mol O   c. 1.06×10^{25} O atoms

14. 6.02×10^{24} Sb atoms

15. a. 1.40×10^3 g   b. 0.10 g

16. a. 52.91% Al, 47.08% O   b. 1.60% H, 22.23% N, 76.17% O
     c. 9.93% C, 58.63% Cl, 31.43% F

17. 74.5% F

18. C_8H_8

19. CrO_3

Chapter 6
Representing Molecules

PRACTICE QUESTIONS

1. 8 bonding and 4 nonbonding
2. P
3. Based on the difference in electronegativity, N—H > C—N > C—H
4. ±0.0206
5. a. 1   b. 2   c. 2
6. a. All atoms have 0 formal charge.   b. All atoms have 0 formal charge.

7.

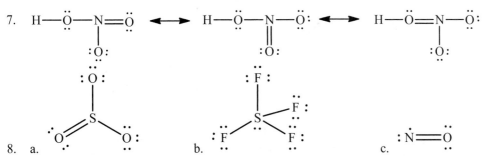

8. a.   b.   c.
9. SF_4 will be most polar due to geometry and bond polarity.

PRACTICE QUIZ

1. O—O < S—O < N—O < Na—O
2. polar covalent
3. Br—Br
4. Se—Cl: polar covalent
   Al—Cl: polar covalent (using electronegativity difference > 2.0)
   K—F: ionic
   Cl—Cl: nonpolar covalent

5.

6. a. 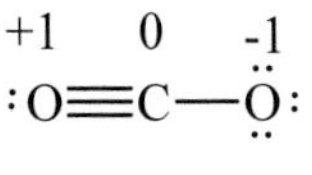 $\overset{+1}{:O}\equiv\overset{0}{C}-\overset{-1}{\ddot{O}}:$ b. $\overset{+1}{\ddot{F}}=\overset{-2}{Be}=\overset{+1}{\ddot{F}}$

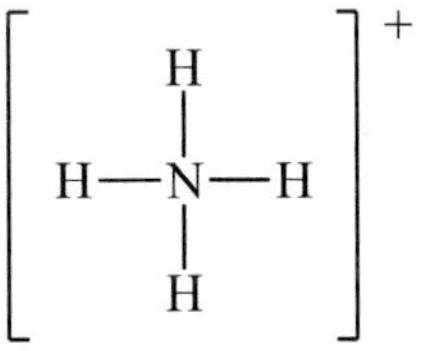

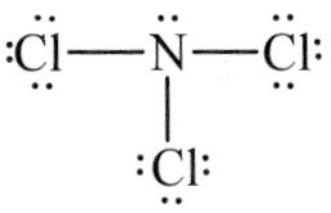

   c. $\ddot{O}=C=\ddot{O}$ $:\ddot{F}-Be-\ddot{F}:$

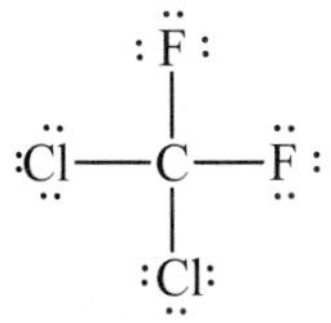

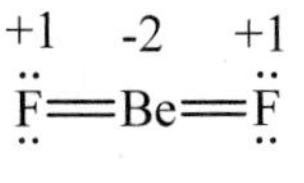

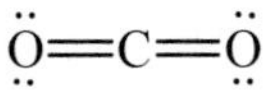

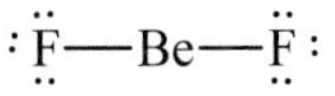

7. No, formal charges merely help us keep track of the electrons involved in bonding in the molecule.

8. $$\left[H-\ddot{O}-C=\ddot{O} \atop :\ddot{O}: \right]^{-} \longleftrightarrow \left[H-\ddot{O}-C-\ddot{O}: \atop :\ddot{O}: \right]^{-} \longleftrightarrow \left[H-\ddot{O}=C-\ddot{O}: \atop :\ddot{O}: \right]^{-}$$

Chapter 7
Molecular Structure and Bonding

PRACTICE QUESTIONS

1. a. linear   b. trigonal pyramidal   c. seesaw-shaped
    d. t-shaped   e. square planar
2. a. bent   b. trigonal planar   c. tetrahedral
3. a. and c.
4. a. $180°$ b. $109.5°$ c. less than $109.5°$
5. a. SH_2 b. PCl_3
6. tetrahedral

7.

8. The two requirements are polar bonds and a molecular geometry in bonds dipoles do not cancel.
9. c
10. c and d
11. a. PCl_3 – dipole-dipole forces and dispersion forces
    b. CO_2 – dispersion forces
    c. Cl_2 – dispersion forces
    d. ICl – dipole-dipole forces and dispersion forces
    e. KCl – ion-ion interactions (ionic bonding)
12. a. CO   b. ICl   c. H_2O d. AsH
13. Each chlorine atom has the same electron configuration: $1s^2 2s^2 2p^6 3s^2 3p_x^2 3p_y^2 3p_z^1$. The $3p_z$ orbitals of each Cl atom are only half full. Therefore, a sigma bond can be formed by overlap of the $3p_z$ orbital from each atom. In HBr, the $1s$ orbital of the H atom overlaps with the $4p_z$ orbital of Br to form a sigma bond.
14. a. $180°$ b. $120°$ c. $109.5°$
15. a. sp   b. sp^2 c. sp^2 d. sp^3d e. sp^2
16. a. sp^3 b. sp^3d c. sp^3d^2
17. octahedral
18. $2\,\sigma$ and $2\,\pi$
19. The antibonding MO; The atomic orbitals from which the bonding molecular orbital was created.
20. The bond order for Be_2 is zero, while for Be_2^+ it is 0.5.
21. Li_2: $(\sigma 1s)^2(\sigma^* 1s)^2(\sigma 2s)^2$ Li_2^+: $(\sigma 1s)^2(\sigma^* 1s)^2(\sigma 2s)^1$ Li_2 has the stronger bond.
22. a and b
23. Compare the electron configurations.

    N_2 $(\sigma 1s)^2((\sigma^* 1s)^2(\sigma 2s)^2((\sigma^* 2s)^2(\pi 2p_y)^2(\pi 2p_z)^2(\sigma 2p_x)^2$

N_2^+ $(\sigma_{1s})^2((\sigma*_{1s})^2(\sigma_{2s})^2((\sigma*_{2s})^2(\pi_{2p_y})^2(\pi_{2p_z})^2(\sigma_{2p_x})^1$

O_2 $(\sigma_{1s})^2((\sigma*_{1s})^2(\sigma_{2s})^2((\sigma*_{2s})^2(\pi_{2p_y})^2(\pi_{2p_z})^2(\sigma_{2p_x})^2(\pi_{2p_y}*)^1(\pi_{2p_z}*)^1$

O_2^+ $(\sigma_{1s})^2((\sigma*_{1s})^2(\sigma_{2s})^2((\sigma*_{2s})^2(\pi_{2p_y})^2(\pi_{2p_z})^2(\sigma_{2p_x})^2(\pi*_{2p_y})^1$

The bond order in N_2 is 3. Loss of 1 electron gives N_2^+. The electron came out of a bonding orbital and so the bond order is less than in N_2, b.o. = 2.5. The bond order in O_2 is 2. Loss of 1 electron gives O_2^+. The electron came out of an antibonding orbital and so the bond order is greater than in O_2, b.o. = 2.5.

24. The central oxygen atom is sp^2 hybridized and forms σ bonds to the two terminal oxygen atoms. The central O atom has an unhybridized $2p_z$ orbital that overlaps "sideways" with the $2p_z$ orbitals of both terminal O atoms.

PRACTICE QUIZ

1. a. Square pyramid   b. linear   c. trigonal planar
   d. bent   e. bent
2. a. sp^3d^2 b. sp c. sp^2 d. sp^2 e. sp^3
3. a. trigonal planar   b. distorted tetrahedral (seesaw)   c. square planar
   d. trigonal planar   e. tetrahedral
4. a. sp^2 b. sp^3d c. sp^3d^2 d. sp^2 e. sp^3
5. CO
6. b. SiH_4
7. a. $< 120°$ b. $90°$ c. $< 90°$ and $120°$
8. HF, CH_3CH_2OH, and CH_3NH_2 are capable of hydrogen bonding.
9. a. Hydrogen bonding must be overcome to vaporize water.
   b. The covalent bond must be broken in order to dissociate H_2.
   c. London dispersion forces must be overcome to boil O_2.
10. CCl_4 has four electron pairs about the central carbon atom, while $AsCl_4^-$ has 5 pairs about the As atom with only 4 Cl atoms attached. The extra pair of unshared electrons prevents the formation of a tetrahedron by $AsCl_4^-$.
11. There are two electron pairs in the valence shell of Be in $BeCl_2$, but there are four pairs in the valence shell of Te in $TeCl_2$. These four pairs are at the corners of a tetrahedron. When two chlorine atoms bond to a Te atom via two of the pairs, the Cl—Te—Cl bond angle is 109.5°.
12. sp^3d, sp^3d^2
13. No
14. HBr
15. b. H_2S
16. b. BCl_3 c. $BeCl_2$
17. c.
18. The O—H bonds are polar. No.

19. a. C_2 $(\sigma_{1s})^2(\sigma*_{1s})^2(\sigma_{2s})^2(\sigma*_{2s})^2(\pi_{2py})^2(\pi_{2pz})^2$
    C_2^{2-} $(\sigma_{1s})^2(\sigma*_{1s})^2(\sigma_{2s})^2(\sigma*_{2s})^2(\pi_{2py})^2(\pi_{2pz})^2(\sigma_{2px})^2$
    b. For C_2 the bond order = 2. For C_2^{2-} the bond order = 3.

20. The bond order in N_2 is 3, while in N_2^+ it is only 2.5. The greater the bond order, the stronger the bond and the shorter the bond distance.

21. Recall that lone-pair electrons take up more space than electrons in a bond. A lone pair in an axial position has three close neighbors in the bonding electron pairs in the equatorial positions. A lone pair in an equatorial position has two close neighbors in the bonding electron pairs in the axial positions. (The neighboring equatorial electron pairs are too far away). This means that a lone pair will have more room and have less repulsion when it is in an equatorial position.

22. The structure for the oxalate anion would show delocalization of one electron pair over each $O - C - O$ "half" of the anion. Those electrons are able to be shared over the pi-system formed by the overlap of the $2p$ orbitals from both carbon and oxygen.

Chapter 8
Chemical Reactions

PRACTICE QUESTIONS

1. $4Fe + 3O_2 \rightarrow 2Fe_2O_3$
2. $2H_2 + CO \rightarrow CH_3OH$
3. The sum of the coefficients on both sides does not have to be the same. The molar coefficients are used to balance the mass and the number of each type of atom in reactants and products.
4. $C_3H_4O_3$
5. 178 amu
6. C_3H_8 and C_4H_4
7. 9.33×10^{23} molecules
8. 28.07 g KOH; 0.504 g H_2
9. 1.79 g O_2
10. 10 g H_2
11. 1.67 mol K_3PO_4
12. 45.0 g $Pb(NO_3)_2$ left
13. a. 100 g Fe (rounded 139.6 g to one significant figure)
    b. 79% yield
14. a. 21.5 g   b. 65.9% yield
15. A is the limiting reactant. Theoretical yield is 3 AB_2.
16. Diagonal effect
17. Group 6 elements have relatively high electron affinity, hence, they will gain 2 electrons to become isoelectronic with the nearest noble gas. They will have a charge of –2 when they combine with metals. Halogens are one electron away from noble gas configuration. They have evem greater electron affinity than group 6. They will have –1 charge.

PRACTICE QUIZ

1. a. $P_4O_{10} + 6\,H_2O \rightarrow 4\,H_3PO_4$
   b. $2\,Ga + 3\,H_2SO_4 \rightarrow Ga_2(SO_4)_3 + 3\,H_2$
   c. $2\,C_4H_{10} + 13\,O_2 \rightarrow 8\,CO_2 + 10\,H_2O$
2. $2\,SO_2 + O_2 \rightarrow 2\,SO_3$

mol SO_2	grams O_2	mol SO_3	grams SO_3
1.50	24.0	1.50	120.
1.25	20.0	1.25	100.
5.21	83.4	5.21	417

3. a. 71.7 g Na_2CO_3 b. 25.9 g Na_2S

4. a. 30.6 g HF   b. 3.30 g CaF_2 remain   c. 85.7% yield
5. 129 g Sb_2S_3
6. 92.7% yield
7. H_2O is excess reactant; 7.5 g H_2O remain after reaction
8. B and Si
9. V is most metallic.
10. a. F   b. O   c. Cl
11. N has an orbital diagram of $[He]2s^22p^3$ while O has an orbital diagram of $[He]2s^22p^4$. It is more difficult to add electrons to orbitals that already contain electrons.
12. +1 and +2, respectively. The have relatively low first ionization energies.
13. CaO would be the most basic; SO_3 would be the most acidic

Chapter 9
Chemical Reactions in Aqueous Solutions

PRACTICE QUESTIONS

1. a, c, d, and e are ionic compounds.
2. b and d
3. a. CH_3CH_2OH and H_2O b. Mg^{2+} and Cl^- ions and water
4. a. $Mg(NO_3)_2(aq) \rightarrow Mg^{2+}(aq) +2\ NO_3^-\ (aq)$ b. $KOH(aq) \rightarrow K^+(aq) + OH^-\ (aq)$
    c. $CaF_2(aq) \rightarrow Ca^{2+}(aq) + 2\ F^-(aq)$
5. a and e.
6. b, c, and e.
7. a. No   b. Yes, iron (III) acetate is insoluble
8. True
9. a. yes, $Mg(OH)_2$ b. yes, AgI   c. no precipitate is formed
10. a. Na^+ and Br^- b. Na^+ and NO_3^-
11. a. $Pb^{2+}(aq) + 2Br^-(aq) \rightarrow PbBr_2(s)$
    b. $Ag^+(aq) + Br^-(aq) \rightarrow AgBr(s)$
    c. $Ba^{2+}(aq) + SO_4^{2-}\ (aq) \rightarrow BaSO_4(s)$
12. Li^+ and NO^{3-} ions floating in solution and solid $CaCO_3$ sinking to the bottom of the test tube.
13. a. $HBr(aq) + H_2O(l) \rightarrow H_3O^+(aq) + Br^-(aq)$
    b. $HC_2H_3O_2(aq) + H_2O(l) \rightarrow H_3O^+(aq) + C_2H_3O_2^-\ (aq)$
    c. $H_2SO_4(aq) + H_2O(l) \rightarrow 2\ H_3O^+(aq) + SO_4^{2-}\ (aq)$
14. a. +4   b. +6   c. −2   d. +1   e. −1
    f. +4   g. +3   h. +6   i. +7
15. $+7 \rightarrow +2$
16. a. Ag is oxidized and N is reduced   b. Fe^{2+} is oxidized and Cr is reduced
17. 0.983 M
18. 4.56 g
19. a. 0.0138 mol b. 0.0138 $M\,Mg^{2+}$ c. 0.0275 $M\,Cl^-$ ion
20. 1.25 M
21. Take 33.3 mL of 5.0 M KOH and dilute to a final volume of 250 mL
22. 1×10^{-4} mol
23. 5.1×10^{-3} M
24. 0.113 mol
25. 0.236 M
26. 20.7%
27. 18.7 mL

28. 9.8 mL
29. 50 mL
30. 0.157 M

PRACTICE QUIZ

1. a. and b. are strong electrolytes   c. and d. are weak electrolytes
   e. is a nonelectrolyte
2. a. $H^+(aq) + OH^-(aq) \rightarrow H_2O(l)$
   b. $CH_3COOH(aq) + OH^-(aq) \rightarrow CH_3COO^-(aq) + H_2O(l)$
   c. $H^+(aq) + OH^-(aq) \rightarrow H_2O(l)$
3. a. $Ba^{2+}(aq) + CO_3^{2-}(aq) \rightarrow BaCO_3(s)$
   b. $Aq^+(aq) + Cl^-(aq) \rightarrow AqCl(s)$
   c. $Pb^{2+}(aq) + S^{2-}(aq) \rightarrow PbS(s)$
4. a. b. c. and d. are insoluble   e. and f. are soluble
5. a. H_2SO_4 and $Cu(OH)_2$ b. HBr and KOH   c. H_3PO_4 and $Ca(OH)_2$
6. a. $Na^+(aq) + Br^-(aq) + Ag^+(aq) + NO_3^-(aq) \rightarrow AgBr(s) + Na^+(aq) + NO_3^-(aq)$
   $Ag^+(aq) + Br^-(aq) \rightarrow AgBr(s)$
   b. $Mg^{2+}(aq) + 2Br^-(aq) + Pb^{2+}(aq) + 2NO_3^-(aq) \rightarrow PbBr_2(s) + Mg^{2+}(aq) + 2NO_3^-(aq)$
   $Pb^{2+}(aq) + 2Br^-(aq) \rightarrow PbBr_2(s)$
7. a. +3   b. −3   c. +1   d. +5
   e. +7   f. +4   g. +7   h. −4
   i. +2   j. +1   k. +2.5
8. a, and c.
9. a. O_2 is the oxidizing agent and S is the reducing agent.
   b. BrO_3^- is the oxidizing agent and I^- is the reducing agent.
   c. NO_3^- is the oxidizing agent and As is the reducing agent.
10. 0.983 M NaOH
11. 3.29 g NaCl
12. 1.1 L
13. 0.060 M KCl
14. 12.0 g $PbSO_4$
15. 7.06×10^{-3} M Pb^{2+}
16. 13 mL of 0.10 M H_2SO_4
17. 1.19 M NaOH
18. 5.0 mL
19. 54.6 mL Ce^{4+} soln
20. 14.6% Fe

Chapter 10
Energy Changes in Chemical Reactions

PRACTICE QUESTIONS

1. a. Melting is endothermic.
   b. Condensation is exothermic.
   c. Sublimation is endothermic.
2. $w = +202$ L·atm
3. $q = 625$ J
4. 116 kJ

5. 37.8 kJ
6. 489 °C
7. 5.77 kJ
8. Iron will reach a higher temperature.
9. 1150 J
10. 2.92 kJ/°C
11. 941 J/°C

12. $\Delta H_f^\circ = -75.0$ kJ/mol

13. −37 kJ/mol
14. a. $CaCl_2$ b. NaI
15. 390 kJ/mol
16. 733 kJ/mol

PRACTICE QUIZ

1. $\Delta E = -350$ J
2. $\Delta E = -1623$ J
3. w = +322 J
4. 162 g
5. −111 kJ/g
6. $\Delta H^\circ = -2.88$ kJ/mol
7. a. 50 kJ is liberated.   b. 198 kJ is liberated
8. $\Delta H_{rxn}^\circ = +366$ kJ/mol
9. 321 kJ
10. 61.6 kJ

11. a. $C_2H_5OH + \dfrac{7}{2}O_2 \rightarrow 2CO_2 + 3H_2O$

     b. $q = 2000(4.18)(1.6) + 950(1.6) = 14{,}900$ J

     c. $\Delta H_{rxn}^\circ = -1370$ kJ

12. $\Delta H_f^\circ (SO_2) = -296$ kJ/mol

13. $\Delta H_f^\circ (CH_4) = -75$ kJ/mol

14. 3340 kJ/mol
15. 20.8°C
16. −44.67 kJ for 1 mol of $KClO_3$ a. exothermic   b. -1.46×10^3 kJ
17. −315.4 kJ/mol
18. 62.6°C
19. a. The enthalpy changes for the reaction is 15.39 kJ, so 1.44 kJ will be absorbed.
     b. 2.23°C
The internal energy of a system contains contributions from both potential and kinetic energy components. Among the kinetic energy components we include translations, rotations, and vibrations of molecules. It also include motions of subatomic particles. All of these components cannot be measured accurately so an absolute value of the internal energy of a system cannot be obtained.
20. −191.2 kJ/mol
21. −54.5 kJ/mol
22. $MgO(s) \rightarrow Mg^{2+}(g) + O^{2-}(g)$ $\Delta H =$ lattice energy
23. a. CaO   b. KCl

Chapter 11
Gases

PRACTICE QUESTIONS

1. 314 m/s
2. 2.8×10^2 m/s
3. O_2
4. a. 0.849 atm   b. 86.0 kPa
5. a. 98 kilopascals   b. 1.45 torr
6. a. 125 torr   b. 96.7 kPa
7. Volume is half of its original value.
8. −124 °C
9. 25 °C
10. 16.9 mL
11. 2.3 L
12. The volume would also decrease.
13. 55.85 L
14. 15.6 L
15. 84.4 mmHg
16. Both samples contain the same number of molecules. The sample of SF_4 has a greater mass.
17. 96.9 atm
18. 1.34 atm
19. 2.15×10^{22} molecules
20. 0.0683 g/L
21. 279 g/mol, $HgCl_2$
22. CF_4
23. 0.54 g H_2 and 0.32 g He
24. The total pressure would increase by 2.5 times.
25. 0.0169 mol H_2, 40.1 g/mol
26. a. 1.30 L   b. 191 g
27. 2.78 L
28. yes

PRACTICE QUIZ

1. a. 0.914 atm   b. 695 torr   c. 9.26×10^4 Pa
2. 1.2 L
3. 17.3 L
4. 5240 L
5. 790 mmHg
6. 0.016 g/L
7. 22.3 g
8. 11.4 L
9. 15.7 g/L
10. 279 g/mol, $HgCl_2$
11. 120 g/mol
12. 0.136 L
13. 0.0289 mol N_2
14. CO: 728 mmHg; $X = 0.968$
15. 0.782 N_2, 0.210 O_2, 0.008 Ar
16. $P_{O_2} = 0.100$ atm $P_{He} = 0.800$ atm
17. a. 5.84 L   b. 3.44 L
18. 13 L
19. 62.5 L

Chapter 12
Liquids and Solids

PRACTICE QUESTIONS

1.

	Higher Surface Tension	Higher Viscocity
a.	XeF_4	XeF_4
b.	BrCl	BrCl
c.	H_2O	H_2O
d.	NH_3	NH_3

2.   a. CH_3OH b. Cl_2 c. CH_3Cl
3. ΔH = 41.7 kJ/mol
4.   At higher temperature, a liquid's surface tension and viscosity are decreased. Its vapor pressure increases. Boiling point doesn't change with temperature; it depends on atmospheric pressure.
5.   352 pm
6.   Dry ice < gold bracelet < table salt < diamond
7.   Yes, it can have a cubic unit cell.
8.   Dry ice < gold bracelet < table salt < diamond
9.   16 g
10. Gas → liquid < liquid → solid, liquid → gas < solid → gas

PRACTICE QUIZ

1.   Surface tension makes it possible to fill a glass of water to a level slightly above the rim.
2. C_2H_5OH
3.   Viscosity is the resistance to flow and is caused by intermolecular forces and the general shape of the molecule. As the temperature of water is raised, more and more H-bonds are broken, and the viscosity decreases.
4.   82.5 kJ/mol
5. ΔH_{vap} = 41.7 kJ/mol
6.   1052 K
7.   100 mmHg
8.   The atmospheric pressure decreases as altitude increases. Therefore the higher one goes the lower the vapor pressure needed to form the bubbles observed during boiling. The lower the vapor pressure needed for boiling, the lower the temperature.
9.   4 atoms, face-centered cubic
10. 361 pm
11. 204 pm
12. 0.154 nm
13. 0.188 kJ/g
14. 388 torr

Chapter 13
Physical Properties of Solutions

PRACTICE QUESTIONS

1.   miscible
2.   Hexane is a nonpolar liquid and water is a polar liquid.
3.   a. soluble   b. soluble   c. insoluble
     d. insoluble   e. soluble   f. soluble

4. Ammonia NH_3 can form hydrogen bonds to water, but hexane cannot form hydrogen bonds..
5. a. $NaCl(s)$ b. $NH_3(g)$ c. $CH_3OH(l)$
6. a. 7.96%   b. 9.81%   c. 1.6%
7. a. 1.48 m b. 0.359 m
8. 2.33 M NaCl
9. 9.95% $AgNO_3$
10. 0.34 m $AgNO_3$
11. 48.5 g KNO_3
12. decreases
13. 1.5×10^{-5} mol/L
14. 31.4 mmHg
15. −0.103 °C
16. −1.8 °C
17. The pasta will cook faster in the boiling water that contains salt because this solution will have a higher boiling point.
18. 237 atm
19. a. 1   b. 1   c. 2   d. 4   e. 2   f. 3
20. 0.75 m solution of $MgCl_2$
21. 127 g/mol
22. 181 g/mol

PRACTICE QUIZ

1. a. HCl   b. $AlCl_3$ c. CH_3Cl d. CH_3OH
2. a. oil   b. C_6H_6 c. I_2
3. 1.04 m urea
4. 24.9 g KI
5. 0.88 M H_2O_2
6. 18 M
7. $X_{CH_3OH} = 0.360$; $X_{H_2O} = 0.640$
8. a. 1.28 M
 b. 16.9% $CuSO_4$, 83.1% H_2O
 c. $X_{CuSO_4} = 0.023$; $X_{H_2O} = 0.977$
 d. 1.29 m
9. 1.9×10^{-8} M CO
10. 520 g
11. 190 g/mol
12. 63.8 g/mol
13. 650 g/mol
14. 0.20 M NaCl
15. 0.154 M NaCl
16. NaBr; $Ca(NO_3)_2$; HCl = ethanol
17. $i = 1.76$
18. $C_6H_6O_4$

Chapter 14
Entropy, Free Energy, and Equilibrium

PRACTICE QUESTIONS

1. a. $\Delta S > 0$ b. $\Delta S < 0$
 c. cannot predict, ΔS will be essentially zero   d. $\Delta S < 0$

2. Think about the system. Let it be the oil or toxic substance. When an oil spill spreads out it is becoming more disordered as its molecules have a much larger area to occupy. Any process that gives rise to a more random distribution of the particles of the substance in space gives rise to an increase in entropy of the substance. In the reverse process, that of cleaning up a dispersed substance, work must be done to bring the substance together into a more concentrated space. This reduces the disorder of the system. This is a nonspontaneous process requiring a continual work input. Keeping the substance contained in the first place takes less work.

3. 357 J/K mol

4. a. Solid to liquid is an endothermic process with an increase in entropy. $\Delta H > 0$ and $\Delta S > 0$. The process will be spontaneous at high temperatures.
    b. Gas to solid is an exothermic process with a decreast in entropy. $\Delta H < 0$ and $\Delta S < 0$. The process will be spontaneous at low temperatures.

5. $\Delta G_{rxn}^{\circ} = 8.7$ kJ

PRACTICE QUIZ

1. b and c
2. a. $H_2O(g)$ b. $CO_2(g)$ c. $Ag^+(g)$
    d. $Cl_2(g)$ e. $2Cl(g)$
3. a. Essentially zero   b. Positive   c. Positive
    d. Negative   e. Negative
4. $\Delta S_{vap} = 81$ J/mol·K
5. $\Delta G_{rxn}^{\circ} = -24.7$ kJ/mol
6. $\Delta S^{\circ} = -110$ J/K·mol
7. a. $\Delta G^{\circ} = 236$ kJ/mol   b. $\Delta G^{\circ} = 318.2$ kJ/mol
8. b. (note: don't forget to double ΔG_f in part a)
9. $\Delta G_{rxn}^{\circ} = -23.3$ kJ/mol
10. $\Delta G_{rxn}^{\circ} = -34.85$ kJ/mol; $K_p = 1.28 \times 10^6$
11. Yes, $\Delta G^{\circ} < 0$
12. a. No, $\Delta G^{\circ} > 0$ b. $T = 7,320$ K
13. $\Delta G = -17.8$ kJ/mol. Spontaneous in the forward direction.
14. a. $\Delta S^{\circ} = 20$ J/K mol   b. 224 J/K mol The reaction is spontaneous at this temperature.
15. -5.56×10^3 kJ/mol
16. -220 J/K
17. -4.84 kJ/mol
18. The reaction is spontaneous at all temperatures.

Chapter 15
Chemical Equilibrium

PRACTICE QUESTIONS

1. $K_c = \dfrac{[NOBr]^2}{[NO]^2[Br_2]}$

2. a. $K_c = [O_2]$ b. $K_c = \dfrac{[Ni(CO)_4]}{[CO]^4}$

3. a. $K_p = 1.4 \times 10^{-5}$ b. $K_p = 270$

4. $K_c = \dfrac{[COCl_2]^2}{[CO_2][Cl_2]^2} = 4.2 \times 10^{11}$

5. $K_c = 1.5 \times 10^{-3}$
6. No, equilibrium will shift in the forward direction (right)
7. $K_c = 0.169$
8. $[NOBr] = 5.85 \times 10^{-3} \ M$
9. $K_c = 0.133$
10. $K_c = 17$
11. $[HI] = 0.53 \ M$, $[H_2] = [I_2] = 0.06 \ M$
12. a. right   b. no shift   c. left   d. left
    e. no shift   f. no shift
13. increase
14. The technician should have known that a catalyst cannot change the position of equilibrium. A catalyst lowers the activation energies of *both* the forward and reverse reactions, and so speeds up the rates of both the forward and reverse reactions. This does not change the equilibrium constant.

PRACTICE QUIZ

1. a. $K_c = \dfrac{[N_2]^2[H_2O]^6}{[NH_3]^4[O_2]^3}$ b. $K_c = \dfrac{[N_2O_4]^2}{[N_2O]^2[O_2]^3}$ c. $K_c = \dfrac{[FClO_2]^2}{[ClO_2]^2[F_2]}$

   d. $K_c = \dfrac{[HBr]^2}{[H_2]}$ e. $K_c = \dfrac{[CO]^2}{[CO_2]}$ f. $K_c = \dfrac{[H_2O]}{[H_2]}$

2. $[HI] = 1.1 \times 10^{-4} \ M$
3. $K_p = 2.2 \times 10^{-5}$, $K_c = 9.0 \times 10^{-7}$
4. $K_c = 0.0314$
5. $K_c = 34$
6. Net reverse reaction.
7. $[HI] = 0.24 \ M$
8. $P_{NO} = 0.017$ atm; $P_{N_2} = 1.99$ atm; $P_{O_2} = 0.39$ atm

9. a. Least extent of reaction   b. Greatest extent of reaction
   c. Intermediate of the three examples
10. b. (note: don't forget to double ΔG°_f in part a)
11. $\Delta G^\circ_{rxn} = -23.3$ kJ/mol
12. $\Delta G^\circ_{rxn} = -34.85$ kJ/mol; $K_p = 1.28 \times 10^6$
13. $\Delta G = -17.8$ kJ/mol. Spontaneous in the forward direction.
14. -5.56×10^3 kJ/mol
15. -4.84 kJ/mol
16. a. The partial pressures of SO_2 and O_2 will increase.
    b. No changes in equilibrium partial pressures.
    c. The partial pressures of SO_2 and O_2 decrease while that of SO_3 increases.
    d. The partial pressures of SO_2 and O_2 decrease while that of SO_3 increases.
    e. The partial pressure of SO_2 increases and that of SO_3 decreases.
17. a. K_c will decrease   b. $[PCl_3]$ decreases
    c. $[PCl_5]$ increases; $[PCl_5]$ and $[Cl_2]$ decrease   d. Cl_2 pressure increases
18. $K_p = 3.0 \times 10^{-8}$ $K_c = 4.7 \times 10^{-6}$
19. a. more solid dissolves (equilibrium shifts to the right
    b. nothing happens (the solution is already saturated)
    c. more solid dissolves (equilibrium shifts to the right)
20. $K_p = 3.2 \times 10^4$

Chapter 16
Acids and Bases

PRACTICE QUESTIONS

1. CO_3^{2-}

2. a. NH_4^+ is the acid on the left and NH_3 is its conjugate base, H_2O is the base on the left and H_3O^+ is its conjugate acid.

   b. HNO_2 is the acid on the left and NO_2^- is its conjugate base, CN^- is the base on the left and HCN is its conjugate acid.

3. a. $H_2PO_4^-$ b. PO_4^{3-}

4. a. HNO_3 b. SiH_4 c. $HOBr$ d. NH_3

5. $[OH^-] = 1.0 \times 10^{-11}\ M$

6. $[OH^-] = 0.0033\ M$ and $[H^+] = 3.0 \times 10^{-12}\ M$

7. $H_2SO_4 < HBr = HNO_3 < HF < KOH < Ba(OH)_2$

8. pure water

9. 3.4%

10. pH = 2.74

11. a. $[OH^-] = 1.3 \times 10^{-4}\ M$ $[H^+] = 7.9 \times 10^{-11}\ M$
    b. weak   c. 1.4×10^{-6}

12. HCN

13. 5.9×10^{-11}

14. 1. $H_3PO_4 \rightleftharpoons H^+ + H_2PO_4^-$ 2. $H_2PO_4^- \rightleftharpoons H^+ + HPO_4^{2-}$

    3. $HPO_4^{2-} \rightleftharpoons H^+ + PO_4^{3-}$

15. 1.21

16. a. pH > 7   b. pH = 7   c. pH < 7

17. Because N atoms in most compounds have an unshared electron pair.

PRACTICE QUIZ

1. 1.3×10^{-12}
2. 1.3×10^{-5}
3. 1.4×10^{-3}
4. 17%
5. Left. $HNO_2(aq) + ClO_4^-\ (aq)$ predominate at equilibrium.
6. HF
7. HNO_2
8. a. and d.
9. 5.6×10^{-10}
10. 10.82
11. Because they have empty orbitals available to accept electron pairs.

Chapter 17
Acid-Base Equilibria and Solubility Equilibria

PRACTICE QUESTIONS

1. a. pH = 2.87   b. pH = 4.05

2. a. Yes, OH^- b. No
3. a. pH = 4.66   b. No
4. a. Slight increase   b. No effect   c. Slight decrease

5. a. initial pH is 9.39 and final is 9.45   b. $NH_4^+ \rightleftharpoons NH_3 + H^+$

    c. $OH^- + NH_4^+ \rightarrow H_2O + NH_3$

6. 6.50
7. pH = 2.12
8. a. $HBr(aq) + NaOH(aq) \rightarrow H_2O(l) + NaBr(aq)$ b. pH = 1.39
9. 4.38
10. 7
11. Sample B > Sample C > Sample A
9. $K_{sp} = 4.9 \times 10^{-9}$
10. $K_{sp} = 1.7 \times 10^{-6}$
11. a. $s = 2.2 \times 10^{-4} M$ b. $[F^-] = 4.4 \times 10^{-4} M$
12. Yes, MgF_2

13. a. $s = 6.9 \times 10^{-7} M$ b. $Ag_3PO_4(s) \rightleftharpoons 3Ag^+ + PO_4^{3-}$
14. $s = 8.0 \times 10^{-9} M$
15. $s = 1.2 \times 10^{-7} M$
16. b, c, and d
17. a. pH increases   b. no change in pH   c. pH increases
18. The solubility reaction of $Zn(OH)_2$ is:

$$Zn(OH)_2 \rightleftharpoons Zn^{2+}(aq) + 2OH^-(aq)$$

The pH of a solution directly affects the OH^- ion concentration. To adjust the pH upward we must add more OH^- ion to a solution. In terms of the solubility equilibrium OH^- ion is a common ion and so the solubility equilibrium is shifted to the left as pH increases. Increasing the pH lowers the solubility of $Zn(OH)_2$.

19. $1.0 \times 10^{-22} M$
20. $s = 3.5 \times 10^{-3} M$

PRACTICE QUIZ

1. pH = 4.96
2. a. pH = 3.77   b. pH = 3.70
3. pH = 9.31
4. b. NH_4Cl and NH_3
5. a. 40.0 mL   b. pH = 2.00   c. pH = 10.89
6. pH = 8.72
7. pH = 4.92
8. b. $Zn(OH)_2$
9. $K_{sp} = 3.2 \times 10^{-10}$
10. $K_{sp} = 1.8 \times 10^{-15}$

Chapter 18
Electrochemistry

PRACTICE QUESTIONS

1. a. $14H^+(aq) + Cr_2O_7^{2-}(aq) + 6Br^-(aq) \rightarrow 2Cr^{3+}(aq) + 3Br_2(g) + 7H_2O(l)$

    b. $2MnO_4^-(aq) + 6I^-(aq) + 4H_2O(l) \rightarrow 2MnO_2(s) + 3I_2(aq) + 8OH^-(aq)$

2. No
3. a. $E^\circ_{cell} = -2.58$ V   b. $E^\circ_{cell} = 0.29$ V
4. Ce^{4+}
5. a. $\Delta G^\circ = 3$ kJ/mol   b. $\Delta G^\circ = -65.6$ kJ/mol
6. $Ni^{2+} + Fe \rightarrow Ni + Fe^{2+}$
7. 9.1×10^{-4} M
8. $E^\circ_{cell} = 0.02$ V; $E_{cell} = 0.02$ V (note: Q=1, ln(1)=0, so $(RT/nF)\ln Q = 0$); $\Delta G = -10$ kJ
9. a. $2Br^-(aq) \rightarrow Br_2(l) + 2e^-$
   b. $2H_2O(l) + 2e^- \rightarrow H_2(g) + 2OH^-(aq)$
10. 63.3 min

PRACTICE QUIZ

1. a. $Al(s) \mid Al^{3+}(aq) \parallel H^+(aq) \mid H_2(g) \mid Pt(s)$

   b. $Pb(s) \mid Pb^{2+}(aq) \parallel Fe^{3+}(aq) \mid Fe(s)$

   c. $Pt(s) \mid H_2(g) \mid H^+(aq) \parallel Cu^{2+}(aq) \mid Cu(s)$

   d. $Pt(s) \mid Br^-(aq) \mid Br_2(l) \parallel I_2(s) \mid I^-(aq) \mid Pt(s)$

   e. $Pt(s) \mid Fe^{2+}(aq), Fe^{3+}(aq) \parallel Sn^{2+}(aq) \mid Sn(s)$

   f. $Pt(s) \mid Fe^{2+}(aq), Fe^{3+}(aq) \parallel Sn^{4+}(aq), Sn^{2+}(aq) \mid Pt(s)$

2. $SO_4^{2-} < O_2 < Ce^{4+} < H_2O_2$
3. a. Yes   b. Yes   c. No
4. $\Delta G^\circ = 3$ kJ/mol; $K = 0.3$
5. $[Fe^{2+}]/[Fe^{3+}] = 0.3$
6. $E = 1.21$ V
7. 0.73 V
8. The cell emf will increase due to a decrease in $[H^+]$. The cell reaction is $Cu^{2+} + H_2 \rightarrow Cu + 2H^+$
9. -1.84 V
10. $[Cu^{2+}] = 3.3 \times 10^{-25}$ M
11. pH = 3.88
12. 0.45 mol e^-
13. 0.35 mol e^-
14. a. 5.7 g Ni   b. 3.8 g Co
15. 320 h
16. a. $H_2O_2 + 2I^- + 2H^+ \rightarrow I_2 + 2H_2O$
    b. $Cr_2O_7^{2-} + 3H_3AsO_3 + 8H^+ \rightarrow 2Cr^{3+} + 3H_3AsO_4 + 4H_2O$
17. $4Cl_2 + 8OH^- \rightarrow ClO_4^- + 7Cl^- + 4H_2O$

Chapter 19
Chemical Kinetics

PRACTICE QUESTIONS

1. rate $= \dfrac{1}{2}\dfrac{\Delta[NH_3]}{\Delta t} = -\dfrac{\Delta[N_2]}{\Delta t} = -\dfrac{1}{3}\dfrac{\Delta[H_2]}{\Delta t}$

2. 0.064 M/s

3. a. $\dfrac{-\Delta[S_2O_3^{2-}]}{\Delta t} = 8.3 \times 10^{-4}$ M/s

   b. $\dfrac{-\Delta[I_2]}{\Delta t} = 4.2 \times 10^{-4}$ M/s

4. 9 times
5. a. rate = k [A] [B]^2 b. k = 4.8 /M^2 s
6. a. k = 6.8 × 10^{-3} /s   b. 102 s
7. 92.2 kJ/mol
8. E_a(rev) = 26.5 kJ/mol
9. a. rate$_1$ = k$_1$ [NO$_2$]^2 rate$_2$ = k$_2$ [NO$_3$] [CO]

   b. when step 1 is rate determining: rate = k$_1$ [NO$_2$]^2 when step 2 is rate determining:

   rate = $\dfrac{k_2 k_1}{k_{-1}}$ [NO$_2$]^2[CO][NO]$^{-1}$

PRACTICE QUIZ

1. Rate = k[B]^2

2. Rate = $-\dfrac{1}{30}\dfrac{\Delta[CH_3OH]}{\Delta t} = -\dfrac{\Delta[B_{10}H_{14}]}{\Delta t} = \dfrac{1}{22}\dfrac{\Delta[H_2]}{\Delta t} = \dfrac{1}{10}\dfrac{\Delta[B(OCH_3)_3]}{\Delta t}$

3. a. [sucrose] = 0.083 M b. $t_{1/2}$ = 3.3 × 10^5 s
4. a. k = 0.106/h   b. 0.252   c. 6.54 h   d. 0.0174 M
5. a. first order   b. 3.7 × 10^{-3} per day
6. a. 2.76 ×10^{-2} /M · min   b. 90 min
7. Twelve-fold
8. 182 kJ/mol
9. k = 2.1 × 10^{-5}/M s   k$_2$/k$_1$ = 1.38
10. 16.8
11. 180 kJ/mol
12. Yes
13. The rate law for the rate determining step is rate = k[NO$_2^-$][O$_2$] The concentration of dissolved O$_2$ is a constant according to Henry's law (Chapter 12). Therefore k[O$_2$] = K (a constant), and rate = K[NO$_2^-$].

14. The rate law for the rate determining step is rate = k[Ni(H$_2$O)$_6^{2+}$] This rate law matches the experimental rate law.

15. The rate law for the rate determining step is rate = k$_2$[I]^2[H$_2$] Obtaining a mathematical substitution for [I] from k$_1$[I$_2$] = k$_{-1}$[I]^2 yields rate = $\dfrac{k_2 k_1}{k_{-1}}$ [H$_2$][I$_2$] This predicted rate law matches the experimental rate law given in the problem.

16. a. The catalyst reacts in one of the first few steps, but then is regenerated in the last step.
    b. If E$_a$ is lowered for the forward reaction it must also be lowered for the reverse.
17. a. Br$_2$ b. 2H$_2$O$_2$(aq) → 2H$_2$O(l) + O$_2$(g)   c. homogeneous
18. 6.25 × 10^{-3} M
19. a. the rate would increase by a factor of 1.5   b. the rate would be 1/8th of its value
    c. the rate would increase by a factor of four
20. E_A = 30.8 kJ/mol

Chapter 20
Nuclear Chemistry

PRACTICE QUESTIONS

1. 15 protons, 17 neutrons, and 15 electrons
2. a. $^{206}_{82}Pb$ b. $^{1}_{0}n$ c. $^{15}_{8}O$
3. a. $^{18}_{9}F \rightarrow \, ^{18}_{8}O + \, ^{0}_{+1}\beta$ b. $^{21}_{9}F \rightarrow \, ^{21}_{10}Ne + \, ^{0}_{-1}\beta$
4. a. $^{4}_{2}He$ b. $^{27}_{13}Al$
5. 1.34×10^{-12} J/nucleon
6. 8 alpha particles and 6 beta particles
7. 3520 y
8. a. $^{27}_{12}Mg$ b. $^{14}_{6}C$
9. $^{93}_{36}Kr$

PRACTICE QUIZ

1. Beta particles and positrons have the same mass, 0.00055 amu. Beta particles have one unit of negative charge while positrons have one unit of positive charge.
2. a. $^{239}_{94}Pu \rightarrow \, ^{4}_{2}\alpha + \, ^{235}_{92}U$

    b. $^{2}_{1}H + \, ^{6}_{3}Li \rightarrow 2 \, ^{4}_{2}He$

    c. $^{90}_{38}Sr \rightarrow \, ^{90}_{39}Y + \, ^{0}_{-1}\beta$

    d. $^{10}_{5}B + \, ^{4}_{2}\alpha \rightarrow \, ^{13}_{7}N + \, ^{1}_{0}n$

    e. $^{56}_{26}Fe + \, ^{1}_{0}n \rightarrow \, ^{57}_{26}Fe$

3. $^{11}_{5}B < \, ^{39}_{20}Ca < \, ^{40}_{20}Ca$. B-11 has an odd number of protons and an odd number of neutrons (rule 2). Ca-39 has an even number of protons and an odd number of neutrons (rule 1). Ca-40 has an even number of protons and an even number of neutrons. It also has a "magic number" of protons and neutrons (rule 2).
4. a. positron emission   b. positron emission   c. beta decay
5. $^{29}_{13}Al$
6. a. $k = 0.132$ /yr   b. $N/N_0 = 0.205$
7. $t_{1/2} = 12.4$ yr
8. $t = 5.1$ yr
9. $t = 1.5 \times 10^{9}$ yr
10. c. $\Delta m = -1.98 \times 10^{-2}$ amu
11. $^{209}_{83}Bi + \, ^{58}_{26}Fe \rightarrow \, ^{266}_{109}X + \, ^{1}_{0}n$
12. $^{238}_{92}U + \, ^{1}_{0}n \rightarrow \, ^{239}_{92}U$

    $^{239}_{92}U \rightarrow \, ^{239}_{93}Np + \, ^{0}_{-1}\beta$

    $^{239}_{93}Np \rightarrow \, ^{239}_{94}Pu + \, ^{0}_{-1}\beta$

13. $^{235}_{92}U + \, ^{1}_{0}n \rightarrow \, ^{137}_{55}Cs + \, ^{96}_{37}Rb + 3 \, ^{1}_{0}n$
14. positron emission
15. 8 alpha particles account for the mass change; 8 alpha and 6 beta particles account for the change in positive charge.
16. 0.10 rad
17. 10 rem

18. 2.3×10^{19} Ra atoms decayed in 10 years; 2.3×10^{19} He atoms have a mass of 1.6×10^{-4} g and occupy a volume of 0.87 cm³ at STP.

Chapter 21
Metallurgy and the Chemistry of Metals

PRACTICE QUESTIONS

1. A mineral is a naturally occurring substance with a characteristic range of chemical composition. An ore is a mineral, or a mixture of minerals, from which a particular metal can be profitably extracted.
2. The electropositive metals can be reduced by a more electropositive metal such as Li, but there is no metal electropositive enough to reduce Li^+ to Li. Electrolytic reduction is the only way to prepare Li. It is possible to reduce other metals with Li. Rather than prepare Li by electrolysis and then use it to reduce aluminum, magnesium, or sodium, it is more convenient simply to prepare these metals in one step by electrolytic reduction.
3. Zone refining is a technique for purifying metals. Figure 20.8 in the text explains how it works.
4. Limestone is used to remove impurities from iron during production. Limestone decomposes to lime (CaO). CaO reacts with SiO_2 and Al_2O_3 impurities from the iron ore forming calcium silicate and calcium aluminate. The mixture of calcium silicate and calcium aluminate is known as slag.
5. $Cr_2O_3(s) + 2Al(s) \rightarrow 2Cr(l) + Al_2O_3(s)$
6. A band is a large number of molecular orbitals that are closely spaced in energy. The valence band is a set of closely spaced MOs that are filled with electrons. The conduction band is a set of closely spaced empty orbitals. An electron can travel freely through the metal since the conduction band is void of electrons.
7. A conductor is capable of conducting an electrical current. In a conductor there is essentially no energy gap between the valence band and the conduction band. Insulators are materials that do not conduct. In an insulator the energy gap between the valence band and the conduction band is much greater than the gap in a conductor. Semiconductors are normally not conductors, but will conduct electricity at higher temperatures, or when combined with small amounts of certain elements.
8. Good conductors of heat and electricity. Low density. Soft enough to cut with a knife. Low melting points.
9. Extremely reactive. React with water to form hydroxides. React with oxygen to form a variety of oxides, peroxides, and superoxides.
10. a. barium   b. barium   c. beryllium   d. barium
11. Magnesium is precipitated from seawater. Calcium is obtained from limestone.
12. Molten cryolite is used as a solvent for alumina, Al_2O_3, in aluminum production. Cryolite melts at 1000 °C as compared to 2050 °C for alumina and therefore it lowers the energy consumption.
13. A fresh surface of aluminum reacts readily with oxygen and forms a surface coating of Al_2O_3. This layer is very impenetrable and does not allow oxygen and water through to continue a reaction with the rest of the aluminum.

PRACTICE QUIZ

1. a. Leaching is the selective dissolution of a metal from an ore. Flotation is a technique used to separate mineral particles from waste clays and silicates called gangue.
   b. Roasting involves heating an ore in the presence of air. The idea is to convert metal carbonates and sulfides to oxides, which can be more conveniently reduced to yield pure metals. Reduction is the gain of electrons by a cation to yield the pure metal.
2. No. Use Appendix 3 of the text to calculate $\Delta G°$. The sign of $\Delta G°$ indicates if the reaction is spontaneous.
3. $Al^{3+}(aq) + 3OH^-(aq) \rightarrow Al(OH)_3(s)$
   $Al(OH)_3(s) + OH^-(aq) \rightarrow Al(OH)_4^-(aq)$

4. An alloy is a mixture of two or more metals, or of metals and nonmetal elements that have metallic properties. An amalgam is an alloy containing mercury.

5. Copper atoms have a half-filled 4s orbital. This means that copper metal has a half-filled valence band. A partially filled valence band is a conduction band.

6. The increasingly random motion of electrons due to a temperature rise opposes motion in one unified direction.

7. $3K + AlCl_3 \rightarrow Al + 3KCl$

8. $Al_2O_3 \cdot 2H_2O$

9. The bauxite is pulverized and digested with sodium hydroxide solution. This converts the aluminum oxide to the aluminate ion AlO_2^- which remains in solution. However treatment with base has no effect on the iron oxide which remains as insoluble $Fe_2O_3(s)$, and is removed by filtration. Aluminum hydroxide is then precipitated by acidification to about pH 6.

10. $Al_2(SO_4)_3$ is used to clarify water through its reaction with $Ca(OH)_2$. The precipitation of $Al(OH)_3$ traps dirt and dust particles.

Chapter 22
Coordination Chemistry

PRACTICE QUESTIONS

1. a. $[Ar]4s^23d^5$ b. $[Ar]3d^5$ c. $[Ar]3d^4$

2. Mn^{3+} will form the strongest metal-ligand bond due to its high positive charge.

3. tribromotrichlorocobaltate(II) ion.

4. a. $[Cu(NH_3)_2(C_2O_4)]$ b. $K[PtNH_3Cl_3]$

5. Both complexes are six-coordinate with a Co ion in the center. The charge on the Co ion in $[CoCl_3Br_3]^{4-}$ is +2 while that for $[Co(NH_3)_3Br_3]$ is +3.

6. 4

7. a. Six d-electrons total. In the low-spin O_2 containing complex there are no unpaired electrons. In the high-spin complex with H_2O there are four unpaired electrons.

    b. The two forms of hemoglobin have different colors due to the differing amount of crystal field splitting (Δ) in the iron complexes. O_2 must be a stronger-field ligand than H_2O. The splitting is much greater in the low-spin O_2-containing complex than in the high-spin H_2O-containing complex. Therefore the color of light absorbed will be different for the two complexes, leading to different observed colors. From the complementary colors you can tell that the O_2 complex absorbs green light and the H_2O complex absorbs orange light.

8. Diamagnetic

PRACTICE QUIZ

1. Fe $[Ar]4s^23d^6$
   Fe $[Ar]3d^6$
   Fe $[Ar]3d^5$

2. $[Pt(NH_3)_3Cl_3]NO_3$

3. +2

4. The coordination number is 6; the oxidation state is +3.

5. The coordination number is 6; the oxidation state is +2.

6. a. tetraamminedichloroplatinum(IV) chloride
   b. hexaaminenickel(II) ion
   c. tetrahydroxochromate(III) ion
   d. hexacyanocobaltate(III) ion
   e. potassium tetracyanocuprate(II)

7. a. $[Al(OH)_4]^-$
   b. $[HgI_4]^{2-}$

c. $K_3[Co(C_2O_4)_2Cl_2]$

d. $[Ni(en)_3]SO_4$

8.

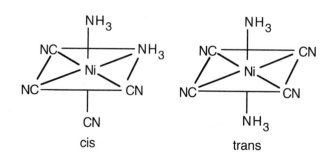

9.

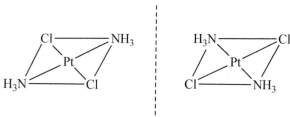

The two structures are not enantiomers.

10. Because when plane-polarized light passes through a solution of a chiral substance, the plane of the polarized light is rotated.

11. a.

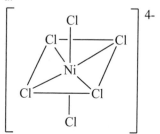

b.

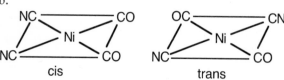

c. Octahedral geometry, Fe^{3+} energy diagram with small crystal-field splitting between the d-orbitals.

d.
$$\begin{array}{ccc} & O & O \\ & \| & \| \\ {}^{-}O{-}C{-}C{-}O^{-} \end{array}$$

e. $[H_3N{-}Ag{-}NH_3]^+$

12. Red

13. Zinc has ten d electrons that completely fill the five d orbitals.

14. $[Ni(en)_3]^{2+}$

Chapter 23
Nonmetallic Elements and Their Compounds

PRACTICE QUESTIONS

1. $-1, +1$

2. Because diamond is a single giant molecule, the melting point of diamond is related to the energy required to break C—C bonds. In diamond each carbon atom is covalently linked to four other carbon atoms. To turn diamond into a liquid about half of these bonds must be broken. This requires extremely high temperatures.

3. CO

4. -5 and -3

5. White phosphorus is very unstable compared to red phosphorus.

6. ozone

7. $-1, -2, 0$

8. K_2O, K_2O_2, and KO_2

9. $S(s) + O_2(g) \rightarrow SO_2(g)$
   $2SO_2(g) + O_2(g) \rightarrow 2SO_3(g)$
   $SO_3(g) + H_2O(l) \rightarrow H_2SO_4(aq)$

10. HCl, HBr, and HI

11. F_2

12. Br_2 can act as both an oxidizing agent and a reducing agent because Br has three oxidation states, namely -1 in Br^- ion, 0 in Br_2, and $+1$ in OBr^-. Therefore, Br_2 can be an oxidizing agent by accepting electrons and being reduced to Br^- ion. And Br_2 can also be a reducing agent by donating electrons and being oxidized to the OBr^- ion.

PRACTICE QUIZ

1. To make ammonia and partially hydrogenated food products.

2. When coal is distilled in the absence of oxygen to release volatile hydrocarbons, the high carbon residue is called coke. Coke is made into graphite by the Acheson process, in which an electric current is passed through coke for several days.

3. a. Create atmospheres devoid of O_2; a coolant.
   b. Ammonia fertilizer
   c. Nitrate fertilizer and explosives

4. A mixture of 3 volumes of concentrated HCl to 1 volume of concentrated HNO_3.

5. NO_2, NO, P_2O_5, P_2O_3

6. It is insoluble.

7. Diammonium phosphate, $(NH_4)_2HPO_4$; superphosphate $Ca(HPO_4)_2 \cdot H_2O$

8. Elemental oxygen, O_2, ozone, O_3; rhombic S_8, monoclinic S_8

9. Peroxide ion: O_2^{2-}; superoxide: O_2^-.

10. $2HF \rightarrow H_2 + F_2$

    $2NaCl + 2H_2O \rightarrow 2NaOH + H_2 + Cl_2$

$$Cl_2 + 2I^- \rightarrow 2Cl^- + I_2$$
11. pale yellow gas.

Chapter 24
Modern Materials

PRACTICE QUESTIONS

1. Initiation, chain growth, and termination.
2. homopolymer
3. Readily accepted by the body is only one requirement of a good biomaterial choice. This material would wear out too readily for a replacement part that requires a surgical placement.
4. Gallium or aluminum (There are other correct responses.)
5. conductor

PRACTICE QUIZ

1. a. $CH_3CH = CHCl$ b. $CH2=CCl2$

2. $HOCH_2CH_2CH_2OH$

3. nitride
4. n-type

NOTES

NOTES

NOTES

NOTES

NOTES

NOTES

NOTES

NOTES